Quick Reference

 **Get ConnectED**

connectED.mcgraw-hill.com

The icons found throughout *Math Connects* provide you with the opportunity to connect the print textbook with online interactive learning.

Investigate

 Animations present an animation of a math concept or graphic novel.

 Vocabulary presents visual representations of math concepts.

 Multilingual eGlossary presents key vocabulary in 13 languages.

Learn

 Personal Tutor presents a teacher explaining step-by-step solutions to problems.

 Virtual Manipulatives provide digital ways to explore concepts.

 Graphing Calculator provides keystrokes other than the TI-83/84 Plus or TI-Nspire used in the textbook.

 Audio recordings provide an opportunity to build oral and listening fluency.

 Foldables provide a unique way to enhance students' study skills.

Practice

 Self-Check Practice allows students to assess their knowledge of foundational skills.

 Worksheets provide additional practice and reteach opportunities.

 Online Assessment checks understanding of concepts and terms.

D1071671

Glencoe McGraw-Hill

Math Connects

Course 3

Authors

Carter • Cuevas • Day • Malloy
Kersaint • Luchin • McClain • Molix-Bailey • Price
Reynosa • Silbey • Vielhaber • Willard

 McGraw Hill Glencoe

The McGraw·Hill Companies

 Glencoe

Copyright © 2012 The McGraw-Hill Companies, Inc. All rights reserved. No part of this
publication may be reproduced or distributed in any form or by any means, or stored
in a database or retrieval system, without the prior written consent of The McGraw-Hill
Companies, Inc., including, but not limited to, network storage or transmission, or
broadcast for distance learning.

Send all inquiries to:
Glencoe/McGraw-Hill
8787 Orion Place
Columbus, OH 43240-4027

ISBN: 978-0-07-895139-8
MHID: 0-07-895139-9   *Math Connects,* Course 3

Printed in the United States of America.

4 5 6 7 8 9 10   DOW   18 17 16 15 14 13 12 11

Contents in Brief

Master the Focal Points

Start Smart A review of concepts from Course 2

End-of-Year Option

Problem-Solving Projects

Online Guide

Get ConnectED
connectED.mcgraw-hill.com

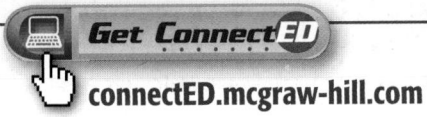

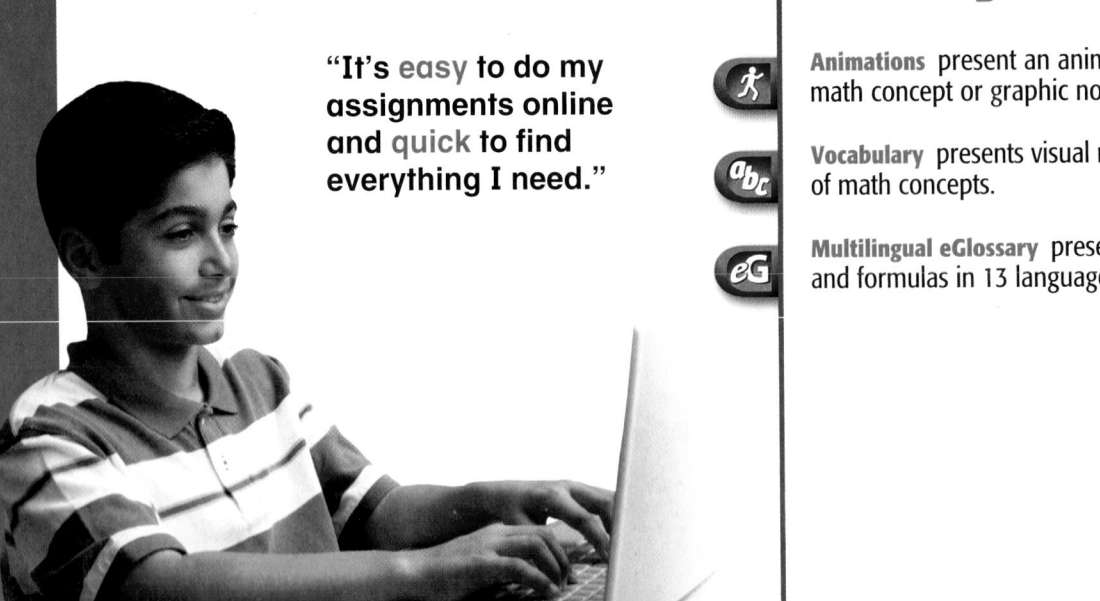

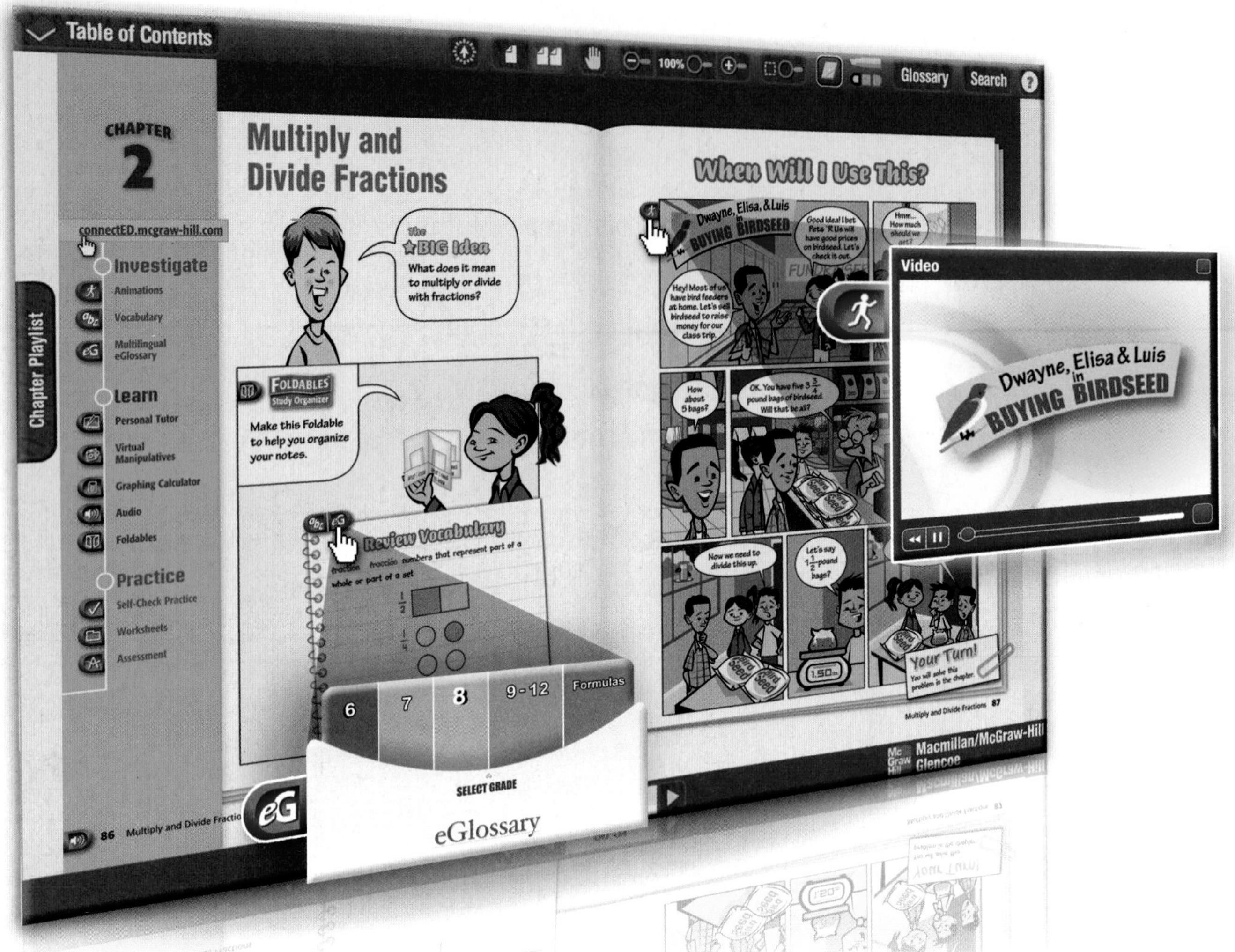

"It's easy to do my assignments online and quick to find everything I need."

Investigate

Animations present an animation of a math concept or graphic novel.

Vocabulary presents visual representations of math concepts.

Multilingual eGlossary presents key vocabulary and formulas in 13 languages.

The icons found throughout Math Connects provide you with the opportunity to connect the print textbook with online interactive learning.

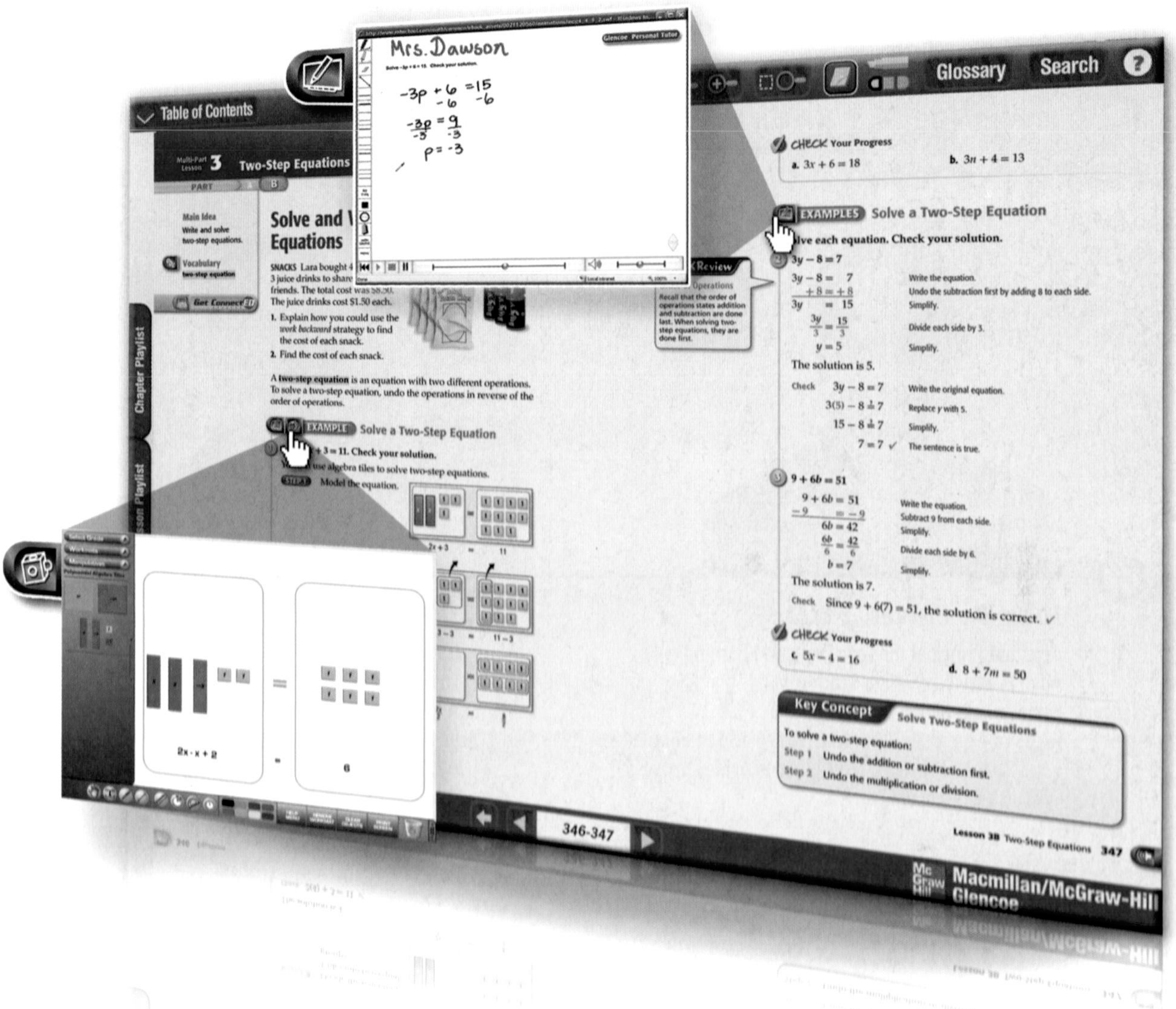

Learn ▶

 Personal Tutor presents a teacher explaining step-by-step solutions to problems.

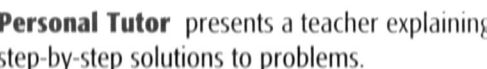 **Virtual Manipulatives** provide digital ways to explore concepts.

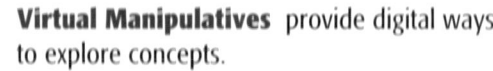 **Graphing Calculator** provides keystrokes other than the TI-83/84 Plus or TI-Nspire used in the textbook.

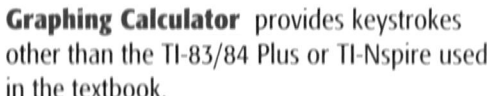 **Audio** recordings provide an opportunity to build oral and listening fluency.

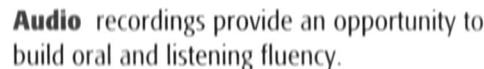 **Foldables** provide a unique way to enhance students' study skills.

Practice ▶

 Self-Check Practice allows students to assess their knowledge of foundational skills.

 Worksheets provide additional practice and reteach opportunities.

 Online Assessment checks understanding of concepts and terms.

Authors

O ur lead authors ensure that the Macmillan/McGraw-Hill and Glencoe/McGraw-Hill mathematics programs are truly vertically aligned by beginning with the end in mind—success in Algebra 1 and beyond. By "backmapping" the content from the high school programs, all of our mathematics programs are well articulated in their scope and sequence.

Lead Authors

John A. Carter, Ph.D.

Assistant Principal for Teaching and Learning
Adlai E. Stevenson High School
Lincolnshire, Illinois
Areas of Expertise: Using technology and manipulatives to visualize concepts; Mathematics Achievement of English-Language Learners

Gilbert J. Cuevas, Ph.D.

Professor of Mathematics Education
Texas State University–San Marcos
San Marcos, Texas
Areas of Expertise: Applying concepts and skills in mathematically rich contexts; Mathematical Representations

Roger Day, Ph.D., NBCT

Mathematics Department Chairperson
Pontiac Township High School
Pontiac, Illinois
Areas of Expertise: Understanding and applying probability and statistics; Mathematics Teacher Education

Carol Malloy, Ph.D.

Associate Professor
University of North Carolina at Chapel Hill
Chapel Hill, North Carolina
Areas of Expertise: Representations and critical thinking; Student Success in Algebra 1

 **Meet the Authors.**

Program Authors

Gladis Kersaint, Ph.D.
Associate Professor of Mathematics
  Education, K–12
University of South Florida
Tampa, Florida

Kay McClain, Ed.D.
Research Professor
Arizona State University
Phoenix, Arizona

Rhonda J. Molix-Bailey
Mathematics Consultant
Mathematics by Design
DeSoto, Texas

Beatrice Moore Luchin
Mathematics Consultant
Houston, Texas

Jack Price, Ed.D.
Professor Emeritus
California State
  Polytechnic University
Pomona, California

Mary Esther Reynosa
Instructional Specialist for Elementary
  Mathematics
Northside Independent School District
San Antonio, Texas

Robyn Silbey
Math Content Coach
Montgomery County Public Schools
Gaithersburg, Maryland

Kathleen Vielhaber
Mathematics Consultant
St. Louis, Missouri

Teri Willard, Ed.D.
Associate Professor
Central Washington University
Ellensburg, Washington

Contributing Author

Dinah Zike FOLDABLES
Educational Consultant
Dinah-Might Activities, Inc.
San Antonio, Texas

Lead Consultant

Viken Hovsepian
Professor of Mathematics
Rio Hondo College
Whittier, California

Consultants and Reviewers

T hese professionals were instrumental in providing valuable input and suggestions for improving the effectiveness of the mathematics instruction.

Consultants

Mathematical Content

Dr. Michaele Chappell
Professor of Mathematics Education
Middle Tennessee State University
Murfreesboro, Tennessee

Melissa D. Young
Mathematics Specialist
Differentiated Accountability Model, Region III
Orlando, Florida

Differentiated Instruction

Jennifer Taylor-Cox, Ph.D.
Educational Consultant
Innovative Instruction: Connecting
  Research and Practice in Education
Severna Park, Maryland

Gifted and Talented

Shelbi K. Cole
Research Assistant
University of Connecticut
Storrs, Connecticut

Problem Solving

Dr. Stephen Krulik
Professor Emeritus – Math Education
Temple University
Philadelphia, Pennsylvania

Reading in the Content Areas

Sue Z. Beers
President / Consultant
Tools for Learning, Inc.
Jewell, Iowa

Reading and Vocabulary

Douglas Fisher
Professor
San Diego State University
San Diego, California

Reviewers

Sheila J. Allen
Secondary Curriculum Coordinator
Medina City Schools
Medina, OH

Kathryn Blizzard Ballin
Mathematics Supervisor
Newark Public Schools
Newark, NJ

Angelee M. Bilbao
Middle School Math Teacher
Discovery School
Glendale, AZ

Christine M. Binkley
8th Grade Math/Algebra
Celina Middle School
Celina, OH

Ronald E. Boggs
Teacher–8th Grade Math and
    Algebra 1
Lakewood Middle School
Hebron, OH

Staci Bolley
Math Teacher
Desert Sky Middle School
Glendale, AZ

Danielle Bouton
District K–12 Coordinator of
    Mathematics and Technology
Schenectady City Schools
Schenectady, NY

Matt Bowser
Math Teacher
Oil City Middle School
Oil City, PA

Thomas Brewer
6th Grade Math Teacher
West Holmes Middle School
Millersburg, OH

Lisa K. Bush
Math Curriculum Specialist–K–12
Deer Valley Unified School District
Phoenix, AZ

Janelle Chisholm Winter
Math Teacher
Palo Verde Middle School
Phoenix, AZ

S. Cox
Math Academic Facilitator
Kirksey Middle School
Rogers, AR

Mary Ellen Dierksheide
Mathematics Teacher
Elmwood Middle School
Bloomdale, OH

Dominick Galimi
Mathematics
School 5
Yonkers, NY

Jacquelyn Gawron
Liaison Instruction Department
Youngstown City Schools
Youngstown, OH

Amber Griffin
Secondary Mathematics Teacher
Kino Jr High School
Mesa, AZ

Chad D. Heuser
Math Department Chair
Elyria High School
Elyria, OH

Jerry Hicks
6th Grade Mathematics Teacher
Queensbury Middle School
Queensbury, NY

Sandra Hughes
Cross Curricular Coach
Gloucester City School District
Gloucester City, NJ

Julene M. Ippolito
Teacher
George Washington Intermediate
    School
New Castle, PA

Tracey Jaehnert
Mathematics Teacher/Dept. Chair
LaGrange Middle School
LaGrangeville, NY

Satish Jagnandan
Administrator for Mathematics and
    Science (K–12)
Mount Vernon City School District
Mount Vernon, NY

Kimberly Knisell
Teacher & Math Department
    Chairperson
Saugerties, NY

Victoria Lautsch
Middle School Math Teacher
Don Mensendick School
Glendale, AZ

Cheryl Ann Lipko
Mathematics Department Chairperson
Mount Pleasant Area Junior and Senior
    High School
Mount Pleasant, PA

Kerri L. Mahan
Middle School Math Teacher
Mount Vernon Middle School
Mount Vernon, OH

William McQuay
K–12 Director of Mathematics
Burnt Hills- Ballston Lake Central
    School District
Burnt Hills, NY

C. Vincent Pané, Ed.D.
Chair-Mathematics and Computer
    Studies
Molloy College
Rockville Centre, NY

Cheryl Peeples
Math Teacher
Tipp Middle School
Tipp City, OH

Steven M. Proehl, NBCT
Math Consultant
Chillicothe City Schools
Chillicothe, OH

Dr. Susan A. Smith
Associate Professor
Molloy College
Rockville Centre, NY

Debra S. Strayer
Middle School Mathematics
Western-Reserve Middle School
Collins, OH

Rebecca W. Sutton
Math Coach
West Jr. High
West Memphis, AR

David Thompson, M.A.
Supervisor of Mathematics
Egg Harbor Township School District
Egg Harbor Township, NJ

Kathleen M. Tucci
Math Coach
New Brighton Area School District
New Brighton, PA

William F. Wales
Director of Mathematics
Niskayuna Central Schools
Niskayuna, NY

Contents

Start Smart

Get ConnectED

connectED.mcgraw-hill.com

Investigate ▷

Learn ▷

Practice ▷

Every chapter and every lesson has a wealth of interactive learning opportunities.

CHAPTER 1
Rational Numbers and Percent

CHAPTER
2 Real Numbers and Monomials

Additional Lessons
Use Lesson 1 **Extend: Scientific Notation using Technology** after Lesson 2-2C.

Contents

connectED.mcgraw-hill.com

Investigate ▷

Learn ▷

Practice ▷

Every chapter and every lesson has a wealth of interactive learning opportunities.

CHAPTER 3 Equations and Inequalities

Additional Lessons

Use Lesson 3 **Solve Equations with Rational Coefficients** after Lesson 3-1B.

CHAPTER 4

Multi-Step Equations and Inequalities

Additional Lessons
Use Lesson 2 **Solve Multi-Step Equations** instead of Lesson 4-2C.

connectED.mcgraw-hill.com

Investigate ▷

Learn ▷

Practice ▷

Every chapter and every lesson has a wealth of interactive learning opportunities.

CHAPTER
5 Expressions and Functions

Additional Lessons

Use Lesson 11 **Extend: Linear and Nonlinear Association** after Lesson 5-4A.

Use Lesson 12 **Two-Way Tables** after Lesson 5-2A.

CHAPTER
6
Linear Functions and Systems of Equations

Additional Lessons

Use Lesson 4 **Extend: Investigating Linear Equations** after Lesson 6-1C.

Use Lesson 5 **Compare Properties of Functions** after Lesson 6-2A.

Use Lesson 6 **Construct Functions** after Lesson 6-1A.

Use Lesson 7 **Qualitative Graphs** after Lesson 6-2C.

Contents

 Get ConnectED

CHAPTER 7 Two- and Three-Dimensional Geometry

connectED.mcgraw-hill.com

Investigate ▷

Learn ▷

Practice ▷

Every chapter and every lesson has a wealth of interactive learning opportunities.

CHAPTER
8 Triangles and Transformations

Additional Lessons

Use Lesson 8 **Congruence and Transformations** after Lesson 8-3B.

Use Lesson 9 **Similarity and Transformations** after Lesson 8-3E.

Use Lesson 10 **Extend: Proofs About the Pythagorean Theorem** after Lesson 8-2B.

Contents

CHAPTER 9 Units of Measure

connectED.mcgraw-hill.com

Investigate ▷

Learn ▷

Practice ▷

Every chapter and every lesson has a wealth of interactive learning opportunities.

CHAPTER
10 Data Analysis and Statistics

ⓘ Contents

connectED.mcgraw-hill.com

Investigate ▷

Learn ▷

Practice ▷

Every chapter and every lesson has a wealth of interactive learning opportunities.

CHAPTER 11 Probability and Combinations

CHAPTER 12 Area and Volume

connectED.mcgraw-hill.com

Investigate ▷

Learn ▷

Practice ▷

Every chapter and every lesson has a wealth of interactive learning opportunities.

Optional Projects

Problem-Solving Projects

Student Handbook

Start Smart

connectED.mcgraw-hill.com

Investigate

 Animations

 Vocabulary

 Multilingual eGlossary

Learn

 Personal Tutor

 Virtual Manipulatives

 Graphing Calculator

 Audio

 Foldables

Practice

 Self-Check Practice

 Worksheets

 Assessment

Are You Ready?

Here are some characters you are going to meet as you move through the book.

Danielle
"Jewelry is my specialty. I love making gifts for my friends. One day, I want to have my own shop."

Brian
"I'm an electronics buff. Everyone comes to me when they have a question about equipment they need answered."

Sarah
"I love learning about foreign places and things. My goal is to visit every continent."

Dion

"My favorite subject is history. I love learning about where we come from and where we are headed."

Hana

"I'm the class clown. I enjoy making people laugh."

Mandar

"Someday I want to have my art displayed in a museum, where everyone can enjoy it."

Jacob

"I am a star on the baseball diamond. Some day, I want to play in the major leagues."

Alma

"One of my hobbies is scrapbooking. I love making fun pages with my photographs."

Roberto

"Some people think I'm a bookworm. I really just love a good story."

Carmen

"I am your typical All-American girl. I love sports, food, and shopping."

Enrique

"I have a huge DVD collection. I could watch a different movie every day of the year."

Jasmine

"I love the water: swimming, diving, skiing—you name it, I've tried it!"

Let's Get Started!

We're going to review a little before you begin Chapter 1.

A Plan for Problem Solving

Log Flumes ▶
The first log flume ride opened in 1963 at Six Flags Over Texas.

RIDES Log flumes are popular attractions at amusement parks. They offer a way to cool off on a hot day. Suppose a log flume ride requires 1 ticket to ride. If 12 tickets cost $18, how much does it cost to ride the log flume ride three times?

In mathematics, you can always use the *four-step plan* to solve a problem.

Understand	• What do you know and what do you need to find?
	• Do you have all the information you need?
	• Is there too much information?
Plan	• Visualize the problem and select a strategy for solving it.
	• Estimate what you think the answer should be.
Solve	• Solve the problem by carrying out your plan.
	• If your plan doesn't work, try another.
Check	• Does your answer fit the facts in the problem?
	• Compare your answer to your estimate.
	• If the answer is not reasonable, make a new plan and start again.

REAL-WORLD EXAMPLE Use the Four-Step Plan

1 **RIDES** A log flume ride requires 1 ticket to ride. If 12 tickets cost $18, how much does it cost to ride the log flume ride three times?

QUICK Review

Unit Rate
A unit rate is a rate with a denominator of 1 unit. Some common unit rates are shown below.
• miles per hour or mph
• price per pound or dollars/lb
• dollars per hour or dollars/h

Understand   You know that 12 tickets cost $18. You need to find the cost per ticket, or the unit rate. Then, multiply the unit rate by three to find how much it costs to ride the log flume ride three times.

Plan   Write the rate as a fraction and then simplify.

Solve

$18 for 12 tickets $= \dfrac{\$18}{12 \text{ tickets}}$ Write the rate as a fraction.

$= \dfrac{\$18 \div 12}{12 \text{ tickets} \div 12}$ Divide the numerator and the denominator by 12.

$= \dfrac{\$1.50}{1 \text{ ticket}}$ Simplify.

It costs $1.50 per ticket. So, it costs $1.50 to ride the log flume ride once. To ride the log flume ride three times, it costs 3 × $1.50 or $4.50.

Study Tip

Check for Reasonableness Always check to be sure your answer is reasonable. If the answer seems unreasonable, solve the problem again.

Check   Check by multiplying. Each ticket costs $1.50. So, 12 tickets cost 12 × $1.50 or $18. So, $4.50 is correct. ✓

 CHECK Your Progress

a. INTERNET A new Web site placed an ad in a newspaper. The table shows the number of visitors to the site, rounded to the nearest thousand, on each of the next five days. If this pattern continues, about how many visitors should the Web site receive on Day 8?

Day	Visitors
1	15,000
2	30,000
3	60,000
4	120,000
5	240,000

b. ROLLER COASTERS The table shows five roller coasters ranked by their longest continuous drop over the course of the ride. Use the formula *distance ÷ rate = time* to find the time it takes to travel the drop. Then, rank the roller coasters by the time.

Rank	Roller Coaster	Drop (ft)	Rate (ft/s)	Time (s)
1	Kingda Ka	418	188	
2	Top Thrill Dragster	400	176	
3	Superman the Escape	$328\frac{1}{12}$	147	
4	Millennium Force	300	136	
5	Goliath	255	125	

Use the four-step plan to solve each problem.

1. **PATTERNS** How many blue squares are needed for the fifth figure in this pattern?

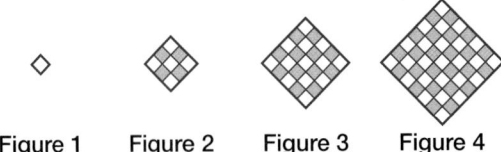

Figure 1   Figure 2   Figure 3   Figure 4

2. **PATTERNS** How many toothpicks are needed to make the 10th figure in the pattern below?

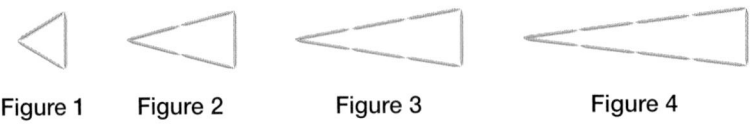

Figure 1   Figure 2   Figure 3   Figure 4

3. **POSTERS** The Math Club purchased 75 posters for $281.25. At this rate, how much would 100 posters cost?

4. **HOCKEY** Jaromir Jagr led the NHL in scoring for five years. The number of points he scored those years is shown in the table.

   a. About how many more points did he score in the last two years than in the first two years?

   b. Estimate the total number of points Jagr scored in those five years.

Year	Points Scored
1995	70
1998	102
1999	127
2000	96
2001	121

Real-World Link
Wayne Gretzky holds the all-time record for career goals scored at 2,857. This is almost 1,000 more goals than the person in second place.

5. **FUEL MILEAGE** A test of a hybrid car resulted in 4,840 miles driven using 88 gallons of gas. At this rate, how many gallons of gas will this vehicle need to travel 1,155 miles?

6. **CLASS TRIP** All of Mr. Bassett's science classes are going to the Natural History Museum. A tour guide is needed for each group of eight students. His classes have 28 students, 35 students, 22 students, 33 students, and 22 students. How many tour guides are needed?

7. **POPULATION** In a recent year, the population of Delaware was 783,600. Its land area is 1,953 square miles. North Carolina's population that year was 8,049,313. Its land area is 48,710 square miles. Which state had a greater number of people per square mile? by how much? Round to the nearest whole number.

DELAWARE

NORTH CAROLINA

Add and Subtract Integers

Digital Music Players
The first digital music player was produced in 1997 and had a storage capacity of 16 megabytes.

Number of Songs Downloaded Per Day

	4	
2		3
Mon	Wed	Fri

MUSIC The graphic above shows the number of songs Felipe downloaded to his MP3 player. It cost $1 for each song. He borrowed money from his brother to pay for the songs. How much money does Felipe owe his brother?

EXAMPLE Add Integers with the Same Sign

1) Find $(-2) + (-4) + (-3)$.

Use a number line.

Step 1 Start at zero.

Step 2 Move 2 units left.

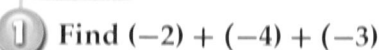

Step 3 From there, move 4 units left and then 3 more units left.

So, $(-2) + (-4) + (-3) = -9$.

Felipe owes his brother $9.

CHECK Your Progress

Add. Use a number line if necessary.

a. $-3 + (-2)$ **b.** $1 + 5$ **c.** $-5 + (-4)$

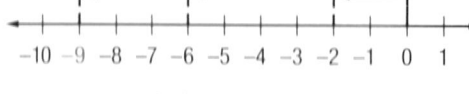

Key Concept · Add Integers with the Same Sign

Words To add integers with the same sign, add their absolute values. The sum has the same sign as the integers.

Examples $5 + 4 = 9$ $\qquad$ $-7 + (-3) = -10$

 EXAMPLE Add Integers with Different Signs

2 **Find $5 + (-2)$.**
Use a number line.

Step 1 Start at zero.

Step 2 Move 5 units right.

Step 3 From there, move 2 units left.

So, $5 + (-2) = 3$.

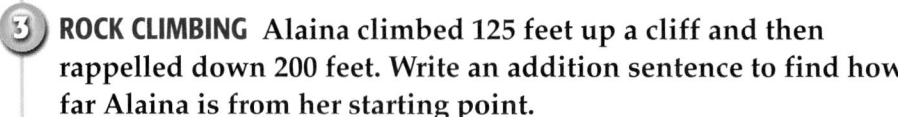

> **Study Tip**
>
> **Adding Integers on a Number Line** Always start at zero. Move right to model a positive integer and left to model a negative integer.

CHECK Your Progress

d. $7 + (-5)$ $\qquad$ **e.** $-6 + 4$ $\qquad$ **f.** $-1 + 8$

Key Concept · Add Integers with Different Signs

Words To add integers with different signs, subtract their absolute values. The sum has the same sign as the integer with the greater absolute value.

Examples $8 + (-3) = 5$ $\qquad$ $-8 + 3 = -5$

 REAL-WORLD EXAMPLE

3 **ROCK CLIMBING** Alaina climbed 125 feet up a cliff and then rappelled down 200 feet. Write an addition sentence to find how far Alaina is from her starting point.

$125 + (-200) = -75$ $\quad$ To find $125 + (-200)$, subtract $|125|$ from $|200|$. The sum is negative because $|-200| > |125|$.

So, Alaina was -75 feet or 75 feet below her starting point.

CHECK Your Progress

g. FINANCIAL LITERACY A checking account has a starting balance of $130. Write an addition sentence to find the balance after writing a check for $58.

You can also subtract integers.

Key Concept — Subtract Integers

Words	To subtract an integer, add its opposite, or additive inverse.
Examples	**Numbers** **Algebra**
	$4 - 7 = 4 + (-7)$ or -3 $a - b = a + (-b)$

 EXAMPLES Subtract Integers

Subtract.

④ $-6 - 8$

$$-6 - 8 = -6 + (-8) \qquad \text{To subtract 8, add } -8.$$
$$= -14 \qquad \text{Add.}$$

⑤ $7 - (-15)$

$$7 - (-15) = 7 + 15 \text{ or } 22 \qquad \text{To subtract } -15, \text{ add } 15.$$

✓ CHECK Your Progress

h. $-5 - 4$ **i.** $-5 - (-19)$ **j.** $-14 - (-2)$

 REAL-WORLD EXAMPLE

⑥ GOLF The table shows Paula Creamer's score for each round of a recent golf tournament. What was the difference between Paula Creamer's score in her best round and her worst round?

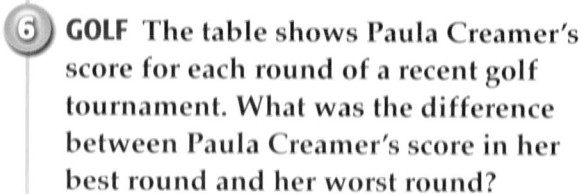

Paula Creamer's Scores	
Round	**Score**
1	−11
2	−6
3	−1
4	+2

In golf, the lowest number is the best score. So, Paula Creamer's best score was −11, and her worst score was +2. Find the difference between these scores.

$$2 - (-11) = 2 + 11 \text{ or } 13 \quad \text{To subtract } -11, \text{ add } 11.$$

The difference in her scores is 13.

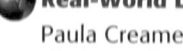 **Real-World Link** · · · · ·
Paula Creamer started playing golf when she was 10 years old. As of June 2009, she was ranked the number three female golfer in the world.

✓ CHECK Your Progress

k. WEATHER The highest recorded temperature in Virginia is 110°F. The lowest recorded temperature in Virginia is −30°F. Find the difference in temperatures.

Add.

1. $14 + 8$
2. $12 + 17$
3. $-14 + (-6)$
4. $-4 + (-5)$
5. $-18 + (-8)$
6. $-3 + (-12)$
7. $-21 + (-13)$
8. $-5 + (-31)$
9. $-7 + (-24)$
10. $10 + (-6)$
11. $7 + (-18)$
12. $-9 + 16$
13. $20 + (-5)$
14. $45 + (-4)$
15. $-15 + 8$
16. $-19 + 2$
17. $-10 + 34$
18. $-17 + 28$

Write an addition expression to describe each situation. Then find each sum and explain its meaning.

19. **AIRPLANES** A pilot descended 350 feet to avoid a thunderstorm, then ascended 400 feet.

20. **MONEY** Gregg earned $27 mowing lawns, then spent $15 on a movie.

21. **STOCK** A stock cost $46 on Wednesday. The price changed −$3 on Thursday and +$5 on Friday. What was the stock worth at the end of the day on Friday?

22. **TIDES** The table shows the change in water levels in feet due to tides.

    a. What is the water level for each day after low tide?

Day	Beginning Water Level	Change Due to Low Tide	Change Due to High Tide
Mon.	20	−22	+24
Tues.	22	−25	+23
Wed.	21	−23	+20

    b. On what day(s) was the final water level the lowest?

Subtract.

23. $8 - 13$
24. $5 - 24$
25. $-4 - 10$
26. $-6 - 3$
27. $7 - (-3)$
28. $2 - (-8)$
29. $-2 - (-6)$
30. $-18 - (-7)$
31. $14 - 8$
32. $17 - 12$
33. $5 - 9$
34. $1 - 8$
35. $-16 - 4$
36. $-15 - 12$
37. $-3 - 14$
38. $-6 - 13$
39. $9 - (-5)$
40. $10 - (-2)$

41. **ELEVATIONS** Find the difference in elevation between the mountain and the valley.

42. **PLANETS** The temperature on Mars can range from 70°F to −225°F. Find the difference in temperatures.

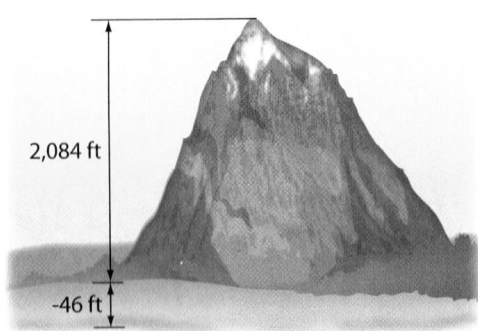

2,084 ft

-46 ft

Multiply and Divide Integers

Air Travel
Commercial air travel began in 1909 with zeppelin service in Germany.

A380
Greener. Cleaner.

PLANES An airplane descends 500 feet each minute to reach the runway. What is the plane's change in elevation after 3 minutes?

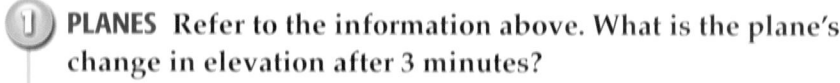 **REAL-WORLD EXAMPLE** **Multiply Integers**

1 **PLANES** Refer to the information above. What is the plane's change in elevation after 3 minutes?

The change in elevation can be expressed as 3(−500). Since multiplication is repeated addition, 3(−500) means that −500 is used as an addend 3 times.

$$3(-500) = -500 + (-500) + (-500)$$
$$= -1,500$$

So, the change in elevation is −1,500 feet.

 CHECK Your Progress

Multiply.

a. 5(−3) **b.** −8(6) **c.** −2(4)

EXAMPLE Multiply Integers with the Same Sign

2 Find $-4(-3)$.

$-4(-3) = 12$ $\qquad$ The factors have the same sign. The product is positive.

 CHECK Your Progress

Multiply.

d. $-3(-7)$ $\qquad$ **e.** $6(4)$ $\qquad$ **f.** $(-5)(-5)$

Read Math

Division In a division sentence like $12 \div 3 = 4$, the number you are dividing, 12, is called the *dividend*. The number you are dividing by, 3, is called the *divisor*. The result is called the *quotient*.

The rules for dividing integers are similar to the rules for multiplying integers.

 EXAMPLES Divide Integers

3 Find $-24 \div 3$.

$-24 \div 3 = -8$ $\qquad$ The signs are different. The quotient is negative.

4 Find $\dfrac{-30}{-15}$.

$\dfrac{-30}{-15} = 2$ $\qquad$ The signs are the same. The quotient is positive.

 CHECK Your Progress

Divide.

g. $-28 \div (-7)$ $\qquad$ **h.** $\dfrac{36}{-2}$ $\qquad$ **i.** $\dfrac{-40}{8}$

Multiply.

1. $7(-8)$ **2.** $8(-9)$ **3.** $-5 \cdot 8$ **4.** $-12 \cdot 7$

5. $-4(9)$ **6.** $-6(8)$ **7.** $-4(-6)$ **8.** $-14(-2)$

9. COOKING The boiling point of water at sea level is 212°F. For every 500 feet above sea level, the boiling point decreases by about 1°F. Find the boiling point of water at an elevation of 2,500 feet above sea level.

10. HEALTH The average person loses 50 to 80 hairs per day to make way for new growth. If you lost 65 hairs per day for 15 days without regrowing any, what would be the change in the number of hairs you have?

11. FARMING The level of a pond receded at the rate of 4 centimeters per day. Find the change in the water level of the pond after 7 days.

Divide.

12. $50 \div (-5)$ **13.** $-60 \div 3$ **14.** $45 \div 9$ **15.** $-34 \div (-2)$

16. $\dfrac{-84}{4}$ **17.** $\dfrac{28}{-7}$ **18.** $\dfrac{-72}{-6}$ **19.** $\dfrac{64}{8}$

20. SWIMMING With the drain open, a pool loses water at a rate of 9 gallons per minute. At that rate, how long will it take to drain 486 gallons of water?

21. AVIATION A weather research airplane began descending from an altitude of 36,000 feet above its base at a rate of 125 feet per minute. How long did it take for the plane to land at its base?

22. MOVIES Predict the number of theater admissions in 2013 if the average change per year following 2007 remains the same as the average change per year from 2005 to 2007. Justify your answer.

U.S. Theater Admissions	
Year	Number of Admissions (millions)
2005	1,415
2007	1,470

Real-World Link
Typically, around 600–700 movies are released in theaters each year.

23. FIND THE ERROR Alma is finding $(-2)(-3)(-4)$. Find her mistake and correct it.

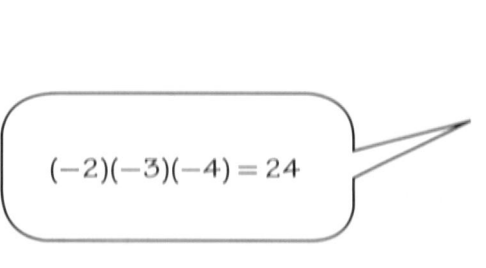

$(-2)(-3)(-4) = 24$

Proportions

Carousels ▶
Binghamton, New York, is considered the Carousel Capital of the World because six original carousels are located in the greater Binghamton area.

CAROUSELS A carousel makes 4 complete turns after 64 seconds. At this rate, how long will it take the carousel to make 10 complete turns?

You can use a proportion to solve the carousel problem.

Key Concept · Proportion

Words A *proportion* is an equation stating that two ratios or rates are equivalent.

Symbols

	Numbers	Algebra
	$\dfrac{1}{2} = \dfrac{3}{6}, \dfrac{8\text{ ft}}{10\text{ s}} = \dfrac{4\text{ ft}}{5\text{ s}}$	$\dfrac{a}{b} = \dfrac{c}{d}$, where $b, d \neq 0$

Consider the following proportion.

$$\frac{a}{b} = \frac{c}{d}$$

$$\frac{a}{\cancel{b}} \cdot \cancel{b}d = \frac{c}{\cancel{d}} \cdot b\cancel{d} \qquad \text{Multiply each side by } bd.$$

$$ad = bc \qquad \text{Simplify.}$$

The products ad and bc are called the cross products of this proportion. The cross products of any proportion are equal. You can use cross products to solve proportions in which one of the quantities is not known.

 **EXAMPLE** Solve a Proportion

① **CAROUSELS** A carousel makes 4 complete turns after 64 seconds. At this rate, how long will it take the carousel to make 10 complete turns?

Write a proportion. Let t represent the time in seconds.

$$\text{turns} \longrightarrow \quad \frac{4}{64} = \frac{10}{t} \quad \longleftarrow \text{turns}$$
$$\text{time} \longrightarrow \qquad\qquad\qquad \longleftarrow \text{time}$$

$\dfrac{4}{64} = \dfrac{10}{t}$ Write a proportion.

$4 \cdot t = 64 \cdot 10$ Find the cross products.

$4t = 640$ Multiply.

$\dfrac{4t}{4} = \dfrac{640}{4}$ Divide each side by 4.

$t = 160$ Simplify.

So, it will take 160 seconds or 2 minutes 40 seconds for the carousel to make 10 complete turns.

 CHECK Your Progress

a. SEWING Melinda bought 16 yards of fabric for $12. At this rate, how much will it cost to buy 24 yards of fabric?

Study Tip

Mental Math
Some proportions can be solved using mental math.

$\dfrac{2.5}{10} = \dfrac{x}{30}$

$\times 3$

$\dfrac{2.5}{10} = \dfrac{7.5}{30}$

$\times 3$

 REAL-WORLD EXAMPLE

② **MEDICINE** For every 18 people who have a sore throat, there are 2 people who actually have strep throat. If 72 patients have sore throats, how many of these would you expect to have strep throat?

Let s represent strep throat.

$\dfrac{2 \text{ strep throats}}{18 \text{ sore throats}} = \dfrac{s}{72 \text{ sore throats}}$ Write a proportion.

$2 \cdot 72 = 18 \cdot s$ Find the cross products.

$144 = 18s$ Multiply.

$8 = s$ Divide each side by 18.

So, if 72 people have sore throats, you would expect 8 people to have strep throat.

 CHECK Your Progress

b. RUNNING Salvador can run 120 meters in 24 seconds. At this rate, how many seconds will it take him to run a 300-meter race?

Solve each proportion.

1. $\dfrac{3}{8} = \dfrac{b}{40}$

2. $\dfrac{x}{12} = \dfrac{12}{4}$

3. $\dfrac{c}{7} = \dfrac{18}{42}$

4. $\dfrac{5}{k} = \dfrac{10}{22}$

5. $\dfrac{3}{8} = \dfrac{n}{4}$

6. $\dfrac{15}{4} = \dfrac{3}{g}$

7. $\dfrac{45}{5} = \dfrac{d}{7}$

8. $\dfrac{30}{a} = \dfrac{8}{20}$

9. $\dfrac{1.6}{m} = \dfrac{2}{3}$

10. $\dfrac{12.4}{8} = \dfrac{t}{10}$

11. $\dfrac{2.5}{4.5} = \dfrac{7.5}{x}$

12. $\dfrac{3.8}{5.2} = \dfrac{7.6}{z}$

13. **GROCERIES** Orange juice is on sale at 3 half-gallons for $5. At this rate, find the cost of 5 half-gallons of orange juice to the nearest cent.

3 half-gallons for $5.00

14. **TRAVEL** Franco drove 203 miles in 3.5 hours. At this rate, how long will it take him to drive another 29 miles to the next town?

15. **SCHOOL** If 4 notebooks weigh 2.8 pounds, how much do 6 of the same notebook weigh?

16. **COOKING** The table shows some cooking measurement equivalents. How many teaspoons are in 1.5 tablespoons?

Cooking Measurement Equivalents
16 tablespoons = 1 cup
2 tablespoons = $\dfrac{1}{8}$ cup
48 teaspoons = 1 cup
3 teaspoons = 1 tablespoon

17. **SCIENCE** The ratio of salt to water in a certain solution is 4 to 15. If the solution contains 6 ounces of water, how many ounces of salt does it contain?

18. **CONCERTS** Serefina purchased 7 tickets for herself and her friends to a concert and paid $164.50. The total cost of tickets to the concert is proportional to the number purchased. How many tickets to the same concert would Serefina have purchased if she paid a total of $94?

19. **MOVIES** After 30 seconds, 720 frames of film have passed through a movie projector. At this rate, what is the approximate running time in minutes of a movie made up of 57,000 frames of film?

Real-World Link Film for a large format projection system passes through the projector at the rate of 330 feet per minute or 5.5 feet per second.

20. **SCHOOL** There are 325 students and 13 teachers at a school. Next school year, the enrollment is expected to increase by 100 students. Write and solve a proportion to find the number of teachers that must be hired so the student-teacher ratio remains the same.

START SMART 5

Perimeter and Area

Football ▶
The end zones of a college football field are 30 feet deep.

FOOTBALL The dimensions of a college football field are 360 feet by 160 feet. What is the perimeter of a college football field?

Perimeter is the distance around a figure. It is measured in linear units. The measure of the surface enclosed by a figure is its area. Area is measured in square units.

Shape	Formula	Model
Rectangle	$P = 2(\ell + w)$ or $P = 2\ell + 2w$ $A = \ell w$	
Square	$P = 4s$ $A = s^2$	
Parallelogram	$P = 2(a + b)$ or $P = 2a + 2b$ $A = bh$	
Triangle	$P = a + b + c$ $A = \frac{1}{2}bh$	

1 **FOOTBALL** The dimensions of a college football field are 360 feet by 160 feet. What is the perimeter of a college football field?

Use the formula for the perimeter of a rectangle.

$P = 2\ell + 2w$	Perimeter of a rectangle
$P = 2 \cdot 360 + 2 \cdot 160$	Replace ℓ with 360 and w with 160.
$P = 720 + 320$	Multiply.
$P = 1,040$	Add.

The perimeter of a college football field is 1,040 feet.

 CHECK Your Progress

a. **GARDENS** A square garden has a side length of 4 meters. What is the perimeter of the garden?

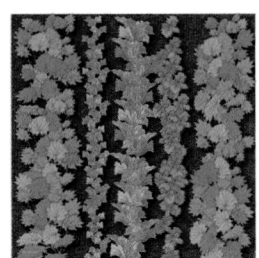

4 m

 EXAMPLES Area

2 **Find the area of the parallelogram.**

$A = bh$	Area of a parallelogram
$A = 12 \cdot 4.5$	Replace b with 12 and h with 4.5.
$A = 54$	Multiply.

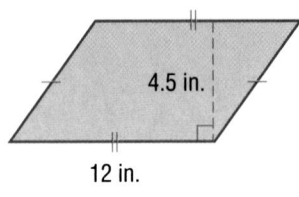

4.5 in.

12 in.

The area of the parallelogram is 54 square inches.

3 **Find the area of the triangle.**

$A = \frac{1}{2}bh$	Area of a triangle
$A = \frac{1}{2} \cdot 9.6 \cdot 5$	Replace b with 9.6 and h with 5.
$A = 24$	Multiply.

5 m

9.6 m

The area of the triangle is 24 square meters.

 CHECK Your Progress

Find the area of each figure.

b.

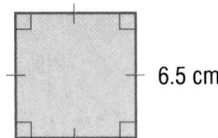

6.5 cm

c.

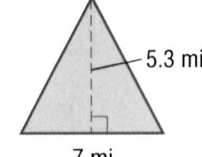

5.3 mi

7 mi

Find the perimeter and area of each figure.

1. 7 ft
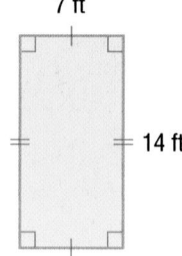
14 ft

2. 3.2 yd
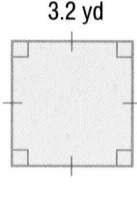

3. 13.3 m
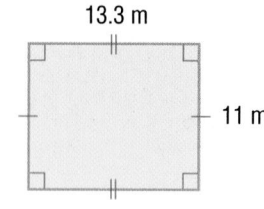
11 m

4.
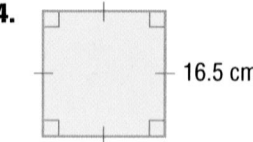
16.5 cm

5. 6 mi
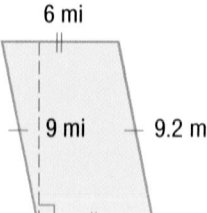
9 mi   9.2 mi

6. 7.5 in.
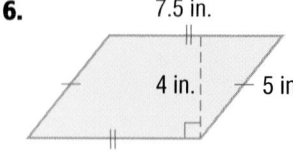
4 in.   5 in.

7.
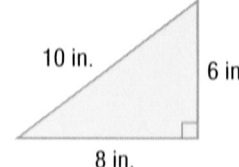
10 in.   6 in.
8 in.

8.
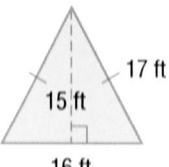
17 ft
15 ft
16 ft

9. 12 m
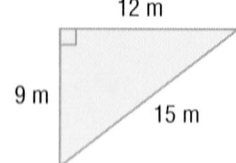
9 m
15 m

10. SAILS A sailboat has a triangular sail with a height of 30 feet and a base of 20.5 feet. What is the area of the sail?

11. HOBBIES Aileen sewed a lace border around a quilt. She used 16 feet of lace. Give two possible dimensions for the quilt.

Real-World Link
The fastest recorded sailboat speed is 38.5 miles per hour.

12. SPORTS The table gives the shape and dimension of the playing field for several sports.

Sport	Shape of Field	Dimensions (ft)
Baseball	Square	90 on each side
Field Hockey	Rectangle	100 by 60
Soccer	Rectangle	360 by 240
Lacrosse	Rectangle	330* by 180

*not including the end zones

a. Which field has the greater perimeter, a regulation soccer field or a lacrosse field?

b. What is the difference between the area of a field hockey field and a regulation soccer field?

c. Is the area of a baseball diamond or a field hockey field greater?

Stem-and-Leaf Plots

Animals ▶
It is believed that the longest living mammal was a bowhead whale that was estimated to have lived for over 200 years.

ANIMALS The chimpanzee has an average life span of about 50 years. How does this compare to other mammals?

A stem-and-leaf plot is used to organize large data sets in a small amount of space. In a stem-and-leaf plot, numerical data are listed in ascending or descending order. The digits in the greatest common place value of the data are used for the *stems*. The digits in the next greatest place value form the *leaves*.

 REAL-WORLD EXAMPLE ▶ **Draw a Stem-and-Leaf Plot**

 ANIMALS The table shows the average life span of several mammals. Display the data in stem-and-leaf plot.

Step 1 Find the least and the greatest number. Then identify the greatest place value digit in each number.

- The least number, 12, has 1 in the tens place.

- The greatest number, 70, has 7 in the tens place.

Average Life Span	
Animal	**Years**
Asian Elephant	40
African Elephant	35
Cat	15
Chimpanzee	50
Cow	15
Dog	13
Gray Whale	70
Hippopotamus	41
Horse	20
Lion	15
Moose	12

Step 2 Draw a vertical line and write the stems from 1 to 7 to the left of the line.

Step 3 Write the leaves to the right of the corresponding stem on the *other* side of the line. For example, for 13, write 3 to the right of 1. Arrange the leaves so they are ordered from least to greatest. Repeat a leaf as often as it occurs. Then include a title and a key.

Average Life Span

Stem	Leaf
1	2 3 5 5 5
2	0
3	5
4	0 1
5	0
6	
7	0

3 | 5 = 35 years

 CHECK Your Progress

a. The number of minutes Caryn spent tutoring for two weeks are shown. Display the data in a stem-and-leaf plot.

Tutoring Time (min)				
73	37	42	85	56
67	45	67	50	49

Because you can see all the data values and the distribution or shape of the data set, stem-and-leaf plots are useful in analyzing data.

 REAL-WORLD EXAMPLE Interpret Data

2 PRESIDENTS The stem-and-leaf plot lists the ages of the U.S. Presidents at the time of their first inaugurations.

Age at Inauguration

Stem	Leaf
4	2 3 6 6 7 7 8 9 9
5	0 1 1 1 1 1 2 2 4 4 4 4 4 5 5 5 5 6 6 6 7 7 7 7 8
6	0 1 1 1 2 4 4 5 8 9

5 | 0 = 50 years

Based on the data, what inference can be made about the ages of the U.S. Presidents at their first inaugurations?

- Most of the data occur in the 50–59 interval.
- The youngest age is 42. The oldest age is 69. The range is 27.
- The median age is 54.5.

 **CHECK Your Progress**

Refer to the stem-and-leaf plot in Example 1.

b. In which interval(s) do most of the ages occur?

c. What is the range of the data?

d. What is the median age?

Two sets of data can be compared using a back-to-back stem-and-leaf plot. The back-to-back stem-and-leaf plot below shows the scores of two basketball teams for the games in one season.

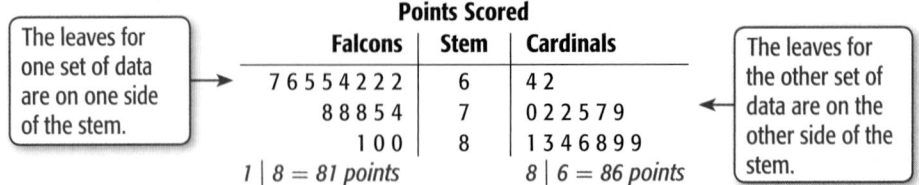

Points Scored

Falcons	Stem	Cardinals
7 6 5 5 4 2 2 2	6	4 2
8 8 8 5 4	7	0 2 2 5 7 9
1 0 0	8	1 3 4 6 8 9 9

$1 \mid 8 = 81$ points $8 \mid 6 = 86$ points

The leaves for one set of data are on one side of the stem. →

← The leaves for the other set of data are on the other side of the stem.

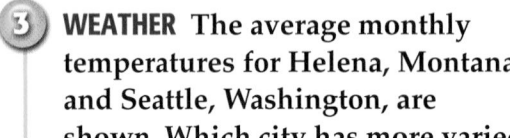 **REAL-WORLD EXAMPLE** **Compare Data**

3 **WEATHER** The average monthly temperatures for Helena, Montana, and Seattle, Washington, are shown. Which city has more varied temperatures? Explain.

The leaves for Helena are spread out while the leaves for Seattle are clustered. So, Helena has the more varied temperatures.

Average Monthly Temperatures

Helena, MT	Stem	Seattle, WA
6 1 0	2	
4 2	3	
5 3	4	1 2 4 6 7
5 3	5	0 4 6
9 7 2	6	1 1 5 6

$3 \mid 5 = 53°$ $6 \mid 1 = 61°$

 CHECK Your Progress

Use the test score data shown in the back-to-back stem-and-leaf plot.

e. Which class had higher test scores? Explain.

f. Which class had more varied test scores? Explain.

Test Scores

3rd Period	Stem	6th Period
9 9 7 4 0	7	3
9 9 5 6 2 1	8	1 2 5 6 6 8 9 9
2 2 2 1 1	9	0 2 2 3 3 3 5 6

$9 \mid 8 = 89\%$ $8 \mid 1 = 81\%$

Practice and Problem Solving

Display each set of data in a stem-and-leaf plot.

1.

Height of 8th Grade Girls (in.)					
67	63	59	61	63	57
58	70	57	59	55	60

2.

Average Weight of Dogs (lb)					
55	7	75	65	18	58
21	13	10	42	27	30

3.

Summer Paralympic Games Participating Countries										
Year	'68	'72	'76	'80	'84	'88	'92	'96	'00	'04	'08
Countries	29	44	42	42	42	61	82	103	128	136	145

4. BASEBALL Use the table that shows the National League home run leaders for several years.

a. What is the average number of home runs hit by a single season home run leader? Round to the nearest tenth.

b. Display the number of home runs in a stem-and-leaf plot.

c. What is the most home runs hit between 1995 and 2009?

d. How many of the season leaders hit fewer than 50 home runs?

e. What is the median number of home runs hit by a single season home run leader?

f. Write a sentence that describes the data.

National League Single Season Home Run Leaders, 1995–2009		
Year	Player	Home Runs
1995	Dante Bichette	40
1996	Andres Galarraga	47
1997	Larry Walker	49
1998	Mark McGwire	70
1999	Mark McGwire	65
2000	Sammy Sosa	50
2001	Barry Bonds	73
2002	Sammy Sosa	49
2003	Jim Thome	47
2004	Adrian Beltre	48
2005	Andruw Jones	51
2006	Ryan Howard	58
2007	Alex Rodriguez	54
2008	Ryan Howard	48
2009	Albert Pujols	47

5. BASKETBALL Use the information shown in the back-to-back stem-and-leaf plot.

a. What is the greatest number of games won by a Big Ten Conference team?

b. What is the least number of games won by a Big East Conference team?

c. How many teams are in the Big East Conference?

d. Compare the median number of games won by each conference.

NCAA Women's Basketball Statistics
Overall Games Won, 2008–2009

Big Ten Conference	Stem	Big East Conference
7	0	
9 1 0 0	1	0 4 7 7 7 9 9
9 5 2 1 1 0	2	0 1 2 3 5 7
	3	4 9

9 | 1 = 19 games   1 | 4 = 14 games

6. NUTRITION Use the information in the table about nutritional content of certain foods to construct a back-to-back stem-and-leaf plot. Then describe an advantage of showing the data in a stem-and-leaf plot instead of a list.

Amount of Protein in Foods (g)									
Dairy Products	26	2	10	7	8	7	6	2	8	9
Legumes, Nuts, and Seeds	5	6	9	14	15	18	39	14	27	9

7. PRESIDENTS The frequency table shows the data from Example 2. Compare and contrast the frequency table and the stem-and-leaf plot.

Age	Frequency
40–49	9
50–59	25
60–69	10

Rational Numbers and Percent

connectED.mcgraw-hill.com

Investigate

 Animations

 Vocabulary

 Multilingual eGlossary

Learn

 Personal Tutor

 Virtual Manipulatives

 Graphing Calculator

 Audio

 Foldables

Practice

 Self-Check Practice

 Worksheets

 Assessment

The ☆BIG Idea

How do the different ways in which rational numbers can be represented help solve real-world problems?

FOLDABLES®
Study Organizer

Make this Foldable to help you organize your notes.

Rational Numbers and Percent
1 Rational Numbers
2 Percents
3 Apply Percents
Vocabulary
Review

Review Vocabulary

percent por ciento a ratio that compares a number to 100

$25\% = 25$ out of 100 or $\dfrac{25}{100}$

Key Vocabulary

English	Español
percent equation	ecuación porcentual
percent of change	porcentaje de cambio
percent proportion	proporción porcentual
rational number	número racional

When Will I Use This?

Are You Ready for the Chapter?

You have two options for checking prerequisite skills for this chapter.

Text Option Take the Quick Check below. Refer to the Quick Review for help.

QUICK Check

Add, subtract, multiply, or divide.

1. $-13 + 4$
2. $28 + (-9)$
3. $-8 - 6$
4. $23 - (-15)$
5. $6(-14)$
6. $-3(-8)$
7. $36 \div (-4)$
8. $-86 \div (-2)$

9. **TEMPERATURE** The high temperature for Saturday was 13°F and the low temperature was −4°F. What was the difference between the high and low temperatures?

10. **FOOTBALL** During a scoring drive, a football team gained or lost yards on each play as shown. What was the average number of yards per play?

Yards Gained or Lost					
+6	−2	+8	0	+23	−4
+5	+12	−4	−3	+18	+1

Solve each proportion.

11. $\dfrac{x}{10} = \dfrac{3}{5}$
12. $\dfrac{4}{9} = \dfrac{14}{b}$
13. $\dfrac{12}{s} = \dfrac{30}{37}$
14. $\dfrac{8}{15} = \dfrac{m}{21}$
15. $\dfrac{d}{5} = \dfrac{18}{45}$
16. $\dfrac{3}{7} = \dfrac{21}{8}$

17. **RECIPES** Rueben's chocolate chip cookie recipe uses 2 eggs for 2 dozen cookies. How many eggs does Rueben need to make 72 cookies?

QUICK Review

EXAMPLE 1

Find $-27 + 13$.

$-27 + 13 = -14$

$|-27| - |13| = |14|$
The sum is negative because $|-27| > |13|$.

EXAMPLE 2

Find $-11 - 8$.

$$-11 - 8 = -11 + (-8)$$
$$= -19$$

To subtract 8, add −8.
$|-11| + |-8| = 19$
Both numbers are negative so the sum is negative.

EXAMPLE 3

Find $-12(7)$.

$-12(7) = -84$

The factors have different signs. The product is negative.

EXAMPLE 4

Solve $\dfrac{w}{12} = \dfrac{5}{6}$.

$$\dfrac{w}{12} = \dfrac{5}{6}$$ Write the proportion.

$6 \cdot w = 12 \cdot 5$ Find cross products.

$6w = 60$ Simplify.

$\dfrac{6w}{6} = \dfrac{60}{6}$ Divide each side by 6.

$w = 10$ Simplify.

 **Online Option** Take the Online Readiness Quiz.

Writing Math

Justify Your Answer

When you justify your answer, you give *reasons* why your answer is correct.

I know my answer is correct, but how do I *justify* it? Can you help?

DVDs The table shows the different plans an online movie rental company offers. Mariah wants to purchase the 2-DVDs-at-a-time plan. This month, the plans are advertised at $\frac{1}{4}$ off. If she has $10.00 to spend, does she have enough? Justify your answer.

Online DVD Rentals	
Plan	**Monthly Price ($)**
3 DVDs at-a-time	14.50
2 DVDs at-a-time	12.00
1 DVD at-a-time	8.00

➡ **Step 1**

Solve the problem.
Find the discount.

$\frac{1}{4}$ of $12 = 0.25 \times 12$ Write $\frac{1}{4}$ as a decimal.

$= 3.00$ The discount is $3.00.

Find the discounted price.
$12.00 - $3.00 = $9.00   Subtract the discount from the regular price.

➡ **Step 2**

Answer the question.
Mariah does have enough money.

➡ **Step 3**

Justify your answer. Always write complete sentences.
Mariah has enough money because $9.00 is less than $10.00.

Practice

1. **SHOPPING** You can buy 3 used CDs at The Music Shoppe for $12.99, or you can buy 5 of the same for $19.99 at Quality Sounds. Which is the better buy? Justify your answer.

2. **TEST** Lexi answered $\frac{3}{4}$ of the test questions correctly. There were 48 questions on the test. Did she get at least 35 questions correct? Justify your answer.

Main Idea

Express rational numbers as decimals and decimals as fractions.

Vocabulary

rational number
terminating decimal
repeating decimal

 Get ConnectED

 8.NS.1

Rational Numbers

MARINE LIFE There are over 360 different species of sharks. Some common shark species are listed below.

Shark Species	Color	Average Length (feet)
Sharpnose shark	brown to green-gray	3
Bonnethead shark	gray or gray-brown	3
Blacknose shark	green-gray	5
Blacktip shark	blue-gray	6
Spinner shark	gray-bronze	6
Sandbar shark	brown or gray	6
Nurse shark	yellow-brown	7
Scalloped hammerhead shark	gray-brown	8
Lemon shark	yellow-gray	9

1. What fraction of the shark species have an average length less than 6 feet?

2. What fraction of the shark species are a shade of blue?

3. What fraction of the shark species are not a shade of gray?

Rational numbers are numbers that can be written as the ratio of two integers, expressed as a fraction. Since -7 can be written as $\frac{-7}{1}$, $2\frac{2}{3}$ can be written as $\frac{8}{3}$, and 9% can be written as $\frac{9}{100}$, -7, $2\frac{2}{3}$, and 9% are rational numbers.

 Key Concept **Rational Numbers**

Words	A rational number is a number that can be written as the ratio of two integers in which the denominator is not zero.	Model
Algebra	$\frac{a}{b}$, where a and b are integers and $b \neq 0$	

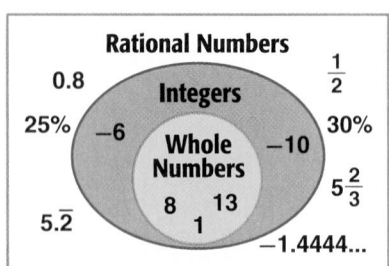

Study Tip

Bar Notation Bar notation is often used to indicate that a digit or group of digits repeats. The bar is placed above the repeating part. To write 8.636363... in bar notation, write 8.$\overline{63}$, not 8.$\overline{6}$ or 8.$\overline{636}$. To write 0.3444... in bar notation, write 0.3$\overline{4}$, not 0.$\overline{34}$.

Any fraction can be expressed as a decimal by dividing the numerator by the denominator. A **terminating decimal**, like 0.625, terminates because the division ends with a remainder of 0. If the division does not end, sometimes a pattern of digits repeats. A **repeating decimal**, like 0.$\overline{3}$, has a pattern in its digits that repeats without end.

 EXAMPLES Write a Fraction as a Decimal

Write each fraction or mixed number as a decimal.

① $\dfrac{5}{8}$

$\dfrac{5}{8}$ means $5 \div 8$.

$$\begin{array}{r} 0.625 \\ 8\overline{)5.000} \\ -48 \\ \hline 20 \\ -16 \\ \hline 40 \\ -40 \\ \hline 0 \end{array}$$ Divide 5 by 8.

② $-1\dfrac{2}{3}$

$-1\dfrac{2}{3}$ can be rewritten as $\dfrac{-5}{3}$.

Divide 5 by 3 and add a negative sign.

The mixed number $-1\dfrac{2}{3}$ can be written as $-1.\overline{6}$.

$$\begin{array}{r} 1.6... \\ 3\overline{)5.0} \\ -3 \\ \hline 20 \\ -18 \\ \hline 2 \end{array}$$

✓ CHECK Your Progress

a. $\dfrac{3}{4}$ **b.** $4\dfrac{13}{25}$ **c.** $-\dfrac{2}{9}$ **d.** $3\dfrac{1}{11}$

Repeating decimals often occur in real-world situations. However, they are usually rounded to a certain place-value position.

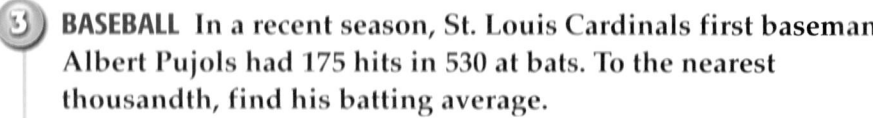 **REAL-WORLD EXAMPLE**

③ **BASEBALL** In a recent season, St. Louis Cardinals first baseman Albert Pujols had 175 hits in 530 at bats. To the nearest thousandth, find his batting average.

To find his batting average, divide the number of hits, 175, by the number of at bats, 530.

175 ⊕ 530 [ENTER] 0.3301886792

Look at the digit to the right of the thousandths place. Since 1 < 5, round down.

Albert Pujols's batting average was 0.330.

✓ CHECK Your Progress

e. **AUTO RACING** In a recent season, NASCAR driver Jimmie Johnson won 6 of the 36 total races held. To the nearest thousandth, find the part of races he won.

Terminating and repeating decimals are also rational numbers because you can write them as fractions.

 EXAMPLES Write Decimals as Fractions

4 **Write 0.45 as a fraction.**

$$0.45 = \frac{45}{100} \quad \text{0.45 is 45 hundredths.}$$

$$= \frac{9}{20} \quad \text{Simplify.}$$

5 **ALGEBRA** Write $0.\overline{5}$ as a fraction in simplest form.

Assign a variable to the value $0.\overline{5}$. Let $N = 0.555...$. Then perform operations on N to determine its fractional value.

$$N = 0.555...$$

$$10(N) = 10(0.555...) \quad \text{Multiply each side by 10 because 1 digit repeats.}$$

$$10N = 5.555... \quad \text{Multiplying by 10 moves the decimal point 1 place to the right.}$$

$$\underline{- N = 0.555...} \quad \text{Subtract } N = 0.555... \text{ to eliminate the repeating part.}$$

$$9N = 5 \quad \text{Simplify.}$$

$$N = \frac{5}{9} \quad \text{Divide each side by 9.}$$

The decimal $0.\overline{5}$ can be written as $\frac{5}{9}$.

CHECK Your Progress

Write each decimal as a fraction or mixed number in simplest form.

f. -0.14 **g.** 8.75 **h.** $0.\overline{27}$ **i.** $-0.\overline{4}$

✓ CHECK Your Understanding

Examples 1 and 2   Write each fraction or mixed number as a decimal.

1. $\frac{4}{5}$ **2.** $\frac{9}{16}$ **3.** $-1\frac{29}{40}$

4. $\frac{5}{9}$ **5.** $4\frac{5}{6}$ **6.** $-7\frac{5}{33}$

Example 3   **7. GOLF** In a recent year, Tiger Woods won 7 of the 16 tournaments he entered. To the nearest thousandth, find his winning average.

Examples 4 and 5   Write each decimal as a fraction or mixed number in simplest form.

8. 0.6 **9** 0.32 **10.** -1.55

11. $-0.\overline{5}$ **12.** $-3.\overline{8}$ **13.** $2.\overline{15}$

Practice and Problem Solving

= **Step-by-Step Solutions** begin on page R1.
Extra Practice begins on page EP2.

Examples 1 and 2 **Write each fraction or mixed number as a decimal.**

14. $\frac{1}{4}$ **15.** $\frac{2}{5}$ **16.** $\frac{7}{80}$ **17.** $\frac{33}{40}$

18. $-\frac{7}{16}$ **19.** $-\frac{5}{32}$ **20.** $2\frac{1}{8}$ **21.** $5\frac{5}{16}$

22. $\frac{4}{33}$ **23** $-\frac{6}{11}$ **24.** $-6\frac{13}{15}$ **25.** $-7\frac{8}{45}$

Example 3 **26. FAMILIES** The table shows statistics about the students at Carter Junior High.

    a. Express the fraction of students with no siblings as a decimal.

    b. Find the decimal equivalent for the number of students with three siblings.

    c. Write the fraction of students with one sibling as a decimal. Round to the nearest thousandth.

    d. Write the fraction of students with two siblings as a decimal. Round to the nearest thousandth.

Number of Siblings	Fraction of Students
None	$\frac{1}{15}$
One	$\frac{1}{3}$
Two	$\frac{5}{12}$
Three	$\frac{1}{6}$
Four or more	$\frac{1}{60}$

Examples 4 and 5 **Write each decimal as a fraction or mixed number in simplest form.**

27. -0.4 **28.** 0.5 **29.** 5.55 **30.** -7.32

31. $0.\overline{2}$ **32.** $-0.\overline{45}$ **33.** $-3.\overline{09}$ **34.** $2.\overline{7}$

WEATHER Write the rainfall amount for each day as a fraction or mixed number.

35 Friday

36. Saturday

37. Sunday

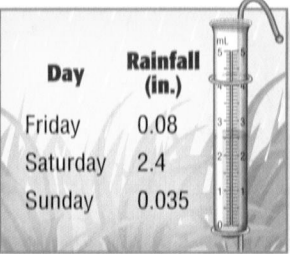

Day	Rainfall (in.)
Friday	0.08
Saturday	2.4
Sunday	0.035

MEASUREMENT Write the length of each insect as a fraction or mixed number and as a decimal.

38.

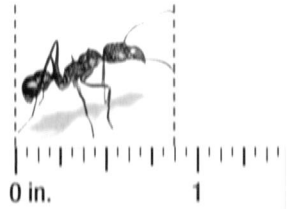

0 in.   1

39.

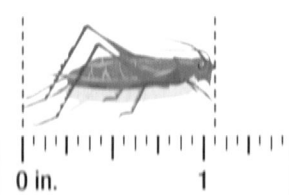

0 in.   1

40. FROZEN YOGURT The table shows three popular flavors according to the results of a survey. What is the decimal value of those who liked vanilla, chocolate, or strawberry? Round to the nearest hundredth.

Flavor	Fraction
Vanilla	$\frac{3}{10}$
Chocolate	$\frac{1}{11}$
Strawberry	$\frac{1}{18}$

41. **OPEN ENDED** Give an example of a repeating decimal where two digits repeat. Explain why your number is a rational number.

42. **Which One Doesn't Belong?** Identify the fraction that does not belong with the other three. Explain your reasoning.

$$\frac{1}{8} \qquad \frac{1}{4} \qquad \frac{1}{6} \qquad \frac{4}{5}$$

43. **CHALLENGE** Explain why any rational number is either a terminating or repeating decimal.

44. **WRITE MATH** Compare 0.1 and $0.\overline{1}$, 0.13 and $0.\overline{13}$, and 0.157 and $0.\overline{157}$ when written as fractions. Make a conjecture about expressing repeating decimals like these as fractions.

Test Practice

45. Which of the following is equivalent to the fraction below?

$$\frac{13}{5}$$

A. 2.4

B. 2.45

C. 2.55

D. 2.6

46. **EXTENDED RESPONSE** The table shows the number of free throws each player made during the last basketball season.

Player	Free Throws Made	Free Throws Attempted
Felisa	18	20
Morgan	13	24
Yasmine	15	22
Gail	10	14

Part A Write the fraction of free throws made in simplest form for each player.

Part B Write each fraction from *Part A* as a decimal. Round to the nearest thousandth if necessary.

Part C Which player has the greatest fraction of free throws made?

47. Which of the following is NOT an example of a rational number?

F. $\frac{-6}{11}$

G. 15

H. 18%

I. 4.23242526...

48. While shopping for a new pair of jeans, Janet notices the sign below.

Which of the following expressions can be used to estimate the total discount on a pair of jeans?

A. 0.033 × $30

B. 0.33 × $30

C. 1.3 × $30

D. 33.3 × $30

Main Idea
Add and subtract rational numbers.

 Vocabulary
like fractions
unlike fractions

Add and Subtract Rational Numbers

APPLES The amount of apples Oleta's family picked is shown.

Person	Amount Picked (baskets)
Oleta	$1\frac{1}{4}$
Mr. Davis	2
Mrs. Davis	$1\frac{3}{4}$
Alvin	$\frac{2}{4}$

1. What is the sum of the whole-number parts of the baskets of apples?

2. How many $\frac{1}{4}$ baskets are there?

3. Can you combine all of the apples into a bushel that holds five baskets? Explain.

Fractions that have the same denominators are called **like fractions**.

Key Concept — Add and Subtract Like Fractions

Words To add or subtract like fractions, add or subtract the numerators and write the result over the denominator.

Examples

Numbers	**Algebra**
$\frac{1}{5} + \frac{3}{5} = \frac{4}{5}$	$\frac{a}{c} + \frac{b}{c} = \frac{a+b}{c}$, where $c \neq 0$
$\frac{7}{8} - \frac{3}{8} = \frac{4}{8}$ or $\frac{1}{2}$	$\frac{a}{c} - \frac{b}{c} = \frac{a-b}{c}$, where $c \neq 0$

You can use the rules for adding integers to determine the sign of the sum of any two signed numbers.

 EXAMPLE Add Like Fractions

1. Find $\frac{5}{8} + \left(-\frac{7}{8}\right)$. Write in simplest form.

$$\frac{5}{8} + \left(-\frac{7}{8}\right) = \frac{5 + (-7)}{8}$$ ◄—Add the numerators.
◄—The denominators are the same.

$$= \frac{-2}{8} \text{ or } -\frac{1}{4}$$ Simplify.

 CHECK Your Progress

a. $\frac{5}{9} + \frac{7}{9}$ **b.** $-\frac{5}{9} + \frac{1}{9}$ **c.** $\frac{-1}{6} + \left(-\frac{5}{6}\right)$

 **EXAMPLE** **Subtract Like Fractions**

2 Find $-\dfrac{8}{9} - \dfrac{7}{9}$. Write in simplest form.

$$-\frac{8}{9} - \frac{7}{9} = -\frac{8}{9} + \left(-\frac{7}{9}\right)$$

$$= \frac{-8 + (-7)}{9} \qquad \text{Subtract the numerators by adding the opposite of 7.}$$

$$= \frac{-15}{9} \text{ or } -1\frac{2}{3} \qquad \text{Rename } \frac{-15}{9} \text{ as } -1\frac{6}{9} \text{ or } -1\frac{2}{3}.$$

CHECK Your Progress

d. $-\dfrac{4}{5} - \dfrac{3}{5}$ **e.** $\dfrac{3}{8} - \dfrac{5}{8}$ **f.** $\dfrac{5}{7} - \left(-\dfrac{4}{7}\right)$

Fractions with unlike denominators are called **unlike fractions**. To add or subtract unlike fractions, rename the fractions using prime factors to find the least common denominator. Then add or subtract as with like fractions.

 EXAMPLES **Add and Subtract Unlike Fractions**

Add or subtract. Write in simplest form.

3 $\dfrac{1}{4} + \left(-\dfrac{2}{3}\right)$

$$\frac{1}{4} + \left(-\frac{2}{3}\right) = \frac{1}{4} \cdot \frac{3}{3} + \left(-\frac{2}{3}\right) \cdot \frac{4}{4} \qquad \text{The LCD is } 3 \cdot 4 \text{ or } 12.$$

$$= \frac{3}{12} + \left(-\frac{8}{12}\right) \qquad \text{Rename using the LCD.}$$

$$= \frac{3 + (-8)}{12} \qquad \text{Add the numerators.}$$

$$= -\frac{5}{12} \qquad \text{Simplify.}$$

4 $-\dfrac{8}{9} - \left(-\dfrac{7}{15}\right)$

$$-\frac{8}{9} - \left(-\frac{7}{15}\right) = -\frac{8}{9} + \frac{7}{15} \qquad \text{To subtract } -\frac{7}{15}, \text{ add } \frac{7}{15}.$$

$$= -\frac{8}{9} \cdot \frac{5}{5} + \frac{7}{15} \cdot \frac{3}{3} \qquad \text{The LCD is } 9 \cdot 5 \text{ or } 45.$$

$$= -\frac{40}{45} + \frac{21}{45} \qquad \text{Rename using the LCD.}$$

$$= \frac{-40 + 21}{45} \qquad \text{Add the numerators.}$$

$$= -\frac{19}{45} \qquad \text{Simplify.}$$

CHECK Your Progress

g. $-\dfrac{5}{6} + \left(-\dfrac{1}{2}\right)$ **h.** $\dfrac{1}{14} - \dfrac{3}{4}$ **i.** $-\dfrac{5}{6} + \dfrac{3}{10}$

> **QUICK Review**
>
> **Least Common Denominator (LCD)**
>
> To find the LCD of two fractions, write the prime factorization of each denominator. Identify common prime factors.
>
> $9 = \boxed{3} \times \boxed{3}$
> $15 = \boxed{3} \times \boxed{5}$
>
> The LCD is the product of each common prime factor and any remaining factors.
>
> LCD = $3 \times 3 \times 5$ or 45

EXAMPLES Add and Subtract Mixed Numbers

Add or subtract. Write in simplest form.

5 $5\frac{7}{9} + 8\frac{4}{9}$

$$5\frac{7}{9} + 8\frac{4}{9} = (5 + 8) + \left(\frac{7}{9} + \frac{4}{9}\right)$$ Add the whole numbers and fractions separately.

$$= 13 + \frac{7 + 4}{9}$$ Add the numerators.

$$= 13\frac{11}{9} \text{ or } 14\frac{2}{9}$$ $\frac{11}{9} = 1\frac{2}{9}$

6 $2\frac{3}{4} - 3\frac{1}{3}$

$$2\frac{3}{4} - 3\frac{1}{3} = \frac{11}{4} - \frac{10}{3}$$ Write as improper fractions.

$$= \frac{33}{12} - \frac{40}{12}$$ $\frac{11}{4} \cdot \frac{3}{3} = \frac{33}{12}$ and $\frac{10}{3} \cdot \frac{4}{4} = \frac{40}{12}$

$$= \frac{33 - 40}{12} \text{ or } \frac{-7}{12}$$ Subtract the numerators. Then simplify.

 CHECK Your Progress

j. $6\frac{4}{7} + 3\frac{5}{7}$ **k.** $9\frac{5}{8} - 3\frac{3}{8}$ **l.** $-8\frac{5}{9} + \left(-6\frac{2}{9}\right)$

m. $3\frac{5}{6} + 2\frac{1}{4}$ **n.** $5\frac{7}{9} - 4\frac{2}{3}$ **o.** $-3\frac{3}{5} + \left(-8\frac{2}{3}\right)$

Sometimes you need to regroup before you can subtract.

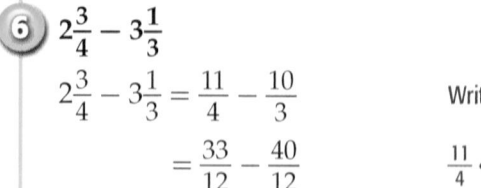 **REAL-WORLD EXAMPLE**

7 **ANIMALS** Horses are measured by a unit called a *handbreadth*, or hand. How much taller is a horse that is $14\frac{1}{4}$ hands tall than one that is $12\frac{3}{4}$ hands tall?

$$\begin{array}{rcl} 14\frac{1}{4} & \rightarrow & 13\frac{5}{4} \\ -12\frac{3}{4} & \rightarrow & -12\frac{3}{4} \\ \hline & & 1\frac{2}{4} \text{ or } 1\frac{1}{2} \end{array}$$

$14\frac{1}{4} = 13 + 1 + \frac{1}{4}$ or $13\frac{5}{4}$

Subtract the whole numbers and fractions separately.

The first horse is $1\frac{1}{2}$ hands taller.

 CHECK Your Progress

p. BAKING A recipe for chocolate cookies calls for $2\frac{3}{4}$ cups of flour. If Alexis has $1\frac{1}{4}$ cups of flour, how much more will she need?

Real-World Link · · · · ·

Dressage is an Olympic sport where a horse and rider must go through a series of tests. These tests require very controlled movements for the horse and require years of training.

Examples 1–6   **Add or subtract. Write in simplest form.**

1. $\frac{2}{5} + \left(-\frac{4}{5}\right)$ 2. $-\frac{3}{4} + \frac{1}{4}$ 3. $-\frac{7}{10} - \frac{9}{10}$ 4. $\frac{3}{8} - \frac{7}{8}$

5. $\frac{3}{4} + \left(-\frac{1}{6}\right)$ 6. $-\frac{5}{8} + \frac{1}{2}$ 7. $\frac{7}{8} - \frac{3}{4}$ 8. $\frac{7}{13} - \frac{2}{9}$

9. $5\frac{4}{9} - 2\frac{2}{9}$ 10. $-1\frac{3}{7} + \left(-2\frac{2}{7}\right)$ ⑪ $-3\frac{2}{5} + 1\frac{5}{6}$ 12. $3\frac{5}{8} - 1\frac{1}{3}$

Example 7   **13. HOMEWORK** Venus wrote a report for her middle school history class in $2\frac{1}{4}$ hours. Her sister Tia is in high school, and she wrote a history paper in $4\frac{3}{4}$ hours. How much longer did it take Tia to write her paper?

Practice and Problem Solving

● = **Step-by-Step Solutions** begin on page R1.
Extra Practice begins on page EP2.

Examples 1–6   **Add or subtract. Write in simplest form.**

14. $-\frac{1}{9} + \frac{4}{9}$ 15. $-\frac{3}{7} + \left(-\frac{2}{7}\right)$ 16. $-\frac{5}{12} + \frac{7}{12}$ 17. $\frac{8}{9} + \left(-\frac{5}{9}\right)$

18. $-\frac{4}{5} - \frac{3}{5}$ 19. $\frac{15}{16} - \frac{9}{16}$ 20. $\frac{1}{12} - \frac{7}{12}$ 21. $\frac{2}{9} - \frac{8}{9}$

22. $\frac{1}{4} + \left(-\frac{7}{12}\right)$ 23. $-\frac{3}{8} + \frac{5}{6}$ 24. $-\frac{6}{7} + \left(-\frac{1}{2}\right)$ 25. $\frac{5}{9} + \left(-\frac{3}{8}\right)$

26. $\frac{1}{3} - \frac{7}{8}$ 27. $\frac{4}{5} - \left(-\frac{2}{15}\right)$ 28. $-\frac{2}{9} - \left(-\frac{3}{11}\right)$ 29. $-\frac{7}{15} - \left(-\frac{12}{25}\right)$

30. ELECTIONS The table shows the fraction of students who voted for Josh or Chuan in the election for class president. What fraction of the students voted for either Josh or Chuan?

Class President	
Candidate	**Fraction of Students**
Josh	$\frac{4}{9}$
Chuan	$\frac{2}{5}$

31. DOGS Omar feeds his dog $\frac{3}{4}$ cup of dog food in the morning, $\frac{2}{3}$ cup in the afternoon, and $\frac{3}{4}$ cup in the evening. How many cups does he feed the dog altogether?

Add or subtract. Write in simplest form.

32. $3\frac{5}{8} + 7\frac{5}{8}$ 33. $8\frac{1}{10} + \left(-2\frac{9}{10}\right)$ 34. $3\frac{1}{5} + \left(-8\frac{1}{2}\right)$ 35. $-15\frac{5}{8} + 11\frac{2}{3}$

36. $-1\frac{5}{6} - 3\frac{5}{6}$ 37. $7 - 5\frac{2}{5}$ 38. $8\frac{3}{7} - \left(-6\frac{1}{2}\right)$ 39. $-8\frac{1}{3} - 4\frac{5}{6}$

Example 7   **40. HOME IMPROVEMENT** Andrew has $42\frac{1}{3}$ feet of molding to use as borders around the windows of his house. If he uses $23\frac{2}{3}$ feet of the molding on the front windows, how much remains for the back windows?

㊶ **WEATHER** One year, Brady's hometown of Powell received about $42\frac{6}{10}$ inches of snow. The following year only $14\frac{3}{10}$ inches of snow fell. What is the difference in the amount of snow between the two years?

Simplify each expression.

42. $-7\frac{4}{5} + 3\frac{1}{5} - \left(2\frac{3}{5}\right)$

43. $-8\frac{1}{8} - \left(-3\frac{5}{6}\right) + 6\frac{3}{4}$

ALGEBRA Evaluate each expression for the given values.

44. $a - b$ if $a = 5\frac{1}{3}$ and $b = -2\frac{1}{3}$

45. $x + y$ if $x = -\frac{5}{12}$ and $y = -\frac{1}{12}$

46. $c - d$ if $c = -\frac{3}{4}$ and $d = -12\frac{7}{8}$

47. $r - s$ if $r = -\frac{5}{8}$ and $s = 2\frac{5}{6}$

Study Tip

Adding Rational Numbers It is helpful to write all rational numbers in the same form before adding. In Exercise 48, write all the numbers as fractions or decimals. Then add.

48. ACTIVITIES Tamara played a computer game for $1\frac{1}{4}$ hours, studied for 2.25 hours, and did some chores for $\frac{1}{2}$ hour. How long did it take Tamara to do these things?

49. HOMEWORK Rob recorded the amount of time he spent on homework last week. Express his total time for the week in terms of hours and minutes.

Day	Time
Mon	$2\frac{1}{6}$ h
Tue	$2\frac{1}{2}$ h
Wed	$1\frac{3}{4}$ h
Thu	$2\frac{5}{12}$ h
Fri	$1\frac{1}{4}$ h

50. PLUMBING A plumber has a pipe that is $64\frac{5}{8}$ inches long. The plumber cuts $2\frac{7}{8}$ inches off the end of the pipe, then cuts off an additional $1\frac{3}{8}$ inches. How long is the remaining pipe after the last cut is made?

MEASUREMENT Find the missing measure for each figure.

51.

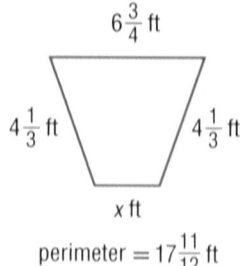

$6\frac{3}{4}$ ft

$4\frac{1}{3}$ ft $4\frac{1}{3}$ ft

x ft

perimeter $= 17\frac{11}{12}$ ft

52.

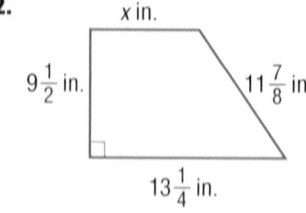

x in.

$9\frac{1}{2}$ in. $11\frac{7}{8}$ in.

$13\frac{1}{4}$ in.

perimeter $= 40\frac{3}{4}$ in.

53. MATH IN THE MEDIA In the life section of a newspaper, in a magazine, or on the Internet, find a recipe. Write a real-world problem in which you would add or subtract the fractions found in the recipe.

H.O.T. Problems

54. OPEN ENDED Write a subtraction problem using unlike fractions with a least common denominator of 12. Find the difference.

55. NUMBER SENSE Without doing the computation, determine whether $\frac{4}{7} + \frac{5}{9}$ is greater than, less than, or equal to 1. Explain.

56. CHALLENGE Suppose a bucket is placed under two faucets. If one faucet is turned on alone, the bucket will be filled in 5 minutes. If the other faucet is turned on alone, the bucket will be filled in 3 minutes. Write the fraction of the bucket that will be filled in 1 minute if both faucets are turned on.

57. WRITE MATH Write a real-world situation that can be solved by adding or subtracting mixed numbers. Then solve the problem.

58. Use the figure shown below.

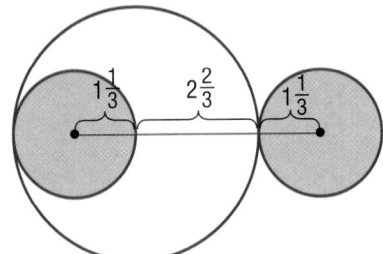

What is the length of the segment connecting the centers of the two smaller circles?

A. $6\frac{1}{3}$ units

B. $5\frac{2}{3}$ units

C. $5\frac{1}{3}$ units

D. $4\frac{5}{3}$ units

59. A recipe for snack mix contains $2\frac{1}{3}$ cups of mixed nuts, $3\frac{1}{2}$ cups of granola, and $\frac{3}{4}$ cup of raisins. What is the total amount of snack mix?

F. $5\frac{2}{3}$ c   H. $6\frac{2}{3}$ c

G. $5\frac{7}{12}$ c   I. $6\frac{7}{12}$ c

60. Which of the following shows the next step using the least common denominator to simplify $\frac{3}{4} - \frac{2}{3}$?

A. $\left(\frac{3}{4} \times \frac{5}{5}\right) - \left(\frac{2}{3} \times \frac{5}{5}\right)$

B. $\left(\frac{3}{4} \times \frac{6}{6}\right) - \left(\frac{2}{3} \times \frac{5}{5}\right)$

C. $\left(\frac{3}{4} \times \frac{3}{3}\right) - \left(\frac{2}{3} \times \frac{4}{4}\right)$

D. $\left(\frac{3}{4} \times \frac{4}{4}\right) - \left(\frac{2}{3} \times \frac{3}{3}\right)$

 **Spiral Review**

Write each fraction or mixed number as a decimal. (Lesson 1A)

61. $\frac{14}{20}$

62. $\frac{5}{6}$

63. $\frac{9}{11}$

64. $\frac{3}{8}$

65. $2\frac{1}{4}$

66. $-3\frac{7}{9}$

67. $4\frac{7}{12}$

68. $1\frac{8}{10}$

69. HOCKEY The sheet of ice that covers a hockey rink is created in two layers. First, an $\frac{1}{8}$-inch layer of ice is made for the lines to be painted on. Then, a $\frac{6}{8}$-inch layer of ice is added on top of the painted layer for a total thickness of $\frac{7}{8}$-inch. Write the total thickness of the ice as a decimal. (Lesson 1A)

Write each decimal as a fraction or mixed number in simplest form. (Lesson 1A)

70. 0.25

71. 1.6

72. −4.35

73. 0.94

74. $1.\overline{6}$

75. $-2.\overline{2}$

76. $0.\overline{7}$

77. $4.\overline{65}$

78. ELECTIONS The results of a recent school election are shown in the table. To the nearest hundredth, find the portion of votes each candidate received. (Lesson 1A)

Candidate	Votes
Carly	243
Christine	311

Main Idea

Multiply rational numbers.

 **Vocabulary**

dimensional analysis

Multiply Rational Numbers

Explore You can use an area model to find $\frac{1}{2}$ of $\frac{3}{4}$. The model also represents the product of $\frac{1}{2}$ and $\frac{3}{4}$.

Step 1 Draw a rectangle with four columns.

Step 2 Divide the rectangle into two rows.

Step 3 Shade a rectangle that is $\frac{1}{2}$ unit by $\frac{3}{4}$ unit blue.

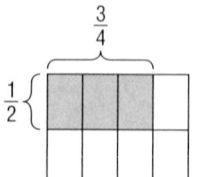

The shaded area represents $\frac{1}{2}$ of $\frac{3}{4}$.

1. What is the product of $\frac{1}{2}$ and $\frac{3}{4}$?

2. Use an area model to find each product.

    a. $\frac{1}{2} \cdot \frac{1}{2}$ **b.** $\frac{2}{5} \cdot \frac{2}{3}$

    c. $\frac{1}{4} \cdot \frac{3}{5}$ **d.** $\frac{2}{3} \cdot \frac{4}{5}$

3. What is the relationship between the numerators of the factors and the numerator of the product?

4. What is the relationship between the denominators of the factors and the denominator of the product?

The area model suggests the following rule for multiplying fractions.

🏃 Key Concept   **Multiply Fractions**

Words   To multiply fractions, multiply the numerators and multiply the denominators.

Examples   **Numbers**   **Algebra**

               $\frac{2}{3} \cdot \frac{4}{5} = \frac{8}{15}$ $\frac{a}{b} \cdot \frac{c}{d} = \frac{ac}{bd}$, where b and $d \neq 0$.

You can use the rules for multiplying integers to determine the sign of the product of any two signed numbers.

EXAMPLES **Multiply Fractions and Mixed Numbers**

Multiply. Write in simplest form.

① $-\frac{5}{6} \cdot \frac{3}{8}$

$$-\frac{5}{6} \cdot \frac{3}{8} = \frac{-5}{\overset{2}{\cancel{6}}} \cdot \frac{\overset{1}{\cancel{3}}}{8} \qquad \text{Divide 6 and 3 by their GCF, 3.}$$

$$= \frac{-5 \cdot 1}{2 \cdot 8} \text{ or } -\frac{5}{16} \qquad \begin{array}{l}\text{Multiply. Then simplify. The fractions have different signs,} \\ \text{so the product is negative.}\end{array}$$

② $4\frac{1}{2} \cdot 2\frac{2}{3}$

$$4\frac{1}{2} \cdot 2\frac{2}{3} = \frac{9}{2} \cdot \frac{8}{3} \qquad \text{Rename } 4\frac{1}{2} \text{ as } \frac{9}{2} \text{ and } 2\frac{2}{3} \text{ as } \frac{8}{3}.$$

$$= \frac{\overset{3}{\cancel{9}}}{\underset{1}{\cancel{2}}} \cdot \frac{\overset{4}{\cancel{8}}}{\underset{1}{\cancel{3}}} \qquad \text{Divide out common factors.}$$

$$= \frac{3 \cdot 4}{1 \cdot 1} \text{ or } 12 \qquad \text{Multiply. Then simplify.}$$

✔ **CHECK Your Progress**

a. $\frac{1}{4} \cdot \frac{2}{3}$ **b.** $\left(-\frac{1}{2}\right)\left(-\frac{6}{7}\right)$ **c.** $2\frac{1}{6} \cdot 1\frac{1}{5}$

Recall that probability can be expressed as a fraction. *Independent events* are two or more events in which the outcome of one does not affect the outcome of the other. To find the probability of independent events, multiply the probability of the first event by the probability of the second event.

 EXAMPLE **Independent Events**

③ **The spinner at the right is spun, and a coin is tossed. What is the probability of spinning an odd number and tossing tails?**

$P(\text{spinning an odd number}) = \frac{4}{8} \text{ or } \frac{1}{2}$

$P(\text{tossing tails}) = \frac{1}{2}$

$P(\text{odd and tails}) = \frac{1}{2} \cdot \frac{1}{2} \text{ or } \frac{1}{4}$

✔ **CHECK Your Progress**

Refer to the situation above to find each probability.

d. $P(\text{less than 4 and heads})$ **e.** $P(\text{prime and tails})$

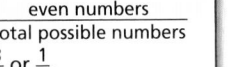

Dimensional analysis is the process of including units of measurement when you compute.

 REAL-WORLD EXAMPLE

④ **AIRCRAFT** Refer to the information at the left. Suppose a VH-71 helicopter is traveling at its cruising speed. Use dimensional analysis to find how far it will travel in $1\frac{3}{4}$ hours.

Words	Distance equals the rate multiplied by the time.
Variable	Let d represent the distance.
Equation	$d = 172$ miles per hour $\cdot 1\frac{3}{4}$ hours

$d = \dfrac{172 \text{ miles}}{1 \text{ hour}} \cdot 1\frac{3}{4}$ hours   Write the equation.

$d = \dfrac{172 \text{ miles}}{1 \text{ hour}} \cdot \dfrac{7}{4} \cdot \dfrac{\text{hours}}{1}$ $1\frac{3}{4} = \dfrac{7}{4}$

$d = \dfrac{\overset{43}{\cancel{172}} \text{ miles}}{\cancel{1 \text{ hour}}} \cdot \dfrac{7}{\underset{1}{\cancel{4}}} \cdot \dfrac{\cancel{\text{hours}}}{1}$ Divide by common factors and units.

$d = 301$ miles

A VH-71 will travel 301 miles in $1\frac{3}{4}$ hours.

 CHECK Your Progress

f. AIRCRAFT The VH-71 has 200 square feet of cabin space. Use dimensional analysis to find the size of its cabin space in square yards. (*Hint:* 1 square yard = 9 square feet)

✓ **CHECK Your Understanding**

Examples 1 and 2   **Multiply. Write in simplest form.**

1. $\dfrac{3}{5} \cdot \dfrac{5}{7}$ **2.** $\dfrac{4}{5} \cdot \dfrac{3}{8}$ **3.** $-\dfrac{1}{8} \cdot \dfrac{4}{9}$

4. $\left(\dfrac{-12}{13}\right)\left(-\dfrac{2}{3}\right)$ **5.** $2\dfrac{1}{2} \cdot 1\dfrac{2}{5}$ **6.** $-6\dfrac{3}{4} \cdot 1\dfrac{7}{9}$

Example 3   **A number cube is rolled and a marble is selected from the bag shown. Find each probability.**

7. P(even and blue)   **8.** P(3 and red)

9 P(greater than 2 and yellow)   **10.** P(odd and green)

Example 4   **11. FRUIT** Terrence bought $2\dfrac{5}{8}$ pounds of grapes that cost \$2 per pound. Use dimensional analysis to find the total cost of the grapes.

= Step-by-Step Solutions begin on page R1.
Extra Practice begins on page EP2.

Examples 1 and 2   **Multiply. Write in simplest form.**

12. $\dfrac{1}{12} \cdot \dfrac{4}{7}$ **13.** $\dfrac{3}{16} \cdot \dfrac{1}{9}$ **14.** $\dfrac{5}{8} \cdot \dfrac{4}{5}$ **15.** $\dfrac{9}{10} \cdot \dfrac{2}{3}$

16. $-\dfrac{9}{10} \cdot \dfrac{2}{3}$ **17.** $\left(-\dfrac{12}{25}\right)\dfrac{15}{32}$ **18.** $\left(-\dfrac{3}{5}\right)\left(-\dfrac{1}{3}\right)$ **19.** $\left(-\dfrac{4}{7}\right)\left(-\dfrac{1}{20}\right)$

20. $3\dfrac{1}{3} \cdot \dfrac{1}{4}$ **㉑** $4\dfrac{1}{4} \cdot 3\dfrac{1}{3}$ **22.** $-3\dfrac{3}{8} \cdot \left(-\dfrac{2}{3}\right)$ **23.** $-\dfrac{5}{6} \cdot \left(-1\dfrac{4}{5}\right)$

Example 3   **The spinner at the right is spun once and the coin is tossed. Find each probability.**

24. *P*(even and heads)

25. *P*(less than 8 and heads)

26. *P*(composite and tails)

27. *P*(factor of 12 and tails)

Example 4   **Use dimensional analysis to solve each problem.**

28. BAKING A recipe calls for $\dfrac{3}{4}$ cup of sugar per batch of cookies. If Gabe wants to make 6 batches of cookies, how many cups of sugar does he need?

29. POPULATION Population density measures how many people live within a certain area. In a certain city, there are about 150,000 people per square mile. How many people live in an area of $2\dfrac{1}{4}$ square miles?

ALGEBRA Evaluate each expression if $r = \dfrac{1}{4}$, $s = \dfrac{2}{5}$, $t = \dfrac{8}{9}$, and $v = -\dfrac{2}{3}$.

30. rs **31.** rt **32.** stv **㉝** rtv

34. GEOGRAPHY There are about 57 million square miles of land on Earth covering seven continents.

    a. What is the approximate land area of Europe?

    b. What is the approximate land area of Asia?

    c. Only about $\dfrac{3}{10}$ of Australia's land area is able to support agriculture. What fraction of Earth's land is this?

Continent	Fraction of Earth's Landmass
Africa	$\dfrac{1}{5}$
Antarctica	$\dfrac{9}{100}$
Asia	$\dfrac{3}{10}$
Australia	$\dfrac{11}{200}$
Europe	$\dfrac{7}{100}$
North America	$\dfrac{33}{200}$
South America	$\dfrac{3}{25}$

Find each product. Write in simplest form.

35. $\dfrac{1}{3} \cdot \left(-\dfrac{3}{8}\right) \cdot \dfrac{4}{5}$ **36.** $\left(-\dfrac{2}{5}\right) \cdot \dfrac{1}{6} \cdot \left(-\dfrac{5}{2}\right)$

37. $3\dfrac{1}{3} \cdot 1\dfrac{1}{2} \cdot 5$ **38.** $10 \cdot 3.78 \cdot \dfrac{1}{5}$

39. $-\dfrac{2}{9} \cdot 0.\overline{3}$ **40.** $-\dfrac{7}{16} \cdot (-2.375)$

41. **FIND THE ERROR** Danielle is finding $2\frac{1}{2} \cdot 3\frac{1}{4}$. Find her mistake and correct it.

$$2\frac{1}{2} \cdot 3\frac{1}{4} = 2 \cdot 3 + \frac{1}{2} \cdot \frac{1}{4}$$
$$= 6 + \frac{1}{8}$$
$$= 6\frac{1}{8}$$

42. **OPEN ENDED** Select two fractions with a product greater than $\frac{1}{2}$ and less than 1. Use a number line to justify your answer.

43. **CHALLENGE** Find the missing fraction. $\frac{3}{4} \cdot \blacksquare = \frac{9}{14}$

44.  **WRITE MATH** Explain why the product of $\frac{1}{2}$ and $\frac{7}{8}$ is less than $\frac{1}{2}$.

Test Practice

45. Which of the following is true when a whole number greater than one is multiplied by a positive fraction less than one?

   A. The product is greater than the whole number.

   B. The product is between the fraction and the whole number.

   C. The product is less than the fraction.

   D. all of the above

46. What is the area of the parallelogram? Use the formula $A = bh$.

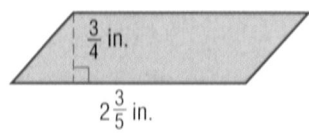

$\frac{3}{4}$ in.

$2\frac{3}{5}$ in.

   F. $\frac{20}{39}$ in^2

   G. $2\frac{3}{10}$ in^2

   H. $1\frac{19}{20}$ in^2

   I. $\frac{4}{5}$ in^2

Spiral Review

Add or subtract. Write in simplest form. (Lesson 1B)

47. $\frac{1}{6} + \frac{1}{7}$

48. $\frac{7}{8} - \frac{1}{6}$

49. $-5\frac{1}{2} - 6\frac{4}{5}$

BIOLOGY Write the weight of each animal as a fraction or mixed number. (Lesson 1A)

50. queen bee

51. hummingbird

52. hamster

ANIMAL	WEIGHT (OUNCES)
QUEEN BEE	0.004
HUMMINGBIRD	0.11
HAMSTER	3.5

Main Idea

Divide rational numbers.

Vocabulary

multiplicative inverses
reciprocals

Divide Rational Numbers

ANIMALS An antelope is one of the fastest animals on Earth. It can run about 60 miles per hour. A squirrel runs one fifth of that speed.

1. Find the value of $60 \div 5$.

2. Find the value of $60 \times \frac{1}{5}$.

3. Compare the values of $60 \div 5$ and $60 \times \frac{1}{5}$.

4. What can you conclude about the relationship between dividing by 5 and multiplying by $\frac{1}{5}$?

Two numbers with a product of 1 are **multiplicative inverses**, or **reciprocals**, of each other. For example, 5 and $\frac{1}{5}$ are multiplicative inverses because $5 \cdot \frac{1}{5} = 1$.

Key Concept **Inverse Property of Multiplication**

Words The product of a number and its multiplicative inverse is 1.

Examples

Numbers	**Algebra**
$\frac{3}{4} \cdot \frac{4}{3} = 1$	$\frac{a}{b} \cdot \frac{b}{a} = 1$, where a and $b \neq 0$

 EXAMPLE **Find a Multiplicative Inverse**

1 Write the multiplicative inverse of $-5\frac{2}{3}$.

$-5\frac{2}{3} = -\frac{17}{3}$ Write $-5\frac{2}{3}$ as an improper fraction.

Since $-\frac{17}{3}\left(-\frac{3}{17}\right) = 1$, the multiplicative inverse of $-5\frac{2}{3}$ is $-\frac{3}{17}$.

 CHECK Your Progress

Write the multiplicative inverse of each number.

a. $-2\frac{1}{3}$ b. $-\frac{5}{8}$ c. 7

Study Tip

Complex Fractions
Recall that a fraction
bar represents division.
So,

$$\frac{a}{b} \div \frac{c}{d} = \frac{\frac{a}{b}}{\frac{c}{d}}.$$

Multiplicative inverses are used in division. Consider $\frac{a}{b} \div \frac{c}{d}$, which can be written as a fraction.

$$\frac{\frac{a}{b}}{\frac{c}{d}} = \frac{\frac{a}{b} \cdot \frac{d}{c}}{\frac{c}{d} \cdot \frac{d}{c}}$$ Multiply the numerator and denominator by $\frac{d}{c}$, the multiplicative inverse of $\frac{c}{d}$.

$$= \frac{\frac{a}{b} \cdot \frac{d}{c}}{1}$$ $\frac{c}{d} \cdot \frac{d}{c} = 1$

$$= \frac{a}{b} \cdot \frac{d}{c}$$

Therefore, $\frac{a}{b} \div \frac{c}{d} = \frac{a}{b} \cdot \frac{d}{c}$.

Key Concept   Divide Fractions

Words   To divide by a fraction, multiply by its multiplicative inverse.

Symbols   **Numbers**

$$\frac{2}{5} \div \frac{3}{4} = \frac{2}{5} \cdot \frac{4}{3}$$

Algebra

$$\frac{a}{b} \div \frac{c}{d} = \frac{a}{b} \cdot \frac{d}{c},\text{ where } b, c \text{ and } d \neq 0$$

EXAMPLES Divide Fractions and Mixed Numbers

Divide. Write in simplest form.

2 $-\frac{4}{5} \div \frac{6}{7}$

$$-\frac{4}{5} \div \frac{6}{7} = -\frac{4}{5} \cdot \frac{7}{6}$$ The multiplicative inverse of $\frac{6}{7}$ is $\frac{7}{6}$.

$$= -\frac{\overset{2}{\cancel{4}}}{5} \cdot \frac{7}{\underset{3}{\cancel{6}}}$$ Divide -4 and 6 by their GCF, 2.

$$= -\frac{14}{15}$$ Multiply.

3 $4\frac{2}{3} \div \left(-3\frac{1}{2}\right)$

$$4\frac{2}{3} \div -3\frac{1}{2} = \frac{14}{3} \div \left(-\frac{7}{2}\right)$$ $4\frac{2}{3} = \frac{14}{3}, -3\frac{1}{2} = -\frac{7}{2}$

$$= \frac{14}{3} \cdot \left(-\frac{2}{7}\right)$$ The multiplicative inverse of $-\frac{7}{2}$ is $-\frac{2}{7}$.

$$= \frac{\overset{2}{\cancel{14}}}{3} \cdot \left(-\frac{2}{\underset{1}{\cancel{7}}}\right)$$ Divide 14 and 7 by their GCF, 7.

$$= -\frac{4}{3} \text{ or } -1\frac{1}{3}$$ Multiply.

CHECK Your Progress

Study Tip

Dividing By a Whole Number
When dividing by a whole
number, rename it as an
improper fraction first.
Then multiply by its
reciprocal.

d. $\frac{3}{4} \div \frac{1}{2}$ **e.** $-\frac{1}{4} \div \frac{7}{8}$ **f.** $2\frac{3}{4} \div \left(-2\frac{1}{5}\right)$ **g.** $-1\frac{1}{2} \div 12$

4 **CRAFTS** Lina's class is making flags for the school's International Day celebration. She needs $1\frac{1}{6}$ feet of paper for the blue portion on each flag. If the class has a 21-foot roll of blue paper, how many flags can she make?

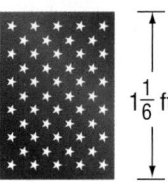

$1\frac{1}{6}$ ft

Divide 21 by $1\frac{1}{6}$.

$21 \div 1\frac{1}{6} = \frac{21}{1} \div \frac{7}{6}$ Write 21 as $\frac{21}{1}$. Write $1\frac{1}{6}$ as $\frac{7}{6}$.

$= \frac{\overset{3}{\cancel{21}}}{1} \cdot \frac{6}{\underset{1}{\cancel{7}}}$ Multiply by the multiplicative inverse of $\frac{7}{6}$, which is $\frac{6}{7}$. Divide 7 and 21 by their GCF, 7.

$= \frac{18}{1}$ or 18   Simplify.

Lina's class can make 18 flags using the 21-foot roll of paper.

✓ **CHECK Your Progress**

h. LUMBER How many $1\frac{1}{2}$-inch-thick boards are in a stack that is 36 inches tall?

✓ **CHECK Your Understanding**

Example 1   **Write the multiplicative inverse of each number.**

1. $\frac{5}{7}$ **2.** -12 **3.** $-2\frac{3}{4}$

Examples 2 and 3   **Divide. Write in simplest form.**

4. $\frac{2}{3} \div \frac{3}{4}$ **5.** $\frac{5}{8} \div \frac{1}{2}$ **6.** $\frac{3}{8} \div \left(-\frac{9}{10}\right)$ **7.** $-\frac{7}{16} \div \left(-\frac{7}{8}\right)$

8. $\frac{4}{5} \div 8$ **9.** $\frac{9}{10} \div 3$ **10.** $-5\frac{5}{6} \div \left(-4\frac{2}{3}\right)$ **11** $-3\frac{7}{12} \div 6\frac{5}{6}$

Example 4   **12. BIRDS** The smallest owl found in the United States is the Elf Owl, which weighs $1\frac{1}{2}$ ounces. One of the largest owls is the Eurasian Eagle Owl, which weighs nearly 10 pounds or 156 ounces. The Eurasian Eagle Owl is how many times as heavy as the Elf Owl?

Elf Owl   Eurasian Eagle Owl

Practice and Problem Solving

= **Step-by-Step Solutions** begin on page R1.
Extra Practice begins on page EP2.

Example 1 Write the multiplicative inverse of each number.

13. $-\dfrac{7}{9}$ **14.** $-\dfrac{5}{8}$ **15** 15

16. 18   **17.** $3\dfrac{2}{5}$ **18.** $4\dfrac{1}{8}$

Example 2 and 3 Divide. Write in simplest form.

19. $\dfrac{2}{5} \div \dfrac{3}{4}$ **20.** $\dfrac{3}{8} \div \dfrac{2}{3}$ **21.** $\dfrac{2}{3} \div \dfrac{5}{6}$ **22.** $\dfrac{2}{5} \div \dfrac{1}{10}$

23. $-\dfrac{4}{5} \div \dfrac{3}{4}$ **24.** $\dfrac{3}{10} \div \left(-\dfrac{2}{3}\right)$ **25.** $-\dfrac{5}{9} \div \left(-\dfrac{2}{3}\right)$ **26.** $-\dfrac{7}{12} \div \left(-\dfrac{5}{6}\right)$

27. $\dfrac{2}{5} \div 4$ **28.** $\dfrac{9}{16} \div 3$ **29.** $\dfrac{4}{5} \div 6$ **30.** $\dfrac{6}{7} \div 4$

31. $3\dfrac{3}{4} \div 2\dfrac{1}{2}$ **32.** $7\dfrac{1}{2} \div 2\dfrac{1}{10}$ **33.** $-12\dfrac{1}{4} \div 4\dfrac{2}{3}$ **34.** $10\dfrac{1}{5} \div \left(-\dfrac{3}{15}\right)$

Example 4 **35. HUMAN BODY** The table shows the composition of a healthy adult male's body. Examples of body cell mass are muscle, body organs, and blood. Examples of supporting tissue are blood, plasma, and bones.

Composition of Human Body	
Component	**Fraction of Body Weight**
Body Cell Mass	$\dfrac{11}{20}$
Supporting Tissue	$\dfrac{3}{10}$
Body Fat	$\dfrac{3}{20}$

    a. How many times more of a healthy adult male's body weight is made up of body cell mass than body fat?

    b. How many times more of a healthy adult male's body weight is made up of body cell mass than supporting tissue?

36. PAINTING It took 3 people $2\dfrac{1}{2}$ hours to paint a large room. How long would it take 5 people to paint a similar room?

37 BIOLOGY How many of the small hummingbirds need to be placed end-to-end to have the same length as the large hummingbird?

$5\dfrac{1}{2}$ cm

22 cm

38. GEOMETRY The circumference C, or distance around a circle, can be approximated using the formula $C = \dfrac{44}{7}r$, where r is the radius of the circle. What is the radius of a circle with a circumference of 53.2 meters? Round to the nearest tenth.

39. BAKING Emily is baking chocolate cupcakes. Each batch of 20 cupcakes requires $\dfrac{2}{3}$ cup of cocoa. If Emily has $3\dfrac{1}{4}$ cups of cocoa, how many full batches of cupcakes will she be able to make and how much cocoa will she have left over?

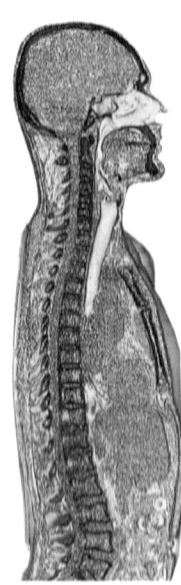

Real-World Link
99% of the mass of the human body is made up of six elements: oxygen, carbon, hydrogen, nitrogen, calcium, and phosphorus.

40. CHALLENGE Give a counterexample to the statement, *The quotient of two fractions between 0 and 1 is never a whole number.*

41. NUMBER SENSE Which is greater: $30 \cdot \frac{3}{4}$ or $30 \div \frac{3}{4}$? Explain.

CHALLENGE Use mental math to find each value.

42. $\frac{43}{594} \cdot \frac{641}{76} \div \frac{641}{594}$

43. $\frac{783}{241} \cdot \frac{241}{783} \div \frac{72}{53}$

44. ✍ **WRITE MATH** Write a real-world problem that can be solved by dividing fractions or mixed numbers. Solve the problem.

Test Practice

45. Some of the ingredients required for one batch of muffins are shown below.

Strawberry Muffins

flour	$\frac{2}{3}$ cup
strawberries	$\frac{3}{4}$ cup

Claudio's father used $1\frac{2}{3}$ cups of flour and $1\frac{7}{8}$ cups of strawberries. How many batches of muffins did he make?

A. 3

C. 2

B. $2\frac{1}{2}$

D. $1\frac{3}{4}$

46. Mr. Jones is doing a science experiment with his class of 20 students. Each student needs $\frac{3}{4}$ cup of vinegar. If he currently has 15 cups of vinegar, which equation could Mr. Jones use to determine if he has enough vinegar for his entire class?

F. $x = 15 \div 20$

H. $x = 20 - (15)$

G. $x = 15 \div \frac{3}{4}$

I. $x = 15(20)$

47. ✏ **GRIDDED RESPONSE** Lucas is storing a set of art books on a shelf that has $11\frac{1}{4}$ inches of space. If each book is $\frac{3}{4}$ inch wide, how many books can be stored on the shelf?

Multiply. Write in simplest form. (Lesson 1C)

48. $\frac{1}{6} \cdot \frac{3}{8}$

49. $\frac{5}{6} \cdot \frac{4}{5}$

50. $1\frac{2}{3} \cdot 4\frac{1}{5}$

51. $\frac{2}{3} \cdot 3\frac{1}{4}$

52. VOLUNTEERING The table shows the number of hours four students volunteered to work at an animal shelter after school. How much time did the students volunteer in all? (Lesson 1B)

53. HEALTH A newborn baby weighs $6\frac{3}{4}$ pounds. Write this weight as a decimal. (Lesson 1A)

Student	Time (h)
Anabel	$2\frac{1}{3}$
Damon	$1\frac{7}{8}$
Jeremiah	$1\frac{5}{6}$
Meghan	$2\frac{1}{4}$

Mid-Chapter Check

1. Write $1\frac{7}{16}$ as a decimal. (Lesson 1A)

2. Write $0.\overline{4}$ as a fraction in simplest form. (Lesson 1A)

3. **MULTIPLE CHOICE** The table gives the durations of spaceflights.

Mission	Year	Duration (h)
Challenger (41–B)	1984	$191\frac{4}{15}$
Discovery (51–A)	1984	$191\frac{3}{4}$
Endeavour (STS–57)	1992	$190\frac{1}{2}$
Discovery (STS–103)	1999	$191\frac{1}{6}$

Which of the following is the combined duration of the space flights? (Lesson 1B)

A. $765\frac{1}{2}$ h

B. $765\frac{7}{16}$ h

C. $764\frac{41}{60}$ h

D. $763\frac{1}{3}$ h

Add or subtract. Write in simplest form. (Lesson 1B)

4. $\frac{1}{5} + \left(-\frac{4}{5}\right)$ 5. $-3\frac{4}{7} - 3\frac{6}{7}$

6. $\frac{5}{12} + \left(-\frac{7}{15}\right)$ 7. $5\frac{3}{5} - 12\frac{1}{2}$

8. **PIZZA** A pizza has 3 toppings with no toppings overlapping. Pepperoni tops $\frac{1}{3}$ of the pizza and mushrooms top $\frac{2}{5}$. The rest is topped with sausage. What fraction is topped with sausage? (Lesson 1B)

Multiply. Write in simplest form. (Lesson 1C)

9. $-\frac{1}{3} \cdot \frac{7}{8}$ 10. $\left(-2\frac{3}{4}\right) \cdot \left(-\frac{1}{5}\right)$

11. $\frac{5}{6} \cdot \frac{3}{5}$ 12. $-2\frac{4}{7} \cdot \left(-3\frac{2}{3}\right)$

13. **WEATHER** The table shows the approximate number of sunny days each year for certain cities. Oklahoma City has about $\frac{3}{5}$ as many sunny days as Phoenix. About how many sunny days each year are there in Oklahoma City? (Lesson 1C)

Sunny Days per Year	
City	Days
Austin, TX	120
Denver, CO	115
Phoenix, AZ	215
Sacramento, CA	195
Santa Fe, NM	175

Divide. Write in simplest form. (Lesson 1D)

14. $\frac{1}{2} \div \left(-\frac{3}{4}\right)$ 15. $\left(-1\frac{1}{3}\right) \div \left(-\frac{1}{4}\right)$

16. $6\frac{1}{6} \div \left(-1\frac{2}{3}\right)$ 17. $8\frac{1}{2} \div 1\frac{7}{10}$

18. **MULTIPLE CHOICE** A board that is $25\frac{1}{2}$ feet long is cut into pieces that are each $1\frac{1}{2}$ feet long. Which of the steps below would give the number of pieces into which the board is cut? (Lesson 1D)

F. Multiply $1\frac{1}{2}$ by $25\frac{1}{2}$.

G. Divide $25\frac{1}{2}$ by $1\frac{1}{2}$.

H. Add $25\frac{1}{2}$ to $1\frac{1}{2}$.

I. Subtract $1\frac{1}{2}$ from $25\frac{1}{2}$.

Problem-Solving Investigation

Main Idea Look for a pattern to solve problems.

P.S.I. TEAM +

Look for a Pattern

DREW: As a member of the Service Club, I need to complete 25 hours of volunteer work each month. After each shift, I total up my hours for the month. My volunteer journal shows $2\frac{2}{3}$, $5\frac{1}{3}$, 8, and $10\frac{2}{3}$ total hours.

YOUR MISSION: Look for a pattern to find after how many days he will have at least 25 hours.

Understand	You know the total hours for the first four days. You want to know how many days it will take to have 25 hours.
Plan	Look for a pattern in the time worked each day. Then continue the pattern to find when he will work at least 25 hours.
Solve	

+1 +1 +1 +1 +1 +1 +1 +1 +1

Shift	1	2	3	4	5	6	7	8	9	10
Total Hours	$2\frac{2}{3}$	$5\frac{1}{3}$	8	$10\frac{2}{3}$	$13\frac{1}{3}$	16	$18\frac{2}{3}$	$21\frac{1}{3}$	24	$26\frac{2}{3}$

$+2\frac{2}{3}$ $+2\frac{2}{3}$ $+2\frac{2}{3}$ $+2\frac{2}{3}$ $+2\frac{2}{3}$ $+2\frac{2}{3}$ $+2\frac{2}{3}$ $+2\frac{2}{3}$ $+2\frac{2}{3}$

Drew will have at least 25 hours after day 10.

Check	Check your pattern to make sure the answer is correct.

Analyze the Strategy

1. Describe how to continue the pattern to find the number of hours Drew could have after 15 days.

2. **WRITE MATH** Write a problem that can be solved by finding a pattern.

Mixed Problem Solving

 = **Step-by-Step Solutions** begin on page R1.
Extra Practice begins on page EP2.

- Look for a pattern.
- Work backward.
- Guess, check, and revise.
- Choose an operation.

Use the *look for a pattern* strategy to solve Exercises 3–5.

3. **PHYSICAL SCIENCE** A ball was dropped from a height of 27 inches. After the first, second, and third bounces, the heights were 18 inches, 12 inches, and 8 inches, respectively. After which bounce will the height of the ball be less than 3 inches?

4. **GEOMETRY** Draw the next two figures in the pattern.

5. **JOBS** Lola is mowing lawns as part of her summer job. The table shows the amount she mows every $\frac{1}{2}$ hour. How many lawns can she mow in 6 hours?

Time (h)	Lawns Mowed
0.5	$\frac{2}{3}$
1	$1\frac{1}{3}$
1.5	2
2	$2\frac{2}{3}$

Use any strategy to solve Exercises 6–12.

6. **MONEY** To attend the class trip, each student will have to pay $7.50 for transportation and $5.00 for food. If there are 360 students in the class, how much money will need to be collected for the trip?

7. **TIME** Carlos and his friends are going out for dinner and a movie. The movie starts at 8:10 P.M. and they want to arrive 20 minutes before it starts. Dinner will take 1 hour 15 minutes, and the total travel time is 55 minutes. At what time should they plan to leave Carlos's house?

8. **ANALYZE TABLES** In computer terminology, a bit is the smallest unit of data. A byte is equal to 8 bits. The table below gives the equivalence for several units of data.

Unit of Data	Equivalence
1 byte	8 bits
1 kilobyte (kB)	1,024 bytes
1 megabyte (MB)	1,024 kilobytes
1 gigabyte (GB)	1,024 megabytes

How many bits are in 1 MB?

9. **PHOTOGRAPHY** Cameras often have multiple shutter speeds. Some common shutter speeds in seconds are $\frac{1}{125}$, $0.0\overline{6}$, $\frac{1}{60}$, 0.125, 0.004, and $\frac{1}{4}$. Which of these shutter speeds is the fastest?

10. **MUSIC THEORY** Musical notes have values that follow a pattern. The first three notes are: whole note, half note, quarter note. Name the next three notes in the pattern.

11. **INSECTS** The longest insect in the world is the stick insect whose length reaches 15 inches. The smallest insect is the fairy fly whose length is only $\frac{1}{100}$ inch. How many times longer is the stick insect than the fairy fly?

12. **ROLLER COASTERS** A roller coaster is 160 feet tall. If a new roller coaster is built that is $2\frac{3}{5}$ times the height of the existing coaster, what is the height of the new roller coaster?

Main Idea

Compare and order rational numbers.

Compare Rational Numbers

GAMES Taino surveyed his classmates about their favorite type of game. The results are shown in the table.

Favorite Types of Games	
Type of Game	**Percent of Favorite**
Board Game	11%
Card Game	9%
Computer or Video Game	45%
Sports Game	17%
None of the above	18%

1. Write each percent as a fraction. Do not simplify the fractions.

2. Write each fraction in Exercise 1 as a decimal.

3. How could you write a percent as a decimal without writing the fraction first?

Remember that *percent* means *per hundred*. You can write percent as a fraction with 100 in the denominator. Similarly, you can write percents as decimals by dividing by 100.

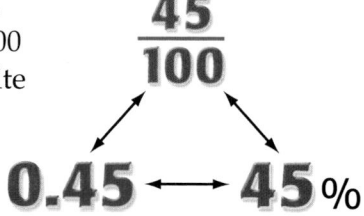

Key Concept — **Percents and Decimals**

Percent ⟶ Decimal

To write a percent as a decimal, divide by 100 and remove the percent symbol.

$$45\% = 45\% = 0.45$$

Decimal ⟶ Percent

To write a decimal as a percent, multiply by 100 and add the percent symbol.

$$0.45 = 0.45 = 45\%$$

EXAMPLES **Percents as Decimals**

Write each percent as a decimal.

1) 35%

$35\% = 35\%$ Divide by 100.

$ = 0.35$ Remove the percent symbol.

2) 115%

$115\% = 115\%$ Divide by 100.

$ = 1.15$ Remove the percent symbol.

Study Tip

Percents and Decimals
To divide by 100, move the decimal point two places to the left.

 CHECK Your Progress

a. 27%   b. 145%   c. 0.2%

 **EXAMPLES** Decimals as Percents

Study Tip

Decimals Greater Than One
Notice that decimals like
1.66 that are greater than
1 are equivalent to percents
greater than 100.

Write each decimal as a percent.

 3 0.2

$0.2 = 0.20$ Multiply by 100.

    $= 20\%$ Add the percent symbol.

4 1.66

$1.66 = 1.66$ Multiply by 100.

     $= 166\%$ Add the percent symbol.

 CHECK Your Progress

d. 0.83   **e.** 1.764   **f.** 0.005

You have learned to write a fraction as a percent by finding an equivalent fraction with a denominator of 100. This method works well if the denominator is a factor of 100. If the denominator is *not* a factor of 100, you can solve a proportion or you can write the fraction as a decimal and then write the decimal as a percent.

 EXAMPLES Fractions as Percents

5 Write $\frac{3}{8}$ as a percent.

Method 1 Use a proportion.

$$\frac{3}{8} = \frac{x}{100}$$
$$3 \cdot 100 = 8 \cdot x$$
$$300 = 8x$$
$$\frac{300}{8} = \frac{8x}{8}$$
$$37.5 = x$$

Method 2 Write as a decimal.

First write as a decimal. Then write as a percent.

$$\frac{3}{8} = 0.375$$
$$= 37.5\%$$

So, $\frac{3}{8} = \frac{37.5}{100}$ or 37.5%.

6 Write $\frac{2}{3}$ as a percent.

$$\frac{2}{3} = 0.6\overline{6}$$
$$= 66.\overline{6}\%$$

$$\begin{array}{r} 0.66... \\ 3\overline{)2.0} \\ -1\,8 \\ \hline 20 \\ -18 \\ \hline 2 \end{array}$$

So, $\frac{2}{3} = 66.\overline{6}\%$.

Study Tip

Percents In real-world
situations, 66.6% will
usually be represented
as $66\frac{2}{3}$% or rounded to
67% or 66.7%.

 CHOOSE Your Method

Write each fraction as a percent.

g. $\frac{7}{25}$ **h.** $\frac{3}{16}$ **i.** $\frac{1}{9}$

EXAMPLE Compare Rational Numbers

7 **TAXES** In a recent survey, 0.6 of the people said they will use their tax refund to pay bills and 7% said they will just spend it. Do more people pay bills or spend their refund?

Since $0.6 = 60\%$ and $60\% > 7\%$, more people plan on using their tax refund for paying bills than for spending.

CHECK Your Progress

j. **VACATION** In a survey, 14% of students said they would travel over spring break and 3 out of 25 said they would watch videos. Is a bigger part of students traveling or watching videos?

EXAMPLE Order Rational Numbers

8 Order 30%, $\dfrac{3}{100}$, $\dfrac{7}{20}$, and 0.33 from least to greatest.

$\dfrac{3}{100} = 3\%$ $\dfrac{7}{20} = \dfrac{35}{100}$ or 35% $0.33 = 33\%$

From least to greatest, the percents are 3%, 30%, 33%, and 35%.

So, from least to greatest, the numbers are $\dfrac{3}{100}$, 30%, 0.33, and $\dfrac{7}{20}$.

CHECK Your Progress

Order each set of numbers from least to greatest.

k. 22%, $\dfrac{1}{10}$, $\dfrac{3}{25}$, 0.25

l. $\dfrac{1}{5}$, 40%, 0.401, $\dfrac{4}{25}$

CHECK Your Understanding

Examples 1 and 2 Write each percent as a decimal.

1. 40%   **2.** 18%   **3.** 0.3%

Examples 3 and 4 Write each decimal as a percent.

4. 0.725   **5.** 1.23   **6.** 0.3

Examples 5 and 6 Write each fraction as a percent.

7 $\dfrac{11}{25}$ **8.** $\dfrac{13}{40}$ **9.** $\dfrac{5}{6}$

Example 7 **10.** **HOMEWORK** At Hancock Middle School, 57% of the eighth-grade students spend at least 30 minutes a day on math homework. Of the seventh-grade students, 0.5 study this long. In which grade do a greater percent of students spend at least 30 minutes a day on math homework?

Example 8 Order each set of numbers from least to greatest.

11. $\dfrac{17}{25}$, 60%, 0.062, $\dfrac{13}{20}$ **12.** 0.99, $\dfrac{9}{10}$, 9%, $\dfrac{19}{20}$

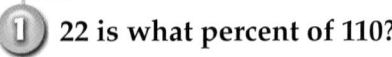

1 **22 is what percent of 110?**

Method 1 **Use the percent proportion.**

$$\begin{matrix} \text{part} \rightarrow \\ \text{whole} \rightarrow \end{matrix} \left. \frac{22}{110} = \frac{n}{100} \right\} \text{percent} \qquad \text{Write the percent proportion.}$$

$$22 \cdot 100 = 110 \cdot n \qquad \text{Find the cross products.}$$

$$2{,}200 = 110n \qquad \text{Multiply.}$$

$$\frac{2{,}200}{110} = \frac{110n}{110} \qquad \text{Divide each side by 110.}$$

$$20 = n \qquad \text{Simplify.}$$

Method 2 **Use the percent equation.**

$$\underbrace{\text{part}} = \underbrace{\text{percent}} \cdot \underbrace{\text{whole}}$$

$$22 = n \cdot 110 \qquad \text{Write the percent equation.}$$

$$\frac{22}{110} = \frac{110n}{110} \qquad \text{Divide each side by 110.}$$

$$0.2 = n \qquad \text{Simplify.}$$

So, 22 is 20% of 110.

Read Math

Percents The *whole* usually follows the word *of*.

2 **What number is 80% of 500?**

Method 1 **Use the percent proportion.**

$$\begin{matrix} \text{part} \rightarrow \\ \text{whole} \rightarrow \end{matrix} \left. \frac{p}{500} = \frac{80}{100} \right\} \text{percent} \qquad \text{Write the percent proportion.}$$

$$p \cdot 100 = 500 \cdot 80 \qquad \text{Find the cross products.}$$

$$100p = 40{,}000 \qquad \text{Multiply.}$$

$$\frac{100p}{100} = \frac{40{,}000}{100} \qquad \text{Divide each side by 100.}$$

$$p = 400 \qquad \text{Simplify.}$$

Study Tip

Percents When using the percent equation, be sure to write the percent in decimal form.

Method 2 **Use the percent equation.**

$$\underbrace{\text{part}} = \underbrace{\text{percent}} \cdot \underbrace{\text{whole}}$$

$$p = 0.8 \cdot 500 \qquad \text{Write the percent equation.}$$

$$p = 400 \qquad \text{Multiply.}$$

So, 400 is 80% of 500.

CHOOSE Your Method

a. 17 is what percent of 68? **b.** 12 is what percent of 50?

c. What number is 35% of 48? **d.** Find 85% of 18.

EXAMPLE The Percent Proportion and Equation

3 **14.4 is 32% of what number?**

Method 1 Use the percent proportion.

$$\begin{array}{l} \text{part}\rightarrow\dfrac{14.4}{w} = \dfrac{32}{100} \end{array}\bigg\}\; \text{percent} \quad \text{Write the percent proportion.}$$

$$14.4 \times 100 = 32 \cdot w \qquad\qquad \text{Find the cross products.}$$

$$1{,}440 = 32w \qquad\qquad \text{Multiply.}$$

$$\dfrac{1{,}440}{32} = \dfrac{32w}{32} \qquad\qquad \text{Divide each side by 32.}$$

$$45 = w \qquad\qquad \text{Simplify.}$$

Method 2 Use the percent equation.

$$\underbrace{\text{part}} = \underbrace{\text{percent}} \cdot \underbrace{\text{whole}}$$

$$14.4 = 0.32 \cdot w \qquad\qquad \text{Write the percent equation.}$$

$$\dfrac{14.4}{0.32} = \dfrac{0.32w}{0.32} \qquad\qquad \text{Divide each side by 0.32.}$$

$$45 = w$$

So, 14.4 is 32% of 45.

CHOOSE Your Method

e. 23.4 is 30% of what number? **f.** 19 is 62.5% of what number?

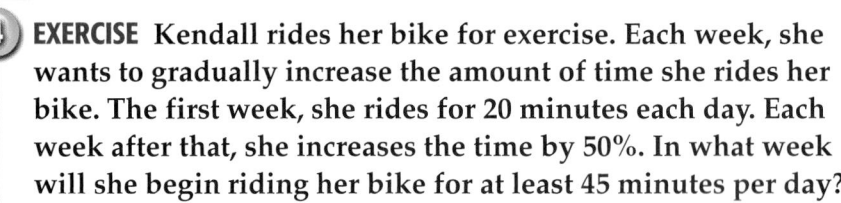 REAL-WORLD EXAMPLE

4 **EXERCISE** Kendall rides her bike for exercise. Each week, she wants to gradually increase the amount of time she rides her bike. The first week, she rides for 20 minutes each day. Each week after that, she increases the time by 50%. In what week will she begin riding her bike for at least 45 minutes per day?

Each week she will increase the time by 50%.

Week 1 20 minutes

Week 2 $20 + 20 \times 0.5 = 20 + 10$ or 30 minutes

Week 3 $30 + 30 \times 0.5 = 30 + 15$ or 45 minutes

In week 3, she will be riding for at least 45 minutes.

Real-World Link · · · · · ·
There is no federal law requiring minors to wear a bicycle helmet. However, 21 states have passed state laws requiring helmets for certain ages.

CHECK Your Progress

g. MONEY Michael deposits $20 into his savings account each month. He wants to increase the amount he deposits by 25% until he is depositing at least $50 per month. In what month will he deposit at least $50?

Concept Summary — The Percent Proportion and Equation

Type	Example	Proportion	Equation
Find the Percent	15 is what percent of 60?	$\frac{15}{60} = \frac{n}{100}$	$15 = n(60)$
Find the Part	What number is 25% of 60?	$\frac{p}{60} = \frac{25}{100}$	$p = 0.25(60)$
Find the Whole	15 is 25% of what number?	$\frac{15}{w} = \frac{25}{100}$	$15 = 0.25w$

 EXAMPLE Percents Greater than 100

 6 is what percent of 5?

$$\begin{array}{l}\text{part} \to \\ \text{whole} \to\end{array} \left. \frac{6}{5} = \frac{n}{100} \right\} \text{percent} \qquad \text{Write the percent proportion.}$$

$$6 \cdot 100 = 5 \cdot n \qquad \text{Find the cross products.}$$

$$600 = 5n \qquad \text{Multiply.}$$

$$\frac{600}{5} = \frac{5n}{5} \qquad \text{Divide each side by 5.}$$

$$120 = n \qquad \text{Simplify.}$$

6 is 120% of 5.

 **CHECK Your Progress**

h. 12 is what percent of 6?   **i.** Find 175% of 18.

 CHECK Your Understanding

Examples 1–3 and 5   Solve each problem using a percent proportion.

 70 is what percent of 280?   **2.** What percent of 49 is 7?

3. Find 118% of 19.   **4.** Find 72% of 200.

5. 151.5 is 75% of what number?   **6.** 126 is 30% of what number?

Examples 1–3 and 5   Solve each problem using a percent equation.

7. 25 is what percent of 625?   **8.** What percent of 800 is 2?

9. Find 85% of 920.   **10.** What number is 4% of 30?

11. 680 is 34% of what number?   **12.** 25% of what number is 10?

Example 4   **13. PROFIT** A dealership sets car prices so that there is a 40% profit. If the dealership paid $5,300 for a car, for how much should they sell the car?

Practice and Problem Solving

 = **Step-by-Step Solutions** begin on page R1.
Extra Practice begins on page EP2.

Examples 1–3 and 5 Solve each problem using a percent proportion.

14. 3 is what percent of 15?

15. 120 is what percent of 360?

16. What is 15% of 60?

17. What is 17% of 350?

18. 18 is 45% of what number?

19. 95 is 95% of what number?

20. 15.12 is what percent of 12?

21. Find 250% of 57.

Examples 1–3 and 5 Solve each problem using a percent equation.

22. Find 60% of 30.

23. What is 40% of 90?

24. What percent of 90 is 36?

25. 45 is what percent of 150?

26. 75 is 50% of what number?

27. 15% of what number is 30?

28. What number is 13% of 52?

29. Find 24% of 84.

Example 4 **30. BRACES** In a recent survey, 34% of kids said they will get dental braces. If 28,800 kids were surveyed, how many will get braces?

31 **TESTS** Michaela is studying for an end of the year test. Each week, she wants to gradually increase her total study time. The first week, she studies for 15 minutes each day. Each week after that, she increases the time by 40%. In what week will she begin studying for at least 40 minutes per day?

Write a percent proportion or equation to solve each problem. Then solve. Round to the nearest tenth if necessary.

32. What is 2.5% of 95?

33. 4 is what percent of 550?

34. 98 is 22.5% of what number?

35. Find 5.8% of 42.

36. What percent of 110 is 1?

37. 57 is 13.5% of what number?

38. SNACKS The card at the right shows a snack recipe.

    a. Chocolate chips make up what percent of the recipe?

    b. The amount of powdered sugar used is what percent of the amount of cereal used?

    c. Which ingredient is 67% of the total recipe?

Kid Chow

$\frac{3}{4}$ cup peanut butter

1 cup chocolate chips

$\frac{1}{4}$ cup butter

8 cups rice squares cereal

2 cups powdered sugar

39 **SPORTS** During the season, a college basketball team won 87.5% of their games. If they played 40 games, how many did they win?

40. MATH IN THE MEDIA In a newspaper or magazine, find an advertised item on sale. What percent of the original price is the sale price?

41. CHALLENGE Choose any two numbers, x and y. Find $x\%$ of y and $y\%$ of x. Will the results always be the same? Explain.

42. ✏ **WRITE MATH** Alonzo scored a 79% on his first test of the quarter. Will a score of 38 out of 45 on the next test help or hurt his grade? Explain your reasoning.

✔ Test Practice

43. A baseball stadium manager expects that 60% of the fans at a game will buy at least $3.00 in concessions. If there are 5,600 fans at a game, which statement does NOT represent the manager's expectation?

   A. 3,360 fans each will buy at least $3.00 in concessions.

   B. 2,240 fans each will buy fewer than $3.00 in concessions.

   C. More than $\frac{1}{2}$ of the fans each will buy at least $3.00 in concessions.

   D. Less than $\frac{2}{5}$ of the fans each will buy fewer than $3.00 in concessions.

44. Student Council has a budget of $5,000. The table shows the portion of the budget that is spent on different items.

Student Council Budget	
Item	**Portion of Budget**
Decorations	35%
Supplies	9%
T-shirts	22%
Activities	?

How much is spent on school activities?

   F. $1,100   **H.** $1,750

   G. $1,700   **I.** $3,300

45. ✏ **GRIDDED RESPONSE** Mr. Dempsey receives a 7% commission for every appliance he sells. If he sells a refrigerator for $1,299, what is his commission in dollars?

Write each percent as a decimal. (Lesson 2B)

46. 45%   **47.** 87.5%   **48.** 0.6%   **49.** $22.\overline{2}\%$

50. SCHEDULES Busses arrive at the station at 11:10 A.M., 11:32 A.M., 11:54 A.M., and 12:16 P.M. If this pattern continues, what time will the next bus arrive at the station? Use the *look for a pattern* strategy. (Lesson 2A)

51. SPORTS The table shows the regular season records of five college baseball teams during a recent season. Which team had the best record? (*Hint:* Divide the number of games won by the number of games played.) (Lesson 1A)

Team	Games Won	Games Played
Illinois State University	25	48
University at Albany	26	58
University of Arkansas	41	65
University of Virginia	49	65
Vanderbilt University	37	64

Main Idea

Apply percents to find discount, markup, and sales tax.

Vocabulary

discount
markup
selling price
sales tax

 Get ConnectED

Discount, Markup, and Sales Tax

 SALES A souvenir shop is selling beach towels for 30% off the original price.

1. Calculate the discount by finding 30% of $25.

2. What is the new cost of a beach towel?

3. Multiply 0.70 and 25. How does the result compare to your answer in Exercise 2?

SALE!
All beach towels
30% off!
Regular price $25

Discount is the amount by which a regular price is reduced. You can find the sale price of an item by subtracting the discount from the original price.

REAL-WORLD EXAMPLE **Find the Sale Price**

1 **MUSIC** The CD Discount Superstore is advertising a 20% off sale. Jonas wants to buy a CD that originally costs $18.50. Find the sale price of the CD.

Method 1 **Find the amount of the discount first.**

The percent is 20% and the whole is 18.50. We need to find the amount of the discount, or the part. Let d represent the amount of discount.

$d = 0.20 \cdot 18.50$ Write the percent equation.

$d = 3.70$ Multiply.

Subtract the amount of the discount from the original price to find the sale price. $18.50 - $3.70 = $14.80.

Method 2 **Find the percent paid first.**

If the amount of the discount is 20%, the percent paid is $100\% - 20\%$ or 80%. Find 80% of $18.50. Let s represent the sale price.

$s = 0.80 \cdot 18.50$ Write the percent equation.

$s = 14.80$ Multiply.

The sale price of the CD is $14.80.

 **CHOOSE** Your Method

Find the sale price of each item to the nearest cent.

a. CD: $14.50, 10% off **b.** sweater: $39.95, 25% off

A store sells an item for more than it paid for that item. The extra money is used to cover the expenses and to make a profit. The increase in the price is called the **markup**. The amount the customer pays is called the **selling price**.

 REAL-WORLD EXAMPLE Find the Selling Price

② **BUSINESS** A bead store buys beads at wholesale prices and then prices them to sell at a 75% markup. If a strand of beads costs the store $9.14, what is the selling price for the strand?

$9.14

Study Tip

Check for Reasonableness To estimate the selling price, think 75% of 9.14 is about $\frac{3}{4}$ of 10 or 7.50. The selling price should be about $9 + $7.50, or $16.50.

 **Method 1** **Find the amount of the markup first.**

The whole is $9.14. The percent is 75%. You need to find the amount of the markup, or the part. Let m represent the amount of the markup.

$m = 0.75 \cdot 9.14$ Multiply the percent equation.
$m \approx 6.86$ Multiply.

Add the markup $6.86 to the store's cost $9.14 to find the selling price. $9.14 + $6.86 = $16.00

Method 2 **Find the total percent first.**

The customer will pay 100% of the store's cost plus an extra 75% of the cost. Find 100% + 75% or 175% of the store's cost. Let p represent the price.

$p = 1.75 \cdot 9.14$ Multiply the percent equation.
$p \approx 16.00$ Multiply.

The selling price of the beads is $16.00.

 CHOOSE Your Method

Find the selling price for each item given the percent of markup.

c. digital camera: $120, 55% markup

d. sunglasses: $7, 30% markup

e. **SHIPPING** Cheng-Yu ordered a book that cost $24 from an online store. Her total with the shipping charge was $27. What was the percent of markup charged for shipping?

Sales tax is an additional amount of money charged on certain goods and services. You can find the total cost of an item by adding the selling price and the sales tax.

 REAL-WORLD EXAMPLE **Find the Total Cost**

3 **GAMES** A board game that costs $25 is on sale for 15% off. The sales tax is 6.25%. What is the total cost of the board game?

Step 1 Find the price of the game after the discount.

Let d represent the total discount.

$\underbrace{\text{part}} = \underbrace{\text{percent}} \cdot \underbrace{\text{whole}}$

d	$=$	$0.15 \cdot 25$	Write the percent equation.
d	$=$	3.75	Multiply.

Subtract the discount from the original price to find the sale price. $\$25 - \$3.75 = \$21.25$.

Step 2 Find the amount of the sales tax.

Let t represent the sales tax.

$\underbrace{\text{part}} = \underbrace{\text{percent}} \cdot \underbrace{\text{whole}}$

t	$=$	$0.0625 \cdot 21.25$	Write the percent equation.
t	$\approx$	1.33	Multiply.

Add the sales tax to the sale price to find the total price. $\$21.25 + \$1.33 = \$22.58$.

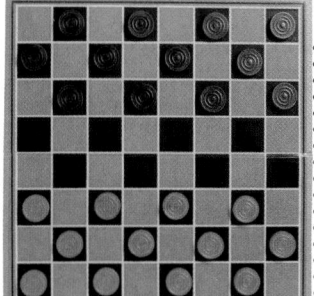

Real-World Link · · · ·
Checkers is one of the oldest and most popular board games. A version of checkers has been played as far back as 3000 B.C.

 CHECK Your Progress

f. BIKES A mountain bike is originally priced at $125. A flyer advertises the bike for 10% off. If there is a 7.85% sales tax, what is the total cost of the mountain bike?

✓ **CHECK Your Understanding**

Example 1   **1. BICYCLES** Find the sale price of a bicycle that is regularly $140 and is on sale for 40% off the original price.

Example 2   **Find the selling price for each item given the percent of markup.**

  2. roller blades: $60, 35% markup   **3.** coat: $87, 33% markup

  4. sunglasses: $25, 40% markup   **5.** snack mix: $7, 51% markup

Example 3   **Find the total cost of each item to the nearest cent.**

  6. DVD player: $75, 25% off, 5% tax   **7** shirt: $19, 10% off, 7.5% tax

  8. car: $8,500, 5% off, 8.25% tax   **9.** trampoline: $130, 15% off, 6.75% tax

Practice and Problem Solving

 = **Step-by-Step Solutions** begin on page R1.
Extra Practice begins on page EP2.

Examples 1 and 3 **Find the sale price or total cost of each item to the nearest cent.**

10. video game: $75, 25% discount
11. mountain bike: $399, 15% discount
12. skateboard: $119.95, 30% discount
13 earrings: $19.50, 35% discount
14. airplane ticket: $275, 6.5% tax
15. coat: $110, 7.85% tax
16. hotel room: $69.95, 5.25% tax
17. MP3 player: $220, 7.15% tax

18. SHOPPING Monica is shopping for clothes at a department store. Before tax, her bill is $95.

a. She has a coupon to receive an additional 25% off her total purchase. What is the total cost of her items before tax?

b. After she receives the discount, how much will her total bill be if there is a 7.95% sales tax?

Example 2 **Find the selling price for each item given the cost to the store and the markup.**

19. computer: $700, 30% markup
20. CD player: $120, 20% markup
21. jeans: $25, 45% markup
22. baseball cap: $12, 48% markup

23 ALGEBRA Students receive a 20% discount off the price of an adult ticket at the theater. If a student ticket is $6.80, find the price of an adult ticket. (*Hint:* Let p represent the part and $p + 6.80$ represent the whole.)

Find the percent of discount given the original price and sale price of each item.

24. coat: original price: $65
    sale price: $42.25
25. shoes: original price: $90
    sale price: $67.50

26. GRAPHIC NOVEL Refer to the graphic novel frame below for Exercises a–c.

a. Find the sale price of the sweater.

b. What is the price of the sweater after the 25% off coupon is used?

c. Jasmine thinks that because the sweater is 50% off and there is a 25% off coupon, the sweater is 75% off. Find the cost of the sweater if it is 75% off. Compare this answer to your answer from part **b.**

27. CHALLENGE Jordyn purchased a sweater which originally cost x dollars. The sale price was 20% off the retail, and she had a coupon which took an additional 15% off the sale price. Sales tax of 6.5% was added to the cost at the end. Write the final cost of the sweater in terms of x.

28. REASONING Should the sales tax *always*, *sometimes*, or *never* be calculated first when dealing with discounts? Explain your reasoning.

29. WRITE MATH Explain the difference between discount and markup.

Test Practice

30. A television was originally priced at $1,250. It is now on sale.

What is the sale price of the television?

A. $875   C. $425

B. $675   D. $375

31. Grace and her two brothers shared the cost of a new video game system equally. The original price of the system was $179. They received a 15% discount off the original price and paid 7.5% sales tax on the discounted price. Find the approximate amount that each paid for the video game system.

F. $51   H. $60

G. $55   I. $66

32. SHORT RESPONSE Pancho paid $36 for a video game that was originally priced at $45. What was the percent of discount?

Solve each problem using a percent equation. (Lesson 2C)

33. Find 40% of 45.

34. What percent of 110 is 44?

35. 60 is 40% of what number?

36. What is 20% of 180?

37. SCHOOL A recent survey asked parents to grade themselves based on their involvement in their children's education. The results are shown at the right. (Lesson 2B)

a. Write the percent of parents who gave themselves an A as a decimal and as a fraction in simplest form.

b. Did more or less than $\frac{2}{5}$ of parents give themselves a B?

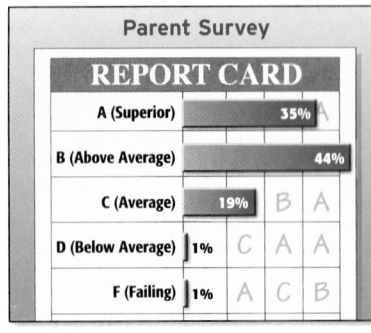

Parent Survey	
REPORT CARD	
A (Superior)	35%
B (Above Average)	44%
C (Average)	19%
D (Below Average)	1%
F (Failing)	1%

Main Idea

Solve problems involving simple and compound interest.

 Vocabulary

interest
simple interest
principal
compound interest

Financial Literacy: Interest

CARS Have you ever dreamed of buying your first car? Suppose you want to buy a used car that costs $4,000. You can pay $400 now and borrow the remaining $3,600. To pay off the loan, you will pay $131.50 each month for the next 36 months.

1. How much money will you pay in all for the car?

2. How much will it cost you to borrow the money for the car?

Interest is the amount of money paid or earned for the use of money. For a savings account, you earn interest from the bank. For a credit card or a loan, you pay interest to the bank. **Simple interest** is paid only on the initial principal of a savings account or loan. To solve problems involving simple interest, use the following formula.

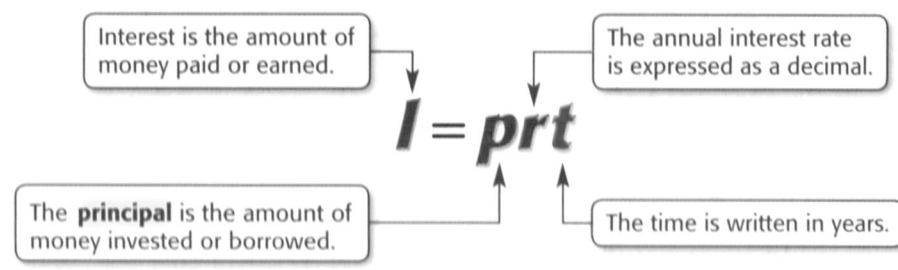

Interest is the amount of money paid or earned.

The annual interest rate is expressed as a decimal.

$$I = prt$$

The **principal** is the amount of money invested or borrowed.

The time is written in years.

 EXAMPLE Find Simple Interest

 Find the simple interest for $500 invested in a savings account at 3.25% for 3 years.

$I = prt$	Write the simple interest formula.
$I = 500 \cdot 0.0325 \cdot 3$	Replace p with 500, r with 0.0325, and t with 3.
$I = 48.75$	

The simple interest is $48.75.

 CHECK **Your Progress**

Find the simple interest to the nearest cent.

a. $400 at 3.67% for 2 years   **b.** $770 at 16% for 6 months

 REAL-WORLD EXAMPLE **Find the Interest Rate**

2 **STUDENT LOANS** Luis makes monthly payments of $290.28 on his loan of $5,000. He plans to pay it off in $1\frac{1}{2}$ years. Find the simple interest rate of his loan.

First find the total that Luis will pay.

$290.28 \cdot 18 = \$5,225.04$ $1\frac{1}{2}$ years = 18 months

He will pay $5,225.04 − $5,000 or $225.04 in interest. So, $I = 225.04$.

$I = prt$	Write the simple interest formula.
$225.04 = 5,000 \cdot r \cdot 1.5$	Replace *I* with 225.04, *p* with 5,000, and *t* with 1.5.
$225.04 = 7,500r$	Simplify.
$\dfrac{225.04}{7,500} = \dfrac{7,500r}{7,500}$	Divide each side by 7,500.
$0.03 \approx r$	

The simple interest rate is about 0.03 or 3%.

✔ CHECK Your Progress

c. **SAVINGS BOND** Louie purchased a $200 savings bond. After 5 years, it is worth $232.50. Find the simple interest rate for his bond.

Compound interest is paid on the initial principal and on interest earned in the past.

 EXAMPLE **Find the Total Amount**

3 How much money is in an account where $1,500 is invested at an interest rate of 5.75% compounded annually for 2 years?

Find the amount of money in the account at the end of the year 1.

$I = prt$	Write the simple interest formula.
$I = 1,500 \cdot 0.0575 \cdot 1$	Substitution
$I = 86.25$	Simplify.
$1,500 + 86.25 = 1,586.25$	Add the amount invested and the interest.

Find the amount of money in the account at the end of year 2.

$I = prt$	Write the simple interest formula.
$I = 1,586.25 \cdot 0.0575 \cdot 1$	Substitution
$I = 91.2094$	Simplify.

So, after 2 years there is $1,586.25 + $91.21 or $1,677.46.

✔ CHECK Your Progress

d. What is the total amount in an account where $875 is invested at an interest rate of 12% compounded annually for 3 years?

Example 1   **Find the simple interest to the nearest cent.**

1. $300 at 7.5% for 5 years

2. $230 at 12% for 8 months

Example 2   **3. CAR SALES** Keisha borrowed $4,000 to buy a car. If her monthly payments are $184.17 for 2 years, what is the simple interest rate for her loan?

Example 3   **Find the total amount in each account to the nearest cent, if the interest is compounded annually.**

4. $660 at 5.25% for 2 years

5. $385 at 12.6% for 4 years

6. Nina invested $1,000 in an account for 3 years. Find the total amount if the interest is compounded annually at a rate of 2.75%.

Practice and Problem Solving

● = **Step-by-Step Solutions** begin on page R1.
Extra Practice begins on page EP2.

Example 1   **Find the simple interest to the nearest cent.**

7. $250 at 6% for 3 years

8. $725 at 4.5% for 4 years

9. $834 at 7.25% for 2 months

10. $3,070 at 8.65% for 24 months

Example 2   **11. BASEBALL CARDS** The prices for a vintage baseball card are given at the right. Determine the simple interest rate for a card purchased as an investment in 1968 and sold in 2011.

Year	Price for Baseball Card
1968	$18.00
2011	$325.80

12. SAVINGS Colin is borrowing money from his parents to purchase a $700 computer. He will pay them $35 per month for two years. Determine the simple interest rate on Colin's loan.

Example 3   **Find the total amount in each account to the nearest cent, if the interest is compounded annually.**

13 $2,250 at 5% for 3 years

14. $5,060 at 7.2% for 2 years

15. $575 at 4.25% for 18 months

16. $950 at 7.85% for 3 years

17 **CARS** Felicia took out a 5-year loan for $15,000 to buy a car. If the interest is compounded annually at a rate of 11%, how much will she pay in all?

Find the simple interest to the nearest cent.

18. $1,000 at $7\frac{1}{2}$% for 30 months

19. $5,200 at $13\frac{1}{5}$% for $1\frac{1}{2}$ years

20. CREDIT CARDS The balance on a credit card was $500. Mrs. Cook paid the minimum monthly payment of $25. The remaining balance was charged a simple interest rate of 18%. If no additional purchases were made, what was the balance the next month?

21. HOUSING The Turners need to borrow $100,000 to purchase a home. The credit union is offering a 30-year mortgage loan at 5.38% interest while the community bank has a 25-year mortgage loan at 6.12% interest. Assuming simple interest, which loan will result in less total interest?

22. **CHALLENGE** What will be the monthly payments on a loan of $25,000 at 9% simple interest so that it will be paid off in 15 years? How much will the total interest be?

23. **OPEN ENDED** Give a principal and interest rate where the amount of simple interest earned in 6 months is more than $100.

24. ✏️ **WRITE MATH** If you have money in a savings account for 8 months, what value for t would you use in the formula $I = prt$ to find the interest you have earned? Explain.

Test Practice

25. Mrs. Owens placed $1,500 in a college savings account with a simple interest rate of 4% when Lauren was born. How much will be in the account in 18 years when Lauren is ready to go to college? Assume no more deposits or withdrawals are made.

    A. $1,080

    B. $2,580

    C. $10,800

    D. $12,300

26. Dave borrowed $4,000 from the bank at an interest rate of 9%. The interest is compounded annually. Suppose he made no payments, approximately how much would he owe at the end of three years?

    F. $1,080

    G. $1,180

    H. $5,080

    I. $5,180

27. **DISCOUNT** A watch that regularly sells for $35 is on sale for $26.95. Find the percent of discount. (Lesson 3A)

28. **COLORS** The table lists the number of each color of candies in a jar. (Lesson 2C)

    a. What percent of the candies are brown?

    b. What percent of the candies are green?

Color	Number
Brown	12
Green	5
Yellow	4
Red	2
Orange	1
Blue	1

29. Order the set of numbers $\frac{1}{6}$, 16%, and 0.016 from least to greatest. (Lesson 2B)

30. **VEGETABLES** Hudson purchased $3\frac{3}{8}$ pounds of vegetables that cost $3 per pound. What was the total cost of the vegetables? (Lesson 1C)

Write each decimal as a fraction or mixed number in simplest form. (Lesson 1A)

31. -5.24

32. $7.\overline{6}$

33. $-12.\overline{75}$

34. 0.625

Main Idea

Find compound interest.

Spreadsheet:
Compound Interest

You can use a spreadsheet to investigate compound interest.

ACTIVITY

Find the value of a $2,000 savings account after four years if the account pays 8% interest compounded semiannually.
8% interest compounded semiannually means that the interest is paid twice a year. The interest rate is 8% ÷ 2 or 4% every 6 months.

Compound Interest   ⬓ ▣ ✕

	A	B	C	D
1	Rate	0.04		
2				
3	Principal	Interest	New Principal	Time (YR)
4	$2000.00	$80.00	$2080.00	0.5
5	$2080.00	$83.20	$2163.20	1
6	$2163.20	$86.53	$2249.73	1.5
7	$2249.73	$89.99	$2339.72	2
8	$2339.72	$93.59	$2433.31	2.5
9	$2433.31	$97.33	$2530.64	3
10	$2530.64	$101.23	$2631.86	3.5
11	$2631.86	$105.27	$2737.14	4
12				

The interest rate is entered as a decimal.

The spreadsheet evaluates the formula A4 × B1.

The interest is added to the principal every 6 months. The spreadsheet evaluates the formula A4 + B4.

◄◄ ◄ ► ►◄ Sheet 1 ╱ Sheet 2 ╱ Sheet 3 ╱

The value of the savings account after four years is $2,737.14.

Practice and Apply

1. Use a spreadsheet to find the value of a savings account if $2,000 is invested for four years at 8% interest compounded quarterly.

2. Suppose you leave $1,000 in each of three bank accounts paying 6% interest per year. One account pays simple interest, one pays interest compounded semiannually, and one pays interest compounded quarterly. Use a spreadsheet to find the amount of money in each account after three years.

3. **MAKE A CONJECTURE** How does the amount of interest change if the compounding occurs more frequently? Explain your reasoning.

Main Idea

Find and use the percent of increase or decrease.

 Vocabulary

percent of change
percent of increase
percent of decrease

 Get ConnectED

Percent of Change

STAMPS Over the years, the price of stamps has increased. Refer to the table that shows the change in stamp prices from 1963 to 1981.

Price of a Stamp	
Effective Date	**Price for the First Ounce (¢)**
January 7, 1963	5
March 2, 1974	10
May 29, 1978	15
November 1, 1981	20

1. How much did the price increase from 1963 to 1974?

2. Write the ratio $\dfrac{\text{amount of increase}}{\text{price in 1963}}$. Then write the ratio as a percent.

3. How much did the price increase from 1974 to 1978? Write the ratio $\dfrac{\text{amount of increase}}{\text{price in 1974}}$. Then write the ratio as a percent.

4. How much did the price increase from 1978 to 1981? Write the ratio $\dfrac{\text{amount of increase}}{\text{price in 1978}}$. Then write the ratio as a percent.

5. **MAKE A CONJECTURE** Why are the amounts of increase the same but the percents different?

The percent that an amount changes from its original amount is called the **percent of change**.

Key Concept   **Percent of Change**

Words   A percent of change is a ratio that compares the change in quantity to the original amount.

Symbols   percent of change $= \dfrac{\text{amount of change}}{\text{original amount}}$

To find the percent of change, do the following:

Step 1   Subtract the original amount from the final amount to find the amount of change.

Step 2   Write the ratio $\dfrac{\text{amount of change}}{\text{original amount}}$ as a decimal.

Step 3   Write the decimal as a percent.

When the percent is positive, the percent of change is a **percent of increase**. When the percent is negative, the percent of change is called a **percent of decrease**.

Study Tip

Percent of Change
When finding percent of change, always use the original amount as the whole.

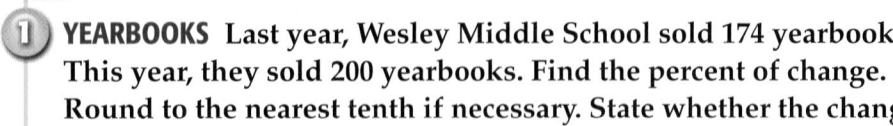

REAL-WORLD EXAMPLES Find Percent of Change

1 **YEARBOOKS** Last year, Wesley Middle School sold 174 yearbooks. This year, they sold 200 yearbooks. Find the percent of change. Round to the nearest tenth if necessary. State whether the change is an *increase* or *decrease*.

Step 1 Subtract to find the amount of change.
$200 - 174$ or 26.   final amount − original amount

Step 2 percent of change $= \dfrac{\text{amount of change}}{\text{original amount}}$ Definition of percent of change

$= \dfrac{26}{174}$ Substitution

≈ 0.1494252 Divide. Use a calculator.

Step 3 $0.1494252 = 14.94252\%$ or 14.9%.

Since the percent of change is positive, it is a percent of increase.

2 **WEATHER** On average, Florida has 54 inches of rainfall per year, but in 2007, it had only 45 inches of rainfall. Find the percent of change. Round to the nearest tenth if necessary. State whether the change is an *increase* or *decrease*.

Step 1 Subtract to find the amount of change.
$45 - 54$ or -9 final amount − original amount

Step 2 percent of change $= \dfrac{\text{amount of change}}{\text{original amount}}$ Definition of percent of change

$= \dfrac{-9}{54}$ Substitution

$\approx -0.1\overline{6}$ Divide. Use a calculator.

Step 3 Express the ratio as a percent.

$-0.1\overline{6} = -16.666...\%$ or -16.7%

Since the percent of change is negative, it is a percent of decrease.

Real-World Link · · · · ·
The National Oceanic and Atmospheric Administration or NOAA is the government agency responsible for monitoring the weather in the United States. The agency is able to monitor storms such as hurricanes from the sky.

CHECK Your Progress

Find each percent of change. Round to the nearest tenth if necessary. State whether the percent of change is an *increase* or a *decrease*.

a. original: 6 hours
new: 10 hours

b. original: 80 water bottles
new: 55 water bottles

Examples 1 and 2 **Find each percent of change. Round to the nearest tenth if necessary. State whether the percent of change is an *increase* or a *decrease*.**

1. original: $40
   new: $32

2. original: 25 CDs
   new: 32 CDs

3. original: 325 miles
   new: 400 miles

4. **BOWLING** Inez bowled a 127 her first game. In her second game, she bowled a 145. Find the percent of change. Round to the nearest tenth if necessary. State whether the change is an *increase* or a *decrease*.

Practice and Problem Solving

= **Step-by-Step Solutions** begin on page R1.
Extra Practice begins on page EP2.

Examples 1 and 2 **Find each percent of change. Round to the nearest tenth if necessary. State whether the percent of change is an *increase* or a *decrease*.**

5. original: 6 tickets
   new: 9 tickets

6. original: 27 guests
   new: 39 guests

7. original: $80
   new: $64

8. original: $560
   new: $420

9. original: 68°F
   new: 51°F

10. original: 150 E-mails
    new: 98 E-mails

11. **TELEVISION** On Tuesday night, 17.8 million households watched a popular television show. On Wednesday night, 16.6 million households watched the same show. Find the percent of decrease in the number of households watching the show from Tuesday to Wednesday.

12. **STOCK** Patrice invested $300 in a particular stock. The amount doubled within a few weeks. Find the percent of increase.

13. **INTERNET** An Internet service provider offers a connection speed that is 35% faster than dial-up. If it takes Brad 8 seconds to connect to the Internet using dial-up, how long would it take using this provider?

14. **GRAPHIC NOVEL** Refer to the graphic novel frame below for Exercises a–b.

a. What is the price of the shirt if Alma uses the 25% off coupon?

b. Find the percent of change from the original price. Is the percent of change a percent of increase or decrease?

H.O.T. Problems

15. **OPEN ENDED** Give an example of a real-world situation where a percent of change occurs.

16. **CHALLENGE** The percent of change over Ethan's last three test scores is 15%. If he scored a 92 on the last test, what are possible scores for the first two tests?

17.  **WRITE MATH** What is the difference between percent of increase and percent of decrease?

Test Practice

18. On Monday, the high temperature was 66°F. On Wednesday, it was 79°F. What was the percent of change?

   A. −19.7%   C. 16.5%

   B. −16.5%   D. 19.7%

19. If the dimensions of the square below are doubled, what is the percent of increase in the area of the square?

   5.5 cm

   5.5 cm

   F. 100%   H. 300%

   G. 200%   I. 400%

20. Which of the following represents the greatest percent of decrease?

   A. A pair of jeans originally priced at $22 on sale for $15.

   B. The number of pages left to read decreased from 281 to 180 pages.

   C. The paper route went from 45 houses to 30 houses.

   D. Attendance decreased from 176 on Friday to 114 on Saturday.

21. **SHORT RESPONSE** A pair of boots costs $130. They are on sale for $110.50. Find the percent of change.

22. **SAVINGS** The Millers opened a savings account for their newborn son with $430. Find the total amount in the account after 3 years if the simple interest rate is 2.5%. (Lesson 3B)

23. **SALES** What is the sale price of a $200 cell phone on sale at 10% off the regular price? (Lesson 3A)

GEOMETRY Find the perimeter of each figure. (Lesson 1B)

24.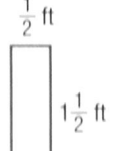
   $\frac{1}{2}$ ft

   $1\frac{1}{2}$ ft

25.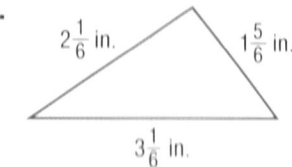
   $2\frac{1}{6}$ in. $1\frac{5}{6}$ in.

   $3\frac{1}{6}$ in.

Feeding the Animals

Are you a curious person who is good with animals and interested in science? If so, then you should consider a career in animal nutrition. An animal nutritionist determines what kinds and amounts of food are best for different animals. Their goal is to make animals healthier through proper nutrition. Animal nutritionists can do research at universities, work for clients on farms and ranches, or work at zoos and aquariums.

21st Century Careers

Are you interested in a career as an animal nutritionist? Take some of the following courses in high school.

- Algebra
- Animal Science and Services
- Biology
- Chemistry
- General Zoology

 Get ConnectED

Weekly Food for Zoo Animals

Animal	Food Needed	Cost ($)
Garden eels, fish	20,000 brine shrimp	190
Koalas	180 pounds of eucalyptus leaves	780
Manatees	840 heads of romaine lettuce	630
	770 pounds of beef	970
Multiple animals	150 pounds of fish	150
	97,000 mealworms	550

Real-World Math

Use the information in the table to solve each problem.

1. If the sales tax on the eucalyptus leaves is 6.75%, what is the total weekly cost?

2. What is the total weekly cost of the mealworms if the sales tax is $5\frac{1}{4}\%$?

3. Romaine lettuce is on sale for 15% off. What is the sale price of 840 heads of the lettuce?

4. If the cost of the brine shrimp is discounted 12%, what is the weekly cost?

5. If the cost of the beef represents a 20% discount, what was the original price?

6. The total weekly food budget for the zoo in the table is about $9,600. What percent of the budget goes to feed the koalas?

Chapter Study Guide and Review

FOLDABLES® Study Organizer

Be sure the following Key Concepts are noted in your Foldable.

Rational Numbers and Percent
1 Rational Numbers
2 Percents
3 Apply Percents
Vocabulary
Review

Key Concepts

Rational Numbers (Lesson 1)
• A rational number is any number that can be expressed in the form $\frac{a}{b}$, where a and b are integers and $b \neq 0$.

Operations with Rational Numbers
(Lesson 1)
• To add and subtract fractions, rename the fractions using the least common denominator. Then add or subtract and simplify, if necessary.
• To multiply fractions, multiply the numerators and multiply the denominators.
• To divide by a fraction, multiply by its multiplicative inverse.

Percents (Lesson 2)
• A percent is a ratio that compares a number to 100.
• The percent proportion is $\frac{\text{part}}{\text{whole}} = \text{percent}$, where the percent is written as a fraction.
• The percent equation is part = percent · whole, where the percent is written as a decimal.

Key Vocabulary

compound interest
dimensional analysis
discount
interest
like fractions
markup
multiplicative inverses
percent equation
percent of change
percent proportion

principal
rational number
reciprocals
repeating decimal
sales tax
selling price
simple interest
terminating decimal
unlike fraction

Vocabulary Check

State whether each sentence is *true* or *false*. If *false*, replace the underlined word or number to make a true sentence.

1. <u>Like</u> fractions have the same denominator.

2. <u>Dimensional analysis</u> is the process of including units of measurement in computation.

3. Numbers that can be written as the ratio of two integers in which the denominator is not zero are called <u>reciprocals</u>.

4. The number 2.75 is a <u>repeating</u> decimal.

5. Two numbers with a product of 1 are <u>multiplicative inverses</u> of each other.

6. A <u>proportion</u> is a ratio that compares a number to 100.

7. <u>Markup</u> is a decrease in the selling price of an item.

8. The <u>interest</u> is the money paid for the use of money.

Multi-Part Lesson Review

Rational Numbers

Rational Numbers (Lesson 1A)

Write each decimal as a fraction or mixed number in simplest form.

9. 0.3 **10.** -7.14 **11.** $4.\overline{3}$ **12.** $-5.\overline{7}$

13. BIOLOGY The average rate of human hair growth is about 0.4 inch per month. Write this decimal as a fraction in simplest form.

EXAMPLE 1   Write 0.28 as a fraction in simplest form.

$0.28 = \dfrac{28}{100}$ 0.28 is 28 hundredths.

$ = \dfrac{7}{25}$ Simplify.

The decimal 0.28 can be written as $\dfrac{7}{25}$.

Add and Subtract Rational Numbers (Lesson 1B)

Add or subtract. Write in simplest form.

14. $\dfrac{5}{11} + \dfrac{6}{11}$ **15.** $\dfrac{1}{8} - \dfrac{7}{8}$ **16.** $-4\dfrac{1}{2} - 6\dfrac{2}{3}$

17. JOBS Kala worked $5\dfrac{3}{20}$ hours on Monday and $2\dfrac{13}{20}$ hours on Tuesday. How much longer did Kala work on Monday than on Tuesday?

EXAMPLE 2   Find $\dfrac{3}{4} + \dfrac{1}{3}$. Write in simplest form.

$\dfrac{3}{4} + \dfrac{1}{3} = \dfrac{9}{12} + \dfrac{4}{12}$ Rename the fractions.

$\phantom{\dfrac{3}{4} + \dfrac{1}{3}} = \dfrac{9 + 4}{12}$ Add the numerators.

$\phantom{\dfrac{3}{4} + \dfrac{1}{3}} = \dfrac{13}{12}$ or $1\dfrac{1}{12}$ Simplify.

Multiply Rational Numbers (Lesson 1C)

Multiply. Write in simplest form.

18. $\dfrac{3}{5} \cdot 1\dfrac{2}{3}$ **19.** $-\dfrac{2}{3} \cdot \left(-\dfrac{2}{3}\right)$ **20.** $\dfrac{1}{2} \cdot \dfrac{10}{11}$

21. COOKING Crystal is making $1\dfrac{1}{2}$ times a recipe. The original recipe calls for $3\dfrac{1}{2}$ cups of milk. How many cups of milk does she need?

EXAMPLE 3   Find $\dfrac{2}{3} \cdot \dfrac{5}{7}$. Write in simplest form.

$\dfrac{2}{3} \cdot \dfrac{5}{7} = \dfrac{2 \cdot 5}{3 \cdot 7}$ Multiply the numerators. Multiply the denominators.

$\phantom{\dfrac{2}{3} \cdot \dfrac{5}{7}} = \dfrac{10}{21}$ Simplify.

Divide Rational Numbers (Lesson 1D)

Divide. Write in simplest form.

22. $\dfrac{7}{9} \div \dfrac{1}{3}$ **23.** $\dfrac{7}{12} \div \left(-\dfrac{2}{3}\right)$ **24.** $-4\dfrac{2}{5} \div (-2)$

25. ALGEBRA Find $a \div b$ if $a = 3\dfrac{1}{2}$ and $b = -\dfrac{7}{8}$.

EXAMPLE 4   Find $-\dfrac{5}{6} \div \dfrac{3}{5}$. Write in simplest form.

$-\dfrac{5}{6} \div \dfrac{3}{5} = -\dfrac{5}{6} \cdot \dfrac{5}{3}$ Multiply by the multiplicative inverse.

$\phantom{-\dfrac{5}{6} \div \dfrac{3}{5}} = -\dfrac{25}{18}$ or $-1\dfrac{7}{18}$ Simplify.

Lesson 2 Percents

PSI: Look for a Pattern (Lesson 2A)

Solve. Use the *look for a pattern* strategy.

26. NUMBERS Find the next two numbers in the sequence 3, 6, 9, 12,… .

27. RUNNING Marcy can run one lap in 65 seconds. Each additional lap takes her 2 seconds longer to run than the previous lap. How many minutes will it take her to run three miles? (1 mile = 4 laps)

28. GEOMETRY What is the total number of rectangles, of any size, in the figure below?

EXAMPLE 5 Raul's phone plan charges a flat monthly rate of $4.95 and $0.06 per minute. If Raul spent a total of $7.35 last month, how many minutes did he use?

Look for a pattern.

Minutes	Charges	Total
0	4.95 + 0(0.06)	$4.95
10	4.95 + 10(0.06)	$5.55
20	4.95 + 20(0.06)	$6.15
30	4.95 + 30(0.06)	$6.75
40	4.95 + 40(0.06)	$7.35

So, Raul used 40 minutes last month.

Compare Rational Numbers (Lesson 2B)

Write each percent as a decimal.

29. 4.3%   **30.** 147%   **31.** 0.7%

Write each decimal as a percent.

32. 0.7   **33.** 0.015   **34.** 2.55

35. CELL PHONES Adam used $\frac{7}{8}$ of his total monthly minutes while Andrea used 88%. Which friend used the greater part of his or her minutes? Explain.

EXAMPLE 6 Write 24% as a decimal.

$24\% = 24\%$ Divide by 100 and remove
$= 0.24$ the percent symbol.

EXAMPLE 7 Write 0.04 as a percent.

$0.04 = 0.04$ Multiply by 100 and add the
$= 4\%$ percent symbol.

Algebra: The Percent Proportion and Equation (Lesson 2C)

Use the percent proportion or equation to solve each problem. Round to the nearest tenth if necessary.

36. 15 is 30% of what number?

37. Find 45% of 18.

38. 75 is what percent of 250?

39. SPORTS Inali made about 81% of his free throws in the game. If he made 13 free throws, how many attempts did he make?

EXAMPLE 8 70 is 25% of what number?

part = percent · whole

$70 = 0.25 \cdot n$ Write the percent equation.

$\frac{70}{0.25} = \frac{0.25n}{0.25}$ Divide each side by 0.25.

$280 = n$ Simplify.

70 is 25% of 280.

Apply Percents

Discount, Markup, and Sales Tax (Lesson 3A)

Find the sale price or total cost of each item to the nearest cent.

40. computer: $975, 15% discount

41. magazine: $4.95, 6.75% tax

Find the selling price for each item given the cost to the store and the markup.

42. sweater: $20, 35% markup

43. hat: $6.50, 60% markup

44. SALES A game that is $55 is on sale for 15% off. If there is a 7.5% sales tax, what is the total cost of the game?

EXAMPLE 9 Find the sale price of a $20 DVD that is on sale for 25% off.

$$\underbrace{part}_{} = \underbrace{percent}_{} \cdot \underbrace{whole}_{}$$

$$
\begin{array}{lll}
d & = & 0.25 \cdot 20 \quad \text{Write the percent equation.} \\
d & = & 5 \quad\quad\quad\; \text{Multiply.}
\end{array}
$$

Subtract the amount of the discount from the original price to find the sale price.

$20 - $5 = $15

So, the DVD is on sale for $15.

Financial Literacy: Interest (Lesson 3B)

Find the total amount in each account to the nearest cent if the interest is compounded annually.

45. $350 at 5% for 3 years

46. $1,500 at 6.25% for 2 years

47. RETIREMENT At age 20, Mark invested $500 into a retirement account with a simple interest rate of 6.5%. He made no more deposits or withdrawals. Find the account value at age 65.

EXAMPLE 10 Find the simple interest for $250 invested at 5.5% for 2 years.

$$
\begin{array}{ll}
I = prt & \text{Simple interest formula} \\
I = 250 \cdot 0.055 \cdot 2 & \text{Write 5.5\% as 0.055.} \\
I = 27.50 & \text{Simplify.}
\end{array}
$$

The simple interest is $27.50.

Percent of Change (Lesson 3D)

Find each percent of change. Round to the nearest tenth if necessary. State whether the percent of change is an *increase* or a *decrease*.

48. original: 10   **49.** original: 37.5
    new: 15   new: 30

50. ANIMALS At birth, a giraffe was 62 inches tall and grew at the highly unusual rate of 0.5 inch per hour. By what percent did the height of the giraffe increase in the first 24 hours?

EXAMPLE 11 Find the percent of change if the original amount is 900 and the new amount is 725. State whether the change is an *increase* or *decrease*.

The amount of change is $725 - 900$ or -175.

$$\text{percent of change} = \frac{\text{amount of change}}{\text{original amount}}$$

$$= \frac{-175}{900}$$

$$\approx -0.194 \text{ or } -19.4\%$$

Since the percent is negative, it is a percent of decrease.

1. FROGS The Gold Frog grows to only 0.375 inch. Write this length as a fraction in simplest form.

Add, subtract, multiply, or divide. Write in simplest form.

2. $-5\frac{1}{4} \cdot \left(-2\frac{1}{3}\right)$　　**3.** $-6 \div \frac{1}{8}$

4. $-\frac{3}{8} + \frac{4}{9}$　　**5.** $\left(-1\frac{7}{8}\right) - \left(-3\frac{1}{4}\right)$

6. ANALYZE TABLES The table shows the time of the back and forth swing of a pendulum and its length. How long is a pendulum with a swing of 5 seconds?

Time of Swing	Length of Pendulum
1 second	1 unit
2 seconds	4 units
3 seconds	9 units
4 seconds	16 units

Express each percent as a decimal.

7. 135%　　**8** 14.6%　　**9.** 0.97%

10. MULTIPLE CHOICE The figure below shows 8 shaded isosceles triangles formed by the diagonals of three adjacent squares.

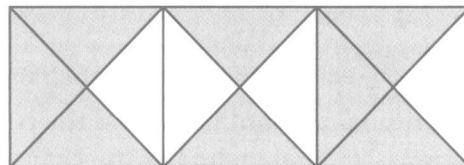

If the total area of the figure is 12 square feet, which statement is true?

A. The shaded area is more than 75% of the area of the figure.

B. The unshaded area is $\frac{2}{3}$ of the area of the figure.

C. The shaded area is 6 square feet.

D. The unshaded area is 4 square feet.

Write a percent proportion and solve each problem. Round to the nearest tenth.

11. What is 2% of 3,600?

12. 62 is 90% of what number?

13. 75 is what percent of 30?

14. BUSINESS A shoe store prices items at a 45% markup rate. If the store purchases an athletic shoe for $40, find the selling price of the shoe.

Find each percent of change and state whether it is an *increase* or a *decrease*. Round to the nearest tenth if necessary.

15. original: $15　　**16.** original: 40 cars
new: $12　　　　　　　new: 55 cars

17. **EXTENDED RESPONSE** The combined monthly income of the Rodriguez family is $3,875. The circle graph shows how the family budgets their money.

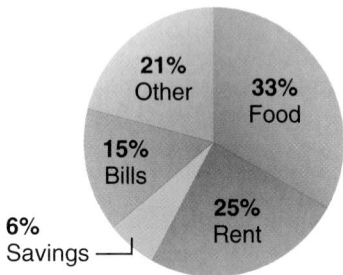

Rodriguez Family Budget

Part A How much money did the family budget for bills and rent?

Part B The family spent $1,575 on food last month. Did they exceed their budget? If so, by how much? If not, how much money did they save?

Part C Two months ago, they only spent 25% of their money on food. They added the leftover money to their savings. How much additional money was added to their savings last month?

Preparing for Standardized Tests

✏️ Gridded Response: Percents

When the answer to a gridded-response question is a percent, do not convert the percent to a decimal. Grid in the percent value without the % symbol.

TEST EXAMPLE

A trampoline that originally cost $1,145 is on sale for $916. What is the percent of discount?

Use a percent proportion.

$$\frac{916}{1,145} = \frac{p}{100} \qquad \text{Write the percent proportion.}$$

$$916 \cdot 100 = 1,145 \cdot p \qquad \text{Find the cross products.}$$

$$91,600 = 1,145p \qquad \text{Simplify.}$$

$$\frac{91,600}{1,145} = \frac{1,145p}{1,145} \qquad \text{Divide each side by 1,145.}$$

$$80 = p \qquad \text{Simplify.}$$

The sale price is 80% of the original price. So, the percent of discount is 100% − 80% or 20%. Grid in 20.

Correct

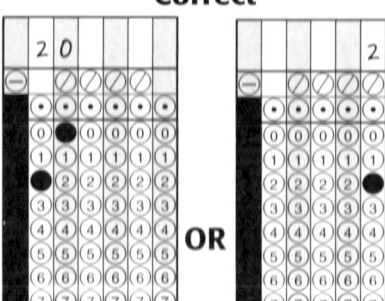 OR

NOT Correct

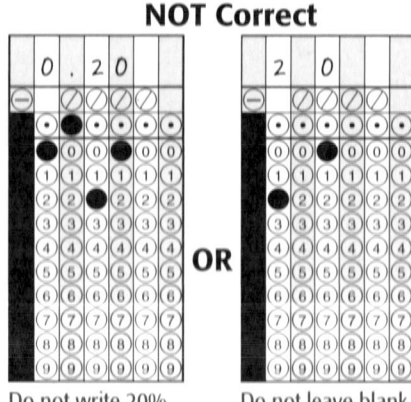 OR

Do not write 20% as 0.20.

Do not leave blank answer boxes in the middle of an answer.

 **Work on It**

Rhonda received 8 text messages on Friday and 14 text messages on Saturday. What was the percent of increase in text messages? Fill in your answer on an answer grid like the ones shown above.

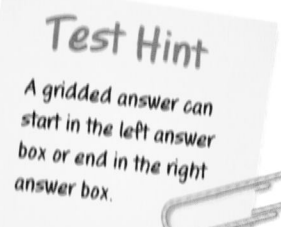

Test Hint

A gridded answer can start in the left answer box or end in the right answer box.

Read each question. Then fill in the correct answer on the answer document provided by your teacher or on a sheet of paper.

1. The inline skates below are on sale for 35% off the regular price.

What is the sale price of the skates?

A. $162   C. $78

B. $94   D. $42

2. ✎ **GRIDDED RESPONSE** The Upton Bobcats won 85% of their games last season. What is 85% written as a fraction in simplest form?

3. A carpenter estimates that it will take one person 54 hours to complete a job. He plans to have three people work on the job for two days. How many hours each day will the workers need to work to complete the job?

F. 8 hours   H. 12 hours

G. 9 hours   I. 18 hours

4. Which fraction is equivalent to the expression below?
$$\frac{3}{5} + \frac{3}{10}$$

A. $\frac{6}{15}$ C. $\frac{9}{50}$

B. $\frac{9}{10}$ D. $\frac{9}{15}$

5. The contents of a jar of mixed nuts are shown below.

Ingredient	Weight (lb)
Peanuts	$2\frac{1}{2}$
Cashews	$1\frac{1}{3}$
Walnuts	$1\frac{5}{6}$

What is the total weight of the contents of the jar?

F. $4\frac{1}{6}$ pounds

G. $4\frac{1}{2}$ pounds

H. $5\frac{2}{3}$ pounds

I. $6\frac{1}{3}$ pounds

6. Martin and his sister agreed to split the cost of a new game. They received a 25% discount on the game and paid 5.5% sales tax. The original price of the game was $30. How much did Martin and his sister each spend on the game?

A. $23.74

B. $22.50

C. $11.87

D. $11.25

7. THINK SOLVE EXPLAIN **SHORT RESPONSE** The table shows the height of three siblings.

Sibling	Height (in.)
Juana	$61\frac{1}{4}$
Maria	$57\frac{3}{4}$
Roberto	$69\frac{1}{8}$

How much taller is Roberto than Juana?

8. Alan is buying a television that is regularly priced at $149.99. It is on sale for $\frac{1}{5}$ off the original price. Which expression can he use to estimate the discount on the television?

F. $0.02 \times \$150$

G. $0.05 \times \$150$

H. $0.5 \times \$150$

I. $0.2 \times \$150$

9. **THINK SOLVE EXPLAIN** **SHORT RESPONSE** Write a fraction that is between $\frac{4}{5}$ and $\frac{5}{6}$.

10. A pattern of percents is shown below.

80% of 62.5 is 50

40% of 125 is 50

20% of 250 is 50

10% of 500 is 50

Which statement **best** describes this pattern of percents?

A. When the percent is halved and the other number is doubled, the answer is 50.

B. When the percent is halved and the other number is halved, the answer is 50.

C. When the percent is increased by 2 and the other number remains the same, the answer is 50.

D. When the percent remains the same and the other number is increased by 2, the answer is 50.

11. **THINK SOLVE EXPLAIN** **SHORT RESPONSE** The widths of a race track are shown below.

Part of Track	Width (feet)
Straightaway	50
Turn	60

What is the percent of increase in the track width from the straightaway to the turn?

12. Eliza purchased a dress off the clearance rack. The original cost for the dress was $35. The dress was marked down 50%, and the sign on the rack said to take an additional 20% off the discounted price. What was the final sale price Eliza paid for the dress?

F. $3.50   **H.** $14.00

G. $10.50   **I.** $17.50

13. **THINK SOLVE EXPLAIN** **EXTENDED RESPONSE** Refer to the figures below.

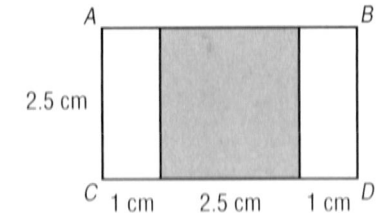

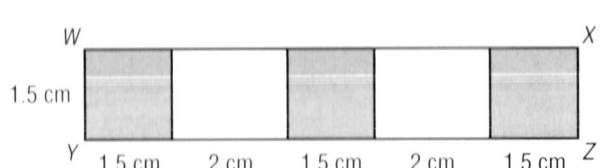

Part A Find the area of the shaded region for each rectangle.

Part B Which rectangle has the greater percent of area that is shaded? Explain.

NEED EXTRA HELP?													
If You Missed Question...	1	2	3	4	5	6	7	8	9	10	11	12	13
Go to Chapter-Lesson...	1-3A	1-2B	1-1D	1-1B	1-1B	1-3A	1-1B	1-3A	1-1A	1-2C	1-3D	1-3A	1-2C

Chapter 1 Test Practice **87**

CHAPTER 2

Real Numbers and Monomials

connectED.mcgraw-hill.com

 Investigate

 Animations

 Vocabulary

 Multilingual eGlossary

 Learn

 Personal Tutor

 Virtual Manipulatives

 Graphing Calculator

 Audio

Foldables

 Practice

 Self-Check Practice

 Worksheets

Assessment

The ☆BIG Idea

How are the different subsets of real numbers related?

FOLDABLES Study Organizer

Make this Foldable to help you organize your notes.

Review Vocabulary

rational number *número racional* any number that can be written as a fraction

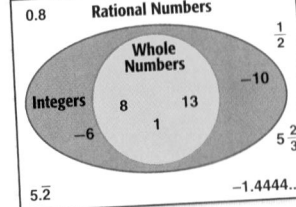

Rational Numbers
0.8
Whole Numbers
$\frac{1}{2}$
Integers
8   13
−6   1
−10
$5\frac{2}{3}$
$5.\overline{2}$
−1.4444...

Key Vocabulary

English	Español
power	potencia
real number	número real
scientific notation	notación científica
square root	raíz cuadrada

When Will I Use This?

Are You Ready for the Chapter?

You have two options for checking prerequisite skills for this chapter.

Text Option Take the Quick Check below. Refer to the Quick Review for help.

QUICK Check

Find each product.

1. $6 \cdot 6 \cdot 6 \cdot 6$

2. $(-3)(-3)(-3)(-3)(-3)$

3. $2 \cdot 2 \cdot 4 \cdot 4 \cdot 4$

4. $(-8)(-8)(5)(5)(-8)$

5. **FUNDRAISER** The students at Hampton Middle School raised $8 \cdot 8 \cdot 2 \cdot 8 \cdot 2$ dollars to help build a new community center. How much money did they raise?

6. **RIVERS** The Nile River is about $2 \times 2 \times 2 \times 2 \times 3 \times 5 \times 5 \times 5$ kilometers long. About how many kilometers long is the Nile?

Find the prime factorization of each number.

7. 36   8. 24

9. 18   10. 100

11. 121   12. −42

13. −64   14. 13

15. **ANIMALS** The table shows the maximum speed for some of the fastest land animals. Find the prime factorization of each speed.

Animal	Maximum Speed (mph)
Cheetah	71
Wildebeest	50
Brown hare	48
Horse	45

QUICK Review

EXAMPLE 1

Find $5 \cdot 4 \cdot 5 \cdot 4 \cdot 5$.

$5 \cdot 4 \cdot 5 \cdot 4 \cdot 5$

$= 4 \cdot 4 \cdot 5 \cdot 5 \cdot 5$ Commutative Property

$= (4 \cdot 4) \cdot (5 \cdot 5 \cdot 5)$ Associative Property

$= 16 \cdot 125$ Multiply.

$= 2{,}000$ Simplify.

EXAMPLE 2

Find $(-2)(-2)(-2)(-2)$.

$(-2)(-2)(-2)(-2) = 16$ Multiply.

EXAMPLE 3

Find the prime factorization of 60.

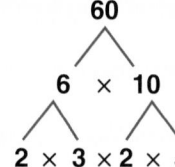

The prime factorization of 60 is $2 \times 2 \times 3 \times 5$.

Online Option Take the Online Readiness Quiz.

Main Idea

Use powers and exponents to write large and small numbers.

 Vocabulary

power
base
exponent

Powers and Exponents

SAVINGS Yogi decided to start saving money by putting a penny in his piggy bank, then doubling the amount he saves each week.

Week	0	1	2	3	4	5	6
Savings	1¢	2¢	4¢	8¢	16¢	32¢	64¢

1. How many 2s are multiplied to find his savings in Week 4? Week 5?

2. How much money will Yogi save in Week 8?

3. When will he have enough to buy a pair of shoes for $80?

A product of repeated factors can be expressed as a **power**, that is, using an exponent and a base.

The **base** is the common factor.

The **exponent** tells how many times the base is used as a factor.

$$\underbrace{2 \cdot 2 \cdot 2 \cdot 2}_{4 \text{ factors}} = 2^4$$

EXAMPLES Write Expressions Using Powers

Write each expression using exponents.

1 $5 \cdot 5 \cdot 5$

The base 5 is a factor 3 times. So, the exponent is 3.
$5 \cdot 5 \cdot 5 = 5^3$

2 $(-3)(-3)(-3)(-3)$

The base -3 is a factor 4 times. So, the exponent is 4.
$(-3)(-3)(-3)(-3) = (-3)^4$

3 $x \cdot x \cdot x \cdot x \cdot x \cdot x$

The base x is a factor 6 times. So, the exponent is 6.
$x \cdot x \cdot x \cdot x \cdot x \cdot x = x^6$

CHECK Your Progress

a. $9 \cdot 9 \cdot 9 \cdot 9 \cdot 9$

b. $\left(\dfrac{3}{4}\right)\left(\dfrac{3}{4}\right)\left(\dfrac{3}{4}\right)\left(\dfrac{3}{4}\right)$

c. $(-6)(-6)(-6)$

d. $c \cdot c \cdot c \cdot c$

 **EXAMPLES** Write Expressions Using Powers

Write each expression using exponents.

4 $2 \cdot 2 \cdot 2 \cdot 3 \cdot 3 \cdot 3 \cdot 3$

$2 \cdot 2 \cdot 2 \cdot 3 \cdot 3 \cdot 3 \cdot 3 = (2 \cdot 2 \cdot 2) \cdot (3 \cdot 3 \cdot 3 \cdot 3)$ Associative Property

$= 2^3 \cdot 3^4$ Definition of exponents

5 $a \cdot b \cdot b \cdot a \cdot b$

$a \cdot b \cdot b \cdot a \cdot b = a \cdot a \cdot b \cdot b \cdot b$ Commutative Property

$= (a \cdot a) \cdot (b \cdot b \cdot b)$ Associative Property

$= a^2 \cdot b^3$ Definition of exponents

CHECK Your Progress

e. $\dfrac{2}{3} \cdot \dfrac{2}{3} \cdot \dfrac{2}{3} \cdot 7 \cdot 7$ **f.** $(-3)(-3)(5)(5)(5)$ **g.** $m \cdot m \cdot n \cdot n \cdot n \cdot m$

Powers are read in a certain way.

Read and Write Powers		
Power	**Words**	**Factors**
3^1	3 to the first power	3
3^2	3 to the second power or 3 squared	$3 \cdot 3$
3^3	3 to the third power or 3 cubed	$3 \cdot 3 \cdot 3$
3^4	3 to the fourth power or 3 to the fourth	$3 \cdot 3 \cdot 3 \cdot 3$
⋮	⋮	⋮
3^n	3 to the nth power or 3 to the nth	$\underbrace{3 \cdot 3 \cdot 3 \cdot \ldots \cdot 3}_{n \text{ factors}}$

 EXAMPLES Evaluate Powers

6 Evaluate $(-3)^3$.

$(-3)^3 = (-3) \cdot (-3) \cdot (-3)$ Write the power as a product.

$= -27$ Multiply.

7 Evaluate $\left(\dfrac{2}{3}\right)^4$.

$\left(\dfrac{2}{3}\right)^4 = \dfrac{2}{3} \cdot \dfrac{2}{3} \cdot \dfrac{2}{3} \cdot \dfrac{2}{3}$ Write the power as a product.

$= \dfrac{16}{81}$ Multiply.

CHECK Your Progress

Evaluate each expression.

h. 4^4 **i.** $(-2)^6$ **j.** $\left(\dfrac{1}{5}\right)^3$

 REAL-WORLD EXAMPLE

⑧ **SKATEBOARDS** The deck of a skateboard has an area of about $2^5 \cdot 7$ square inches. What is the area of the skateboard deck?

$$2^5 \cdot 7 = 2 \cdot 2 \cdot 2 \cdot 2 \cdot 2 \cdot 7 \qquad \text{Write the power as a product.}$$
$$= (2 \cdot 2 \cdot 2 \cdot 2 \cdot 2) \cdot 7 \qquad \text{Associative Property}$$
$$= 32 \cdot 7 \text{ or } 224 \qquad \text{Multiply.}$$

The deck of the skateboard has an area of about 224 square inches.

 CHECK Your Progress

k. **BASKETBALL** A high school basketball court has an area of $2^3 \cdot 3 \cdot 5^2 \cdot 7$ square feet. What is the area of a high school basketball court?

Powers are included in the rules for order of operations.

Key Concept / Order of Operations

1. Perform all operations within grouping symbols first; start with the innermost grouping symbols.
2. Evaluate all powers before other operations.
3. Multiply and divide in order from left to right.
4. Add and subtract in order from left to right.

 EXAMPLES Evaluate Algebraic Expressions

Evaluate each expression if $a = 3$ and $b = 5$.

⑨ $a^2 + b^4$

$$a^2 + b^4 = 3^2 + 5^4 \qquad \text{Replace } a \text{ with 3 and } b \text{ with 5.}$$
$$= (3 \cdot 3) + (5 \cdot 5 \cdot 5 \cdot 5) \qquad \text{Write the powers as products.}$$
$$= 9 + 625 \text{ or } 634 \qquad \text{Add.}$$

⑩ $(a - b)^2$

$$(a - b)^2 = (3 - 5)^2 \qquad \text{Replace } a \text{ with 3 and } b \text{ with 5.}$$
$$= (-2)^2 \qquad \text{Perform operations in the parentheses first.}$$
$$= (-2) \cdot (-2) \text{ or } 4 \qquad \text{Write the powers as products. Then simplify.}$$

 CHECK Your Progress

Evaluate each expression if $c = -4$ and $d = 9$.

l. $c^3 + d^2$ m. $(c + d)^3$ n. $d^3 - (c^2 - 2)$

Examples 1–5 Write each expression using exponents.

1. $4 \cdot 4 \cdot 4 \cdot 4$

2. $(-11)(-11)(-11)$

3. $a \cdot a \cdot a \cdot a$

4. $2 \cdot 2 \cdot 2 \cdot 3 \cdot 3 \cdot 3$

5. $r \cdot s \cdot r \cdot r \cdot s \cdot s \cdot r \cdot r$

6. $\frac{1}{2} \cdot p \cdot k \cdot \frac{1}{2} \cdot p \cdot p \cdot k$

Examples 6 and 7 Evaluate each expression.

7 2^6

8. 6^3

9. $(-4)^4$

10. $(-3)^5$

11. $\left(\frac{1}{7}\right)^3$

12. $\left(\frac{1}{2}\right)^4$

13. $\left(\frac{3}{4}\right)^4$

14. $\left(\frac{4}{5}\right)^2$

Example 8 15. **ANIMALS** The table shows the average weights of some endangered mammals. What is the weight of each animal?

Animal	Weight (lb)
Black bear	$2 \cdot 5^2 \cdot 7$
Key deer	$3 \cdot 5^2$
Panther	$2^3 \cdot 3 \cdot 5$

Examples 9 and 10 **ALGEBRA** Evaluate each expression if $x = 2$ and $y = 10$.

16. $x^2 + y^4$

17. $3x^3 \cdot y^2$

18. $\dfrac{y^2}{x^6}$

19. $(x^2 + y)^3$

Practice and Problem Solving

= Step-by-Step Solutions begin on page R1.
Extra Practice begins on page EP2.

Examples 1–5 Write each expression using exponents.

20. $(-5)(-5)(-5)(-5)$

21. $\left(\frac{5}{6}\right)\left(\frac{5}{6}\right)\left(\frac{5}{6}\right)$

22. $m \cdot m \cdot m \cdot m \cdot m$

23. $3 \cdot 3 \cdot 5 \cdot q \cdot q \cdot q$

24. $s \cdot (-7) \cdot s \cdot (-7) \cdot (-7)$

25. $4 \cdot b \cdot b \cdot 4 \cdot b \cdot b$

26. $n \cdot \frac{1}{4} \cdot p \cdot n \cdot \frac{1}{4}$

27. $d \cdot 2 \cdot 2 \cdot d \cdot k \cdot d \cdot k$

28. $2 \cdot 7 \cdot a \cdot 7 \cdot a \cdot 2 \cdot a$

Examples 6 and 7 Evaluate each expression.

29. 2^3

30. $(-9)^4$

31. $\left(\frac{1}{3}\right)^4$

32. $\left(\frac{5}{7}\right)^3$

33. $4^2 + 2^3$

34. $8^2 - 5^2$

35. $3^3 \cdot 4^2$

36. $(-6)^3 \div 3^5$

Example 8 37. **TEXT MESSAGING** In the United States, nearly $8 \cdot 10^9$ text messages are sent every month. About how many text messages is this?

38. **TRAVEL** Interstate 70 stretches almost $2^3 \cdot 5^2 \cdot 11$ miles across the United States. About how many miles long is Interstate 70?

Examples 9 and 10 **ALGEBRA** Evaluate each expression.

39 $g^5 - h^3$ if $g = 2$ and $h = 7$

40. $x^3 + y^4$, if $x = -3$ and $y = 4$

41. $a^2 \cdot b^6$ if $a = \frac{1}{2}$ and $b = 2$

42. $k^4 \cdot m$, if $k = 3$ and $m = \frac{5}{6}$

43. $(r - s)^3 + r^2$ if $r = -3$ and $s = -4$

44. $(c^3 + d^4)^2 - (c + d)^3$, if $c = -1$ and $d = 2$

45 **PLANETS** The table shows planetary distances from the Sun. Write your answers in standard form.

Planetary Distances from the Sun	
Planet	**Distance (mi)**
Mercury	$3.6 \cdot 10^7$
Venus	$6.7 \cdot 10^7$
Earth	$9.3 \cdot 10^7$
Mars	$1.42 \cdot 10^8$
Jupiter	$4.84 \cdot 10^8$
Saturn	$8.87 \cdot 10^8$
Uranus	$1.8 \cdot 10^9$
Neptune	$2.8 \cdot 10^9$

a. How far is Earth from the Sun?

b. How far is Saturn from the Sun?

c. How far is Neptune from the Sun?

d. How much farther is Neptune than Saturn from the Sun?

Replace each ● with <, >, or = to make a true statement.

46. $(6 - 2)^2 + 3 \cdot 4 \ ● \ 5^2$ **47.** $5 + 7^2 + 3^3 \ ● \ 3^4$ **48.** $\left(\frac{1}{2}\right)^4 \ ● \ \left(\frac{1}{4}\right)^2$

49. $(-4)^3 \ ● \ 4^3$ **50.** $6^4 \ ● \ (5 + 7)^2 + 12^2$ **51.** $3^3 + 1 + 6^2 \ ● \ (-8)^2$

52. 🔄 **MULTIPLE REPRESENTATIONS** A square has a side length of s inches.

a. **TABLES** Copy and complete the table showing the side length, perimeter, and area of the square.

Side Length (in.)	Perimeter (in.)	Area (in²)
1	4	1
2	■	■
⋮	⋮	⋮
10	■	■

b. **GRAPHS** Graph the ordered pairs (side length, perimeter) and (side length, area) on the same coordinate plane. Then connect the points for each set.

c. **WORDS** Compare and contrast the graphs of the perimeter and area of the square. Which graph is linear?

53. **GRAPHIC NOVEL** Refer to the graphic novel frame below for Exercises a–c.

The metric system is based on powers of 10. For example, one kilometer is equal to 1,000 meters or 10^3 meters. Write each measurement in meters as a power of 10.

a. megameter (1,000,000 meters)

b. gigameter (1,000,000,000 meters)

c. petameter (1,000,000,000,000,000 meters)

H.O.T. Problems

54. OPEN ENDED Write an expression with an exponent that has a value between 0 and 1.

55. CHALLENGE Complete the following pattern.

$$3^4 = 81, 3^3 = 27, 3^2 = 9, 3^1 = 3, 3^0 = \blacksquare, 3^{-1} = \blacksquare, 3^{-2} = \blacksquare, 3^{-3} = \blacksquare$$

56. 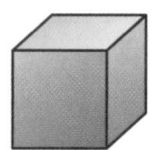 **WRITE MATH** What is the role of the order of operations when evaluating expressions?

Test Practice

57. To find the volume of a cube, multiply its base, its height, and its width.

6 in.

What is the volume of the cube expressed as a power?

A. 6^2 in^3 **C.** 6^4 in^3

B. 6^3 in^3 **D.** 6^6 in^3

58. Which expression is equivalent to the expression below?

$$2^3 \cdot 3^4$$

F. $3 \cdot 3 \cdot 4 \cdot 4 \cdot 4$

G. $2 \cdot 2 \cdot 2 \cdot 3 \cdot 3 \cdot 3$

H. $2 \cdot 2 \cdot 2 \cdot 3 \cdot 3 \cdot 3 \cdot 3$

I. $6 \cdot 12$

59.  **GRIDDED RESPONSE** The volume of an ice cube in cubic millimeters is represented by the term 11^3. What is 11^3 in standard form?

More About Exponents

A scientific calculator performs operations using the order of operations. Use the caret △ key for exponents greater than 2.

EXAMPLES

Evaluate each expression.

1) $(-5)^3$

Keystrokes: ⦅ (–) 5 ⦆ △ 3 ENTER −125

2) $4^2 + 6^3$

Keystrokes: 4 x^2 + 6 △ 3 ENTER 232

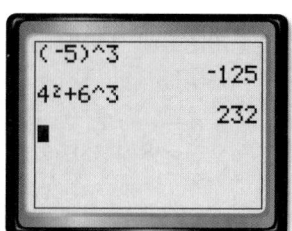

Evaluate each expression.

60. 10^5 **61.** $-(11^2)$ **62.** $2^3 \cdot 3^4$ **63.** $(8 + 15)^2$

64. MAKE A CONJECTURE Evaluate -10^2 and $(-10)^2$ using a scientific calculator. Compare the answers and explain any differences.

Main Idea

Simplify real number expressions by multiplying and dividing monomials.

 **Vocabulary**

monomial

CCSS 8.EE.1

Multiply and Divide Monomials

MEASUREMENT The edge of a dime is approximately 1 millimeter thick. The table shows how other metric measurements of length are related to the millimeter.

Unit of Length	Times Longer than a Millimeter	Written Using Powers
Millimeter	1	10^0
Centimeter	$1 \times 10 = 10$	10^1
Decimeter	$10 \times 10 = 100$	$10^1 \times 10^1 = 10^2$
Meter	$100 \times 10 = 1{,}000$	$10^1 \times 10^2 = 10^3$
Dekameter	$1{,}000 \times 10 = 10{,}000$	$10^1 \times 10^3 = 10^4$
Hectometer	$10{,}000 \times 10 = 100{,}000$	$10^1 \times 10^4 = 10^5$
Kilometer	$100{,}000 \times 10 = 1{,}000{,}000$	$10^1 \times 10^5 = 10^6$

1. Examine the exponents of the factors and the exponents of the products in the last column. What do you observe?

2. A *megameter* is $100{,}000{,}000 \times 10$ or $1{,}000{,}000{,}000$ times longer than a millimeter. Extend the pattern to write this number using powers.

A **monomial** is a number, a variable, or a product of a number and one or more variables. Exponents are used to show repeated multiplication. You can use this fact to find a rule for multiplying monomials.

$$\overbrace{}^{\text{2 factors}} \qquad \overbrace{}^{\text{4 factors}}$$
$$3^2 \cdot 3^4 = (3 \cdot 3) \cdot (3 \cdot 3 \cdot 3 \cdot 3 \cdot) \text{ or } 3^6$$
$$\underbrace{}_{\text{6 factors}}$$

Notice that the sum of the original exponents is the exponent in the final product. This relationship is stated in the following rule.

Key Concept — Product of Powers

Words To multiply powers with the same base, add their exponents.

Examples

Numbers	**Algebra**
$2^4 \cdot 2^3 = 2^{4+3}$ or 2^7	$a^m \cdot a^n = a^{m+n}$

Simplify. Express using exponents.

① $5^2 \cdot 5$

$5^2 \cdot 5 = 5^2 \cdot 5^1$ $5 = 5^1$

$ = 5^{2+1}$ The common base is 5.

$ = 5^3$ Add the exponents.

Check $5^2 \cdot 5 = (5 \cdot 5) \cdot 5$

$ = 5 \cdot 5 \cdot 5$

$ = 5^3$ ✔

② $c^3 \cdot c^5$

$c^3 \cdot c^5 = c^{3+5}$ The common base is c.

$ = c^8$ Add the exponents.

③ $-3x^2 \cdot 4x^5$

$-3x^2(4x^5) = (-3 \cdot 4)(x^2 \cdot x^5)$ Commutative and Associative Properties

$ = (-12)(x^{2+5})$ The common base is x.

$ = -12x^7$ Add the exponents.

✓ **CHECK Your Progress**

a. $9^3 \cdot 9^2$ **b.** $a^3 \cdot a^2$ **c.** $-2m(-8m^5)$

There is also a Law of Exponents for dividing powers with the same base.

$$\frac{5^7}{5^4} = \frac{\overbrace{5 \cdot 5 \cdot 5 \cdot 5 \cdot 5 \cdot 5 \cdot 5}^{\text{7 factors}}}{\underbrace{5 \cdot 5 \cdot 5 \cdot 5}_{\text{4 factors}}} \text{ or } 5^3$$

Notice that the difference of the original exponents is the exponent in the final quotient. This relationship is stated in the following rule.

🏃 **Key Concept**   **Quotient of Powers**

Words   To divide powers with the same base, subtract their exponents.

Examples

	Numbers	**Algebra**
	$\dfrac{3^7}{3^3} = 3^{7-3}$ or 3^4	$\dfrac{a^m}{a^n} = a^{m-n}$, where $a \neq 0$

 **EXAMPLES** Divide Powers

Simplify. Express using exponents.

 $\dfrac{4^8}{4^2}$

$\dfrac{4^8}{4^2} = 4^{8-2}$ The common base is 4.

$= 4^6$ Simplify.

 $\dfrac{n^9}{n^4}$

$\dfrac{n^9}{n^4} = n^{9-4}$ The common base is n.

$= n^5$ Simplify.

CHECK Your Progress

d. $\dfrac{5^7}{5^4}$ **e.** $\dfrac{x^{10}}{x^3}$ **f.** $\dfrac{12w^5}{2w}$

 REAL-WORLD EXAMPLE

 SHORELINES Hawaii's total shoreline is about 2^{10} miles long. New Hampshire's shoreline is about 2^7 miles long. About how many times longer is Hawaii's shoreline than New Hampshire's?

To find how many times longer, divide 2^{10} by 2^7.

$\dfrac{2^{10}}{2^7} = 2^{10-7}$ or 2^3 Quotient of Powers

Hawaii's shoreline is about 2^3 or 8 times longer.

CHECK Your Progress

g. SOUND The loudness of a vacuum cleaner is 10^4 times as intense as the loudness of a mosquito buzzing, while the loudness of a jack hammer is 10^9 times as intense. How many times as intense is the loudness of a jack hammer than that of a vacuum cleaner?

Vocabulary Link

Everyday Use

Simplify to make less complex, easier

Math Use

Simplify to perform all possible operations in an expression

 EXAMPLE Simplify Expressions

 Simplify $\dfrac{2^5 \cdot 3^5 \cdot 5^2}{2^2 \cdot 3^4 \cdot 5}$.

$\dfrac{2^5 \cdot 3^5 \cdot 5^2}{2^2 \cdot 3^4 \cdot 5} = \left(\dfrac{2^5}{2^2}\right)\left(\dfrac{3^5}{3^4}\right)\left(\dfrac{5^2}{5}\right)$ Group by common base.

$= 2^3 \cdot 3^1 \cdot 5^1$ Subtract the exponents.

$= 8 \cdot 3 \cdot 5$ $2^3 = 8$

$= 120$ Simplify.

CHECK Your Progress

Simplify.

h. $\dfrac{3^4 \cdot 5^2 \cdot 7^5}{3^2 \cdot 5 \cdot 7^3}$ **i.** $\dfrac{5^6 \cdot 7^4 \cdot 8^3}{5^4 \cdot 7^2 \cdot 8^2}$ **j.** $\dfrac{(-2)^5 \cdot 3^4 \cdot 5^7}{(-2)^2 \cdot 3 \cdot 5^4}$

Examples 1–5 **Simplify. Express using exponents.**

1. $4^5 \cdot 4^3$

2. $5^2 \cdot 5^5$

3. $r^7 \cdot r^3$

4. $n^2 \cdot n^9$

5. $-2a(3a^4)$

6. $5^2 x^2 y^4 \cdot 5^3 x y^3$

7. $\dfrac{7^6}{7}$

8. $\dfrac{2^{13}}{2^9}$

9. $\dfrac{y^8}{y^5}$

10. $\dfrac{z^2}{z}$

11. $\dfrac{9c^7}{3c^2}$

12. $\dfrac{24k^9}{6k^6}$

Example 6 **13. LANGUAGES** The table shows the number of people worldwide that speak certain languages. How many times as many people speak French than Sicilian?

Language	Total (millions)
French	2^6
Sicilian	2^2

Example 7 **Simplify.**

14. $\dfrac{2^2 \cdot 3^3 \cdot 4^5}{2 \cdot 3 \cdot 4^4}$

15. $\dfrac{6^2 \cdot 7^4 \cdot 8^3}{6 \cdot 7^2 \cdot 8^2}$

16. $\dfrac{(-3)^4 \cdot (-4)^3 \cdot 5^2}{(-3)^2 \cdot (-4) \cdot 5}$

Practice and Problem Solving

● = Step-by-Step Solutions begin on page R1.
Extra Practice begins on page EP2.

Examples 1–5 **Simplify. Express using exponents.**

17. $(-6)^8 \cdot (-6)^5$

18. $2g^2 \cdot 7g^6$

19. $(3x^8)(5x)$

20. $-4a^5(6a^5)$

21. $(8w^4)(-w^7)$

22. $(-p)(-9p^2)$

23. $-5y^3(-8y^6)$

24. $(-7a^4bc^3)(5ab^4c^2)$

25. $\dfrac{8^{15}}{8^4}$

26. $\dfrac{2^9}{2}$

27. $\dfrac{h^7}{h^6}$

28. $\dfrac{g^{18}}{g^6}$

29. $\dfrac{36d^{10}}{6d^5}$

30. $\dfrac{16t^4}{8t}$

31. $\dfrac{x^6 y^{14}}{x^4 y^9}$

32. $\dfrac{3^4 x^4}{3x^2}$

Example 6 **33. COMPUTERS** The processing speed of a certain computer is 10^{11} instructions per second. Another computer has a processing speed that is 10^3 times as fast. How many instructions per second can the faster computer process?

34. SEATING The table shows the seating capacity of two different facilities. About how many times as great is the capacity of Madison Square Garden than a movie theater?

Place	Seating Capacity
Movie theater	3^5
Madison Square Garden	3^9

Example 7 **Simplify.**

35. $\dfrac{4^5 \cdot 5^3 \cdot 6^2}{4^4 \cdot 5^2 \cdot 6}$

36. $\dfrac{6^3 \cdot 6^6 \cdot 6^4}{6^2 \cdot 6^3 \cdot 6^3}$

37. $\dfrac{(-2)^5 \cdot (-3)^4 \cdot (-5)^3}{(-2)^3 \cdot (-3) \cdot (-5)^2}$

38. MEASUREMENT Refer to the information in the table.

   a. How many times as great is one quadrillion than one million?

   b. One quintillion is one trillion times as great as what number?

Power of Ten	U.S. Name
10^3	one thousand
10^6	one million
10^9	one billion
10^{12}	one trillion
10^{15}	one quadrillion
10^{18}	one quintillion

39. **OPEN ENDED** Write a multiplication expression with a product of 5^{13}.

40. **NUMBER SENSE** Is $\dfrac{3^{100}}{3^{99}}$ *greater than, less than,* or *equal to* 3? Explain your reasoning.

41. **CHALLENGE** What is twice 2^{30}? Write using exponents.

42. **REASONING** The figure at the right is composed of a circle and a square. The circle touches the square at the midpoints of the four sides.

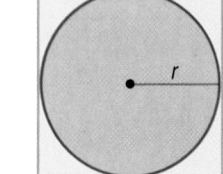

   a. What is the length of one side of the square?

   b. The formula $A = \pi r^2$ is used to find the area of a circle. The formula $A = (2r)^2$ can be used to find the area of the square. Find the ratio of the area of the circle to the area of the square in simplest form.

   c. Copy and complete the table shown.

   d. What can you conclude about the relationship between the areas of the circle and the square?

Radius (units)	Area of Circle (units²)	Area of Square (units²)	Ratio $\left(\dfrac{\text{Area of circle}}{\text{Area of square}}\right)$
r	πr^2	$(2r)^2$ or $4r^2$	■
$2r$	■	■	■
$4r$	■	■	■
r^2	■	■	■

43. **WRITE MATH** Explain why the Quotient of Powers Rule cannot be used to simplify the expression $\dfrac{x^5}{y^2}$.

Test Practice

44. Which expression is equivalent to $8x^2y \cdot 8yz^2$?

   A. $64x^2y^2z^2$

   B. $64x^2yz^2$

   C. $16x^2y^2z^2$

   D. $384x^2y^2z^2$

45. Which of the following is equivalent to $\left(-\dfrac{2}{3}\right)^3$?

   F. $-\dfrac{6}{9}$　　　　**H.** $\dfrac{8}{27}$

   G. $-\dfrac{8}{27}$　　　　**I.** $\dfrac{6}{9}$

46. **SHORT RESPONSE** What is the area of the rectangle below?

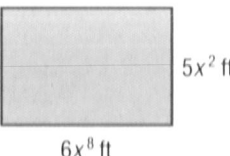

$5x^2$ ft

$6x^8$ ft

47. One meter is 10^3 times longer than one millimeter. One kilometer is 10^6 times longer than one millimeter. How many times longer is one kilometer than one meter?

   A. 10^9　　　　**C.** 10^3

   B. 10^6　　　　**D.** 10

Main Idea
Use laws of exponents to find powers of monomials.

 **Get ConnectED**

CCSS 8.EE.1

Powers of Monomials

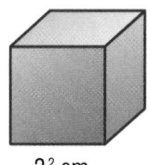

MEASUREMENT Suppose the side length of a cube is 2^2 centimeters.

2^2 cm

1. Write a multiplication expression for the volume of the cube.

2. Simplify the expression. Write as a single power of 2.

3. Using 2^2 as the base, write the multiplication expression $2^2 \cdot 2^2 \cdot 2^2$ using an exponent.

4. Explain why $(2^2)^3 = 2^6$.

You can use the rule for finding the *product* of powers to discover the rule for finding the *power* of a power.

$$\overbrace{(6^4)^5 = (6^4)(6^4)(6^4)(6^4)(6^4)}^{\text{5 factors}}$$
$$= 6^{4+4+4+4+4} \longleftarrow \text{Apply the rule for the product of powers.}$$
$$= 6^{20}$$

Notice that the product of the original exponents, 4 and 5, is the final power 20. This relationship is stated in the following Law of Exponents.

Key Concept — Power of a Power

Words   To find the power of a power, multiply the exponents.

Examples

Numbers

$(5^2)^3 = 5^{2 \cdot 3}$ or 5^6

Algebra

$(a^m)^n = a^{m \cdot n}$

EXAMPLES   Find the Power of a Power

1 Simplify $(8^4)^3$.

$(8^4)^3 = 8^{4 \cdot 3}$ Power of a Power

$= 8^{12}$ Simplify.

2 Simplify $(k^7)^5$.

$(k^7)^5 = k^{7 \cdot 5}$ Power of a Power

$= k^{35}$ Simplify.

CHECK Your Progress

Simplify. Express using exponents.

a. $(2^5)^2$

b. $(w^4)^6$

c. $[(3^2)^3]^2$

Extend the power of a *power* rule to find the power of a *product*.

$$
(3a^4)^5 = \overbrace{(3a^4)(3a^4)(3a^4)(3a^4)(3a^4)}^{\text{5 factors}}
$$

$$
= 3 \cdot 3 \cdot 3 \cdot 3 \cdot 3 \cdot a^4 \cdot a^4 \cdot a^4 \cdot a^4 \cdot a^4 \qquad \text{Associative and Commutative Properties of Multiplication}
$$

$$
= 3^5 \cdot (a^4)^5 \qquad \text{Write using powers.}
$$

$$
= 243 \cdot a^{20} \text{ or } 243a^{20} \qquad \text{Apply the rule for power of a power.}
$$

This example suggests the following Law of Exponents.

Key Concept — Power of a Product

Words   To find the power of a product, find the power of each factor and multiply.

Examples

Numbers	Algebra
$(6x^2)^3 = (6)^3 \cdot (x^2)^3$ or $216x^6$	$(ab)^m = a^m b^m$

EXAMPLES Power of a Product

③ Simplify $(4p^3)^4$.

$$(4p^3)^4 = 4^4 \cdot p^{3 \cdot 4}$$

$$= 256p^{12} \quad \text{Simplify.}$$

④ Simplify $(-2m^7 n^6)^5$.

$$(-2m^7 n^6)^5 = (-2)^5 m^{7 \cdot 5} n^{6 \cdot 5}$$

$$= -32m^{35} n^{30} \quad \text{Simplify.}$$

CHECK Your Progress

Simplify.

d. $(8b^9)^2$ **e.** $(6x^5 y^{11})^4$ **f.** $(-5w^2 z^8)^3$

REAL-WORLD EXAMPLE

⑤ GEOMETRY Express the area of the square as a monomial.

$$A = s^2 \qquad \text{Area of a square}$$

$$A = (7a^4 b)^2 \qquad \text{Replace } s \text{ with } 7a^4 b.$$

$$A = 7^2 (a^4)^2 (b^1)^2 \qquad \text{Power of a Product}$$

$$A = 49a^8 b^2 \qquad \text{Simplify.}$$

The area of the square is $49a^8 b^2$ square units.

$7a^4 b$

CHECK Your Progress

g. GEOMETRY Find the volume of a cube with sides of length $8x^3 y^5$. Express as a monomial.

Examples 1–4 **Simplify.**

1. $(3^2)^5$
2. $(h^6)^4$
3. $[(2^3)^2]^3$
4. $(7w^7)^3$
5. $(5g^8k^{12})^4$
6. $(-6r^5s^9)^2$

Example 5 **(7) MEASUREMENT** Express the volume of the cube at the right as a monomial.

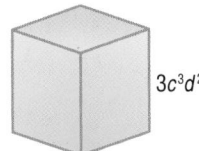

$3c^3d^2$

Practice and Problem Solving

 = **Step-by-Step Solutions** begin on page R1.
Extra Practice begins on page EP2.

Examples 1–4 **Simplify.**

8. $(4^2)^3$
9. $(2^2)^7$
10. $(5^3)^3$
11. $(3^4)^2$
12. $(d^7)^6$
13. $(m^8)^5$
14. $(h^4)^9$
15. $(z^{11})^5$
16. $[(3^2)^2]^2$
17. $[(4^3)^2]^2$
18. $[(5^2)^2]^2$
19. $[(2^3)^3]^2$
20. $(5j^6)^4$
21. $(8v^9)^5$
22. $(11c^4)^3$
23. $(14y)^4$
24. $(6a^2b^6)^3$
25. $(2m^5n^{11})^6$
26. $(-3w^3z^8)^5$
27. $(-5r^4s^{12})^4$

Example 5 **GEOMETRY** Express the area of each square below as a monomial.

28.

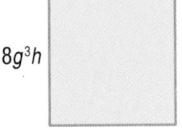

$8g^3h$

29.

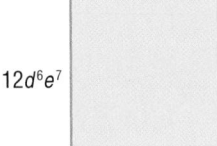

$12d^6e^7$

GEOMETRY Express the volume of each cube below as a monomial.

30.

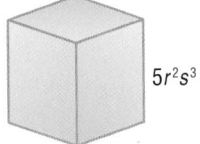

$5r^2s^3$

31.

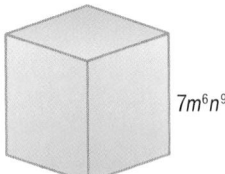

$7m^6n^9$

Simplify.

32. $(0.5k^5)^2$
33. $(0.3p^7)^3$
34. $\left(\frac{1}{4}w^5z^3\right)^2$
35. $\left(\frac{3}{5}a^6b^9\right)^2$
36. $(3x^2)^4(5x^6)^2$
37. $(-2v^7)^3(-4v^2)^4$

38. **PHYSICS** A ball is dropped from the top of a building. The expression $4.9x^2$ gives the distance in meters the ball has fallen after x seconds. Write and simplify an expression that gives the distance in meters the ball has fallen after x^2 seconds. Then write and simplify an expression that gives the distance the ball has fallen after x^3 seconds.

39. **BACTERIA** A certain culture of bacteria doubles in population every hour. At 1 P.M., there are 5 cells. The expression $5(2^x)$ gives the number of bacteria that are present x hours after 1 P.M. Simplify the expressions $[5(2^x)]^2$ and $[5(2^x)]^3$.

40. REASONING The table gives the area and volume of a square and cube, respectively, with side lengths shown.

a. Copy and complete the table.

b. Describe how the area and volume are each affected if the side length is doubled. Then describe how they are each affected if the side length is tripled.

Side Length (units)	Area of Square (units²)	Volume of Cube (units³)
x	x^2	x^3
$2x$		
$3x$		
x^2		
x^3		

c. Describe how the area and volume are each affected if the side length is squared. Describe how they are each affected if the side length is cubed.

CHALLENGE Solve each equation for x.

41. $(7^x)^3 = 7^{15}$

42. $(-2m^3n^4)^x = -8m^9n^{12}$

43.  **WRITE MATH** Compare and contrast how you would correctly simplify the expressions $(2a^3)(4a^6)$ and $(2a^3)^6$.

Test Practice

44. Which expression is equivalent to $(10^4)^8$?

A. 10^2

B. 10^4

C. 10^{12}

D. 10^{32}

45. Which expression has the same value as $81h^8k^6$?

F. $(9h^6k^4)^2$

G. $(9h^4k^3)^2$

H. $(6h^5k^3)^3$

I. $(3h^2k)^6$

46. What is the volume of the cube shown below?

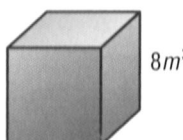

A. $8m^3$

B. $16m^5$

C. $64m^9$

D. $512m^9$

Simplify. Express using exponents. (Lesson 1B)

47. $6^4 \cdot 6^7$

48. $18^3 \cdot 18^5$

49. $(-3x^{11})(-6x^3)$

50. $(-9a^4)(2a^7)$

51. WATERFALLS The table shows the heights of some United States waterfalls. What is the height of each waterfall?

(Lesson 1A)

Waterfall	Height (ft)
Bridalveil (California)	$2^2 \cdot 5 \cdot 31$
Fall Creek (Tennessee)	2^8
Shoshone (Idaho)	$2^2 \cdot 53$

PART   A   B   C   **D**

Problem-Solving Investigation

Main Idea Solve problems by acting it out.

P.S.I. TEAM +

Act It Out

SETH: I received an E-mail about a concert. I forwarded the E-mail to two of my friends. They each forwarded it to two more friends, and so on. I wonder how many E-mails were sent at the 4th stage?

YOUR MISSION: Act it out to find the number of E-mails sent at the 4th stage.

Understand	You know that each person at each stage sends the E-mail to two people. You can use counters to represent the trail of E-mails sent.
Plan	Use red counters to represent the E-mails you sent. Use yellow counters to show the E-mails sent at the 2nd stage. Continue the pattern.
Solve	1st stage 2nd stage 3rd stage 4th stage There are 16 counters in the 4th row. So, 16 E-mails were sent during the 4th stage.
Check	The number of E-mails at each stage is a power of 2. So, find 2^4. Since $2^4 = 16$, the answer is correct. ✔

Analyze the Strategy

1. How did using the *act it out* strategy make it easier to solve the problem?

2. At what stage in the problem would it become more difficult to use the *act it out* strategy? Explain.

3. **WRITE MATH** Write a problem that could be solved by acting it out. Then use the strategy to solve the problem.

Mixed Problem Solving

 = **Step-by-Step Solutions** begin on page R1.
Extra Practice begins on page EP2.

- Act it out.
- Work backward.
- Look for a pattern.
- Choose an operation.

Use the *act it out* strategy to solve Exercises 4–6.

4. **MONEY** Shiro bought an apple juice and a bag of pretzels for $4.55. If he paid the cashier with a $5 bill, in how many different ways can he receive his change if the cashier only gives him quarters, dimes, and nickels?

5. **FITNESS** The length of a basketball court is 84 feet. Kareem runs 20 feet forward and then 8 feet back. How many more times will he have to do this until he reaches the end of the basketball court?

6. **PHOTOGRAPHS** Malcolm is taking a picture of the French Club's five officers. The club secretary will always stand on the left and the treasurer will always stand on the right. How many different ways can he arrange the officers in a single row for the picture?

Use any strategy to solve Exercises 7–13.

7. **MEASUREMENT** Mrs. Lopez is designing her garden in the shape of a rectangle. The area of her garden is 2 times greater than the area of the rectangle shown. Write the area of Mrs. Lopez's garden in simplest form.

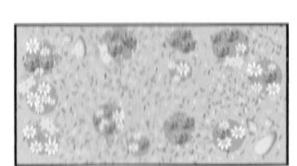

$8s^2$ ft

$4s^3$ ft

8. **ALGEBRA** Complete the pattern.

    100, 81, 64, ■, 36, ■.

9. **MONEY** Caroline received money for a birthday gift. She loaned $5 to her sister Mara and spent half of the remaining money. The next day she received $10 from her uncle. After spending $9 at the movies, she still had $11 left. How much money did she receive for her birthday?

10. **UNIFORMS** Nick has to wear a uniform to school. He can wear either navy blue, black, or khaki pants with a green, white, or yellow shirt. How many uniform combinations can Nick wear?

11. **PATTERNS** Find the next term in the pattern below.

    $$2^2 = 4, 4^2 = 16, 16^2 = 256, \ldots$$

12. **THEME PARKS** Refer to the graphic below. How does the attendance of the Magic Kingdom compare with the attendance of Disney's Animal Kingdom?

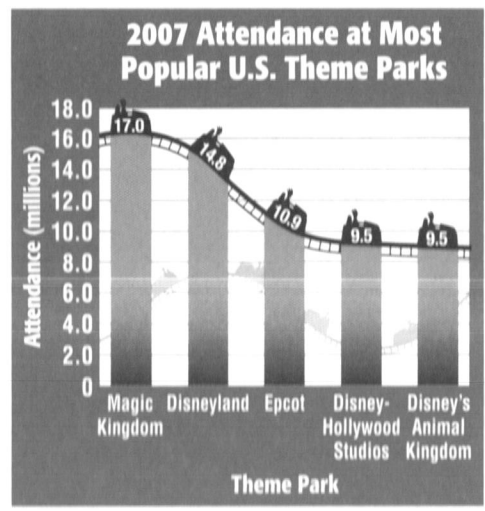

13. **PARTIES** Abby sent a text message to three friends inviting each of them to a party. Each of those friends sent the message to three more friends. How many people received the text message at the fourth stage?

Main Idea

Write and evaluate expressions using negative exponents.

8.EE.1

Negative Exponents

INSECTS The table shows the approximate wing beats per minute for certain insects.

Insect	Wing Beats per Minute
house fly	10,000
small butterfly	100

1. Write a ratio in simplest form that compares the number of wing beats for a butterfly to a housefly.

2. Write the ratio as a fraction with an exponent in the denominator and as a decimal.

You can use exponents to represent very small numbers. Consider the pattern in the powers of 10.

Negative powers are the result of repeated division. The pattern suggests the following rules.

Exponential Form	Standard Form
10^2	100
10^1	10
10^0	1
10^{-1}	$\frac{1}{10}$
10^{-2}	$\frac{1}{100}$

$100 \div 10 = 10$
$10 \div 10 = 1$
$1 \div 10 = \frac{1}{10}$ or $\frac{1}{10^1}$
$\frac{1}{10} \div 10 = \frac{1}{100}$ or $\frac{1}{10^2}$

Key Concept — Zero and Negative Exponents

Words Any nonzero number to the zero power is 1. Any nonzero number to the negative n power is the multiplicative inverse of its nth power.

Examples

Numbers

$5^0 = 1$

$7^{-3} = \frac{1}{7} \cdot \frac{1}{7} \cdot \frac{1}{7}$ or $\frac{1}{7^3}$

Algebra

$x^0 = 1, x \neq 0$

$x^{-n} = \frac{1}{x^n}, x \neq 0$

 **EXAMPLES** Write Expressions using Positive Exponents

Study Tip

Negative Exponents
Remember that 6^{-3} is equal to $\frac{1}{6^3}$, not −216 or −18.

Write each expression using a positive exponent.

① 6^{-3}

$6^{-3} = \frac{1}{6^3}$ Definition of negative exponent

② a^{-5}

$a^{-5} = \frac{1}{a^5}$ Definition of negative exponent

 CHECK Your Progress

a. 7^{-2} **b.** 5^0 **c.** b^{-4} **d.** m^{-3}

 **EXAMPLES** ### Evaluate Expressions with Negative Exponents

Evaluate each expression.

 3 2^{-5}

$2^{-5} = \dfrac{1}{2^5}$ Definition of negative exponent

$= \dfrac{1}{32}$ $2^5 = 2 \cdot 2 \cdot 2 \cdot 2 \cdot 2$ or 32

 4 $(-4)^{-3}$

$(-4)^{-3} = \dfrac{1}{(-4)^3}$ Definition of negative exponent

$= \dfrac{1}{-64}$ $(-4)^3 = (-4)(-4)(-4)$ or -64

✔️ **CHECK Your Progress**

   e. 3^{-3} **f.** 4^{-2} **g.** 2^{-3} **h.** $(-6)^{-4}$

 EXAMPLES ### Write Expressions using Negative Exponents

Write each fraction as an expression using a negative exponent.

5 $\dfrac{1}{5^2}$

$\dfrac{1}{5^2} = 5^{-2}$ Definition of negative exponent

6 $\dfrac{1}{36}$

$\dfrac{1}{36} = \dfrac{1}{6^2}$ Definition of exponent

$= 6^{-2}$ Definition of negative exponent

✔️ **CHECK Your Progress**

   i. $\dfrac{1}{8^3}$ **j.** $\dfrac{1}{c^5}$ **k.** $\dfrac{1}{4}$ **l.** $\dfrac{1}{27}$

Negative exponents are often used in science to express values that are very small. Most often, the number is a power of 10.

REAL-WORLD EXAMPLE ### Use Negative Exponents

7 **BIOLOGY** One human hair is about 0.001 inch in diameter. Write the decimal as a power of 10.

$0.001 = \dfrac{1}{1,000}$ Write the decimal as a fraction.

$= \dfrac{1}{10^3}$ $1,000 = 10^3$

$= 10^{-3}$ Definition of negative exponent

A human hair is 10^{-3} inch thick.

 CHECK Your Progress

   m. **SCIENCE** A water molecule is about 0.0000000001 meter long. Write the decimal as a power of 10.

The Product of Powers and the Quotient of Powers rules can be used to multiply and divide powers with negative exponents.

EXAMPLES **Multiply and Divide with Negative Exponents**

Simplify each expression. Express using positive exponents.

8 $x^4 \cdot x^{-2}$

$x^4 \cdot x^{-2} = x^{4 + (-2)}$ Product of Powers

$\qquad = x^2$ Add the exponents.

9 $\dfrac{w^{-1}}{w^{-4}}$

$\dfrac{w^{-1}}{w^{-4}} = w^{-1 - (-4)}$ Quotient of Powers

$\qquad = w^{(-1) + 4}$ or w^3 Subtract the exponents.

CHECK Your Progress

n. $3^8 \cdot 3^{-2}$ **o.** $n^9 \cdot n^{-4}$ **p.** $\dfrac{11^{-2}}{11^8}$ **q.** $\dfrac{b^{-4}}{b^{-7}}$

✓ CHECK Your Understanding

Examples 1 and 2   **Write each expression using a positive exponent.**

1. 2^{-4} **2.** 4^{-3} **3.** a^{-4} **4.** g^{-7}

Examples 3 and 4   **Evaluate each expression.**

5. 3^{-6} **6.** 5^{-2} **7.** $(-2)^{-6}$ **8.** $(-4)^{-3}$

Examples 5 and 6   **Write each fraction as an expression using a negative exponent.**

9. $\dfrac{1}{3^4}$ **10.** $\dfrac{1}{m^5}$ **11.** $\dfrac{1}{16}$ **12.** $\dfrac{1}{49}$

Example 7   **13. ANIMALS** An American green tree frog tadpole is about 0.00001 kilometer in length when it hatches. Write this decimal as a power of 10.

Examples 8 and 9   **Simplify. Express using positive exponents.**

14. $5^{-1} \cdot 5^{-2}$ **15** $3^{-3} \cdot 3^{-2}$ **16.** $r^{-7} \cdot r^3$ **17.** $m^{-4} \cdot m^{-3}$

18. $\dfrac{15^{-6}}{15^2}$ **19.** $\dfrac{12^{-3}}{12^{-5}}$ **20.** $\dfrac{h^5}{h^{-5}}$ **21.** $\dfrac{p^1}{p^{-4}}$

Practice and Problem Solving

 = **Step-by-Step Solutions** begin on page R1.
Extra Practice begins on page EP2.

Examples 1 and 2   Write each expression using a positive exponent.

22. 6^{-8} **23.** 7^{-10} **24.** $(-3)^{-5}$ **25.** $(-5)^{-4}$

26. s^{-9} **27.** g^{-7} **28.** t^{-11} **29.** w^{-13}

Examples 3 and 4   Evaluate each expression.

30. 3^{-5} **31.** 7^{-3} **32.** 2^{-4} **33.** 12^{-3}

34. $(-5)^{-4}$ **35.** $(-9)^{-3}$ **36.** $(-10)^{-4}$ **37.** $(-8)^{-5}$

Examples 5 and 6   Write each fraction as an expression using a negative exponent.

38. $\dfrac{1}{9^8}$ **39.** $\dfrac{1}{12^4}$ **40.** $\dfrac{1}{(-4)^5}$ **41.** $\dfrac{1}{(-5)^7}$

42. $\dfrac{1}{256}$ **43.** $\dfrac{1}{125}$ **44.** $\dfrac{1}{216}$ **45.** $\dfrac{1}{1,024}$

Example 7   **46. MEASUREMENT** The table shows different metric measurements. Write each decimal as a power of 10.

47. SCIENCE An atom is a small unit of matter. A small atom measures about 0.0000000001 meter. Write the decimal as a power of 10.

Measurement	Value
Decimeter	0.1
Centimeter	0.01
Millimeter	0.001
Micrometer	0.000001

Examples 8 and 9   Simplify. Express using positive exponents.

48. $4^{-2} \cdot 4^3$ **49.** $2^{-3} \cdot 2^{-4}$ **50.** $x^6 \cdot x^{-3}$ **51** $y^{-1} \cdot y^4$

52. $z^{-2} \cdot z^{-3}$ **53.** $s^{-5} \cdot s^{-2}$ **54.** $m^2 n^{-1} \cdot m^{-3} n^3$ **55.** $-3ab \cdot 4a^{-3}b^3$

56. $\dfrac{2^{-9}}{2^4}$ **57.** $\dfrac{3^{-1}}{3^{-5}}$ **58.** $\dfrac{b^{-7}}{b^5}$ **59.** $\dfrac{a^{-4}}{a^{-6}}$

60. $\dfrac{w^{-6}}{w^{-2}}$ **61.** $\dfrac{y^{-6}}{y^{-10}}$ **62.** $\dfrac{x^4}{x^{-2}}$ **63.** $\dfrac{z^{-4}}{z^{-8}}$

64. ANIMALS A common flea that is 2^{-4} inch long can jump about 2^3 inches high. About how many times its body size can a flea jump?

65 **MEDICINE** The mass of a molecule of penicillin is 10^{-18} kilogram and the mass of a molecule of insulin is 10^{-23} kilogram. How many times greater is the mass of a molecule of penicillin than the mass of a molecule of insulin?

66. MASS The table shows the average mass of different objects.

    a. How many times greater is the mass of a paper clip than a drop of water?

    b. About how many blood cells would fit in one drop of water?

    c. A box of paper clips holds approximately 1 kilogram of paper clips. How many paper clips are in the box?

Object	Mass (kg)
Paper clip	10^{-3}
Drop of water	10^{-6}
Blood cell	10^{-15}

Find each missing exponent.

67. $\dfrac{17^{\bullet}}{17^4} = 17^8$ **68.** $\dfrac{k^6}{k^{\bullet}} = k^2$ **69.** $\dfrac{5^{\bullet}}{5^{-9}} = 5^3$ **70.** $\dfrac{p^{-1}}{p^{\bullet}} = p^{10}$

71. **NUMBER SENSE** Without evaluating, order 11^{-3}, 11^2, and 11^0 from least to greatest. Explain your reasoning.

72. **OPEN ENDED** Write an expression with a negative exponent with a value between 0 and $\frac{1}{2}$.

73. **CHALLENGE** Select several fractions between 0 and 1. Find the value of each fraction after it is raised to the –1 power. Explain the relationship between the –1 power and the original fraction.

74. **WRITE MATH** Explain the difference between the expressions $(-4)^2$ and 4^{-2}.

Test Practice

75. Which of the following shows the expressions $6^3, 6^0, 6^{-1}, 6^{-2}$, and 6^1 in order from least to greatest?

   A. $6^{-2}, 6^{-1}, 6^0, 6^1, 6^3$

   B. $6^{-1}, 6^0, 6^1, 6^{-2}, 6^3$

   C. $6^3, 6^{-2}, 6^{-1}, 6^1, 6^0$

   D. $6^3, 6^1, 6^0, 6^{-1}, 6^{-2}$

76. **GRIDDED RESPONSE** Evaluate 3^{-4}. Write your answer as a fraction.

77. A blood cell has a diameter of about 5^{-5} inches.

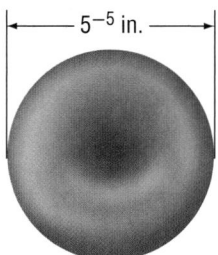

5^{-5} in.

Write 5^{-5} using positive exponents.

   F. 5^5

   G. $\frac{1}{5^{-5}}$

   H. $\frac{5^5}{1}$

   I. $\frac{1}{5^5}$

Spiral Review

78. **PIZZA** A pizza parlor has thin crust and thick crust, 2 different cheeses, and 4 toppings. Use the *act it out* strategy to determine how many different one-cheese and one-topping pizzas can be ordered. (Lesson 1D)

Simplify. (Lesson 1C)

79. $(6^3)^5$

80. $(n^7)^2$

81. $(2a^3b^2)^4$

82. $(-4p^{11}q)^3$

83. **ELEVATION** The table shows the approximate elevation of two cities. How many times as great is Anchorage's elevation than Honolulu's? (Lesson 1B)

City	Elevation (ft)
Anchorage, Alaska	2^7
Honolulu, Hawaii	2^4

= Step-by-Step Solutions begin on page R1.
Extra Practice begins on page EP2.

Examples 1 and 2 Write each expression using a positive exponent.

22. 6^{-8} **23.** 7^{-10} **24.** $(-3)^{-5}$ **25.** $(-5)^{-4}$

26. s^{-9} **27.** g^{-7} **28.** t^{-11} **29.** w^{-13}

Examples 3 and 4 Evaluate each expression.

30. 3^{-5} **31.** 7^{-3} **32.** 2^{-4} **33.** 12^{-3}

34. $(-5)^{-4}$ **35.** $(-9)^{-3}$ **36.** $(-10)^{-4}$ **37.** $(-8)^{-5}$

Examples 5 and 6 Write each fraction as an expression using a negative exponent.

38. $\dfrac{1}{9^8}$ **39.** $\dfrac{1}{12^4}$ **40.** $\dfrac{1}{(-4)^5}$ **41.** $\dfrac{1}{(-5)^7}$

42. $\dfrac{1}{256}$ **43.** $\dfrac{1}{125}$ **44.** $\dfrac{1}{216}$ **45.** $\dfrac{1}{1,024}$

Example 7 **46. MEASUREMENT** The table shows different metric measurements. Write each decimal as a power of 10.

Measurement	Value
Decimeter	0.1
Centimeter	0.01
Millimeter	0.001
Micrometer	0.000001

47. SCIENCE An atom is a small unit of matter. A small atom measures about 0.0000000001 meter. Write the decimal as a power of 10.

Examples 8 and 9 Simplify. Express using positive exponents.

48. $4^{-2} \cdot 4^3$ **49.** $2^{-3} \cdot 2^{-4}$ **50.** $x^6 \cdot x^{-3}$ **51** $y^{-1} \cdot y^4$

52. $z^{-2} \cdot z^{-3}$ **53.** $s^{-5} \cdot s^{-2}$ **54.** $m^2 n^{-1} \cdot m^{-3} n^3$ **55.** $-3ab \cdot 4a^{-3}b^3$

56. $\dfrac{2^{-9}}{2^4}$ **57.** $\dfrac{3^{-1}}{3^{-5}}$ **58.** $\dfrac{b^{-7}}{b^5}$ **59.** $\dfrac{a^{-4}}{a^{-6}}$

60. $\dfrac{w^{-6}}{w^{-2}}$ **61.** $\dfrac{y^{-6}}{y^{-10}}$ **62.** $\dfrac{x^4}{x^{-2}}$ **63.** $\dfrac{z^{-4}}{z^{-8}}$

64. ANIMALS A common flea that is 2^{-4} inch long can jump about 2^3 inches high. About how many times its body size can a flea jump?

65 **MEDICINE** The mass of a molecule of penicillin is 10^{-18} kilogram and the mass of a molecule of insulin is 10^{-23} kilogram. How many times greater is the mass of a molecule of penicillin than the mass of a molecule of insulin?

66. MASS The table shows the average mass of different objects.

Object	Mass (kg)
Paper clip	10^{-3}
Drop of water	10^{-6}
Blood cell	10^{-15}

  a. How many times greater is the mass of a paper clip than a drop of water?

  b. About how many blood cells would fit in one drop of water?

  c. A box of paper clips holds approximately 1 kilogram of paper clips. How many paper clips are in the box?

Find each missing exponent.

67. $\dfrac{17^{\bullet}}{17^4} = 17^8$ **68.** $\dfrac{k^6}{k^{\bullet}} = k^2$ **69.** $\dfrac{5^{\bullet}}{5^{-9}} = 5^3$ **70.** $\dfrac{p^{-1}}{p^{\bullet}} = p^{10}$

71. **NUMBER SENSE** Without evaluating, order 11^{-3}, 11^2, and 11^0 from least to greatest. Explain your reasoning.

72. **OPEN ENDED** Write an expression with a negative exponent with a value between 0 and $\frac{1}{2}$.

73. **CHALLENGE** Select several fractions between 0 and 1. Find the value of each fraction after it is raised to the –1 power. Explain the relationship between the –1 power and the original fraction.

74. **WRITE MATH** Explain the difference between the expressions $(-4)^2$ and 4^{-2}.

Test Practice

75. Which of the following shows the expressions $6^3, 6^0, 6^{-1}, 6^{-2}$, and 6^1 in order from least to greatest?

   A. $6^{-2}, 6^{-1}, 6^0, 6^1, 6^3$

   B. $6^{-1}, 6^0, 6^1, 6^{-2}, 6^3$

   C. $6^3, 6^{-2}, 6^{-1}, 6^1, 6^0$

   D. $6^3, 6^1, 6^0, 6^{-1}, 6^{-2}$

76. **GRIDDED RESPONSE** Evaluate 3^{-4}. Write your answer as a fraction.

77. A blood cell has a diameter of about 5^{-5} inches.

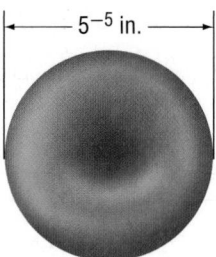

5^{-5} in.

Write 5^{-5} using positive exponents.

   F. 5^5

   G. $\frac{1}{5^{-5}}$

   H. $\frac{5^5}{1}$

   I. $\frac{1}{5^5}$

Spiral Review

78. **PIZZA** A pizza parlor has thin crust and thick crust, 2 different cheeses, and 4 toppings. Use the *act it out* strategy to determine how many different one-cheese and one-topping pizzas can be ordered. (Lesson 1D)

Simplify. (Lesson 1C)

79. $(6^3)^5$

80. $(n^7)^2$

81. $(2a^3b^2)^4$

82. $(-4p^{11}q)^3$

83. **ELEVATION** The table shows the approximate elevation of two cities. How many times as great is Anchorage's elevation than Honolulu's? (Lesson 1B)

City	Elevation (ft)
Anchorage, Alaska	2^7
Honolulu, Hawaii	2^4

Main Idea
Use scientific notation to write large and small numbers.

 Vocabulary
scientific notation

 Get Connect**ED**

 **8.EE.3**

Scientific Notation

Explore Copy and complete each table below.

Expression	Product
$8.7 \times 10^1 = 8.7 \times 10$	87
$8.7 \times 10^2 = 8.7 \times 100$	
$8.7 \times 10^3 = 8.7 \times \blacksquare$	

Expression	Product
$8.7 \times 10^{-1} = 8.7 \times \frac{1}{10}$	0.87
$8.7 \times 10^{-2} = 8.7 \times \frac{1}{100}$	
$8.7 \times 10^{-3} = 8.7 \times \blacksquare$	

1. If 8.7 is multiplied by a positive power of 10, what relationship exists between the decimal point's new position and the exponent?

2. When 8.7 is multiplied by a negative power of 10, how does the new position of the decimal point relate to the negative exponent?

Scientific notation is a compact way of writing numbers with absolute values that are very large or very small.

One factor is greater than or equal to 1, but less than 10. → 8.7×10^{-4} ← The power of 10 is written in exponential form.

Multiplying a factor by a positive power of 10 moves the decimal point right. Multiplying a factor by a negative power of 10 moves the decimal point left.

EXAMPLES **Express Numbers in Standard Form**

Write each number in standard form.

1 5.34×10^4

$5.34 \times 10^4 = 53,400.$ The decimal point moves 4 places right.

2 3.27×10^{-3}

$3.27 \times 10^{-3} = 0.00327$ The decimal point moves 3 places left.

CHECK Your Progress

a. 7.42×10^5 **b.** 6.1×10^{-2} **c.** 3.714×10^2

Use these rules to express a number in scientific notation.

- If the number is greater than or equal to 1, the power of ten is positive.
- If the number is between 0 and 1, the power of ten is negative.

 EXAMPLES Write Numbers in Scientific Notation

Write each number in scientific notation.

3 3,725,000

$3{,}725{,}000 = 3.725 \times 1{,}000{,}000$ The decimal point moves 6 places.

$\qquad\qquad\ = 3.725 \times 10^6$ Since 3,725,000 > 1, the exponent is positive.

4 0.000316

$0.000316 = 3.16 \times 0.0001$ The decimal point moves 4 places.

$\qquad\quad\ = 3.16 \times 10^{-4}$ Since 0 < 0.000316 < 1, the exponent is negative.

CHECK Your Progress

d. 14,140,000   **e.** 0.00876   **f.** 0.114

REAL-WORLD EXAMPLE

5 **TOURISM** Refer to the table at the right. Order the countries according to the amount of money visitors spent in the United States from greatest to least.

Dollars Spent by International Visitors in the U.S

Country	Dollars Spent
Canada	1.03×10^7
India	1.83×10^6
Mexico	7.15×10^6
United Kingdom	1.06×10^7

Canada and United Kingdom   Mexico and India

Step 1 $\begin{Bmatrix} 1.06 \times 10^7 \\ 1.03 \times 10^7 \end{Bmatrix} > \begin{Bmatrix} 7.15 \times 10^6 \\ 1.83 \times 10^6 \end{Bmatrix}$

Step 2 1.06 > 1.03   7.15 > 1.83

United Kingdom   Canada   Mexico   India

Top U.S. Cities Visited by Overseas Travelers

U.S. City	Number of Visitors
Boston	7.21×10^5
Las Vegas	1.3×10^6
Los Angeles	2.2×10^6
Metro D.C. area	9.01×10^5
New York	4×10^6
Orlando	1.8×10^6
San Francisco	1.6×10^6

 Real-World Link . . . .
The table lists seven of the top U.S. cities visited by overseas travelers in a recent year.

 CHECK Your Progress

g. **TRAVEL** Refer to the information at the left. Order the cities according to the number of visitors from least to greatest.

Examples 1 and 2   **Write each number in standard form.**

     1. 7.32×10^4 **2.** 9.931×10^5

     3. 4.55×10^{-1} **4.** 6.02×10^{-4}

Examples 3 and 4   **Write each number in scientific notation.**

     5. 277,000   **6.** 8,785,000,000

     7. 0.00004955   **8.** 0.524

Example 5   **9. MUSIC** The table lists the total value of music shipments for four years. List the years from least to greatest dollar amount.

Year	Music Shipments ($)
1	1.22×10^{10}
2	1.12×10^{10}
3	9.87×10^9
4	7.99×10^9

Practice and Problem Solving

● = **Step-by-Step Solutions** begin on page R1.
Extra Practice begins on page EP2.

Examples 1 and 2   **Write each number in standard form.**

     10. 2.08×10^2 **11.** 3.16×10^3 **12.** 7.113×10^7 **13.** 4.265×10^6

     14. 7.8×10^{-3} **15.** 1.1×10^{-4} **16.** 8.73×10^{-4} **17.** 2.52×10^{-5}

Examples 3 and 4   **Write each number in scientific notation.**

     18. 6,700   **19.** 43,000   **20.** 52,300,000   **21.** 147,000,000

     22. 0.037   **23.** 0.0072   **24.** 0.00000707   **25.** 0.0000901

Example 5   **26. CHEMISTRY** The table shows the mass in grams of one atom of each of several elements. List the elements in order from the least mass to greatest mass per atom.

27 GEOGRAPHY The areas of the world's oceans are listed in the table. Order the oceans according to their area from least to greatest.

Element	Mass per Atom
Carbon	1.995×10^{-23} g
Gold	3.272×10^{-22} g
Hydrogen	1.674×10^{-24} g
Oxygen	2.658×10^{-23} g
Silver	1.792×10^{-22} g

World's Oceans	
Ocean	Area (mi²)
Atlantic	2.96×10^7
Arctic	5.43×10^6
Indian	2.65×10^7
Pacific	6×10^7
Southern	7.85×10^6

Arrange these numbers in increasing order.

     28. $216{,}000{,}000,\ 2.2 \times 10^3,\ 3.1 \times 10^7,\ 310{,}000$

     29. $-4.56 \times 10^{-3},\ 4.56 \times 10^2,\ -4.56 \times 10^2,\ 4.56 \times 10^{-2}$

30. ASTRONOMY A light year is used to measure distances in the solar system. A light year is 5,865,696,000,000 miles.

    a. Write the number of miles in a light year in scientific notation.

    b. The star Sirius is about 8.6 light years away from Earth. Use scientific notation to write the distance in miles.

31. MATH IN THE MEDIA Find examples of large and small numbers in a newspaper or magazine or on the Internet. Write each number in scientific notation.

32. DINOSAURS The giganotosaurus weighed about 1.6×10^4 pounds. The microceratops weighed about 1.1×10^1 pounds. How many times heavier was the giganotosaurus than the microceratops? Write your answer in standard form. Round to the nearest tenth.

Replace each ● with <, >, or = to make a true statement.

33 6.25×10^3 ● 6.3×10^3 **34.** 5.49×10^6 ● $5,490,000$

35. $678,000$ ● 6.78×10^6 **36.** 3.14×10^{-4} ● 0.00314

37. GRAPHIC NOVEL Refer to the graphic novel frame below for Exercises a–b.

    a. Find Jacob's and Sarah's heights in nanometers.

    b. Write each height using scientific notation.

H.O.T. Problems

38. NUMBER SENSE Determine whether 1.2×10^5 or 1.2×10^6 is closer to one million. Explain.

39. CHALLENGE Compute and express each value in scientific notation.

    a. $\dfrac{(130,000)(0.0057)}{0.0004}$ **b.** $\dfrac{(90,000)(0.0016)}{(200,000)(30,000)(0.00012)}$

40. ✍ **WRITE MATH** When are numbers written in scientific notation?

41. The average width of a strand of a spider web is 7×10^{-6} meter. What is this length expressed in standard notation?

    A. 7,000,000 m
    C. 0.00007 m

    B. 700,000 m
    D. 0.000007 m

42.  **GRIDDED RESPONSE** By the year 2050, the world population is expected to reach 10 billion people. When ten billion is written in scientific notation, what is the exponent of the power of ten?

43. The thermosphere layer of the atmosphere is between 90 thousand and 110 thousand meters above sea level. What is 110 thousand written in scientific notation?

    F. 1.1×10^5
    H. 1.1×10^{-4}

    G. 1.1×10^4
    I. 1.1×10^{-5}

44. The attendance records for four Major League baseball teams for a recent year are shown below.

Team	Attendance
Florida Marlins	6.76×10^5
Los Angeles Angels	1.87×10^6
Pittsburgh Pirates	9.68×10^5
St. Louis Cardinals	1.98×10^6

Which team had the greatest attendance?

A. Florida Marlins

B. Los Angeles Angels

C. Pittsburgh Pirates

D. St. Louis Cardinals

Spiral Review

Write each expression using a positive exponent. (Lesson 2A)

45. 5^{-4}

46. 9^{-3}

47. 7^{-2}

48. 4^{-3}

49. **FOOD** Ariana wants to get an ice cream cone with two different flavors of ice cream. How many two-scoop ice cream cones can be created from the list of flavors shown? Use the *act it out* strategy. (Lesson 1D)

Ice Cream Flavors		
Vanilla	Chocolate	Strawberry
Chocolate Chip	Cookie Dough	Neopolitan
Peanut Butter	Rocky Road	Banana Split
Raspberry	Butter Pecan	Mint

GEOMETRY Express the area of each square below as a monomial. (Lesson 1C)

50.  $6a^3b$

51.  $3g^5h$

Simplify. Express using exponents. (Lesson 1B)

52. $3a^4 \cdot 12a^2$

53. $(5x)^2 \cdot 2x^5$

54. $\dfrac{3^9}{3^2}$

Write each expression using exponents. (Lesson 1A)

55. $6 \cdot 6 \cdot 6$

56. $2 \cdot 3 \cdot 3 \cdot 2 \cdot 2 \cdot 2$

57. $s \cdot t \cdot t \cdot s \cdot s \cdot t \cdot s$

Main Idea

Compute with numbers written in scientific notation.

 **8.EE.3, 8.EE.4**

Compute with Scientific Notation

E-MAIL Every day, nearly 130 billion spam E-mails are sent worldwide! How many is that each year? The numbers are too large even for your calculator.

1. Express 130 billion in scientific notation.

2. Round 365 to the nearest hundred and express it in scientific notation.

3. Write a multiplication expression using the numbers in Exercises 1 and 2 to find the total number of spam E-mails sent each year.

You can use the Product of Powers and Quotient of Powers properties to multiply and divide numbers written in scientific notation.

 **EXAMPLES** Multiplication and Division with Scientific Notation

Evaluate each expression. Express the result in scientific notation.

① $(4.2 \times 10^3)(1.6 \times 10^4)$

$(4.2 \times 10^3)(1.6 \times 10^4) = (4.2 \times 1.6)(10^3 \times 10^4)$ — Commutative and Associative Properties

$= (6.72)(10^3 \times 10^4)$ — Multiply 4.2 by 1.6.

$= 6.72 \times 10^{3+4}$ — Product of Powers

$= 6.72 \times 10^7$ — Add the exponents.

② $\dfrac{1.449 \times 10^6}{2.1 \times 10^3}$

$\dfrac{1.449 \times 10^6}{2.1 \times 10^3} = \left(\dfrac{1.449}{2.1}\right)\left(\dfrac{10^6}{10^3}\right)$ — Associative Property

$= (0.69)\left(\dfrac{10^6}{10^3}\right)$ — Divide 1.449 by 2.1.

$= 0.69 \times 10^{6-3}$ — Quotient of Powers

$= 0.69 \times 10^3$ — Subtract the exponents.

$= 0.69 \times 10^3$ — Write 0.69×10^3 in scientific notation.

$= 6.9 \times 10^2$ — Since the decimal point moved 1 place to the right, subtract 1 from the exponent.

CHECK Your Progress

a. $(8.4 \times 10^2)(2.5 \times 10^6)$

b. $\dfrac{9.72 \times 10^7}{3.6 \times 10^3}$

3 **SPACE** The average distance from Earth to the Sun is 1.46×10^8 kilometers. The average distance from Earth to the Moon is 3.84×10^5 kilometers. About how many times as great is the distance from Earth to the Sun than to the Moon?

Find $\dfrac{1.46 \times 10^8}{3.84 \times 10^5}$.

$$\dfrac{1.46 \times 10^8}{3.84 \times 10^5} = \left(\dfrac{1.46}{3.84}\right)\left(\dfrac{10^8}{10^5}\right) \qquad \text{Associative Property}$$

$$\approx 0.38 \times 10^3 \qquad \text{Simplify.}$$

$$\approx 3.8 \times 10^2 \qquad \text{Write } 0.38 \times 10^3 \text{ in scientific notation.}$$

The distance from Earth to the Sun is about 3.8×10^2 or 380 times as great as the distance from Earth to the Moon.

✓ **CHECK Your Progress**

c. **GEOGRAPHY** Refer to the information at the left. About how many times as great is the area covered by Lake Superior than Lake Ontario?

Real-World Link · · ·
Superior, the largest Great Lake, covers an area of 3.17×10^4 square miles. The smallest Great Lake, Ontario, covers an area of 7.34×10^3 square miles.

When adding or subtracting decimals in standard form, it is necessary to line up the place values. In scientific notation, the place value is represented by the exponent.

 **EXAMPLES** **Addition and Subtraction with Scientific Notation**

Evaluate each expression. Express the result in scientific notation.

4 $(6.89 \times 10^4) + (9.24 \times 10^5)$

$(6.89 \times 10^4) + (9.24 \times 10^5)$

$= (6.89 \times 10^4) + (92.4 \times 10^4) \qquad \text{Write } 9.24 \times 10^5 \text{ as } 92.4 \times 10^4.$

$= (6.89 + 92.4) \times 10^4 \qquad \text{Distributive Property}$

$= 99.29 \times 10^4 \qquad \text{Add 6.89 and 92.4.}$

$= 9.929 \times 10^5 \qquad \text{Write } 99.29 \times 10^4 \text{ in scientific notation.}$

5 $(7.83 \times 10^8) - (1.161 \times 10^7)$

$(7.83 \times 10^8) - (1.161 \times 10^7)$

$= (78.3 \times 10^7) - (1.161 \times 10^7) \qquad \text{Write } 7.83 \times 10^8 \text{ as } 78.3 \times 10^7.$

$= (78.3 - 1.161) \times 10^7 \qquad \text{Distributive Property}$

$= 77.139 \times 10^7 \qquad \text{Subtract 1.161 from 78.3.}$

$= 7.7139 \times 10^8 \qquad \text{Write } 77.139 \times 10^7 \text{ in scientific notation.}$

✓ **CHECK Your Progress**

d. $(8.41 \times 10^3) + (9.71 \times 10^4)$ e. $(1.263 \times 10^9) - (1.525 \times 10^7)$

Examples 1 and 2 **Evaluate each expression. Express the result in scientific notation.**

1. $(2.6 \times 10^5)(1.9 \times 10^2)$

2. $(5.3 \times 10^4)(9 \times 10^2)$

3. $(3.7 \times 10^{-2})(1.2 \times 10^3)$

4. $(3.3 \times 10^3)(2.1 \times 10^{-5})$

5. $\dfrac{8.37 \times 10^8}{2.7 \times 10^3}$

6. $\dfrac{8.04 \times 10^5}{6.7 \times 10^2}$

7. $\dfrac{9.72 \times 10^{-9}}{1.8 \times 10^5}$

8. $\dfrac{4.64 \times 10^{-4}}{2.9 \times 10^{-6}}$

Example 3 **9. TEXT MESSAGING** In 2005, 8.1×10^{10} text messages were sent in the United States. By 2007, the number of annual text messages had risen to 3.63×10^{11}. About how many times as great was the number of text messages in 2007 than 2005?

Examples 4 and 5 **Evaluate each expression. Express the result in scientific notation.**

10. $(5.4 \times 10^3) + (6.8 \times 10^5)$

11. $(8.9 \times 10^9) + (4.2 \times 10^6)$

12. $(1.35 \times 10^6) - (1.17 \times 10^5)$

13. $(9.64 \times 10^8) - (5.29 \times 10^6)$

Practice and Problem Solving

 = **Step-by-Step Solutions** begin on page R1.
Extra Practice begins on page EP2.

Examples 1 and 2 **Evaluate each expression. Express the result in scientific notation.**

14. $(8.5 \times 10^3)(1.1 \times 10^1)$

15. $(3.9 \times 10^2)(2.3 \times 10^6)$

16. $(6.45 \times 10^5)(1.2 \times 10^3)$

17. $(4.18 \times 10^{-4})(9 \times 10^{-6})$

18. $(1.26 \times 10^{-7})(5 \times 10^5)$

19. $(9.75 \times 10^3)(8.4 \times 10^{-6})$

20. $\dfrac{8.32 \times 10^7}{1.3 \times 10^5}$

21. $\dfrac{9.45 \times 10^{10}}{1.5 \times 10^6}$

22. $\dfrac{4.2 \times 10^8}{1.68 \times 10^2}$

23. $\dfrac{9 \times 10^{-11}}{2.4 \times 10^8}$

24. $\dfrac{3.24 \times 10^{-4}}{8.1 \times 10^{-7}}$

25 $\dfrac{1.14 \times 10^6}{4.8 \times 10^{-3}}$

Example 3 **26. SCIENCE** Neurons are cells in the nervous system that process and transmit information. An average neuron is about 5×10^{-6} meter in diameter. A standard table tennis ball is 4×10^{-2} meter in diameter. About how many times as great is the diameter of a ball than a neuron?

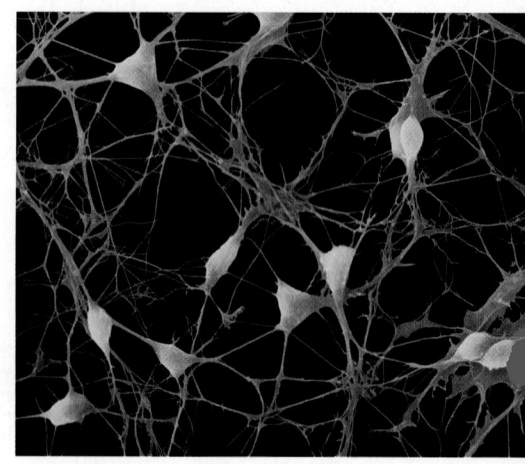

27. ASTRONOMY The Sun burns about 4.4×10^6 tons of hydrogen per second. How much hydrogen does the Sun burn in one year? (*Hint:* one year $\approx 3.16 \times 10^7$ seconds)

Examples 4 and 5 **Evaluate each expression. Express the result in scientific notation.**

28. $(7.3 \times 10^5) + (2.4 \times 10^6)$

29. $(9.5 \times 10^{11}) + (6.3 \times 10^9)$

30. $(1.357 \times 10^9) + (5.9 \times 10^5)$

31. $(8.64 \times 10^6) + (1.334 \times 10^{10})$

32. $(1.21 \times 10^5) - (9.5 \times 10^3)$

33. $(1.03 \times 10^9) - (4.7 \times 10^7)$

34. $(1.54 \times 10^{12}) - (6.94 \times 10^{10})$

35. $(8.71 \times 10^4) - (6.34 \times 10^1)$

36. **MEASUREMENT** A circular swimming pool holds 1.22×10^6 cubic inches of water. It is being filled at a rate of 1.5×10^3 cubic inches per minute. About how long will it take to fill the swimming pool?

37. **PARKS** Central Park in New York City is rectangular in shape and measures approximately 1.37×10^4 feet by 2.64×10^2 feet. If one acre is equal to 4.356×10^4 square feet, how many acres does Central Park cover? Round to the nearest hundredth.

MEASUREMENT Find the missing measure for each figure. Express the result in scientific notation.

38.

$A = \blacksquare$ m^2

8.3×10^{-4} m

2.5×10^{-3} m

39.

$P = 5 \times 10^4$ in.

s in.

H.O.T. Problems

40. **FIND THE ERROR** Enrique is finding $\dfrac{6.63 \times 10^{-6}}{5.1 \times 10^{-2}}$. Find his mistake and correct it.

$$\frac{6.63 \times 10^{-6}}{5.1 \times 10^{-2}} = \left(\frac{6.63}{5.1}\right)\left(\frac{10^{-6}}{10^{-2}}\right)$$
$$= 1.3 \times 10^{-6-2}$$
$$= 1.3 \times 10^{-8}$$

41. **Which One Doesn't Belong?** Identify the expression that does not belong with the other three. Explain your reasoning.

| 14.28×10^9 | $(3.4 \times 10^6)(4.2 \times 10^3)$ | 1.4×10^9 | $(3.4)(4.2) \times 10^{(6+3)}$ |

42. **WRITE MATH** Explain how to estimate the sum of (4.215×10^{-2}) and (3.2×10^{-4}).

43. A music download Web site announced that over 4×10^9 songs were downloaded by 5×10^7 registered users. What is the average number of downloads per user?

   A. 8×10^{-1}

   B. 1.25×10^{-2}

   C. 1.25×10^2

   D. 8×10^1

44. There are approximately 45 hundred species of mammals on Earth and 2.8×10^4 species of fish. What is the difference in the number of species?

   F. 6.2×10^0

   G. 2.35×10^4

   H. 1.6×10^{-1}

   I. 3.25×10^4

45. The rectangle has an area of 9.14×10^{-7} square kilometers.

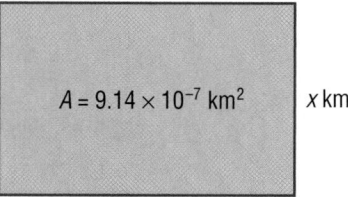

$A = 9.14 \times 10^{-7}$ km² x km

1.656×10^{-3} km

What is the approximate length of the missing side?

   A. 2.74×10^{-6}

   B. 5.52×10^{-4}

   C. 1.656×10^{-3}

   D. 1.51×10^{11}

Spiral Review

46. LANGUAGES It is estimated that over 836 million people speak Mandarin Chinese. Write this number in scientific notation. (Lesson 2B)

Write each expression using a positive exponent. (Lesson 2A)

47. 5^{-4} **48.** 6^{-3} **49.** 3^{-5} **50.** 8^{-2}

GEOMETRY Express the volume of each cube as a monomial. (Lesson 1C)

51.
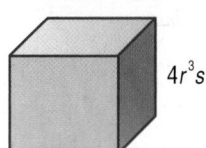
$4r^3s$

52.
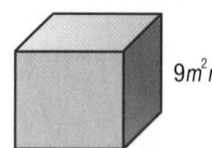
$9m^2n^4$

53. GEOGRAPHY The table shows the approximate land area of two counties. How many times as great is Alamosa County's land area than Grand Isle County's? (Lesson 1B)

County	Land Area (square miles)
Alamosa County, Colorado	3^6
Grand Isle County, Vermont	3^4

54. SPORTS The total points scored in a recent season for three players is shown in the table. How many total points did each player score? (Lesson 1A)

Player	Total Points
LeBron James	$2^8 \cdot 3^2$
Dwight Howard	$2^3 \cdot 7 \cdot 29$
Carmelo Anthony	$2^5 \cdot 47$

Mid-Chapter Check

Write each expression using exponents.
(Lesson 1A)

1. $3 \cdot 3 \cdot 3 \cdot 3$

2. $(-7)(-7)(-7)$

3. $2 \cdot 5 \cdot 5 \cdot 2 \cdot 2$

4. $a \cdot b \cdot a \cdot a \cdot b$

5. **SPORTS** The table shows the mass of different sports objects. Find the mass of each object. (Lesson 1A)

Object	Mass (g)
Softball	14^2
Bat	3^6
Glove	5^4

Simplify. Express using exponents. (Lesson 1B)

6. $10^4 \cdot 10^7$

7. $3^3 \cdot 3^5 \cdot 3^2$

8. $2^3 a^7 \cdot 2a^3$

9. $(3^2 xyz^2)(3^5 x^3 yz^3)$

10. **MULTIPLE CHOICE** Which expression below has the same value as $5m^2$? (Lesson 1A)

   A. $5m$

   B. $5 \cdot m \cdot m$

   C. $5 \cdot 5 \cdot m \cdot m$

   D. $5 \cdot m \cdot m \cdot m$

Simplify. Express using exponents. (Lesson 1B)

11. $\dfrac{9^5}{9^3}$

12. $\dfrac{k^{15}}{k^6}$

13. $\dfrac{24y^4}{4y^2}$

14. $\dfrac{45g^7}{3g^3}$

15. **AGE** Angelina is 2^3 years old. Her grandfather is 2^3 times her age. How old is her grandfather? (Lesson 1B)

Simplify. Express using exponents. (Lesson 1C)

16. $\left(3^3\right)^4$

17. $\left(5^2\right)^3$

18. $\left(a^7\right)^2$

19. $\left(x^4\right)^5$

20. **BOOKS** Jackie has four different textbooks that she wants to place on a shelf. Use the *act it out* strategy to determine how many different ways she can organize the books. (Lesson 1D)

Write each fraction as an expression using a negative exponent. (Lesson 2A)

21. $\dfrac{1}{4^3}$

22. $\dfrac{1}{5^5}$

23. $\dfrac{1}{(-8)^2}$

24. **MULTIPLE CHOICE** Which of the following shows 0.0000035 in scientific notation? (Lesson 2B)

   F. 3.5×10^6

   G. 3.5×10^5

   H. 3.5×10^{-5}

   I. 3.5×10^{-6}

25. **LIFE SCIENCE** A Petri dish contains 2.53×10^{11} bacteria. Write the number of bacteria in standard form. (Lesson 2B)

26. **GEOGRAPHY** The table shows the length of certain states' coastlines.

State	Coastline (mi)
Alaska	6.64×10^3
California	8.4×10^2
Florida	1.35×10^3
Louisiana	3.97×10^2

   a. Order the states from least to greatest coastline. (Lesson 2B)

   b. How many times longer is the coastline of Alaska than Louisiana? (Lesson 2C)

27. **POPULATION** The population of Groveton is 7.78×10^3. The population of Putnam is 1.68×10^6. About how many more people live in Putnam than Groveton? (Lesson 2C)

Main Idea
Find square roots and cube roots.

 **Vocabulary**

perfect square
square root
radical sign
perfect cube
cube root

 Get ConnectED

 8.EE.2

Roots

  Explore   Continue the pattern of coins until you reach 5 coins on each side.

1. Make a table showing the number of coins on a side and the total number of coins in the square arrangement.

2. Suppose a square arrangement has 36 coins. How many coins are on a side?

3. Suppose you have 420 coins. How many coins are on one side of the largest square you can construct? Explain.

Numbers such as 1, 4, 9, 16, and 25 are called **perfect squares** because they are squares of integers. Squaring a number and finding a square root are inverse operations.

Key Concept   Square Root

Words   A **square root** of a number is one of its two equal factors.

Symbols   If $x^2 = y$, then x is a square root of y.

A **radical sign**, $\sqrt{}$, is used to indicate a positive square root. Every positive number has *both* a negative and a positive square root.

 EXAMPLES   Find Square Roots

Find each square root.

① $\sqrt{64}$

$\sqrt{64} = 8$ Find the positive square root of 64; $8^2 = 64$.

② $\pm\sqrt{1.21}$

$\sqrt{1.21} = \pm 1.1$ Find both square roots of 1.21; $1.1^2 = 1.21$.

③ $-\sqrt{\dfrac{25}{36}}$

$-\sqrt{\dfrac{25}{36}} = -\dfrac{5}{6}$ Find the negative square root of $\sqrt{\dfrac{25}{36}}$; $\left(-\dfrac{5}{6}\right)^2 = \dfrac{25}{26}$.

④ $\sqrt{-16}$

There is no real square root because no number times itself is equal to -16.

✔ CHECK Your Progress

a. $\sqrt{\dfrac{9}{16}}$ **b.** $\pm\sqrt{0.81}$ **c.** $-\sqrt{49}$ **d.** $\sqrt{-100}$

By the definition of a square root, if $n^2 = a$, then $n = \pm\sqrt{a}$. You can use this relationship to solve equations that involve squares.

 EXAMPLE Use Square Roots to Solve an Equation

⑤ ALGEBRA Solve $t^2 = 169$. Check your solution(s).

$t^2 = 169$ Write the equation.

$t = \pm\sqrt{169}$ Definition of square root

$t = 13$ and -13 Check $13 \cdot 13 = 169$ and $(-13)(-13) = 169$ ✓

The equation has two solutions, 13 and -13.

CHECK Your Progress

Solve each equation. Check your solution(s).

e. $289 = a^2$ **f.** $m^2 = 0.09$ **g.** $y^2 = \dfrac{4}{25}$

In most real-world situations, a negative square root does not make sense. Only the positive or *principal* square root is considered.

REAL-WORLD EXAMPLE

⑥ HISTORY The square base of the Great Pyramid covers about 562,500 square feet. Determine the length of each side of the base.

Words	Area is equal to the square of the length of a side.
Variable	Let *s* represent the length of a side.
Equation	$s^2 = 562{,}500$

$s^2 = 562{,}500$ Write the equation.

$s = \pm\sqrt{562{,}500}$ Definition of square root

To find $\sqrt{562{,}500}$, find two equal factors of 562,500.

$562{,}500 = 2 \cdot 2 \cdot 3 \cdot 3 \cdot 5 \cdot 5 \cdot 5 \cdot 5 \cdot 5 \cdot 5$ Find the prime factors.

$\phantom{562{,}500} = (2 \cdot 3 \cdot 5 \cdot 5 \cdot 5)(2 \cdot 3 \cdot 5 \cdot 5 \cdot 5)$ Regroup into two equal factors.

So, $s = 2 \cdot 3 \cdot 5 \cdot 5 \cdot 5$ or 750.

Since distance cannot be negative, the length of each side is 750 feet.

CHECK Your Progress

h. CONCERTS A concert crew needs to set up 900 chairs on the floor level. If the chairs are placed in a square arrangement, how many should be in each row?

Real-World Link · · · · ·
The Great Pyramid of Khufu is the largest of the ancient pyramids.

Numbers such as 8, 27, and 64 are **perfect cubes** because they are the cubes of integers.

$$8 = 2 \cdot 2 \cdot 2 \text{ or } 2^3 \qquad 27 = 3 \cdot 3 \cdot 3 \text{ or } 3^3 \qquad 64 = 4 \cdot 4 \cdot 4 \text{ or } 4^3$$

> **Key Concept**   Cube Roots
>
> **Words**   A **cube root** of a number is one of its three equal factors.
>
> **Symbols**   If $x^3 = y$, then x is the cube root of y.

The symbol $\sqrt[3]{}$ is used to indicate a cube root of a number.

 EXAMPLES Find Cube Roots

Find each cube root.

7 $\sqrt[3]{125}$

$\sqrt[3]{125} = 5$ $5^3 = 5 \cdot 5 \cdot 5$ or 125

8 $\sqrt[3]{-27}$

$\sqrt[3]{-27} = -3$ $(-3)^3 = (-3) \cdot (-3) \cdot (-3)$ or -27

 CHECK Your Progress

Find each cube root.

i. $\sqrt[3]{729}$ **j.** $\sqrt[3]{-64}$ **k.** $\sqrt[3]{1,000}$

✓ CHECK Your Understanding

Examples 1–4   Find each square root.

1. $\sqrt{25}$ **2.** $\sqrt{0.64}$ **3.** $-\sqrt{1.69}$ **4.** $-\sqrt{\dfrac{16}{81}}$

5. $\pm\sqrt{100}$ **6.** $\pm\sqrt{\dfrac{49}{144}}$ **7.** $\sqrt{-\dfrac{25}{64}}$ **8.** $\sqrt{-1.44}$

Example 5   **ALGEBRA** Solve each equation. Check your solution(s).

9. $p^2 = 36$ **10.** $t^2 = \dfrac{1}{9}$ **11.** $6.25 = r^2$

Example 6   **12. GAMES** A checkerboard is a large square that is made up of 32 small red squares and 32 small black squares. How many small squares are along one side of a checkerboard?

Examples 7 and 8   Find each cube root.

13. $\sqrt[3]{216}$ **14.** $\sqrt[3]{4,913}$ **15.** $\sqrt[3]{-8}$ **16.** $\sqrt[3]{-125}$

Practice and Problem Solving

= **Step-by-Step Solutions** begin on page R1.
Extra Practice begins on page EP2.

Examples 1–4 Find each square root.

17. $\sqrt{16}$
18. $-\sqrt{81}$
19. $-\sqrt{484}$
20. $\sqrt{-36}$

21. $\sqrt{\dfrac{121}{324}}$
22. $-\sqrt{\dfrac{64}{225}}$
23. $\pm\sqrt{\dfrac{9}{49}}$
24. $-\sqrt{\dfrac{16}{25}}$

25. $-\sqrt{2.56}$
26. $\pm\sqrt{1.44}$
27. $\sqrt{-0.25}$
28. $\pm\sqrt{0.0196}$

Example 5 **ALGEBRA** Solve each equation. Check your solution(s).

29. $v^2 = 81$
30. $b^2 = 100$
31. $144 = s^2$
32. $225 = y^2$

33. $w^2 = \dfrac{36}{100}$
34. $\dfrac{9}{64} = c^2$
35. $0.0169 = d^2$
36. $a^2 = 1.21$

Example 6 **37** **PHOTOGRAPHY** A group of 169 students needs to be seated in a square formation for a yearbook photo. How many students should be in each row?

38. **BAND** A marching band wants to form a square in the middle of the field. If there are 225 members in the band, how many should be in each row?

Examples 7 and 8 Find each cube root.

39. $\sqrt[3]{1,331}$
40. $\sqrt[3]{-216}$
41. $\sqrt[3]{2,197}$
42. $\sqrt[3]{-512}$

43. $\sqrt[3]{1,728}$
44. $\sqrt[3]{-1,000}$
45. $\sqrt[3]{3,375}$
46. $\sqrt[3]{-343}$

ALGEBRA Solve each equation. Check your solution(s).

47. $\sqrt{x} = 5$
48. $\sqrt{y} = 20$
49. $\sqrt{z} = 10.5$

MEASUREMENT The formula for the perimeter of a square is $P = 4s$, where s is the length of a side. Find the perimeter of each square.

50.

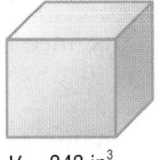

Area = 121 square inches

51.

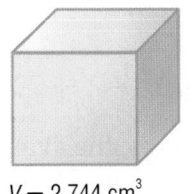

Area = 25 square feet

52.

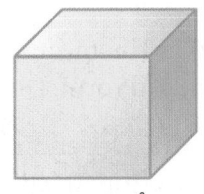

Area = 36 square meters

GEOMETRY The volume of each cube is given. Find the side length of each cube.

53.

$V = 343\ \text{in}^3$

54.

$V = 2,744\ \text{cm}^3$

55.

$V = 4,096\ \text{ft}^3$

56. **PUZZLES** A puzzle cube is shown. The volume of the cube is 512 cubic centimeters. What are the dimensions of the puzzle cube?

57. CHALLENGE Find each value.

a. $(\sqrt{36})^2$ **b.** $\left(\sqrt{\dfrac{25}{81}}\right)^2$ **c.** $(\sqrt{199})^2$ **d.** $(\sqrt{x})^2$

58. NUMBER SENSE Explain why $\sqrt{64}$ has a positive and a negative value.

59. 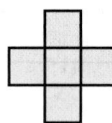 **WRITE MATH** Read the cartoon. Then find $\sqrt{1,296}$. Create a cartoon of your own that uses the square root of a perfect square.

Test Practice

60. The area of each square is 16 square units.

Find the perimeter of the figure.

A. 16 units   **C.** 40 units

B. 32 units   **D.** 48 units

61. Mr. Freeman's farm has a square cornfield. Find the area of the cornfield if the sides are measured in whole numbers.

F. 164,000 ft²

G. 170,150 ft²

H. 170,586 ft²

I. 174,724 ft²

Evaluate each expression. Express the result in scientific notation. (Lesson 2C)

62. $(9.7 \times 10^5)(4.2 \times 10^3)$ **63.** $(2.1 \times 10^4)(6.8 \times 10^2)$ **64.** $(8.4 \times 10^{-3})(1.2 \times 10^{-6})$

65. SPACE The radius of the Sun is 6.96×10^8 meters. Write this distance in standard form. (Lesson 2B)

66. INSECTS Depending on the species, the average length of a ladybug ranges from 0.001 meter to 0.01 meter. Write the length of the ladybug shown as a power of 10. (Lesson 2A)

Evaluate each expression. (Lesson 1A)

67. $5 \cdot 2^3 \cdot 7$

68. $2^2 \cdot 7 \cdot 10^4$

69. $2^4 \cdot 6^2 \cdot (-3)^3$

70. $4^3 \cdot 5^2 \cdot (-2)^3$

0.01 m

Explore

Main Idea

Estimate square roots of non-perfect squares.

Get ConnectED

CCSS 8.NS.2

Roots of Non-Perfect Squares

CRAFTS Mindi is making a quilting piece from a square pattern. The pattern is on dotted paper as shown in the activity below. She needs to figure out the side length of the square to cut the correct size quilting piece.

ACTIVITIES

STEP 1 On dot paper, copy and cut out a square like the one shown below. The area of section A is $\frac{1}{2}(2 \cdot 2)$ or 2 square units. So, the shaded square has an area of 8 square units.

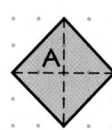

STEP 2 Draw a number line on your dot paper so that 1 unit equals the distance between dots.

Analyze the Results

1. Place your square on the number line. Between what two consecutive whole numbers is $\sqrt{8}$, the side length of the square, located?

2. Between what two perfect squares is 8 located?

3. Estimate the length of a side of the square. Verify your estimate by using a calculator to compute the value of $\sqrt{8}$.

4. How does the area of a square relate to the square of a number?

Practice and Apply

Estimate the side length of each square using the method shown above.

5.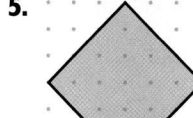

6.

7.

Main Idea

Use roots to estimate solutions.

CCSS 8.NS.2

Estimate Roots

TREES When a healthy apple becomes ripe, it will fall from the tree. Suppose an apple is growing on a tree 25 feet above the ground.

1. What is the square root of 25?

2. The formula $t = \dfrac{\sqrt{h}}{4}$ can be used to find the time t in seconds it will take for an object to fall from a certain height h in feet. How long will it take for the apple to hit the ground?

In the previous lesson, you found that $\sqrt{8}$ is not a whole number because 8 is not a perfect square.

The number line below shows that $\sqrt{8}$ is between 2 and 3. Since 8 is closer to 9 than 4, the best positive whole number estimate for $\sqrt{8}$ is 3.

$$
\begin{array}{c}
\sqrt{4} \qquad\qquad \sqrt{8}\ \ \sqrt{9} \\
\underset{2\quad 2.25\quad 2.5\quad 2.75\quad 3}{\longmapsto\!\!+\!\!+\!\!+\!\!\bullet\!\!\longrightarrow}
\end{array}
$$

EXAMPLES **Estimate Square Roots**

1 **Estimate $\sqrt{83}$ to the nearest whole number.**

- The largest perfect square less than 83 is 81. $\sqrt{81} = 9$
- The smallest perfect square greater than 83 is 100. $\sqrt{100} = 10$

Plot each square root on a number line. Then estimate $\sqrt{83}$.

$$
\begin{array}{c}
\sqrt{81}\ \ \sqrt{83} \qquad\qquad \sqrt{100} \\
\underset{9\quad 9.25\quad 9.5\quad 9.75\quad 10}{\longmapsto\!\!\bullet\!\!+\!\!+\!\!+\!\!\longrightarrow}
\end{array}
$$

$$81 < \ 83 < 100 \qquad \text{Write an inequality.}$$
$$9^2 < \ 83 < 10^2 \qquad 81 = 9^2 \text{ and } 100 = 10^2$$
$$\sqrt{9^2} < \sqrt{83} < \sqrt{10^2} \qquad \text{Find the square root of each number.}$$
$$9 < \sqrt{83} < 10 \qquad \text{Simplify.}$$

So, $\sqrt{83}$ is between 9 and 10. Since $\sqrt{83}$ is closer to $\sqrt{81}$ than $\sqrt{100}$, the best whole number estimate for $\sqrt{83}$ is 9.

② **Estimate** $\sqrt{21.5}$ **to the nearest whole number.**

- The largest perfect square less than 21.5 is 16. $\sqrt{16} = 4$
- The smallest perfect square greater than 21.5 is 25. $\sqrt{25} = 5$

$16 < \ \ 21.5 < 25$ Write an inequality.

$4^2 < \ \ 21.5 < 5^2$ $16 = 4^2$ and $25 = 5^2$

$\sqrt{4^2} < \sqrt{21.5} < \sqrt{5^2}$ Find the square root of each number.

$4 < \sqrt{21.5} < 5$ Simplify.

So, $\sqrt{21.5}$ is between 4 and 5. Since 21.5 is closer to 25 than 16, the best whole number estimate for $\sqrt{21.5}$ is 5.

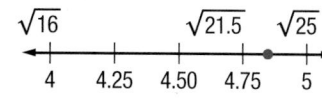

Read Math

Inequalities
$16 < 21.5 < 25$ is read *16 is less than 21.5 is less than 25* or *21.5 is between 16 and 25.*

✓ CHECK Your Progress

Estimate to the nearest whole number.

a. $\sqrt{35}$ **b.** $\sqrt{44.8}$ **c.** $\sqrt{170}$

🏃 ✏ REAL-WORLD EXAMPLE

③ **NATURE** The *golden rectangle* is found frequently in the nautilus shell. The length of the longer side divided by the length of the shorter side is equal to $\dfrac{1 + \sqrt{5}}{2}$. **Estimate this value.**

$1 + \sqrt{5}$

2

First estimate the value of $\sqrt{5}$.

$4 < \ \ 5 < 9$ 4 and 9 are the closest perfect squares.

$2^2 < \ \ 5 < 3^2$ $4 = 2^2$ and $9 = 3^2$

$\sqrt{2^2} < \sqrt{5} < \sqrt{3^2}$ Find the square root of each number.

$2 < \sqrt{5} < 3$ Simplify.

Since 5 is closer to 4 than 9, the best whole number estimate for $\sqrt{5}$ is 2. Use this value to evaluate the expression.

$\dfrac{1 + \sqrt{5}}{2} \approx \dfrac{1 + 2}{2}$ or 1.5

✓ CHECK Your Progress

d. BASEBALL In Little League, the bases are squares with sides of 14 inches. The expression $\sqrt{(s^2 + s^2)}$ represents the distance *diagonally across* a square of side length s. Estimate the diagonal distance across a base to the nearest inch.

🌐 Real-World Link . . . .
Major league baseball has specific requirements for the size of first, second, and third base. They are to be 15 inches square, and no more than 5 inches thick nor less than 3 inches thick.

 EXAMPLE **Estimate Cube Roots Mentally**

 Estimate $\sqrt[3]{320}$ to the nearest integer.

Use the *guess, check, and revise* strategy.

Base, b	b^3	Check
5	$5^3 = 125$	less than 320
6	$6^3 = 216$	less than 320
7	$7^3 = 343$	greater than 320

$$\sqrt[3]{125} \qquad \sqrt[3]{216} \qquad \sqrt[3]{320}\ \sqrt[3]{343}$$

5   6   7

Six and 7 are consecutive integers, and 320 is between 216 and 343. So, $\sqrt[3]{320}$ is between 6 and 7. Since 320 is closer to 343 than to 216, $\sqrt[3]{320}$ is closer to 7 than to 6.

CHECK Your Progress

e. $\sqrt[3]{62}$ **f.** $\sqrt[3]{25}$

You can also estimate cube roots mentally by using perfect cubes. The first ten positive perfect cubes are shown.

$1 = 1^3$	$216 = 6^3$
$8 = 2^3$	$343 = 7^3$
$27 = 3^3$	$512 = 8^3$
$64 = 4^3$	$729 = 9^3$
$125 = 5^3$	$1{,}000 = 10^3$

CHECK Your Understanding

Examples 1 and 2   **Estimate to the nearest whole number.**

     1. $\sqrt{28}$ **2.** $\sqrt{60}$ **3.** $\sqrt{135}$

     4. $\sqrt{13.5}$ **5.** $\sqrt{38.7}$ **6.** $\sqrt{79.2}$

Example 3   **7. SCIENCE** The number of swings back and forth of a pendulum of length L in inches each minute is $\dfrac{375}{\sqrt{L}}$. About how many swings will a 40-inch pendulum make each minute?

Example 4   **Estimate each cube root to the nearest integer. Do not use a calculator.**

     8. $\sqrt[3]{51}$ **9.** $\sqrt[3]{14}$ **10.** $\sqrt[3]{200}$

     11. $\sqrt[3]{145}$ **12.** $\sqrt[3]{95}$ **13.** $\sqrt[3]{360}$

Practice and Problem Solving

● = **Step-by-Step Solutions** begin on page R1.
Extra Practice begins on page EP2.

Examples 1 and 2   **Estimate to the nearest whole number.**

14. $\sqrt{44}$ 15. $\sqrt{23}$ 16. $\sqrt{125}$ 17. $\sqrt{197}$

18. $\sqrt{15.6}$ 19. $\sqrt{23.5}$ 20. $\sqrt{85.1}$ 21. $\sqrt{38.4}$

Example 3   22. **GEOMETRY** The radius of a circle with area A is approximately $\sqrt{\dfrac{A}{3}}$.
If a pizza has an area of 78 square inches, estimate its radius.

23 **PHYSICS** The formula $t = \dfrac{\sqrt{h}}{4}$ represents the time t in seconds that it takes an object to fall from a height of h feet. If a rock falls from a height of 125 feet, estimate how long it will take to reach the ground.

Example 4   **Estimate each cube root to the nearest integer. Do not use a calculator.**

24. $\sqrt[3]{199}$ 25. $\sqrt[3]{22}$ 26. $\sqrt[3]{59}$ 27. $\sqrt[3]{34}$

28. $\sqrt[3]{802}$ 29. $\sqrt[3]{989}$ 30. $\sqrt[3]{430}$ 31. $\sqrt[3]{275}$

Estimate to the nearest whole number.

32. $\sqrt{5\dfrac{1}{5}}$ 33. $\sqrt{21\dfrac{7}{10}}$ 34. $\sqrt{17\dfrac{3}{4}}$

Order from least to greatest.

35. $7, 9, \sqrt{50}, \sqrt{85}$ 36. $\sqrt[3]{105}, 7, 5, \sqrt{38}$ 37. $\sqrt{62}, 6, \sqrt{34}, 8$

ALGEBRA Estimate the solution of each equation to the nearest integer.

38. $y^2 = 55$ 39 $d^2 = 95$ 40. $p^2 = 6.8$

GEOMETRY The volume of each cube is given. Estimate the side length of the cube to the nearest integer. Use the formula $V = s^3$.

41.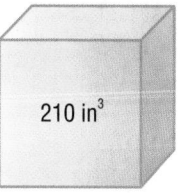
210 in³

42.
520 cm³

43. **STORAGE** Amanda purchased a storage cube that has a volume of 4 cubic feet. She wants to put it on a bookshelf that is 12 inches tall. Will the cube fit? Explain.

44. **HOME IMPROVEMENT** Jacob is buying the grass seed shown at the right. Estimate the side length of the largest square Jacob could seed if he purchases 5 bags.

45. **NUMBER SENSE** Without a calculator, determine which is greater, $\sqrt{94}$ or 10. Explain your reasoning.

46. OPEN ENDED Find two numbers that have square roots between 7 and 8. One number should have a square root closer to 7 and the other number should have a square root closer to 8. Justify your answer.

47. FIND THE ERROR Jasmine is estimating $\sqrt{200}$. Find her mistake and correct it.

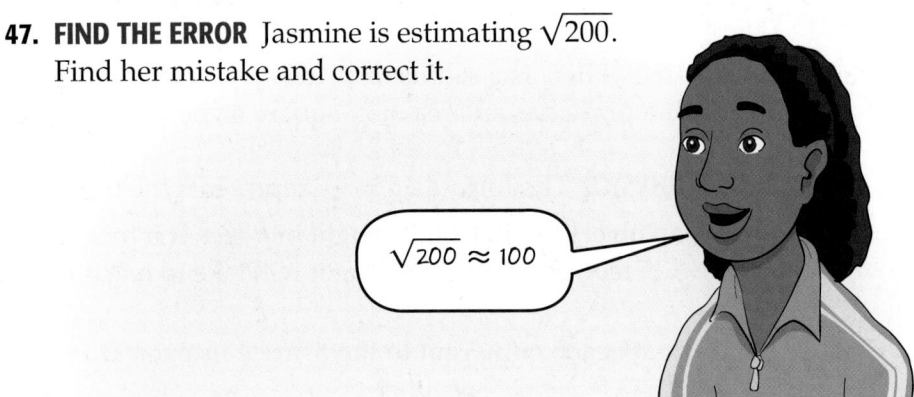

$\sqrt{200} \approx 100$

48. CHALLENGE If $x^4 = y$, then x is the fourth root of y. Explain how to estimate the fourth root of 30. Find the fourth root of 30 to the nearest whole number.

49. ✏ **WRITE MATH** Explain how to graph $\sqrt{78}$ on a number line.

✔ Test Practice

50. THINK SOLVE EXPLAIN **SHORT RESPONSE** After an accident, officials use the formula below to estimate the speed the car was traveling based on the length of the car's skid marks.

$$s = \sqrt{24m}$$

In the formula, s represents the speed in miles per hour and m is the length of the skid marks in feet. If a car leaves a skid mark of 50 feet, what was its approximate speed? Show all work necessary to justify your answer.

51. Leah found the side of a square to be $\sqrt{30}$ inches. Which point is closest to $\sqrt{30}$ on the number line?

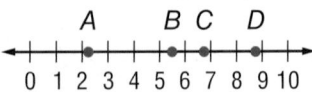

```
        A       B C  D
  +--+--+--+--+--+-+--+-+--+--+
  0  1  2  3  4  5 6  7 8  9 10
```

A. point A

B. point B

C. point C

D. point D

Spiral Review

52. ALGEBRA Find a number that, when squared, equals 8,100. (Lesson 3A)

Evaluate each expression. Express the result in scientific notation. (Lesson 2C)

53. $(5.9 \times 10^3) + (6.7 \times 10^4)$

54. $(8.1 \times 10^9) - (2.4 \times 10^7)$

Write each fraction as an expression using negative exponents. (Lesson 2A)

55. $\dfrac{1}{125}$

56. $\dfrac{1}{p^4}$

57. $\dfrac{1}{1,296}$

Main Idea

Compare mathematical expressions involving real numbers.

 Vocabulary

irrational number
real number

 **8.NS.1, 8.NS.2**

Compare Real Numbers

SPORTS Major League baseball has rules for the dimensions of the baseball diamond.

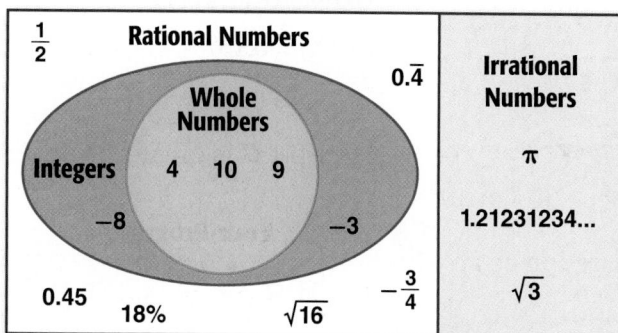

90 ft

√16,200 ft

60.5 ft

1. The distance from the pitching mound to home plate is 60.5 feet. Is 60.5 a rational number? Explain.

2. The distance from first base to second base is 90 feet. Is 90 a rational number? Explain.

3. The distance from home plate to second base is √16,200 feet. Can this square root be written as a rational number? Explain.

A calculator gives a decimal value of 127.2792206 for √16,200. Although this continues on and on, it does not repeat. Since the decimal does not terminate or repeat it is *not* a rational number.

Numbers that are not rational are called **irrational numbers**. The square root of any number that is not a perfect square number is irrational.

Key Concept | Irrational Numbers

Words | An irrational number is a number that cannot be expressed as the ratio $\frac{a}{b}$, where a and b are integers and $b \neq 0$.

Examples | $\sqrt{2} \approx 1.414213562...$ $-\sqrt{3} \approx -1.732050807...$

The set of rational numbers and the set of irrational numbers together make up the set of **real numbers**. Study the Venn diagram below.

Real Numbers

$\frac{1}{2}$ **Rational Numbers**

$0.\overline{4}$ **Irrational Numbers**

Whole Numbers

Integers   4   10   9

π

−8   −3

1.21231234...

0.45

18%   √16 $-\frac{3}{4}$ $\sqrt{3}$

EXAMPLES Classify Numbers

Name all sets of numbers to which each real number belongs.

1 0.252525... The decimal ends in a repeating pattern. It is a rational number because it is equivalent to $\frac{25}{99}$.

> **Study Tip**
>
> **Classifying Numbers**
> Always simplify numbers before classifying them.

2 $\sqrt{36}$ Since $\sqrt{36} = 6$, it is a whole number, an integer, and a rational number.

3 $-\sqrt{7}$ $-\sqrt{7} \approx -2.645751311...$ Since the decimal does not terminate or repeat, it is an irrational number.

CHECK Your Progress

   a. $\sqrt{10}$ **b.** $-2\frac{2}{5}$ **c.** $\sqrt{100}$

EXAMPLES Compare Real Numbers

Replace each ● with <, >, or = to make a true statement.

4 $\sqrt{7}$ ● $2\frac{2}{3}$

Write each number as a decimal.

$\sqrt{7} \approx 2.645751311...$

$2\frac{2}{3} = 2.666666666...$

Since 2.645751311...is less than 2.66666666..., $\sqrt{7} < 2\frac{2}{3}$.

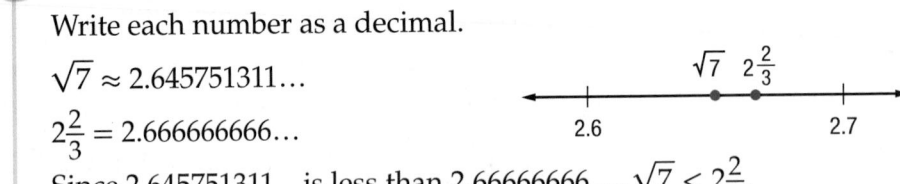

> **Study Tip**
>
> **Calculator Use**
> To find $\sqrt{7}$ on a calculator, use the keystrokes [2nd] [x²] 7 [ENTER] 2.645751311

5 $1.\overline{5}$ ● $\sqrt{2.25}$

Write $\sqrt{2.25}$ as a decimal.

$\sqrt{2.25} = 1.5$

$1.\overline{5} = 1.555555555...$

Since 1.555555555...is greater than 1.5, $1.\overline{5} > \sqrt{2.25}$.

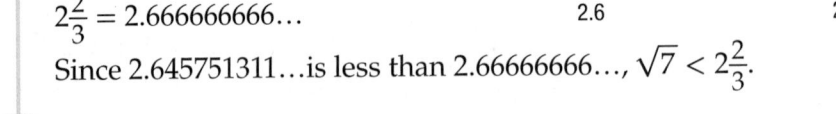

6 15.7% ● $\sqrt{0.02}$

Write each number as a decimal.

15.7% = 0.157

$\sqrt{0.02} \approx 0.141$

Since 0.157 is greater than 0.141, $15.7\% > \sqrt{0.02}$.

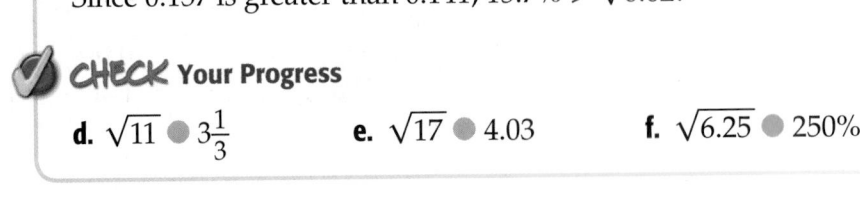

CHECK Your Progress

   d. $\sqrt{11}$ ● $3\frac{1}{3}$ **e.** $\sqrt{17}$ ● 4.03   **f.** $\sqrt{6.25}$ ● 250%

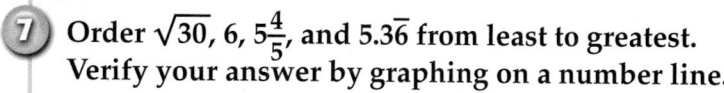

EXAMPLE Order Real Numbers

⑦ Order $\sqrt{30}$, 6, $5\frac{4}{5}$, and $5.3\overline{6}$ from least to greatest.
Verify your answer by graphing on a number line.

Write each number as a decimal. Then order the decimals.

$\sqrt{30} \approx 5.48$

$6 = 6.00$

$5\frac{4}{5} = 5.80$

$5.3\overline{6} \approx 5.37$

From least to greatest, the order is $5.3\overline{6}$, $\sqrt{30}$, $5\frac{4}{5}$, and 6.

CHECK Your Progress

g. Order -7, $-\sqrt{60}$, $-7\frac{7}{10}$, and $-\frac{66}{9}$ from least to greatest. Verify your answer by graphing on a number line.

Real-World Link · · · ·
One Bayfront Plaza is a skyscraper sheduled for completion in 2015. It will be 1,049 feet tall with 70 floors, making it the tallest building in Miami.

REAL-WORLD EXAMPLE

⑧ **SKYSCRAPERS** On a clear day, the number of miles a person can see to the horizon is about 1.23 times the square root of his or her distance from the ground, in feet. Suppose Frida is at the top of the Four Seasons Hotel and Kia is at the top of the Wachovia Financial Center. How much farther can Frida see than Kia?

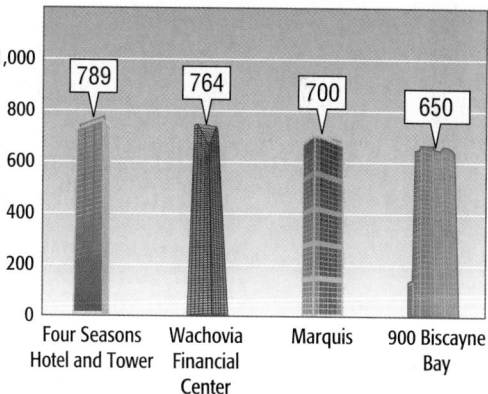

Use a calculator to approximate the distance each person can see.

Frida: $1.23 \sqrt{789} \approx 34.55$ Kia: $1.23 \sqrt{764} \approx 34.0$

Frida can see about $34.55 - 34.0$ or 0.55 miles farther than Kia.

CHECK Your Progress

h. MEASUREMENT How much greater is the perimeter of a square with area 250 square meters than a square with area 125 square meters?

Examples 1–3 **Name all sets of numbers to which each real number belongs.**

    1. 0.050505...   **2.** $-\sqrt{64}$ **3.** $\sqrt{17}$ **4.** $-3\frac{1}{4}$

Examples 4–6 **Replace each ● with <, >, or = to make a true statement.**

    5. $\sqrt{15}$ ● 3.5   **6.** $\sqrt{2.25}$ ● 150%   **7** 2.$\overline{21}$ ● $\sqrt{5.2}$

Example 7 **8.** Order $\sqrt{5}$, 220%, 2.25, and 2.$\overline{2}$ from least to greatest. Verify your answer by graphing on a number line.

Example 8 **9. AREA** The formula $A = \sqrt{s(s-a)(s-b)(s-c)}$ can be used to find the area A of a triangle. The variables a, b, and c are the side measures and s is one half the perimeter. Use the formula to find the area of the triangle at the right.

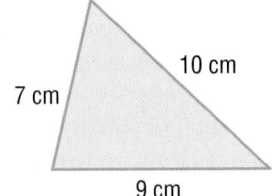

7 cm   10 cm   9 cm

Practice and Problem Solving

● = **Step-by-Step Solutions** begin on page R1.
Extra Practice begins on page EP2.

Examples 1–3 **Name all sets of numbers to which each real number belongs.**

    10. 14   **11.** $\frac{2}{3}$ **12.** $-\sqrt{16}$ **13.** $-\sqrt{20}$

    14. 4.83   **15.** 7.$\overline{2}$ **16.** $-\sqrt[3]{90}$ **17.** $\frac{12}{4}$

Examples 4–6 **Replace each ● with <, >, or = to make a true statement.**

    18. $\sqrt{10}$ ● 3.2   **19.** $\sqrt{12}$ ● 3.5   **20.** $6\frac{1}{3}$ ● $\sqrt[3]{240}$

    21. 240% ● $\sqrt{5.76}$ **22.** $5\frac{1}{6}$ ● 5.1$\overline{6}$ **23.** $\sqrt{6.2}$ ● 2.$\overline{4}$

Example 7 **Order each set of numbers from least to greatest. Verify your answer by graphing on a number line.**

    24. −415%, $-\sqrt{17}$, −4.$\overline{1}$, −4.01   **25** $\sqrt{5}$, $\sqrt{6}$, 2.5, 2.55, $\frac{7}{3}$

Example 8 **26. ROADS** The equation $s = \sqrt{30fd}$ can be used to find a car's speed s in miles per hour given the length d in feet of a skid mark and the friction factor f of the road. Police measured a skid mark of 90 feet on a dry concrete road. If the speed limit is 35 mph, was the car speeding? Explain.

Friction Factor		
Road	Concrete	Tar
Wet	0.4	0.5
Dry	0.8	1.0

27. HEALTH The surface area in square meters of the human body can be found using the expression $\sqrt{\dfrac{hm}{3,600}}$ where h is the height in centimeters and m is the mass in kilograms. Find the surface area of a 15-year-old boy with a height of 183 centimeters and a mass of 74 kilograms.

28. ALGEBRA In the sequence 4, 12, ■, 108, 324, the missing number can be found by simplifying $\sqrt{ab}$ where a and b are the numbers on either side of the missing number. Find the missing number.

H.O.T. Problems

29. OPEN ENDED Give a counterexample for the statement *all square roots are irrational numbers*. Explain your reasoning.

NUMBER SENSE Replace each ● with <, >, or = to make a true statement.

30. $3 + \sqrt{7}$ ● 6

31. $4 - \sqrt{10}$ ● $\sqrt{2}$

32. 13 ● $8 + \sqrt{20}$

CHALLENGE Tell whether the following statements are *always, sometimes,* or *never* true. If a statement is not always true, explain.

33. Integers are rational numbers.

34. Rational numbers are integers.

35. The product of a rational number and an irrational number is irrational.

36. **WRITE MATH** The set of natural numbers consists of the counting numbers 1, 2, 3, . . ., . Describe the relationship between the set of natural numbers and other subsets of the real number system.

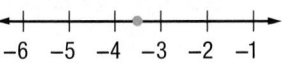
Test Practice

37. Which of the following is an irrational number?

  A. -6

  B. $\frac{2}{3}$

  C. $\sqrt{9}$

  D. $\sqrt{3}$

38. **SHORT RESPONSE** Which of the two real numbers below is greater?

$$\sqrt{3} \qquad \frac{1}{3}$$

39. Which number represents the point graphed on the number line?

```
<---+---+---+--•-+---+---+--->
   -6  -5  -4  -3  -2  -1
```

  F. $-\sqrt{12}$

  G. $-\sqrt{10}$

  H. $-\sqrt{15}$

  I. $-\sqrt{8}$

40. Order 7, $\sqrt{53}$, $\sqrt{32}$, and 6 from least to greatest. (Lesson 3C)

ALGEBRA Solve each equation. (Lesson 3A)

41. $t^2 = 25$

42. $y^2 = \frac{1}{49}$

43. $0.64 = a^2$

Evaluate each expression. Express the result in scientific notation. (Lesson 2C)

44. $(7.2 \times 10^4)(1.1 \times 10^{-6})$

45. $(3.6 \times 10^3) + (5.7 \times 10^5)$

46. POPULATION The table shows the approximate population of several countries. Order the countries from the greatest population to the least population. (Lesson 2B)

Country	Population
China	1.3×10^9
India	1.2×10^9
Indonesia	2.3×10^8
United States	3.1×10^8

Problem Solving in Engineering

Relying on Robots

Are you mechanically inclined? Do you like to find new ways to solve problems? If so, a career as a robotics engineer is something you should consider. Robotics engineers design and build robots to perform tasks that are difficult, dangerous, or tedious for humans. For example, a robotic insect was developed based on a real insect. Its purpose was to travel over water surfaces, take measurements, and monitor water quality.

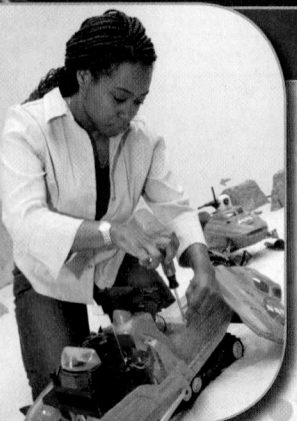

21ˢᵗ Century Careers

Are you interested in a career as a robotics engineer? Take some of the following courses in high school.

- Calculus
- Electro-Mechanical Systems
- Fundamentals of Robotics
- Physics

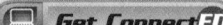
Get ConnectED

Robotic Insect Characteristics

Mass	3.5×10^{-4} kg
Length	0.09 m
Leg Diameter	0.2 mm
Speed	180 mm/s

Real-World Math

Use the information in the table to solve each problem.

1. Write the mass of the robot in standard form.

2. Write the length of the robot in scientific notation.

3. Write the leg diameter of the robot in scientific notation.

4. What is the mass in milligrams? Write in standard form.

5. Real insects called water striders can travel 8.3 times faster than the robot. Write the speed of water striders in scientific notation.

Chapter Study Guide and Review

FOLDABLES Study Organizer

Be sure the following Key Concepts are noted in your Foldable.

Powers and Exponents | Multiply and Divide Monomials | Powers of Monomials | Negative Exponents | Scientific Notation | Square Roots and Cube Roots | Estimate Roots | Compare Real Numbers

Key Concepts

Monomials (Lesson 1)
- To multiply powers with the same base, add their exponents.
- To divide powers with the same base, subtract their exponents.
- To find the power of a power, multiply the exponents.
- To find the power of a product, find the power of each factor and multiply.

Scientific Notation (Lesson 2)
- A number is expressed in scientific notation when it is written as the product of a factor and a power of 10. The factor must be greater than or equal to 1 and less than 10.

Square Roots, Cube Roots, and Irrational Numbers (Lesson 3)
- A square root of a number is one of its two equal factors.
- A cube root of a number is one of its three equal factors.
- An irrational number is a number that cannot be expressed as $\frac{a}{b}$, where a and b are integers and $b \neq 0$.

Key Vocabulary

base
cube root
exponent
irrational number
monomial
perfect cube
perfect square
power
radical sign
real number
scientific notation
square root

Vocabulary Check

State whether each statement is *true* or *false*. If *false*, replace the underlined word or number to make a true sentence.

1. The number 4.05×10^8 is written in <u>bar notation</u>.

2. The <u>base</u> tells how many times a number is used as a factor.

3. The number 5^4 is a <u>power</u>.

4. The symbol that is used to indicate a square root is the <u>radical sign</u>.

5. The number <u>11</u> is a perfect square.

6. A <u>real number</u> is a number that cannot be expressed as the quotient of two integers.

7. To divide powers with the same base, <u>subtract</u> their exponents.

8. The Product of Powers states that when multiplying powers with the same base, <u>multiply</u> their exponents.

Multi-Part Lesson Review

Laws of Exponents

Powers and Exponents (Lesson 1A)

Write each expression using exponents.

9. $x \cdot x \cdot x \cdot x \cdot y$ 10. $4 \cdot 4 \cdot 9 \cdot 9$

11. **PHONE TREES** To close school the principal calls six teachers, who in turn call six more. If each of those teachers call six more, how many calls will be made by the teachers in this last group?

EXAMPLE 1 Write $3 \cdot 3 \cdot 3 \cdot 7 \cdot 7$ using exponents.

$$3 \cdot 3 \cdot 3 \cdot 7 \cdot 7 = 3^3 \cdot 7^2$$

Multiply and Divide Monomials (Lesson 1B)

Simplify. Express using exponents.

12. $x^6 \cdot x^2$ 13. $-9y^2(-4y^9)$ 14. $\dfrac{n^5}{n}$

15. **MEASUREMENT** The area of the den is 3^4 square feet. The area of the kitchen is 3^3 square feet. How many times larger is the den than the kitchen?

EXAMPLE 2 Simplify $4^2 \cdot 4^5$. Express using exponents.

$$4^2 \cdot 4^5 = 4^{2+5} \quad \text{The common base is 4.}$$
$$= 4^7 \quad \text{Add the exponents.}$$

Powers of Monomials (Lesson 1C)

Simplify.

16. $(9^2)^3$ 17. $(5y^5)^4$ 18. $[(p^2)^3]^2$

19. **GEOMETRY** Find the volume of a cube with side length $5x^2t^4$ as a monomial.

EXAMPLE 3 Simplify $(7^3)^5$.

$$(7^3)^5 = 7^{3 \cdot 5} \quad \text{Power of a Power}$$
$$= 7^{15} \quad \text{Simplify.}$$

PSI: Act it Out (Lesson 1D)

Solve. Use the *act it out* strategy.

20. **READING** In English class, each student must select 4 short stories from a list of 5 short stories to read. How many different combinations of short stories could a student read?

21. **CARPENTRY** Jaime has $14\frac{1}{4}$ feet of lumber. She uses $2\frac{7}{8}$ feet for a shelf. Does Jaime have enough lumber for 4 more shelves?

EXAMPLE 4 The Spirit Club is making a banner using three sheets of paper. How many different banners can they make using their school colors of black, orange, and white?

Use three index cards labeled black, orange, and white to model the different banners.

There are six different combinations they can make.

Scientific Notation

Negative Exponents (Lesson 2A)

Write each expression using a positive exponent.

22. 3^{-5} **23.** 2^{-7} **24.** 5^{-3}

Simplify. Express using positive exponents.

25. $6^{-2} \cdot 6^5$ **26.** $4^{-4} \cdot 4^{-3}$ **27.** $9^{-3} \cdot 9^8$

28. SEAHORSES A seahorse swims at a maximum rate of 0.0001 mile per minute. Write the decimal as a power of 10.

EXAMPLE 5 Write 4^{-4} using a positive exponent.

$4^{-4} = \dfrac{1}{4^4}$ Definition of negative exponent

EXAMPLE 6 Simplify $5^{-3} \cdot 5^{-6}$. Express using positive exponents.

$$5^{-3} \cdot 5^{-6} = 5^{-3 + (-6)} \quad \text{Product of Powers}$$
$$= 5^{-9} \quad \text{Add the exponents.}$$
$$= \dfrac{1}{5^9} \quad \text{Definition of negative exponent}$$

Scientific Notation (Lesson 2B)

Write each number in standard form.

29. 3.2×10^{-3} **30.** 6.71×10^4

Write each number in scientific notation.

31. 0.000064   **32.** 87,500,000

33. ANIMALS The smallest mammal is Kitti's hog-nosed bat weighing about 4.375×10^{-3} pound. Write this weight in standard form.

EXAMPLE 7 Write 3.21×10^{-6} in standard form.

$3.21 \times 10^{-6} = 0.00000321$ Move the decimal point 6 places to the left.

EXAMPLE 8 Write 0.004 in scientific notation.

$0.004 = 4 \times 0.001$ The decimal point moves 3 places.

$= 4 \times 10^{-3}$ Since $0 < 0.004 < 1$, the exponent is negative.

Compute with Scientific Notation (Lesson 2C)

Evaluate each expression. Express the result in scientific notation.

34. $(3.4 \times 10^4)(2.8 \times 10^5)$

35. $\dfrac{2.97 \times 10^{-5}}{0.4 \times 10^3}$

36. $(9.6 \times 10^6) + (2.4 \times 10^4)$

37. $(11.2 \times 10^2) - (9.5 \times 10^{-1})$

38. RIVERS The Guadalupe River is 2.56×10^2 miles long. The Amazon River is 4.096×10^3 miles long. How many times longer is the Amazon River than the Guadalupe River?

EXAMPLE 9 Evaluate $(2.6 \times 10^3)(4.9 \times 10^7)$. Express the result in scientific notation.

$(2.6 \times 10^3)(4.9 \times 10^7)$

$= (2.6 \times 4.9)(10^3 \times 10^7)$ Commutative and Associative Properties

$= (12.74)(10^3 \times 10^7)$ Multiply 2.6 by 4.9.

$= 12.74 \times 10^{3+7}$ Product of Powers

$= 12.74 \times 10^{10}$ Add the exponents.

$= 1.274 \times 10^{11}$ Write in scientific notation.

Lesson 3 **Square Roots and Cube Roots**

Roots (Lesson 3A)

Find each square root or cube root.

39. $\sqrt{81}$　　　　**40.** $\pm\sqrt{121}$

41. $-\sqrt{64}$　　　**42.** $\sqrt[3]{-729}$

43. SEWING A quilter made 256 small squares for a large quilt. If the quilt is shaped like a square, how many small squares will she use on each side?

44. MEASUREMENT What is the length of a side of the square?

Area = 225 m²

EXAMPLE 10 Find $\sqrt{36}$.

$\sqrt{36} = 6$　Find the positive square root of 36; $6^2 = 36$.

EXAMPLE 11 Find $-\sqrt{169}$.

$-\sqrt{169} = -13$　Find the negative square root of $-\sqrt{169}$; $(-13)^2 = 169$.

EXAMPLE 12 Find $\pm\sqrt{2.56}$.

$\pm\sqrt{2.56} = \pm1.6$　Find both square roots of 2.56; $1.6^2 = 2.56$.

Estimate Roots (Lesson 3C)

Estimate to the nearest integer.

45. $\sqrt{32}$　　　　**46.** $\sqrt{42}$

47. $\sqrt{230}$　　　**48.** $\sqrt{96}$

49. $\sqrt[3]{150}$　　**50.** $\sqrt[3]{311}$

51. $\sqrt[3]{50}$　　　**52.** $\sqrt[3]{224}$

53. ALGEBRA Estimate the solution of $b^2 = 60$.

EXAMPLE 13 Estimate $\sqrt{135}$ to the nearest whole number.

$121 < 135 < 144$　Write an inequality.
$11^2 < 135 < 12^2$　$121 = 11^2$ and $144 = 12^2$
$11 < \sqrt{135} < 12$　Take the square root of each number.

Since 135 is closer to 144 than to 121, the best whole number estimate is 12.

Compare Real Numbers (Lesson 3D)

Name all sets of numbers to which each real number belongs.

54. $-\sqrt{19}$　　**55.** $0.\overline{3}$

56. 7.43　　　**57.** -12

58. $\sqrt{32}$　　　**59.** 101

60. MEASUREMENT The area of a square vegetable garden is 360 square meters. To the nearest hundredth meter, what is the perimeter of the garden?

61. Order the numbers $2.\overline{2}$, $2\frac{1}{5}$, 2.25, and $\sqrt{5}$ from least to greatest.

EXAMPLE 14 Name all sets of numbers to which $-\sqrt{33}$ belongs.

$-\sqrt{33} \approx -5.744562647$

Since the decimal does not terminate or repeat, it is an irrational number.

Write each expression using exponents.

1. $(-5)(-5)(4)(4)(4)$

2. $x \cdot y \cdot x \cdot x \cdot y \cdot x$

3. $4 \cdot 4 \cdot a \cdot a \cdot b \cdot 3 \cdot 4 \cdot 3 \cdot a$

4. **EXTREME SPORTS** Recently, San Antonio, Texas, hosted the first ever summer Global X Games. Team USA won the gold medal count with a total of $7^2 \cdot 2^2$ points. Evaluate the number of points won by Team USA.

5. **MULTIPLE CHOICE** Simplify the algebraic expression $(3x^3y^2)(7x^3y)$.

A. $21x^9y^2$

C. $21x^6y^3$

B. $21x^6y^2$

D. $21x^6y^6$

Simplify. Express using exponents.

6. $15^3 \cdot 15^5$

7. $-5m^6(-9m^8)$

8. $\dfrac{3^{15}}{3^7}$

9. $\dfrac{-40w^8}{8w}$

10. **GEOMETRY** Express the area of the square below as a monomial.

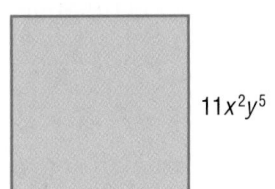

$11x^2y^5$

Simplify. Express using exponents.

11. $(3^3)^2$

12. $(4^2x^3)^4$

13. $[(x^2)^4]^3$

14. $(-2b^3)^2(4b^2)^2$

15. Write 8.83×10^{-7} in standard form.

16. **INTERNET** There are 5.79×10^8 Internet users in Asia and 2.48×10^8 Internet users in North America. How many more Internet users are there in Asia than North America? Write in scientific notation.

17. **SHOES** A tennis shoe comes in men's and women's sizes and in blue, black, or white. Use the *act it out* strategy to determine how many different pairs of shoes are available.

Find each square root or cube root.

18. $\pm\sqrt{81}$

19. $\sqrt[3]{-1{,}331}$

20. $-\sqrt{\dfrac{1}{25}}$

21. **PICTURES** The area of a square picture frame is 529 square centimeters. How long is each side of the frame?

Estimate to the nearest integer.

22. $\sqrt{90}$

23. $\sqrt[3]{12}$

24. $\sqrt{226}$

25. **MEASUREMENT** The radius of a circle with area A is approximately $\sqrt{\dfrac{A}{3}}$. If a pie has an area of 42 square inches, estimate its radius.

Replace each ● with $<$, $>$, or $=$ to make a true statement.

26. $\sqrt{15}$ ● 4.1

27. 6.5 ● $\sqrt{45}$

28. **THINK SOLVE EXPLAIN** **EXTENDED RESPONSE** The table gives the approximate diameter in miles for several planets.

Planet	Diameter
Mercury	3.032×10^3
Saturn	7.4948×10^4
Neptune	3.0603×10^4
Earth	7.926×10^3

Part A Order the planets from least to greatest diameters.

Part B How many times greater is the diameter of Saturn than Mercury?

Part C How much longer is Neptune's diameter than Earth's?

Preparing for Standardized Tests

Multiple Choice: Eliminate Unreasonable Answers

Sometimes answer choices are not reasonable for the information given in a multiple-choice question. These choices can be eliminated.

TEST EXAMPLE

Earth moves in an orbit that is 585 million miles around the sun. What is 585 million written in scientific notation?

A. 5.85×10^8

B. 5.85×10^5

C. 5.85×10^{-5}

D. 5.85×10^{-8}

Negative exponents used in scientific notation indicate very small numbers. Since 585 million is not a small number, choices C and D can be eliminated automatically.

$$585 \text{ million} = 585,000,000$$
$$= 5.85 \times 100,000,000$$
$$= 5.85 \times 10^8$$

The correct answer is A.

 **Work on It**

About 18.7 million ounces of blood flow through your kidneys in one year. What is 18.7 million written in scientific notation?

F. 1.87×10^6

G. 1.87×10^{-6}

H. 1.87×10^7

I. 1.87×10^{-7}

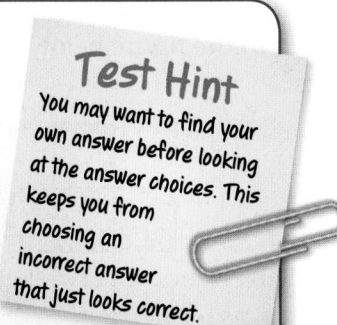

Test Hint

You may want to find your own answer before looking at the answer choices. This keeps you from choosing an incorrect answer that just looks correct.

Read each question. Then fill in the correct answer on the answer document provided by your teacher or on a sheet of paper.

1. The distance from Earth to the Sun is 92,900,000 miles. What is this number in scientific notation?

   A. 92.9×10^6 **C.** 9.29×10^6

   B. 9.29×10^7 **D.** 929×10^5

2. ✎ **GRIDDED RESPONSE** Ms. Leigh wants to organize the desks in the study hall into a square. If she has 64 desks, how many should be in each row?

3. What is the simple interest for $450 invested in a savings account at 2.25% for 2 years?

   F. $18   **H.** $202.50

   G. $20.25   **I.** $470.25

4. Between which two whole numbers is $\sqrt{66}$ located on a number line?

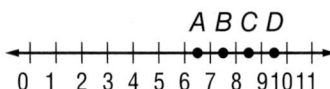

   A. 6 and 7   **C.** 8 and 9

   B. 7 and 8   **D.** 9 and 10

5. **THINK SOLVE EXPLAIN** **SHORT RESPONSE** The area of each square in the figure below is 25 square units.

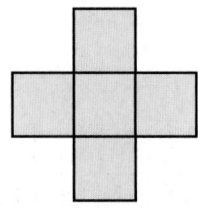

What is the perimeter of the figure?

6. What percent of the area of the rectangle is shaded?

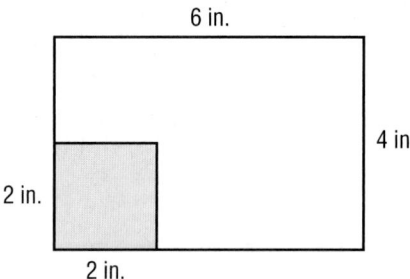

   F. 15%

   G. $16\frac{2}{3}\%$

   H. 20%

   I. $25\frac{1}{2}\%$

7. **THINK SOLVE EXPLAIN** **SHORT RESPONSE** The area of a rectangle is $30m^{11}$ square feet. If the length of the rectangle is $6m^4$ feet, what is the width of the rectangle?

8. The table shows the area in square miles of certain states.

State	Area (mi²)
California	1.64×10^5
Ohio	4.48×10^4
Oregon	9.84×10^4
Vermont	9.62×10^3

Which state has the greatest area?

   A. California   **C.** Oregon

   B. Ohio   **D.** Vermont

9. The mass of a paper clip is 9.4×10^{-4} kilogram. What is this mass in standard form?

   F. 0.000000094 kg   **H.** 0.000094 kg

   G. 0.0000094 kg   **I.** 0.00094 kg

10. Shane and his two brothers equally shared the cost of a new computer game with a list price of $35. They received a 25% discount on the video game and paid 5.5% sales tax on the discounted price. Find the approximate amount that each of the brothers paid toward the cost of the game.

   A. $14.77   **C.** $9.23

   B. $11.73   **D.** $8.42

11. Which of the following is equivalent to $(-3)^{-3}$?

   F. -9 **H.** $\frac{1}{27}$

   G. $\frac{1}{-27}$ **I.** 9

12. ✍ **GRIDDED RESPONSE** In soccer practice one week, Lena blocked 51 shots on goal. This was 85% of the shots on goal. How many shots on goal were there that week?

13. Which of the following is closest to point *A* on the number line below?

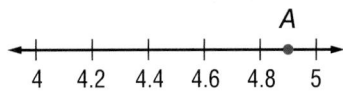

   A. $\frac{1}{2}$ **C.** $\sqrt{24}$

   B. $\sqrt{20}$ **D.** 480%

14. Part of a recipe for one batch of macaroni and cheese is shown below.

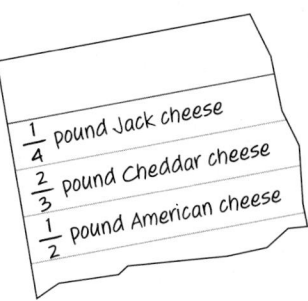

$\frac{1}{4}$ pound Jack cheese
$\frac{2}{3}$ pound Cheddar cheese
$\frac{1}{2}$ pound American cheese

Javier is making two batches of the recipe for a party. How much total cheese will Javier need?

   F. $1\frac{5}{12}$ pounds

   G. $2\frac{1}{6}$ pounds

   H. $2\frac{5}{6}$ pounds

   I. $3\frac{5}{12}$ pounds

15. 📋 **EXTENDED RESPONSE** The container for a child's set of blocks is 9 inches by 9 inches by 9 inches. The blocks measure 3 inches by 3 inches by 3 inches.

   Part A Describe how to determine the number of blocks needed to fill the container.

   Part B Write and simplify an expression to solve the problem.

   Part C How many blocks will it take?

NEED EXTRA HELP?															
If You Missed Question...	1	2	3	4	5	6	7	8	9	10	11	12	13	14	15
Go to Chapter-Lesson...	2-2B	2-3A	1-3B	2-3C	2-3A	1-2C	2-1B	2-2B	2-2B	1-3A	2-2A	1-2C	2-3D	1-1B	2-1A

Equations and Inequalities

connectED.mcgraw-hill.com

Investigate

 Animations

 Vocabulary

 Multilingual eGlossary

Learn

 Personal Tutor

 Virtual Manipulatives

 Graphing Calculator

 Audio

 Foldables

Practice

 Self-Check Practice

 Worksheets

 Assessment

The **★ BIG Idea**

How does the solution of an equation differ from the solution of an inequality?

FOLDABLES
Study Organizer

Make this Foldable to help you organize your notes.

Review Vocabulary

variable variable a symbol, usually a letter, used to represent an unknown number

$$5 - 2 \cdot x$$

variable

Key Vocabulary

English	Español
compound inequality	desigualdad compuesta
inequality	desigualdad
two-step equation	ecuación de dos pasos
two-step inequality	desigualdad de dos pasos

When Will I Use This?

Are You Ready for the Chapter?

You have two options for checking prerequisite skills for this chapter.

Text Option Take the Quick Check below. Refer to the Quick Review for help.

QUICK Check

Add, subtract, multiply, or divide.

1. $64 + (-13)$
2. $35 + 156$
3. $200 - 48$
4. $59 - (-26)$
5. $4(-4)$
6. $-3(5)$
7. $72 \div (-9)$
8. $36 \div 3$

9. **TRAVEL** The Perez family drove for 4 hours at 65 miles per hour. How far did they drive?

10. **SHOPPING** Mrs. Wilson spent the amounts shown on school clothes for her children. How much did she spend altogether?

Store	Amount Spent ($)
A	80
B	72
C	69
D	61

Determine whether each statement is *true* or *false*.

11. $10 > 4$
12. $3 < -3$
13. $-8 < -7$
14. $-1 > 0$
15. $\frac{1}{2} \geq \frac{12}{24}$
16. $\frac{4}{5} < \frac{5}{6}$

17. **WEATHER** The temperature in Sioux City, Iowa, was $-7°F$ while the temperature in Des Moines, Iowa, was $-5°F$. Which city was warmer? Explain.

QUICK Review

EXAMPLE 1

Find $13 - (-8)$.

$13 - (-8) = 13 + 8$ To subtract -8, add 8.
         $= 21$ Add.

EXAMPLE 2

Find $-12(7)$.

$-12(7) = -84$ The factors have different signs. The product is negative.

EXAMPLE 3

Find $-18 \div (-6)$.

$-18 \div (-6) = 3$ The signs are the same. The quotient is positive.

EXAMPLE 4

Determine whether the statement $-2 > 1$ is *true* or *false*.

Plot the points on a number line.

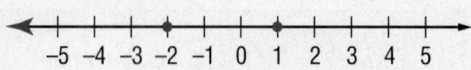

Since -2 is to the left of 1, $-2 < 1$. The statement is false.

Online Option Take the Online Readiness Quiz.

 # Studying Math

Use a Flowchart

A *flowchart* is like a map that tells you how to get from the beginning of a problem to the end.

Try using a flowchart when you take notes to map out the steps you should follow.

Flowchart Symbols	
◇	A diamond contains a question. You need to stop and make a decision.
▭	A rectangle tells you what to do.
⬭	An oval indicates the beginning or end.

Here is a flowchart for comparing two integers. Just follow the arrows.

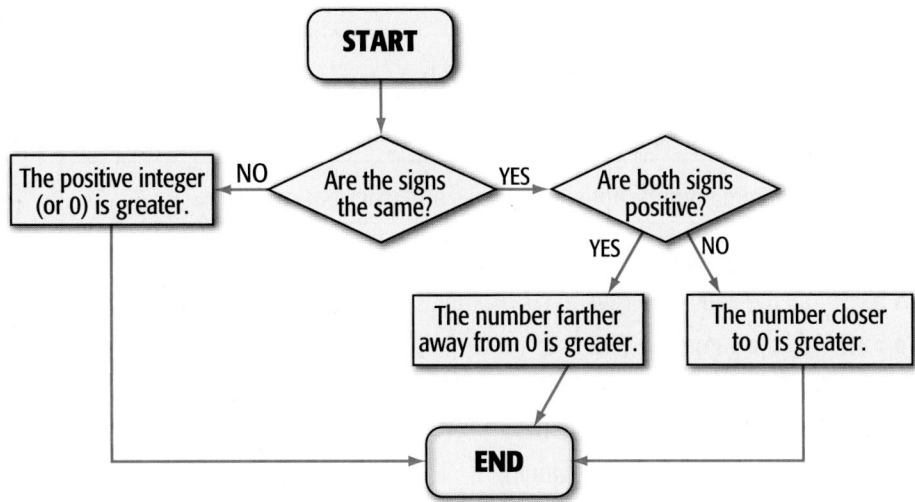

Practice

Make a flowchart for each kind of problem.

1. adding two integers (See Start Smart 2.)

2. graphing inequalities (See Lesson 3A.)

Problem-Solving Investigation

Main Idea Solve problems by working backward.

P.S.I. TEAM +

Work Backward

ALEX: Alisa and I traded video games. I gave Alisa one fourth of my video games in exchange for 6 video games. Then I sold 3 video games and gave 2 video games to my brother. I ended up with 16 video games.

YOUR MISSION: Work backward to find how many video games Alex started with.

Understand	You know how many video games Alex now has. You know how many video games he gave away, sold, or traded. You need to determine how many video games he started with.	
Plan	Start with the ending number of video games and work backward.	
Solve	Alex has 16 video games **Go back** Add the games he gave to his brother. **Go back** Add the games he sold. **Go back** Subtract the games he traded with Alisa. **Go back** This number is $\frac{3}{4}$ of his games. Alex had 20 video games at the beginning.	$+2$ ⟍ 16 → 18 $+3$ ⟍ 21 -6 ⟍ 15 $\times\frac{4}{3}$ ⟍ 20
Check	Start with 20. Perform operations in reverse order.	

Analyze the Strategy

1. Tell why the *work backward* strategy is the best way to solve this problem.

2. Explain how you can check a solution when you solve by working backward.

3. **WRITE MATH** Write a problem that can be solved by working backward. Then write the steps you would take to find the solution to your problem.

= **Step-by-Step Solutions** begin on page R1.
Extra Practice begins on page EP2.

- Work backward.
- Look for a pattern.
- Choose an operation.

Use the *work backward* strategy to solve Exercises 4–6.

4. **MONEY** Aurora collected money for a charity. Jacy made the first donation. Guillermo's donation was twice Jacy's donation. Rosa's mother contributed triple what Aurora had collected so far. Now Aurora has $120. How much did Jacy donate?

5. **SCHEDULE** Nyoko needs to be at school at 7:45 A.M. It takes her 25 minutes to walk to school, 25 minutes to eat breakfast, and 35 minutes to get dressed. What time should Nyoko get up to be at school 5 minutes early?

6. **SHOPPING** Janelle has $75. She buys jeans that are on sale for half price and then uses an in-store coupon for $10 off. After paying $1.80 in sales tax, she receives $37.20 in change. What was the original price of the jeans?

Use any strategy to solve Exercises 7–11.

7. **FINANCIAL LITERACY** Examine the graph below.

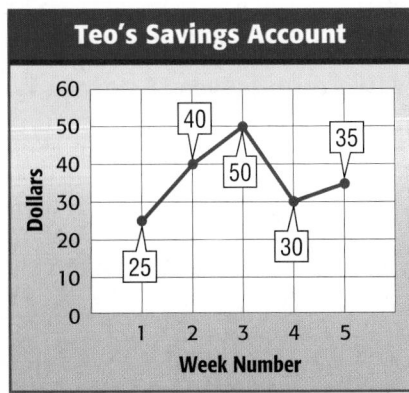

Teo's Savings Account

The activity in Teo's savings account is reflected in the graph. After week 5, he continues to save at the same rate as between weeks 2 and 3. In what week will his balance be at least $100?

8. **ANALYZE TABLES** The table gives the average weekly television viewing time in hours:minutes for teens and children.

Group	Total per Week
Teens (ages 12–17)	19:19
Children (ages 2–11)	21:00

How many more minutes each week do children spend watching television than teens do?

9. **FURNITURE** Ms. Ruiz makes an initial payment of $150 when buying a sofa. She pays the remaining cost of the sofa over 12 months, at no additional charge. If her monthly payment is $37.50, what was the original price of the sofa?

10. **ANALYZE TABLES** The table gives information about two different airplanes.

Airplane	Top Speed (mph)	Flight Length (mi)	Operating Cost per Hour
B747-400	534	3,960	$8,443
B727-200	430	644	$4,075

How much greater is the operating cost of a B747-400 than a B727-200 if each plane flies at its top speed for its maximum length of flight?

11. **MONEY** At the end of the month, Mr. Copley had $1,475 in his checking account. His statement showed the following transactions.

Item	Amount
Deposit	$150.00
Check #132	$45.00
Withdrawal	$100.00
Check #133	$18.50
Deposit	$250.00

What was his balance at the beginning of the month?

Main Idea
Write algebraic equations from verbal sentences and problem situations.

 Vocabulary
equation
defining a variable

Write Equations

UNIFORMS Cheerleading uniforms cost $32 each. The Weberstown booster club wants to purchase uniforms for the team.

1. What is the relationship between the number of uniforms and the total cost?

2. Write an expression to represent the total cost for *n* uniforms.

Number of Uniforms	Total Cost
1	$32
3	$96
5	$160

An **equation** is a mathematical sentence stating that two quantities are equal. An important skill in algebra is modeling situations using equations.

①
WORDS
Describe the situation. Use only the most important words.

②
VARIABLE
Assign a variable to represent the unknown quantity. This is called defining a variable.

③
EQUATION
Translate your verbal model into an algebraic equation.

To translate your verbal model, look for common words or phrases that suggest one of the four operations.

EXAMPLE Write an Algebraic Equation

 GAMES Eduardo had a score of −150 points in the first round of a game. His final score after two rounds was 75 points. Write an equation to find his second round score.

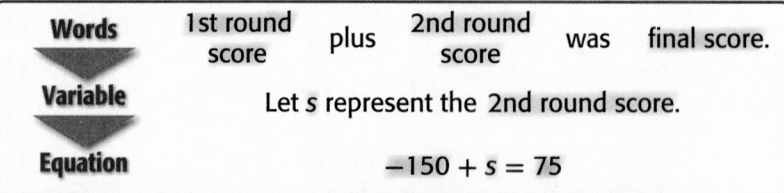

| Words | 1st round score | plus | 2nd round score | was | final score. |

Variable: Let *s* represent the 2nd round score.

Equation: $-150 + s = 75$

CHECK Your Progress

a. The winning time of 27 seconds was 2 seconds shorter than Tina's. Define a variable. Then write an equation to model the situation.

REAL-WORLD EXAMPLE

 2 **CARS** For the average consumer, a hybrid SUV would cost $1,473 less per year in gas to drive than a regular SUV. Use the information at the left to write an equation that could be used to find the yearly gas cost for a regular SUV.

Words	Gas cost for a hybrid SUV	is	$1,473 less than	gas cost for a regular SUV.
Variable	Let g represent gas cost for a regular SUV.			
Equation	$1,386 = g - 1,473$			

CHECK Your Progress

b. DANCE The change in attendance from last year's spring dance was −45 students. The attendance this year was 128 students. Define a variable. Then write an equation that could be used to find the attendance last year.

You can also write an equation with two variables to express the relationship between two unknown quantities.

REAL-WORLD EXAMPLE

 3 **BATS** The number of pounds of insects a bat can eat is 2.5 times its own body weight. Given b, a bat's body weight in pounds, write an equation that can be used to find p, the pounds of insects it can eat.

Words	Pounds of insects eaten	is	2.5 times	body weight.
Variable	Let p represent pounds eaten and b represent body weight.			
Equation	$p = 2.5b$			

CHECK Your Progress

c. VOTES A state's number of electoral votes is 2 more than its number of Representatives. Given r, a state's number of Representatives, write an equation that can be used to find e, the state's number of electoral votes.

 Real-World Link . . . .
Hybrid cars operate on a combination of gasoline and electricity. The electricity is generated by the brakes of a car and stored in a battery. The estimated gas costs for a hybrid SUV is $1,386 per year.

Example 1   **Define a variable. Then write an equation to model each situation.**

   1. Toya's score of 20 points was four times Corey's score.

   2. The total was $28 after a $4 tip was added to the bill.

Example 2   **Define a variable. Then write an equation that could be used to solve each problem.**

   3. SUBMARINES A submarine dove 75 feet below its original depth. If the submarine's new depth is −600 feet, what was its original depth?

   4. TESTING The total time given to take a state test is equally divided among the 3 subjects tested. If the time given for each subject test is 45 minutes, how many minutes long is the entire test?

Example 3   **5. AGE** Mateo is 4 years younger than his sister Rita. Given m, Mateo's age, write an equation that can be used to find r, Rita's age.

Practice and Problem Solving

● = **Step-by-Step Solutions** begin on page R1.
Extra Practice begins on page EP2.

Example 1   **Define a variable. Then write an equation to model each situation.**

   6. After dropping 12°C, the temperature outside was −5°C.

   7. Jamal's score of 82 was 5 points less than the class average.

   8. At 30 meters per second, a cheetah's top speed is three times that of the fastest recorded human speed.

     An archaeological site is excavated to a level of −75 centimeters over several days for an average dirt removal of 15 centimeters each day.

   10. A class of 24 students was separated into equal-sized teams, resulting in 6 students per team.

   11. When the money was divided among the four grade levels, each grade received $235.

Example 2   **Define a variable. Then write an equation that could be used to solve each problem.**

   12. PETS Nikki's cat is 5 pounds heavier than her sister's cat. If Nikki's cat weighs 9 pounds, how much does her sister's cat weigh?

   13. MEASUREMENT The length of a triangle's base is one-fourth its height. If the base is 15 meters long, what is the height of the triangle?

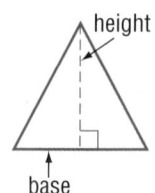

height

base

14. **CREDIT** For charging the cost of 4 equally priced shirts, Antonio's father's credit card statement shows an entry of −$74. What would the statement have shown for a charge of just one shirt?

15. **GOLF** The table shows some of the top 20 leaders in a golf tournament after the first round. If the 6th place participant is 5 strokes behind the leader, what was the leader's score after the first round?

6.	Poole	−3
7.	Shaw	−2
8.	Kendrick	−2
9.	Rodriguez	1

Example 3   **Write an equation that could be used to express the relationship between the two quantities.**

16. **HEALTH** Your heart rate r in beats per minute is the number of times your heart beats h in 15 seconds multiplied by 4. Given h, write an equation to find r.

17. **CARS** Ashley's car travels 24 miles per gallon of gas. Given d, the distance the car travels, write an equation to find g, the gallons of gas used.

18. **FRAMING** A mat for a picture frame should be cut so that its width is $\frac{1}{8}$ inch less than the frame's opening. Given p, the width of the frame's opening, write an equation to find m, the width of the mat.

19. **MEASUREMENT** A seam allowance indicates that the total length of fabric needed is $\frac{1}{2}$ inch more than that measured. Given t, the total length of fabric needed, write an equation to find m, the length measured.

Real-World Link · · · · ·
The earliest year a musical group can be inducted into the Rock and Roll Hall of Fame is 25 years after the year its first album debuted.

20. **MUSIC** Refer to the information at the left. If an artist was inducted in 2011, write an equation that could be used to find the latest year the artist's first album could have debuted.

Write an equation to model the relationship between the quantities in each table.

21.

Yards, y	Feet, f
1	3
2	6
3	9
4	12
y	f

22.

Centimeters, c	Meters, m
200	2
300	3
400	4
500	5
c	m

23. **MAPS** The scale on a map indicates that 1 inch on the map represents an actual distance of 20 miles. Create a table of values showing the number of miles represented by 1, 2, 3, 4, and m inches on the map. Given m, a distance on the map, write an equation to find a, the actual distance.

24. OPEN ENDED Write a real-world situation to represent the equation $n + 12 = 25$.

CHALLENGE Consider the sequence 2, 4, 6, 8, ...

25. Express the relationship between a number in this sequence and its position using words. For example, 6 is the third number in this sequence.

26. Define two variables and write an equation to express this relationship.

27. Describe how this relationship would change, using words and a new equation, if the sequence were changed to 0, 2, 4, 6, 8, ...

28. ✏ **WRITE MATH** Analyze the meaning of the equations $b = 2h$ and $h = 2b$ if b represents the base of a rectangle and h its height. Then draw a rectangle that demonstrates each relationship.

✔ Test Practice

29. The length of an actual car is 87 times the corresponding length of a model of the car. Given a, an actual length of the car, which equation can be used to find m, the corresponding model length?

A. $a = 87 + m$ **C.** $a = 87 \cdot m$

B. $a = 87 - m$ **D.** $a = 87 \div m$

30. The sides of each square are 1 unit long. Which equation can be used to represent the area of the figure that contains x squares?

  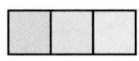

Figure 1   Figure 2   Figure 3

F. $A = x + 4$ **H.** $A = x + 2$

G. $A = x + 1$ **I.** $A = x$

More About Equations

A verbal problem is one way to represent an equation.

 EXAMPLE

Write a problem based on the given equation.

b = **number of bags of apples;** $5b$ = **cost of b bags of apples;** $5b = 20$

Since the cost $5b$ is part of the equation, the problem should relate to money.

Sample problem
Rene was at the Farmers Market and spent a total of $20. He bought some bags of apples that cost $5 each. How many bags of apples did Rene buy?

Write a problem based on the given equation.

31. g = the number of guests at a party; $\frac{g}{7} = 4$ **32.** a = Megan's age; $a - 5 = 12$

33. s = number of swimmers on a team; $4s = 60$ **34.** b = cost of a bracelet; $b + 0.9 = 15.90$

Main Idea

Solve equations using the Subtraction and Addition Properties of Equality.

 Vocabulary

inverse operations

 **8.EE.7**

Solve Addition and Subtraction Equations

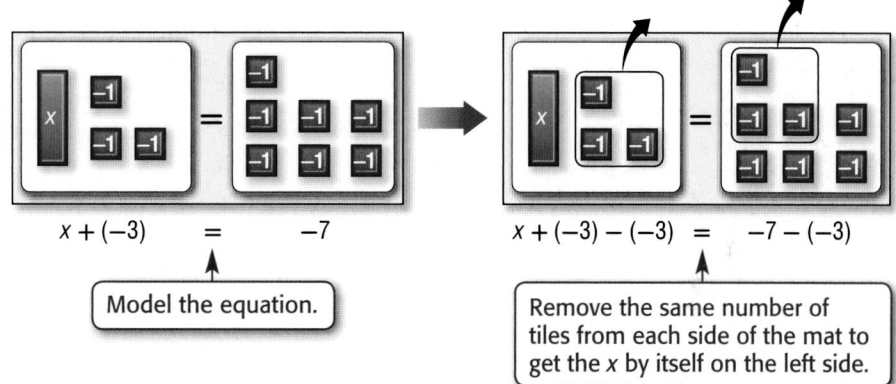 **Explore** When you solve an equation, you are finding the values of the variable that make the equation true. These values are called the solutions of the equation. You can use algebra tiles and an equation mat to solve $x + (-3) = -7$.

$$x + (-3) = -7 \qquad x + (-3) - (-3) = -7 - (-3)$$

Model the equation.

Remove the same number of tiles from each side of the mat to get the x by itself on the left side.

The number of tiles remaining on the right side of the mat represents the value of x. So, -4 is the solution of the equation $x + (-3) = -7$.

Solve each equation using algebra tiles.

1. $x + 1 = 4$ **2.** $x + 3 = 7$ **3.** $x + (-4) = -5$

4. Explain how you would find a value of x that makes $x + (-3) = -8$ true without using algebra tiles.

In the example above, you solved the equation $x + (-3) = -7$ by *removing*, or subtracting, the same number of negative counters from each side of the mat. This suggests the **Subtraction Property of Equality**.

Key Concept   Subtraction Property of Equality

Words   If you subtract the same number from each side of an equation, the two sides remain equal.

Examples   **Numbers**   **Algebra**

$$7 = 7 \qquad\qquad x + 4 = 6$$

$$7 - 3 = 7 - 3 \qquad \underline{-4 = -4}$$

$$4 = 4 \qquad\qquad x \quad = 2$$

You can use this property to solve any addition equation. Remember to check your solution by substituting it into the original equation.

 EXAMPLE Solve an Addition Equation

1 Solve $x + 5 = 3$. Check your solution.

> **Method 1** **Use the vertical method.**
>
> | | |
> |---|---|
> | $x + 5 = \quad 3$ | Write the equation. |
> | $\underline{\quad -5 = -5}$ | Subtraction Property of Equality |
> | $x \quad\quad = -2$ | |

> **Method 2** **Use the horizontal method.**
>
> | | |
> |---|---|
> | $x + 5 = 3$ | Write the equation. |
> | $x + 5 - 5 = 3 - 5$ | Subtraction Property of Equality |
> | $x = -2$ | |

Study Tip

Isolating the Variable
When trying to decide which value to subtract from each side of an addition equation, remember that your goal is to get the variable by itself on one side of the equation. This is called isolating the variable.

The solution is -2.

Check	$x + 5 = 3$	Write the original equation.
	$-2 + 5 \overset{?}{=} 3$	Replace x with -2. Is this sentence true?
	$3 = 3 \checkmark$	The sentence is true.

CHOOSE Your Method

Solve each equation. Check your solution.
a. $a + 6 = 2$ **b.** $y + 3 = -8$ **c.** $5 = n + 4$

Addition and subtraction are called **inverse operations** because they "undo" each other. For this reason, you can use the **Addition Property of Equality** to solve subtraction equations like $x - 7 = -5$.

Key Concept   **Addition Property of Equality**

Words   If you add the same number to each side of an equation, the two sides remain equal.

Examples

Numbers	**Algebra**
$7 = 7$	$x - 5 = \quad 6$
$7 + 3 = 7 + 3$	$\underline{\quad +5 = +5}$
$10 = 10$	$x \quad\quad = 11$

---

2 **MEASUREMENT** Two angles are supplementary if the sum of their measures is 180°. The two angles shown are supplementary. Write and solve an equation to find the measure of angle X.

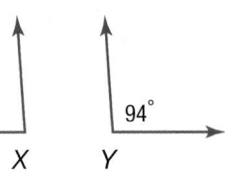

Let x represent the measure of angle X.

So, $x + 94 = 180$.

$$\begin{aligned} x + 94 &= 180 && \text{Write the equation.} \\ -94 &= -94 && \text{Subtraction Property of Equality} \\ x &= 86 && \text{Simplify.} \end{aligned}$$

The measure of angle X is 86°.

✔ **CHECK Your Progress**

d. READING A novel is ranked 7th on a best-seller list. This is up 8 places from its position last week. Write and solve an equation to determine the novel's ranking last week.

✎ **EXAMPLE** Solve a Subtraction Equation

3 Solve $-6 = y - 7$.

Study Tip

Position of the Variable
You could also begin solving Example 3 by rewriting the equation so that the variable is on the left side of the equation.

$-6 = y - 7$
↓
$y - 7 = -6$

Method 1 · Use the vertical method.

$$\begin{aligned} -6 &= y - 7 && \text{Write the equation.} \\ +7 &= \quad +7 && \text{Addition Property of Equality} \\ 1 &= y && -6 + 7 = 1 \text{ and } -7 + 7 = 0 \end{aligned}$$

Method 2 Use the horizontal method.

$$\begin{aligned} -6 &= y - 7 && \text{Write the equation.} \\ -6 + 7 &= y - 7 + 7 && \text{Addition Property of Equality} \\ 1 &= y && -6 + 7 = 1 \text{ and } -7 + 7 = 0 \end{aligned}$$

The solution is 1. Check the solution.

✔ **CHOOSE Your Method**

Solve each equation.

e. $x - 8 = -3$ **f.** $b - 4 = -10$ **g.** $7 = p - 12$

Example 1 Solve each equation. Check your solution.

1. $a + 4 = 10$ **2.** $2 = z + 7$ **3.** $x + 9 = -3$

Example 2 **4. RUGS** The length of a rectangular rug is 12 inches shorter than its width. If the length is 30 inches, write and solve an equation to find the width.

Example 3 Solve each equation. Check your solution.

5. $y - 2 = 5$ **6.** $n - 5 = -6$ **7.** $-8 = d - 11$

Practice and Problem Solving

= Step-by-Step Solutions begin on page R1.
Extra Practice begins on page EP2.

Examples 1 and 3 Solve each equation. Check your solution.

8. $x + 5 = 18$ **9.** $n + 3 = 20$ **10.** $9 = p + 11$

11. $1 = a + 7$ **12.** $y + 12 = -3$ **13.** $w + 8 = -6$

14. $m - 15 = 3$ **15.** $b - 9 = -8$ **16.** $g - 2 = -13$

17. $-16 = t - 6$ **18.** $-4 = r - 20$ **19.** $k - 14 = -7$

Example 2 **20. MEASUREMENT** Two angles are supplementary if the sum of their measures is 180°. The two angles shown are supplementary. Write and solve an equation to find the measure of angle B.

65°

A   B

21. BANKING After you withdraw $50 from your savings account, the balance is $124. Write and solve an equation to find your starting balance.

22. TEMPERATURE On one day in Fairfield, Montana, the temperature dropped 84°F from noon to midnight. If the temperature at midnight was -21°F, write and solve an equation to determine the noon temperature that day.

23 TREES Before planting a tree, Sanjay digs a hole with a depth 18 inches below ground level. Once planted, the top of the tree is 54 inches above ground. Write and solve an equation to find the height of the tree that Sanjay planted.

24. BASKETBALL The table shows the points leaders for a recent WNBA season.

a. Lauren Jackson averaged 2.5 points per game less than Seimone Augustus. Write and solve an equation to find Augustus' average points scored per game.

b. Sheryl Swoopes averaged 6.4 fewer points per game than Seimone Augustus. Write and solve an equation to find how many points Swoopes averaged per game.

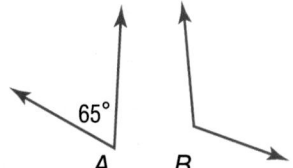

WNBA Regular Season Points Leaders	
Player	**AVG**
Diana Taurasi	25.3
Seimone Augustus	p
Lisa Leslie	20
Lauren Jackson	19.4

25. **OPEN ENDED** Write one addition equation and one subtraction equation that each have –3 as a solution.

26. **Which One Doesn't Belong?** Identify the equation that does not belong with the other three. Explain your reasoning.

| $y + 4 = 2$ | $x + 6 = 4$ | $t + 5 = -3$ | $1 + m = -1$ |

27. **CHALLENGE** Solve $|x| + 5 = 7$. Explain your reasoning.

28. **WRITE MATH** Write a problem about a real-world situation that can be answered by solving the equation $x + 60 = 20$. Then solve the equation. Explain the meaning of its solution in the context of your problem.

Test Practice

29. A scuba diver dives 125 feet below the surface of the water, then ascends 42 feet. What is her depth?

   A. +167 feet   C. −83 feet

   B. +83 feet   D. −167 feet

30. **SHORT RESPONSE** Two angles are complementary if the sum of their measures is 90°. The two angles shown are complementary. Write and solve an equation to find the measure of angle 1.

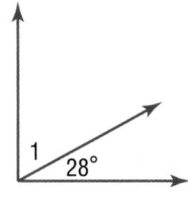

31. The bill for dinner came to $15.68 and included a $3.25 tip. Which equation could be used to find the cost of dinner before the tip?

   F. $x - 3.25 = 15.68$

   G. $x + 3.25 = 15.68$

   H. $x + 15.68 = 3.25$

   I. $x - 15.68 = 3.25$

32. What is the solution of the equation $-25 = t - 5$?

   A. 30

   B. 20

   C. −20

   D. −30

Spiral Review

ALGEBRA Define a variable. Then write an equation to model each situation. (Lesson 1B)

33. Lindsay, 59 inches tall, is 5 inches shorter than her sister.

34. After cutting the recipe in half, Ricardo needed 3 cups of flour.

35. **NUMBER SENSE** A number is halved. Then five is subtracted from the quotient, and 9 is multiplied by the difference. Finally, 2 is added to the product. If the ending number is 47, what was the beginning number? Use the *work backward* strategy. (Lesson 1A)

Main Idea

Solve equations by using the Division and Multiplication Properties of Equality.

 Get ConnectED

 CCSS 8.EE.7

Solve Multiplication and Division Equations

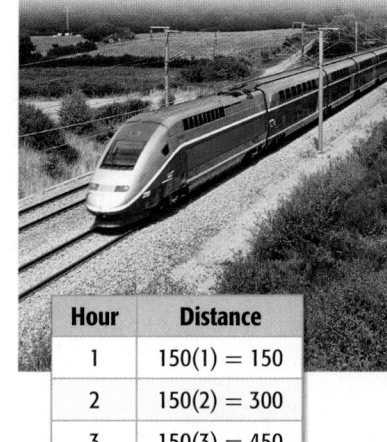

TRANSPORTATION High speed trains are very popular in Europe and Asia and can travel at speeds of 150 miles per hour or more. That's almost three times the speed of a car on the highway!

1. If h represents the number of hours the train has traveled, write a multiplication equation you could use to find how long it would take the train to travel 675 miles.

Hour	Distance
1	$150(1) = 150$
2	$150(2) = 300$
3	$150(3) = 450$
⋮	⋮
h	?

The equation $150h = 675$ models the relationship described above. To undo the multiplication of 150, divide each side of the equation by 150.

 EXAMPLE Solve a Multiplication Equation

 Solve $150h = 675$.

$150h = 675$ Write the equation.

$\dfrac{150h}{150} = \dfrac{675}{150}$ Divide each side of the equation by 150.

$h = 4.5$ $150 \div 150 = 1$ and $675 \div 150 = 4.5$

CHECK Your Progress

Solve each equation. Check your solution.

a. $8x = 72$ **b.** $-4n = 28$ **c.** $-12 = -6k$

In Example 1, you used the **Division Property of Equality**.

Key Concept   **Division Property of Equality**

Words   If you divide each side of an equation by the same nonzero number, the two sides remain equal.

Examples

Numbers	Algebra
$12 = 12$	$5x = -60$
$\dfrac{12}{4} = \dfrac{12}{4}$	$\dfrac{5x}{5} = \dfrac{-60}{5}$
$3 = 3$	$x = -12$

You can also use the **Multiplication Property of Equality** to solve equations.

> ### Key Concept · Multiplication Property of Equality
>
> **Words** If you multiply each side of an equation by the same number, the two sides remain equal.
>
> **Examples**
>
> | **Numbers** | **Algebra** |
> |---|---|
> | $5 = 5$ | $\dfrac{x}{2} = 8$ |
> | $5(-4) = 5(-4)$ | $\dfrac{x}{2}(2) = 8(2)$ |
> | $-20 = -20$ | $x = 16$ |

 EXAMPLE Solve a Division Equation

Read Math

Division Expressions
Remember, $\dfrac{a}{-3}$ means *a divided by −3.*

② Solve $\dfrac{a}{-3} = -7$.

$$\dfrac{a}{-3} = -7 \qquad \text{Write the equation.}$$

$$\dfrac{a}{-3}(-3) = -7(-3) \qquad \text{Multiplication Property of Equality}$$

$$a = 21 \qquad -7 \cdot (-3) = 21$$

CHECK Your Progress

Solve each equation.

d. $\dfrac{y}{-4} = -8$ **e.** $\dfrac{m}{5} = -9$ **f.** $30 = \dfrac{b}{-2}$

 REAL-WORLD EXAMPLE

Real-World Link · · ·
There are nearly 5,000 species of lizards. One of the most commonly known lizards is the chameleon, which has the ability to change color to match its environment.

③ **REPTILES** An adult lizard is about 5 times as long as a hatchling. If an adult lizard is 11 centimeters long, about how long is a hatchling?

Let g represent the length of a hatchling.

So, $11 = 5g$.

$$11 = 5g \qquad \text{Write the equation.}$$

$$\dfrac{11}{5} = \dfrac{5g}{5} \qquad \text{Division Property of Equality}$$

$$2.2 = g \qquad 11 \div 5 = 2.2$$

A lizard hatchling is about 2.2 centimeters long.

CHECK Your Progress

g. METEOROLOGY The recorded amount of precipitation is one-tenth the amount of fallen snow. If Redfield received 3.6 inches of precipitation in one week, how many inches of snow fell?

Examples 1 and 2 Solve each equation. Check your solution.

1. $5b = 40$ **2.** $-7k = 14$ **3.** $-18 = -3n$

4. $\dfrac{p}{9} = 9$ **5.** $\dfrac{a}{12} = -3$ **6.** $22 = \dfrac{m}{-2}$

Example 3 **7. FUNDRAISING** King Middle School is selling cards as a fundraiser and keeps $\dfrac{3}{4}$ of the proceeds. Write and solve a multiplication equation to find how much they need to sell in order to raise $1,350.

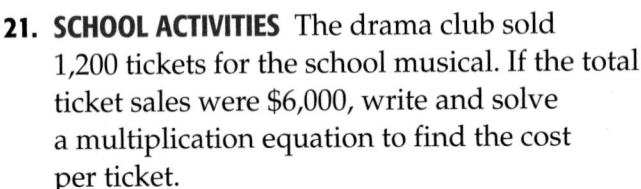

Practice and Problem Solving

● = **Step-by-Step Solutions** begin on page R1.
Extra Practice begins on page EP2.

Examples 1 and 2 Solve each equation. Check your solution.

8. $4c = 44$ **9** $9b = 72$ **10.** $34 = -2x$

11. $36 = -18y$ **12.** $-32 = 8d$ **13.** $-35 = 5n$

14. $\dfrac{m}{7} = 10$ **15.** $\dfrac{u}{9} = 6$ **16.** $\dfrac{h}{-3} = 33$

17. $20 = \dfrac{q}{-5}$ **18.** $-8 = \dfrac{c}{12}$ **19.** $\dfrac{r}{24} = -3$

Example 3 **20. MAMMALS** A koala eats an average of 2.5 pounds of leaves from eucalyptus trees each day. If a small tree has 30 pounds of leaves on it, write and solve a multiplication equation to find how many days the leaves will last.

21. SCHOOL ACTIVITIES The drama club sold 1,200 tickets for the school musical. If the total ticket sales were $6,000, write and solve a multiplication equation to find the cost per ticket.

MEASUREMENT Refer to the table. Write and solve an equation to find each quantity.

22. the number of cups in 96 teaspoons

23. the number of cups in 64 tablespoons

24. the number of gallons in 24 quarts

25. the number of gallons in 120 pints

26. the number of gallons in 128 fluid ounces

Customary System Conversions (capacity)
1 tablespoon = 3 teaspoons
1 cup = 16 tablespoons
1 cup = 8 fluid ounces
1 pint = 2 cups
1 quart = 2 pints
1 gallon = 4 quarts

Solve each equation.

27. $7 = \dfrac{-56}{z}$ **28.** $\dfrac{10}{x} = -5$ **29.** $\dfrac{-126}{a} = -21$ **30.** $-17 = \dfrac{136}{g}$

31 **DISTANCE** If an object is traveling at a rate of speed r, then the distance d the object travels after a time t is given by the distance formula $d = rt$. A car travels 245 miles at an average speed of 70 miles per hour. How long did the car travel?

32. GRAPHIC NOVEL Refer to the graphic novel frame below for Exercises a–c.

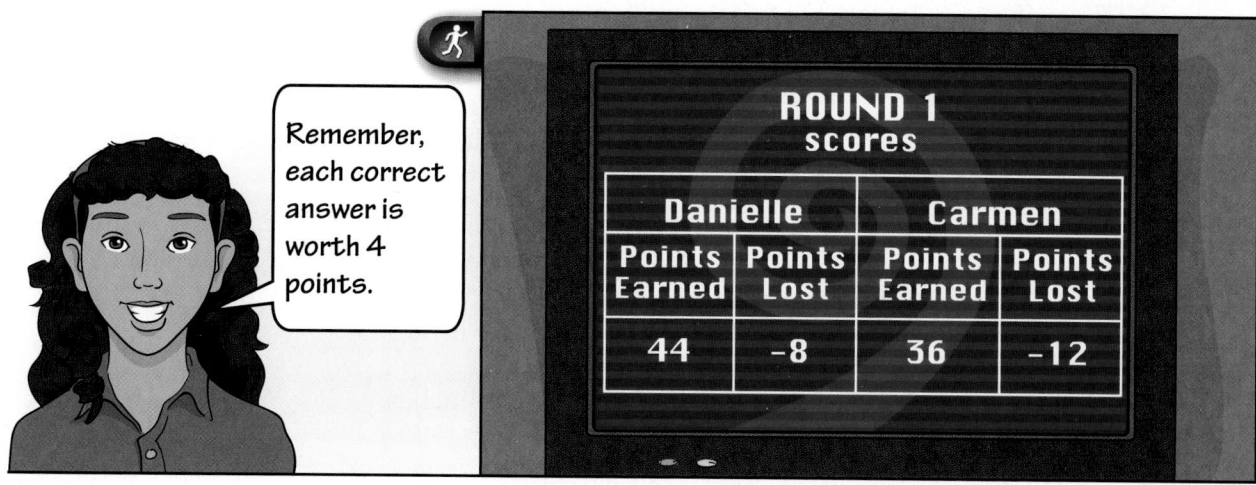

Remember, each correct answer is worth 4 points.

ROUND 1
scores

Danielle		Carmen	
Points Earned	Points Lost	Points Earned	Points Lost
44	-8	36	-12

a. Write an equation that can be used to find the number of questions each person answered correctly.

b. Write an equation that can be used to find the number of questions each person answered incorrectly.

c. How many total points has Danielle earned? Carmen earned?

33 RECREATION Refer to the table. Write and solve a multiplication equation to determine how many times more visitors t visit the Golden Gate National Recreation Area than the Great Smoky Mountains National Park. Round your answer to the nearest tenth.

Most Popular National Parks	
Park	Visitors (millions)
Blue Ridge	18.3
Golden Gate	13.9
Great Smoky Mountains	9.4

MEASUREMENT Find the area of each rectangle.

34.

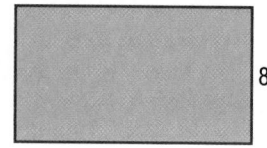

$8\frac{3}{4}$ in.

Perimeter $= 38\frac{1}{2}$ in.

35.

4.84 m

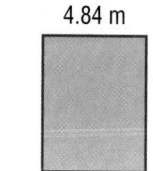

Perimeter $= 22.83$ m

H.O.T. Problems

36. OPEN ENDED Describe a real-world situation in which you would use a division equation to solve a problem. Then write your equation.

37. CHALLENGE The amount of work W, measured in foot-pounds, needed to move an object d feet using a force of F pounds, is given by the equation $W = Fd$. Rewrite this equation so that it expresses the value of d in terms of W and F.

38. WRITE MATH Explain how to solve $-4a = 84$. Be sure to state which property you use and why you used it.

39. Tonya paid $2.24 for 4 granola bars. All 4 granola bars were the same price. How much did each granola bar cost?

A. $0.52   C. $1.24

B. $0.56   D. $1.56

40. Jason paid half of what Prem paid for new shoes. If Jason paid $38 for his shoes, which equation can be used to find the amount s that Prem paid for his shoes?

F. $s = \dfrac{32}{8}$

G. $s = 38 - \dfrac{1}{2}$

H. $s = 2(38)$

I. $s + \dfrac{1}{2} = 38$

41. The area of the triangle is $13\frac{1}{2}$ square inches. What is the height of the triangle? Use the formula $A = \frac{1}{2}bh$.

A. $3\frac{3}{8}$ inches

B. $4\frac{1}{2}$ inches

C. $5\frac{7}{8}$ inches

D. $6\frac{3}{4}$ inches

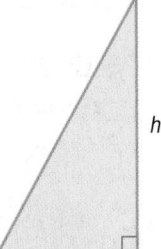

4 in.

42. **THINK SOLVE EXPLAIN** **SHORT RESPONSE** Write and solve an equation to represent the phrase *the product of a number and nine is forty-nine.*

Spiral Review

43. **SPORTS** At a recent track meet, Lucia tried to break her school's high jump record, but fell short by 4 inches. Write and solve an equation to find the height of Lucia's jump. (Lesson 1C)

ALGEBRA Define a variable. Then write an equation to model each situation. (Lesson 1B)

44. Eight feet longer than she jumped is 15 feet.

45. The temperature fell 28°F from 6 A.M. to 17°F at 11 A.M.

46. Three friends shared a $9 parking fee equally.

47. In 13 years, Malik will be 26 years old.

48. The basketball score of 59 was 3 more points than last week's game.

49. When slices of pizza were shared among 5 people, each person ate 3 slices.

School record 58 inches

4 inches

x inches

50. **MONEY** Audrey played a round of golf, rode the go-karts once, and bought 100 arcade tokens. At the end of the day, she had $12.60 left. How much money did Audrey take to the fun center? Use the *work backward* strategy. (Lesson 1A)

Item	Cost
100 Tokens	$20.00
Bumper Boats	$4.95
Go-Kart	$6.45
Golf	$5.95

Two-Step Equations

Main Idea

Use a bar diagram to write and solve two-step equations.

CCSS 8.EE.7

POSTCARDS Miranda bought two large postcards and four small postcards at a souvenir shop. Each small postcard cost $0.50. If Miranda spent $5.00 on postcards, what is the cost of one large postcard?

ACTIVITY

STEP 1 Make a bar diagram that represents the total number of postcards and the total cost. Label the parts.

$5					
large	large	small	small	small	small
?	?	$0.50	$0.50	$0.50	$0.50

STEP 2 Write an equation that represents the bar diagram. The cost of a large postcard is the unknown, so it is represented by the variable p.

$$2p \quad + \quad \$2 \quad = \quad \$5$$

STEP 3 Find the cost of the large postcards by working backward.

$3		$2			
large	large	small	small	small	small
?	?	$0.50	$0.50	$0.50	$0.50

The cost of one large postcard is $3 ÷ 2 or $1.50.

Analyze the Results

1. Suppose 4 medium postcards and 4 small postcards cost $5. Draw a bar diagram and write an equation to find the cost of one medium postcard. Then solve.

2. **FOOD** Reynaldo bought eight packages of hot dog buns and four packages of hamburger buns. Each package of hot dog buns cost $2.

   a. Suppose Reynaldo spent a total of $22. Draw a bar diagram and write an equation to find the cost of one package of hamburger buns.

   b. Use the diagram from part **a** to find the cost of one package of hamburger buns.

Main Idea

Solve two-step equations.

Vocabulary

two-step equation
coefficient

CCSS 8.EE.7

Solve Two-Step Equations

PETS Mario bought two small bags of dog treats and three bags of cat food. His total was $7, but the receipt did not show the individual price of the dog treats.

1. Explain how you could use the work backward strategy to find the cost of each bag of dog treats.

2. Find the cost of each bag of dog treats.

The solution to this problem can also be found by solving the two-step equation $2x + 3 = 7$, where x is the cost per bag of dog treats.

A **two-step equation** contains two operations. In the equation $2x + 3 = 7$, x is multiplied by 2 and then 3 is added. To solve two-step equations, undo each operation in reverse order.

EXAMPLE **Solve Two-Step Equations**

 Solve $2x + 3 = 7$.

Method 1 Use a model.

Remove three 1-tiles from each mat.

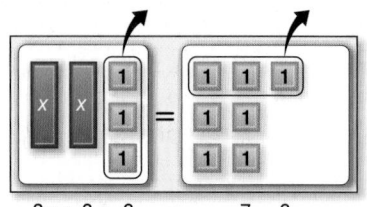

$$2x + 3 - 3 = 7 - 3$$

Separate the remaining tiles into 2 equal groups.

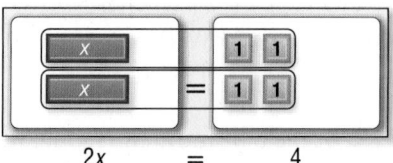

$$2x = 4$$

There are two 1-tiles in each group, so $x = 2$.

Method 2 Use symbols.

Use the Subtraction Property of Equality.

$$\begin{aligned} 2x + 3 &= 7 & \text{Write the equation.} \\ -3 &= -3 & \text{Subtraction Property} \\ 2x &= 4 & \text{of Equality} \end{aligned}$$

Use the Division Property of Equality.

$$\begin{aligned} 2x &= 4 \\ \frac{2x}{2} &= \frac{4}{2} & \text{Division Property of Equality} \\ x &= 2 & \text{Simplify.} \end{aligned}$$

Using either method, the solution is 2.

 **EXAMPLE** **Solve Two-Step Equations**

2 Solve $25 = \frac{1}{4}n - 3$.

$25 = \frac{1}{4}n - 3$	Write the equation.
$\underline{+3 = \quad +3}$	Addition Property of Equality
$28 = \frac{1}{4}n$	Simplify.
$4 \cdot 28 = 4 \cdot \frac{1}{4}n$	Multiplication Property of Equality
$112 = n$	

The solution is 112.

QUICK Review

Multiplying Fractions

$4 \cdot \frac{1}{4} = \frac{4}{1} \cdot \frac{1}{4}$

$= \frac{\overset{1}{\cancel{4}}}{1} \cdot \frac{1}{\underset{1}{\cancel{4}}}$

$= 1$

 **CHOOSE Your Method**

a. $3x + 2 = 20$ **b.** $5 + 2n = -1$ **c.** $-1 = \frac{1}{2}a + 9$

The numerical factor of a term that contains a variable is called the **coefficient** of the variable.

Some two-step equations have a term with a negative coefficient.

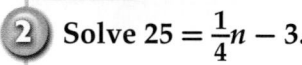 **EXAMPLE** **Equations with Negative Coefficients**

3 Solve $6 - 3x = 21$.

$6 - 3x = 21$	Write the equation.
$6 + (-3x) = 21$	Rewrite the left side as addition.
$\underline{-6 \qquad = -6}$	Subtraction Property of Equality
$-3x = 15$	Simplify.
$\dfrac{-3x}{-3} = \dfrac{15}{-3}$	Division Property of Equality
$x = -5$	Simplify.

The solution is -5.

Check	$6 - 3x = 21$	Write the equation.
	$6 - 3(-5) \overset{?}{=} 21$	Replace x with -5.
	$6 - (-15) \overset{?}{=} 21$	Multiply.
	$6 + 15 \overset{?}{=} 21$	To subtract a negative number, add its opposite.
	$21 = 21$ ✓	The sentence is true.

Study Tip

Common Error A common mistake when solving the equation in Example 3 is to divide each side by 3 instead of -3. Remember that you are dividing by the coefficient of the variable, which in this instance is a negative number.

CHECK Your Progress

d. $10 - \frac{2}{3}p = 52$ **e.** $-19 = -3x + 2$ **f.** $\frac{n}{-3} - 2 = -18$

You can use two-step equations to solve many real-world problems.

REAL-WORLD EXAMPLE

④ TEMPERATURE The formula $F = 1.8C + 32$ can be used to convert between degrees Fahrenheit (°F) and degrees Celsius (°C). The table shows the lowest recorded temperature in three cities. Solve the equation $-27 = 1.8C + 32$ to convert Chicago's lowest recorded temperature to degrees Celsius.

Lowest Recorded Temperature (°F)	
City	Temperature
Buffalo, New York	−20
Chicago, Illinois	−27
Norfolk, Virginia	−3

$$-27 = 1.8C + 32 \qquad \text{Write the equation.}$$

$$\underline{-32 = \qquad -32} \qquad \text{Subtraction Property of Equality}$$

$$-59 = 1.8C \qquad \text{Simplify.}$$

$$\frac{-59}{1.8} = \frac{1.8C}{1.8} \qquad \text{Division Property of Equality}$$

$$-32.8 \approx C \qquad \text{Simplify. Check the solution.}$$

So, Chicago's lowest recorded temperature is about −32.8 degrees Celsius.

CHECK Your Progress

g. CLASS TRIP Simone has $150 for a class trip that costs $320. She needs to save money for the next 10 weeks in order to pay for the trip. Solve the equation $10d + 150 = 320$ to find the amount of money per week Simone must save.

CHECK Your Understanding

Examples 1–3   **Solve each equation. Check your solution.**

1. $6x + 5 = 29$

2. $-2 = 9m - 11$

3. $10 = \frac{a}{4} + 3$

4. $\frac{2}{3}x - 5 = 7$

5. $3 - 5y = -37$

6. $\frac{c}{-2} - 4 = 3$

7. $6 - 10k = 16$

8. $4 - 2d = 12$

9. $1 = 4\frac{1}{2} - 1\frac{1}{3}p$

Example 4   **10. ELECTRONICS** Mr. Sampson bought a television for $816, and he pays $34 a month on the balance. The current balance owed is $272. Solve the equation $272 = 816 - 34m$ to determine the number of monthly payments he has made.

11 MOVIES Cassidy went to the movies with some of her friends. The tickets cost $6.50 each, and they spent $17.50 on snacks. The total amount paid was $63.00. Solve the equation $63 = 6.50p + 17.50$ to determine how many people went to the movies.

Practice and Problem Solving

● = **Step-by-Step Solutions** begin on page R1
Extra Practice begins on page EP2.

Examples 1–3   **Solve each equation. Check your solution.**

12. $2h + 9 = 21$ **13.** $11 = 2b + 17$ **14.** $5 = 4a - 7$

15. $-17 = 6p - 5$ **16.** $2g - 3 = -19$ **17.** $16 = 5x - 9$

18. $13 = \frac{g}{3} + 4$ **19.** $5 + \frac{y}{8} = -3$ **20.** $3 - 8c = 35$

21. $13 - 3d = -8$ **22.** $-\frac{1}{2}x - 7 = -11$ **23** $15 - \frac{w}{4} = 28$

24. $-\frac{2}{3}m - 4 = 10$ **25.** $12 - \frac{3}{5}p = -27$ **26.** $-3 - 6x = 9$

27. $-5y - 25 = 25$ **28.** $42 + 8x + 13 = 7$ **29.** $-7w + 17 - 30 = 71$

Example 4   **30. VACATION** Four friends decide to go to the aquarium together. Each person pays x dollars to get in and $10 for the shark exhibit. The total cost is $64. Solve $4x + 4(10) = 64$ to find how much each person pays to get into the aquarium.

31. GIFTS Larina received a $50 gift card to an online store. She wants to purchase some bracelets that cost $8 each. There will be a $10 overnight delivery fee. Solve $8n + 10 = 50$ to find the number of bracelets she can purchase.

32. GAMES Brent had $26 when he went to the fair. After playing 7 games, he had $15.50 left. Solve $15.50 = 26 - 7p$ to find the price for each game.

33. SPORTS LaTasha paid $75 to join a summer golf program. The course where she plays charges $30 per round. Since she is a student, she receives a $10 discount per round. If LaTasha spent $375, use the equation $375 = 20g + 75$ to find how many rounds of golf LaTasha played.

Solve each equation. Check your solution.

34. $\dfrac{a - 4}{5} = 12$ **35.** $\dfrac{n + 3}{8} = -4$ **36.** $\dfrac{6 + z}{10} = -2$

37 **HOME IMPROVEMENT** If Mr. Arenth wants to put new carpeting in the room shown, how many square feet should he order?

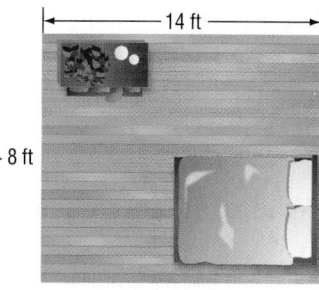
14 ft

$6c - 8$ ft

$5 + 3c$ ft

38. ANIMALS Solve $4x + 12 = 171$. If x stands for the number of animals in a pet store, can it be a solution? Explain.

39. GEOMETRY Write an equation to represent the length of $\overline{AB}$. Then find the value of x.

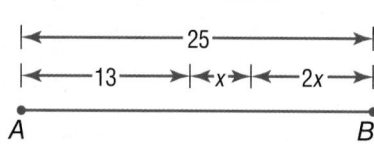
25
13   x $2x$
A B

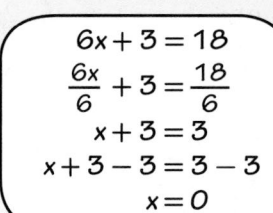

40. FIND THE ERROR Sarah is solving the equation $6x + 3 = 18$. Find her mistake and correct it.

$$6x + 3 = 18$$
$$\frac{6x}{6} + 3 = \frac{18}{6}$$
$$x + 3 = 3$$
$$x + 3 - 3 = 3 - 3$$
$$x = 0$$

41. CHALLENGE Solve $(x + 5)(x + 5) = 49$. (*Hint:* There are two solutions.)

42. 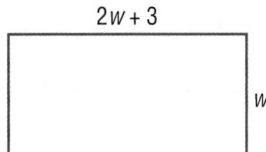 **WRITE MATH** Explain how you can use the *work backward* problem-solving strategy to solve a two-step equation.

✓ Test Practice

43. The width of the rectangle below can be found by solving the equation $6w + 6 = 36$.

$2w + 3$

w

Perimeter = 36 units

What is the width of the rectangle?

A. 4 units   **C.** 6 units

B. 5 units   **D.** 7 units

44. What is the value of m if $-6m + 4 = -32$?

F. 6

G. $4\frac{2}{3}$

H. $2\frac{1}{3}$

I. -6

45. ▨ **GRIDDED RESPONSE** What value of y makes the equation true?

$$\frac{y}{4} - 7 = 3$$

**Spiral Review**

46. The product of two integers is 72. If one integer is -18, what is the other integer? (Lesson 1D)

ALGEBRA Solve each equation. Check your solution. (Lesson 1C)

47. $t + 17 = -5$ **48.** $a - 5 = 14$ **49.** $5 = 9 + x$ **50.** $m - 5 = -14$

Write an equation to model the relationship in each table. (Lesson 1B)

51.

Kilometers, k	Miles, m
1	0.62
2	1.24
3	1.86
4	2.48

52.

Inches, n	Centimeters, c
1	2.54
2	5.08
3	7.62
4	10.16

Main Idea

Write two-step equations that represent real-world situations.

Write Two-Step Equations

SUMMER CAMP You want to attend a two-week robotics day camp this summer that costs $700. Your parents have agreed to pay the deposit of $400 if you pay the rest in weekly payments of $15.

Payments	Amount Paid
0	$400 + 15(0) = 400$
1	$400 + 15(1) = 415$
2	$400 + 15(2) = 430$
3	$400 + 15(3) = 445$
⋮	⋮

1. Let n represent the number of payments. Write an expression that represents the amount of the camp fee paid after n payments.

2. Write and solve an equation to find the number of payments you will have to make in order to pay off the balance of the fees.

3. What type of equation did you write for Exercise 2? Explain.

You have already learned how to write verbal sentences as one-step equations. Some verbal sentences translate into two-step equations.

Words	The sum of 400 and 15 times a number is 700.
Variable	Let n represent the number.
Equation	$400 + 15n = 700$

EXAMPLES Translate Sentences into Equations

Translate each sentence into an equation.

Sentence	Equation
1 Eight less than three times a number is −23.	$3n - 8 = -23$
2 Thirteen is 7 more than twice a number.	$13 = 2n + 7$
3 The quotient of a number and 4, decreased by 1, is equal to 5.	$\frac{n}{4} - 1 = 5$

CHECK Your Progress

a. Fifteen equals three more than six times a number.

b. Ten increased by the quotient of a number and 6 is 5.

c. The difference between 12 and twice a number is 18.

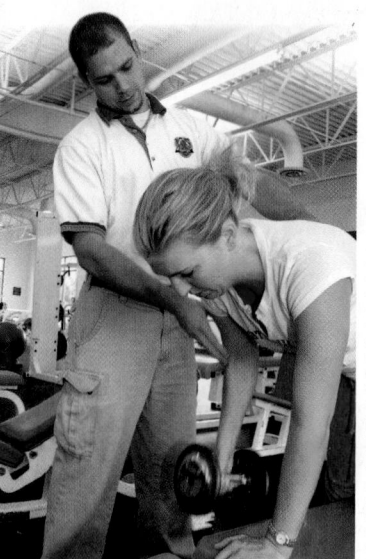

④ **PERSONAL TRAINING** A personal trainer buys a weight bench for $500 and w weights for $25 each. The total cost of the purchase is $850. How many weights were purchased?

Words	Bench	plus	$25 per weight	equals	$850.
Variable	Let w represent the number of weights.				
Expression	500	+	25 ·	w	= 850

$$500 + 25w = 850 \quad \text{Write the equation.}$$
$$-500 \qquad = -500 \quad \text{Subtraction Property of Equality}$$
$$25w = 350 \quad \text{Simplify.}$$
$$\frac{25w}{25} = \frac{350}{25} \quad \text{Division Property of Equality}$$
$$w = 14$$

So, 14 weights were purchased.

Real-World Link · · ·
Personal trainers help people set individual fitness goals. Statistics have projected that by the year 2016, there will be nearly 300,000 personal trainers in the U.S. This is a 27% increase from recent times.

⑤ **DINING** Your and your friend's lunch cost $19. Your lunch cost $3 more than your friend's. How much was your friend's lunch?

Words	Your friend's lunch	plus	your lunch	equals	$19.
Variable	Let f represent the cost of your friend's lunch.				
Equation	f	+	$f + 3$	=	19

$$f + f + 3 = 19 \quad \text{Write the equation.}$$
$$2f + 3 = 19 \quad f + f = 2f$$
$$\underline{-3 = -3} \quad \text{Subtraction Property of Equality}$$
$$2f = 16 \quad \text{Simplify.}$$
$$\frac{2f}{2} = \frac{16}{2} \quad \text{Division Property of Equality}$$
$$f = 8$$

Your friend spent $8.

✓ **CHECK Your Progress**

d. METEOROLOGY Suppose the current temperature is 54°F. It is expected to rise 2°F each hour for the next several hours. In how many hours will the temperature be 78°F?

e. MEASUREMENT The perimeter of a rectangle is 40 inches. The width is 8 inches shorter than the length. Find the dimensions of the rectangle.

Practice and Problem Solving

= **Step-by-Step Solutions** begin on page R1
Extra Practice begins on page EP2.

Examples 1–3   **Solve each equation. Check your solution.**

12. $2h + 9 = 21$ **13.** $11 = 2b + 17$ **14.** $5 = 4a - 7$

15. $-17 = 6p - 5$ **16.** $2g - 3 = -19$ **17.** $16 = 5x - 9$

18. $13 = \frac{g}{3} + 4$ **19.** $5 + \frac{y}{8} = -3$ **20.** $3 - 8c = 35$

21. $13 - 3d = -8$ **22.** $-\frac{1}{2}x - 7 = -11$ **23** $15 - \frac{w}{4} = 28$

24. $-\frac{2}{3}m - 4 = 10$ **25.** $12 - \frac{3}{5}p = -27$ **26.** $-3 - 6x = 9$

27. $-5y - 25 = 25$ **28.** $42 + 8x + 13 = 7$ **29.** $-7w + 17 - 30 = 71$

Example 4   **30. VACATION** Four friends decide to go to the aquarium together. Each person pays x dollars to get in and $10 for the shark exhibit. The total cost is $64. Solve $4x + 4(10) = 64$ to find how much each person pays to get into the aquarium.

31. GIFTS Larina received a $50 gift card to an online store. She wants to purchase some bracelets that cost $8 each. There will be a $10 overnight delivery fee. Solve $8n + 10 = 50$ to find the number of bracelets she can purchase.

32. GAMES Brent had $26 when he went to the fair. After playing 7 games, he had $15.50 left. Solve $15.50 = 26 - 7p$ to find the price for each game.

33. SPORTS LaTasha paid $75 to join a summer golf program. The course where she plays charges $30 per round. Since she is a student, she receives a $10 discount per round. If LaTasha spent $375, use the equation $375 = 20g + 75$ to find how many rounds of golf LaTasha played.

Solve each equation. Check your solution.

34. $\frac{a - 4}{5} = 12$ **35.** $\frac{n + 3}{8} = -4$ **36.** $\frac{6 + z}{10} = -2$

37 **HOME IMPROVEMENT** If Mr. Arenth wants to put new carpeting in the room shown, how many square feet should he order?

38. ANIMALS Solve $4x + 12 = 171$. If x stands for the number of animals in a pet store, can it be a solution? Explain.

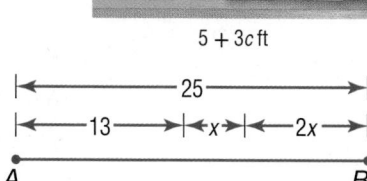

14 ft

$6c - 8$ ft

$5 + 3c$ ft

39. GEOMETRY Write an equation to represent the length of $\overline{AB}$. Then find the value of x.

25

13   x $2x$

A B

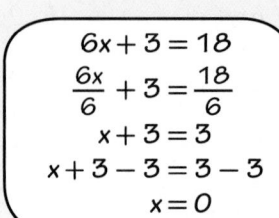 **H.O.T. Problems**

40. FIND THE ERROR Sarah is solving the equation
$6x + 3 = 18$. Find her mistake and correct it.

$$6x + 3 = 18$$
$$\frac{6x}{6} + 3 = \frac{18}{6}$$
$$x + 3 = 3$$
$$x + 3 - 3 = 3 - 3$$
$$x = 0$$

41. CHALLENGE Solve $(x + 5)(x + 5) = 49$. (*Hint:* There are two solutions.)

42. ✍ **WRITE MATH** Explain how you can use the *work backward* problem-solving strategy to solve a two-step equation.

✔ **Test Practice**

43. The width of the rectangle below can be found by solving the equation $6w + 6 = 36$.

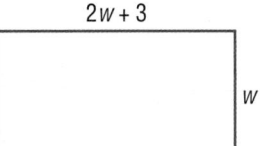

2w + 3

w

Perimeter = 36 units

What is the width of the rectangle?

A. 4 units   **C.** 6 units

B. 5 units   **D.** 7 units

44. What is the value of m if $-6m + 4 = -32$?

F. 6

G. $4\frac{2}{3}$

H. $2\frac{1}{3}$

I. -6

45. ▤✎ **GRIDDED RESPONSE** What value of y makes the equation true?

$$\frac{y}{4} - 7 = 3$$

46. The product of two integers is 72. If one integer is -18, what is the other integer? (Lesson 1D)

ALGEBRA Solve each equation. Check your solution. (Lesson 1C)

47. $t + 17 = -5$ **48.** $a - 5 = 14$ **49.** $5 = 9 + x$ **50.** $m - 5 = -14$

Write an equation to model the relationship in each table. (Lesson 1B)

51.

Kilometers, k	Miles, m
1	0.62
2	1.24
3	1.86
4	2.48

52.

Inches, n	Centimeters, c
1	2.54
2	5.08
3	7.62
4	10.16

Examples 1–3 **Translate each sentence into an equation.**

1. One more than three times a number is 7.

2. Seven less than twice a number is −1.

3. The quotient of a number and 5, less 10, is 3.

Examples 4 and 5 **Write and solve an equation to solve each problem.**

4. **MOVIES** You already owe $4.32 in overdue rental fees and are returning a movie that is 4 days late. Now you owe $6.48. How much is a daily fine for an overdue movie?

5. **SHOPPING** Marty paid $121 for shoes and clothes. He paid $45 more for clothes than he did for shoes. How much did Marty pay for the shoes?

Practice and Problem Solving

● = **Step-by-Step Solutions** begin on page R1.
Extra Practice begins on page EP2.

Examples 1–3 **Translate each sentence into an equation.**

6. Four less than five times a number is equal to 11.

7. Fifteen more than twice a number is 9.

8. Eight more than four times a number is −12.

9 Six less than seven times a number is equal to −20.

Examples 4 and 5 **Write and solve an equation to solve each problem.**

10. **MUSIC** Amy has saved $725 for a new guitar and lessons. Her guitar costs $475, and guitar lessons are $25 per hour. Determine how many hours of lessons she can afford.

11. **BOOKS** You buy 3 books that each cost the same amount and a magazine, all for $55.99. You know that the magazine costs $1.99. How much does each book cost?

12. **CELL PHONES** Refer to the poster at the right. If your bill for one month was $113.74, find the number of minutes you used.

13. **TICKETS** It costs $13 for admission to an amusement park, plus $1.50 for each ride. If you have a total of $35.50 to spend, what is the greatest number of rides you can go on?

14. **MONUMENTS** From ground level to the tip of the torch, the Statue of Liberty and its pedestal are 92.99 meters high. The pedestal is 0.89 meter higher than the statue. How high is the Statue of Liberty?

15. ANIMALS Refer to the information at the left.

   a. The top speed of a peregrine falcon is 20 miles per hour less than three times the top speed of a cheetah. What is the cheetah's top speed?

   b. A sailfish can swim up to 1 mile per hour less than one fifth the top speed of a peregrine falcon. Find the top speed that a sailfish can swim.

   c. The peregrine falcon can reach speeds about 14 miles per hour more than 7 times the speed of the fastest human. What is the approximate top speed of the fastest human?

16. BASEBALL Trey went to the batting cages to practice hitting. He rented a helmet for $4 and paid $0.75 for each group of 20 pitches. If he spent a total of $7 at the batting cages, how many groups of pitches did he pay for?

17 SNOWBOARDING Elsie would like to take snowboarding lessons at Powder Mountain. She has saved $550 for lessons and a junior season pass. How many more semi-private lessons than private lessons can she take?

Powder Mountain Ski Resort Snowboarding Lessons	
Semi-Private	$45/lesson
Private	$60/lesson
Junior Season Pass	$315

18. ALGEBRA Three consecutive even integers can be represented by n, $n + 2$, and $n + 4$. If the sum of three consecutive even integers is 36, what are the integers?

19. FINANCIAL LITERACY Hunter and Amado are each trying to save $600 for a summer trip. Hunter started with $150 and earns $7.50 per hour working at a grocery store. Amado has nothing saved, but he earns $12 per hour painting houses.

   a. Make a conjecture about who will take longer to save enough money for the trip. Justify your reasoning.

   b. Write and solve two equations to check your conjecture.

Write a problem that could be solved using each equation.

20. $4x + 20 = 70$ **21.** $2x - 6 = 25$

H.O.T. Problems

22. OPEN ENDED If 12 less than 4 times a number is 8, the number is 5. Write a different sentence where the unknown number is also 5.

23. CHALLENGE The ages of three siblings combined is 27. The oldest is twice the age of the youngest. The middle child is 3 years older than the youngest. Write and solve an equation to find the ages of each sibling.

24. **WRITE MATH** Write about a real-world situation that can be solved using a two-step equation. Then write the equation and solve the problem.

25. A company employs 72 workers. It plans to increase the number of employees by 6 per month until it has twice its current workforce. Which equation can be used to determine m, the number of months it will take for the number of employees to double?

   A. $6m + 72m = 144$

   B. $2m + 72 = 144$

   C. $2(6m + 72) = 144$

   D. $6m + 72 = 144$

26. What is the value of x in the following equation?

$$-3x + 4 = -23$$

   F. -9

   G. $-6\frac{1}{3}$

   H. $6\frac{1}{3}$

   I. 9

27. Kimberly needs $45 to go to the amusement park. She has $13. She earns $8 per hour working at her job. The equation $8h + 13 = 45$ shows this relationship. How many hours does Kimberly need to work to earn enough money to go to the park?

   A. 8   **C.** 6

   B. 7   **D.** 4

28. THINK SOLVE EXPLAIN **SHORT RESPONSE** The table shows the number of baseball cards in two baseball collections.

Person	Cards
Marcus	m
James	$2m + 6$

If Marcus and James have 120 cards altogether, write and solve an equation that could be used to find the number of cards in Marcus' collection.

Solve each equation. Check your solution. (Lesson 2B)

29. $5x + 2 = 17$

30. $-7b + 13 = 27$

31. $-6 = \frac{n}{8} + 1$

32. $-15 = -4p + 9$

ALGEBRA **Solve each equation. Check your solution.** (Lesson 1D)

33. $\frac{y}{7} = 22$

34. $4p = -60$

35. $20 = \frac{t}{15}$

36. $81 = -3d$

37. $\frac{a}{6} = -108$

38. $-4n = -96$

39. FOOTBALL In a recent NFL game, the Green Bay Packers scored 14 points less than the Tennessee Titans. Write and solve an equation to find the total points the Tennessee Titans scored.
(Lesson 1C)

Preseason Week 4	
Team	**Total Points**
Packers	17
Titans	p

Define a variable. Then write an equation to model each situation.
(Lesson 1B)

40. After collecting $15 for charity, Eduardo had a total of $50.

41. A group of 200 students were separated into equal-size groups to ride the bus. Fifty people rode each bus.

Mid-Chapter Check

1. **TIME** Lo arrived home at 5:45 P.M. from the store. At the store, she spent a half hour trying on clothes, 15 minutes looking at shoes, and 10 minutes waiting in line. If it took her 35 minutes to walk home, what time did she arrive at the store? Use the *work backward* strategy. (Lesson 1A)

Write an equation to model each situation.
(Lesson 1B)

2. Ten increased by a number is −8.

3. The difference of −5 and a number is 12.

4. The product of a number and 4 is 32.

Solve each equation. Check your solution.
(Lesson 1C)

5. $7 + a = 15$

6. $23 = d + 44$

7. $28 = n - 14$

8. $t - 22 = -31$

9. **STOCKS** The changes in the price of a certain stock each day are shown.

Day	Change
Monday	−$2.25
Tuesday	+$0.50
Wednesday	+$1.50
Thursday	+$0.75

The overall change for the week was −$0.50. Write and solve an equation to find the change in price on Friday. (Lesson 1C)

Solve each equation. Check your solution.
(Lesson 1D)

10. $42 = -14x$

11. $144 = 18a$

12. $\frac{n}{3} = 7$

13. $-6 = \frac{t}{9}$

Solve each equation. Check your solution.
(Lesson 2B)

14. $3m + 5 = 14$

15. $-2k + 7 = -3$

16. $11 = \frac{1}{3}a + 2$

17. $-15 = -7 - p$

18. **MULTIPLE CHOICE** Jane charges $5 per hour for babysitting. Last week, she earned a total of $55. Which of the following equations could be used to find the number of hours h she babysat last week? (Lesson 1D)

A. $h \div 55 = 5$

B. $5h = 55$

C. $5 + h = 55$

D. $55 - h = 5$

19. **EXERCISE** Brandi rode her bike the same distance on Tuesday and Thursday, and 20 miles on Saturday for a total of 50 miles for the week. Solve the equation $2m + 20 = 50$ to find the distance Brandi rode on Tuesday and Thursday. (Lesson 2B)

20. **MULTIPLE CHOICE** A diagram of a room is shown.

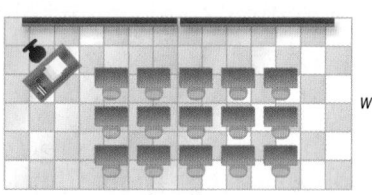

If the perimeter of the room is 78 feet, find the value of w. (Lesson 2B)

F. 12 ft   H. 25 ft

G. 15 ft   I. 27 ft

21. **TEXT MESSAGES** A cell phone company charges $45 a month for service and $0.12 extra for each text message. Ms. Barnes was charged $49.32 last month. Write and solve an equation to find the number of text messages she sent. (Lesson 2C)

Main Idea

Write and graph inequalities.

Vocabulary

inequality

Graph Inequalities

POSTAL SERVICE Iko wants to mail square invitations to a birthday party. Square envelopes must be 5 inches by 5 inches or *greater*. She will pay $0.44 in postage for every invitation that weighs 1 ounce or *less*.

First-Class Mail Rates

Weight not over (ounces)	Rate
1	$0.44
2	$0.61
3	$0.78
3.5	$0.95

1. List three envelope sizes that Iko can use.

2. How much will it cost to mail an invitation that weighs 2.5 ounces?

An **inequality** is a mathematical sentence that contains > or <. When used to compare a variable and a number, inequalities can describe a range of values. The symbols ≤ and ≥ combine < and > with part of the equals sign.

Inequalities				
Words	• is less than • is fewer than	• is greater than • is more than • exceeds	• is less than or equal to • is no more than • is at most	• is greater than or equal to • is no less than • is at least
Symbols	<	>	≤	≥

 **EXAMPLES** Write Inequalities

Write an inequality for each sentence.

1 A suitcase must weigh less than 40 pounds.

Let w = suitcase's weight.
$w < 40$

2 You must be older than 12 years old to play.

Let a = person's age.
$a > 12$

Read Math

Inequality Symbols

≤ less than or equal to

≥ greater than or equal to

3 You must be at least 48 inches tall to ride the roller coaster.

Let h = person's height.
$h \geq 48$

4 You must be 12 years of age or younger to order from the children's menu.

Let a = person's age.
$a \leq 12$

a. You must be older than 17 to see the movie.

b. You must be 18 years or older to vote.

c. A fuel tank holds at most 16 gallons of gasoline.

Inequalities with variables are open sentences. When the variable is replaced with a number, the inequality becomes either true or false.

EXAMPLES **Determine the Truth of an Inequality**

State whether each inequality is *true* or *false* for the given value.

(5) $a + 2 > 8$, $a = 5$

$a + 2 > 8$ Write the inequality.

$5 + 2 \overset{?}{>} 8$ Replace a with 5.

$\quad 7 \not> 8$ Simplify.

Since 7 is not greater than 8, $7 > 8$ is false.

(6) $10 \leq 7 - x$, $x = -3$

$10 \leq 7 - x$ Write the inequality.

$10 \overset{?}{\leq} 7 - (-3)$ Replace x with -3.

$10 \leq 10$ Simplify.

While $10 < 10$ is false, $10 = 10$ is true, so $10 \leq 10$ is true.

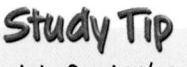 **Study Tip**

Symbols Read $7 \not> 8$ as 7 is not greater than 8.

 **CHECK Your Progress**

d. $n - 6 < 15$, $n = 18$ e. $-3p \geq 24$, $p = 8$ f. $-2 > 5y - 7$, $y = 1$

Inequalities can be graphed on a number line. An open or closed dot is used to indicate where they begin. An arrow to the left or to the right is used to show that they continue in the indicated direction.

EXAMPLES **Graph an Inequality**

Graph each inequality on a number line.

(7) $n < 3$

Place an open dot at 3. Then draw a line and an arrow to the left.

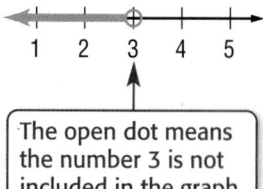

The open dot means the number 3 is not included in the graph.

(8) $n \geq 3$

Place a closed dot at 3. Then draw a line and an arrow to the right.

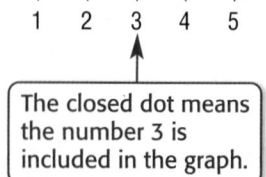

The closed dot means the number 3 is included in the graph.

 **CHECK Your Progress**

g. $x > 2$ h. $x < 1$ i. $x \leq 5$ j. $x \geq -4$

Examples 1–4   Write an inequality for each sentence.

1. Your speed must be 55 miles per hour or less.

2. The game is recommended for ages greater than 6.

Examples 5 and 6   State whether each inequality is *true* or *false* for the given value.

3. $x - 11 < 9, x = 20$ **4.** $42 \geq 6a, a = 8$ **5.** $\frac{n}{3} + 1 \leq 6, n = 15$

Examples 7 and 8   Graph each inequality on a number line.

6. $n > 4$ **7** $p \leq 2$ **8.** $x \geq 0$ **9.** $a < 7$

Practice and Problem Solving

 = **Step-by-Step Solutions** begin on page R1.
Extra Practice begins on page EP2.

Examples 1–4   Write an inequality for each sentence.

10. To be a U.S. senator, you must be 30 years of age or older.

11. For a group of 10 or more, an 18% tip is already included.

12. The maximum occupancy must be less than 512 people.

13. The phone costs no more than $25.

14. You must spend more than $50 to receive a discount.

15. The heavyweight division is greater than 200 pounds.

Examples 5 and 6   State whether each inequality is *true* or *false* for the given value.

16. $12 + a < 20, a = 9$ **17** $15 - k > 6, k = 8$ **18.** $-3y < 21, y = 8$

19. $32 \leq 2x, x = 16$ **20.** $\frac{n}{4} \geq 5, n = 12$ **21.** $\frac{-18}{x} > 9, x = -2$

Examples 7 and 8   Graph each inequality on a number line.

22. $x > 6$ **23.** $a > 0$ **24.** $y < 8$ **25.** $h < 2$

26. $w \leq 3$ **27.** $p \geq 7$ **28.** $1 \leq n$ **29.** $4 \geq d$

30. SPORTS The graph shows the number of children ages 5–14 recently treated in U.S. emergency rooms.

a. In which sport(s) were more than 150,000 children injured?

b. In which sport(s) were at least 75,000 children injured?

c. Of the sports listed, which have fewer than 125,000 injuries?

d. Write an inequality comparing the number treated for soccer-related injuries with those treated for football-related injuries.

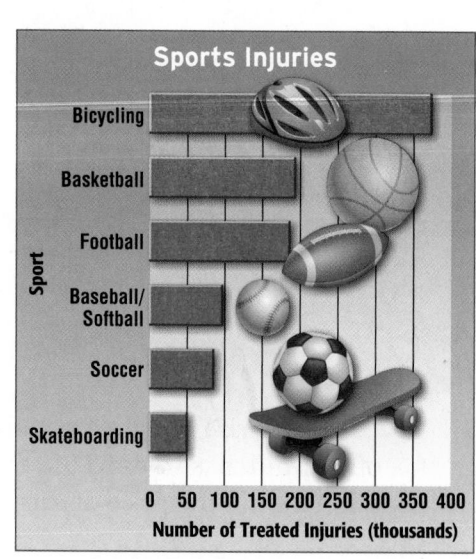

Sports Injuries

Bicycling

Basketball

Football

Baseball/
Softball

Soccer

Skateboarding

Sport

0 50 100 150 200 250 300 350 400
Number of Treated Injuries (thousands)

31. FIND THE ERROR Roberto is writing an inequality for the expression *at least 2 hours of homework*. Find his mistake and correct it.

$h \leq 2$

32. CHALLENGE If $x = 3$, is the following inequality *true* or *false*? Explain.

$$\frac{108}{12} + x \geq 15 - 4x + 9$$

33. ✏️ **WRITE MATH** If $a < b$ and $b < c$, what is true about the relationship between a and c? Explain your reasoning and give examples using both positive and negative values for a, b, and c.

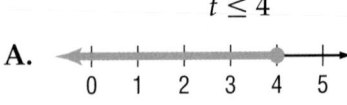

Test Practice

34. Conner can spend no more than 4 hours at the swimming pool today. Which graph represents the time that Conner can spend at the pool?

$t \leq 4$

A.
```
  <──┼──┼──┼──┼──●──┼──>
     0  1  2  3  4  5
```

B.
```
  <──┼──┼──┼──┼──⊕──┼──>
     0  1  2  3  4  5
```

C.
```
  <──┼──●──┼──┼──┼──┼──>
     3  4  5  6  7  8
```

D.
```
  <──┼──⊕──┼──┼──┼──┼──>
     3  4  5  6  7  8
```

35. Which inequality matches the sentence below?

Members must be 18 years of age or older.

F. $m > 18$

G. $m \geq 18$

H. $m < 18$

I. $m \leq 18$

36. 📋 **SHORT RESPONSE** Graph the inequality $x < -8$ on a number line.

Spiral Review

37. SHOPPING Marisa bought 4 paperback books, each at the same price. The tax on her purchase was $2.35, and the total was $34.15. Write and solve an equation to find the price of each book. (Lesson 2C)

ALGEBRA Solve each equation. (Lesson 2B)

38. $9 + 5y = 19$

39. $-6 = 4 + 2x$

40. $8 - k = 17$

41. $2 = 18 - 4d$

Main Idea

Solve and graph one-step inequalities by using the Addition or Subtraction Properties of Inequality.

Solve Inequalities by Addition or Subtraction

WEATHER The temperature at 7 A.M. in three Florida cities is shown. The temperature in Miami is less than the temperature in Jacksonville.

Temperature (°F)	
Jacksonville	82°
Miami	78°
Tampa	76°

1. If the temperature rises 5° in both cities, will this still be true?

2. Would it be colder in Tampa or Jacksonville if the temperature in both cities dropped 10°? Explain.

The examples above demonstrate properties of inequalities.

Key Concept   **Properties of Inequality**

Words	When you add or subtract the same number from each side of an inequality, the inequality remains true.
Symbols	For all numbers, a, b, and c,
	1. if $a > b$, then $a + c > b + c$ and $a - c > b - c$.
	2. if $a < b$, then $a + c < b + c$ and $a - c < b - c$.
Examples	$2 > -3$ $3 < 8$
	$2 + 5 > -3 + 5$ $3 - 4 < 8 - 4$
	$7 > 2$ ✓ $-1 < 4$ ✓

These properties are also true for $a \geq b$ and $a \leq b$.

Solving an inequality means finding values that make the inequality true.

Study Tip

Checking Solutions To check the solution of an inequality, replace the variable with a value that satisfies the solution. In Example 1, replace n with 22.
$n - 8 < 15$
$22 - 8 < 15$
$14 \overset{?}{<} 15$
Since $14 < 15$, the statement is true.

EXAMPLES Solve Inequalities

1 Solve $n - 8 < 15$. Graph the solution set on a number line.

$$n - 8 < \quad 15 \qquad \text{Write the inequality.}$$
$$\underline{+8 \quad +8} \qquad \text{Addition Property of Inequality}$$
$$n < 23 \qquad \text{Simplify.}$$

Graph the solution set.

<--+---+---+---+---+---+--⊕--+---+---+---+---+---+-->
  18 19 20 21 22 23 24 25 26 27 28

Draw an open dot at 23 with an arrow to the left.

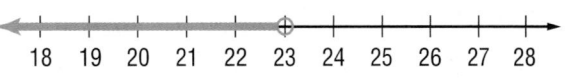

2 **Solve $-4 \geq a + 7$. Graph the solution set on a number line.**

$$-4 \geq a + 7 \qquad \text{Write the inequality.}$$

$$\underline{-7 \qquad -7} \qquad \text{Subtraction Property of Inequality}$$

$$-11 \geq a \text{ or } a \leq -11 \qquad \text{Simplify.}$$

The solution is $a \leq -11$.

Study Tip

Equivalent Inequalities If −11 is greater than or equal to a, then a is less than or equal to −11.

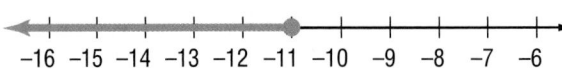

Draw a closed dot at −11 with an arrow to the left.

CHECK Your Progress

Solve each inequality. Graph the solution set on a number line.

a. $t + 3 > 12$ **b.** $n + \dfrac{1}{2} \geq 4$ **c.** $y - 1.5 < 2$

Real-World Link

The Airbus A380 is one of the largest airplanes in the world.

 REAL-WORLD EXAMPLE

3 **PLANES** The Airbus A380 can seat up to 853 passengers. Suppose there are currently 632 passengers boarded on the airplane. Write and solve an inequality to determine how many more people are able to board.

The phrase *up to* means *less than or equal to*. Let $p =$ the number of passengers left to board.

Estimate $850 - 650 = 200$

Current passengers	plus	passengers left to board	is less than or equal to	853 total passengers.
↓	↓	↓	↓	↓
632	+	p	≤	853

$$632 + p \leq 853 \qquad \text{Write the inequality.}$$

$$\underline{-632 \qquad\quad -632} \qquad \text{Subtraction Property of Inequality}$$

$$p \leq 221 \qquad \text{Simplify.}$$

Check for Reasonableness $200 \approx 221$ ✓

At most, 221 people are able to board.

CHECK Your Progress

d. **WEATHER** An F1 tornado has wind speeds that are at least 73 miles per hour. An F2 tornado has wind speeds that are at least 113 miles per hour. Write and solve an inequality to determine how much the winds of an F1 tornado need to increase to become at least an F2 tornado.

Examples 1 and 2   **Solve each inequality. Graph the solution set on a number line.**

1. $b + 5 > 9$ **2.** $12 + n \leq 4$ **3.** $x - 4 < 10$

Example 3   **4. WEATHER** Concord, New Hampshire, receives an average of 37 inches of precipitation per year, and there has been 13 inches so far this year. Write and solve an inequality to determine how much more precipitation Concord can get and stay at or below the average.

Practice and Problem Solving

 = **Step-by-Step Solutions** begin on page R1.
Extra Practice begins on page EP2.

Examples 1 and 2   **Solve each inequality. Graph the solution set on a number line.**

5 $5 + x \leq 18$ **6.** $10 + n \geq -2$ **7.** $-4 < k + 6$

8. $3 < y + 8$ **9.** $c + 10 < 9$ **10.** $g - 4 \geq 13$

11. $2 + m \geq 3.5$ **12.** $q + 0.8 \leq -0.5$ **13.** $v - 6 > 2.7$

14. $p - 4.8 > -6$ **15.** $d + 1\frac{2}{3} \leq \frac{1}{2}$ **16.** $5 > f + 1\frac{1}{4}$

Example 3   **17 BASKETBALL** Amos, who is 15 years old, is thinking about joining the City Basketball League. Write and solve an inequality to determine how many years until he is able to join.

Join the City Basketball League

Division A (ages 18–23)
Division B (ages 24–29)
Division C (ages 30 and up)

18. ANIMALS Hippos weigh up to 5,300 pounds. Write and solve an inequality that describes how much weight a young hippo could gain if its current weight is 2,200 pounds.

19. 🔀 MULTIPLE REPRESENTATIONS Keisha wants to compare the weights of one cup of different substances. On one side of a balance she has 4.5 ounces of flour. On the other side of the balance, she has 8 ounces of water.

a. PICTORIAL Draw a picture to represent the situation.

b. ALGEBRAIC Write an inequality to represent the balance.

c. NUMERICAL Suppose Keisha added a second cup of each substance to the scale. Write an inequality to represent the balance.

d. ALGEBRAIC Write an inequality to represent the situation if Keisha adds x cups of each substance to each side of the balance.

20. MATH IN THE MEDIA Find two related numbers in a newspaper or magazine, on television, or on the Internet. Write a real-world problem in which you would solve an inequality.

CHALLENGE Determine whether each equation or inequality has no solution, one solution, or more than one solution.

21. $y - y = 0$ **22.** $x + 4 = 9$ **23.** $x + 4 > 9$ **24.** $y > y + 1$

25. OPEN ENDED Write two different inequalities that each have the solution $x < 9$. One inequality should be solved using addition properties, and the other should be solved using subtraction properties.

26.  **WRITE MATH** Write a word problem that has the solution $w \leq 200$.

Test Practice

27. [THINK SOLVE EXPLAIN] **SHORT RESPONSE** Kali has $80 to go clothes shopping. She spends $25 on a new shirt. Write an inequality that represents how much money she has left to spend on clothes.

28. The solution set of which inequality is represented in the number line below?

$$\xleftarrow{\quad\begin{array}{cccccccccc} + & + & + & + & \bullet & + & + & + & + \\ 4 & 5 & 6 & 7 & 8 & 9 & 10 & 11 & 12 \end{array}\quad}\rightarrow$$

A. $x + 6 \leq 14$ **C.** $x + 6 < 14$
B. $x + 6 \geq 14$ **D.** $x + 6 > 14$

29. Which of the following shows the solution set of the inequality $w + 4 < 35$?

F. $\xleftarrow{\quad\begin{array}{ccccccccc} + & + & + & + & \bullet & + & + & + & + \\ 27 & 28 & 29 & 30 & 31 & 32 & 33 & 34 & 35 \end{array}\quad}\rightarrow$

G. $\xleftarrow{\quad\begin{array}{ccccccccc} + & + & + & \bullet & + & + & + & + & + \\ 27 & 28 & 29 & 30 & 31 & 32 & 33 & 34 & 35 \end{array}\quad}\rightarrow$

H. $\xleftarrow{\quad\begin{array}{ccccccccc} + & + & + & \oplus & + & + & + & + & + \\ 27 & 28 & 29 & 30 & 31 & 32 & 33 & 34 & 35 \end{array}\quad}\rightarrow$

I. $\xleftarrow{\quad\begin{array}{ccccccccc} + & + & \oplus & + & + & + & + & + & + \\ 27 & 28 & 29 & 30 & 31 & 32 & 33 & 34 & 35 \end{array}\quad}\rightarrow$

State whether each inequality is *true* or *false* for the given value. (Lesson 3A)

30. $18 - n > 4, n = 11$ **31.** $13 + x < 21, x = 8$ **32.** $34 \leq 5p, p = 7$

33. FOOTBALL In football, a touchdown with an extra point is worth 7 points and a field goal is worth 3 points. The winning team scored 27 points. The score consisted of two field goals, and the rest were touchdowns with extra points. Write and solve an equation to determine how many touchdowns the winning team scored. (Lesson 2C)

Solve each equation. Check your solution. (Lesson 2B)

34. $11 = 3x + 2$ **35.** $5 = -\dfrac{x}{2} - 1$ **36.** $20 = -2x + 4$

37. MONEY The table at the right shows how much Julianne earned per paycheck. She saved a total of $244.72. Write and solve an equation to find the amount she spent. (Lesson 1C)

Paycheck	Total Earnings ($)
1	272
2	298
3	304

Main Idea

Solve and graph one-step inequalities by using the Multiplication or Division Properties of Inequality.

Solve Inequalities by Multiplication or Division

COINS Lamar, Oscar, and Nick put the money from their pockets on the table. Lamar has more money than Nick. Will this still be true if each boy spends half of his money?

Name	Money	Amount
Lamar	1 dollar bill, 2 quarters, 2 dimes	$1.70
Oscar	1 dollar bill, 3 quarters, 1 dime, 1 nickel	$1.90
Nick	5 quarters, 1 dime, 1 nickel	$1.40

1. Divide each side of the inequality $1.70 > 1.40$ by 2. Is the resulting inequality *true* or *false*?

2. Oscar and Lamar each tripled their money by doing lawn work. Who has more now?

The example above demonstrates additional properties of inequality.

Key Concept · Properties of Inequality

Words When you multiply or divide each side of an inequality by a positive number, the inequality remains true.

Symbols For all numbers a, b, and c, where $c > 0$,

1. if $a > b$, then $ac > bc$ and $\frac{a}{c} > \frac{b}{c}$.

2. if $a < b$, then $ac < bc$ and $\frac{a}{c} < \frac{b}{c}$.

Examples

$$5 < 8 \qquad\qquad 2 > -10$$
$$4(5) < 4(8) \qquad \frac{2}{2} > \frac{-10}{2}$$
$$20 < 32 \qquad\qquad 1 > -5$$

These properties are also true for $a \geq b$ and $a \leq b$.

Study Tip

Checking Solutions You can check the solution in Example 1 by substituting numbers greater than -6 into the inequality and testing it to verify that it holds true.

✏️ 📷 EXAMPLES Solve Inequalities

① Solve $7y > -42$. Graph the solution set on a number line.

$7y > -42$ Write the inequality.

$\dfrac{7y}{7} > \dfrac{-42}{7}$ Division Property of Inequality

$y > -6$ Simplify.

```
 ←――――――――――○――――――――――→
  -11 -10 -9 -8 -7 -6 -5 -4 -3 -2 -1
```

Draw an open dot at -6 with an arrow to the right.

2 Solve $\frac{1}{3}x \leq 8$. Graph the solution set on a number line.

$$\frac{1}{3}x \leq 8 \qquad \text{Write the inequality.}$$

$$3\left(\frac{1}{3}x\right) \leq 3(8) \qquad \text{Multiplication Property of Inequality}$$

$$x \leq 24 \qquad \text{Simplify.}$$

The solution is $x \leq 24$.

Graph the solution set.

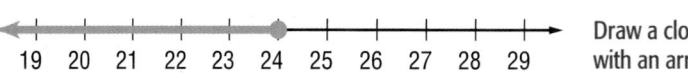

Draw a closed dot at 24 with an arrow to the left.

CHECK Your Progress

Solve each inequality. Graph the solution set on a number line.

a. $3a \geq 45$ **b.** $\frac{n}{4} < -16$ **c.** $81 \leq 9p$

What happens when each side of an inequality is multiplied or divided by a negative number?

Graph 3 and 5 on a number line.

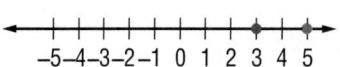

Since 3 is to the left of 5, $3 < 5$.

Multiply each number by -1.

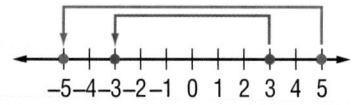

Since -3 is to the right of -5, $-3 > -5$.

The numbers being compared switched positions as a result of being multiplied by a negative number. In other words, their order reversed.

Study Tip

Common Error Do not reverse the inequality symbol just because there is a negative sign in the inequality, as in $7y < -42$. Only reverse the inequality symbol when you multiply or divide each side by a negative number.

Key Concept — Properties of Inequality

Words   When you multiply or divide each side of an inequality by a negative number, the direction of the inequality symbol must be reversed for the inequality to remain true.

Symbols   For all numbers a, b, and c, where $c < 0$,

1. if $a > b$, then $ac < bc$ and $\frac{a}{c} < \frac{b}{c}$.

2. if $a < b$, then $ac > bc$ and $\frac{a}{c} > \frac{b}{c}$.

Examples

$$8 > 5$$
$$-1(8) < -1(5) \qquad \text{Reverse the inequality symbols.}$$
$$-8 < -5$$

$$-3 < 9$$
$$\frac{-3}{-3} > \frac{9}{-3}$$
$$1 > -3$$

These properties are also true for $a \geq b$ and $a \leq b$.

192   Equations and Inequalities

Solve each inequality. Graph the solution set on a number line.

3 $\dfrac{a}{-2} \geq 8$

$\dfrac{a}{-2} \geq 8$ Write the inequality.

$-2\left(\dfrac{a}{-2}\right) \leq -2(8)$ Multiplication Property of Inequality; reverse inequality symbol.

$a \leq -16$ Simplify.

```
<---+---+---+---+---+---●---+---+---+---+---+--->
  -21 -20 -19 -18 -17 -16 -15 -14 -13 -12 -11
```

4 $-24 > -6n$

$-24 > -6n$ Write the inequality.

$\dfrac{-24}{-6} < \dfrac{-6n}{-6}$ Division Property of Inequality; reverse inequality symbol.

$4 < n$ or $n > 4$ Simplify.

```
+---+---+---+---+---○---+---+---+---+---+
-1   0   1   2   3   4   5   6   7   8   9
```

CHECK Your Progress

d. $\dfrac{c}{-7} < -14$ **e.** $-5d \geq 30$ **f.** $-3 \leq \dfrac{w}{-8}$

Real-World Link · · · ·

More than 400,000 people participate each year in marathons. The most popular marathon is the New York Marathon.

REAL-WORLD EXAMPLE

5 **MARATHONS** Dyani is training for a marathon by running no less than 45 kilometers per week. She runs at an average rate of 12 kilometers per hour. Write and solve an inequality to find the minimum number of hours she should run. Interpret the solution.

The phrase *no less than* means *greater than or equal to*. Let $h =$ the number of hours she runs in one week.

$12h \geq 45$ Write the inequality.

$\dfrac{12h}{12} \geq \dfrac{45}{12}$ Division Property of Inequality

$h \geq 3.75$ Simplify.

Dyani should run for at least 3.75 hours, or 3 hours 45 minutes, per week.

CHECK Your Progress

g. **GRADES** Each of the 20 questions on a math test is worth 3 points. Write and solve an inequality to find how many questions you must answer correctly to earn a score of at least 45 points. Interpret the solution.

Examples 1–4 **Solve each inequality. Graph the solution set on a number line.**

1. $3x > 12$ **2.** $\frac{1}{2} > \frac{1}{3}y$ **3.** $8x \leq -72$ **4.** $\frac{h}{4} \geq -6$

5 $-4y > 32$ **6.** $-56 \leq -7p$ **7.** $\frac{g}{-2} < -7$ **8.** $\frac{d}{-3} \geq -3$

Example 5 **9. MUSIC** Monique wants to practice the piano at least 6 hours each week. If she averages 1.5 hours per day, write and solve an inequality to find how many days she will have to practice to have at least 6 hours of practice per week. Interpret the solution.

Practice and Problem Solving

 = **Step-by-Step Solutions** begin on page R1.
Extra Practice begins on page EP2.

Examples 1–4 **Solve each inequality. Graph the solution set on a number line.**

10. $5x < 15$ **11.** $9n \leq 45$ **12.** $14k \geq -84$

13. $-12 > 3g$ **14.** $-100 \leq 50p$ **15.** $2y < -22$

16. $\frac{x}{9} \leq -3$ **17.** $\frac{n}{7} < -14$ **18.** $\frac{m}{-2} < -7$

19. $\frac{t}{-5} \leq -2$ **20.** $-8 \leq \frac{y}{0.2}$ **21.** $\frac{-1}{2}k > -10$

Example 5 **22. GYM MEMBERSHIP** A gym charges $5 each time you go, or you can buy a yearly membership for $190. Write and solve an inequality to find how many times a person should use the gym so that a yearly membership is less expensive than paying each time. Interpret the solution.

23 WORK Max charges $6 an hour to rake leaves. He is saving money for shoes that cost $89. Write and solve an inequality to find how many whole hours Max must work to buy the shoes. Interpret the solution.

24. GRAPHIC NOVEL Refer to the graphic novel frame below for Exercises a–b.

a. In order to win the game, a player must earn 50 points. How many more points does each player need to earn 50 points?

b. Write and solve an inequality to determine how many questions each player needs to answer correctly to earn a total of 50 points.

25. OPEN ENDED Write an inequality for the following sentence and then solve. Then name three numbers that are possible solutions. Explain.
The quotient of a number and −6 increased by 5 is at most 9.

26. CHALLENGE The product of an integer and −12 is less than −132. What is the least integer that meets this condition?

27. Which One Doesn't Belong? Identify the inequality that does not belong with the other three. Explain your reasoning.

| $-2m < 6$ | $-10 > -5x$ | $4a < -16$ | $-b > 12$ |

28.  **WRITE MATH** Explain when you should reverse the inequality symbol when solving an inequality.

✔ Test Practice

29. Which is a possible value of x if the area of the trapezoid is less than 256 square feet?

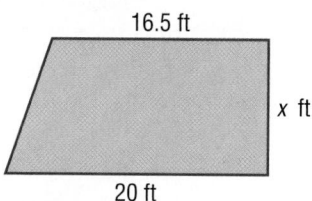

16.5 ft

x ft

20 ft

    A. 14   **C.** 16

    B. 15   **D.** 17

30. If $3n > 18$, then n could be which of the following values?

    F. 2   **H.** 6

    G. 4   **I.** 8

31. Which of the following shows the solution set of $\frac{x}{-4} \geq -3$?

A.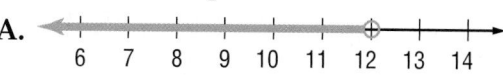

B.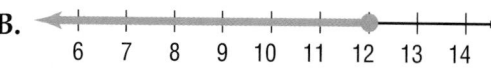

C.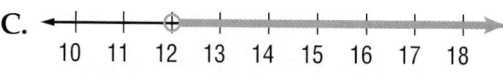

D.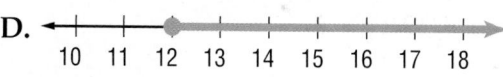

32. **THINK SOLVE EXPLAIN** **SHORT RESPONSE** Solve the inequality $\frac{-x}{5} \leq -6$.

Spiral Review

Solve each inequality. Graph the solution set on a number line. (Lesson 3B)

33. $y + 7 < 9$ **34.** $a - 5 < 2$ **35.** $j - 9 \geq -12$ **36.** $-14 > 8 + n$

Write an inequality for each sentence. (Lesson 3A)

37. A minimum speed on a certain highway is 45 miles per hour.

38. A hummingbird's wings can beat up to 200 times per second.

Sensational Sites

Are you someone who is artistic and loves computers? You might want to think about a career designing Web sites. Web designers combine their artistic abilities with technological skills to design Web sites that are well-organized, attractive, and easily accessible. Web designers must understand basic design principles and have strong technical knowledge of programming languages and computer design tools.

21st Century Careers

Are you interested in a career as a Web designer? Take some of the following courses in high school.

- Geometry
- Web Design
- Web Authoring-HTML
- Web Page Programming

 Get ConnectED

Web Design Tips

TIP 1	Black text on a light yellow or white background provides the best viewing conditions. Using light text on a dark background reduces the readability by 10 percent or more.
TIP 2	A large background image is at least 30 kilobytes and small seamless images are no more than 3 kilobytes. So, use the smaller images to produce faster Web page display.
TIP 3	To make pages easier to read, limit the width of the text display area so that it is less than 600 pixels.
TIP 4	Images on a Web page should have a resolution that is no higher than 72 pixels per inch.

Real-World Math

Use the tips in the table to solve each problem.

1. Refer to Tip 1. Write an inequality to describe the reduction in readability r when a Web site has a dark background.

2. Write inequalities to describe the size of a background image b and the size of a small seamless image s described in Tip 2.

3. Write an inequality to describe the text display area a in Tip 3. Then graph the inequality.

4. Write an inequality to describe the recommended image resolution r in Tip 4. Then graph the inequality.

5. A designer is creating a Web page using the small images that are described in Tip 2. She wants the total size of the images to be less than 18 kilobytes. Write and solve an inequality to determine how many images she can include on the page. Then graph the solution set on a number line.

Solve Two-Step Inequalities

NEWSPAPERS Kaitlyn is placing an ad in the newspaper for an arts and crafts show. The cost of placing an ad is shown in the table.

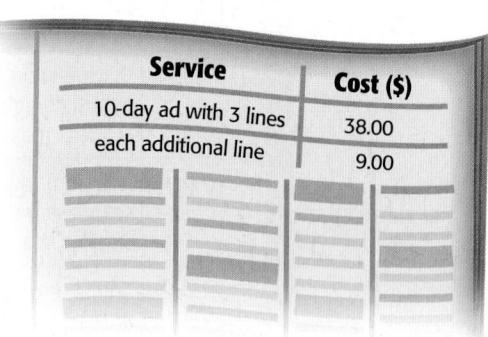

Service	Cost ($)
10-day ad with 3 lines	38.00
each additional line	9.00

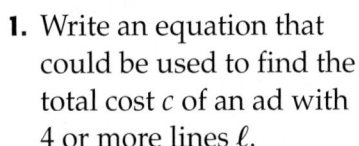

1. Write an equation that could be used to find the total cost c of an ad with 4 or more lines ℓ.

2. How much will it cost to place the ad if it is 5 lines long?

3. Suppose Kaitlyn can spend only $60 on the ad. Does she have enough money to place the ad?

A **two-step inequality** is an inequality that contains two operations. To solve a two-step inequality, use inverse operations to undo each operation in reverse order of the order of operations.

EXAMPLE Solve a Two-Step Inequality

① **Solve $3x + 4 \geq 16$. Graph the solution set on a number line.**

$3x + 4 \geq 16$ Write the inequality.

$3x + 4 \geq 16$
$\underline{-4 \quad -4}$ Subtraction Property of Inequality

$3x \geq 12$ Simplify.

$\dfrac{3x}{3} \geq \dfrac{12}{3}$ Division Property of Inequality

$x \geq 4$ Simplify.

Graph the solution set.

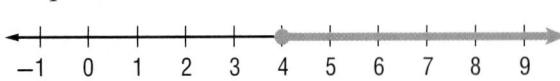

Draw a closed dot at 4 with an arrow to the right.

✓ **CHECK Your Progress**

Solve each inequality. Graph the solution set on a number line.

a. $2x + 8 > 24$ **b.** $5 + 4x < 33$ **c.** $9 + 2x > 15$

EXAMPLE Solve a Two-Step Inequality

2 Solve $7 - 2x > 11$. Graph the solution set on a number line.

$$7 - 2x > 11 \quad \text{Write the inequality.}$$

$$\underline{-7 \qquad -7} \quad \text{Subtraction Property of Inequality}$$

$$-2x > 4 \quad \text{Simplify.}$$

$$\frac{-2x}{-2} < \frac{4}{-2} \quad \text{Division Property of Inequality; reverse inequality symbol.}$$

$$x < -2 \quad \text{Simplify. Check your solution.}$$

Graph the solution set.

Draw an open dot at -2 with an arrow to the left.

CHECK Your Progress

Solve each inequality. Graph the solution set on a number line.

d. $\frac{x}{2} + 9 \geq 5$ **e.** $16 - 4x > 20$ **f.** $8 - \frac{x}{3} \leq 7$

● **Real-World Link** · · · ·
Ichiro Suzuki holds the major league record for most hits in a single season with 262 hits.

 REAL-WORLD EXAMPLE Two-Step Inequalities

3 **BASEBALL** Halfway through baseball season, Stewart has 34 hits. He averages 2 hits per game. Write and solve an inequality to find how many more games it will take for Stewart to have at least 61 hits, the school record. Interpret the solution.

Words	The number of hits plus two hits per game is at least 61.
Variable	Let g represent the number of games he needs to play.
Inequality	$34 \quad + \quad 2g \quad \geq \quad 61.$

$$34 + 2g \geq 61 \quad \text{Write the inequality.}$$

$$\underline{-34 \qquad \quad -34} \quad \text{Subtraction Property of Inequality}$$

$$2g \geq 27 \quad \text{Simplify.}$$

$$\frac{2g}{2} \geq \frac{27}{2} \quad \text{Division Property of Inequality}$$

$$g \geq 13.5 \quad \text{Simplify.}$$

Stewart should have at least 61 hits after 14 more games.

CHECK Your Progress

g. **DVDs** Joan has $250. DVDs cost $18.95 each. Write and solve an inequality to find how many DVDs she can buy and still have at least $50. Interpret the solution.

Examples 1 and 2 Solve each inequality. Graph the solution set on a number line.

1. $4x + 8 > 40$ **2.** $9x - 2 < 34$ **3.** $5x - 7 \geq 43$

4. $\frac{x}{3} + 10 < 12$ **⑤** $11 \leq 7 + \frac{x}{5}$ **6.** $-3 - x > 4$

Example 3 **7. FINANCIAL LITERACY** A rental car company charges $45 plus an additional $0.20 per mile to rent a car. If Mr. Lawrence does not want to spend more than $100 for his rental car, write and solve an inequality to find how many miles he can drive and not spend more than $100. Interpret the solution.

Practice and Problem Solving

 = **Step-by-Step Solutions** begin on page R1.
Extra Practice begins on page EP2.

Examples 1 and 2 Solve each inequality. Graph the solution set on a number line.

8. $6x + 14 \geq 20$ **9.** $5x + 6 > 71$ **10.** $6x - 3 > 33$

11. $16 > 2x + 4$ **12.** $4x - 15 \leq 5$ **13.** $4x - 13 < 11$

14. $\frac{x}{13} + 3 \geq 4$ **15.** $\frac{x}{5} - 2 > 1$ **16.** $\frac{x}{4} - 8 \leq 16$

17. $9 \leq \frac{x}{14} + 6$ **18.** $18.6 \geq \frac{x}{3} - 1.4$ **19.** $7.5 > 4.5 + \frac{x}{4}$

20. $-7 \leq \frac{x}{10} - 12$ **21.** $-73 \geq 15 + 11x$ **22.** $-1 \geq \frac{x}{-10} - 6$

23. $-3x - 11 \geq 34$ **24.** $-20 > -2x + 4$ **25.** $143 \leq -12x - 1$

Example 3 **26. ENTERTAINMENT** Tyler needs at least $205 for a new video game system. He has already saved $30. He earns $7 an hour at his part-time job. Write and solve an inequality to find how many hours he will need to work to buy the system. Interpret the solution.

㉗ BABYSITTING Catie is starting a babysitting business. She spent $26 to make signs and flyers to advertise. She charges an initial fee of $5 and then $3 for each hour of service. Write and solve an inequality to find the number of hours she will have to babysit to make a profit. Interpret the solution.

Solve each inequality. Graph the solution set on a number line.

28. $6y > 15 + y$ **29.** $-5g + 5 \geq -7 - 2g$ **30.** $10 - 3x \geq 25 + 2x$

Write an inequality for each sentence. Then solve the inequality and graph the solution set on a number line.

31. Three times a number increased by four is less than -62.

32. The quotient of a number and -5 increased by one is at most 7.

33. The quotient of a number and 3 minus two is at least -12.

34. The product of -2 and a number minus six is greater than -18.

35. OPEN ENDED Write a real-world example that could be solved by using the inequality $4x + 8 \geq 32$. Then solve the inequality.

36. CHALLENGE In five games, you score 16, 12, 15, 13, and 17 points. How many points must you score in the sixth game to have an average of at least 15 points?

37. ✏ WRITE MATH Solve $2x + 8 > 18$ and $2x + 8 \leq 18$. How are the inequalities and solutions similar? How are they different?

38. CHALLENGE Solve $-x + 6 > -(2x + 4)$. Then graph the solution set on a number line.

Test Practice

39. You want to purchase a necklace for $325. You have already saved $115 and can set aside $22 a week. Which inequality can be used to find the number of weeks it will take to save at least $325?

   A. $22w + 115 \geq 325$

   B. $22w + 115 \leq 325$

   C. $22 + 115w \leq 325$

   D. $22w + 115 < 325$

40. Which inequality represents *six less than three times a number is at least fifteen*?

   F. $3n - 6 \leq 15$

   G. $3n - 6 \geq 15$

   H. $3n - 6 < 15$

   I. $3n - 6 > 15$

41. Which of the following inequalities has the solution set shown below?

$$\xleftarrow{\hspace{3cm}} \mid \mid \mid \mid \mid \mid \mid \mid \xrightarrow{\hspace{2cm}}$$
$$-11\ -10\ -9\ -8\ -7\ -6\ -5\ -4\ -3$$

   A. $-2x - 5 < 7$ **C.** $-2x - 5 \leq 7$

   B. $-2x - 5 > 7$ **D.** $-2x - 5 \geq 7$

42. ☐ EXTENDED RESPONSE Dante has 60 baseball cards. This is at least six more than three times as many cards as Anna.

   Part A Write an inequality to represent the situation.

   Part B Solve the inequality from part a. Interpret the solution.

   Part C Graph the solution set on a number line.

43. STATISTICS In a recent year, the Boston Marathon had more than 2,600,000 spectators along its 26-mile route. Write and solve an inequality to find the average number of spectators per mile. Interpret the solution. (Lesson 3C)

Solve each inequality. Graph the solution set on a number line. (Lesson 3B)

44. $6 + x \leq 16$ **45.** $5 + n \geq -4$ **46.** $-3 < k + 8$ **47.** $9 < y + 12$

Main Idea
Write and graph compound inequalities in one variable.

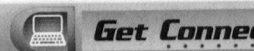

Vocabulary
compound inequality
intersection
union

Get ConnectED

Compound Inequalities

BIKES Jacob is saving money to buy a new mountain bike. He knows he will have to spend at least $150 and no more than $275 on the bike.

1. What is the least amount that Jacob will spend? the most?

2. Write two inequalities to represent the situation.

This situation can be written using a compound inequality. A **compound inequality** is two inequalities connected by the words *and* or *or*. In the above example, Jacob will spend at least $150 *and* no more than $275.

Compound inequalities containing *and* can be combined and written without using *and*. For example, $m \geq 150$ and $m \leq 275$ can be written as $150 \leq m \leq 275$.

REAL-WORLD EXAMPLE Write Compound Inequalities

1. **PLANTS** Rebecca bought some flowers to plant in her backyard. The description of the plant states that the flowers will grow to be at least 8 inches tall and no more than 15 inches tall. Write a compound inequality to represent this situation.

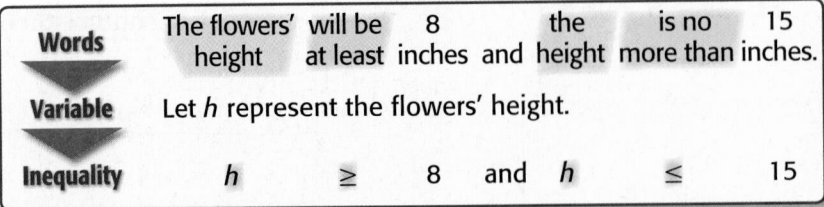

Words	The flowers' height	will be at least	8 inches	and	the height	is no more than	15 inches.
Variable	Let h represent the flowers' height.						
Inequality	h	$\geq$	8	and	h	$\leq$	15

Since $h \geq 8$ and $h \leq 15$, the compound inequality is $8 \leq h \leq 15$.

 **Your Progress**

a. **DRIVING** On a highway, you must drive less than 70 miles per hour and at least 45 miles per hour. Write a compound inequality to represent this situation.

Read Math

Inequalities The compound inequality $8 \leq h \leq 15$ is read *8 is less than or equal to h, which is less than or equal to 15.*

The graph of a compound inequality containing the word *and* is the intersection of the graphs of the two inequalities. The **intersection** is where both graphs overlap.

  **EXAMPLES** Graph an Intersection

Study Tip

Inequalities In Example 2, the open dot above *9* indicates that it is *not* included in the solution set. The closed dot above *4* indicates that it is included in the solution set.

② **Graph the solution set of $b < 9$ and $b \geq 4$.**

Graph both inequalities to find the intersection.

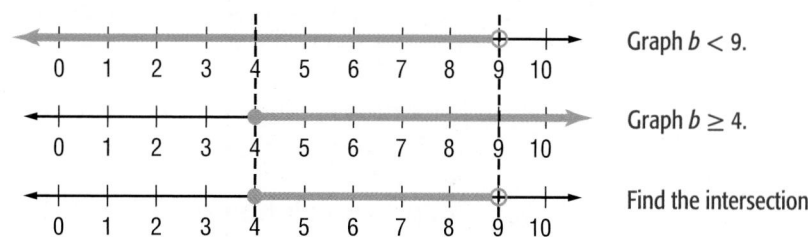

Graph $b < 9$.

Graph $b \geq 4$.

Find the intersection.

The intersection is all real numbers greater than or equal to 4 but less than 9.

✓ **CHECK Your Progress**

Graph the solution set of each compound inequality.

b. $m < 7$ and $m \geq 2$ **c.** $g > 5$ and $g \leq 12$

Some compound inequalities contain the word *or*. Compound inequalities containing the word *or* cannot be combined into one statement.

 **REAL-WORLD EXAMPLE** Write Compound Inequalities

③ **GAMES** In order to win a board game, Reggie must spin a number less than three or greater than seven. Write a compound inequality to represent this situation.

Let n represent the number.

So, the compound inequality is $n < 3$ or $n > 7$.

✓ **CHECK Your Progress**

d. ADMISSION An amusement park charges a reduced admission rate for people under 10 years of age or 60 years of age or older. Write a compound inequality to represent this situation.

Vocabulary Link

Everyday Use

Union a number of people associated together for a common purpose

Math Use

Union the combination of the elements of two or more sets

The graph of a compound inequality containing the word *or* is the union of the graphs of the two inequalities. The **union** is everything shown in both graphs.

 **EXAMPLES** Graph a Union

④ **Graph the solution set of $n \geq 4$ or $n \leq -2$.**

Graph both inequalities to find the union.

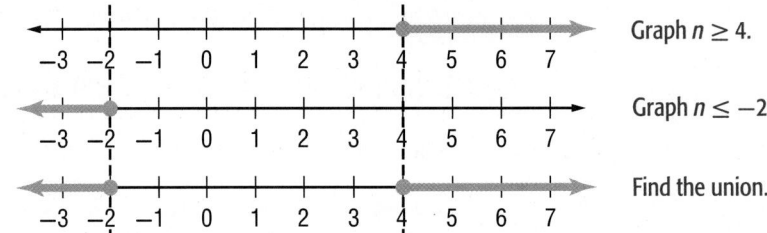

Graph $n \geq 4$.

Graph $n \leq -2$.

Find the union.

The union is all real numbers less than or equal to –2 or greater than or equal to 4.

⑤ **Graph the solution set of $c < 12$ or $c \leq 8$.**

Graph both simple inequalities to find the union.

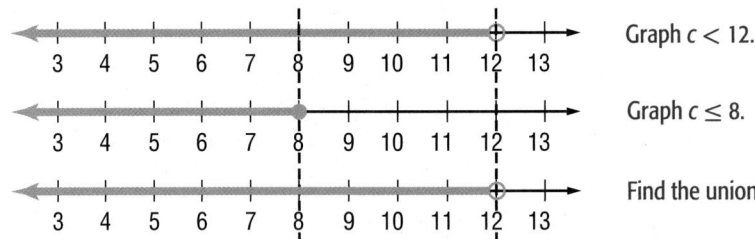

Graph $c < 12$.

Graph $c \leq 8$.

Find the union.

The union is all real numbers less than 12.

CHECK Your Progress

Graph the solution set of each compound inequality.

e. $d < 15$ or $d \geq 19$ **f.** $t > 3$ or $t > 6$

CHECK Your Understanding

Examples 1 and 3   ① **CARS** Tire pressure in a car is measured in pounds per square inch (psi). Standard tire pressure should be at least 28 psi but no more than 35 psi. Write a compound inequality to represent this situation.

2. PETS Alejandro wants to buy an iguana. He notices the card at the right that states the appropriate temperature range for the iguana's cage. Write a compound inequality to show the dangerous temperatures for the cage.

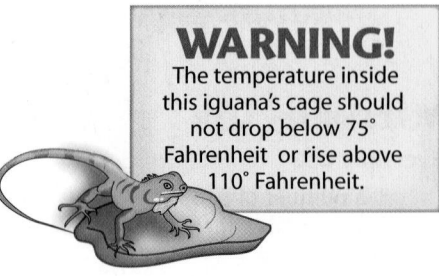

WARNING!
The temperature inside this iguana's cage should not drop below 75° Fahrenheit or rise above 110° Fahrenheit.

Examples 2, 4, and 5   **Graph the solution set of each inequality.**

3. $a \leq 13$ and $a > 8$ **4.** $m < -5$ and $m > -11$ **5.** $s \geq -2$ and $s < 6$

6. $b \geq 5$ or $b < 2$ **7.** $c < -4$ or $c \geq -1$ **8.** $r < 9$ or $r > 14$

Practice and Problem Solving

● = **Step-by-Step Solutions** begin on page R1.
Extra Practice begins on page EP2.

Examples 1 and 3

9. GAMES If a student in Mrs. Hansen's class can guess the exact number of jelly beans in a jar, they win the jar. She tells the students that there are more than 150 but less than 275 jelly beans in the jar. Write a compound inequality to represent this situation.

HURRICANES The Saffir-Simpson Hurricane Scale assigns a rating to a hurricane based on the hurricane's present wind speed.

Saffir-Simpson Hurricane Sale	
Category	**Wind Speed (mph)**
1	74-95
2	96-110
3	111-130
4	131-155
5	155 +

10. Write a compound inequality representing the wind speed of a category 3 hurricane.

11. Write a compound inequality representing the wind speed of a category 4 or 5 hurricane.

12. RIDES The sign for a roller coaster at the fair states that persons under 48 inches tall or over 72 inches tall cannot ride. Write a compound inequality showing the heights of people that cannot ride the roller coaster.

Examples 2, 4, and 5

Graph the solution set of each inequality.

13 $d > 8$ and $d < 21$ **14.** $f < 12$ and $f > 8$ **15.** $w > -7$ and $w \leq 0$

16. $h \leq 2$ and $h \geq -11$ **17.** $x \geq -15$ and $x < -2$ **18.** $n \leq -4$ and $n \geq -10$

19. $t \leq 12$ or $t < 9$ **20.** $k < 5$ or $k > 13$ **21.** $b > -2$ or $b > 5$

22. $p \geq 7$ or $p \leq -1$ **23.** $g \geq -6$ or $g > -1$ **24.** $r > -4$ or $r \leq -9$

Write a compound inequality for each graph.

25.

```
    ◀─┼──●──┼──┼──┼──┼──┼──┼──⊕──┼──▶
     −7 −6 −5 −4 −3 −2 −1  0  1  2  3
```

26.

```
    ◀─┼──┼──┼──⊕──┼──┼──┼──┼──●──┼──▶
      12 13 14 15 16 17 18 19 20 21 22
```

27.

```
    ◀─┼──┼──┼──●──┼──┼──┼──●──┼──┼──┼──▶
     −15 −14 −13 −12 −11 −10 −9 −8 −7 −6 −5
```

28.

```
    ◀─┼──┼──┼──┼──┼──⊕──┼──┼──┼──┼──┼──▶
     −2 −1  0  1  2  3  4  5  6  7  8
```

Write a compound inequality to represent each situation. Then graph the solution set on a number line.

29. Twenty-five is greater than a number, which is greater than or equal to 19.

30. A number is less than 15 but greater than or equal to 7.

31 BIOLOGY The table at the right shows the average weight for two types of endangered sea turtles.

    a. Write and graph a compound inequality for the average weight of each turtle.

    b. What is the union of the two graphs? the intersection?

Turtle	Average Weight
Loggerhead	200–350 lb
Green	300–350 lb

CHALLENGE Solve each compound inequality. Then graph the solution set.

32. $m - 6 \geq -1$ and $m + 4 < 16$

33. $p - 8 > -12$ and $p - 2 \leq 5$

34. $-2r - 3 \leq -13$ and $\frac{r}{3} < 6$

35. $3s + 4 \leq 7$ and $2s - 3 \geq -15$

36. OPEN ENDED Give an example of a compound inequality for which the solution set includes all numbers.

37.  **WRITE MATH** In your own words, write a couple of sentences that explain the difference between intersection and union.

Test Practice

38. The average length of a bald eagle is about 29 to 42 inches long, as shown in the inequality below.

$$29 \leq \ell \leq 42$$

Which is the correct graph?

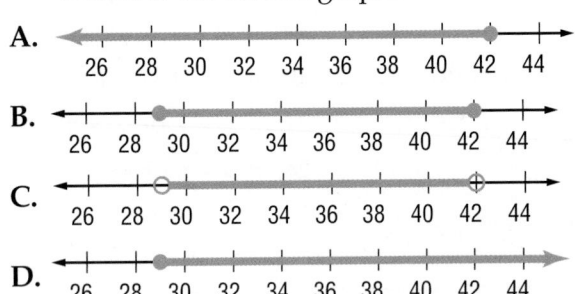

39. The weather forecaster predicts that the temperature would increase by no more than 10 degrees from today's high temperature to tomorrow's high temperature. If today's high temperature was 65°F, which of the inequalities below represents a possible high temperature range for tomorrow?

F. $65 \leq t \leq 70$

G. $65 \geq t \geq 75$

H. $65 \leq t \leq 75$

I. $65 \leq t \leq 80$

Spiral Review

Solve each inequality. Graph the solution set on a number line. (Lesson 4A)

40. $4h - 7 \leq 13$

41. $-3m + 5 > 17$

42. $-\frac{n}{4} + 3 \leq -2$

43. BABYSITTING You want to buy a pair of $42 skate shoes with the money that you make from babysitting. If you charge $5.25 an hour, write and solve an inequality to find how many whole hours you must babysit to buy the shoes.
(Lesson 3C)

44. FINANCIAL LITERACY The table shows the activity in Kirsten's bank account. (Lesson 1C)

a. Write an equation that represents the balance.

b. Kirsten owes the cable company $65. What is the minimum amount she must deposit in order to pay the bill?

Description	Amount
Beginning Balance	435
Gas Company	−75
Electric Company	−75
Phone Company	−100
Deposit	d
Rent	−200

Chapter Study Guide and Review

Be sure the following Key Concepts are noted in your Foldable.

Equations

Inequalities

Key Concepts

Solving Equations (Lessons 1 and 2)
- If you add or subtract the same number to/from each side of an equation, the two sides remain equal.
- If you multiply or divide each side of an equation by the same nonzero number, the two sides remain equal.
- To solve a two-step equation, undo each operation in reverse order.

Inequalities (Lesson 3)
- When used to compare a variable and a number, inequalities can describe a range of values.

Inequality Properties (Lesson 3)
- When you add or subtract the same number from each side of an inequality, the inequality remains the same.
- When you multiply or divide each side of an inequality by a positive number, the inequality remains true.
- When you multiply or divide each side of an inequality by a negative number, the direction of the symbol must be reversed for the inequality to be true.

Key Vocabulary

coefficient

compound inequality

defining a variable

equation

inequality

intersection

inverse operations

two-step equation

two-step inequality

union

Vocabulary Check

State whether each sentence is *true* or *false*. If *false*, replace the underlined word or number to make a true sentence.

1. Operations that "undo" each other are called <u>order of operations</u>.

2. A <u>two-step equation</u> is an equation that contains two operations.

3. The <u>intersection</u> of a compound inequality is where the graphs of both inequalities overlap.

4. An <u>inequality</u> is a mathematical statement stating that two quantities are equal.

5. A <u>compound inequality</u> is when two inequalities are connected by the words *and* or *or*.

6. Addition and <u>multiplication</u> are inverse operations.

7. When each side of an inequality is multiplied by a <u>negative</u> number, the inequality symbol must be reversed.

Multi-Part Lesson Review

Lesson 1 **One-Step Equations**

PSI: Work Backward (Lesson 1A)

Solve. Use the *work backward* strategy.

8. TRAVEL Malik's flight to Phoenix departs at 7:15 P.M. It takes 30 minutes to drive to the airport from his home, and it is recommended that he arrive at the airport 2 hours prior to departure. What time should Malik leave his house?

9. TICKETS After Candace purchased tickets to the play for herself and her two brothers, ticket sales totaled $147. If tickets were $5.25 each, how many tickets were sold before her purchase?

EXAMPLE 1 Fourteen years ago, Samuel's parents had their oldest child, Isabel. Six years later, Julia was born. If Samuel was born last year, how many years older than Samuel is Julia?

Since Samuel was born last year, he must be one year old. Since Isabel was born fourteen years ago, she must be fourteen years old. Since Julia was born six years after Isabel, she must be eight years old. This means that Julia is seven years older than Samuel.

Write Equations (Lesson 1B)

10. SPORTS An athlete's long jump attempt measured 670 centimeters. This was 5 centimeters less than her best jump. Define a variable. Then write an equation that could be used to find the measure of her best jump.

11. ALGEBRA Claire uses a copier to reduce the length of an image so it is $\frac{1}{4}$ of its original size. Given ℓ, the length of the image, write an equation to find the length n of the new image.

EXAMPLE 2 Tennessee became a state 4 years after Kentucky. Tennessee became a state in 1796. Write an equation that could be used to find the year Kentucky became a state.

Words Tennessee's year is 4 years after Kentucky's year.

Variable Let y represent Kentucky's year.

Equation $1796 = y + 4$

Solve Addition and Subtraction Equations (Lesson 1C)

Solve each equation. Check your solution.

12. $n + 40 = 90$ **13.** $x - 3 = 10$

14. $c - 30 = -18$ **15.** $9 = a + 31$

16. MOVIES Thirty-two people arrived at the movie during the previews. There were 150 people at the movie when it started. Write and solve an equation to find how many were there before the previews.

EXAMPLE 3 Solve $5 + k = 18$.

$5 + k = 18$ Write the equation.
$\underline{-5 \quad = -5}$ Subtraction Property of Equality
$k = 13$ Simplify.

One-Step Equations (continued)

Solve Multiplication and Division Equations (Lesson 1D)

Solve each equation. Check your solution.

17. $15x = -75$ **18.** $-4x = 52$

19. $\frac{s}{7} = 42$ **20.** $\frac{y}{-10} = -15$

21. MONEY Laura borrowed $168 from her father to buy clothes. She plans to pay $28 a month toward this debt. Write and solve an equation to find how many months it will take to repay her father.

22. CARS Mr. Mitchell bought 12 quarts of motor oil for $36. Write and solve an equation to find the cost of each quart of motor oil.

EXAMPLE 4 Solve $60 = 5t$.

$60 = 5t$ Write the equation.

$\frac{60}{5} = \frac{5t}{5}$ Division Property of Equality

$12 = t$ Simplify.

EXAMPLE 5 Solve $\frac{m}{-2} = 8$.

$\frac{m}{-2} = 8$ Write the equation.

$\frac{m}{-2}(-2) = 8(-2)$ Multiplication Property of Equality

$m = -16$ Simplify.

Two-Step Equations

Solve Two-Step Equations (Lesson 2B)

Solve each equation. Check your solution.

23. $2x + 5 = 17$ **24.** $4 = -3y - 2$

25. $\frac{c}{5} + 2 = 42$ **26.** $39 = 7a + 11$

27. ZOO Four adults spend $37 for admission and $3 for parking at the zoo. Solve the equation $4a + 3 = 40$ to find the cost of admission per person.

EXAMPLE 6 Solve $5h + 8 = -12$.

$5h + 8 = -12$ Write the equation.

$\underline{ -8 = -8}$ Subtraction Property of Equality

$5h = -20$ Simplify.

$\frac{5h}{5} = \frac{-20}{5}$ Division Property of Equality

$h = -4$ Simplify.

Write Two-Step Equations (Lesson 2C)

Translate each sentence into an equation. Then solve.

28. Six more than twice a number is -4.

29. The quotient of a number and 8, less 2, is 5.

30. MEDICINE Dr. Miles recommended that Jerome take 8 tablets on the first day and then 4 tablets each day until the prescription was used. The prescription contained 28 tablets. Write and solve an equation to find how many days Jerome will be taking tablets after the first day.

EXAMPLE 7 Translate the following sentence into an equation. Then solve.

6 less than 4 times a number is equal to 10.

6 less than	4 times a number	is	10.
	$4n - 6$	$=$	10

$4n - 6 = 10$ Write the equation.

$\underline{ +6 = +6}$ Addition Property of Equality

$4n = 16$ Simplify.

$\frac{4n}{4} = \frac{16}{4}$ Division Property of Equality

$n = 4$ Simplify.

Lesson 3 **One-Step Inequalities**

Graph Inequalities (Lesson 3A)

Write an inequality for each sentence.

31. Participants must be at least 12 years old to play.

32. No more than 15 people are at the party.

Graph each inequality on a number line.

33. $t < 2$ **34.** $g \geq 92$

35. NUTRITION A food can be labeled low fat only if it has no more than 3 grams of fat per serving. Write an inequality to describe low-fat foods.

EXAMPLE 8   All movie tickets are $9 and less. Write an inequality for this situation.

Let $t =$ the cost of a ticket.

$t \leq 9$

EXAMPLE 9   Graph the inequality $a < -4$ on a number line.

Place an open dot at -4. Then draw a line and an arrow to the left.

$$\begin{array}{cccccccccccc} & -9 & -8 & -7 & -6 & -5 & -4 & -3 & -2 & -1 & 0 & 1 \end{array}$$

Solve Inequalities by Addition or Subtraction (Lesson 3B)

Solve each inequality. Graph the solution set on a number line.

36. $b - 9 \geq 8$ **37.** $15 > 3 + n$

38. $t + \dfrac{1}{2} < 4$ **39.** $-12 < k - 3$

40. MOVING A moving company is loading a 920-pound piano into a service elevator. The elevator can carry a maximum of 1,800 pounds. Write and solve an inequality to determine how much additional weight the elevator can carry. Interpret the solution.

EXAMPLE 10   Solve $x - 7 > 3$. Graph the solution set on a number line.

$$x - 7 > \quad 3 \qquad \text{Write the inequality.}$$
$$\underline{+ 7 \qquad + 7} \qquad \text{Addition Property of Inequality}$$
$$x > 10 \qquad \text{Simplify.}$$

Graph the solution set.

Place an open dot at 10. Then draw a line and an arrow to the right.

$$\begin{array}{cccccccccccc} 5 & 6 & 7 & 8 & 9 & 10 & 11 & 12 & 13 & 14 & 15 \end{array}$$

Solve Inequalities by Multiplication or Division (Lesson 3C)

Solve each inequality. Graph the solution set on a number line.

41. $\dfrac{n}{4} < 6$ **42.** $\dfrac{k}{2} \leq 3$

43. $5x > 15$ **44.** $-56 \geq 8y$

45. GOLF Aubrey wants to spend less than $38.50 on new golf balls. Each box costs $11. What is the maximum number of boxes of golf balls that she can buy?

EXAMPLE 11   Solve $-2n \geq 26$. Graph the solution set on a number line.

$$-2n \geq 26 \qquad \text{Write the inequality.}$$
$$\dfrac{-2n}{-2} \leq \dfrac{26}{-2} \qquad \text{Division Property of Inequality}$$
$$n \leq -13 \qquad \text{Simplify.}$$

Place a closed dot at -13. Then draw a line and an arrow to the left.

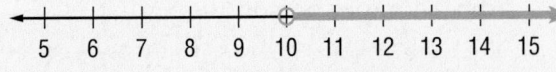

$$\begin{array}{cccccccccccc} -18 & -17 & -16 & -15 & -14 & -13 & -12 & -11 & -10 & -9 & -8 \end{array}$$

Two-Step Inequalities

Solve Two-Step Inequalities (Lesson 4A)

Solve each inequality. Graph the solution set on a number line.

46. $3n + 5 \geq 14$ **47.** $-4b - 8 > 12$

48. $\dfrac{x}{-6} + 7 \leq 11$ **49.** $7 - \dfrac{1}{5}y \leq 3$

50. $4w + 8 \leq -64$ **51.** $-6s + 21 > 15$

52. RENTALS Susan is renting a floating trampoline for a party. There is a delivery fee of $10 plus $3 per hour. Susan can spend at most $60. Write and solve an inequality to find how many hours she can rent the trampoline. Interpret the solution.

EXAMPLE 12 Solve $-3g - 5 > 16$. Graph the solution set on a number line.

$$
\begin{array}{ll}
-3g - 5 > 16 & \text{Write the inequality.} \\
\underline{+5 \ \ +5} & \text{Addition Property of Inequality} \\
-3g > 21 & \text{Simplify.} \\
\dfrac{-3g}{-3} < \dfrac{21}{-3} & \text{Division Property of Inequality} \\
g < -7 & \text{Simplify.}
\end{array}
$$

Graph the solution set.
Place an open dot at −7. Then draw a line and an arrow to the left.

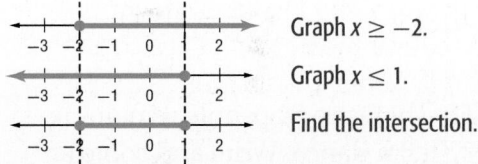

Compound Inequalities (Lesson 4B)

Graph the solution set of each inequality.

53. $d \leq 18$ and $d > 10$

54. $g > 5$ and $g < 10$

55. $m \leq -5$ and $m \geq -16$

56. $b < -4$ and $b \geq -12$

57. $x < 4$ or $x > 11$ **58.** $v \geq -6$ or $v \geq 0$

59. $w > 9$ or $w \geq 13$ **60.** $s < 12$ or $s > 20$

61. SOCCER An international standard soccer ball must weigh at least 14 ounces but no more than 16 ounces. Write a compound inequality to represent this situation.

62. MONEY A discounted rate for an admission ticket to an amusement park costs $25 for people under the age of 5 or over the age of 60. Write a compound inequality showing the ages that qualify for a discounted ticket.

EXAMPLE 13 Graph the solution set of $x \geq -2$ and $x \leq 1$.

Graph both inequalities to find the intersection.

Graph $x \geq -2$.

Graph $x \leq 1$.

Find the intersection.

The intersection is all real numbers greater than or equal to −2 and less than or equal to 1.

1. **JEANS** A store set the price it sold a pair of jeans for by tripling the amount it had paid for the jeans. After a month, the jeans were marked down by $5. Two weeks later, the price was divided in half. Finally, the price was reduced by $3, down to $14.99. How much did the store pay for the jeans?

2. **MULTIPLE CHOICE** A circle's radius is half its diameter. Which of the following equations could be used to find the radius r of a circle with diameter d?

   A. $2d = r$ **C.** $\frac{r}{2} = d$

   B. $d - 2 = r$ **D.** $\frac{d}{2} = r$

Solve each equation. Check your solution.

3. $x + 15 = -3$ 4. $-7 = a - 11$

5. $\frac{n}{-2} = 16$ 6. $-96 = 8y$

7. $3n + 18 = 6$ 8. $\frac{k}{2} - 11 = 5$

9. **DANCES** There are 54 people remaining after 37 left a dance. Write and solve an equation to find the number of people who were originally at the dance.

Translate each sentence into an equation. Then find each number.

10. Three more than twice a number is 15.

11. The quotient of a number and 6 plus 3 is 11.

Write an inequality for each sentence. Then graph the inequality on a number line.

12. A recordable DVD can hold at most 4.7 gigabytes of data.

13. Your score must be over 55,400 points to have the new high score.

Solve each inequality. Graph the solution set on a number line.

14. $-4 + x < 12$

15. $t - 15 \geq 20$

16. $-4 > \frac{c}{9}$

17. $6s \leq 36$

18. $8k + 3 \leq -5$

19. $7 - \frac{n}{3} < 4$

20. **PETS** A cage for a leopard gecko needs to maintain a temperature from 82°F to 88°F. Write a compound inequality to represent the situation. Then graph the solution set on a number line.

21. THINK SOLVE EXPLAIN **EXTENDED RESPONSE** A square pool containing a fountain is being designed for the city park. The pool needs to have a perimeter that is no more than 80 feet.

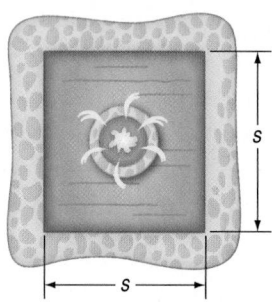

   Part A Write and solve an inequality to represent the situation. Interpret the solution.

   Part B Find the perimeter of 3 pools that would meet these guidelines.

   Part C Suppose the area of the pool must be between 200 and 400 square feet. Write a compound inequality to represent the situation.

   Part D What is a possible side length for the pool now?

Preparing for Standardized Tests

 Extended-Response: Scoring Points

In order to receive all possible points for an extended-response question, an answer must be correct and all work must be shown.

TEST EXAMPLE

A pet store sold 12 kittens for $150 each and 9 puppies at p dollars each. The store earned $4,275 on the sale of the kittens and puppies. Write and solve an equation that could be used to find the price p of one puppy.

Twelve kittens sold for $150 each. Their total cost was 12 × $150 or $1,800.

The 9 puppies were all sold at the same price, p. Their total cost can be written as 9 × p or 9p.

An equation that could be used to find the price p of each puppy is 1,800 + 9p = 4,275.

To find the price of the puppy, I can solve the equation for p.

$$
\begin{array}{rcl}
1,800 + 9p &=& 4,275 \\
-1,800 &=& -1,800 \\
\hline
9p &=& 2,475 \\
\dfrac{9p}{9} &=& \dfrac{2,475}{9} \\
p &=& 275
\end{array}
$$

The price of each puppy was $275.

> All the steps and calculations are shown. The reasoning is clearly stated.

A full-credit solution scores all possible points. A solution that is incomplete, does not include an explanation, or is partially correct receives partial credit.

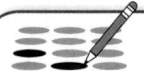 **Work on It**

It took Hannah 24 minutes to ride her bike 3 miles. She then rides her bike for 5 more miles at a different speed. Her entire trip lasts 54 minutes. Write and solve an equation that could be used to find the number of minutes m that it took her to ride each of the last five miles.

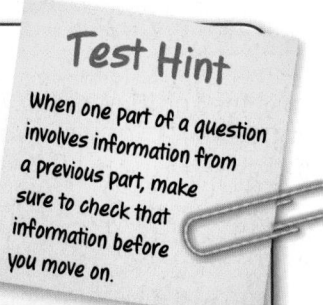

Test Hint

When one part of a question involves information from a previous part, make sure to check that information before you move on.

✓ Test Practice

Read each question. Then fill in the correct answer on the answer document provided by your teacher or on a sheet of paper.

1. Which of the following equations matches the description below?

> *Six more than the quotient of a number and three is 14.*

A. $14 = \frac{x}{3} + 6$

B. $6 = 14 + \frac{x}{3}$

C. $14 = \frac{x + 6}{3}$

D. $6 = \frac{x + 14}{3}$

2. In the inequality $3x + \$5,000 \leq \$80,000$, x represents the salary of an employee at a factory. Which phrase most accurately describes the employee's salary?

F. less than $25,000

G. more than $25,000

H. at least $25,000

I. at most $25,000

3. ✏️ **GRIDDED RESPONSE** The volume of a cube is given below. What is the length in feet of one side of the cube?

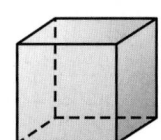

$V = 27 \text{ ft}^3$

4. What is the solution of the inequality $4n - 8 \leq 40$?

A. $n \leq 8$

B. $n \leq 12$

C. $n \geq 8$

D. $n \geq 12$

5. Clarence surveyed the students in his class about their favorite cafeteria food. The table shows the results of the survey. What percent of the students voted for pizza?

Favorite Cafeteria Food				
Food	Chili	Pizza	Chicken	Soup
Votes	3	12	6	3

F. 12%

G. 12.5%

H. 25%

I. 50%

6. 📝 THINK SOLVE EXPLAIN **SHORT RESPONSE** The sum of a number and 6 is 23. Write an equation to represent this situation.

7. The average weight of an adult male manatee is about 900 to 1,200 pounds, as shown in the inequality below.

$$900 \leq w \leq 1,200$$

Which is the correct graph of this inequality?

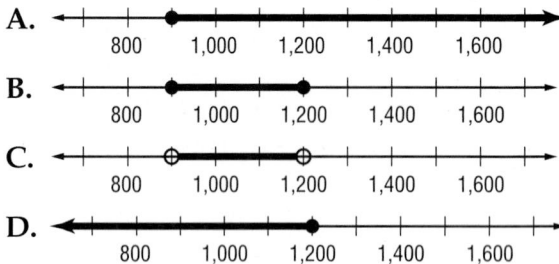

8. Zach, Luke, and Charlie ordered a large pizza for $11.99, breadsticks for $2.99, and chicken wings for $5.99. If the three friends agree to split the cost of the food evenly, about how much will each friend pay?

F. $20.79

G. $7.93

H. $7.32

I. $6.99

9. Stu bought the following items at the store.

Item	Cost ($)
Shoes	45
T-shirt	18
Hat	25
Socks	10

He had a coupon to save 25%. If the sales tax is 6.75%, how much did he pay for all four items?

A. $104.62
C. $78.46
B. $98
D. $73.50

10. After reading the salon prices listed below, Alex chose Special #1. She is going to tip her hairdresser 20% of the total cost. What is the total amount Alex will spend on her haircut?

Hair Salon Prices			
Trim	$12	**Special #1**	
Haircut	$19	Haircut, style, and	
Shampoo	$4	shampoo	$25
Style	$4	**Special #2**	
Highlights	$55	Haircut, style,	
		shampoo, and	
Perm	$50	highlights	$75

F. $25

G. $30

H. $32.40

I. $90

11. What is the sum of the fractions below in simplest form?

$$\frac{2}{3} + \frac{4}{5}$$

A. $\frac{7}{15}$
C. 1
B. $\frac{6}{8}$
D. $\frac{22}{15}$

12. **THINK SOLVE EXPLAIN** **SHORT RESPONSE** Silvia earns $8 per hour working at a landscaping company. She wants to earn at least $1,200 this summer. Write and solve an inequality to represent this situation.

13. Which inequality has the solution set shown below?

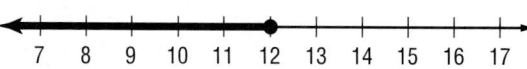

F. $3x \geq 36$
H. $3x > 36$
G. $3x \leq 36$
I. $3x < 36$

14. **THINK SOLVE EXPLAIN** **EXTENDED RESPONSE** Jocelyn and Shawn need to save at least $70 each to go to the amusement park. Jocelyn's parents gave her $20 and then she saves $10 a week. Shawn saves $10 a week.

Part A Write and solve an inequality to find the number of weeks it will take Jocelyn to save at least $70.

Part B Write and solve an inequality to find the number of weeks it will take Shawn to save at least $70.

Part C Graph the solution set for the inequalities from parts **a** and **b**.

NEED EXTRA HELP?														
If You Missed Question...	1	2	3	4	5	6	7	8	9	10	11	12	13	14
Go to Chapter-Lesson...	3-2B	3-4A	2-3A	3-4A	1-2B	3-1B	3-4B	1-1D	1-3A	1-2C	1-1B	3-3C	3-3C	3-4A

CHAPTER 4

Multi-Step Equations and Inequalities

connectED.mcgraw-hill.com

Investigate

 Animations

 Vocabulary

 Multilingual eGlossary

Learn

 Personal Tutor

 Virtual Manipulatives

 Graphing Calculator

 Audio

 Foldables

Practice

 Self-Check Practice

 Worksheets

 Assessment

The ★BIG Idea

Which properties of mathematics are used to solve multi-step equations and inequalities?

 FOLDABLES Study Organizer

Make this Foldable to help you organize your notes.

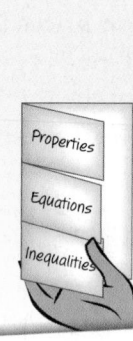

Properties

Equations

Inequalities

Review Vocabulary

inequality   desigualdad   a mathematical sentence that contains <, >, ≤, or ≥

-5 is less than 4.
$$-5 < 4$$

4 is greater than -5.
$$4 > -5$$

Key Vocabulary

English	Español
coefficient	coeficiente
Distributive Property	propiedad distributiva
property	propiedad
simplest form	forma reducida

When Will I Use This?

Are You Ready for the Chapter?

You have two options for checking prerequisite skills for this chapter.

Text Option Take the Quick Check below. Refer to the Quick Review for help.

QUICK Check

Solve each equation. Check your solution.

1. $n + 8 = -9$ **2.** $4 = m + 19$

3. $-4 + a = 15$ **4.** $z - 6 = -10$

5. $3c = -18$ **6.** $-42 = -6b$

7. $\dfrac{w}{4} = -8$ **8.** $12 = \dfrac{r}{-7}$

9. $0.25d = 130$ **10.** $48r = 12$

11. $0.4m = 22$ **12.** $0.02n = 9$

13. MARBLES Barry has 18 more marbles than Heidi. If Barry has 92 marbles, write and solve an equation to determine the number of marbles Heidi has.

Solve each inequality. Graph the solution set on a number line.

14. $a - 9 < 14$ **15.** $b + (-5) < -12$

16. $-7 + c \geq 47$ **17.** $d + 15 > -8$

18. $6e < 78$ **19.** $-18 < 3f$

20. $-5m \leq 125$ **21.** $8n > -40$

22. $\dfrac{g}{-3} > 24$ **23.** $\dfrac{h}{6} \leq -5$

24. $-7 < \dfrac{r}{5}$ **25.** $\dfrac{s}{-4} \geq -10$

26. ART Suppose an artist had more than 67 paintings and sold 34 of them at an art show. Write and solve an inequality that describes how many paintings the artist has left to sell.

QUICK Review

EXAMPLE 1

Solve $44 = k - 7$.

$44 = k - 7$ Write the equation.

$\underline{+7 = +7}$ Addition Property of Equality

$51 = k$

EXAMPLE 2

Solve $18 + m = 10$.

$18 + m = 10$ Write the equation.

$\underline{-18 = -18}$ Subtraction Property of Equality

$m = -8$

EXAMPLE 3

Solve $-4x \geq 12$. Graph the solution set on a number line.

$-4x \geq 12$ Write the inequality.

$\dfrac{-4x}{-4} \leq \dfrac{12}{-4}$ Division Property of Inequality

$x \leq -3$ Simplify.

Graph the solution set.

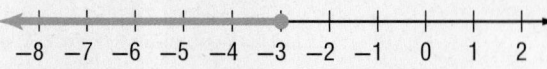

Draw a closed dot at -3 with an arrow to the left.

Online Option Take the Online Readiness Quiz.

 # Reading Math

Topic Sentences

Topic sentences are usually found near the beginning of the paragraph and are followed by supporting details. Here's the beginning of a paragraph about Mrs. Garcia's math class.

> A topic sentence is a sentence that expresses the main idea in a paragraph.

Topic sentence

> Mrs. Garcia's math class was doing research about wild horses living on public lands. They found that there are about 30,000 wild horses living in Nevada, 4,000 living in Wyoming, and 2,000 living in California.

In a word problem, the "topic sentence" is usually found near the end. It is the sentence or question that tells you what you need to find. Here's the same information, written as a word problem.

> Mrs. Garcia's math class was doing research about wild horses living on public lands. They found that there are about 30,000 wild horses living in Nevada, 4,000 living in Wyoming, and 2,000 living in California. Is the number of wild horses living on public lands in Nevada, Wyoming, and California greater than 35,000?

Topic sentence

When you start to solve a word problem, follow these steps.

STEP 1 Skim through the problem, looking for the "topic sentence."

STEP 2 Go back and read the problem more carefully, looking for the supporting details you need to solve the problem.

Practice

Refer to Lesson 2D. For each exercise below, write the "topic sentence." Do not solve the problem.

1. Exercise 20

2. Exercise 21

3. Exercise 22

4. Exercise 23

Main Idea

Identify and use mathematical properties to simplify algebraic expressions.

 Vocabulary

property
counterexample
simplify

Properties

WORK Mrs. Fuentes drives back and forth to work along the same route Monday through Friday.

1. She drives 25 miles to get to work from home. What is the distance she drives from her work to her home?

2. Does the distance change when she travels from home to work or from work to home?

A **property** is a statement that is true for any number. In the example above, Mrs. Fuentes is a *commuter* because she drives back and forth to work. The order in which she drives does not change the distance she drives. This is an example of the **Commutative Property**.

🗝 Key Concept · Commutative Properties

Words	The order in which numbers are added or multiplied does not change the sum or product.	
	Addition	**Multiplication**
Symbols	$a + b = b + a$	$a \cdot b = b \cdot a$
Examples	$6 + 1 = 1 + 6$	$7 \cdot 3 = 3 \cdot 7$

Sometimes it is easier to regroup numbers and use mental math to add or multiply. The **Associative Property** allows you to regroup numbers without changing the value.

🗝 Key Concept · Associative Properties

Words	The way in which numbers are grouped when they are added or multiplied does not change the sum or product.	
	Addition	**Multiplication**
Symbols	$a + (b + c) = (a + b) + c$	$a \cdot (b \cdot c) = (a \cdot b) \cdot c$
Examples	$2 + (3 + 8) = (2 + 3) + 8$	$3 \cdot (4 \cdot 5) = (3 \cdot 4) \cdot 5$

The following properties are also true for any numbers.

Key Concept   Number Properties

Property	Words	Symbols	Examples
Additive Identity	When 0 is added to any number, the sum is the number.	$a + 0 = a$ $0 + a = a$	$9 + 0 = 9$ $0 + 9 = 9$
Multiplicative Identity	When any number is multiplied by 1, the product is the number.	$a \cdot 1 = a$ $1 \cdot a = a$	$5 \cdot 1 = 5$ $1 \cdot 5 = 5$
Multiplicative Property of Zero	When any number is multiplied by 0, the product is 0.	$a \cdot 0 = 0$ $0 \cdot a = 0$	$8 \cdot 0 = 0$ $0 \cdot 8 = 0$

EXAMPLE   Identify Properties

① Name the property shown by the statement $2 \cdot (5 \cdot n) = (2 \cdot 5) \cdot n$.

The order of the numbers and variable did not change, but their grouping did. This is the Associative Property of Multiplication.

CHECK Your Progress

a. $42 + x + y = 42 + y + x$ **b.** $3x + 0 = 3x$

Vocabulary Link

Everyday Use

Conjecture a guess made without sufficient evidence

Math Use

Conjecture an informed guess based on known information

You may wonder if any of the properties apply to subtraction or division. If you can find a **counterexample**, an example that shows that a conjecture is false, the property does not apply.

EXAMPLE   Find a Counterexample

② State whether the following conjecture is *true* or *false*. If *false*, provide a counterexample.

Division of whole numbers is commutative.

Write two division expressions using the Commutative Property.

$15 \div 3 \overset{?}{=} 3 \div 15$ State the conjecture.

$5 \neq \frac{1}{5}$ Divide.

The conjecture is false. We found a counterexample. That is, $15 \div 3 \neq 3 \div 15$. So, division is *not* commutative.

CHECK Your Progress

c. *The difference of two different whole numbers is always less than either of the two numbers.*

3 **SHOPPING** Alana wants to buy a sweater that costs $38, a pair of sunglasses that costs $14, a pair of jeans that costs $22, and a T-shirt that costs $16. Use mental math to find the total cost of the items before tax.

Write an expression for the total cost. You can rearrange the numbers using the properties of math. Look for sums that are multiples of ten.

$38 + 14 + 22 + 16$

$= 38 + 22 + 14 + 16$	Commutative Property
$= (38 + 22) + (14 + 16)$	Associative Property
$= 60 + 30$	Add.
$= 90$	Simplify.

The total cost of the items is $90.

Real-World Link · · · ·
Teenagers contribute about $180 billion annually to the sales industry.

 CHECK Your Progress

d. CALLS Lance made four phone calls from his cell phone today. The calls lasted 4.7, 9.4, 2.3, and 10.6 minutes. Use mental math to find the total amount of time he spent on the phone.

You can use the properties you learned in this lesson to simplify algebraic expressions. To **simplify** an expression is to perform all possible operations. When you justify each step with a property, you are using *deductive reasoning* and are proving the statement to be true.

 EXAMPLES Simplify Algebraic Expressions

Simplify each expression. Justify each step.

4 $(7 + g) + 5$

$(7 + g) + 5 = (g + 7) + 5$	Commutative Property of Addition
$= g + (7 + 5)$	Associative Property of Addition
$= g + 12$	Simplify.

5 $(m \cdot 11) \cdot m$

$(m \cdot 11) \cdot m = (11 \cdot m) \cdot m$	Commutative Property of Multiplication
$= 11 \cdot (m \cdot m)$	Associative Property of Multiplication
$= 11m^2$	Simplify.

 CHECK Your Progress

e. $12 + (12x + 13)$ **f.** $9 + (7d + 8)$ **g.** $4 \cdot (3c \cdot 2)$

Example 1 Name the property shown by each statement.

1. $3m \cdot 0 \cdot 5m = 0$ **2.** $7c + 0 = 7c$

Example 2 **3.** State whether the following conjecture is *true* or *false*. If *false*, provide a counterexample.

Subtraction of whole numbers is associative.

Example 3 **4.** **ATTENDANCE** The number of students in each of Mr. Hernandez's classes on Monday were 22, 31, 27, 29, and 18. Use mental math to find the total number of students in all of his classes on Monday. Explain.

Examples 4 and 5 Simplify each expression. Justify each step.

5. $9c + (8 + 3c)$ **6.** $(6b \cdot 4) \cdot 3b$ **7.** $5 \cdot (7h \cdot 4)$

Practice and Problem Solving

 = Step-by-Step Solutions begin on page R1.
Extra Practice begins on page EP2.

Name the property shown by each statement.

Example 1 **8.** $a + (b + 12) = (b + 12) + a$ **9.** $(5 + x) + 0 = 5 + x$

10. $16 + (c + 17) = (16 + c) + 17$ **11.** $d \cdot e \cdot 0 = 0$

12. $9(ab) = (9a)b$ **13.** $y \cdot 7 = 7y$

Example 2 State whether each conjecture is *true* or *false*. If *false*, provide a counterexample.

14. Division of whole numbers is associative.

15. Subtraction of whole numbers is commutative.

Example 3 **16.** **RELAYS** The times for each leg of a relay for four runners are shown. Use mental math to find the total time for the relay team. Explain.

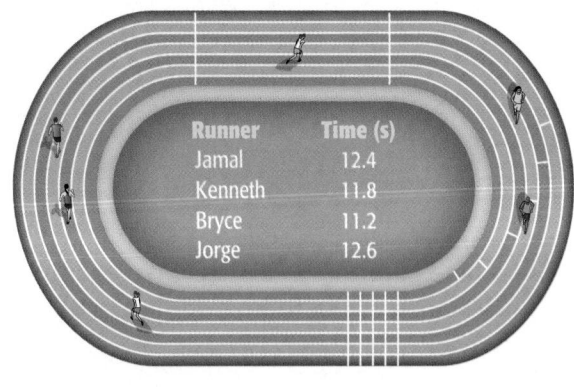

Runner	Time (s)
Jamal	12.4
Kenneth	11.8
Bryce	11.2
Jorge	12.6

17 **MONEY** At dinner, Darien ordered a soda for $2.75, a sandwich for $8.50, and a dessert for $3.85. If the tax on his meal was $1.15, use mental math to find the total amount of his dinner bill. Explain.

Examples 4 and 5 Simplify each expression. Justify each step.

18. $15 + (12 + 8a)$ **19.** $(22 + 19b) + 7$ **20.** $18 + (5 + 6m)$

21. $11s(4)$ **22.** $(5n \cdot 9) \cdot 2n$ **23.** $3x \cdot (7 \cdot x)$

24. Simplify the expression $(7 + 47 + 3)[5 \cdot (2 \cdot 3)]$. Use properties to justify each step.

25. OPEN ENDED Write about something you do every day that is commutative. Then write about another situation that is not commutative.

26. CHALLENGE Does the Associative Property *always, sometimes,* or *never* hold for subtraction? Explain your reasoning using examples and counterexamples.

27. FIND THE ERROR Brian is simplifying $4 \cdot (5 \cdot m)$. Find his mistake and correct it.

$$4 \cdot (5 \cdot m) = 20 \cdot 5m$$
$$= 100m$$

28. **WRITE MATH** Explain the difference between the Commutative and Associative Properties.

Test Practice

29. Which equation is an example of the Commutative Property?

   A. $4 \cdot 1 = 4$

   B. $16 + 0 = 16$

   C. $w + (3 + 2) = w + (2 + 3)$

   D. $d(9 \cdot f) = (d \cdot 9)f$

30. The equation $15 + 0 = 15$ is an example of which of the following properties?

   F. Multiplicative Property of Zero

   G. Multiplicative Identity

   H. Additive Identity

   I. Associative Property

31. **THINK SOLVE EXPLAIN** **EXTENDED RESPONSE** The table shows the cost of different items at a bakery.

Item	Cost ($)
Doughnut	2.29
Muffin	2.50
Cookie	2.21
Roll	1.15

Part A Write a numerical expression to find the total cost of a doughnut, muffin, and cookie.

Part B Use mental math to find the total cost. Justify each step.

32. **THINK SOLVE EXPLAIN** **SHORT RESPONSE** Simplify the expression below. Show and justify each step.

$$14 + (4p + 46) + 0$$

Main Idea

Apply the Distributive Property to rewrite algebraic expressions.

 Vocabulary

equivalent expressions

 **Get ConnectED**

The Distributive Property

SUPPLIES Jordan buys three notebooks that cost $5 each and three packages of pens for $6 each.

1. What does the expression $3 \cdot 5 + 3 \cdot 6$ represent?

2. What does the expression $3(5 + 6)$ represent?

3. Evaluate both expressions. What do you notice?

An expression like $3(5 + 6)$ can be rewritten as $3 \cdot 5 + 3 \cdot 6$ using the **Distributive Property**.

Key Concept — Distributive Property

Words	To multiply a sum or difference by a number, multiply each term inside the parentheses by the number outside the parentheses.
Symbols	$a(b + c) = ab + ac$ $a(b - c) = ab - ac$
Examples	$4(6 + 2) = 4 \cdot 6 + 4 \cdot 2$ $3(7 - 5) = 3 \cdot 7 - 3 \cdot 5$

EXAMPLES Evaluate Numerical Expressions

Use the Distributive Property to evaluate each expression.

1 $8(9 + 4)$

$$8(9 + 4) = 8 \cdot 9 + 8 \cdot 4 \qquad \text{Distributive Property}$$
$$= 72 + 32 \text{ or } 104 \qquad \text{Multiply. Then add.}$$

2 $(5 - 3)15$

$$(5 - 3)15 = 5 \cdot 15 - 3 \cdot 15 \qquad \text{Distributive Property}$$
$$= 75 - 45 \text{ or } 30 \qquad \text{Multiply. Then subtract.}$$

 CHECK Your Progress

a. $5(9 + 11)$ **b.** $7(10 - 5)$ **c.** $(12 - 8)9$

You can use the Distributive Property to help find products mentally. For example, you can find $9 \cdot 22$ mentally by evaluating $9(20 + 2)$.

 **REAL-WORLD EXAMPLE**
Use the Distributive Property

3 **EQUIPMENT** Mr. Ito needs to buy batting helmets for the baseball team. The helmets he plans to buy are $19.95 each. Find the total cost if Mr. Ito needs to buy 9 batting helmets for the team.

Rename $19.95 as $20.00 − $0.05. Then use the Distributive Property to find the total cost mentally.

$9(\$20.00 - \$0.05) = 9(\$20.00) - 9(\$0.05)$ Distributive Property

$\qquad\qquad\qquad = \$180 - \$0.45$ Multiply.

$\qquad\qquad\qquad = \$179.55$ Subtract.

The total cost of the helmets is $179.55.

✓ CHECK Your Progress

d. RENTALS The Mathis family is having their family reunion at Oleta River State Park. They can rent mountain bikes for $37.50 each. Find the total cost for the family to rent 20 bikes. Justify your answer by using the Distributive Property.

You can model the Distributive Property with algebraic expressions using algebra tiles. The expression $2(x + 2)$ is modeled below.

Model $x + 2$ using algebra tiles.

Double the amount of tiles to represent $2(x + 2)$.

Rearrange the tiles by grouping together the ones with the same shapes.

$2(x + 2) = 2(x) + 2(2)$ Distributive Property
$\qquad\qquad = 2x + 4$ Multiply.

The expressions $2(x + 2)$ and $2x + 4$ are **equivalent expressions** because no matter what x is, these expressions have the same value.

Real-World Link · · · ·
The Oleta River State Park is a large urban park. It is known primarily for its nearly 20 miles of off-road biking trails.

EXAMPLES Simplify Algebraic Expressions

Use the Distributive Property to rewrite each expression.

4 $4(x + 7)$

$$4(x + 7) = 4(x) + 4(7) \qquad \text{Distributive Property}$$
$$= 4x + 28 \qquad \text{Simplify.}$$

5 $6(p - 5)$

$$6(p - 5) = 6[p + (-5)] \qquad \text{Rewrite } p - 5 \text{ as } p + (-5).$$
$$= 6(p) + 6(-5) \qquad \text{Distributive Property}$$
$$= 6p + (-30) \qquad \text{Simplify.}$$
$$= 6p - 30 \qquad \text{Definition of subtraction}$$

6 $-2(x - 8)$

$$-2(x - 8) = -2[x + (-8)] \qquad \text{Rewrite } x - 8 \text{ as } x + (-8).$$
$$= -2(x) + -2(-8) \qquad \text{Distributive Property}$$
$$= -2x + 16 \qquad \text{Simplify.}$$

CHECK Your Progress

e. $6(a + 4)$ **f.** $(n + 3)8$ **g.** $3(y - 10)$ **h.** $-7(w - 4)$

✓ CHECK Your Understanding

Examples 1 and 2   **Use the Distributive Property to evaluate each expression.**

1. $7(5 + 4)$ **2.** $(8 + 11)(-3)$

3. $9(10 - 6)$ **4.** $(11 - 5)(-4)$

Example 3   **5**   **FOOD** Amelia bought roast beef from the deli for $6.85 per pound. Find the total cost if Amelia bought 4 pounds of roast beef. Justify your answer by using the Distributive Property.

6. ADMISSION The table shows the number of seniors, adults, and children going on a group trip to an aquarium. The tickets are $14.95 per person. Find the total cost of the tickets. Justify your answer by using the Distributive Property.

Type of Ticket	Tickets Purchased
Senior	7
Adult	11
Child	12

Examples 4–6   **Use the Distributive Property to rewrite each expression.**

7. $5(x + 4)$ **8.** $2(n + 7)$ **9.** $(y + 6)3$

10. $2(p - 3)$ **11.** $6(4 - k)$ **12.** $-6(g - 2)$

Practice and Problem Solving

● = **Step-by-Step Solutions** begin on page R1.
Extra Practice begins on page EP2.

Examples 1 and 2 **Use the Distributive Property to evaluate each expression.**

13. 3(5 + 6)   **14.** 5(6 + 4)   **15.** (3 + 6)(−8)

16. (6 + 4)(−12)   **17.** 4(8 − 7)   **18.** 4(11 − 5)

19. −6(9 − 4)   **20.** (5 − 7)(−3)   **21.** (12 − 4)(−5)

Example 3 **22.** **INSECTS** A housefly can fly at a speed of about 6.4 feet per second. At this rate, how far can a housefly travel in 25 seconds? Justify your answer by using the Distributive Property.

23. **CRAFTS** Theresa is planning on making a fleece blanket for her nephew. She learns that the fabric she wants to use is $7.99 per yard. Find the total cost of 4 yards of fabric. Justify your answer by using the Distributive Property.

Examples 4–6 **Use the Distributive Property to rewrite each expression.**

24. 3(x + 8)   **25.** −8(a + 1)   **26.** (b + 8)5

27. (p + 7)(−2)   **28.** 4(x − 6)   **29.** 6(5 − q)

30. −8(c − 8)   **31.** −3(5 − b)   **32.** (d + 2)(−7)

33. **ACTIVITIES** The prices for different activities at a a fun center are shown in the table.

a. Write two equivalent expressions for the total cost if 3 people participate in each activity.

b. What is the total cost of all the activities for all three people?

Activity	Cost ($)
Go-karts	7
Miniature golf	11
Laser tag	8

34. **MOVIES** The table shows the different prices of items at a movie theater.

a. Suppose Mina and two of her friends go to the movies. Write an expression that could be used to find the total cost for them to go to the movies and buy one of each item.

Movie Theater Prices

$7.50 Ticket $3.25 Drink $4.50 Popcorn $2.25 Box of candy

b. What is the total cost for all three people?

Use the Distributive Property to rewrite each expression.

35. 3(2y + 1)   **36.** −4(3x + 5)   **37.** −6(12 − 8n)

38. 4(x − y)   **39.** −2(3a − 2b)   **40.** (−2 − n)(−7)

41. 5x(y − z)   **42.** −6a(2b + 5c)   **43** −4m(3n − 6p)

44. **ANIMALS** A local club is showing animals at the fair. The club is showing 4 horses, 5 cows, and some chickens. If the animals have a total of 52 legs, how many chickens are being shown?

MEASUREMENT Write two equivalent expressions for the area of each figure.

45.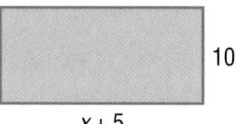
10
$x + 5$

46.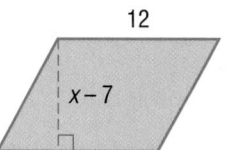
12
$x - 7$

47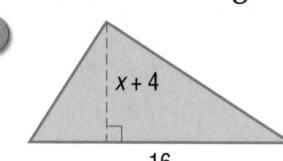
$x + 4$
16

48. SCHOOL You are ordering T-shirts with your school's mascot printed on them. Each T-shirt costs $4.75. The printer charges a setup fee of $30 and $2.50 to print each shirt. Write two expressions that you could use to represent the total cost of printing n T-shirts.

Find each product mentally. Justify your answer.

49. $9 \cdot 35$ **50.** $8 \cdot 28$ **51.** $112 \cdot 6$ **52.** $85 \cdot 8$

53. $4 \cdot 122$ **54.** $12 \cdot 64$ **55.** $108 \cdot 7$ **56.** $264 \cdot 9$

57. GRAPHIC NOVEL Refer to the graphic novel frame below for Exercises a–b.

   a. Write an equation that requires the use of the Distributive Property to represent the number of messages Jacob and Roberto can each send with a $50 budget.

   b. Solve each equation from part **a** to find the number of text messages each person can still send.

H.O.T. Problems

58. OPEN ENDED Write an expression using three terms that can be simplified to $12a + 18b - 6c$.

59. NUMBER SENSE Use the Distributive Property to rewrite the expression $7bx + 7by$ as an equivalent expression.

60. CHALLENGE Use the Distributive Property to write an equivalent expression for the expression $(a + b)(2 + y)$.

61. 📝 **WRITE MATH** Describe how the formula to find the perimeter of a rectangle is an application of the Distributive Property.

62. Which of the following expressions is equivalent to the expression below?

$$5a + 5b$$

  A. $5ab$

  B. $5(a + b)$

  C. $5a + b$

  D. $a + 5b$

63. Which property is demonstrated in the equation below?

$$4x + 32 = 4(x + 8)$$

  F. Associative Property of Addition

  G. Commutative Property of Addition

  H. Distributive Property

  I. Multiplicative Identity

64. Celeste is going to summer camp. The table below shows the cost of items she will need to purchase with the camp logo.

Item	Cost ($)
T-shirt	8.00
Shorts	4.50
Socks	2.25

Celeste needs to buy four of each item. Which expression below *cannot* be used to find the total cost of the items?

  A. $4(14.75)$

  B. $4(8) + 4(4.50) + 4(2.25)$

  C. $4(8.00) + 4.50 + 2.25$

  D. $4(8.00 + 4.50 + 2.25)$

More About Properties

If you choose any two whole numbers and add them together, the sum is always a whole number. So, the set of whole numbers {0, 1, 2, 3, 4, …} is said to be *closed* under addition. The examples below demonstrate the **Closure Property for Addition**.

$$6 + 4 = 10 \qquad 15 + 3 = 18 \qquad 7 + 25 = 32$$

State whether each statement is *true* or *false*. If *false*, provide a counterexample.

65. The set of whole numbers is closed under subtraction.

66. The set of whole numbers is closed under multiplication.

67. The set of whole numbers is closed under division.

68. Consider the set {0, 1}.

  a. Is the set closed under addition? If not, provide a counterexample.

  b. Is the set closed under multiplication? If not, provide a counterexample.

  c. Is the set closed under subtraction? If not, provide a counterexample.

  d. Is the set closed under division? If not, provide a counterexample.

69. Consider the set of even numbers. Is the set closed under addition? Explain your reasoning.

Main Idea

Simplify algebraic expressions.

 Vocabulary

term
like terms
constant
simplest form

Get Connect ED

Simplify Algebraic Expressions

Explore You can use algebra tiles to simplify the algebraic expression $2x + 3 + x - 2$.

Model the expression using algebra tiles. Then group like tiles together and remove zero pairs.

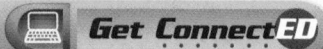

$$2x + 3 + x + (-2) \qquad 3x + 1$$

1. Which properties allow you to group like tiles together?

2. Which property allows you to remove zero pairs?

When addition or subtraction signs separate an algebraic expression into parts, each part is called a **term**. Recall that the numerical factor of a term that contains a variable is called the coefficient of the variable.

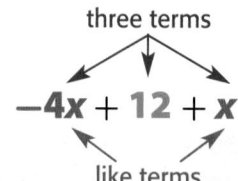

three terms

$$-4x + 12 + x$$

like terms

Like terms contain the same variables to the same powers. For example, $3x^2$ and $-7x^2$ are like terms. So are $8xy^2$ and $12xy^2$. But $10x^2z$ and $22xz^2$ are *not* like terms. A term without a variable is called a **constant**. Constant terms are also like terms.

EXAMPLE Identify Parts of an Expression

1. Identify the terms, like terms, coefficients, and constants in the expression $6n - 7n - 4 + n$.

$6n - 7n - 4 + n = 6n + (-7n) + (-4) + 1n$ Rewrite the expression.

- Terms: $6n, -7n, -4, n$ • Like terms: $6n, -7n, n$

- Coefficients: $6, -7, 1$ • Constants: -4

CHECK Your Progress

Identify the terms, like terms, coefficients, and constants in each expression.

a. $9y - 4 - 11y + 7$ **b.** $3x + 2 - 10 - 3x$

An algebraic expression is in **simplest form** if it has no like terms and no parentheses. Use the Distributive Property to combine like terms.

 EXAMPLES Simplify Algebraic Expressions

Write each expression in simplest form.

(2) $4y + y$

$4y$ and y are like terms.

$$4y + y = 4y + 1y \qquad \text{Identity Property; } y = 1y$$
$$= (4 + 1)y \text{ or } 5y \quad \text{Distributive Property; simplify.}$$

(3) $7x - 2 - 7x + 6$

$7x$ and $-7x$ are like terms. -2 and 6 are also like terms.

$$7x - 2 - 7x + 6 = 7x + (-2) + (-7x) + 6 \quad \text{Definition of subtraction}$$
$$= 7x + (-7x) + (-2) + 6 \quad \text{Commutative Property}$$
$$= [7 + (-7)]x + (-2) + 6 \quad \text{Distributive Property}$$
$$= 0x + 4 \quad \text{Simplify.}$$
$$= 0 + 4 \text{ or } 4 \quad 0x = 0 \cdot x \text{ or } 0$$

Study Tip

Equivalent Expressions To check whether $4y + y$ and $5y$ are equivalent expressions, substitute any value for y and see whether the expressions have the same value.

✓ **CHECK Your Progress**

   c. $4z - z$ **d.** $6 - 3n + 3n$ **e.** $2g - 3 + 11 - 8g$

 REAL-WORLD EXAMPLE

(4) **CONCERTS** At a concert, you buy some souvenir T-shirts for $12.00 each and the same number of CDs for $7.50 each. Write an expression in simplest form that represents the total amount spent.

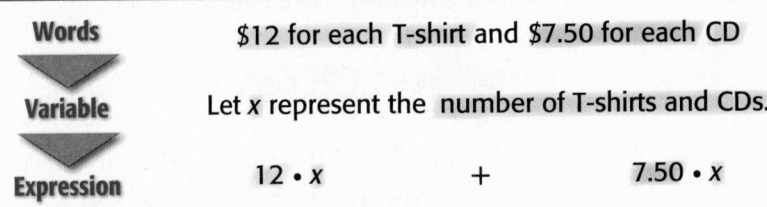

Words	$12 for each T-shirt and $7.50 for each CD
Variable	Let x represent the number of T-shirts and CDs.
Expression	$12 \cdot x$ $+$ $7.50 \cdot x$

$$12x + 7.50x = (12 + 7.50)x \quad \text{Distributive Property}$$
$$= 19.50x \quad \text{Simplify.}$$

The expression $19.50x$ represents the total amount spent.

Real-World Link · · · · ·

In a recent year, the top 10 music moneymakers made a collective $973 million. The majority of this money comes from concert sales.

✓ **CHECK Your Progress**

   f. **MONEY** You have some money. Your friend has $50 less than you. Write an expression in simplest form that represents the total amount of money you and your friend have.

Example 1 Identify the terms, like terms, coefficients, and constants in each expression.

1. $5n - 2n - 3 + n$ **2.** $8a + 4 - 6a - 5a$ **3.** $7 - 3d - 8 + d$

Examples 2 and 3 Write each expression in simplest form.

4. $8n + n$ **5.** $7n + 5 - 7n$ **6.** $4p - 7 + 6p + 10$

Example 4   **7. CONCESSIONS** You go to watch a basketball game and buy 3 bottles of water that each cost x dollars and a large bag of peanuts for \$4.50. Write an expression in simplest form that represents the total amount of money you spent.

Practice and Problem Solving

 = **Step-by-Step Solutions** begin on page R1.
Extra Practice begins on page EP2.

Example 1 Identify the terms, like terms, coefficients, and constants in each expression.

8. $2 + 3a + 9a$ **9.** $7 - 5x + 1$ **10.** $4 + 5y - 6y + y$

11. $n + 4n - 7n - 1$ **12.** $-3d + 8 - d - 2$ **13.** $9 - z + 3 - 2z$

Examples 2 and 3 Write each expression in simplest form.

14. $n + 5n$ **15.** $12c - c$ **16.** $5x + 4 + 9x$

17. $2 + 3d + d$ **18.** $-3r + 7 - 3r - 12$ **19.** $-4j - 1 - 4j + 6$

Example 4 Write an expression in simplest form that represents the total amount in each situation.

20. BOWLING You rent x pairs of shoes for \$2 each. You buy the same number of drinks for \$1.50 each. You also pay \$9 for a bowling lane.

21 TELEVISION You watch x minutes of television on Monday, the same amount on Wednesday, and 30 minutes on Friday.

22. MAGAZINES You subscribe to m different magazines. Your friend subscribes to 2 fewer than you.

23. BIRTHDAYS Today is your friend's birthday. She is y years old. Her brother is 5 years younger.

24. GOVERNMENT In 2009, in the Texas Legislature, there were 119 more members in the House of Representatives than in the Senate. If there were m members in the Senate, write an expression in simplest form to represent the total number of members in the Texas Legislature.

25. MATH IN THE MEDIA Refer to a newspaper or magazine. Choose some data and write a real-world problem for which you could write and simplify an algebraic expression.

ALGEBRA Write a real-world verbal expression for each algebraic expression.

26. $3x + 15$ **27.** $6a - 14$ **28.** $7.50y + 9$

29 **MONEY** Elian and his friends paid a total of $7 for tickets to the school football game. While at the game, they bought 5 hot dogs at x dollars each, 4 boxes of popcorn at y dollars each, and 2 pretzels at z dollars each.

     a. Write an expression to show the total cost of admission and the snacks.

     b. Hot dogs cost $4, popcorn cost $3, and pretzels cost $2. What was the total cost for admission and snacks?

MEASUREMENT Write an expression in simplest form for the perimeter of each figure.

30. 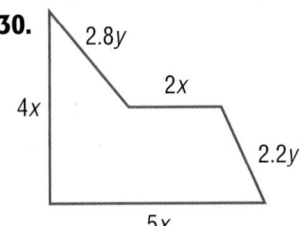   **31.**   **32.**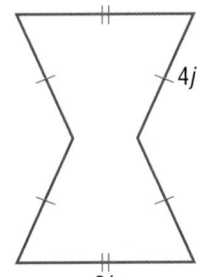

Simplify each expression.

33. $3(4x - 5) + 4(2x + 6)$ **34.** $-8(2a - 3b) - 5(6b - 4a)$

35. $10(5g + 2h - 3) - 4(3g - 4h + 2)$ **36.** $12r + 7(3r - 5s) - 9(6r + 3s) - 2s$

37. $-5(8m - 4n + 6) + 6(9m + 7)$ **38.** $-6(4y - 8z) + 8(3y - 6z)$

39. **SCHOOL** You spent m minutes studying on Monday. On Tuesday, you studied 15 more minutes than you did on Monday. Wednesday, you studied 30 minutes less than you did on Tuesday. You studied twice as long on Thursday as you did on Monday. On Friday, you studied 20 minutes less than you did on Thursday. Write an expression in simplest form to represent the number of minutes you studied in all.

H.O.T. Problems

40. **OPEN ENDED** Write an expression that has three terms and simplifies to $4x - 7$. Identify the coefficient(s) and constant(s) in your expression.

41. **Which One Doesn't Belong?** Identify the expression that is not equivalent to the other three. Explain your reasoning.

     $x - 2 + 3x$ $4(x - 2)$ $7 + 4x - 9$ $4x - 2$

42. **CHALLENGE** Simplify the expression $8x^2 - 2x + 12x - 3$. Show that your answer is true for $x = 2$.

43. **WRITE MATH** Is $2(x - 1) + 3(x - 1) = 5(x - 1)$ a true statement? If so, justify your answer using mathematical properties. If not, give a counterexample.

44. Samir has c cards in his baseball card collection. On his birthday, he received 20 more cards than the number of cards already in his collection. Which expression represents the total number of cards now in his collection?

    A. $c + 20$ **C.** $2c + 20$

    B. $c - 20$ **D.** $2c - 20$

45. Which of the following expressions is $7a - 3(2a - 4)$ in simplest form?

    F. $a - 12$ **H.** $13a - 12$

    G. $a + 12$ **I.** $13a + 12$

46. THINK SOLVE EXPLAIN **SHORT RESPONSE** Simplify the expression below.

$$5(3x + 4y) - 6(2x + 5y)$$

47. The table shows the number of tickets needed and the number of times Patricia participated in different activities at a carnival.

Activity	Tickets	Times Completed
Ring toss	2	a
Dunk tank	4	b
Balloon pop	3	a
Trampoline	5	b

Which expression represents the total number of tickets she used?

    A. 14

    B. $a + b$

    C. $2a + 2b$

    D. $5a + 9b$

 **Simplifying Algebraic Expressions** . . . .

You can also simplify algebraic expressions that contain exponents.

 **EXAMPLE**

Write the expression $x^2 + 2x + x^2$ in simplest form.

Use algebra tiles to model the expression. The tile $\boxed{x^2}$ represents the term x^2.

$$\underbrace{\boxed{x^2}}_{x^2} + \underbrace{\boxed{x}\boxed{x}}_{2x} + \underbrace{\boxed{x^2}}_{x^2} \longrightarrow \underbrace{\boxed{x^2}\boxed{x^2}}_{2x^2} + \underbrace{\boxed{x}\boxed{x}}_{2x}$$

So, $x^2 + 2x + x^2 = 2x^2 + 2x$.

Simplify each expression. Use models if needed.

48. $y^2 + 3 + y^2$ **49.** $5 + p^2 + p^2$

50. $2s^2 + 3s + 4s^2 + 9s$ **51.** $6x + x^2 + 6x + 7x^2$

52. $-5y^2 + 3y - 12y - 8y^2$ **53.** $-m + 10m^2 + m^2 + 7m$

Problem-Solving Investigation

Main Idea Solve a simpler problem.

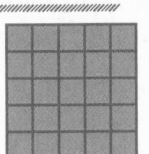

P.S.I. TEAM ✛

Solve a Simpler Problem

GINA: It looks like the figure is made of 25 squares. But, I think there are more squares than that.

YOUR MISSION: Solve a simpler problem to find how many squares of any size are in the figure.

Understand	You know that the figure is a 5 × 5 grid, so the possible sizes for squares are 1 × 1, 2 × 2, 3 × 3, 4 × 4, and 5 × 5. You want to find the total number of squares.
Plan	Count the number of squares when the figure is a 2 × 2 grid and a 3 × 3 grid.
Solve	There are four 1 × 1 squares, and one 2 × 2 square in a 2 × 2 grid. There are 4 + 1 or 5 different squares. 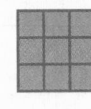 There are nine 1 × 1 squares, four 2 × 2 squares, and one 3 × 3 square. There are 9 + 4 + 1 or 14 different squares.

Make a conjecture with a 4 × 4 grid, then look for a pattern.

Number of Small Squares	1	4	9	16	25
Number of Squares of Any Size	1	5	14	30	55

+4   +9   +16   +25

So, a 5 × 5 grid has 55 squares.

Check	Check your pattern carefully to make sure the answer is correct.

Analyze the Strategy

1. Explain why it was helpful for Gina to solve a simpler problem.

2. **WRITE MATH** Write about a situation in which you might need to solve a simpler problem. Then solve.

- Solve a simpler problem.
- Look for a pattern.
- Work backward.
- Choose an operation.

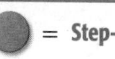 = **Step-by-Step Solutions** begin on page R1.
Extra Practice begins on page EP2.

Use the *solve a simpler problem* strategy to solve Exercises 3–6.

3. **CARPENTRY** Working separately, three carpenters can each make three chairs in three days. How many chairs can 7 carpenters working at the same rate make in 30 days?

4. **TABLES** The school cafeteria has 15 square tables that can each be pushed together to form one long table for class parties. Each square table can seat only one person on each side. How many people can be seated at the combined tables?

5. **PROGRAMS** The school needs 250 programs for the band concert. They can purchase them in packages of 30 or 80 from the printer. How many of each package should they buy?

6. **CRAFTS** Levon needs to cut a long straw into 25 smaller pieces. How many cuts will he need to make?

Use any strategy to solve Exercises 7–13.

7. **VOLUNTEER** Five students can each volunteer for five hours in five days. At this rate, how many hours can 11 students volunteer in 15 days?

8. **SOUVENIRS** Corinne bought 3 hats, postcards, and keychains for her friends.

Item	Cost ($)
Hat	8.50
Keychain	2.25
Postcard	1.75

Write and solve an expression to find how much she spent altogether.

9. **ANALYZE GRAPHS** The graph represents a survey of students about their chocolate preferences. About what percent of students preferred dark chocolate?

10. **RESTAURANTS** Consuelo had dinner with some friends and ordered a meal for $9.95, a drink for $1.25, and dessert for $2.75. A sales tax of 6.5% is added to her bill. If she leaves a tip of 18% after the sales tax, what is the total cost of her meal?

11. **PIZZA** What is the largest number of pieces that can be cut from one pizza using five straight cuts?

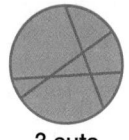 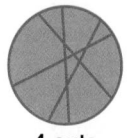

3 cuts   4 cuts

12. **SCHOOL SUPPLIES** Susie wishes to buy 4 of each of the items shown below from the school store. If there is no tax, is $11 enough to pay for Susie's school supplies? Explain.

Item	Cost
Pens	$1.75
Pencils	$1.09
Folder	$0.55

13 **PHONES** A cellular phone company charges $25 per month plus $0.03 per minute. If Cora's monthly bill is $35.38, how many minutes did she use the phone?

Mid-Chapter Check

Name the property shown by each statement. (Lesson 1A)

1. $1(3 \cdot 5) = 3 \cdot 5$

2. $10(9 \cdot 7) = (10 \cdot 9)7$

3. $x + (y + 8) = (y + 8) + x$

4. $(12 + m) + 0 = 12 + m$

5. MONEY Marco spent $12, $23, $18, and $17 at different stores. Use mental math to find the total amount of money that he spent. Explain your reasoning. (Lesson 1A)

Use the Distributive Property to rewrite each expression. (Lesson 1B)

6. $3(x + 2)$

7. $-2(a - 3)$

8. $5(3c - 7)$

9. $-4(2n + 3)$

10. MULTIPLE CHOICE Which of the following is equivalent to $a(3 + b)$? (Lesson 1B)

A. $3a + b$

C. $3a + 3b$

B. $3ab$

D. $3a + ab$

Use the Distributive Property to evaluate each expression. (Lesson 1B)

11. $4(3 + 5)$

12. $(9 + 8)(-7)$

13. $12(8 - 4)$

14. $(10 - 7)(-3)$

15. SNACKS A tourist group had three options of a bagged lunch to select from. The table shows the number of each option ordered by the group. Each lunch costs $8. Find the total cost of the lunches. Justify your answer by using the Distributive Property. (Lesson 1B)

Type of Lunch	Number Ordered
Chicken	12
Beef	15
Veggie	9

Write each expression in simplest form. (Lesson 1C)

16. $2a - 13a$

17. $6b + 5 - 6b$

18. $2m + 5 - 8m$

19. $7x + 2 - 8x$

20. Identify the terms, like terms, coefficients, and constants in the expression $5 - 4x + x - 3$. (Lesson 1C)

21. MULTIPLE CHOICE At the movie theater, Mr. Dawson spent $30 on tickets, bought x bags of popcorn for $5.50 each, and x drinks for $3.75 each. Which of the following expressions represents the total amount of money Mr. Dawson spent? (Lesson 1C)

F. $30 + 2x$

G. $30 + 9.25x$

H. $39.25 + 2x$

I. $39.25 + 9.25x$

22. SPORTS At a football game, you bought x hot dogs for $2.50 each and x candy bars for $1.50 each. If admission cost $8.50, write an expression in simplest form that represents the total spent at the game. (Lesson 1C)

23. BAKING Three bakers can each bake three cakes in three hours. How many cakes can 8 bakers bake working at the same rate in 20 hours? Use the *solve a simpler problem* strategy. (Lesson 1D)

24. DANCE Balloons come in packages of 15 or 35. Vera needs 195 balloons for the spring dance. How many packages of each size should she buy? Use the *solve a simpler problem* strategy. (Lesson 1D)

Explore

Main Idea

Solve equations with variables on each side using algebra tiles.

Get ConnectED

CCSS 8.EE.7, 8.EE.7b

Equations with Variables on Each Side

You can use algebra tiles to solve equations that have variables on each side of the equation.

ACTIVITY

1 **Use algebra tiles to solve** $4x + 2 = 2x + 8$.

STEP 1 Model the equation.

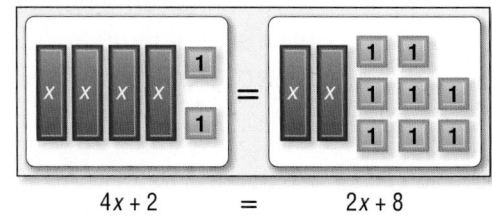
$4x + 2 \qquad = \qquad 2x + 8$

STEP 2 Remove the same number of x-tiles from each side of the mat until there are x-tiles on only one side.

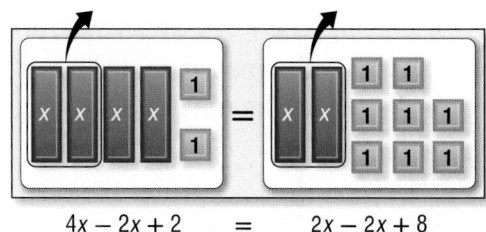
$4x - 2x + 2 \qquad = \qquad 2x - 2x + 8$

STEP 3 Remove the same number of 1-tiles from each side of the mat until the x-tiles are by themselves on one side.

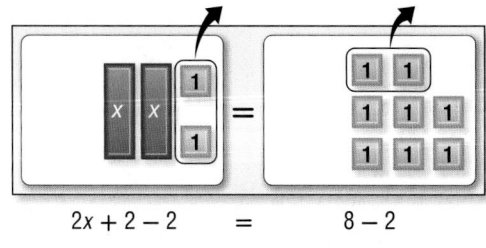
$2x + 2 - 2 \qquad = \qquad 8 - 2$

STEP 4 Separate the tiles into two equal groups.

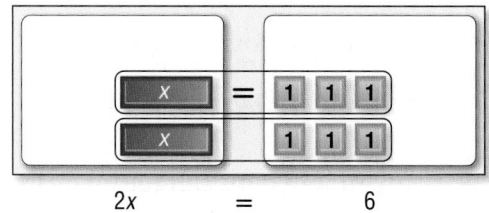
$2x \qquad = \qquad 6$

So, $x = 3$. Since $4(3) + 2 = 2(3) + 8$, the solution is correct.

2 Use algebra tiles to solve $3x + 3 = 2x - 3$.

STEP 1 Model the equation.

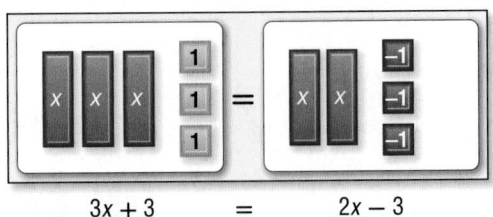

$$3x + 3 \quad = \quad 2x - 3$$

STEP 2 Remove the same number of *x*-tiles from each side of the mat until there is an *x*-tile by itself on one side.

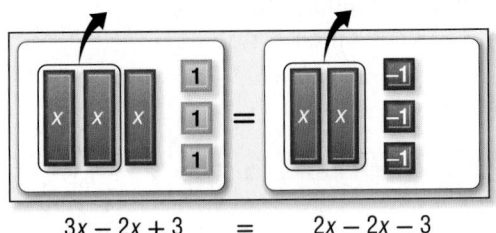

$$3x - 2x + 3 \quad = \quad 2x - 2x - 3$$

STEP 3 To isolate the *x*-tile, it is not possible to remove the same number of 1-tiles from each side of the mat. Add three -1-tiles to each side of the mat.

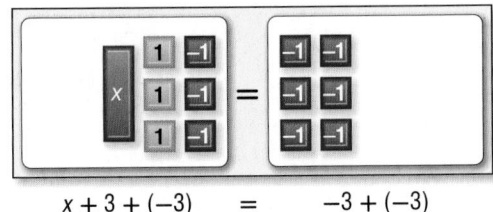

$$x + 3 + (-3) \quad = \quad -3 + (-3)$$

STEP 4 Remove the zero pairs from the left side. There are six -1-tiles on the right side of the mat. The *x*-tile is isolated on the left side of the mat.

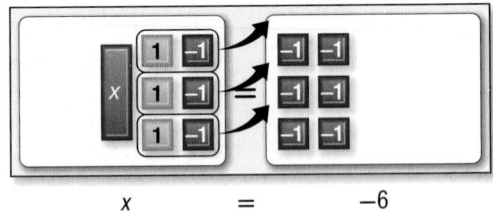

$$x \quad = \quad -6$$

So, $x = -6$. Since $3(-6) + 3 = 2(-6) - 3$, the solution is correct.

Practice and Apply

Use algebra tiles to solve each equation.

1. $x + 2 = 2x + 1$ **2.** $2x + 7 = 3x + 4$ **3.** $2x - 5 = x - 7$ **4.** $8 + x = 3x$

5. $x + 6 = 3x - 2$ **6.** $3x + 3 = x - 5$ **7.** $2x + 1 = x - 7$ **8.** $2x + 5 = 4x - 1$

Analyze the Results

9. Explain why you can remove an *x*-tile from each side of the mat.

10. Solve $x + 4 = 3x - 4$ by removing 1-tiles first. Then solve the equation by removing *x*-tiles first. Does it matter whether you remove *x*-tiles or 1-tiles first? Is one way more convenient? Explain.

Main Idea
Solve equations with variables on each side.

CCSS 8.EE.7, 8.EE.7b

Solve Equations with Variables on Each Side

FUNDRAISING Noah and Tanner are both selling gift wrap for a school fundraiser. Noah has already sold 8 packages of gift wrap before Tanner starts. Tanner sells an average of 5 packages of gift wrap per day, and Noah sells an average of 4 packages per day.

Time (days)	Noah's Sales	Tanner's Sales
0	8 + 4(0) = 8	5(0) = 0
1	8 + 4(1) = 12	5(1) = 5
2	8 + 4(2) = 16	5(2) = 10
3	8 + 4(3) = 20	5(3) = 15
⋮	⋮	⋮

1. Copy the table. Continue filling in rows to find how many days until Tanner and Noah sell the same number of packages.

2. Write an expression for Noah's gift wrap sales after d days.

3. Write an expression for Tanner's gift wrap sales after d days.

4. On which day will Tanner's sales pass Noah's sales?

5. Write an equation that could be used to find how many days it will take until Tanner and Noah sell the same number of packages.

Some equations, like $8 + 4d = 5d$, have variables on each side of the equals sign. Use the properties of equality to write an equivalent equation with the variables on one side of the equals sign. Then solve the equation.

EXAMPLES Equations with Variables on Each Side

1. **Solve $8 + 4d = 5d$. Check your solution.**

   $8 + 4d = \quad 5d$ Write the equation.

   $\underline{-4d = -4d}$ Subtraction Property of Equality

   $8 = d$ Simplify by combining like terms.

   > Subtract 4d from the left side of the equation to isolate the variable.

   > Subtract 4d from the right side of the equation to keep it balanced.

   To check your solution, replace d with 8 in the original equation.

   Check $\quad 8 + 4d = 5d$ Write the original equation.

   $\qquad 8 + 4(8) \stackrel{?}{=} 5(8)$ Replace d with 8.

   $\qquad\qquad 40 = 40$ ✓ The sentence is true.

   The solution is 8.

2 Solve $6n - 1 = 4n - 5$.

$$
\begin{array}{ll}
6n - 1 = \quad 4n - 5 & \text{Write the equation.} \\
\underline{-4n \quad\quad = -4n} & \text{Subtraction Property of Equality} \\
2n - 1 = -5 & \text{Simplify.} \\
\underline{+1 = +1} & \text{Addition Property of Equality} \\
2n = -4 & \text{Simplify.} \\
n = -2 & \text{Mentally divide each side by 2.}
\end{array}
$$

QUICK Review

Properties of Equality
The Properties of Equality state that when the same operation is performed on both sides of an equation, the two sides remain equal.

CHECK Your Progress

Solve each equation. Check your solution.

a. $8a = 5a + 21$ **b.** $3x - 7 = 8x + 23$ **c.** $7g - 12 = 3 + \frac{7}{3}g$

 REAL-WORLD EXAMPLE

3 **CELL PHONES** A cellular phone company charges $24.95 per month plus $0.10 per minute for calls. Another company charges $19.95 per month plus $0.20 per minute. For how many minutes is the monthly cost of both providers the same?

Words	$24.95 per month plus $0.10 per minute	equals	$19.95 per month plus $0.20 per minute.
Variable		Let *m* represent the minutes.	
Equation		$24.95 + 0.10m = 19.95 + 0.20m$	

$$
\begin{array}{ll}
24.95 + 0.10m = \quad 19.95 + 0.20m & \text{Write the equation.} \\
\underline{-0.10m = \quad\quad\quad -0.10m} & \text{Subtraction Property of Equality} \\
24.95 = \quad 19.95 + 0.10m & \text{Simplify.} \\
\underline{-19.95 = -19.95} & \text{Subtraction Property of Equality} \\
5 = 0.10m & \text{Simplify.} \\
\dfrac{5}{0.10} = \dfrac{0.10m}{0.10} & \text{Division Property of Equality} \\
50 = m
\end{array}
$$

The monthly cost is the same for 50 minutes of calls.

CHECK Your Progress

Real-World Link
Congress established the first official United States flag on June 14, 1777.

d. FLAGS The length of a flag is 0.3 foot less than twice its width. If the perimeter is 14.4 feet longer than the width, find the dimensions of the flag.

Examples 1 and 2   **Solve each equation. Check your solution.**

1. $5n + 9 = 2n$ **2.** $3k + 14 = k$

3. $10x = 3x - 28$ **4.** $7y - 8 = 6y + 1$

5. $2a + 21 = 8a - 9$ **6.** $-4p - 3 = 2 + p$

Example 3   **7. CAR RENTAL** EZ Car Rental charges $40 a day plus $0.25 per mile. Ace Rent-A-Car charges $25 a day plus $0.45 per mile. What number of miles results in the same cost for one day?

Practice and Problem Solving

⬤ = **Step-by-Step Solutions** begin on page R1.
Extra Practice begins on page EP2.

Examples 1 and 2   **Solve each equation. Check your solution.**

8. $7a + 10 = 2a$ **9.** $11x = 24 + 8x$

10. $9g - 14 = 2g$ **11.** $m - 18 = 3m$

12. $5p + 2 = 4p - 1$ **13** $8y - 3 = 6y + 17$

14. $15 - 3n = n - 1$ **15.** $3 - 10b = 2b - 9$

16. $-6f + 13 = 2f - 11$ **17.** $2z - 31 = -9z + 24$

18. $2.5h - 15 = 4h$ **19.** $21.6 - d = 5d$

Example 3   **Define a variable, write an equation, and solve to find each number.**

20. Eighteen less than three times a number is twice the number.

21. Eleven more than four times a number equals the number less 7.

22. Fifteen more than twice a number is eight more than three times the number.

23. Nine fewer than half a number is five more than four times the number.

24. TICKETS The table shows ticket prices for the local minor league baseball team for fan club members and non-members. For how many tickets is the cost the same for club members and non-members?

Ticket Prices		
	Club Members	Non-Club Members
Membership Fee (one-time)	$30	none
Ticket Price	$3	$6

25 BASKETBALL Will averages 18 points a game and is the all-time scoring leader on his team with 483 points. Tom averages 21 points a game and is currently second on the all-time scorers list with 462 points. If both players continue to play at the same rate, how many more games will it take until Tom and Will have scored the same number of total points?

MEASUREMENT Write an equation to find the value of x so that each pair of polygons has the same perimeter. Then solve.

26.

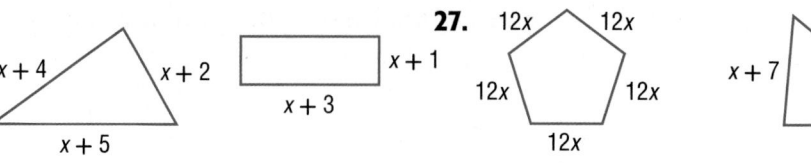

27.

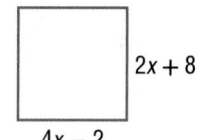

Solve each equation. Check your solution.

28. $8.5x - 4.3 = 3.3x + 3.5$

29. $9.7x + 8.2 = 5.9x - 7$

30. $12.4x + 3.35 = 15.7x - 14.8$

31. $-9.5x + 16.4 = -8.7x + 18.6$

32. $15.4x - 13.7 = 14x - 17.2$

33 $-19.7x - 12.4 = -8.5x + 15.6$

34. **🔁 MULTIPLE REPRESENTATIONS** Refer to the square at the right.

$2x + 8$

$4x - 2$

a. **WORDS** Explain a method you could use to find the value of x.

b. **SYMBOLS** Write an equation to find the side length of the square.

c. **ALGEBRA** What is the side length of the square?

H.O.T. Problems

35. **CHALLENGE** The cheerleaders are selling school sweatshirts at a local fall festival. The fee for a booth is $10 plus 7% of their sales. The sweatshirts are being sold for $15, and they each cost $9 to make. Write and solve an equation to find how many sweatshirts they must sell to break even.

36. **FIND THE ERROR** Alma is solving the equation $4a - 5 = 2a - 3$. Find her mistake and correct it.

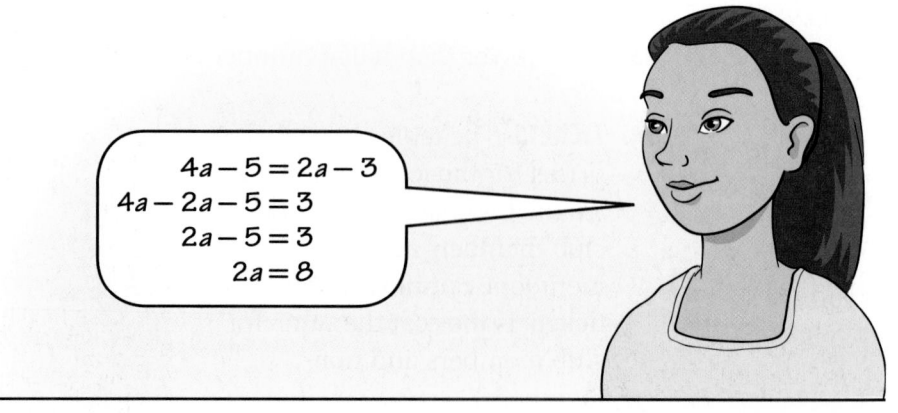

37. **OPEN ENDED** Write a word problem that can be solved using the equation $5x = 3x + 20$.

38. **CHALLENGE** Find the area of the rectangle at the right.

$2x + 17$

$4x - 1$

$6x + 9$

39. **✏️ WRITE MATH** Explain how to solve the equation $2 - 4x = 6x - 8$.

40. Carpet cleaner A charges $28.25 plus $18 a room. Carpet cleaner B charges $19.85 plus $32 a room. Which equation can be used to find the number of rooms for which the total cost of both carpet cleaners is the same?

A. $28.25x + 18 = 19.85x + 32$

B. $28.25 + 32x = 19.85 + 18x$

C. $28.25 + 18x = 19.85 + 32x$

D. $(28.25 + 18)x = (19.85 + 32)x$

41. What is the solution of the following equation?

$$5x + 7 = -3x - 9$$

F. -2 **H.** 2

G. 1   **I.** 8

42. Find the value of x so that the polygons have the same perimeter.

A. 4   **C.** 2

B. 3   **D.** 1

43. Which of the following equations has a solution of 5?

F. $-12x - 6 = -10x + 4$

G. $12x - 6 = 10x + 4$

H. $12x + 6 = 10x - 4$

I. $12x - 6 = 10x - 4$

Spiral Review

44. PRIZES Every 12th person that enters a store is given a coupon for 50% off one item. How many coupons will be given away if 200 people enter the store? Use the *solve a simpler problem* strategy. (Lesson 1D)

Write each expression in simplest form. (Lesson 1C)

45. $5x + 6 - x$

46. $8 - 3n + 3n$

47. $7a - 7a - 9$

48. $3 - 4y + 9y$

49. $10 + 8a + 6a$

50. $2p - 5 + 9p + 2$

51. MEASUREMENT Write an expression in simplest form for the perimeter of the figure. (Lesson 1C)

Use the Distributive Property to evaluate each expression. (Lesson 1B)

52. $8(4 + 3)$

53. $7(9 - 4)$

54. $(9 + 2)(-6)$

55. $(12 - 8)(-7)$

56. $-5(9 - 8)$

57. $-8(10 - 5)$

58. VOLUNTEERS The number of students in each of the eighth grade homerooms that volunteer in the office are shown in the table. Use mental math to find the total number of students who volunteered. Explain. (Lesson 1A)

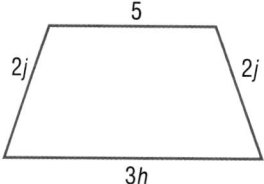

Office Volunteers	
Homeroom	Number of Students
A	6
B	5
C	4
D	8

Main Idea

Use Properties of Equality to solve multi-step equations.

 CCSS 8.EE.7, 8.EE.7b

Solve Multi-Step Equations

 **FOOD** An all-you-can-eat buffet costs $15 per person.

1. Write an equation that can be used to find the total cost c for any number of people p.

2. In order to have dessert, each person must pay an additional d dollars. Write an equation that can be used to find the total cost c for any number of people p to eat and have dessert.

3. Suppose the total cost for 5 people to eat and have dessert is $90. Write an equation to show the total cost of the buffet if all 5 people order dessert.

To find the cost of the dessert in the above example, you can solve the equation $5(15 + d) = 90$. First, you can use the Distributive Property to remove the grouping symbols. Then solve the equation using Properties of Equality.

 **EXAMPLE** Solve Multi-Step Equations

1 Solve $5(15 + d) = 90$.

$5(15 + d) = 90$	Write the equation.
$75 + 5d = 90$	Distributive Property
$\underline{-75 \qquad = -75}$	Subtraction Property of Equality
$5d = 15$	Simplify.
$\dfrac{5d}{5} = \dfrac{15}{5}$	Division Property of Equality
$d = 3$	Simplify.

CHECK Your Progress

Solve each equation. Check your solution.

a. $-3(9 + x) = 33$

b. $5(a - 7) = 24$

c. $2(g + 8) = 4(g - 3)$

d. $-6(n + 9) = 4(5n - 7)$

2 **MONEY** At the fair, Hunter bought 3 snacks and 10 ride tickets. Each ride ticket costs $1.50 less than a snack. If he spent a total of $24.00, what was the cost of each snack?

Use a bar diagram.

snack	snack	snack

$24

$--s--|--s--|--s--$

ticket	ticket	ticket	ticket	ticket	ticket	ticket	ticket	ticket	ticket

$|s-1.5|s-1.5|s-1.5|s-1.5|s-1.5|s-1.5|s-1.5|s-1.5|s-1.5|s-1.5|$

Write an equation to represent the bar model.

$24 = 3s + 10(s - 1.5)$	Write the equation.
$24 = 3s + 10s - 15$	Distributive Property
$24 = 13s - 15$	Simplify.
$\underline{+15 = \qquad +15}$	Addition Property of Equality
$39 = 13s$	Simplify.
$\dfrac{39}{13} = \dfrac{13s}{13}$	Division Property of Equality
$3 = s$	Simplify.

The cost of each snack was $3.

CHECK Your Progress

e. **PETS** Deandra's dog weighs fifteen pounds more than Ruby's dog. Jennifer's dog weighs twice the amount of Deandra's dog. If the dogs weigh 91 pounds altogether, how many pounds does Deandra's dog weigh?

✓ **CHECK Your Understanding**

Example 1   Solve each equation. Check your solution.

1. $5(a - 4) = 30$ **2.** $-8(w - 6) = 32$

3. $4(t - 9) = 6(t + 7)$ **4.** $5(2d + 8) = 7(2d + 8)$

5. $9(g - 10) - 4g = 8g + 27$ **6.** $6(r - 4) = 2(r - 8) + 3r$

7. $12(x + 3) = 4(x + 12) + 6x$ **8.** $8z - 22 = 3(3z + 11) - 6z$

Example 2   9 **CHARITY** Mr. Richards's class is holding a canned food drive for charity. Juliet collected 10 more cans than Rosana. Santiago collected twice as many cans as Juliet. If they collected 130 cans altogether, how many cans did Juliet collect?

Practice and Problem Solving

● = **Step-by-Step Solutions** begin on page R1.
Extra Practice begins on page EP2.

Example 1 **Solve each equation. Check your solution.**

10. $9(j - 4) = 81$

11. $-12(k + 4) = 60$

12. $-5(3m + 6) = -3(4m - 2)$

13. $8(3a + 6) = 9(2a - 4)$

14. $\frac{1}{2}r + 2\left(\frac{3}{4}r - 1\right) = \frac{1}{4}r + 6$

15 $\frac{1}{3}h - 4\left(\frac{2}{3}h - 3\right) = \frac{2}{3}h - 6$

16. $8(4q - 5) - 7q = 20q - 10$

17. $8(t + 2) - 3(t - 4) = 6(t - 7) + 8$

18. $-7(k + 9) = 9(k - 5) - 14k$

19. $-10y + 18 = -3(5y - 7) - 8$

20. $10p - 2(3p - 6) = 4(5p - 6) - 10p$ **21.** $7(c - 9) = 9(2c - 8) + 4c$

Example 2 **22. PARTIES** The school has budgeted $2,000 for an end-of-year party at the local park. The cost to rent the park shelter is $150. How much can the student council spend per student on food if each of the 225 students receives a $3.50 gift?

23 SCHOOL The table shows the number of students in each homeroom.

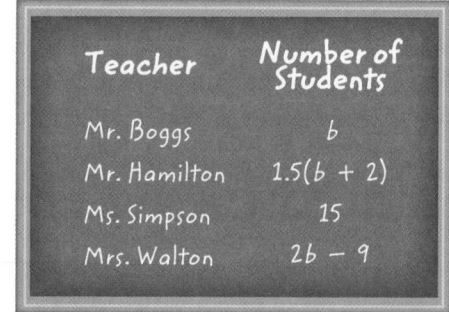

a. Write an equation to find the number of students in Mr. Boggs's homeroom if the total number of students is 90.

b. Solve the equation from part **a** to find the number of students in Mr. Boggs's homeroom.

Teacher	Number of Students
Mr. Boggs	b
Mr. Hamilton	$1.5(b + 2)$
Ms. Simpson	15
Mrs. Walton	$2b - 9$

24. GRAPHIC NOVEL Refer to the graphic novel frame below for Exercises a–b.

a. Write an equation that can be used to determine the number of text messages Jacob and Roberto can send for their plans to cost the same.

b. Solve the equation from part **a** to find the number of text messages each person can send for their costs to be the same.

25. REASONING Does a multi-step equation *always, sometimes,* or *never* have a solution? Explain your reasoning.

26. CHALLENGE The perimeter of a rectangle is $8(2x + 1)$ inches. If the length of the sides of the rectangle are $3x + 4$ inches and $4x + 3$ inches, what is the length of each side of the rectangle?

27. 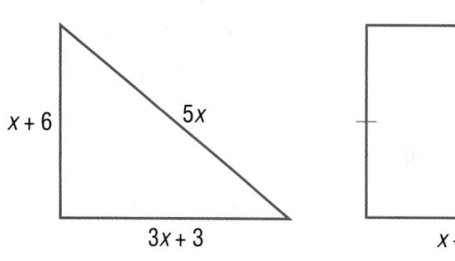 **WRITE MATH** Write about a real-world situation that could be represented by the equation $5(x + 2) = 5x$.

Test Practice

28. The Yeoman family spent a total of $26.75 on lunch. They bought 5 drinks and 3 sandwiches. Each drink costs $2.50 less than a sandwich. Which of the following equations could be used to find the cost of each sandwich?

A. $\$26.75 = 5(\$2.50) + 3s$

B. $\$26.75 = 3(\$2.50) + 5s$

C. $\$26.75 = 5s + 3(s + 2.50)$

D. $\$26.75 = 3s + 5(s - \$2.50)$

29. What value of x makes the perimeters of the figures below equal?

$x + 6$ $5x$

$3x + 3$

$x + 6$

F. 2   **H.** 4

G. 3   **I.** 5

Spiral Review

30. RENTALS Suppose you can rent a car for either $35 a day plus $0.40 a mile or for $20 a day plus $0.55 per mile. Write and solve an equation to find the number of miles that result in the same cost for one day. (Lesson 2B)

31. MONEY While shopping, Ophelia spent $215. Of that, she spent 72% on clothes. About how much money was not spent on clothes? Use the *solve a simpler problem* strategy. (Lesson 1D)

32. MEASUREMENT Write an expression in simplest form for the perimeter of the figure at the right. (Lesson 1C)

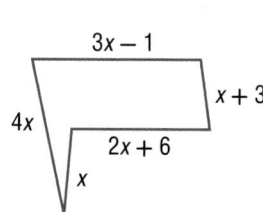

$3x - 1$

$x + 3$

$4x$

$2x + 6$

x

Name the property shown by each statement. (Lessons 1A and 1B)

33. $a + (a + 0) = a + a$

34. $-5(2x + 7) = -10x - 35$

35. $16 + (4x + 12) = (4x + 12) + 16$

36. $5a \cdot 0 = 0$

Main Idea

Use properties of inequality to solve multi-step inequalities.

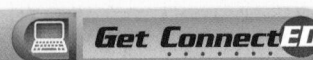

Solve Multi-Step Inequalities

TICKETS The Collins family can spend at most $125 on a trip to a safari park.

1. There are five people in the Collins family. If admission to the park is d dollars per person, write an inequality to represent the amount the Collins can spend on admission.

2. The park sells bags of food to feed the giraffes that cost $9 less than admission. Suppose the three children each buy a bag of food. Write an inequality to represent this situation.

You can use the inequality $5d + 3(d - 9) \leq 125$ to find the most the family can pay for an admission. You can solve the inequality using properties of inequality.

EXAMPLE Solve Multi-Step Inequalities

1. Solve $5d + 3(d - 9) \leq 125$. Graph the solution set on a number line.

$5d + 3(d - 9) \leq 125$	Write the inequality.
$5d + 3d - 27 \leq 125$	Distributive Property
$8d - 27 \leq 125$	Simplify.
$\underline{+\ 27\quad +27}$	Addition Property of Inequality
$8d \leq 152$	Simplify.
$\dfrac{8d}{8} \leq \dfrac{152}{8}$	Division Property of Inequality
$d \leq 19$	Simplify.

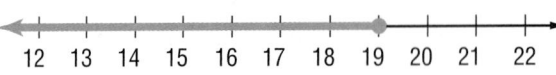

Graph the solution set on a number line. Use a closed dot because 19 is included.

CHECK Your Progress

Solve each inequality. Graph the solution on a number line.

a. $8(m + 6) \leq 16$

b. $-5(p - 7) > 15$

c. $3(3r - 4) > -3(r - 12)$

d. $-4(6s - 6) \leq -9(3s - 2)$

 **REAL-WORLD EXAMPLE** **Solve Multi-Step Inequalities**

② **SCHOOL** Louis has test scores of 90, 87, 70, 97, and 94. He would like to finish with an average score of at least 88. Write and solve an inequality to find the minimum score Louis must earn to achieve his goal.

Words	The total points earned divided by the number of tests is at least 88.
Variable	Let t represent the unknown test score.
Inequality	$\dfrac{90 + 87 + 70 + 97 + 94 + t}{6} \geq 88$

$\dfrac{90 + 87 + 70 + 97 + 94 + t}{6} \geq 88$ Write the inequality.

$\dfrac{438 + t}{6} \geq 88$ Add.

$\dfrac{438 + t}{6} \cdot 6 \geq 88 \cdot 6$ Multiplication Property of Inequality

$438 + t \geq 528$ Simplify.

$\underline{ -438 -438 }$ Subtraction Property of Inequality

$t \geq 90$ Simplify.

So, Louis must earn a score of 90 on the next test to average at least 88 points.

✓ CHECK Your Progress

e. PARTIES Ava can spend no more than $200 on her birthday party. The cost to rent a party room is $50. Write and solve an inequality to determine how much she can spend per guest on food if each of her 15 party guests receives a $4 gift bag.

 **✓ CHECK Your Understanding**

Example 1   Solve each inequality. Graph the solution set on a number line.

1. $7(v + 5) \geq 56$ **2.** $-12(c - 6) < 120$

3. $48 \geq 3(w + 7)$ **4.** $-3(2b + 4) > 4(b + 7)$

5. $-5(3e - 2) < -2(6e + 7)$ **6.** $8(4s - 7) \leq 6(2s + 4)$

Example 2   **⑦ PICTURES** Elise can spend no more than $10 to print 40 pictures from a photo printing Web site. Each picture costs c cents to print plus an additional $0.03 printing fee per picture. If shipping is $3.50, write and solve an inequality to find the most a single print can cost.

Practice and Problem Solving

= **Step-by-Step Solutions** begin on page R1.
Extra Practice begins on page EP2.

Example 1   **Solve each inequality. Graph the solution set on a number line.**

8. $-9(f - 6) \leq 117$ 9. $8(m + 5) > -32$

10. $6(p - 8) \geq 78$ **11** $-12(g + 8) > 24$

12. $-5(c + 15) \leq 75$ 13. $3(4b + 8) < 132$

14. $8(2j - 6) > 2(4j - 4)$ 15. $7(3g - 9) \geq 6(5g - 6)$

16. $-4(5m + 9) \leq -2(8m + 6)$ 17. $9(3x - 8) < 3(7x - 8)$

18. $10(2n - 5) \geq 2(7n + 11)$ 19. $3(6a - 8) < 4(5a - 4)$

Example 2   20. **BOWLING** Over the last two weeks, Diego had bowling scores of 120, 134, 162, 115, and 125. He would like to have an average score of at least 130. Write and solve an inequality to find the minimum score Diego needs to achieve his goal.

21. **COMPUTERS** A computer technician tells Audrey that it will cost no more than $150 to fix her computer. If the cost of the computer parts is $80, and the technician charges $25 per hour, how many hours is the technician planning to work on the computer?

22. **DONATIONS** The table shows the number of food products the five eighth-grade homerooms donated for a food drive. If their average number of donations was at least 72 items per room, what is the least number of items Homeroom 103 donated?

Homeroom	Donations
101	58
102	80
103	?
104	64
105	90

23 **MEASUREMENT** A triangle has side lengths of $(x + 4)$, $(4x - 8)$, and $(2x + 8)$ units. If the perimeter of the triangle is at least 88 units, what is the minimum length of each side of the triangle?

H.O.T. Problems

24. **OPEN ENDED** Write a multi-step inequality with a solution of $k \geq -5$.

25. **Which One Doesn't Belong?** Identify the inequality that does not belong with the other three. Explain your reasoning.

| $3(4x - 2) \geq 30$ | $3(-3x + 9) \leq 0$ | $2(2x + 9) \geq 6$ | $-5(x - 4) \leq 5$ |

26. **CHALLENGE** Can the inequality $12 \geq -4(x + 5)$ be solved without multiplying or dividing either side by a negative number? Explain your reasoning.

27. [E] **WRITE MATH** Is it possible for there to be no solution to an inequality? Justify your reasoning with an example.

28. Monday through Thursday, Michaela spent 75, 60, 95, and 55 minutes practicing the piano. Suppose she wants her average practice time to be at least 70 minutes for Monday through Friday. Which inequality represents the time she must practice on Friday?

A. $x \geq 70$ **C.** $x \geq 65$

B. $x \leq 70$ **D.** $x \leq 65$

29. Which of the following is the solution set of the inequality below?

$$3(x - 6) \geq 2(3x + 12)$$

F. $x \leq -9$ **H.** $x \leq -14$

G. $x \geq -9$ **I.** $x \geq -14$

30. Which of the following shows the solution set of the inequality $4(x + 6) < 6(x - 3)$?

A. ———————————●———————————→
   18   19   20   21   22   23   24

B. ———————————●————————————→
   18   19   20   21   22   23   24

C. ———————————○————————————→
   18   19   20   21   22   23   24

D. ———————————○————————————→
   18   19   20   21   22   23   24

Solve each equation. Check your solution. (Lesson 2C)

31. $6(a + 3) = 42$

32. $-7(b - 8) = 105$

33. $5(2d + 8) = 7(2d + 8)$

34. MEASUREMENT Write an equation that can be used to find the value of x so that the polygons have the same perimeter. Then solve. (Lesson 2B)

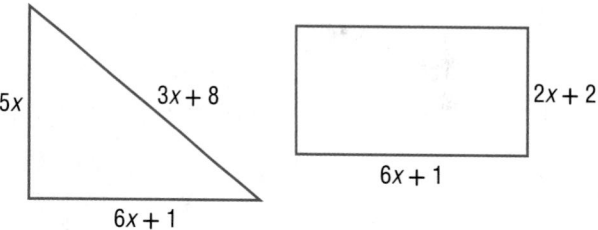

Identify the terms, like terms, coefficients, and constants in each expression. (Lesson 1C)

35. $6n - 3n - 4 + n$

36. $8 + 6d - 9 + d$

37. $4 + 8k + 9k$

38. $10 - 9x + 2$

39. $j + 15j - 8j - 3$

40. $-7d + 12 - d + 5$

41. BAKE SALE The prices for different items at a bake sale are shown in the table. (Lesson 1B)

Item	Price per Package
Brownies	$2.75
Cookies	$1.00
Cupcakes	$3.50

   a. Write two equivalent expressions for the total cost if Emily buys two of each package.

   b. What is Emily's total cost if she buys two of each package?

Name the property shown by each statement. (Lesson 1A)

42. $8x + 0 = 8x$

43. $x + (y + 6) = (y + 6) + x$ **44.** $a \cdot b \cdot 0 = 0$

45. $z \cdot 20 = 20z$

46. $(j + k) + 13 = j + (k + 13)$ **47.** $5(mn) = (5m)n$

Problem Solving in Design

It's Great to Skate!

If you love the sport of skateboarding, and you are creative and have strong math skills, you should think about a career designing skateboards. A skateboard designer applies engineering principles and artistic ability to design high-performance skateboards that are both strong and safe. To have a career in skateboard design, you should study physics and mathematics and have a good understanding of skateboarding.

21st Century Careers

Are you interested in becoming a skateboard designer? Take some of the following courses in high school.

- Digital Design
- Geometry
- Physics
- Trigonometry

Get Connect ED

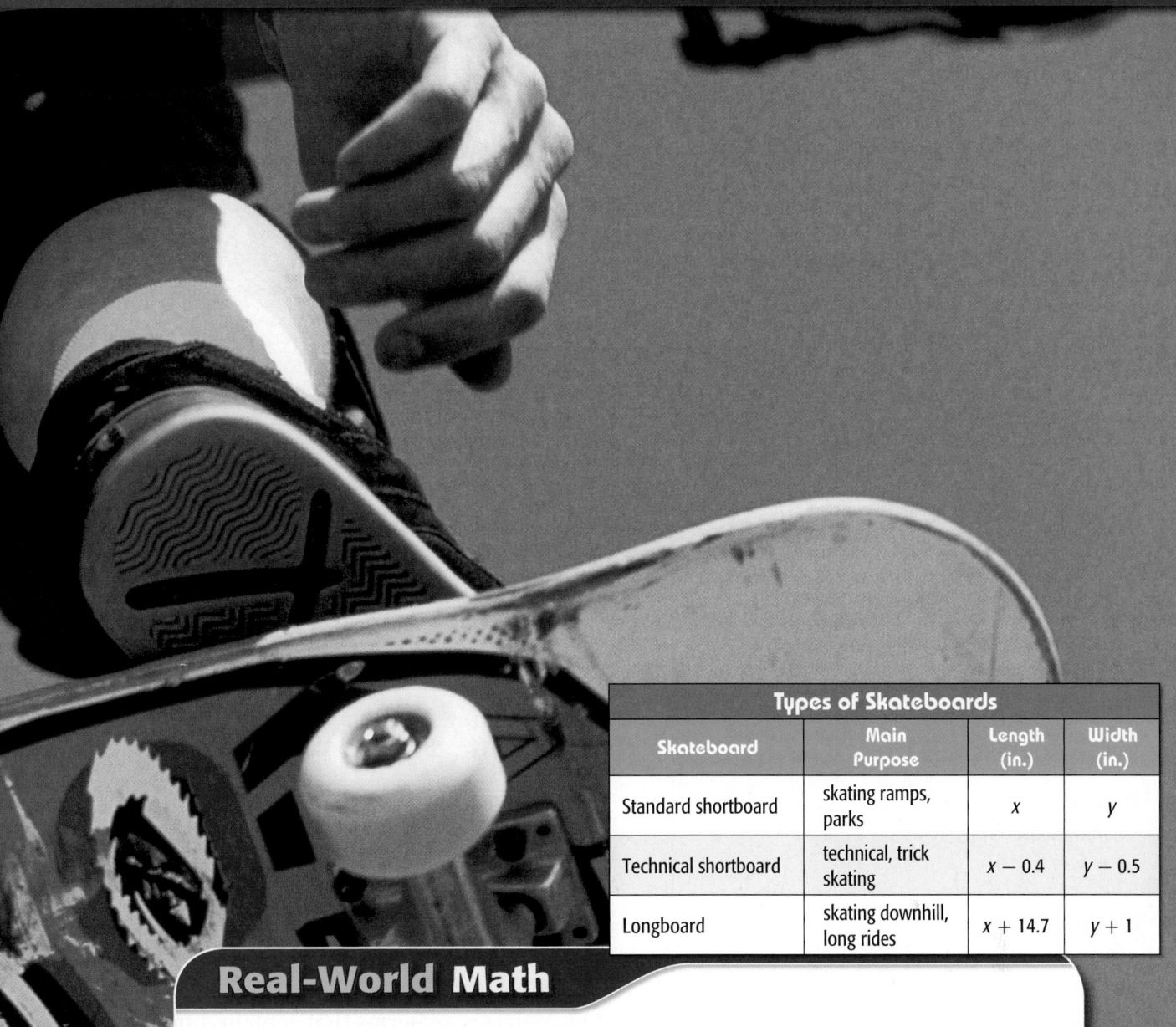

Types of Skateboards			
Skateboard	Main Purpose	Length (in.)	Width (in.)
Standard shortboard	skating ramps, parks	x	y
Technical shortboard	technical, trick skating	$x - 0.4$	$y - 0.5$
Longboard	skating downhill, long rides	$x + 14.7$	$y + 1$

Real-World Math

Use the information in the table to solve each problem.

1. Write an expression to represent the total width of two standard shortboards and a technical shortboard.

2. Identify the terms, like terms, coefficients, and constants in the expression that you wrote in Exercise 1.

3. Write an expression for the total length of two longboards. Then write the expression in simplest form.

4. Write an expression to represent the total width of five longboards. Then use the Distributive Property to rewrite the expression.

5. Write an expression to represent the total length of three technical shortboards. Then use the Distributive Property to rewrite the expression.

Chapter Study Guide and Review

FOLDABLES®
Study Organizer

Be sure the following Key Concepts are noted in your Foldable.

Properties
Equations
Inequalities

Key Concepts

Properties (Lesson 1)

• Commutative Property
$$4 + 5 = 5 + 4$$
$$2 \cdot 7 = 7 \cdot 2$$

• Associative Property
$$10 + (8 + 3) = (10 + 8) + 3$$
$$6 \cdot (5 \cdot 9) = (6 \cdot 5) \cdot 9$$

• Identity Property
$$3 + 0 = 3$$
$$3 \cdot 1 = 3$$

• Distributive Property
$$a(b + c) = ab + ac$$

Multi-Step Equations and Inequalities
(Lesson 2)

• To solve equations that have variables on each side of the equals sign, use the properties of equality to write an equivalent equation with the variables on one side of the equals sign. Then solve the equation.

• To solve multi-step equations and inequalities, first use the Distributive Property to remove any grouping symbols. Then solve the equation or inequality.

Key Vocabulary

constant

counterexample

equivalent expressions

like terms

property

simplest form

simplify

term

Vocabulary Check

State whether each sentence is *true* or *false*. If *false*, replace the underlined word(s) to make a true sentence.

1. Like terms are terms that contain <u>different</u> variables.

2. A <u>constant</u> is a term without a variable.

3. A <u>counterexample</u> is a statement that is true for any number.

4. When addition or subtraction signs separate an algebraic expression into parts, each part is called a <u>term</u>.

5. An algebraic expression is in <u>simplest form</u> if it has no like terms and no parentheses.

6. When you use the Distributive Property to combine like terms, you are <u>simplifying the expression</u>.

7. A <u>property</u> is an example that shows that a conjecture is false.

Multi-Part Lesson Review

Properties of Mathematics

Properties (Lesson 1A)

Name the property shown by each statement.

8. $5 \cdot 0 = 0$

9. $a + b + c = a + (b + c)$

10. $m \cdot 5 = 5m$

11. $1(7 + a) = 7 + a$

12. HOMEWORK Graham earned 8, 13, 7, 12, and 9 points on his last few homework assignments. Use mental math to find his total number of points. Explain.

> **EXAMPLE 1** Name the property shown by the statement $6 \cdot (4 \cdot a) = (6 \cdot 4) \cdot a$.
>
> The order of the numbers and variable did not change, but their grouping did. This is the Associative Property of Multiplication.

The Distributive Property (Lesson 1B)

Use the Distributive Property to evaluate each expression.

13. $3(8 + 7)$ **14.** $(8 + 12)(-5)$

15. $9(3 + 9)$ **16.** $(11 - 4)(-3)$

Use the Distributive Property to rewrite each expression.

17. $4(a + 13)$ **18.** $(n - 5)(-7)$

19. $(r - 9)8$ **20.** $-6(b + 10)$

21. PICTURES Carlita had some pictures printed for $0.20 per print. Find the total cost if Carlita had 65 pictures printed. Justify your answer by using the Distributive Property.

> **EXAMPLE 2** Use the Distributive Property to rewrite $-8(x - 9)$.
>
> | | |
> |---|---|
> | $-8(x - 9)$ | Write the expression. |
> | $= -8[x + (-9)]$ | Rewrite $x - 9$ as $x + (-9)$. |
> | $= -8(x) + (-8)(-9)$ | Distributive Property |
> | $= -8x + 72$ | Simplify. |

Simplify Algebraic Expressions (Lesson 1C)

Write each expression in simplest form.

22. $p + 6p$ **23.** $6b - 3 + 7b + 5$

24. $2m - 5 + 3m - 4m + 2$

25. $8s + 2s + 12 - 3s - 10 + (-2) - 6s$

26. SOCCER Pan scored n goals. Leo scored 5 fewer than Pan. Write an expression in simplest form to represent the total number of goals scored.

> **EXAMPLE 3** Write $5y + 3 + 2y + 4$ in simplest form.
>
> $5y$ and $2y$ are like terms. 3 and 4 are also like terms.
>
> | | |
> |---|---|
> | $5y + 3 + 2y + 4$ | |
> | $= 5y + 2y + 3 + 4$ | Commutative Property |
> | $= (5 + 2)y + 3 + 4$ | Distributive Property |
> | $= 7y + 7$ | Simplify. |

Lesson 1 ## Properties of Mathematics (continued)

PSI: Solve a Simpler Problem (Lesson 1D)

Solve. Use the *solve a simpler problem* strategy.

27. **GEOGRAPHY** The total area of Arizona is 14,006 square miles. Of that, about 42% of the land is desert. About how many square miles of Arizona's land is *not* covered by desert?

28. **BIOLOGY** An average person blinks their eyes about 20 times per minute. Estimate the number of times a person will blink in one year.

EXAMPLE 4 A total of 450 students were surveyed. If 60% of the students voted to hold a carnival, find the number of students who voted for the carnival.

Find 10% of 450 and use the result to find 60% of 450.

10% of 450 = 45; so 60% of 450 is 6×45 or 270. So, 270 students voted for the carnival.

Lesson 2 ## Multi-Step Equations and Inequalities

Solve Equations with Variables on Each Side (Lesson 2B)

Solve each equation. Check your solution.

29. $11x = 20x + 18$

30. $7b - 3 = -2b + 24$

31. **GEOGRAPHY** The coastline of California is 46 miles longer than twice the length of Louisiana's coastline. It is also 443 miles longer than Louisiana's coastline. Find the lengths of the coastlines of California and Louisiana.

32. **MEASUREMENT** Write an equation to find the value of x so that the polygons below have the same perimeter. Then solve.

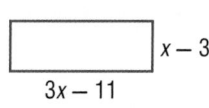

$2x - 7$
$2x - 5$
$x - 1$

$x - 3$
$3x - 11$

EXAMPLE 5 Solve $-7x + 5 = x - 19$.

$-7x + 5 =$	$x - 19$	Write the equation.
$+ 7x$	$= + 7x$	Addition Property of Equality
$5 =$	$8x - 19$	Simplify.
$+ 19 =$	$+ 19$	Addition Property of Equality
$24 = 8x$		Simplify.
$\dfrac{24}{8} = \dfrac{8x}{8}$		Division Property of Equality
$3 = x$		Simplify.

The solution is 3.

Solve Multi-Step Equations (Lesson 2C)

Solve each equation. Check your solution.

33. $-8(t + 9) = 24$

34. $4(3k - 6) = 6(3k + 5)$

35. MONEY Mr. and Mrs. Hawkins have budgeted $500 for Marion's graduation party. The cost to rent the room is $150. How much can they spend per person on food if each of the 30 guests receives a $2.50 picture?

EXAMPLE 6 Solve $3(4x - 12) = 24$.

$3(4x - 12) = 24$	Write the equation.
$12x - 36 = 24$	Distributive Property
$\underline{+ 36 = + 36}$	Addition Property of Equality
$12x = 60$	Simplify.
$\dfrac{12x}{12} = \dfrac{60}{12}$	Division Property of Equality
$x = 5$	Simplify.

The solution is 5.

Solve Multi-Step Inequalities (Lesson 2D)

Solve each inequality. Graph the solution set on a number line.

36. $3x + 7 \geq 2x$

37. $7p - 6 < 4p$

38. $3y - 5 \leq 5y + 7$

39. $6(2a - 6) > 4(4a + 4)$

40. CLUBS Over the year, 32, 28, 41, 17, 35, and 40 people have attended French Club meetings. Write and solve an inequality to find the minimum number of people that need to attend the next meeting so that the average attendance is at least 30 people.

EXAMPLE 7 Solve $4(y - 4) > 2(5y + 4)$.

Graph the solution set on a number line.

$4(y - 4) > 2(5y + 4)$	Write the inequality.
$4y - 16 > 10y + 8$	Distributive Property
$\underline{+16 +16}$	Addition Property of Inequality
$4y > 10y + 24$	Simplify.
$\underline{-10y -10y}$	Subtraction Property of Inequality
$-6y > 24$	Simplify.
$\dfrac{-6y}{-6} < \dfrac{24}{-6}$	Division Property of Inequality
$y < -4$	Simplify.

Graph the solution set on a number line. Use an open dot because -4 is not included.

Name the property shown by each statement.

1. $8(cd) = (8c)d$

2. $x \cdot 4 = 4x$

3. $(a + b) + 0 = a + b$

4. $(15 + r) + 12 = 12 + (15 + r)$

Use the Distributive Property to rewrite each expression.

5. $-7(x - 10)$

6. $8(2y + 5)$

7. $-4(2c - 5)$

8. $9(4a - 10)$

Write each expression in simplest form.

9. $9a - a + 15 - 10a - 6$

10. $2x + 17x$

11. $8b + 7 - 6b + 4b - 7 + b$

12. $10y - 8y - 10y + 8y + 12$

13. **MULTIPLE CHOICE** The perimeter of the rectangle is 44 inches.

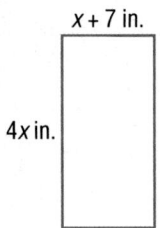

$x + 7$ in.

$4x$ in.

What is the area of the rectangle?

A. 22 in^2

C. 392 in^2

B. 120 in^2

D. 440 in^2

14. **FINANCIAL LITERACY** First Bank charges $4.50 per month for a basic checking account plus $0.15 for each check written. Citizen's Bank charges a flat fee of $9. How many checks would you have to write each month in order for the cost to be the same at both banks?

15. **MUSICAL** Russell sold tickets to the school musical. He had 12 bills worth $175 for the tickets sold. If all the money was in $5 bills, $10 bills, and $20 bills, how many of each bill did he have? Use the *solve a simpler problem* strategy.

Solve each equation. Check your solution.

16. $-23 = 3p + 5 + p$ 17. $-3a - 2 = 2a + 3$

18. **SKATEBOARDING** A skate park charges $6 each time you skate. They also offer a membership for a one-time fee of $24 plus $2 for each time you skate. Write and solve an equation to determine how many times you would have to skate to break even when purchasing the membership.

Solve each inequality. Graph the solution set on a number line.

19. $x + 5 < 4x + 26$ 20. $3d \le 25 - 2d$

21. $-2g + 15 \ge 45 - 8g$

22. THINK SOLVE EXPLAIN **EXTENDED RESPONSE** The members of the local rec center are planning to go to an amusement park. The admission rates for two different parks are shown in the table. As few as 10 people or as many as 25 people will go on the trip.

Park	Admission Cost ($)	
	< 15 people	≥ 15 people
Fun World	$37n$	$37n$
Coaster City	$40n$	$30n + 75$

Part A Write an expression to find the total cost for any number of people to visit each park.

Part B Find the total cost for each possible group size if they go to each park.

Part C Write a recommendation that details which park they should go to based on the number of people they expect to attend. Justify your answer.

Preparing for Standardized Tests

 **Extended Response: Show Your Work or Explain in Words**

When answering extended-response questions, you can either show all the steps that you used to find the solution, or explain in words the steps that you used.

TEST EXAMPLE

JR's Landscaping Service charges $225 for a design layout and $48.50 an hour for working on a yard. Great Landscapes charges $405 for a design layout and $25.90 an hour.

Part A Write an expression to represent the total cost of hiring each landscaping company for h hours.

JR's Landscaping Service $\underline{225 + 48.50h}$
Great Landscapes $\underline{405 + 25.90h}$

Part B For how many hours of yard work would the landscaping companies cost the same? Show your work or explain in words how you got your answer.

Show Your Work	Explain in Words
$225 + 48.50h = 405 + 25.90h$ $-225 \qquad\qquad = -225$ $\overline{\quad 48.50h = \quad 180 + 25.90h}$ $-25.90h = \qquad\quad -25.90h$ $\overline{\quad 22.60h = \quad 180}$ $\dfrac{22.60h}{22.60} = \dfrac{180}{22.60}$ $h \approx 7.96$ For 8 hours of work, the cost is the same.	I first set the expressions equal to each other. Then I solve the equation $225 + 48.50h = 405 + 25.90h$ for h. To do this, I subtract 225 and 25.90h from each side to get $22.60h = 180$. Then I divide each side by 22.60. My final answer is $h \approx 7.96$. So, for about 8 hours of work, the cost is the same.

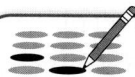 **Work on It**

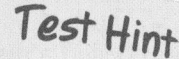 **Test Hint**
Be sure to keep your work or explanation inside the work space provided.

For a fundraiser, Hailey raises $27, plus $1.50 for each lap in the pool that she swims. Tariana raises $18, plus $3 for each lap that she swims.

Part A Write an expression to represent the total amount raised by each swimmer after swimming n laps.

Part B After how many laps would the two swimmers have raised the same amount of money? Show your work or give an explanation.

Read each question. Then fill in the correct answer on the answer document provided by your teacher or on a sheet of paper.

1. Which property is illustrated by the equation below?

$$5(x + 2) = 5x + 10$$

A. Associative Property of Addition

B. Commutative Property of Addition

C. Distributive Property

D. Multiplicative Identity

2. The table shows the fraction of votes won by Janie and Jamal in the Student Council elections. What fraction of the votes did Marissa receive?

Candidate	Fraction of Votes
Jamal	$\frac{2}{3}$
Janie	$\frac{1}{5}$
Marissa	x

F. $\frac{2}{15}$

G. $\frac{1}{3}$

H. $\frac{4}{5}$

I. $\frac{13}{15}$

3. THINK SOLVE EXPLAIN **SHORT RESPONSE** Simplify the expression shown below.

$$(3m^3n^2)(6m^4n)$$

4. Damien purchased a new digital camera for $499 and a printer for $299 including tax. If he plans to pay the total amount in 6 equal monthly payments, what is a reasonable estimate of the amount he will pay each month?

A. $66.50

B. $133.00

C. $155.00

D. $165.00

5. ≡≡✍ **GRIDDED RESPONSE** Write the length of the pencil in inches as a decimal.

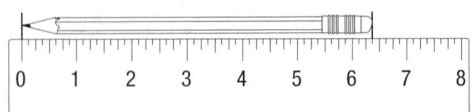

6. Ryan solved the problem below and then graphed the solution on a number line.

Ten less than three fifths of a number is at least 41.

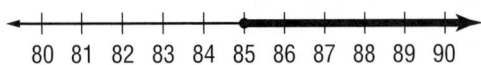

Which of the following most appropriately describes the unknown number?

F. less than 85

G. at most 85

H. more than 85

I. 85 or more

7. Which expression is equivalent to $x(2 + y)$?

A. $2x + y$

B. $2xy$

C. $2x + 2y$

D. $2x + xy$

8. ≡≡✍ **GRIDDED RESPONSE** To approximate the radius r of a circle, you can use the formula $r = \sqrt{\dfrac{A}{3.14}}$, where A is the area of the circle. Find the radius in feet of the circle below. Round to the nearest tenth.

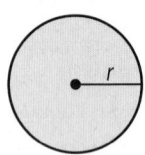

$A = 60$ ft²

9. [THINK SOLVE EXPLAIN] **SHORT RESPONSE** What value of x makes the polygons below have the same perimeter?

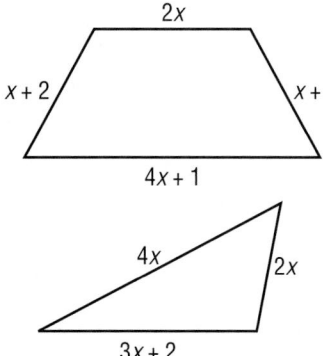

10. What is the value of m in the equation below?

$$4m + 7 = -3m + 49$$

F. -6

G. 6

H. 12

I. 42

11. A pizzeria sells large pizzas for $11.50, medium pizzas for $8.75, and small pizzas for $6.50. Suppose a scout group orders 3 large pizzas, 2 medium pizzas, and 2 small pizzas. Which equation can be used to find the total cost of the pizzas?

A. $t = (3 + 2 + 2)(11.50 + 8.75 + 6.50)$

B. $t = (3)(11.50) + 2(8.75) + 2(6.50)$

C. $t = (3 + 2 + 2)\left(\dfrac{11.50 + 8.75 + 6.50}{3}\right)$

D. $t = (3)(11.50) + 8.75 + 2(6.50)$

12. A circle with an area of approximately 3.14 square meters is to be cut from the board shown. What percent of the area of the total board will be left after the cut?

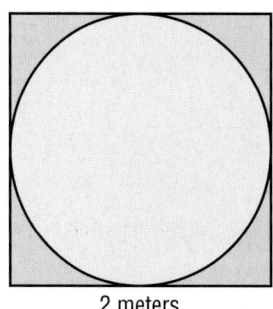

2 meters

F. 21.5%

G. 36%

H. 64%

I. 78.5%

13. [THINK SOLVE EXPLAIN] **EXTENDED RESPONSE** Lucy earned the scores shown on her first four tests.

Test	Score
1	85
2	88
3	92
4	90

Part A Lucy wants to have an average test score of at least 90 points on five tests. Write an inequality to represent this situation.

Part B Solve the inequality from *Part A*. Justify each step using Properties of Inequality.

Part C Suppose only 100 points are possible on the next test. Is it possible for Lucy to achieve an average of 95 points? Explain.

NEED EXTRA HELP?

If You Missed Question...	1	2	3	4	5	6	7	8	9	10	11	12	13
Go to Chapter-Lesson...	4-1B	1-1B	2-1B	1-1D	1-1A	3-4A	4-1B	2-3D	4-2B	4-2B	3-1B	1-2C	4-2D

Chapter 4 Test Practice **263**

CHAPTER 5

Expressions and Functions

connectED.mcgraw-hill.com

Investigate

 Animations

 Vocabulary

 Multilingual eGlossary

Learn

 Personal Tutor

 Virtual Manipulatives

 Graphing Calculator

 Audio

 Foldables

Practice

 Self-Check Practice

 Worksheets

 Assessment

The ☆BIG Idea

What is the relationship among tables, graphs, and equations in modeling a given situation?

 FOLDABLES Study Organizer

Make this Foldable to help you organize your notes.

Multi-Part Lesson	Vocabulary	Need to know	Things I know
1			
2			
3			
4			

Review Vocabulary

expression **expresión** A combination of variables, numbers, and at least one operation.

$$3x + 4$$

variable → $3x$
number → 4
operation → $+$

Key Vocabulary

English	Español
algebraic expression	expresión algebraica
domain	dominio
function	función
range	rango
relation	relación

When Will I Use This?

Are You Ready for the Chapter?

You have two options for checking prerequisite skills for this chapter.

Text Option Take the Quick Check below. Refer to the Quick Review for help.

QUICK Check

Name the ordered pair for each point.

1. R
2. S
3. T
4. U
5. V
6. W

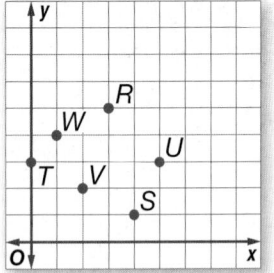

Graph each point on a coordinate plane.

7. $A(3, 4)$ 8. $B(2, 1)$ 9. $C(0, 2)$

10. $D(4, 3)$ 11. $E(6, 0)$ 12. $F(2, 5)$

13. **WALKING** From his cabin, Derek walked 4 miles north and 2 miles east, where he rested. If the origin represents the cabin, graph the point representing Derek's resting point.

Evaluate each expression if $x = 6$.

14. $3x$ 15. $4x - 9$

16. $2x + 8$ 17. $5 + x$

18. $\dfrac{x}{2}$ 19. $\dfrac{3x}{9}$

20. **PROFIT** The weekly profit of a certain company is $48x - 875$, where x represents the number of units sold. Find the weekly profit if the company sells 37 units.

QUICK Review

EXAMPLE 1

Graph $P(1, 2)$, $Q(3, 4)$, and $R(4, 0)$ on a coordinate plane.

Start at the origin. The first number in each ordered pair is the x-coordinate. The second number in each ordered pair is the y-coordinate.

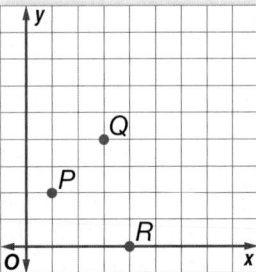

EXAMPLE 2

Evaluate $6x - 1$ if $x = 4$.

$$6x - 1 = 6(4) - 1 \quad \text{Replace } x \text{ with 4.}$$
$$= 24 - 1 \quad \text{Multiply 6 by 4.}$$
$$= 23 \quad \text{Subtract.}$$

Online Option Take the Online Readiness Quiz.

 # Reading Math

Reading For Understanding

One way to make a word problem easier to understand is to rewrite it using fewer words. Here's an example.

STEP 1 Read the problem and identify the important words and numbers.

CELL PHONES There is a wide range of cell phone plans available for students. With Janelle's plan, she pays $15 per month for 200 minutes, plus $0.10 per minute once she talks for more than 200 minutes. Suppose Janelle can spend $20 each month for her cell phone. How many more minutes can she talk?

STEP 2 Simplify the problem. Keep all of the important words and numbers, but use fewer of them.

The total monthly cost is $15 for 200 minutes plus $0.10 times the number of minutes over 200. How many minutes can she talk for $20?

STEP 3 Simplify it again. Use a variable for the unknown.

The cost of *m* minutes at $0.10 per minute plus $15 is $20.

Practice

Use the method above to rewrite each problem.

1. **MONEY** Akira is saving money to buy a scooter that costs $125. He has already saved $80 and plans to save an additional $5 each week. In how many weeks will he have enough money for the scooter?

2. **SHOPPING** Joaquin wants to buy some DVDs that are each on sale for $10 plus a CD that costs $15. How many DVDs can he buy if he has $75 to spend?

Problem-Solving Investigation

Main Idea Solve problems by making a table.

✎ P.S.I. TEAM ✚

Make a Table

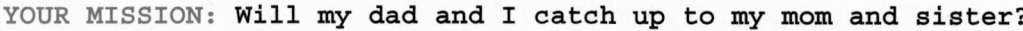

EMILIO: My family is moving from Washington, D.C., to Jacksonville, Florida, a distance of about 720 miles. My mom and sister left at 7:00 this morning and my mom drives an average of 45 miles per hour with stops. My dad and I are leaving at 8:00, but he drives an average of 60 miles per hour with stops.

YOUR MISSION: Will my dad and I catch up to my mom and sister?

Understand	You know the times they left and their rates. You need to know if Emilio and his dad will catch up to his mom and sister.
Plan	Make a table that shows how many miles each driver has driven.

Solve

Hours Since 7:00 A.M.	Distance Traveled (mi)	
	Emilio's Mom	Emilio's Dad
0	0	0
1	45	0
2	90	60
3	135	120
4	180	180

At 11:00 A.M., Emilio and Emilio's dad will catch up to his mom and sister.

Check	45 mph × 4 h = 180 mi 60 mph × 3 h = 180 mi
	The distances are equal. ✔

Analyze the Strategy

1. Describe another method you can use to solve this problem.

2. Describe two types of information you have seen recorded in a table.

3. ✏ **WRITE MATH** Write a problem that can be solved using a table. Then solve the problem.

Mixed Problem Solving

 = **Step-by-Step Solutions** begin on page R1.
Extra Practice begins on page EP2.

- Make a table.
- Use logical reasoning.
- Guess, check, and revise.
- Choose an operation.

Use the *make a table* strategy to solve Exercises 4 and 5.

4. **GAMES** Ross wants to rent a karaoke machine for a family reunion. The prices to rent the machine from two different companies are shown.

Company	Deposit	Cost per Day
Mike's Music	$5	$1.25
Karaoke Korner	$4	$1.50

For how many days must he rent the machine for the cost from each place to be the same?

5. **CONCERTS** An auditorium has 300 seats. At the beginning of a concert, 100 people were in the auditorium. Every 10 minutes, 15 people leave and 40 people sit down. If this continues, how long will it take for the auditorium to be full?

Use any strategy to solve Exercises 6–12.

6. **PHOTOGRAPHY** How many ways are there to arrange five French club members for a yearbook photo if the president and vice president must be seated in front with the other three members behind them?

7. **COUSINS** William, Scott, Sophia, and Christina are all cousins that live in different states: Ohio, Idaho, Colorado, and Arizona. William and Scott visited Sophia in Ohio. Christina likes to snowboard in her hometown of Denver. Scott lives the farthest south. Who lives in Idaho?

8. **FOOD** In 1984, a 12-ounce box of cold cereal cost $0.89. In 2008, an 18-ounce box of the same cereal sold for $3.29. How much more was the cost of the cereal per ounce in 2008 than in 1984? Explain.

9. **TELEVISION** The first sitcom in the United States aired in 1949. If it aired 30 new episodes per year, how many new episodes did it air until the sitcom went off the air in 1956?

10. **PLANTS** The table shows the height of a giant bamboo plant.

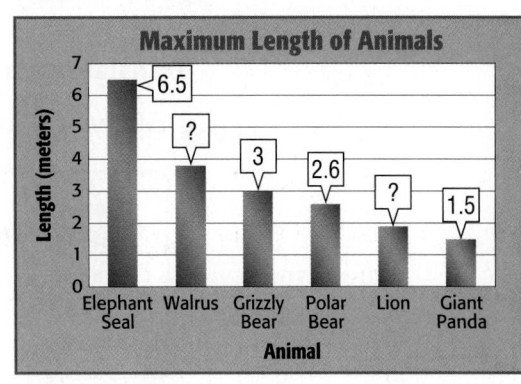

Number of Days	Total Growth (ft)
5	?
6	?
7	10
8	13.5
9	17

Assuming the bamboo grew at a steady rate, what was the height of the bamboo on the fifth day?

11. **E-MAIL** About 97 billion E-mails are sent daily worldwide. Over 40 billion of them are spam messages. At this rate, how many spam messages are sent yearly in a non-leap year?

12. **ANIMALS** The graph shows the maximum length of several animals. The maximum length of a walrus is twice the maximum length of a lion, which is 0.4 meter longer than the maximum length of a giant panda. Find the maximum length of a walrus.

Main Idea

Translate verbal phrases into algebraic expressions. Evaluate algebraic expressions.

 Vocabulary

variable
algebra
algebraic expression

 **Get ConnectED**

Variables and Expressions

PIZZA Mark's Pizza is offering a special deal. For a limited time, each pizza is only $4 and the delivery charge is $3. The total cost for 1, 2, 3, or 4 pizzas is shown in the table.

Number of Pizzas	Total Cost ($)
1	7
2	11
3	15
4	19

1. How could you find the total cost of the pizzas without the delivery charge?

2. How could you find the total cost of all the pizzas with the delivery charge?

3. Suppose you need to order 30 pizzas for a party. How much will the pizzas cost with delivery? Describe two different methods you could use to find the cost for any number of pizzas.

A **variable** is a symbol, usually a letter, used to represent a number. You can use the variable p to represent the number of pizzas ordered.

delivery charge → $3 + 4 \times p$ ← number of pizzas

expression for total cost of all pizzas

Algebra is a branch of mathematics that involves expressions with variables. The expression $3 + 4 \times p$ is called an **algebraic expression** because it contains at least one variable and at least one operation.

To translate verbal phrases into algebraic expressions, choose a variable to represent the unknown quantity in the problem.

EXAMPLES Translate Phrases into Algebraic Expressions

Translate each phrase into an algebraic expression.

1 **fifteen dollars more than Devon has**

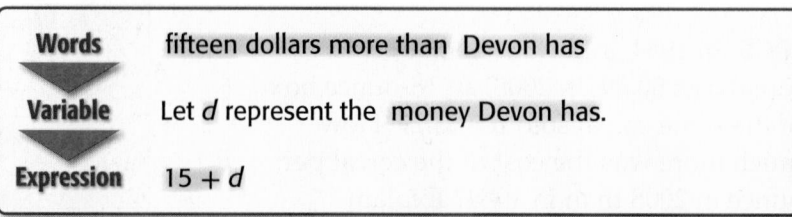

Words	fifteen dollars more than Devon has
Variable	Let d represent the money Devon has.
Expression	$15 + d$

2 **Kylie had 30 pencils and gave 4 to each of her friends.**

Words	had thirty and gave four to each friend.
▼	
Variable	Let f represent the number of friends.
▼	
Expression	$30 - 4f$

 CHECK Your Progress

Translate each phrase into an algebraic expression.

a. twenty-four more cookies than Hallie made
b. four less than three times the number of students

Recall, to evaluate algebraic expressions, first replace the variable or variables with the known values to produce a numerical expression. Then find the value of the numerical expression using the order of operations.

EXAMPLES Evaluate Algebraic Expressions

3 Evaluate $(a - b) + 2 \cdot (-1)$ if $a = -5$ and $b = 4$.

$(a - b) + 2 \cdot (-1) = (-5 - 4) + 2 \cdot (-1)$ Replace a with -5 and b with 4.

$= (-9) + 2 \cdot (-1)$ Perform operations in the parentheses first.

$= -9 + (-2)$ or -11 Multiply 2 and -1. Then add.

4 Evaluate $6(x - y) \div 3$ if $x = 7$ and $y = 4$.

$6(x - y) \div 3 = 6(7 - 4) \div 3$ Replace x with 7 and y with 4.

$= 6(3) \div 3$ Perform operations in the parentheses first.

$= 18 \div 3$ or 6 Multiply 6 and 3. Then divide.

QUICK Review

Absolute Value

$|-4| = 4 \quad |4| = 4$

5 Evaluate $\dfrac{3m + |n|}{2n + 8}$ if $m = 9$ and $n = -5$.

$\dfrac{3m + |n|}{2n + 8} = \dfrac{3(9) + |-5|}{2(-5) + 8}$ Replace m with 9 and n with -5.

$= \dfrac{27 + 5}{-10 + 8}$ Simplify.

$= \dfrac{32}{-2}$ or -16 Evaluate the numerator and denominator. Then divide.

 CHECK Your Progress

Evaluate each expression if $c = -3$ and $d = 7$.

c. $6c + 4 - 3d$ **d.** $4(d - c) + 1$ **e.** $\dfrac{7 + 3d}{4c - 2}$

REAL-WORLD EXAMPLE Use Expressions to Solve Problems

6 **GAMES** Games Galore rents video game systems and games. They charge a fee of $15 to rent a system plus $6 per game.

a. Write an expression that can be used to find the total cost to rent a system and any number of games.

Words	fifteen dollar system fee plus six dollars per game
Variable	Let g represent the number of games.
Expression	$15 + 6g$

The expression is $15 + 6g$.

b. Suppose Ravi wants to rent a system and five games. What will be the total cost?

$15 + 6g = 15 + 6(5)$ Replace g with 5.

$= 15 + 30$ or 45 Multiply. Then add.

The total cost will be $45.

CHECK Your Progress

DVDs Jenifer received 8 DVDs free when she joined a DVD mail-order club. She must buy 2 DVDs per month after that.

f. Write an expression to find the total number of DVDs she will have after any number of months.

g. Find the number of DVDs she will have if she keeps the membership for 15 months.

✓ CHECK Your Understanding

Examples 1 and 2 Translate each phrase into an algebraic expression.

1. twelve times as many people

2. fifteen less than twice Mary's points

Examples 3–5 Evaluate each expression if $a = 2$, $b = 7$, and $c = 4$.

3 $3\left(a + \dfrac{c}{a}\right) - b$ **4.** $4(a + b - c)$ **5.** $\dfrac{bc}{2}$ **6.** $\dfrac{c}{a - 4}$

Example 6 **7. SHIPPING** Antoine is shipping a birthday gift to his grandmother. Ship Express charges $4.25 for shipping plus an additional $0.15 per pound.

a. Write an expression to find the total cost of shipping the package for any number of pounds.

b. Find the total cost to ship the package if it weighs 6.5 pounds.

Practice and Problem Solving
= Step-by-Step Solutions begin on page R1.
Extra Practice begins on page EP2.

Examples 1 and 2 **Translate each phrase into an algebraic expression.**

8. twenty more dollars than three times the amount Amalia has

9. the number of pieces of candy divided equally among 6 friends

10. Kevin has 16 baseball cards and gets 5 cards from each of his friends

11. thirty-five cars less than twice the number in the parking lot

Examples 3–5 **Evaluate each expression if $w = 2$, $x = 6$, $y = 4$, and $z = -5$.**

12. $2x + y$ **13.** $3z - 2w$ **14.** $9 + 7x - y$ **15.** $12 + |z - x|$

16. $x(wx)$ **17.** $z(x + z) - 2$ **18.** $\dfrac{x - 3}{|2z + 1|}$ **19.** $\dfrac{wz}{y + 6}$

Example 6 **20. MEASUREMENT** One inch is equal to about 2.54 centimeters.

  a. Write an expression to find the centimeters in any number of inches.

  b. How many centimeters are in 32 inches?

21. RAFTING The rates to rent a raft from two different rental companies are shown.

  a. Write an expression to find the total cost to rent a raft for any number of hours from each company.

  b. Find the total cost to rent for 8 hours from each company.

Total Cost ($)		
Time (h)	Ryan's Rafts	Water Raft Rentals
1	15.00	6.00
2	17.25	9.00
3	19.50	12.00
4	21.75	15.00
5	24.00	18.00

22. GRAPHIC NOVEL Refer to the graphic novel frame below for Exercises a–b.

  a. Write an expression to find the cost to print and ship any number of pictures.

  b. How much would it cost for Brian to print and ship 75 pictures? 100 pictures?

23. **REASONING** Suppose you evaluate the expression $-16 + (n - 3)$ for various integer values of n. For what values of n will the expression be positive? negative? zero? Explain your reasoning.

CHALLENGE Write an expression to find the total number of counters in any figure. Then evaluate the expression for the tenth figure.

24.

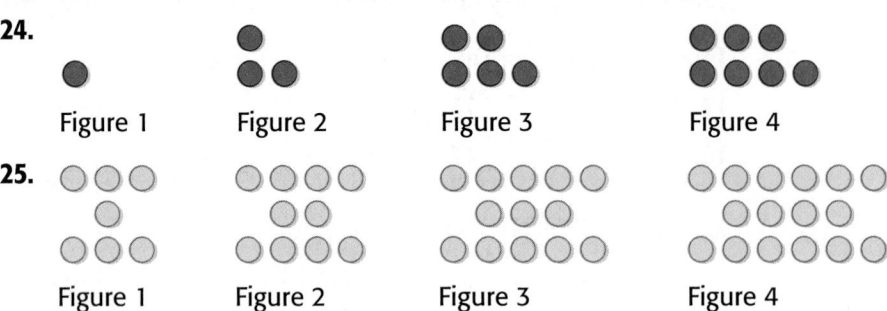

Figure 1   Figure 2   Figure 3   Figure 4

25.

Figure 1   Figure 2   Figure 3   Figure 4

26.  **WRITE MATH** Write a real-world problem for the expression $16 + 7x$.

Test Practice

27. Which of the following phrases is equivalent to $4 + 9x$?

   A. four more than nine players

   B. nine more than four players

   C. four times more than nine players

   D. four more than nine times the players

28. What is the value of the expression $\frac{a - b}{2c}$ if $a = 5$, $b = -3$, and $c = -2$?

   F. -4 **H.** 2

   G. -2 **I.** 4

29. Samantha has saved $100 to buy a new laptop. Each month, she saves $15 towards the laptop. Which expression could be used to find the total amount saved after m months?

   A. $15m$

   B. $100 + 15m$

   C. $100m + 15$

   D. $100m + 15m$

30. The expression s^2 can be used to find the area of a square, where s is the length of a side of the square. If the area of a square is 144 square centimeters, what is the perimeter of the square?

   F. 3 cm   **H.** 48 cm

   G. 16 cm   **I.** 412 cm

31. **THINK SOLVE EXPLAIN** **EXTENDED RESPONSE** The cost to rent a jet ski from a marina is $20 plus $5 per hour.

   Part A Write an expression to find the total cost for any number of hours.

   Part B Copy and complete the table to show how much it will cost for 2, 3, 4, or 5 hours.

Hours	Total Cost ($)
2	▪
3	▪
4	▪
5	▪

Main Idea

Graph ordered pairs on the coordinate plane and use the coordinate plane to represent relations.

 **Vocabulary**

coordinate plane
origin
y-axis
x-axis
quadrants
ordered pair
x-coordinate
y-coordinate
relation
domain
range

Get Connect ED

Ordered Pairs and Relations

MAPS A map of the Lincoln city park is shown.

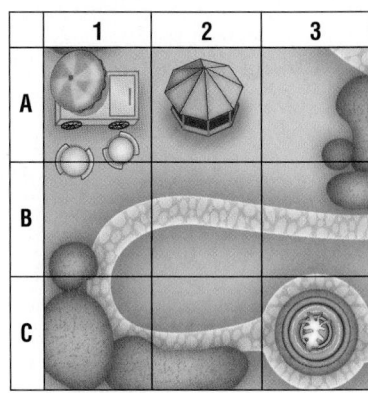

1. The map guide says that a fountain is located in section C3. Describe a method you would use to locate the fountain on the map.

2. A snack vendor is located 2 squares west and 2 squares north of the fountain. How could you use the map coordinates to tell a friend where the snack vendor is located?

You can locate a point by using a coordinate system similar to the grid of the park. It is called a **coordinate plane**.

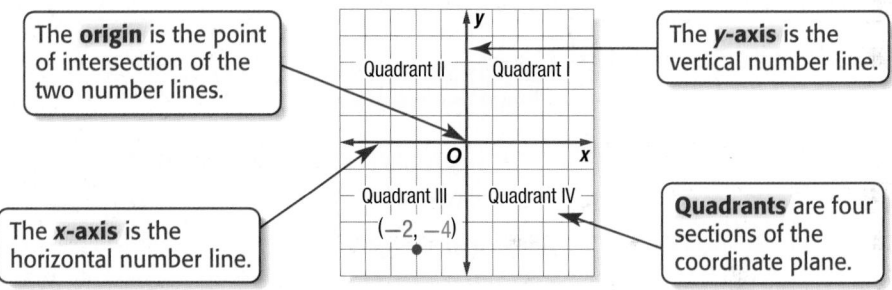

The **origin** is the point of intersection of the two number lines.

The **y-axis** is the vertical number line.

The **x-axis** is the horizontal number line.

Quadrants are four sections of the coordinate plane.

Any point on the coordinate plane can be graphed by using an **ordered pair** of numbers. The first number in the ordered pair is the **x-coordinate**. The second number is the **y-coordinate**.

 EXAMPLES Name an Ordered Pair

1) **Name the ordered pair for point P.**

- Start at the origin.
- Move right to find the x-coordinate of point P, which is $3\frac{1}{2}$.
- Move up to find the y-coordinate, which is 2.

So, the ordered pair for point P is $\left(3\frac{1}{2}, 2\right)$.

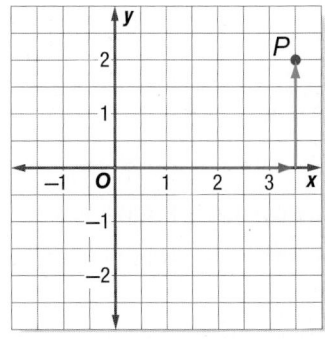

2 Name the ordered pair for point Q.

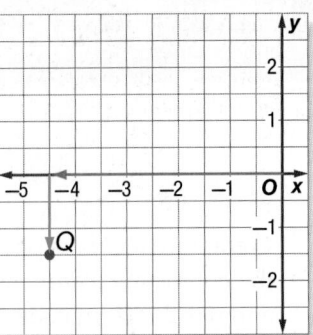

- Start at the origin.

- Move left to find the x-coordinate of point Q, which is $-4\frac{1}{2}$.

- Move down to find the y-coordinate, which is $-1\frac{1}{2}$.

So, the ordered pair for point Q is $\left(-4\frac{1}{2}, -1\frac{1}{2}\right)$.

✓ **CHECK Your Progress**

Name the ordered pair for each point.

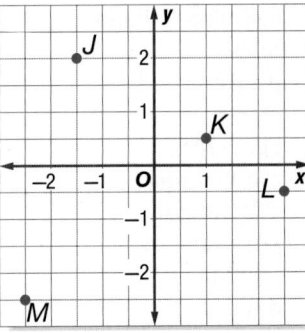

a. J

b. K

c. L

d. M

📝 **EXAMPLES** Graph Ordered Pairs

Graph each ordered pair on a coordinate plane.

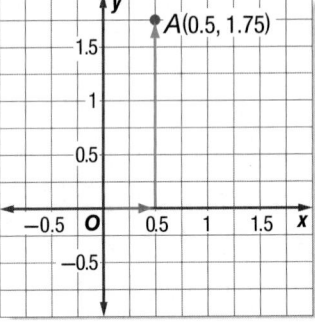

3 $A(0.5, 1.75)$

- Start at the origin and move 0.5 unit to the right. Then move up 1.75 units.

- Draw a dot and label it $A(0.5, 1.75)$.

4 $B\left(-2, -3\frac{1}{4}\right)$

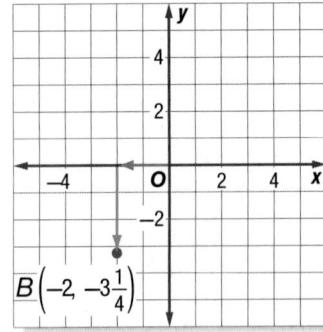

- Start at the origin and move 2 units to the left. Then move down $3\frac{1}{4}$ or 3.25 units.

- Draw a dot and label it $B\left(-2, -3\frac{1}{4}\right)$.

✓ **CHECK Your Progress**

e. $R\left(2\frac{1}{4}, 3\frac{1}{2}\right)$ **f.** $S(-1.5, 3)$ **g.** $T\left(-\frac{1}{2}, -3\frac{3}{4}\right)$

Study Tip

Graphing The grid below shows the correct location for each type of point.

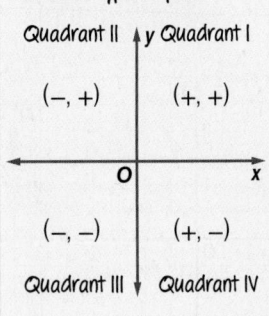

Quadrant II ↑y Quadrant I

$(-, +)$ $(+, +)$

O ————————→ x

$(-, -)$ $(+, -)$

Quadrant III ↓ Quadrant IV

A **relation** is any set of ordered pairs. Relations can be represented as a table and as a graph. The **domain** of the relation is the set of x-coordinates. The **range** of the relation is the set of y-coordinates.

Key Concept   Relations

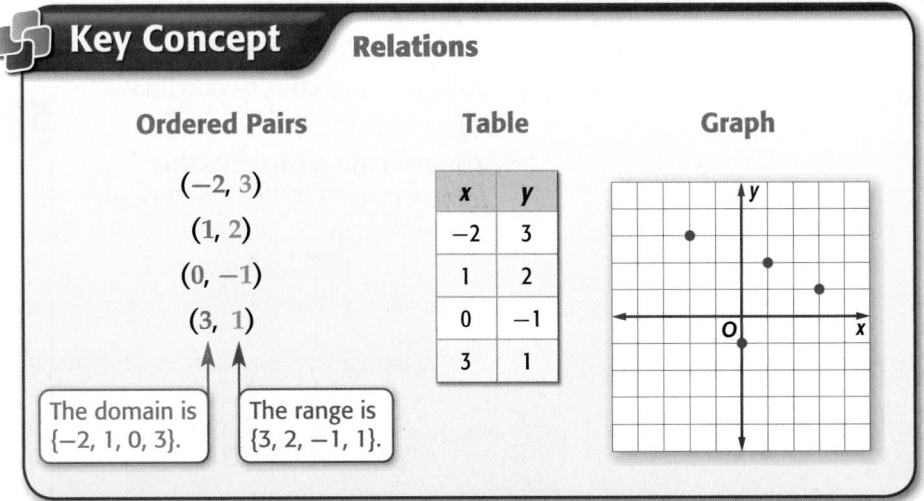

Ordered Pairs	Table	Graph
(−2, 3) (1, 2) (0, −1) (3, 1)		

x	y
−2	3
1	2
0	−1
3	1

The domain is {−2, 1, 0, 3}.

The range is {3, 2, −1, 1}.

EXAMPLE   **Relations as Tables and Graphs**

5 Express the relation {(2, 6), (−4, −8), (−3, 6), (0, −4)} as a table and a graph. Then state the domain and range.

Place the ordered pairs in a table with x-coordinates in the first column and the y-coordinates in the second column.

x	y
2	6
−4	−8
−3	6
0	−4

Graph the ordered pairs on a coordinate plane.

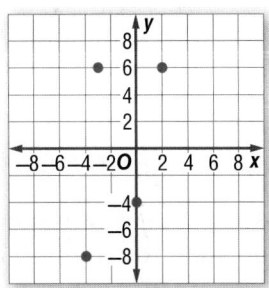

The domain is {−4, −3, 0, 2}. The range is {−8, −4, 6}.

CHECK Your Progress

Express each relation as a table and a graph. Then state the domain and range.

h. {(−5, 2), (3, −1), (6, 2), (1, 7)}   **i.** {(−4, 3), (2, −6), (−4, 5), (−3, −6)}

6 **PARKING** It costs $3 per hour to park at the Wild Wood Amusement Park.

a. Make a table of ordered pairs in which the *x*-coordinate represents the hours and the *y*-coordinate represents the total cost for 3, 4, 5, and 6 hours.

x	y
3	9
4	12
5	15
6	18

b. Graph the ordered pairs.

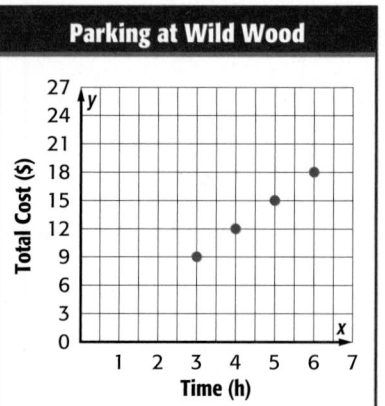

CHECK Your Progress

MOVIES A movie rental store charges $3.95 per movie rental.

j. Make a table of ordered pairs in which the *x*-coordinate represents the number of movies rented and the *y*-coordinate represents the total cost for 1, 2, 3, or 4 movies.

k. Graph the ordered pairs.

CHECK Your Understanding

Examples 1 and 2 **Name the ordered pair for each point.**

1. *A*

2. *B*

3. *C*

4. *D*

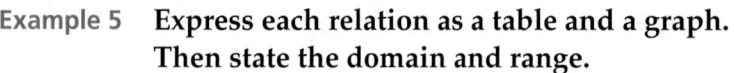

Examples 3 and 4 **Graph each ordered pair on a coordinate plane.**

5. $J\left(\dfrac{1}{4}, 3\dfrac{1}{2}\right)$

6. $K\left(-1, -2\dfrac{3}{4}\right)$

7. $L(4.5, -2.25)$

8. $M(-2.5, 2.5)$

Example 5 **Express each relation as a table and a graph. Then state the domain and range.**

9 $\{(-4, 3), (2, 1), (0, 3), (-3, -2)\}$

10. $\{(5, 3), (-4, 1), (2, -5), (3, -4)\}$

Example 6 **11. ACTIVITIES** At a vacation resort, you can rent a personal watercraft for $20 per hour.

a. Make a table of ordered pairs in which the *x*-coordinate represents the number of hours and the *y*-coordinate represents the total cost for 1, 2, 3, or 4 hours.

b. Graph the ordered pairs.

Practice and Problem Solving

● = Step-by-Step Solutions begin on page R1.
Extra Practice begins on page EP2.

Examples 1 and 2   Name the ordered pair for each point.

12. P

13. Q

14. R

15. S

16. T

17. U

18. V

19 W

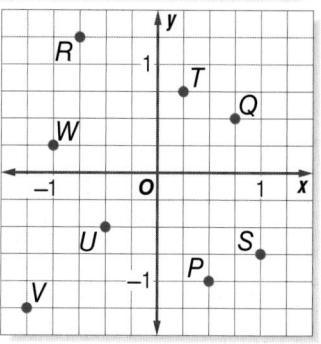

Examples 3 and 4   Graph each ordered pair on a coordinate plane.

20. $E\left(\dfrac{3}{4}, 2\dfrac{1}{4}\right)$ **21.** $F\left(\dfrac{2}{5}, 1\dfrac{1}{2}\right)$ **22.** $G\left(-3, 4\dfrac{2}{3}\right)$

23. $H\left(-2\dfrac{1}{4}, 3\dfrac{4}{5}\right)$ **24.** $J(4.3, -3.1)$ **25.** $K(-3.75, -0.5)$

Example 5   Express each relation as a table and a graph. Then state the domain and range.

26. $\{(8, 5), (-6, -9), (2, 5), (0, -8)\}$ **27.** $\{(9, 4), (5, -7), (-3, -4), (-8, 7)\}$

28. $\left\{\left(2\dfrac{1}{2}, -1\dfrac{1}{2}\right), \left(2, \dfrac{1}{2}\right), \left(-1, 2\dfrac{1}{2}\right),\right.$ **29.** $\{(-1.25, 3.75), (2.5, -1.75), (3, -1),$
$\left.(-2, -2)\right\}$ $(-1.5, -3.25)\}$

Example 6   **30. MANUFACTURING** A company can manufacture 825 small cars per day.

     a. Make a table of ordered pairs in which the x-coordinate represents the number of days and the y-coordinate represents the total number of cars produced in 1, 2, 3, 4, and 5 days.

     b. Graph the ordered pairs.

31. PARADES An annual parade follows a 5.5-mile route at a speed of 2.5 miles per hour.

     a. Make a table of ordered pairs in which the x-coordinate represents the minutes and the y-coordinate represents the total distance traveled for 30, 60, 90, and 120 minutes.

     b. Graph the ordered pairs.

32. GEOMETRY Graph the points $A(1, 4)$, $B(-4, 4)$, and $C(-4, -1)$ on a coordinate plane. What are the coordinates of point D if points A, B, C, and D form a square?

33 **MULTIPLE REPRESENTATIONS** Refer to the table at the right.

     a. WORDS Describe the pattern, if any, in the table.

     b. NUMBERS Write the ordered pairs (x, y).

     c. GRAPHS Graph the ordered pairs on a coordinate plane.

     d. WORDS Describe the graph. How is it different from the other graphs in this lesson?

x	y
1	1
2	4
3	9
4	16
5	25

34. FIND THE ERROR Hana is finding the domain of the relation {(2, 3), (−4, 2), (0, −4), (1, 5)}. Find her mistake and correct it.

Domain: {-4, 2, 3, 5}

35. ✏️ **WRITE MATH** Use the Internet or another source to find information that could be written as ordered pairs. Then describe how you would make a table and a graph to represent the information.

Test Practice

36. Elias plotted his four favorite stores on a coordinate plane.

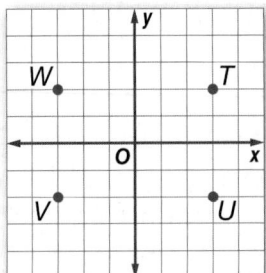

Which point on the grid **best** represents the location of the video store V?

A. (3, 2) **C.** (−3, −2)

B. (3, −2) **D.** (−3, 2)

37. What is the range of the relation {(1, 4), (3, 0), (5, 5), (7, 4)}?

F. {(1, 4), (3, 0), (5, 5), (7, 4)}

G. {1, 3, 5, 7}

H. {0, 4, 5}

I. {1, 4}

38. Point A is located 5 units right and two units up from point $B(−2, 4)$. What are the coordinates of point A?

A. (7, 6) **C.** (−7, 2)

B. (3, 6) **D.** (3, 2)

Spiral Review

39. ALGEBRA Evaluate the expression $\frac{1}{2}ab$ if $a = 7$ and $b = 4$. (Lesson 1B)

Translate each phrase into an algebraic expression. (Lesson 1B)

40. sixteen more tokens than Samantha has

41. three less than two times the number of busses

42. PLATES Julian needs to buy paper plates for a party. If he needs 110 plates, which package from the table should he buy to pay the least amount? Use the *make a table* strategy. (Lesson 1A)

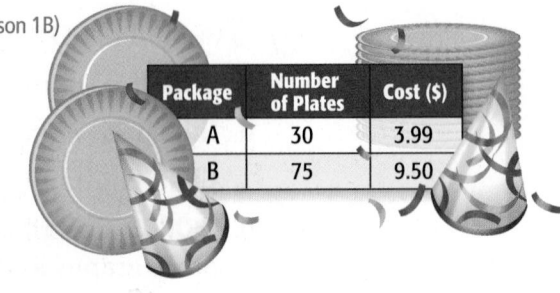

Package	Number of Plates	Cost ($)
A	30	3.99
B	75	9.50

Patterns

Main Idea

Find a rule for a given pattern.

PATTERNS Keenan created the pattern shown below using toothpicks. What will be the perimeter of Figure 10?

ACTIVITY

STEP 1 Study the pattern. If one toothpick is one unit, Figure 1 has a perimeter of 6 units, Figure 2 has a perimeter of 8 units, and Figure 3 has a perimeter of 10 units.

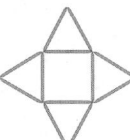

Figure 1 **Figure 2** **Figure 3**

STEP 2 Make the next two figures in the pattern.

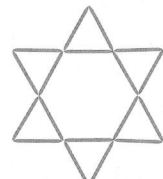

Figure 4 **Figure 5**

STEP 3 Copy and complete the table.

Figure Number	1	2	3	4	5	6	7
Perimeter	6	8	10	■	■	■	■

STEP 4 Describe a rule you could use to find the perimeter for any figure. The perimeter increases by 2 for each figure. So, multiply the figure number by 2. Then add 4.

The perimeter of the tenth figure is $2 \times 10 + 4$ or 24 units.

Analyze the Results

1. Use the rule to find the perimeter of Figures 15 and 25.

2. Suppose Keenan wanted to find the number of toothpicks needed to make Figure 10. Find a pattern and write a rule to find the number of toothpicks in any figure.

Main Idea

Translate information in tables to expressions.

 Vocabulary

sequence
term
arithmetic sequence
common difference
geometric sequence

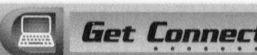 Get Connect**ED**

Analyze Tables

PARTIES Abigail is planning to have her birthday party at a skating rink. Each rink charges a party fee plus an additional charge for each guest.

Number of Guests	Total Cost ($)	
	Skate World	**Roller Heaven**
1	50	45
2	53	50
3	56	55
4	59	59
5	62	63
6	65	67
7	68	70

1. For each skating rink, describe the pattern in the total cost if one exists.

2. If possible, write an algebraic expression for the total cost for each skating rink. If an expression cannot be written, explain why.

In the example above, the cost at Skate World forms a sequence. A **sequence** is an ordered list of numbers. Each number in the list is called a **term**. An **arithmetic sequence** is a sequence of numbers in which the difference between any two consecutive terms is the same.

$$+1 \quad +1 \quad +1 \quad +1 \quad +1 \quad +1$$

Term Number (n)	1	2	3	4	5	6	7
Term	50	53	56	59	62	65	68

$$+3 \quad +3 \quad +3 \quad +3 \quad +3 \quad +3$$

The difference is called the **common difference**.

EXAMPLES Use a Table to Analyze a Sequence

Use a table to write an expression that can be used to find the nth term of each sequence. Then use the expression to find the next three terms.

 4, 8, 12, 16, …

Make a table to analyze the pattern. The terms have a common difference of 4. Also, each term is 4 times its term number.

$$+1 \quad +1 \quad +1$$

Term Number (n)	1	2	3	4
Term	4	8	12	16

$$+4 \quad +4 \quad +4$$

An expression that can be used to find the nth term is $4n$.

The next three terms are $4(5)$ or 20, $4(6)$ or 24, and $4(7)$ or 28.

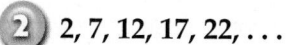

2 **2, 7, 12, 17, 22, . . .**

The terms have a common difference of 5. Start with 5n. Then subtract 3 to get the term. So, the expression to find the nth term is 5n − 3.

	+1	+1	+1	+1
Term Number (n)	1	2	3	4	5
Term	2	7	12	17	22
	+5	+5	+5	+5

The next three terms are 5(6) − 3 or 27, 5(7) − 3 or 32, and 5(8) − 3 or 37.

 CHECK Your Progress

Use a table to write an expression that can be used to find the nth term of each sequence. Then use the expression to find the next three terms.

a. −2, −4, −6, −8, −10, ...   **b.** 8, 12, 16, 20, 24, ...

c. −9, −4, 1, 6, 11, ...   **d.** $\frac{1}{2}$, 1, $1\frac{1}{2}$, 2, $2\frac{1}{2}$, ...

Real-World Link · · ·
Approximately 45% of teens use a cell phone and 33% use text messaging.

REAL-WORLD EXAMPLES

TEXT MESSAGING The table shows the monthly cost of sending text messages.

3 Write an expression that can be used to find the total cost for t text messages.

The common difference between the costs is 0.10. Start with 0.10t. Then add $10 to get the total cost.

The expression is 0.10t + 10.

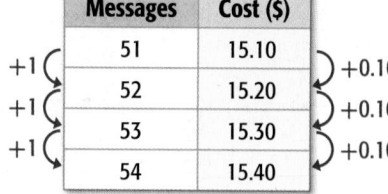

Messages	Cost ($)
51	15.10
52	15.20
53	15.30
54	15.40

4 How much would it cost to send 60 text messages?

Evaluate the expression when t = 60.

$0.10t + 10 = 0.10(60) + 10$ Replace t with 60.

$= 6 + 10$ Multiply 0.10 and 60.

$= 16$ Simplify.

It will cost $16 to send 60 text messages.

 CHECK Your Progress

MONEY The table shows how much money Sydney earns at her job.

Number of Hours	Money Earned ($)
3	19.50
4	26.00
5	32.50
6	39.00

e. Write an expression that can be used to find her total earnings in h hours.

f. How much would she earn if she worked for 1 hour?

Examples 1 and 2

Write an expression that can be used to find the *n*th term of each sequence. Then use the expression to find the next three terms.

1.

Term Number (*n*)	1	2	3	4
Term	3	6	9	12

2.

Term Number (*n*)	1	2	3	4
Term	−20	−16	−12	−8

3. $\frac{1}{6}, \frac{1}{3}, \frac{1}{2}, \frac{2}{3}, \ldots$

4. $1\frac{2}{3}, 2\frac{1}{3}, 3, 3\frac{2}{3}, \ldots$

Examples 3 and 4

5. FITNESS Each week, Luther increases the amount of time he spends jogging per day.

   a. Write an expression that can be used to find how many minutes he will jog per day in *w* weeks.

   b. How many minutes will he spend jogging each day during his tenth week of jogging?

Week	Time Jogging (min)
1	8
2	16
3	24
4	32

Practice and Problem Solving

> = Step-by-Step Solutions begin on page R1.
> Extra Practice begins on page EP2.

Examples 1 and 2

Write an expression that can be used to find the *n*th term of each sequence. Then use the expression to find the next three terms.

6.

Term Number (*n*)	1	2	3	4
Term	2	4	6	8

7.

Term Number (*n*)	1	2	3	4
Term	12	24	36	48

8.

Term Number (*n*)	1	2	3	4
Term	5	9	13	17

9.

Term Number (*n*)	1	2	3	4
Term	1	4	7	10

10. 8, 2, −4, −10, …

11. 25, 23, 21, 19, …

12. 3, 10, 17, 24, …

13. $\frac{1}{10}, \frac{1}{5}, \frac{3}{10}, \frac{2}{5}, \ldots$

14. $\frac{2}{5}, \frac{4}{5}, 1\frac{1}{5}, 1\frac{3}{5}, \ldots$

15. $2, 1\frac{1}{4}, \frac{1}{2}, -\frac{1}{4}, \ldots$

Examples 3 and 4

16. MEMBERSHIPS Mrs. Perry wants to buy a membership to the zoo. There are different options available, depending on the number of people she wants included on the membership.

   a. Write an expression that can be used to find how much the membership will cost for *p* people.

   b. How much will the membership cost if she buys a membership for 10 people?

Number of People	Cost ($)
5	65
6	80
7	95
8	110

17. MONEY An online movie rental company charges an annual rate of $125 for their service. However, after renting 100 movies, an additional fee per movie is charged.

Number of Movies	Cost ($)
101	126.50
102	128.00
103	129.50
104	131.00
105	132.50

    a. Write an expression that can be used to find how much it will cost to rent *m* movies.

    b. How much will you pay if you rent 130 movies?

18. SKIING A ski resort offers a one-day lift pass for $40 and a yearly lift pass for $400.

Number of Visits	1	2	3	4	5
Total Cost with One-Day Passes ($)	40	80	■	■	■
Total Cost with Yearly Passes ($)	400	400	■	■	■

    a. Is the sequence formed by the total cost with one-day passes arithmetic? yearly passes? Explain your reasoning.

    b. How many times would a person have to go skiing to make the yearly pass a better buy?

19 PATTERNS The figures at the right were made using algebra tiles.

    a. Make a table relating the figure number to the number of tiles in each figure.

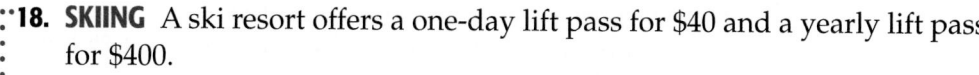

Figure 1   Figure 2   Figure 3

    b. Write an expression to find the number of tiles in any figure *f*.

    c. How many tiles will be in Figure 18?

    d. Is the number of tiles in each figure proportional to the number of the figure? Explain.

20. MULTIPLE REPRESENTATIONS The pattern below was formed using square pattern blocks.

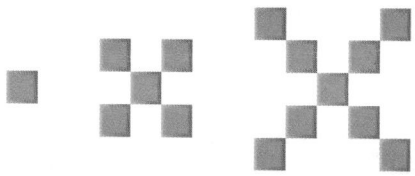

Figure 1   Figure 2   Figure 3

    a. MODELS Use pattern blocks to create the next two figures in the pattern.

    b. TABLES Make a table to show the number of blocks used to create Figures 1 through 5. Then write an expression that could be used to find the number of squares in any figure.

    c. GRAPHS Write the data as a set of ordered pairs (figure number, number of blocks). Graph the ordered pairs on a coordinate plane. Then describe the graph.

21. **Which One Doesn't Belong?** Each set of ordered pairs represents a term number and term in a pattern. Identify the ordered pair that does not belong with the other three. Explain your reasoning.

| (1, 4) | (4, 13) | (3, 10) | (2, 5) |

22. **CHALLENGE** Write an expression that can be used to find the nth term of the sequence shown.

Position	1	3	5	7
Term	8	14	20	26

23. **WRITE MATH** Write a real-world situation involving a sequence. Describe the sequence. Then, write an algebraic expression to find any number in the sequence.

Test Practice

24. Which expression can be used to find the value of the term in the nth position?

Position	Value of Term
1	0.6
2	1.2
3	1.8
4	2.4
5	3.0
n	?

A. $n - 0.4$ **C.** $\frac{3}{5}n$

B. $\frac{n}{5}$ **D.** $n + 0.6$

25. **GRIDDED RESPONSE** Mr. Counts is using lengths of rope to make nets. The table shows how many nets can be made from different lengths of rope.

Length of Rope (ft)	Number of Nets
6	2
12	4
18	6
24	8

How many nets can be made from 36 feet of rope?

More About Sequences

Geometric sequences are sequences in which each term after the first is found by *multiplying* the previous term by the same number. In the example below, each term is found by multiplying the previous term by 4.

Term Number (n)	1	2	3	4
Term	1	4	16	64

Determine whether each sequence is *arithmetic* or *geometric*. Explain your reasoning. Then write the next three terms in the sequence.

26. 6, 12, 24, 48, … **27.** 4, 7, 10, 13, … **28.** 27, 9, 3, 1, …

Main Idea

Translate information in graphs to expressions.

Vocabulary
linear

Analyze Graphs

MARATHONS A marathon is a race of 26.2 miles. In 2007, Haile Gebrselassie set a world record time of 2 hours 4 minutes and 26 seconds. The table shows the distance he traveled at his average rate.

Time (s)	Distance (ft)
1	18.5
2	37.0
3	55.5
4	74.0
5	92.5

1. Describe the change of the values in the distance column. Do the data form an arithmetic sequence?

2. Write the data as ordered pairs (time, distance). Then graph the ordered pairs on a coordinate plane. Describe the graph.

Any set of ordered pairs is called a relation. A graph can help to find a pattern, if one exists, in the relation. The graph of the relation in the example above is considered **linear** because the points fall in a straight line.

REAL-WORLD EXAMPLES

Translate a Graph into an Expression

CRAFTS A craft store is having a sale. If you buy 1 yard of a fabric at full price, then each yard after that is on sale. The total cost of different amounts of fabric are shown in the graph.

1. Write an algebraic expression to represent the data in the graph.

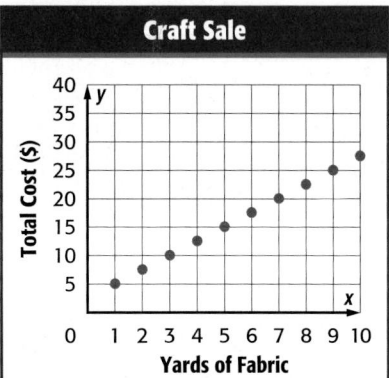

Write the ordered pairs in a table.

	+1	+1	+1	+1

Yards of Fabric (*f*)	1	2	3	4	5
Total Cost ($)	5	7.50	10	12.50	15

+2.50 +2.50 +2.50 +2.50

The common difference of 2.50 implies that the expression is 2.50*f*. But you need to add 2.50 to get the total cost. The expression is 2.50*f* + 2.50.

Use the expression to find the cost of 7.5 yards of fabric.

Evaluate the expression when $f = 7.5$.

$2.50f + 2.50 = 2.50(7.5) + 2.50$ Replace f with 7.5.

$= 18.75 + 2.5$ Multiply 2.50 and 7.5.

$= 21.25$ Simplify.

It will cost $21.25 to buy 7.5 yards of fabric.

✓ CHECK Your Progress

TRAINS An electric train set runs on a track that is 25 meters long. The graph shows how many times the electric train goes around its track each minute.

a. Write an algebraic expression to represent the data in the graph.

b. Use the expression to find how many times the train would circle the track in one hour.

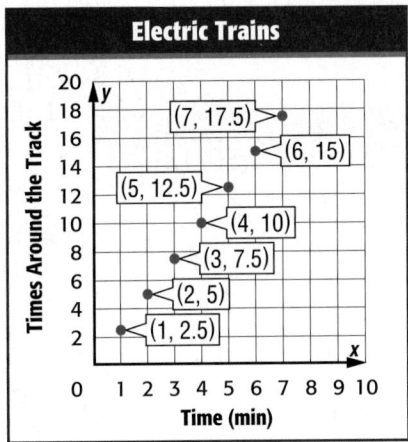

DRIVING The graph shows Bruce's distance from home on a car trip.

c. Write an algebraic expression to represent the data in the graph.

d. Use the expression to find Bruce's distance from home after driving 2.5 hours.

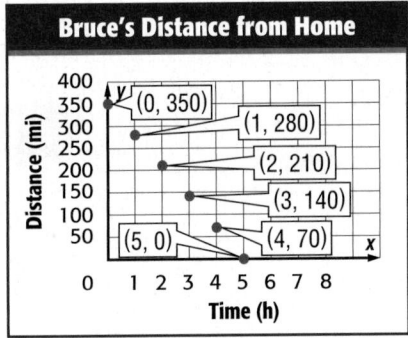

 Key Concept   Tables and Graphs

Words   Each term is two less than twice the term number.

Expression $2n - 2$

Table

Term	1	2	3	4
Term Number	0	2	4	6

Graph

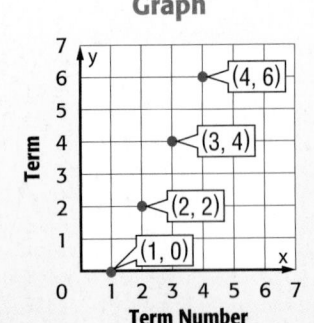

Example 1 **Write an algebraic expression to represent the data in each graph.**

1.
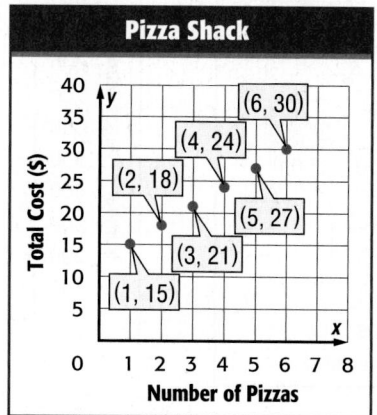

Pizza Shack

(6, 30)
(4, 24)
(2, 18)
(5, 27)
(3, 21)
(1, 15)

Total Cost ($)
Number of Pizzas

2.
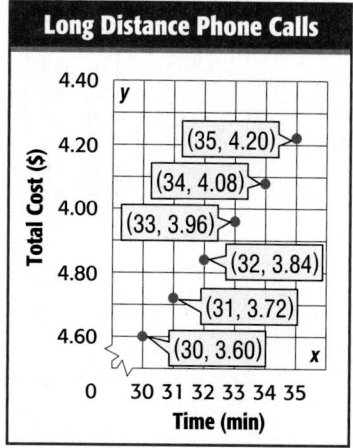

Long Distance Phone Calls

(35, 4.20)
(34, 4.08)
(33, 3.96)
(32, 3.84)
(31, 3.72)
(30, 3.60)

Total Cost ($)
Time (min)

Examples 1 and 2 **3. GAMES** The graph shows the number of tokens you receive for each dollar spent at the Play More Arcade.

  a. Write an algebraic expression to represent the data in the graph.

  b. Use the expression to find how many tokens you would receive for $30.

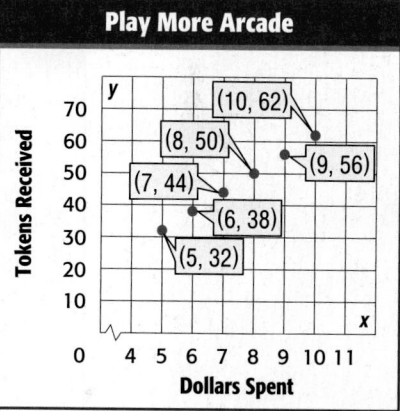

Play More Arcade

(10, 62)
(8, 50)
(9, 56)
(7, 44)
(6, 38)
(5, 32)

Tokens Received
Dollars Spent

Practice and Problem Solving

● = **Step-by-Step Solutions** begin on page R1.
Extra Practice begins on page EP2.

Example 1 **Write an algebraic expression to represent the data in each graph.**

4.
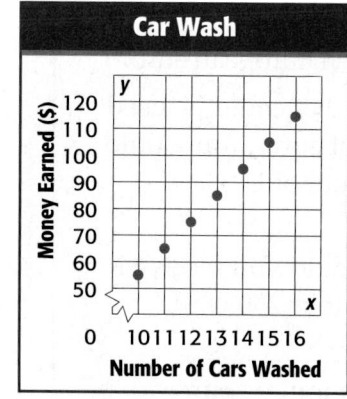

Car Wash

Money Earned ($)
Number of Cars Washed

5.
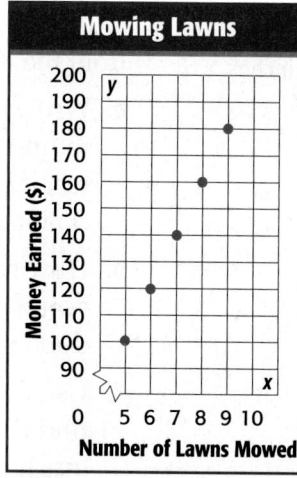

Mowing Lawns

Money Earned ($)
Number of Lawns Mowed

Example 1 Write an algebraic expression to represent the data in each graph.

6.

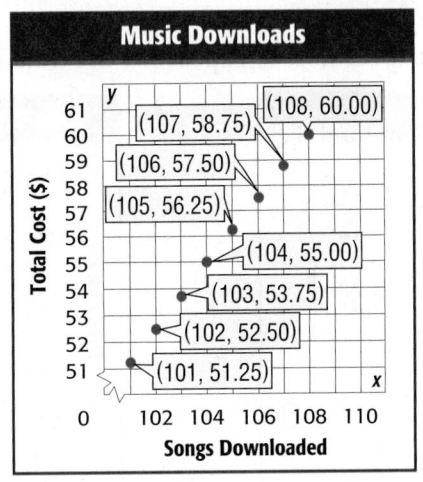

Music Downloads

(108, 60.00)
(107, 58.75)
(106, 57.50)
(105, 56.25)
(104, 55.00)
(103, 53.75)
(102, 52.50)
(101, 51.25)

Total Cost ($)
Songs Downloaded

7.

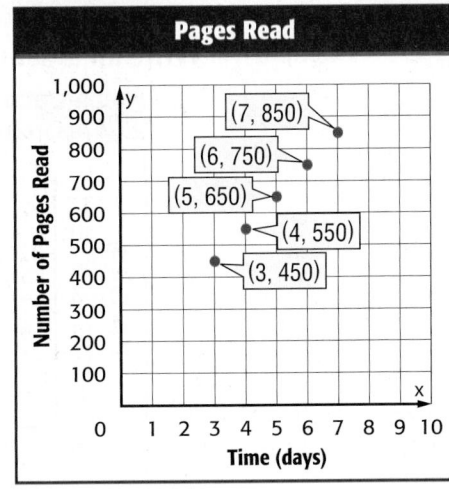

Pages Read

(7, 850)
(6, 750)
(5, 650)
(4, 550)
(3, 450)

Number of Pages Read
Time (days)

Examples 1 and 2

8. ANIMALS The graph shows the pounds of food elephants in captivity eat per day.

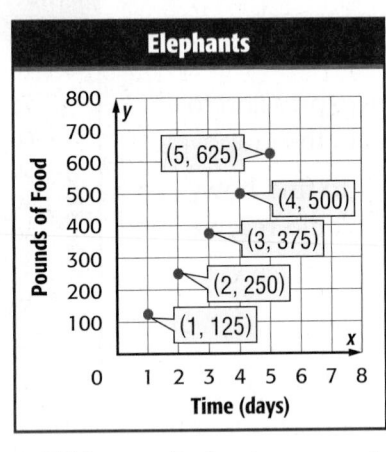

Elephants

(5, 625)
(4, 500)
(3, 375)
(2, 250)
(1, 125)

Pounds of Food
Time (days)

a. Write an algebraic expression to represent the data in the graph.

b. Use the expression to find how much food an elephant would eat in two weeks.

9 TIRES The graph shows the number of times per minute a tire on a car rotates.

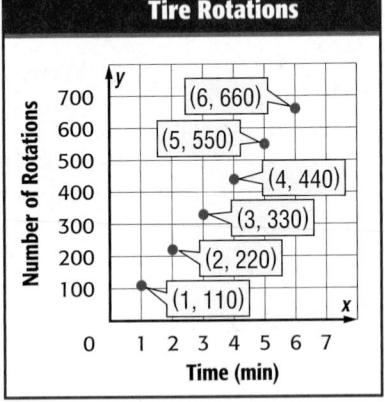

Tire Rotations

(6, 660)
(5, 550)
(4, 440)
(3, 330)
(2, 220)
(1, 110)

Number of Rotations
Time (min)

a. Write an algebraic expression to represent the data in the graph.

b. Use the expression to find how many rotations a tire will make in 1.5 hours.

10. MONEY The graph shows the exchange rate from the U.S. dollar to the European euro.

a. Write an expression that could be used to find the amount of euros you would receive for any amount of U.S. dollars.

b. How many euros would you receive if you wanted to exchange 250 U.S. dollars? Explain.

c. Use the Internet or another source to find the current exchange rate. How much is 250 U.S. dollars currently worth in European euros?

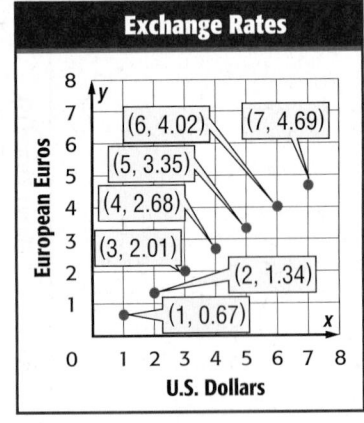

Exchange Rates

(6, 4.02) (7, 4.69)
(5, 3.35)
(4, 2.68)
(3, 2.01)
(2, 1.34)
(1, 0.67)

European Euros
U.S. Dollars

11. FIND THE ERROR Mandar is finding an expression to represent the graph. Find his mistake and correct it.

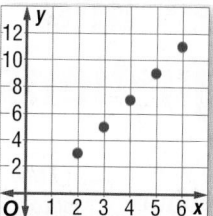

The expression is $2n + 2$.

12. OPEN ENDED Make a graph and a scenario to represent the expression $4n - 2$.

13. WRITE MATH Is it easier to write an expression based directly on a graph or to translate data into a table first? Explain your reasoning.

Test Practice

14. Which expression **best** represents the information in the graph below?

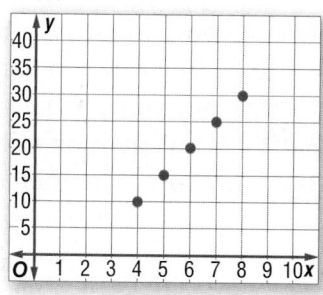

A. $5n$

B. $5n + 10$

C. $5n - 10$

D. 5

15. GRIDDED RESPONSE The miles Mr. Johnson drove are shown.

Mr. Johnson's Travels

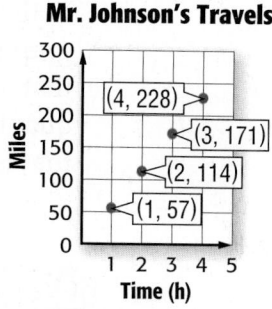

If the pattern continues, how many miles will he have driven after six hours?

Spiral Review

16. SAVINGS Orlando has $20 in his savings account. Each week, he adds $5 and does not take out any money. (Lesson 2B)

a. Write an expression that can be used to find his total savings in w weeks.

b. How much money will he have saved after 7 weeks?

Number of Weeks	Money Saved ($)
1	25
2	30
3	35
4	40

Graph each ordered pair on a coordinate plane. (Lesson 1C)

17. $A(-4, 2)$

18. $B(3, -1)$

19. $C(0, -3)$

20. $D(1, 4)$

Main Idea

Translate tables and graphs into linear equations.

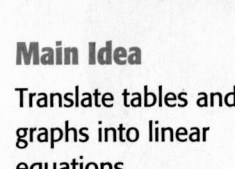

Vocabulary

linear equation

Translate Tables and Graphs into Equations

SPACE The fastest spacecraft, New Horizons, travels at a rate of about 12 miles per second. The table shows the total distance d that the craft covers in certain periods of time t.

Time (seconds)	Distance (miles)
1	12
2	24
3	36
4	48
5	60

1. Describe the shape of the graph that would be formed by the ordered pairs in the table.

2. Write an algebraic expression for the distance in miles for any number of seconds s.

In the previous lessons, you wrote an expression from a table or graph. When two expressions are equal, they form an equation. Recall that an equation is a mathematical sentence stating that two quantities are equal. A **linear equation** is an equation with a graph that is a straight line. Some equations contain more than one variable.

REAL-WORLD EXAMPLES

Translate a Table into an Equation

MEASUREMENT The table shows the liters in quarts of liquid.

(1) **Write an equation to find the number of liters in any number of quarts. Describe the relationship in words.**

Let ℓ represent the liters and q represent the quarts. The equation is $\ell = 0.95q$.

There are 0.95 liters in every quart.

Quarts, q	Liters, ℓ
1	0.95
2	1.9
3	2.85
4	3.8
5	4.75

$+0.95$
$+0.95$
$+0.95$
$+0.95$

(2) **About how many liters are in 8 quarts?**

$\ell = 0.95q$ Write the equation.

$\ell = 0.95(8)$ Replace q with 8.

$\ell = 7.6$ Multiply.

There are about 7.6 liters in 8 quarts.

TICKETS The total cost of tickets to the school play are shown in the table.

Number of Tickets, t	Total Cost ($), c
1	4.50
2	9.00
3	13.50
4	18.00

a. Write an equation to find the total cost of any number of tickets. Describe the relationship in words.

b. Use the equation to find the cost of 15 tickets.

 REAL-WORLD EXAMPLES **Translate a Graph into an Equation**

RUNNING The total number of miles Marlon ran are shown.

3 Write an equation to find the number of miles run after any number of days.

Let m represent the miles and d represent the days.

The equation is $m = 3.5d$.

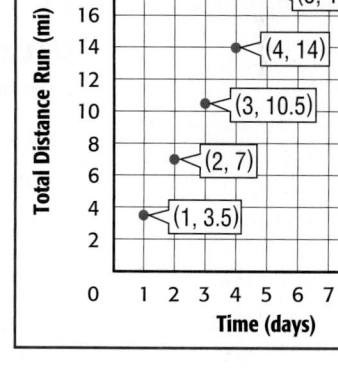

Total Distance Run

4 How many miles will Marlon run after 2 weeks?

$m = 3.5d$ Write the equation.

$m = 3.5(14)$ Replace d with 14.

$m = 49$ Multiply.

Marlon will run 49 miles in 2 weeks.

 CHECK Your Progress

RECYCLING The number of trees saved by recycling paper is shown.

c. Write an equation to find the total number of trees t that can be saved for any number of tons of paper p.

d. Use the equation to find how many trees could be saved if 500 tons of paper are recycled.

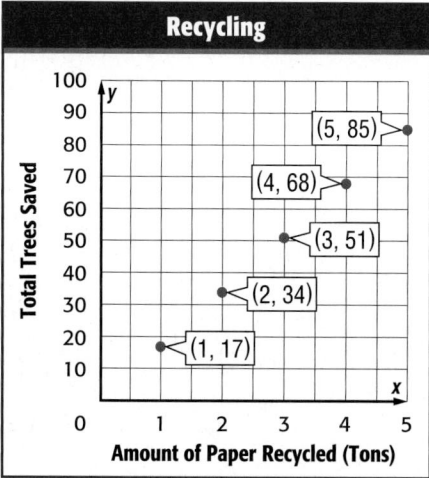

Recycling

Real-World Link · · · · ·
In a recent year, about 350 pounds of paper were recycled for every man, woman, and child in the United States.

On the previous page, you learned that the New Horizons spacecraft travels at a rate of 12 miles per second. You can use words, tables, graphs, and equations to represent this linear situation.

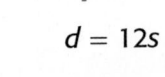

Key Concept   Representing Equations

Words

Distance traveled is equal to 12 miles per second times the number of seconds.

Equation

$d = 12s$

Table

Time (seconds)	Distance (miles)
1	12
2	24
3	36
4	48
5	60

Graph

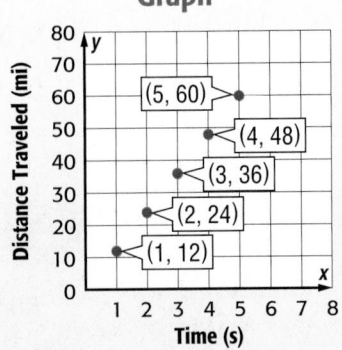

REAL-WORLD EXAMPLES   **Represent Equations**

SPORTS Chloe competes in jump rope competitions. Her average rate is 225 jumps per minute.

(5) Write an equation to find the number of jumps in any number of minutes.

Let j represent the number of jumps and m represent the minutes.

The equation is $j = 225m$.

(6) Make a table to find the number of jumps in 1, 2, 3, 4, or 5 minutes. Then graph the ordered pairs.

m	$225m$	j
1	225(1)	225
2	225(2)	450
3	225(3)	675
4	225(4)	900
5	225(5)	1,125

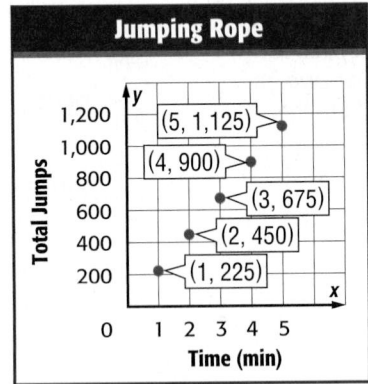

 CHECK Your Progress

MONEY Paul earns $7.50 an hour working at a grocery store.

e. Write an equation to find Paul's money earned m for any number of hours h.

f. Make a table to find his earnings if he works 5, 6, 7, or 8 hours. Then graph the ordered pairs.

Real-World Link · · · ·
New York City was the first school district to make Double Dutch an official school sport.

294   Expressions and Functions

Examples 1 and 2

1. TEXT MESSAGING The table shows the average number of text messages that Brad sends per day.

Number of Days, d	Total Messages, m
1	50
2	100
3	150
4	200

a. Write an equation to find the total number of messages sent in any number of days. Describe the relationship in words.

b. Use the equation to determine how many text messages Brad would send in 30 days.

Examples 3 and 4

2. FINANCIAL LITERACY The graph shows the amount of money the Rockwell family budgets for food each month.

a. Write an equation to find the total amount of money c budgeted in any number of months m.

b. Use the equation to determine how much money the Rockwell family should budget for 12 months.

Food Budget

(4, 2,500)
(3, 1,875)
(2, 1,250)
(1, 625)

Total Budgeted ($)

Number of Months

Examples 5 and 6

3. MOVIES A store receives an average of 7 new movies per week.

a. Write an equation to find the number of new movies m in any number of weeks w.

b. Make a table to find the number of new movies received in 4, 5, 6, or 7 weeks. Then graph the ordered pairs.

Practice and Problem Solving

● = **Step-by-Step Solutions** begin on page R1.
Extra Practice begins on page EP2.

Examples 1 and 2

4. MEASUREMENT The table shows the number of square inches per square foot.

Square Feet, f	Square Inches, i
1	144
2	288
3	432
4	576

a. Write an equation to find the number of square inches in any number of square feet. Describe the relationship in words.

b. Use the equation to determine how many square inches are in 15 square feet.

5 CRAFTS The number of baskets a company produces each day is shown in the table.

Number of Days, d	Total Baskets, b
1	45
2	90
3	135
4	180

a. Write an equation to find the total number of baskets crafted in any number of days. Describe the relationship in words.

b. Use the equation to determine how many baskets the company makes in one non-leap year.

6. INSECTS A type of dragonfly is the fastest insect. The graph shows how far the dragonfly can travel.

   a. Write an equation to find how far the dragonfly can travel d in any number of seconds s.

   b. Use the equation to determine how far the dragonfly can travel in one minute.

Dragonflies

Distance Traveled (ft) vs Time (s)
- (5, 115)
- (4, 92)
- (3, 69)
- (2, 46)
- (1, 23)

7. BOOKS A library charges a late return fee of $3.50 plus $0.15 per day that a book is returned late.

   a. Write an equation to find the total late fee f for any number of days late d.

   b. Make a table to find the total fee if a book is 10, 15, 20, or 25 days late. Then graph the ordered pairs.

8. 🔄 **MULTIPLE REPRESENTATIONS** The fastest times for swimming the English Channel for men and women are shown in the table at the right.

Swimmer	Average Rate (feet per minute)
Petar Stoychev	265
Yvetta Hlaváčová	249

   a. TABLES Make a table of ordered pairs in which the x-coordinate represents the time and the y-coordinate represents the total distance swum in 1-10 minutes.

   b. WORDS Explain whether a graph of each set of ordered pairs would be a straight line or a curve.

   c. GRAPHS Graph each set of ordered pairs on a coordinate plane.

   d. ALGEBRA Write an equation for each swimmer to find the number of feet swam d in any number of minutes t.

   e. NUMBERS If Petar Stoychev swam the Channel in 6 hours, 57 minutes, and 50 seconds, approximately how wide in miles is the English Channel? (*Hint:* 1 mi = 5,280 ft)

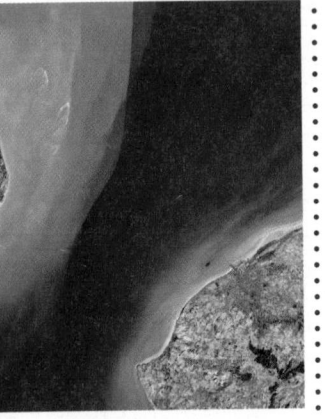

Real-World Link
The English Channel is the portion of the Atlantic Ocean that separates England and Northern France. Nearly 1,000 people have swum across the English Channel.

9 MONEY Kara and Mandy are each saving money for a trip to Washington, D.C. The table shows how much they save per week.

   a. Copy and complete the table.

   b. Graph the ordered pairs (week number, Kara's Savings) and (week number, Mandy's Savings) on the same coordinate plane.

Week Number	Kara's Savings ($)	Mandy's Savings ($)
1	20	50
2	35	60
3	50	70
⋮	⋮	⋮
10	■	■

   c. Compare the graphs of each savings plan. Explain any differences.

   d. Describe each girl's savings plan.

   e. Write an equation to represent each girl's savings plan.

   f. What does the point (7, 110) represent on the graph?

10. OPEN ENDED Write an equation with two variables that represents a real-world situation. Then make a table and graph to match your equation.

11. CHALLENGE The table shows the areas of circles with radii from 1 through 3 feet. Recall that π has a value of about 3. Write an equation in two variables to represent the relation in the table.

Radius (ft), r	Area (ft²), A
1	π
2	4π
3	9π

12. ✏ **WRITE MATH** Write about a real-world situation that can be represented by the equation $c = 9.5s + 10$.

Test Practice

13. John was swimming lengths across the pool. The table shows the time it took him to complete different lengths.

Number of Lengths (n)	Time (t) (minutes)
3	5.25
5	8.75
7	12.25
9	15.75

Based on the table, which equation can be used to approximate the time t it will take for John to complete any number of lengths n?

A. $n = 1.75t$ **C.** $n = 3.5t$

B. $t = 1.75n$ **D.** $t = 3.5n$

14. ✏ **GRIDDED RESPONSE** The graph represents the total cost to send a text message.

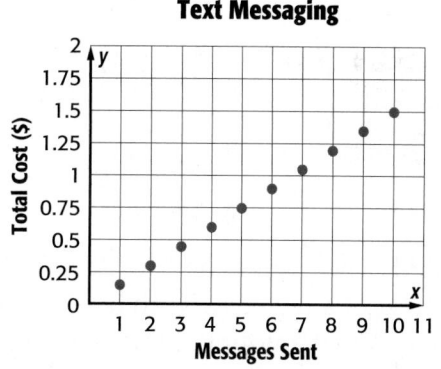

What will be the total cost in dollars if Javier sends 70 text messages?

 Spiral Review

15. MEASUREMENT The graph at the right shows the approximate number of grams in one ounce. (Lesson 2C)

  a. Write an algebraic expression to represent the data in the graph.

  b. Use the expression to find the number of grams in 150 ounces.

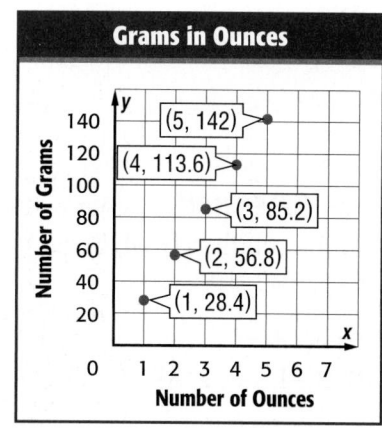

Write an expression that can be used to find the nth term of each sequence. Then use the expression to find the next three terms. (Lesson 2B)

16. 2, 6, 10, 14, 18

17. −6, −4, −2, 0, 2

Mid-Chapter Check

1. **PARTIES** Dasan is throwing a surprise party for his parents. The rates of two different banquet halls are shown in the table.

Banquet Hall	Rental Fee	Cost per Guest
Wonderland	$200	$2.50
Haven	$150	$3.75

Use the *make a table* strategy to determine how many guests must attend the party for the cost from each place to be the same. (Lesson 1A)

Evaluate each expression if $x = 3$, $y = 6$, and $z = 2$. (Lesson 1B)

2. $2x + 3y + 2z$

3. $\dfrac{xy}{z} - 4z$

4. $-3x - 4yz$

5. $xyz - 2x$

6. **GARDENS** The expression $2\ell + 2w$ gives the perimeter of a rectangle with length ℓ and width w. What amount of fencing would Mr. Nagawaka need in order to fence a garden that is 12 feet long and 9 feet wide? (Lesson 1B)

Name the ordered pair for each point. (Lesson 1C)

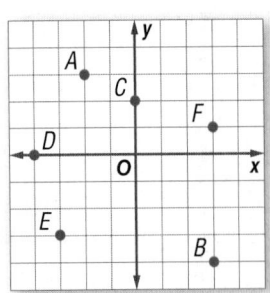

7. A
8. B
9. C
10. D
11. E
12. F

13. **MULTIPLE CHOICE** What is the domain of the relation $\{(-4, 5), (2, 0), (3, -2), (-1, 0)\}$? (Lesson 1C)

A. $\{-4, 2\}$

B. $\{-4, -1, 2, 3\}$

C. $\{-2, 0, 5\}$

D. $\{(-4, 5), (2, 0), (3, -2), (-1, 0)\}$

Write an expression that can be used to find the nth term of each sequence. Then use the expression to find the next three terms. (Lesson 2B)

14.

Term Number (n)	1	2	3	4	5
Term	13	17	21	25	29

15.

Term Number (n)	1	2	3	4	5
Term	-7	-16	-25	-34	-43

16.

Term Number (n)	1	2	3	4	5
Term	2	$3\frac{1}{3}$	$4\frac{2}{3}$	6	$7\frac{1}{3}$

17. **MULTIPLE CHOICE** Which expression **best** represents the information in the graph below? (Lesson 2C)

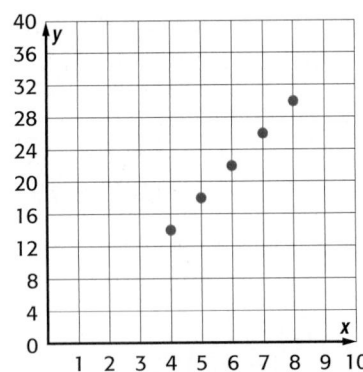

F. $4n$

G. $4n + 2$

H. $4n - 2$

I. 4

18. **MONEY** There are 20 nickels in one dollar. (Lesson 2D)

a. Write an equation to find the number of nickels n in any number of dollars d.

b. Make a table to find the number of nickels in 5, 10, 15, or 20 dollars. Then graph the ordered pairs.

Main Idea

Determine whether a relation is a function.

CCSS 8.F.1

Relations and Functions

A *function* is a special relation in which each member of the domain is paired with *exactly* one member in the range.

Suppose three students select a favorite color. The mappings below show some possible results.

Relation 1 is a function.	Relation 2 is a function.	Relation 3 is not a function.
Domain Range	Domain Range	Domain Range
Amber → Blue Tim → Red Kaela → Green	Alonso → Blue Malina → Blue Chen → Green	Elena → Blue Haley → Red Tony → Green

Relation 3 is *not* a function because Elena chose two favorite colors, blue and red.

ACTIVITY

Mr. Morgan asked his students how many pets they have. Some of the student responses are shown below.

Copy and complete the mapping diagram below.

Student Number	1	3	6
Number of Pets	2	5	7

Domain   Range

1
3
6

Analyze the Results

1. Describe why the relation in the Activity is or is not a function. Explain your reasoning in terms of the ordered pairs.

2. In the Activity above, suppose Student 8 has 2 pets. Make a mapping diagram of this new situation. Is this relation a function? Explain.

3. Suppose Student 1 got a new kitten after Mr. Morgan recorded the data. Make a mapping diagram to include the updated information.

Student Number	1	1	3	6
Number of Pets	2	3	5	7

PART   A   **B**   C

Main Idea

Complete function tables.

Vocabulary

function
function table
independent variable
dependent variable

8.F.1

Functions

ENTERTAINMENT Suppose you can buy DVDs for $15 each.

1. Copy and complete the table at the right.

2. If 6 DVDs are purchased, what is the total cost?

3. Explain how to find the total cost of 9 DVDs.

DVDs	Cost ($)
1	15
2	30
3	
4	
5	

The total cost depends on, or is a function of, the number of DVDs purchased. A **function** is a relation in which each member of the domain (input value) is paired with exactly one member of the range (output value). Functions are often written as equations.

f(x) is read *the function of x*, or more simply, *f of x*. It is the *output*. It is the *range*.

$$f(x) = 15x$$

The *input x* is any real number. It is the domain.

To find the value of a function for a certain number, substitute the number for the variable x.

EXAMPLE   Find a Function Value

 Find $f(-3)$ if $f(x) = 2x + 1$.

$f(x) = 2x + 1$ Write the function.

$f(-3) = 2(-3) + 1$ Substitute -3 for x into the function rule.

$f(-3) = -6 + 1$ or -5 Simplify.

So, $f(-3) = -5$.

CHECK Your Progress

Find each function value.

a. $f(2)$ if $f(x) = x - 4$ **b.** $f(6)$ if $f(x) = 2x - 8$

You can organize the input, rule, and output into a **function table**. The variable for the domain is called the **independent variable** because it can be any number. The variable for the range is called the **dependent variable** because it depends on the domain.

 **EXAMPLE** Make a Function Table

 Choose four values for x to make a function table for $f(x) = x + 5$. Then state the domain and range of the function.

Domain	Rule	Range
x	$f(x) = x + 5$	$f(x)$
-2	$-2 + 5$	3
-1	$-1 + 5$	4
0	$0 + 5$	5
1	$1 + 5$	6

Substitute each domain value x into the function rule. Then simplify to find the range value.

The domain is $\{-2, -1, 0, 1\}$. The range is $\{3, 4, 5, 6\}$.

CHECK Your Progress

Choose four values for x to make a function table for each function. Then state the domain and range of the function.

c. $f(x) = x - 7$ **d.** $f(x) = 4x$ **e.** $f(x) = 2x + 3$

A function can also describe the relationship between two quantities.

 REAL-WORLD EXAMPLES

FOOD There are approximately 770 peanuts in a 16.3-ounce jar of peanut butter. The total number of peanuts p in any number of jars of peanut butter j can be represented by the function $p(j) = 770j$.

3 Identify the independent and dependent variables.

Since the total number of peanuts depends on the number of jars of peanut butter, the number of peanuts p is the dependent variable and the jars of peanut butter j is the independent variable.

4 What values of the domain and range make sense for this situation? Explain.

Only whole numbers make sense for the domain because you cannot buy a fraction of a jar. The range values depend on the domain values, so the range will be multiples of 770.

Study Tip

Functions The total sales depends on the number of stamps sold. In other words, the total sales is a function of the number of stamps.

 CHECK Your Progress

SALES A scrapbooking store is selling rubber stamps for $4.95 each. The total sales s for any number of stamps n can be represented by the function $s(n) = 4.95n$.

f. Identify the independent and dependent variables.

g. What values of the domain and range make sense for this situation? Explain.

5 **DOGS** A veterinarian needs to give medication to a dog. The dosage is 5 milligrams for every 1 pound of weight. Write a function to represent the amount of medication $m(p)$ needed for p pounds of weight. Then determine how much to give to a dog that weighs 33 pounds.

Words	Amount of medication	equals	5 times	the number of pounds.
Function	$m(p)$	$=$	$5 \cdot$	p

The function $m(p) = 5p$ represents the situation.

To find the amount needed for 33 pounds, substitute 33 for p.

$m(p) = 5p$ Write the function.

$m(p) = 5(33)$ or 165   Substitute 33 for p.

The veterinarian should give the dog 165 milligrams of medication.

Real-World Link · · · ·
About 70% of all veterinarians work in a small animal practice. They focus on taking care of pets like dogs and cats.

✓ **CHECK Your Progress**

h. **HOME REPAIR** An air conditioner repair service charges $60 for a service call plus $30 per hour for labor. Write a function to represent the charge $c(h)$ for a service call with h hours of labor. How much would the charge be if there are 3 hours of labor?

✓ **CHECK Your Understanding**

Example 1   **Find each function value.**

1. $f(4)$ if $f(x) = x - 6$ **2.** $f(-2)$ if $f(x) = 4x + 1$

Example 2   **Choose four values for x to make a function table for each function. Then state the domain and range of the function.**

 $f(x) = 8 - x$ **4.** $f(x) = 5x + 1$ **5.** $f(x) = 3x - 2$

Examples 3 and 4   **6.** **BALLOONS** A hot air balloon can hold 90,000 cubic feet of air. It is being inflated at a rate of 6,000 cubic feet per minute. The total cubic feet of air a for any number of minutes t can be represented by the function $a(t) = 6,000t$.

    a. Identify the independent and dependent variables.

    b. What values of the domain and range make sense for this situation? Explain.

Example 5   **7.** **TRAVEL** On a highway, a car travels an average of 55 miles in one hour. Write a function to represent the distance $d(t)$ a car can travel in t hours. At this rate, how far can the car travel in 5 hours?

Practice and Problem Solving

● = **Step-by-Step Solutions** begin on page R1.
Extra Practice begins on page EP2.

Example 1 Find each function value.

8. $f(7)$ if $f(x) = 5x$ **9.** $f(9)$ if $f(x) = x + 13$ **10.** $f(4)$ if $f(x) = 3x - 1$

11. $f(5)$ if $f(x) = 2x + 5$ **12.** $f(-5)$ if $f(x) = 4x - 1$ **13.** $f(-12)$ if $f(x) = 2x + 15$

Example 2 Choose four values for x to make a function table for each function. Then state the domain and range of the function.

14. $f(x) = 6x - 4$ **15.** $f(x) = 5 - 2x$ **16.** $f(x) = 7 + 3x$

17. $f(x) = x - 9$ **18.** $f(x) = 7x$ **19.** $f(x) = 4x + 3$

Examples 3 and 4 **20. BASKETBALL** In a recent 82-game season, Dwight Howard of the Orlando Magic averaged 20.7 points per game. His approximate total points p for any number of games g can be represented by the function $p(g) = 20.7g$.

   a. Identify the independent and dependent variables.

   b. What values of the domain and range make sense for this situation? Explain.

21 **PORTRAITS** A photographer takes an average of 15 pictures per session. The total number of pictures p taken in any number of sessions s can be represented by the function $p(s) = 15s$.

   a. Identify the independent and dependent variables.

   b. What values of the domain and range make sense for this situation? Explain.

Example 5 **22. SPORTS** Tyree's bowling score is handicapped by 30 points, meaning that he receives an additional 30 points on his final score. Write a function that can be used to represent Tyree's final score $s(b)$ given his base score b. What is his adjusted score if he bowled 185?

23. MUSIC Leon belongs to a music club that charges a monthly fee of $5, plus $0.50 per song that he downloads. Write a function to represent the amount of money $m(s)$ he would pay in one month to download s songs. What is the cost if he downloads 30 songs?

Find each function value.

24. $f\left(\dfrac{5}{6}\right)$ if $f(x) = 2x + \dfrac{1}{3}$ **25.** $f\left(\dfrac{5}{8}\right)$ if $f(x) = 4x - \dfrac{1}{4}$

26. BIKING After 1 hour, a cyclist had ridden 12 miles. If she then continued riding at an average rate of 8 miles per hour, how long did it take her to ride 60 miles?

27. SCUBA DIVING The table shows the water pressure encountered by a diver. Write a function to represent the pressure $p(d)$ encountered at a depth of d feet. What would the pressure be at a depth of 175 feet? Round to the nearest tenth.

Depth (ft)	Pressure (lb/in²)
0	14.7
33	29.4
66	44.1
99	58.8
132	73.5

Determine whether each relation is a function. Explain.

28. {(2, 7), (2, −4), (2, 0), (2, 12)}   29 {(−12, −7), (−8, 5), (8, 5), (12, 7)}

30. ⊞ **MULTIPLE REPRESENTATIONS** Bonnie's Bead Warehouse sells beads for jewelry-making at discount prices. You pay a $10 membership fee and then $2 for every bag of beads you select.

   a. ALGEBRA Write a function to represent the situation.

   b. TABLES Make a function table to find the cost of 0, 1, 2, 3, 4, and 5 bags.

   c. GRAPHS Graph the ordered pairs from the function table.

31. **GRAPHIC NOVEL** Refer to the graphic novel frame below for Exercises a–c.

   a. Write a function to represent the total cost *c* of printing and shipping any number of pictures *p*.

   b. Make a function table to find the total cost of printing and shipping 25, 50, 75, and 100 pictures.

   c. Graph the ordered pairs on a coordinate plane. Can you determine how many pictures Brian can ship for $25?

H.O.T. Problems

32. **OPEN ENDED** If $f(-3) = -8$, write a function rule and find the function values for zero, a negative, and a positive value of *x*.

33. **CHALLENGE** Write the function rule for each function table.

   a.
x	f(x)
−3	−30
−1	−10
2	20
6	60

   b.
x	f(x)
−5	−9
−1	−5
3	−1
7	3

   c.
x	y
−2	−3
1	3
3	7
5	11

   d.
x	y
−2	−5
1	1
3	5
5	9

34. 🖹 **WRITE MATH** What are the similarities and differences among the terms domain, range, independent variable, and dependent variable?

35. The Cracker Crumbs Company has been making different varieties of crackers for many years. Currently, their most popular cracker sells for $4.80 per box. The table below shows the cost of these crackers in various years.

Year	Cost ($)
1965	?
1975	?
1985	2.55
1995	3.30
2005	4.05
2015	4.80

Assuming the price changed at a steady rate, what was the price of the crackers in 1965?

A. $0.30

B. $1.05

C. $1.80

D. $2.55

36. Stephanie received a $25 gift certificate to an online music store. If the cost of purchasing a song is $0.95, which table **best** describes $b(s)$, the balance remaining after she buys s songs?

F.

s	b(s)
1	$24.10
2	$23.20
4	$21.40
6	$19.60
8	$17.80

H.

s	b(s)
2	$23.10
4	$21.20
5	$20.25
8	$17.40
10	$15.50

G.

s	b(s)
0	$25.00
3	$22.00
6	$19.00
9	$16.00
12	$13.00

I.

s	b(s)
5	$20.05
10	$15.10
15	$10.15
20	$5.20
25	$0.25

 Spiral Review

37. RUNNING Cheryl is training for a marathon. She runs about 85 miles per week. (Lesson 2D)

a. Write an equation to find the total miles m run in any number of weeks w.

b. Make a table to find the total miles ran in 3, 4, 5, or 6 weeks. Then graph the ordered pairs.

38. SALES The graph shows the total cost of key chains at a souvenir shop. (Lesson 2D)

a. Write an algebraic equation to represent the data in the graph.

b. Use the equation to find the total cost of 15 key chains.

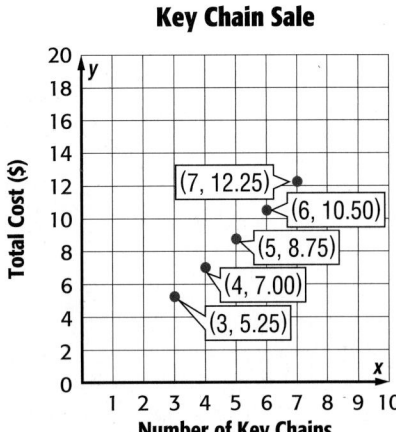

Key Chain Sale

Evaluate each expression if $p = 5$ and $q = 12$. (Lesson 1B)

39. $\dfrac{3p - 6}{8 - p}$

40. $\dfrac{4q}{q + 2(p + 1)}$

41. $\dfrac{q \cdot q}{4p - 2}$

42. $\dfrac{pq}{p + 5}$

43. $\dfrac{p + 3}{|q - 20|}$

44. $\dfrac{4p + q}{p + 11}$

Main Idea

Represent linear functions using function tables and graphs. Determine whether a set of data is continuous or discrete.

Vocabulary

linear function
continuous data
discrete data

 Get Connect**ED**

 CCSS 8.F.1, 8.F.3, 8.F.4

Linear Functions

AVIATION The Lockheed SR-71 Blackbird has a top speed of 36.6 miles per minute. If x represents the minutes traveled at this speed, the function rule for the distance traveled is $y = 36.6x$.

1. Copy and complete the function table.

2. Graph the ordered pairs (x, y) on a coordinate plane. What do you notice?

Input	Rule	Output	(Input, Output)
x	36.6x	y	(x, y)
1	36.6(1)	36.6	(1, 36.6)
2	36.6(2)		
3			
4			

Sometimes functions are written using two variables. One variable, usually x, represents the domain and the other, usually y, represents the range. When a function is written in this form it is an equation.

Like equations, functions can be represented in words, in a table, with a graph, and as ordered pairs.

REAL-WORLD EXAMPLE   Graph a Function

1 **SCHOOL SUPPLIES** The school store sells book covers for $2 each and notebooks for $1. Toni has $5 to spend on x book covers and y notebooks. Graph the equation $y = 5 - 2x$ to find how many book covers and notebooks Toni can buy.

Step 1 Choose values for x and substitute them to find y.

x	5 − 2x	y
0	5 − 2(0)	5
1	5 − 2(1)	3
2	5 − 2(2)	1
3	5 − 2(3)	−1

Step 2 Graph the ordered pairs (x, y).

She cannot buy negative amounts. So she can buy 0 covers and 5 notebooks, 1 cover and 3 notebooks, or 2 covers and 1 notebook.

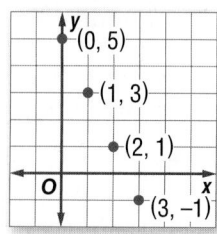

 CHECK Your Progress

a. **DECORATING** A repeating pattern is made using a set of 6 triangular tiles x and 1 hexagonal tile y. Graph the function $y = 35 - 6x$ to find the number of each tile needed if 35 tiles are used.

2 Graph $y = x + 2$.

- Select any four values for the domain x. Substitute these values for x to find the value of y.

- Graph each ordered pair. Draw a line that passes through each point.

x	x + 2	y	(x, y)
0	0 + 2	2	(0, 2)
1	1 + 2	3	(1, 3)
2	2 + 2	4	(2, 4)
3	3 + 2	5	(3, 5)

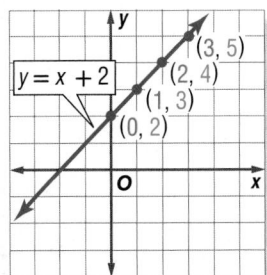

Study Tip

Solutions The solutions of an equation are ordered pairs that make an equation representing the function true.

The line is the complete graph of the function. The ordered pair corresponding to any point on the line is a solution of the equation $y = x + 2$.

Check It appears that $(-2, 0)$ is also a solution. Check this by substitution.

$y = x + 2$ Write the function.

$0 \stackrel{?}{=} -2 + 2$ Replace x with −2 and y with 0.

$0 = 0$ ✓ Simplify.

✅ **CHECK Your Progress**

Graph each function.

b. $y = x - 5$ **c.** $y = -2x$ **d.** $y = 2x + 1$

A **linear function** is a function in which the graph of the solutions forms a line. Therefore, $y = x + 2$ is a *linear function*.

🧩 **Key Concept**   **Representing Functions**

Words   The value of y is one less than the corresponding value of x.

Equation $y = x - 1$ **Ordered Pairs**   (0, −1), (1, 0), (2, 1), (3, 2)

Table

x	y
0	−1
1	0
2	1
3	2

Graph

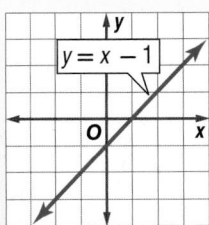

A function can be considered continuous or discrete. **Continuous data** can take on any value, so there is no space between data values for a given domain. **Discrete data** have space between possible data values. Graphs of continuous data are represented by solid lines and graphs of discrete data are represented by dots.

You can determine if data that model real-world situations are discrete or continuous by considering whether all numbers are reasonable as part of the domain.

Continuous Data	Discrete Data
the number of ounces in a glass	the number of glasses in a cupboard
the weight of each chocolate chip	the number of chocolate chips in a bag

EXAMPLES Continuous and Discrete Data

SHOPPING Each person that enters a store receives a coupon for $5 off his or her entire purchase.

3 **Write a function to represent the situation.**

Let y represent the total value of the coupons and x represent the number of people. The function is $y = 5x$.

4 **Make a function table to find the total value of the coupons given out to 5, 10, 15, and 20 customers.**

x	$5x$	y
5	5(5)	25
10	5(10)	50
15	5(15)	75
20	5(20)	100

5 **Graph the function. Is the function continuous or discrete? Explain.**

Use the ordered pairs from the function table to graph the function.

There can only be a whole number amount of customers. The function is discrete. So, the points are not connected.

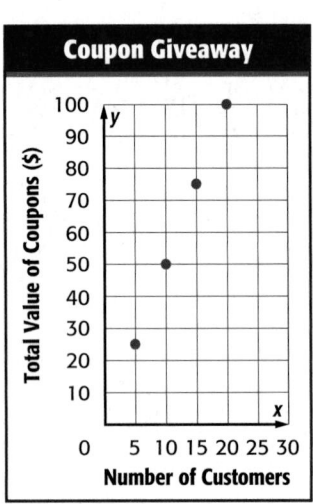

Coupon Giveaway

CHECK Your Progress

NUTS A store sells assorted nuts for $5.95 per pound.

e. Write a function to represent the situation.

f. Make a function table to find the total cost of 1, 2, 3, 4, or 5 pounds of nuts.

g. Graph the function. Is the function continuous or discrete? Explain.

Example 1   **1. GARDENING** Marigolds x come in containers with 4 flowers and daisies y come individually. Graph the function $y = 15 - 4x$ to find the number of containers of marigolds and daisies you can get if you want 15 flowers.

Example 2   **Graph each function.**

2. $y = x + 5$ **3.** $y = 3x - 2$ **4.** $y = -2x + 1$

Examples 3–5   **5** **TELEVISION** A satellite cable company charges an installation fee of $50 plus an additional $35.95 per month for service.

    a. Write a function to represent the situation.

    b. Make a function table to find the total cost for 1, 2, 3, 4, or 5 months.

    c. Graph the function. Is the function continuous or discrete? Explain.

Practice and Problem Solving

● = **Step-by-Step Solutions** begin on page R1.
Extra Practice begins on page EP2.

Example 1   **6. PETS** Fancy goldfish x cost $3 each and common goldfish y cost $1 each. Graph the function $y = 20 - 3x$ to determine how many of each type of goldfish Tasha can buy for $20.

    7. CLOTHES A store sells T-shirts x in packs of 5 and regular shirts y individually. Graph the function $y = 10 - 5x$ to determine the number of each type of shirt Bethany can have if she buys 10 shirts.

Example 2   **Graph each function.**

8. $y = 4x$ **9.** $y = -3x$ **10.** $y = x - 3$ **11.** $y = x + 1$

12. $y = 3x - 7$ **13.** $y = 2x + 3$ **14.** $y = \frac{1}{3}x + 1$ **15.** $y = \frac{1}{2}x - 3$

Examples 3–5   **16. FINANCIAL LITERACY** Manuel is saving money for college. He already has $250. He plans to save another $50 per month.

    a. Write a function to represent the situation.

    b. Make a function table to find his total savings after 2, 4, 6, 8, and 10 months.

    c. Graph the function. Is the function continuous or discrete? Explain.

17. RENTALS The table shows the cost to rent different items.

    a. Write a function to represent each situation.

    b. Make a function table to find the total cost to rent each item for 2, 3, 4, or 5 hours.

Item	Deposit ($)	Cost per Hour ($)
Mountain bike	15	4.25
Scooter	25	2.50

    c. Graph the functions on the same coordinate plane. Are the functions continuous or discrete? Explain.

    d. Will the mountain bike or the scooter cost more to rent for 8 hours?

18. **TEMPERATURE** The formula $F = 1.8C + 32$ compares temperatures in degrees Celsius C to temperatures in degrees Fahrenheit F. Find four ordered pairs (C, F) that are solutions of the equation. Then graph the function.

19. **MEASUREMENT** The equation $y = 1.09x$ describes the approximate number of yards y in x meters.

    a. Would negative values of x have any meaning in this situation? Explain.

    b. Graph the function.

    c. About how many meters is a 40-yard race?

20. **ELEVATION** If the temperature is 80°F at sea level, the function $t = 80 - 3.6h$ describes the temperature t at a height of h thousand feet above sea level.

    a. Graph the temperature function.

    b. What is the temperature at each peak on a day that is 80°F at sea level?

Eastern U.S. Mountains	
Mountain	**Elevation (ft)**
Mount Mitchell, NC	6,684
Mount Rogers, VA	5,729
Black Mountain, KY	4,139
Sassafras Mountain, SC	3,560

21. **MONEY** Drake is saving money to buy a new computer for $1,200. He already has $450 and plans to save $30 a week. The function $f(x) = 30x + 450$ represents the amount Drake has saved after x weeks. Graph the function to determine the number of weeks it will take Drake to save enough money to buy the computer.

H.O.T. Problems

22. **OPEN ENDED** Draw a graph of a linear function. Name three solutions of the function.

23. **Which One Doesn't Belong?** Identify the ordered pair that is not a solution of $y = -4x + 3$. Explain your reasoning.

| (2, 5) | (0, 3) | (−1, 7) | (1, −1) |

24. **CHALLENGE** Name the coordinates of four points that satisfy each function. Then give the function rule.

    a.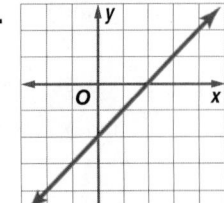

    b.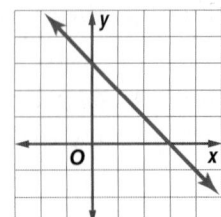

25. **WRITE MATH** Explain how a function table can be used to graph a function.

26. Which line graphed below **best** represents the table of values for the ordered pairs (x, y)?

x	−4	0	4	8
y	−2	−1	0	1

A.
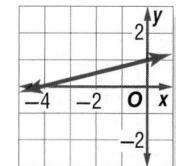

C.
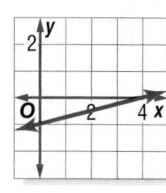

B.

D.

27.  **SHORT RESPONSE** Does the graph of the equation $y = 3x - 4$ go through the origin? If not, give the point where the line crosses the y-axis.

28. A graph of the equation $y = 5x - 1$ is below.

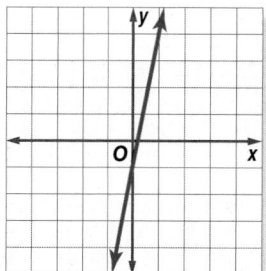

Which table of ordered pairs contains only points on this line?

F.

x	−2	−1	0	1
y	−9	−4	1	6

G.

x	−3	−2	−1	0
y	−8	−7	−6	−5

H.

x	0	1	2	3
y	−1	0	1	2

I.

x	−1	0	1	2
y	−6	−1	4	9

Spiral Review

Find each function value. (Lesson 3B)

29. $f(6)$ if $f(x) = 7x - 3$

30. $f(-5)$ if $f(x) = 3x + 15$

31. $f(3)$ if $f(x) = 2x - 7$

32. HOTELS The table shows the rate a hotel charges per day for a room. (Lesson 2D)

   a. Write an equation to find the total cost to stay in the hotel for any number of days.

   b. Use the equation to determine how much it would cost to stay for 9 days.

Number of Days	Total Charge ($)
1	65
2	130
3	195
4	260

33. Write an expression that can be used to find the nth term of the arithmetic sequence 15, 30, 45, 60, … . Then write the next three terms. (Lesson 2B)

34. TICKETS If Linda buys movie tickets online, the tickets are $6 each plus a service fee of $5. If she buys them at the theater, they are $8 each. For how many tickets is the online price the better buy? Use the *make a table* strategy. (Lesson 1A)

Focusing on Recovery

Are you a compassionate person? Do you have a strong desire to help others? If so, a career as a physical therapist might be a good choice for you. Physical therapists help restore function, improve mobility, and relieve pain of patients suffering from injuries or disease. One of their jobs is to teach exercises or recommend activities to help patients regain balance, flexibility, endurance, and strength.

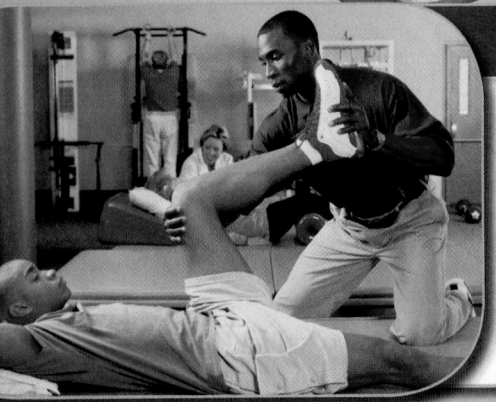

21st Century Careers

Are you interested in a career as a physical therapist? Take some of the following courses in high school.

- Algebra
- Biology
- Chemistry
- Introduction To Physical Therapy
- Physics

Get ConnectED

Endurance Exercise: Inline Skating	
Time (min)	Distance (mi)
15	2.25
30	4.5
45	6.75
60	9.0

Real-World Math

Use the information above to solve each problem.

1. The function $t(r) = 12r$, where r is the number of repetitions, represents the total time $t(r)$ in seconds to complete a flexibility exercise. Find $t(8)$. Then interpret the solution.

2. Refer to the information in Exercise 1. Make a function table to find the time it will take to complete 1, 2, 5, and 10 repetitions.

3. Write a function to represent the distance d in miles an inline skater will travel in t minutes.

4. Refer to the function that you wrote in Exercise 3. How far will an inline skater travel after 80 minutes?

5. Graph the function from Exercise 3. Then use the graph to estimate the distance an inline skater will travel after 90 minutes.

Linear and Nonlinear Functions

 **FOOTBALL**

The table shows the approximate height and horizontal distance traveled by a football thrown at an angle of 30° with an initial velocity of 30 yards per second.

Time (s)	Height (yd)	Length (yd)
0.00	0	0
0.50	6.2	13.0
1.00	9.7	26.0
1.50	10.5	39.0
2.00	8.7	52.0
2.50	4.2	65.0

1. Did the football travel the same height each half-second? Justify your answer.

2. Did the football travel the same length each half-second? Justify your answer.

3. Graph the ordered pairs (time, height) and (time, length) on separate grids. Connect the points with a straight line or smooth curve. Then compare the graphs.

In the previous lesson, you learned that linear functions have graphs that are straight lines. This is because the rate of change between any two data points is a constant. **Nonlinear functions** are functions whose rates of change are not constant. Therefore, their graphs are not straight lines.

 EXAMPLES Identify Functions Using Tables

Determine whether each table represents a *linear* or *nonlinear* function. Explain.

 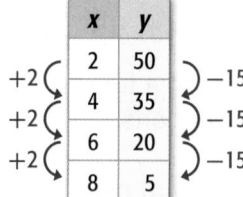

x	y
2	50
4	35
6	20
8	5

+2 each, −15 each

As *x* increases by 2, *y* decreases by 15 each time. The rate of change is constant, so this function is linear.

Check

Graph the points on a coordinate plane.

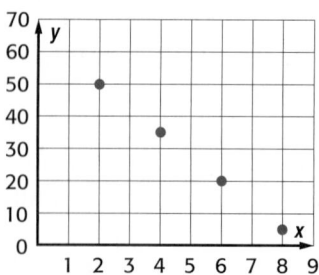

The points fall in a line. The function is linear. ✓

②

x	y
1	1
4	16
7	49
10	100

+3 (1→4), +3 (4→7), +3 (7→10)

+15 (1→16), +33 (16→49), +51 (49→100)

As *x* increases by 3, *y* increases by a greater amount each time. The rate of change is not a constant, so this function is nonlinear.

Check

Graph the points on a coordinate plane.

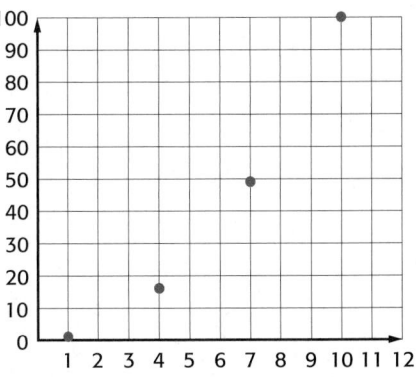

The points do not fall in a line. The function is nonlinear. ✓

CHECK Your Progress

Determine whether each table represents a *linear* or *nonlinear* function. Explain.

a.

x	0	5	10	15
y	20	16	12	8

b.

x	0	2	4	6
y	0	2	8	18

REAL-WORLD EXAMPLE

③ TIGERS Use the table to determine whether the minimum number of Calories a tiger cub should eat is a linear function of its age in weeks.

Age (weeks)	Minimum Calorie Intake
1	825
2	1,000
3	1,185
4	1,320
5	1,420

Examine the differences between the number of Calories for each week.

$1,000 - 825 = 175$ $1,185 - 1,000 = 185$
$1,320 - 1,185 = 135$ $1,420 - 1,320 = 100$

The difference in Calories is not the same. Therefore, this function is nonlinear.

Check Graph the data to verify the ordered pairs do not lie on a straight line.

Real-World Link

An adult Siberian tiger can weigh as much as 660 pounds and eats about 10,000 Calories per day.

CHECK Your Progress

c. **TICKETS** Tickets to the school dance cost $5 per student. Are the ticket sales a linear function of the number of tickets sold? Explain.

Number of Tickets Sold	1	2	3
Ticket Sales	$5	$10	$15

4 **E-MAILS** Jonathan sent an E-mail to two of his friends. Each of those friends forwarded the E-mail to two more friends. Does this situation represent a linear or nonlinear function? Explain.

Make a table to show the number of E-mails sent each time starting with Jonathan (Level 1).

Level	1	2	3	4	5
E-mails Sent	2	4	8	16	32

Graph the function. The data do not lie on a straight line. This function is nonlinear.

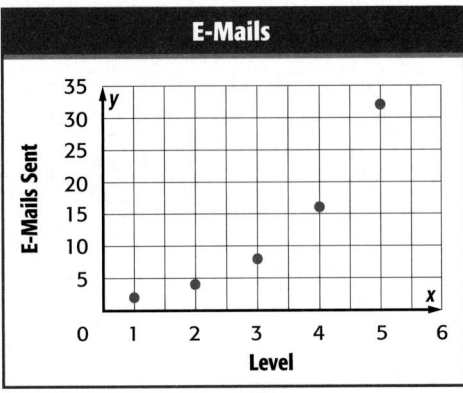

CHECK Your Progress

d. MEASUREMENT A square has a side length of *s* inches. The area of the square is a function of the side length. Does this situation represent a linear or nonlinear function? Explain.

CHECK Your Understanding

Examples 1 and 2   **Determine whether each table represents a *linear* or *nonlinear* function. Explain.**

1.

x	0	1	2	3
y	1	3	6	10

2.

x	0	3	6	9
y	−3	9	21	33

 3.

x	−2	0	2	4
y	−1	0	1	2

4.

x	0	1	2	3
y	0	1	1.5	1.75

Example 3   **5. MEASUREMENT** The table shows the measures of the sides of several rectangles. Are the widths of the rectangles a linear function of the lengths? Explain.

Length (in.)	1	4	8	10
Width (in.)	64	16	8	6.4

Example 4   **6. MEASUREMENT** A cube has a side length of *s* meters. The volume of the cube is represented by the expression s^3. The volume of the cube is a function of the side length. Does this situation represent a linear or nonlinear function? Explain.

Practice and Problem Solving

● = **Step-by-Step Solutions** begin on page R1.
Extra Practice begins on page EP2.

Examples 1 and 2 Determine whether each table represents a *linear* or *nonlinear* function. Explain.

7.

x	3	6	9	12
y	12	10	8	6

8.

x	1	2	3	4
y	1	4	9	16

9.

x	5	10	15	20
y	13	28	43	58

10.

x	1	3	5	7
y	−2	−18	−50	−98

11.

x	2	4	6	8
y	10	12	16	24

12.

x	4	8	12	16
y	3	0	−3	−6

Example 3 **13** **TRAVEL** The Guzman family drove from Anderson to Myrtle Beach. Use the table to determine whether the distance driven is a linear function of the hours traveled. Explain.

Time (h)	1	2	3	4
Distance (mi)	65	130	195	260

14. BUILDINGS The table shows the height of several buildings in Chicago. Use the table to determine whether the height of the building is a linear function of the number of stories. Explain.

Building	Stories	Height (ft)
Harris Bank III	35	510
One Financial Place	40	515
Kluczynski Federal Building	45	545
Mid Continental Plaza	50	582
North Harbor Tower	55	556

Example 4 **15. TIME** There are 3,600 seconds in one hour. The total seconds is a function of the hours. Does this situation represent a linear or nonlinear function? Explain.

16. SPORTS A football is placed on the ground to kick a field goal. The height of the ball is a function of the time in seconds. Does the path the football follows after being kicked represent a linear or nonlinear function? Explain.

Graph each function by making a table of ordered pairs. Determine whether each function is *linear* or *nonlinear*. Explain.

17. $y = -x + 1$

18. $y = \dfrac{-9}{x}$

19. $y = \dfrac{3x}{2}$

20. FINANCIAL LITERACY The graph shows the amount of money Charlotte has in a savings account with a 5.75% interest rate that is compounded monthly after several years. Would you describe the graph as linear or nonlinear? Explain.

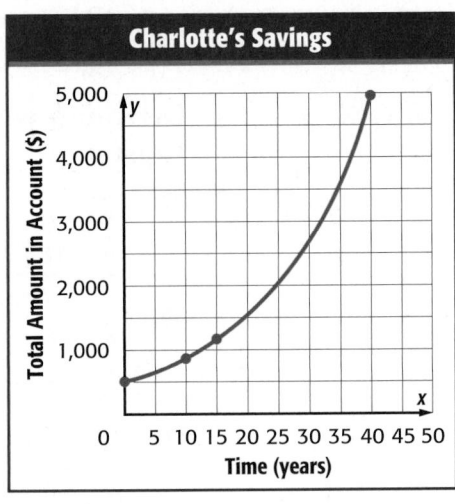

Charlotte's Savings

21 **MEASUREMENT** Make a graph showing the area of a square as a function of its perimeter. Explain whether the function is linear.

22. GRAPHING Water is poured at a constant rate into the vase at the right. Draw a graph of the water level as a function of time. Is the water level a linear or nonlinear function of time? Explain.

23. MATH IN THE MEDIA Find examples of graphs in a newspaper or on the Internet. Explain whether each graph represents a linear or nonlinear function.

24. MULTIPLE REPRESENTATIONS Recall that the circumference of a circle is equal to two times π times its radius and that the area of a circle is equal to π times the square of the radius.

a. TABLES Copy and complete the table showing the circumference and area of circles with radius r.

b. GRAPHS Graph the ordered pairs (radius, circumference) and (radius, area) for each function on the same coordinate plane.

Radius r	Circumference $2 \cdot \pi \cdot r$	Area πr^2
1	$2 \cdot \pi \cdot 1 \approx 6.28$	$\pi \cdot 1^2 \approx 3.14$
2	■	■
3	■	■
4	■	■
5	■	■

c. WORDS Is the circumference of a circle a linear or nonlinear function of its radius? the area? Explain your reasoning.

H.O.T. Problems

25. CHALLENGE Does the graph at the right represent a linear function? Explain.

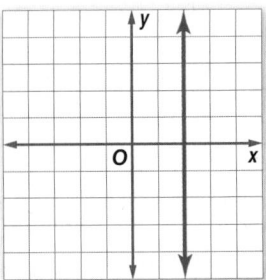

26. OPEN ENDED Give an example of a situation that can be represented by a linear function.

27. WRITE MATH Describe two methods for determining whether a function is linear or nonlinear.

 **Test Practice**

28. Which of the following tables is NOT an example of a linear function?

A.
x	y
2	15
4	20
6	25
8	30

C.
x	y
−3	−6
−7	−1
−11	4
−15	9

B.
x	y
9	1.25
11.5	2
14	2.75
16.5	3.5

D.
x	y
10	5
13	7
16	10
19	14

29. **EXTENDED RESPONSE** Jung has $200 in a safe. Each month, he adds another $10 to the safe. Miguel opens a savings account with a $200 deposit and earns 10% interest each month on the total amount of money in the bank.

Part A Make a table showing the money they have each saved after 7 months.

Part B Graph each person's savings on the same coordinate plane.

Part C Explain why one function is linear and the other is not.

Graph each function. (Lesson 3C)

30. $y = 5x$ **31.** $y = x - 2$ **32.** $y = 2x - 1$ **33.** $y = 3x + 2$

Find each function value. (Lesson 3B)

34. $f(5)$ if $f(x) = 3x + 4$ **35.** $f(-3)$ if $f(x) = 2x - 8$ **36.** $f(7)$ if $f(x) = 9x - 24$

37. PHONES The table shows the average number of phone calls Riley makes per day. (Lesson 2D)

Number of Days, d	Total Phone Calls, c
1	5
2	10
3	15
4	20

  a. Write an equation to find the total number of phone calls made in any number of days. Describe the relationship in words.

  b. Use the equation to determine how many phone calls Riley would make in 1 week.

Name the ordered pair for each point. (Lesson 1C)

38. Q **39.** R **40.** S

41. T **42.** U **43.** V

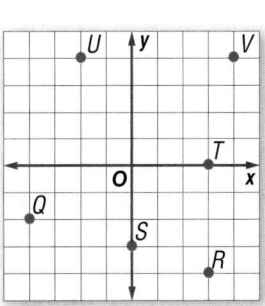

44. TEMPERATURE The formula $F = \frac{9}{5}C + 32$ is used to convert degrees Celsius to degrees Fahrenheit. (Lesson 1B)

  a. Find the degrees Fahrenheit if it is 30°C outside.

  b. A local newscaster announces that the temperature is currently ten degrees Celsius. What is the temperature in degrees Fahrenheit?

Main Idea

Graph quadratic functions.

 Vocabulary

quadratic function
cubic function

 Get ConnectED

 8.F.3

Graph Quadratic Functions

 Explore You know that the area A of a square is equal to the length of a side s squared, or $A = s^2$.

s	s^2	(s, A)
0	0	(0, 0)
1	1	(1, 1)
2		
3		
4		
5		
6		

Step 1 Copy and complete the table.

Step 2 Graph the ordered pairs from the table. Connect them with a smooth curve.

1. Describe the graph. Is the relationship between the side length and the area of a square linear or nonlinear? Explain.

A **quadratic function**, like $A = s^2$, is a function in which the greatest power of the variable is 2. Its graph is U-shaped, opening upward or downward. The graph opens upward if the coefficient of the variable that is squared is *positive*, downward if it is *negative*.

EXAMPLES Graph Quadratic Functions

① Graph $y = x^2$.

To graph a quadratic function, make a table of values, plot the ordered pairs, and connect the points with a smooth curve.

x	x^2	y	(x, y)
−2	$(-2)^2 = 4$	4	(−2, 4)
−1	$(-1)^2 = 1$	1	(−1, 1)
0	$(0)^2 = 0$	0	(0, 0)
1	$(1)^2 = 1$	1	(1, 1)
2	$(2)^2 = 4$	4	(2, 4)

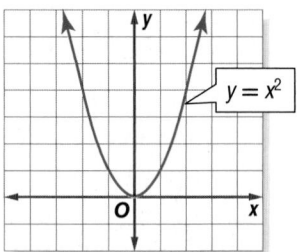

② Graph $y = -x^2 + 4$.

x	$-x^2 + 4$	y	(x, y)
−2	$-(-2)^2 + 4 = 0$	0	(−2, 0)
−1	$-(-1)^2 + 4 = 3$	3	(−1, 3)
0	$-(0)^2 + 4 = 4$	4	(0, 4)
1	$-(1)^2 + 4 = 3$	3	(1, 3)
2	$-(2)^2 + 4 = 0$	0	(2, 0)

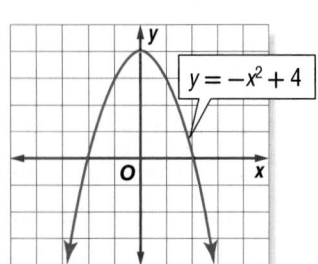

Graph each function.

a. $y = 6x^2$ **b.** $y = x^2 - 2$ **c.** $y = -2x^2 - 1$

REAL-WORLD EXAMPLE

③ BALLOONS The function $h = 0.66d^2$ represents the distance d in miles you can see from a height of h feet. Graph this function. Then use your graph and the information at the left to estimate how far you could see from the first hot air balloon.

Distance cannot be negative, so use only positive values of d.

Real-World Link · · · ·

In 1783, one of the first hot air balloons to take flight rose about 1,000 feet in the air.

d	$h = 0.66d^2$	(d, h)
0	$0.66(0)^2 = 0$	(0, 0)
10	$0.66(10)^2 = 66$	(10, 66)
20	$0.66(20)^2 = 264$	(20, 264)
25	$0.66(25)^2 = 412.5$	(25, 412.5)
30	$0.66(30)^2 = 594$	(30, 594)
35	$0.66(35)^2 = 808.5$	(35, 808.5)
40	$0.66(40)^2 = 1,056$	(40, 1,056)

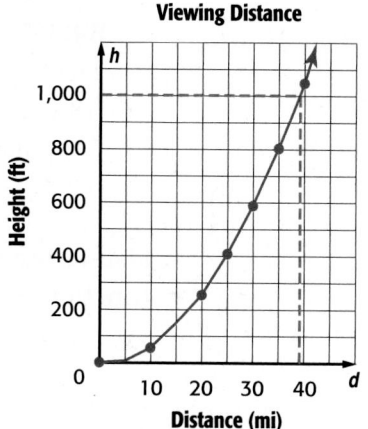

At a height of 1,000 feet, you could see approximately 39 miles.

 CHECK Your Progress

d. TOWERS The outdoor observation deck of the Space Needle in Seattle, Washington, is 520 feet above ground level. Estimate how far you could see from the observation deck.

✓ CHECK Your Understanding

Examples 1 and 2   Graph each function.

1. $y = 3x^2$ **2.** $y = -5x^2$

3. $y = -4x^2$ **4.** $y = -x^2 + 1$

5. $y = x^2 - 3$ **6.** $y = -x^2 + 2$

Example 3   **⑦ CARS** The function $d = 0.006s^2$ represents the braking distance d in meters of a car traveling at a speed s in kilometers per second. Graph this function. Then use your graph to estimate the speed of the car if its braking distance is 12 meters.

Practice and Problem Solving

= Step-by-Step Solutions begin on page R1.
Extra Practice begins on page EP2.

Examples 1 and 2   **Graph each function.**

8. $y = 4x^2$

9. $y = 5x^2$

10. $y = -3x^2$

11. $y = -6x^2$

12. $y = x^2 + 6$

13. $y = x^2 - 4$

14. $y = -x^2 + 2$

15. $y = -x^2 - 5$

16. $y = 2x^2 - 1$

17. $y = 2x^2 + 3$

18. $y = -4x^2 - 1$

19. $y = -3x^2 + 2$

Example 3   **20. CARNIVAL RIDES** The function $a = 0.2v^2$ models the acceleration of a carnival ride, where a is the acceleration toward the center of the ride in meters per second every second and v is the velocity in meters per second. Graph this function. Then use your graph to estimate the velocity of the ride at an acceleration of 1 meter per second every second.

21 **BRIDGES** A penny is dropped from a height of 196 feet off a bridge. The function $d = -16t^2 + 196$ models the distance d in feet the penny is from the surface of the water at time t seconds. Graph this function. Then use your graph to estimate the time it will take for the penny to reach the water.

Determine whether each equation represents a linear or nonlinear function. Explain.

22. $y = 3x$

23. $y = 2x^2$

24. $y = -3x^2$

25. $y = -6x$

26. $5x + y = 7$

27. $7x^2 + y = 24$

28. $x + x^2 = y$

29. $x + y = x^2$

30. $xy = 24$

31. CRAFTS Annika is making a fabric memo board.

a. Write a function that represents the area A of the memo board.

b. Graph the function.

c. If the width of the memo board is 8 inches, what is its area?

x in.

$(x + 4)$ in.

H.O.T. Problems

CHALLENGE The graphs of quadratic functions may have exactly one highest point, called a *maximum*, or exactly one lowest point, called a *minimum*. Graph each quadratic equation. Determine whether each graph has a maximum or a minimum. If so, give the coordinates of each point.

32. $y = 2x^2 + 1$

33. $y = -x^2 + 5$

34. $y = x^2 - 3$

35. OPEN ENDED Write and graph a quadratic function that opens upward and has its minimum at $(0, -3.5)$.

36. **WRITE MATH** Write a quadratic function of the form $y = ax^2 + c$ and explain how to graph it.

37. Which graph represents the function $y = -0.5x^2 - 2$?

A.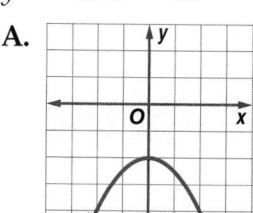

C.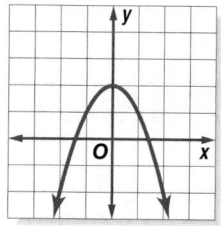

B.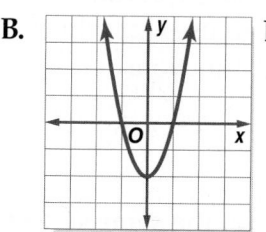

D.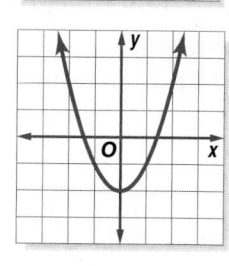

38. Which equation represents a linear function?

F. $y = 2x^2$

G. $y = x^2 + 3$

H. $y = 5x - 1$

I. $y = 10 + x^2$

More About **Nonlinear Functions**

The formula for the volume of a cube, $V = s^3$, is a cubic function. A **cubic function** is a nonlinear function in which the greatest power of the variable is 3. You can make a table of values to graph a cubic function.

 EXAMPLE

Graph $y = x^3$.

Make a table of values, graph the ordered pairs, and draw a smooth curve that connects the points.

x	$y = x^3$	(x, y)
−1.5	$(-1.5)^3 \approx -3.4$	(−1.5, −3.4)
−1	$(-1)^3 = -1$	(−1, −1)
0	$(0)^3 = 0$	(0, 0)
1	$(1)^3 = 1$	(1, 1)
1.5	$(1.5)^3 \approx 3.4$	(1.5, 3.4)

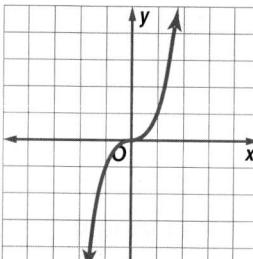

Graph each function.

39. $y = x^3 - 1$

40. $y = x^3 + 4$

41. $y = -4x^3$

42. PACKAGING A packaging company wants to manufacture a cardboard box with a square base with a side length of x feet and a height of $(x - 2)$ feet. Write the function for the volume V of the box. Graph the function. Then estimate the dimensions of the box that would have a volume of about 1 cubic foot.

Extend

Main Idea

Use a graphing calculator to graph families of nonlinear functions.

 Vocabulary

exponential function

 Get ConnectED

 CCSS 8.F.3

Graphing Technology: Families of Nonlinear Functions

Families of nonlinear functions share a common characteristic based on a parent function. The parent function, or simplest function, of a family of quadratic functions is $y = x^2$. You can use a graphing calculator to investigate families of quadratic functions.

ACTIVITY

1. **Graph $y = x^2$, $y = x^2 + 5$, and $y = x^2 - 3$ on the same screen.**

   STEP 1 Clear any existing equations from the Y= list by pressing [Y=] [CLEAR].

   STEP 2 Enter each equation. Press [X,T,θ,n] [x²] [ENTER], [X,T,θ,n] [x²] [+] 5 [ENTER], and [X,T,θ,n] [x²] [−] 3 [ENTER].

   STEP 3 Graph the equations in the standard viewing window. Press [ZOOM] 6.

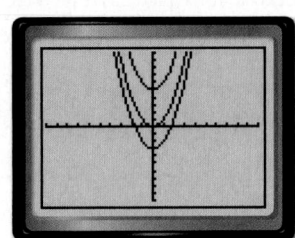

Analyze the Results

1. Compare and contrast the three equations you graphed.

2. Describe how the graphs of the three equations are related.

3. **MAKE A CONJECTURE** How does changing the value of c in the equation $y = x^2 + c$ affect the graph?

4. Use a graphing calculator to graph $y = 0.5x^2$, $y = x^2$, and $y = 2x^2$.

5. Compare and contrast the three equations you graphed in Exercise 4.

6. Describe how the graphs of the three equations are related.

7. **MAKE A CONJECTURE** How does changing the value of a in the equation $y = ax^2$ affect the graph?

8. Without graphing, determine whether the graph of $y = \frac{1}{3}x^2$ or the graph of $y = 3x^2$ is narrower. Explain.

A function such as $y = 4^x$ is an exponential function. An **exponential function** is a nonlinear function in which the base is a constant and the exponent is an independent variable, x. Exponential functions are also *nonquadratic*.

ACTIVITY

2 **BIOLOGY A certain type of bacteria doubles every hour. The function $y = 2^x$ represents the total number of bacteria y at the end of every hour x. Graph the function. Then find the number of bacteria at the end of 5 hours.**

STEP 1 Clear any existing equations from the Y= list by pressing Y= CLEAR.

STEP 2 Enter the equation. Press Y= 2 ∧ X,T,θ,n.

STEP 3 Graph the equation in the standard viewing window. Press ZOOM 6.

STEP 4 Use the TABLE feature. Press 2nd [GRAPH]. The y value that corresponds to the x value of 5 is 32.

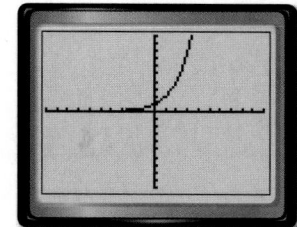

So, there are 32 bacteria at the end of 5 hours.

Analyze the Results

9. Graph $y = \left(\frac{1}{2}\right)^x$ in the same window as the equation from the Activity. What do you notice about this graph?

10. Use a graphing calculator to graph $y = 5^x$ and $y = 0.2^x$ in the same window.

11. Compare and contrast the graphs.

12. **MAKE A CONJECTURE** Without graphing, predict which graph slopes up from left to right: $y = 3^x + 1$ or $y = 50\left(\frac{1}{3}\right)^x$. Explain.

13. Without graphing, determine whether the graph of $y = 6^x$ or the graph of $y = 2^x$ is steeper. Explain.

14. **FINANCIAL LITERACY** Miley's parents started a savings account when she was born. They initially deposited $100 into the account, which earns an annual interest rate of 3%. The balance y in the account can be represented by the function $y = 100(1.03)^x$, where x is the number of years.

   a. Graph the function.

   b. Determine how much money is in the account after 13 years.

Chapter Study Guide and Review

FOLDABLES® Study Organizer

Be sure the following Key Concepts are noted in your Foldable.

Key Concepts

Expressions (Lessons 1 and 2)

• An algebraic expression contains at least one variable and at least one operation.

• You can translate words, tables, and graphs to expressions or equations.

Relations and Functions (Lesson 3)

• A relation is a set of ordered pairs. A function is a special relation in which each member of the domain is paired with exactly one member in the range.

• A function is a relationship in which one value is dependent upon another.

• Functions can be represented by words, equations, tables, ordered pairs, and graphs.

Nonlinear Functions (Lesson 4)

• Linear functions are functions in which the graph of the solutions forms a line.

• To graph a linear function, make a function table and plot ordered pairs.

• Nonlinear functions are functions whose graphs are not straight lines.

Key Vocabulary

algebra	nonlinear function
algebraic expression	ordered pair
arithmetic sequence	origin
common difference	quadrants
continuous data	range
coordinate plane	relation
dependent variable	sequence
discrete data	term
domain	variable
function	*x*-axis
function table	*x*-coordinate
independent variable	*y*-axis
linear	*y*-coordinate
linear function	

Vocabulary Check

Choose the correct term to complete each sentence.

1. The (domain, range) is the set of input values of a function.

2. A (common difference, sequence) is an ordered list of numbers.

3. A(n) (dependent, independent) variable is the variable for the output of a function.

4. Data is (continuous, discrete) if it can take on any value.

5. A relationship where one thing depends on another is called a (function, linear relationship).

6. The number that corresponds to a point is called its (domain, coordinate).

7. A (relation, function) is any set of ordered pairs.

Multi-Part Lesson Review

Lesson 1 Expressions

PSI: Make a Table (Lesson 1A)

Solve. Use the *make a table* strategy.

8. CARS The rates for two different rental companies are shown below. At how many miles will the cost to rent a car be the same for both companies?

Company	Deposit ($)	Charge per Mile ($)
ABC Rentals	25	0.20
Rent and Go	30	0.15

9. RIDES Fifty people are standing in line for a ride. Every three minutes, 20 people get on the ride, but 35 more people get in line. After one hour, how many people are in line?

EXAMPLE 1 At the grand opening of a store, every twelfth customer receives a coupon. The coupons have a value of 10%, 20%, or 30% off the total purchase. If 130 people visited the store, how many 30% off coupons were distributed?

Customer	Coupon	Customer	Coupon
12	10%	72	30%
24	20%	84	10%
36	30%	96	20%
48	10%	108	30%
60	20%	120	10%

Three 30% off coupons were distributed.

Variables and Expressions (Lesson 1B)

10. ALGEBRA Evaluate $a(b + 4)$ if $a = 6$ and $b = 2$.

11. WEATHER The time s in seconds between seeing lightning and hearing thunder can be used to estimate a storm's distance in miles. Use the expression $\frac{s}{5}$ to determine how far away a storm is if this time is 15 seconds.

EXAMPLE 2 Evaluate $x + yx - z$ if $x = 4$, $y = 2$, and $z = 1$.

$x + yx - z$	Write the expression.
$= 4 + (2)(4) - 1$	$x = 4, y = 2,$ and $z = 1$
$= 4 + 8 - 1$	Multiply.
$= 11$	Add and subtract from left to right.

Ordered Pairs and Relations (Lesson 1C)

Graph each ordered pair on a coordinate plane.

12. $A\left(-2\frac{1}{4}, 1\frac{3}{4}\right)$ **13.** $B\left(1\frac{1}{2}, -\frac{1}{4}\right)$

Express each relation as a table and a graph. Then state the domain and range.

14. $\{(1, -1), (2, -2), (3, -3), (4, -4)\}$

15. $\{(7, 2), (-3, 9), (0, 4), (-3, 2), (7, 8)\}$

EXAMPLE 3 Graph $X(2, -1)$, $Y(-3, 2)$, and $Z(0, -3)$ on a coordinate plane.

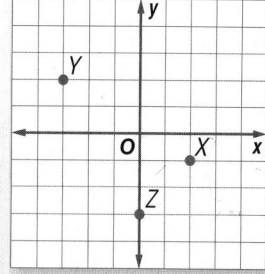

Start at the origin. The first number in each ordered pair is the x-coordinate. The second number in each ordered pair is the y-coordinate.

Translate Among Words, Tables, Graphs, and Equations

Analyze Tables (Lesson 2B)

16. Write an expression that can be used to find the nth term of the sequence below. Then use the expression to find the next three terms.

Term Number (n)	1	2	3	4	5
Term	−5	−1	3	7	11

EXAMPLE 4 Write an expression that can be used to find the nth term of the sequence 5, 10, 15, 20, 25, ... Then write the next three terms.

Term Number (n)	1	2	3	4	5
Term	5	10	15	20	25

The expression is $5n$. The next three terms are 5(6) or 30, 5(7) or 35, and 5(8) or 40.

Analyze Graphs (Lesson 2C)

17. PARKING The graph shows the total cost of parking in a parking garage.

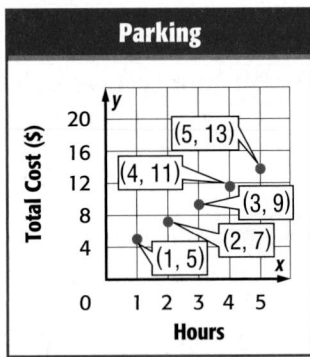

a. Write an algebraic expression to represent the data in the graph.

b. Use the expression to find the total cost to park for eight hours.

EXAMPLE 5 The graph shows the cost of tickets to the school play. Write an algebraic expression to represent the data in the graph.

The common difference is $4.50. So, the expression is $4.50t.

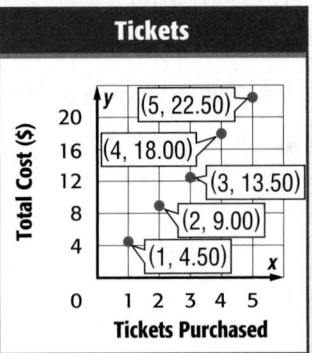

Translate Tables and Graphs into Equations (Lesson 2D)

18. WORK Gloria charges $15 per night to babysit. She also charges an additional $5 per child.

a. Write an equation to find the total charge c to babysit any number of children n.

b. Make a table to find the total charge to babysit 1, 2, 3, or 4 children. Then graph the ordered pairs.

EXAMPLE 6 The table shows the average number of minutes Kendra talks on her phone each day. Write an equation to find the total minutes in any number of days.

The common difference is 35. So, the equation is $m = 35d$.

Days, d	Total Minutes, m
1	35
2	70
3	105
4	140

Functions (Lesson 3B)

Find each function value.

19. $f(3)$ if $f(x) = 3x + 1$

20. $f(-11)$ if $f(x) = -2x$

21. $f(2)$ if $f(x) = \frac{1}{2}x - 4$

22. Make a function table for $f(x) = 3x - 4$. Then state the domain and the range.

EXAMPLE 7 Find $f(-2)$ if $f(x) = 2x - 1$.

$$f(x) = 2x - 1$$
$$f(-2) = 2(-2) - 1$$
$$f(-2) = -4 - 1$$
$$f(-2) = -5$$

Linear Functions (Lesson 3C)

Graph each function.

23. $y = -2x + 1$

24. $y = \frac{1}{2}x - 2$

25. $y = 4x + 5$

26. $y = -3x - 4$

EXAMPLE 8 Graph $y = 3 - x$.

x	3 − x	y
−1	3 − (−1)	4
0	3 − 0	3
2	3 − 2	1
3	3 − 3	0

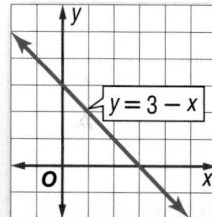

Linear and Nonlinear Functions (Lesson 4A)

Determine whether each table represents a *linear* or *nonlinear* function. Explain.

27.
x	2	3	4	5
y	98	147	199	248

28.
x	1	2	3	4
y	15	30	45	60

EXAMPLE 9 Determine whether the table represents a *linear* or *nonlinear* function.

As x increases by 1, y increases by 2. The rate of change is constant, so this function is linear.

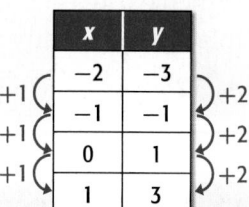

Graph Quadratic Functions (Lesson 4B)

Graph each function.

29. $y = -2x^2 + 1$ 30. $y = 3x^2 - 1$

31. **CLIFFS** The function $d = -4.9t^2 + 43$ models the distance d in meters a rock is from the surface of the water t seconds after being dropped from a 43-meter-tall cliff. Graph this function.

EXAMPLE 10 Graph $y = -x^2 - 1$.

x	$y = -x^2 - 1$	(x, y)
−2	$-(-2)^2 - 1$	(−2, −5)
−1	$-(-1)^2 - 1$	(−1, −2)
0	$-(0)^2 - 1$	(0, −1)
1	$-(1)^2 - 1$	(1, −2)

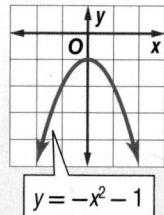

1. MONEY Piper and Michelle are both saving money to buy a new digital music player. Piper already has $75 and plans on saving $10 per month. Michelle only has $25 but plans on saving $20 per month. How many months will pass before they will have saved the same amount of money?

Evaluate each expression if $a = 3$, $b = 2$, and $c = -5$.

2. $(2c + b) \div b - 3$ **3.** $4a(a) - 5a - 12$

Graph each point on a coordinate plane.

4. $W\left(-2\frac{1}{2}, -3\frac{1}{2}\right)$ **5.** $X\left(-4, 1\frac{1}{2}\right)$

6. $Y(1, -3)$ **7.** $Z(2, 0)$

8. MULTIPLE CHOICE The table shows the total cost of a monthly gym membership.

Months	Cost ($)
1	125
2	200
3	275
4	350

Which expression can be used to find the total cost after any number of months?

A. $125n$ **C.** $125n + 75$

B. $75n$ **D.** $75n + 50$

9. SWIMMING The graph shows the number of laps Rafael swims every minute. Write an algebraic expression to represent the data in the graph.

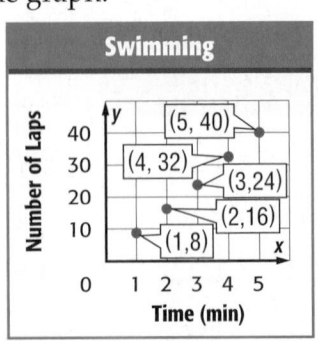

10. KEYBOARDING Megan types 35 words per minute.

a. Write an equation to find the number of words w typed in any number of minutes m.

b. Make a table to find the total words typed in 15, 20, 25, and 30 minutes. Then graph the ordered pairs.

c. What are the domain and range of the function?

Find each function value.

11. $f(3)$ if $f(x) = -2x + 6$ **12.** $f(-2)$ if $f(x) = \frac{x}{2} + 5$

Graph each function.

13. $y = -2x + 5$ **14.** $y = 4x^2 - 1$

15. Does the table represent a *linear* or *nonlinear* function? Explain your reasoning.

x	-1	0	1	2
y	1	0	1	4

16. THINK SOLVE EXPLAIN **EXTENDED RESPONSE** The width of a rectangular garden will be one foot longer than the length.

Part A Copy and complete the table showing the perimeter P of a garden with length ℓ.

Length, ℓ	Perimeter, P
10	42
11	46
12	■
13	■
14	■
15	■

Part B Graph the ordered pairs (length, perimeter) on a coordinate grid.

Part C Write an equation to represent the relationship in the table.

Part D Is the function continuous or discrete? Explain.

Part E What is the length of a garden with a perimeter of 82 feet?

Preparing for Standardized Tests

Multiple Choice: Use the Answer Choices

Sometimes you can use the answer choices to help you answer a multiple-choice question. In the example below, you can use substitution to determine the correct answer choice.

TEST EXAMPLE

The table shows how far Trevor travels while water skiing. Based on the linear relation in the table, which equation can be used to approximate d, the distance in feet that Trevor travels in t seconds?

Water Skiing	
Time in Seconds, t	Distance in Feet, d
4	105.6
6	158.4
8	211.2

A. $t = 26.4d$

B. $d = 26.4t$

C. $t = 0.04d$

D. $d = 0.04t$

A shortcut method for solving the problem is to choose values from the table and substitute them into the equations given in the answer choices.

$t = 26.4d$ Equation A

$4 \stackrel{?}{=} 26.4(105.6)$ Replace t with 4 and d with 105.6.

$4 \neq 2787.84$ ✗

$d = 26.4t$ Equation B

$105.6 \stackrel{?}{=} 26.4(4)$ Replace t with 4 and d with 105.6.

$105.6 = 105.6$ ✓

The correct equation is $d = 26.4t$. So, the correct answer is B.

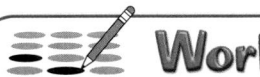 Work on It

The graph represents the speed of a tennis ball. Which equation can be used to approximate the distance of the tennis ball d in meters after t seconds?

F. $d = 0.02t$

G. $t = 0.02d$

H. $d = 56t$

I. $t = 56d$

Speed of Tennis Ball

(3, 168)

(2, 112)

Distance (m)

150

100

50

0   1   2   3

Time (s)

Test Hint

Once you find the correct answer, you do not need to substitute values in the remaining answer choices.

✓ Test Practice

Read each question. Then fill in the correct answer on the answer document provided by your teacher or on a sheet of paper.

1. Beth's monthly charge for Internet access c is represented by the function $c = 12 + 2.50h$, where h represents the number of hours of usage during a month. What is the total charge for a month in which Beth used the Internet for 9 hours?

   A. $39.95

   B. $34.50

   C. $27.00

   D. $22.50

2. The graph of the line $y = -2x + 1$ is shown below. Which table of ordered pairs contains only points on this line?

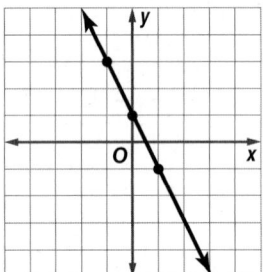

   F.
x	−2	−1	0
y	5	3	−1

   G.
x	−2	−1	0
y	3	1	−1

   H.
x	−1	0	1
y	−3	−1	1

   I.
x	−1	0	1
y	3	1	−1

3. **THINK SOLVE EXPLAIN** **SHORT RESPONSE** What is the domain of the relation $\{(-1, 4), (4, 6), (-3, -7), (2, -1)\}$?

4. The table shows the number of shoppers in a store on each of the first four days after its grand opening.

Day	Shoppers
1	250
2	310
3	370
4	430

   Suppose the pattern continues. Which expression can be used to find the number of shoppers on any day?

   A. $250d$ C. $60d$

   B. $250d + 60$ D. $60d + 190$

5. ✏ **GRIDDED RESPONSE** What is the sum of the fractions below?
   $$\frac{1}{6} + \frac{3}{4}$$

6. A refrigerator costs $560. The refrigerator is on sale for 30% off the regular price. What is the total cost of the refrigerator after the discount?

   F. $728   H. $175

   G. $392   I. $168

7. Lorraine plotted her friends' houses on a coordinate plane.

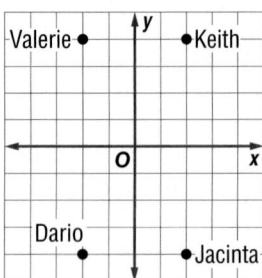

   Which point on the grid **best** represents the location of Dario's house?

   A. $(2, 4)$ C. $(-2, -4)$

   B. $(-2, 4)$ D. $(2, -4)$

8. Andrew purchased a coat for $67.20 that regularly sells for $84.00. What was the percent discount that Andrew received?

F. 16.8%   **H.** 25%

G. 20%   **I.** 80%

9. ☐THINK SOLVE EXPLAIN **SHORT RESPONSE** Landon is charged for each text message that he sends. The graph shows the total cost of sending text messages.

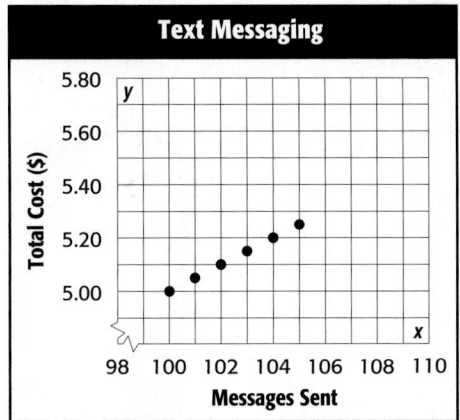

Text Messaging

Write an expression that can be used to find the total cost of sending any number of text messages.

10. Which equation represents a linear function?

A. $y = -x^2 - 4$

B. $-3x^2 + 1 = y$

C. $y = x^2$

D. $x + 2 = y$

11. Which of the following is the value of $f(-4)$ in the function below?

$$f(x) = -2x - 3$$

F. −11   **H.** 5

G. −5   **I.** 11

12. There are four children in the Velez family. Alano is $1\frac{1}{2}$ times as tall as Lupe, and 6 inches taller than Olivia. Nelia is 56 inches tall, which is 2 inches taller than Olivia. How tall is Lupe?

A. 40 in.   **C.** 54 in.

B. 50 in.   **D.** 56 in.

13. ☐THINK SOLVE EXPLAIN **EXTENDED RESPONSE** Video Mania is having a sale on DVDs. If you buy one DVD for $5, the cost of any additional DVDs is $2. The table shows the total cost c of any number of DVDs d.

Number of DVDs	Total Cost ($)
1	5
2	7
3	9
⋮	⋮
10	■

Part A Copy and complete the table.

Part B Graph the ordered pairs (DVDs, total cost) on a coordinate plane.

Part C Write an equation to find the total cost of any number of DVDs.

Part D How much would it cost to purchase 15 DVDs?

NEED EXTRA HELP?												
If You Missed Question...	1	2	3	4	5	6	7	8	9	10	11	12	13
Go to Chapter-Lesson...	5-3B	5-3C	5-1C	5-2B	1-1B	1-3A	5-1C	1-3A	5-2C	5-4B	5-3B	1-1D	5-2D

Linear Functions and Systems of Equations

connectED.mcgraw-hill.com

Investigate

 Animations

 Vocabulary

 Multilingual eGlossary

Learn

 Personal Tutor

 Virtual Manipulatives

 Graphing Calculator

 Audio

 Foldables

Practice

 Self-Check Practice

 Worksheets

 Assessment

The ☆BIG Idea

What types of real-life situations can be represented by a linear equation or a system of linear equations?

 FOLDABLES® Study Organizer

Make this Foldable to help you organize your notes.

Intercepts

Review Vocabulary

function función a relation in which each member of the domain is paired with exactly one member of the range

Range

$f(x) = 3x + 4$

Domain

Key Vocabulary

English	Español
slope	pendiente
slope-intercept form	forma pendiente
	intersección
system of equations	sistema de ecuaciones
y-intercept	intersección y

When Will I Use This?

Your Turn!
You will solve this problem in the chapter.

Are You Ready for the Chapter?

You have two options for checking prerequisite skills for this chapter.

Text Option Take the Quick Check below. Refer to the Quick Review for help.

QUICK Check

Subtract.

1. $5 - (-4)$
2. $10 - 8$
3. $-4 - 3$
4. $-6 - (-2)$
5. $12 - 6$
6. $-5 - (-3)$
7. $-8 - 7$
8. $4 - (6)$
9. $-10 - (-4)$
10. $-9 - 5$

11. **WEATHER** At noon, the outside temperature was $-5°F$. By 5:00 P.M., the temperature had dropped $8°F$. What was the temperature at 5:00 P.M.?

12. **CHEMISTRY** The melting point of mercury is about $-39°C$ and the melting point of aluminum is about $660°C$. Find the difference between these two temperatures.

Evaluate each expression.

13. $\dfrac{6-2}{5+5}$
14. $\dfrac{7-4}{8-4}$
15. $\dfrac{3-1}{1+9}$
16. $\dfrac{5+7}{8-6}$
17. $\dfrac{2-4}{3+2}$
18. $\dfrac{1-5}{8-2}$
19. $\dfrac{4-8}{4-6}$
20. $\dfrac{5-8}{4-7}$

21. **ALGEBRA** Evaluate the expression $\dfrac{b-a}{c+a}$ if $a = -4$, $b = 6$, and $c = -8$.

QUICK Review

EXAMPLE 1

Find $-15 - 8$.

$$\begin{aligned} -15 - 8 &= -15 + (-8) \quad \text{To subtract 8, add } -8. \\ &= -23 \quad \text{Add.} \end{aligned}$$

EXAMPLE 2

Find $-8 - (-4)$.

$$\begin{aligned} -8 - (-4) &= -8 + 4 \quad \text{To subtract } -4, \text{ add 4.} \\ &= -4 \quad \text{Add.} \end{aligned}$$

EXAMPLE 3

Evaluate $\dfrac{11+4}{9-4}$.

$$\begin{aligned} \dfrac{11+4}{9-4} &= \dfrac{15}{5} \quad \begin{array}{l}\text{Simplify the numerator} \\ \text{and denominator.}\end{array} \\ &= 3 \quad \text{Simplify.} \end{aligned}$$

Online Option Take the Online Readiness Quiz.

Main Idea

Identify proportional and nonproportional linear relationships by finding a constant rate of change.

 Vocabulary

linear relationship
constant rate of change

Get ConnectED

 CCSS 8.EE.5, 8.F.4

Constant Rate of Change

MUSIC Vineeta can download two songs from the Internet each minute. This is shown in the table and in the graph.

Time (minutes)	0	1	2	3	4
Number of Songs	0	2	4	6	8

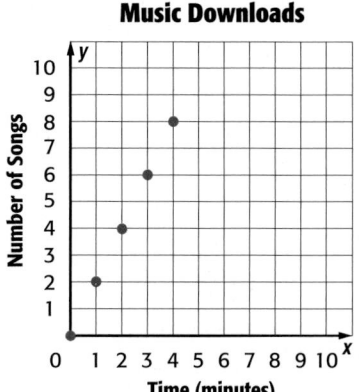

Music Downloads

1. Pick several pairs of points. Find the rate of change between them. What is true of these rates?

Relationships that have straight-line graphs, like the one above, are called **linear relationships**. Notice that as the number of songs increases by 2, the time in minutes increases by 1.

$$+2 \quad +2 \quad +2 \quad +2$$

Number of Songs	0	2	4	6	8
Time (minutes)	0	1	2	3	4

$$+1 \quad +1 \quad +1 \quad +1$$

Rate of Change
$\frac{2}{1} = 2$ songs per minute

The rate of change between any two points in a linear relationship is the same or *constant*. A linear relationship has a **constant rate of change**.

 EXAMPLE Identify Linear Relationships

1 **FINANCIAL LITERACY** The balance in an account after several transactions is shown. Is the relationship between the balance and number of transactions linear? If so, find the constant rate of change. If not, explain your reasoning.

Number of Transactions	Balance ($)
3	170
6	140
9	110
12	80

	Number of Transactions	Balance ($)	
+3	3	170	−30
+3	6	140	−30
+3	9	110	−30
	12	80	

As the number of transactions increases by 3, the balance in the account decreases by $30.

(continued on the next page)

Since the rate of change is constant, this is a linear relationship. The constant rate of change is $\frac{-30}{3}$ or $-\$10$ per transaction. This means that each transaction involved a $10 *withdrawal*.

 CHECK Your Progress

Determine whether the relationship between the two quantities described in each table is linear. If so, find the constant rate of change. If not, explain your reasoning.

a.

Cooling Water	
Time (min)	Temperature (°F)
5	95
10	90
15	86
20	82

b.

Wrapping Paper	
Number of Rolls	Total Cost ($)
2	8.50
4	17.00
6	25.50
8	34.00

📝 **EXAMPLE** Find A Constant Rate of Change

② **FOOD** Find the constant rate of change for the cost of each personal pizza ordered. Interpret its meaning.

Choose any two points on the line and find the rate of change between them.

$(1, 3) \longrightarrow$ 1 pizza, $3

$(3, 9) \longrightarrow$ 3 pizzas, $9

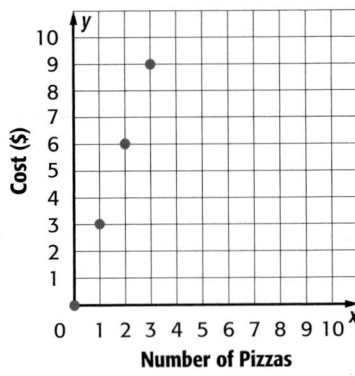

Pizza Cost

$\dfrac{\text{change in cost}}{\text{change in number}} = \dfrac{\$(9-3)}{(3-1)\text{ pizzas}}$ The cost changed from $9 to $3 while the number changed from 3 pizzas to 1 pizza.

$= \dfrac{\$6}{2\text{ pizzas}}$ Subtract to find the change in the cost and number of pizzas.

$= \dfrac{\$3}{1\text{ pizza}}$ Express this rate as a unit rate.

The cost is $3 for every 1 pizza ordered.

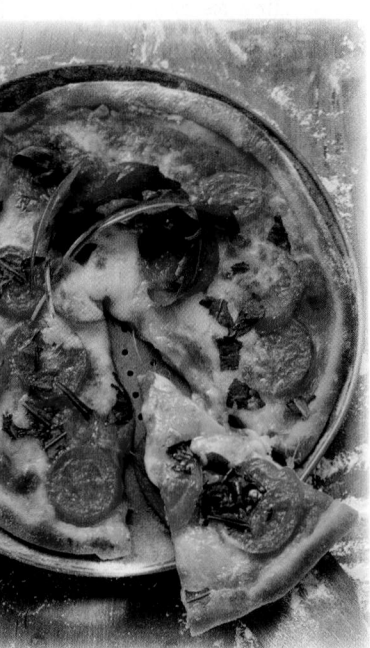

Real-World Link · · · · ·
Pizza Margherita was actually named after Queen Margherita of Italy in 1889. Very simple pizzas may have been eaten as early as 500 B.C.

 CHECK Your Progress

c. **SERVICE PROJECT** Find the constant rate of change for the time it takes to complete a trash pickup project for each number of volunteers in the graph. Interpret its meaning.

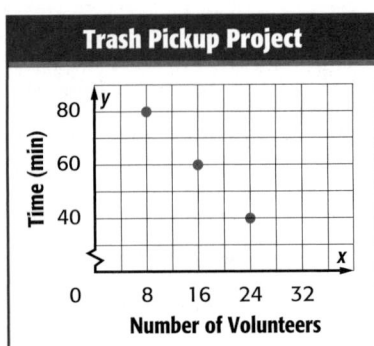

Trash Pickup Project

REAL-WORLD EXAMPLE

3 **TEMPERATURE** Use the graph to
determine if there is a proportional
linear relationship between a
temperature in degrees Fahrenheit
and a temperature in degrees
Celsius. Explain your reasoning.

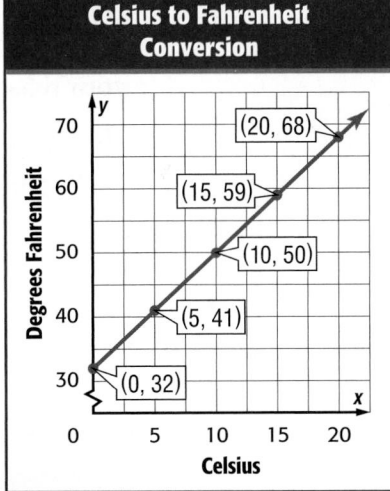

**Celsius to Fahrenheit
Conversion**

Since the graph of the data forms a
line, the relationship between the
two scales is linear. This can also be
seen in the table of values created
using the points on the graph.

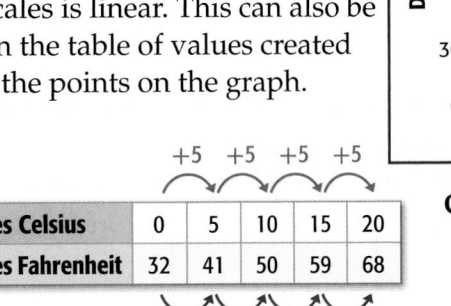

Degrees Celsius	0	5	10	15	20
Degrees Fahrenheit	32	41	50	59	68

Constant Rate of Change

$$\frac{\text{change in }°\text{F}}{\text{change in }°\text{C}} = \frac{9}{5}$$

To determine if the two scales are proportional, express the
relationship between the degrees for several columns as a ratio.

$$\frac{\text{degrees Fahrenheit}}{\text{degrees Celsius}} \rightarrow \frac{41}{5} = 8.2 \qquad \frac{50}{10} = 5 \qquad \frac{59}{15} \approx 3.9$$

Since the ratios are not the same, the relationship between degrees
Fahrenheit degrees Celsius is *not* proportional.

 CHECK Your Progress

d. MEASUREMENT Use the graph
to determine if there is a
proportional linear relationship
between the weight of an
object in pounds and the mass
of the object in kilograms.
Explain your reasoning.

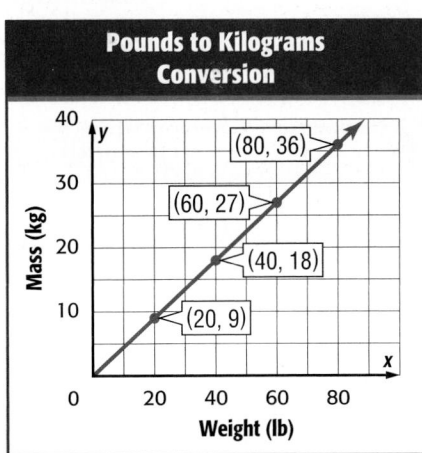

**Pounds to Kilograms
Conversion**

Key Concept   **Proportional Linear Relationships**

Words   Two quantities *a* and *b* have a proportional linear relationship
if they have a constant ratio and a constant rate of change.

Symbols $\frac{b}{a}$ is constant and $\frac{\text{change in } b}{\text{change in } a}$ is constant.

Example 1 Determine whether the relationship between the two quantities described in each table is linear. If so, find the constant rate of change. If not, explain your reasoning.

1.

Volume of Cube	
Side Length (cm)	Volume (cm³)
2	8
3	27
4	64
5	125

2.

Paint Needed for Chairs	
Chairs	Cans of Paint
5	6
10	12
15	18
20	24

Example 2 Find the constant rate of change for each graph and interpret its meaning.

3.
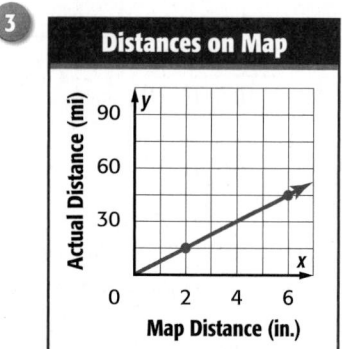

4.
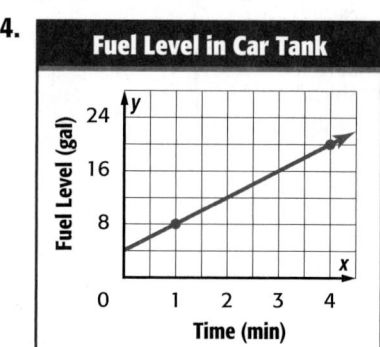

Example 3 Determine whether a proportional linear relationship exists between the two quantities shown in each of the indicated graphs. Explain your reasoning.

5. Exercise 3

6. Exercise 4

Practice and Problem Solving

● = **Step-by-Step Solutions** begin on page R1.
Extra Practice begins on page EP2.

Example 1 Determine whether the relationship between the two quantities described in each table is linear. If so, find the constant rate of change. If not, explain your reasoning.

7.

Cost of Electricity to Run Personal Computer	
Time (h)	Cost (¢)
5	15
8	24
12	36
24	72

8.

Total Number of Customers Helped at Jewelry Store	
Time (h)	Total Helped
1	12
2	24
3	36
4	60

9.

Distance Traveled by Falling Object				
Distance (m)	4.9	19.6	44.1	78.4
Time (s)	1	2	3	4

10.

Italian Dressing Recipe				
Oil (c)	2	4	6	8
Vinegar (c)	$\frac{3}{4}$	$1\frac{1}{2}$	$2\frac{1}{4}$	3

Example 2 **Find the constant rate of change for each graph and interpret its meaning.**

11.
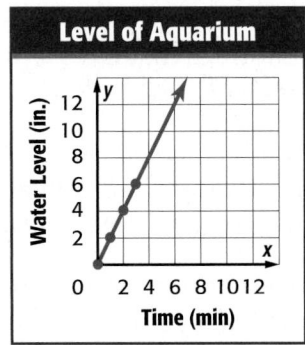

12.
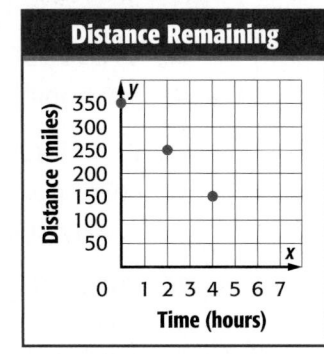

Real-World Link · · · ·
Nearly 12 million U.S.
homes have an
aquarium.

13.

Aircraft Altitude

14.
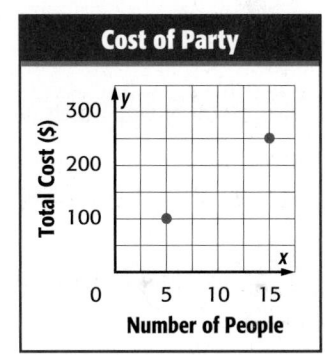

15.

Sale Price

16.

Cost of Party

Example 3 **Determine whether a proportional relationship exists between the two quantities shown in each of the indicated graphs. Explain your reasoning.**

17. Exercise 11   18. Exercise 12   19. Exercise 13

20. Exercise 14   21. Exercise 15   22. Exercise 16

23 **CELL PHONES** Both Tyrell and Miriam have cell phone plans. Their costs for several minutes are shown.

a. Who is spending more money each minute? Explain your reasoning.

b. Whose plan is proportional to the number of minutes the phone is used? Explain.

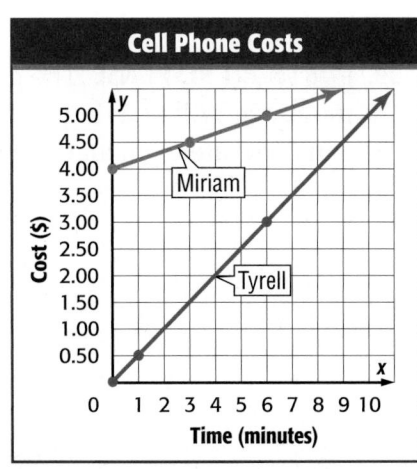

24. **OPEN ENDED** Graph two quantities that have a proportional linear relationship. Justify your answer.

25. **CHALLENGE** Examine the graphs in Exercises 11–16, as well as your corresponding answers in Exercises 17–22. What point do all of the graphs that represent proportional linear relationships have in common?

26. ✍ **WRITE MATH** Write a real-world problem in which you would need to find a constant rate of change. Then solve your problem. Is the relationship described in your problem proportional? Explain.

Test Practice

27. Tickets to the school play are $2.50 each. Which table contains values that fit this situation if c represents the total cost for t tickets?

A.

Cost of Play Tickets ($)				
t	1	2	3	4
c	2.50	3.25	4.00	4.75

B.

Cost of Play Tickets ($)				
t	1	2	3	4
c	3.50	6.00	8.50	11.00

C.

Cost of Play Tickets ($)				
t	1	2	3	4
c	3.50	4.00	4.50	5.00

D.

Cost of Play Tickets ($)				
t	1	2	3	4
c	2.50	5.00	7.50	10.00

28. The graph shows the distance Bianca traveled on her 2-hour bike ride.

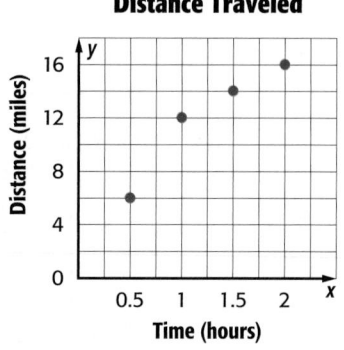

Distance Traveled

Which of the following is true?

F. She traveled at a constant speed of 12 miles per hour for the entire ride.

G. She traveled at a constant speed of 8 miles per hour for the last hour.

H. She traveled at a constant speed of 4 miles per hour for the last hour.

I. She traveled at a constant speed of 8 miles per hour for the entire ride.

29. 📊 **SHORT RESPONSE** The graph shows the amount of money in Will's savings account each week. Find the constant rate of change. Then interpret its meaning.

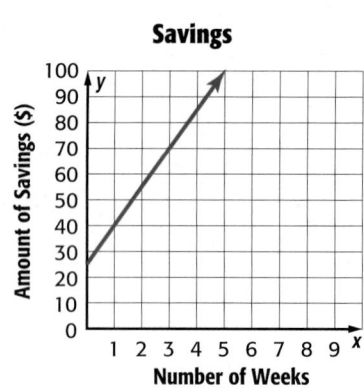

Savings

Main Idea

Find rates of change.

CCSS 8.EE.5, 8.F.4

Graphing Technology:
Rate of Change

Recall that a rate of change is a rate that describes how one quantity changes in relation to another.

📱 ACTIVITY

SCHOOL At the school store, three-ring notebooks are sold for $5 each. The equation $y = 5x$ can be used to find the total cost y of any number of notebooks x. Find the rate of change.

STEP 1 Enter the equation. Press [Y=] 5 [X,T,θ,n].

STEP 2 Graph the equation in the standard viewing window. Press [Zoom] 6.

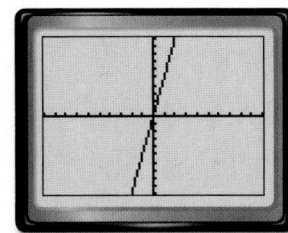

STEP 3 Choose any two points on the line to find the rate of change. Use the points (0, 0) and (1, 5).

$$\frac{\text{change in total cost}}{\text{change in number of notebooks}} = \frac{\$(5-0)}{(1-0) \text{ notebooks}}$$

$$= \frac{\$5}{1 \text{ notebook}}$$

The rate of change is 5.

Analyze the Results

1. School T-shirts are sold for $10 each and packages of markers are sold for $2.50 each.

   a. Write an equation that can be used to find the total cost y of each item x.

   b. Graph the equations in the same window as the equation from the Activity. Find each rate of change.

2. Look at the graphs of all three lines. What do you notice about the steepness of the lines?

3. **MAKE A CONJECTURE** Without graphing, predict which graph has a steeper line: $y = 3x$ or $y = \frac{1}{3}x$. Explain.

Main Idea

Find the slope of a line.

 Vocabulary

slope
rise
run
qualitative graph

 *Get Connect**ED***

 CCSS 8.EE.5, 8,F,4

Slope

45 ft

30 ft

SAFETY A ladder truck uses a moveable ladder to reach upper levels of houses and buildings.

1. The rate of change of the ladder compares the height it is raised to the distance of its base from the building. Write this rate as a fraction in simplest form.

2. Find the rate of change of a ladder that has been raised 100 feet and has a base of 50 feet from the building.

The term *slope* is used to describe the steepness of a straight line. **Slope** is the ratio of the **rise**, or vertical change, to the **run**, or horizontal change. In linear functions, no matter which two points you choose, the slope, or rate of change, of the line is always constant.

$$\textbf{slope} = \frac{\textbf{rise}}{\textbf{run}}$$ ← vertical change between any two points

← horizontal change between the same two points

 REAL-WORLD EXAMPLE

1 **EXERCISE** **Find the slope of the treadmill.**

$$\text{slope} = \frac{\text{rise}}{\text{run}} \qquad \text{Definition of slope}$$

$$= \frac{10 \text{ in.}}{48 \text{ in.}} \qquad \begin{array}{l}\text{rise} = 10 \text{ in.,} \\ \text{run} = 48 \text{ in.}\end{array}$$

$$= \frac{5}{24} \qquad \text{Simplify.}$$

The slope of the treadmill is $\frac{5}{24}$.

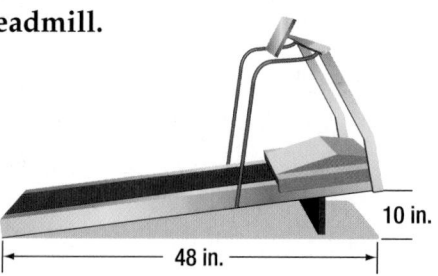

10 in.

48 in.

 CHECK Your Progress

a. **HIKING** A hiking trail rises 6 feet for every horizontal change of 100 feet. What is the slope of the hiking trail?

Vocabulary Link

Everyday Use

Slope any slanting surface, hill side

Math Use

Slope the steepness of a straight line defined as the ratio $\frac{\text{rise}}{\text{run}}$

Since slope is a rate of change, it can be positive (slanting upward) or negative (slanting downward).

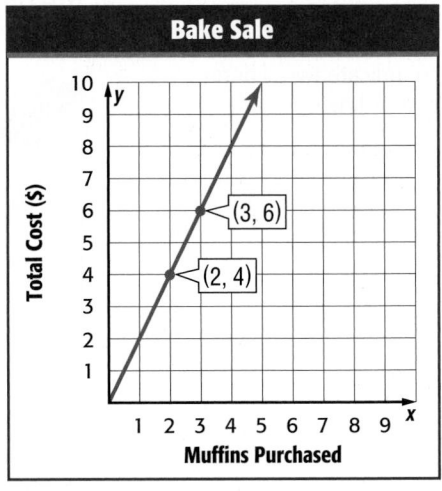

Study Tip

Translating Rise and Run

up	→ positive
down	→ negative
right	→ positive
left	→ negative

EXAMPLES Find Slope Using a Graph or Table

2 **FOOD** The graph shows the cost of muffins at a bake sale. Find the slope of the line.

Choose two points on the line. The vertical change is 2 units and the horizontal change is 1 unit.

$$\text{slope} = \frac{\text{rise}}{\text{run}} \quad \text{Definition of slope}$$

$$= \frac{2}{1} \quad \text{rise} = 2, \text{run} = 1$$

The slope of the line is $\frac{2}{1}$ or 2.

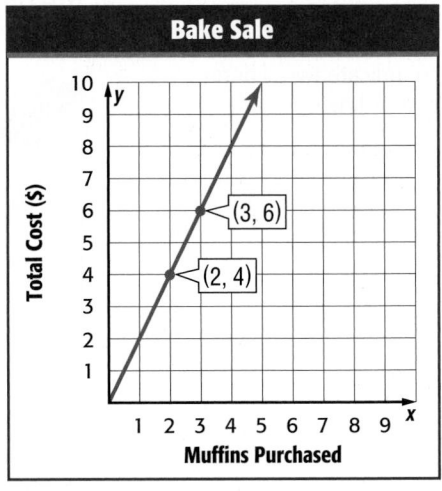

3 **READING** The table shows the number of pages Garrett has left to read after a certain number of minutes. Find the slope of the line. Then graph it.

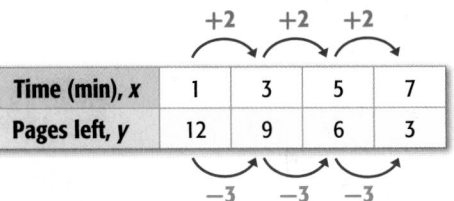

Choose two points from the table to find the changes in the x- and y-values.

$$\text{slope} = \frac{\text{change in } y}{\text{change in } x}$$

$$= \frac{9 - 12}{3 - 1}$$

$$= \frac{-3}{2} \text{ or } -\frac{3}{2}$$

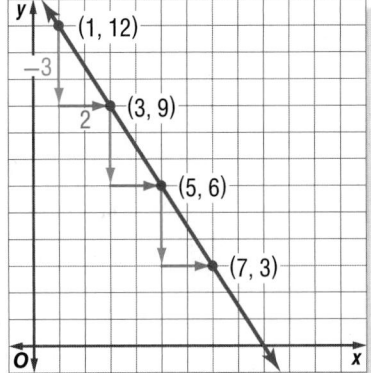

Study Tip

Slope You can choose any two points to calculate slope. Whichever y-value you use first, be sure to use the corresponding x-value first.

CHECK Your Progress

Find the slope of each line.

b.

c.

d.

x	−6	−2	2	6
y	−2	−1	0	1

e.

x	−4	0	4	8
y	−1	−2	−3	−4

 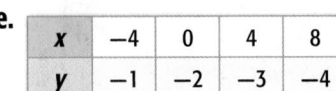

Subscripts x_1 is read
x sub one and x_2 is read
x sub two. They are used
to indicate two different
x-coordinates.

You can also find the slope of a line by using the coordinates of any two points on the line. One point can be represented by (x_1, y_1) and the other by (x_2, y_2).

Key Concept — Slope Formula

Words		Model
The slope m of a line passing through points (x_1, y_1) and (x_2, y_2) is the ratio of the difference in the y-coordinates to the corresponding difference in the x-coordinates.		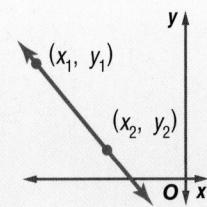

Symbols $m = \dfrac{y_2 - y_1}{x_2 - x_1}$, where $x_2 \neq x_1$

EXAMPLES   Find Slope Using Coordinates

Find the slope of the line that passes through each pair of points.

4 $C(-1, -4), D(2, 2)$

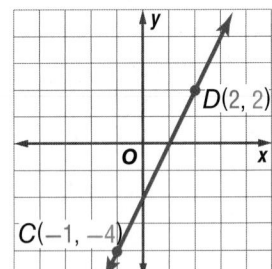

$m = \dfrac{y_2 - y_1}{x_2 - x_1}$ Slope formula

$m = \dfrac{2 - (-4)}{2 - (-1)}$ $(x_1, y_1) = (-1, -4)$
                                  $(x_2, y_2) = (2, 2)$

$m = \dfrac{6}{3}$ or 2   Simplify.

Check   When going from left to right, the graph of the line slants upward. This is correct for positive slope.

Study Tip

Using the Slope Formula
• It does not matter which point you define as (x_1, y_1) and (x_2, y_2).
• However, the coordinates of both points must be used in the same order.

To check Example 5, let $(x_1, y_1) = (-4, 3)$ and $(x_2, y_2) = (1, 2)$. Then find the slope.

5 $R(1, 2), S(-4, 3)$

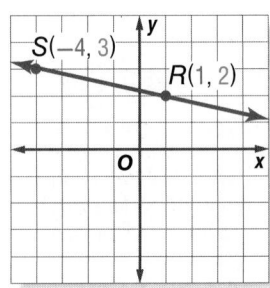

$m = \dfrac{y_2 - y_1}{x_2 - x_1}$ Slope formula

$m = \dfrac{3 - 2}{-4 - 1}$ $(x_1, y_1) = (1, 2)$
                             $(x_2, y_2) = (-4, 3)$

$m = \dfrac{1}{-5}$ or $-\dfrac{1}{5}$ Simplify.

Check   When going from left to right, the graph of the line slants downward. This is correct for negative slope.

CHECK Your Progress

f. $A(2, 2), B(5, 3)$ **g.** $C(-2, 1), D(0, -3)$ **h.** $J(-7, -4), K(-3, -2)$

Example 1 **1. BUILDINGS** Find the slope of the storage shed's roof.

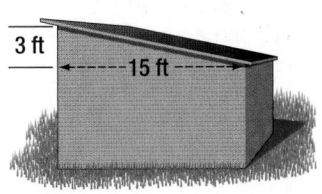

Examples 2 and 3 Find the slope of each line.

2.

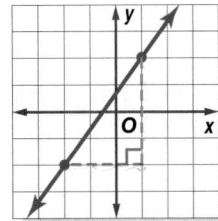

3.

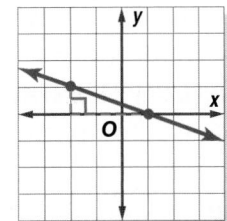

4.

x	0	1	2	3
y	1	3	5	7

Examples 4 and 5 Find the slope of the line that passes through each pair of points.

5. $A(-3, -2)$, $B(5, 4)$

6. $C(-4, 2)$, $D(1, 5)$

7. $E(-6, 5)$, $F(3, -3)$

8. $G(1, 5)$, $H(4, -3)$

Practice and Problem Solving

● = Step-by-Step Solutions begin on page R1.
Extra Practice begins on page EP2.

Example 1 **⑨ SKIING** Find the slope of a ski run that descends 15 feet for every horizontal change of 24 feet.

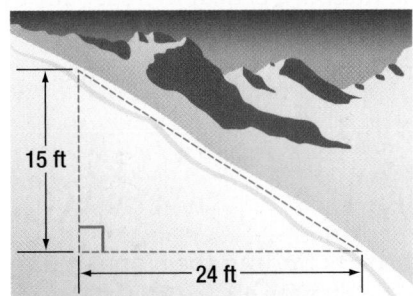

10. ROADS Find the slope of a road that rises 12 feet for every horizontal change of 100 feet.

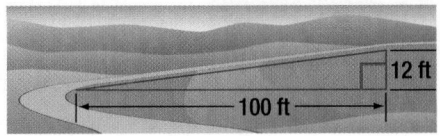

Example 2 Find the slope of each line.

11.

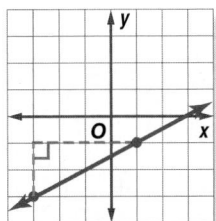

12.

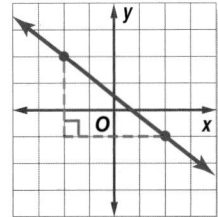

13.

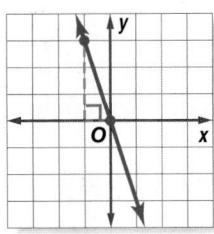

14.

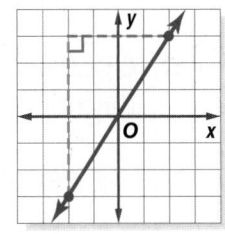

Example 3 The points given in each table lie on a line. Find the slope of the line. Then graph the line.

15.

x	0	2	4	6
y	9	4	−1	−6

16.

x	−3	3	9	15
y	−3	1	5	9

Examples 4 and 5 Find the slope of the line that passes through each pair of points.

17. $A(0, 1)$, $B(2, 7)$ **18.** $C(2, 5)$, $D(3, 1)$ **19.** $E(1, 2)$, $F(4, 7)$

20. $G(−6, −1)$, $H(4, 1)$ **21.** $J(−9, 3)$, $K(2, 1)$ **22.** $M(−2, 3)$, $N(7, −4)$

23. **MULTIPLE REPRESENTATIONS** For working 3 hours, Sofia earns $30.60. For working 5 hours, she earns $51. For working 6 hours, she earns $61.20.

   a. GRAPHS Graph the information with hours on the horizontal axis and money earned on the vertical axis. Draw a line through the points.

   b. NUMBERS What is the slope of the line?

   c. WORDS What does the slope of the line represent?

24. GEOMETRY Two lines that are parallel have the same slope. Determine whether quadrilateral $ABCD$ is a parallelogram. Justify your reasoning.

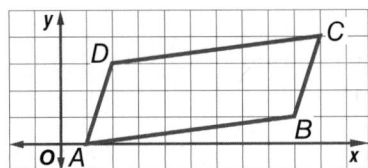

25. DISABILITIES Wheelchair ramps for access to public buildings are allowed a maximum of one inch of vertical increase for every one foot of horizontal distance. Would a ramp that is 10 feet long and 8 inches tall meet this guideline? Explain your reasoning.

Real-World Link · · · ·

The Americans with Disabilities Act (ADA) was signed on July 26, 1990. It is a law that prohibits discrimination based on disabilities.

H.O.T. Problems

26. FIND THE ERROR Jacob is finding the slope of the line that passes through $X(0, 2)$ and $Y(4, 3)$. Find his mistake and correct it.

$$m = \frac{3 - 2}{0 - 4}$$
$$m = \frac{1}{-4} \text{ or } -\frac{1}{4}$$

27. CHALLENGE Find the slope of the straight line that is the graph of the function expressing the circumference of a circle as a function of the radius.

28. WRITE MATH In any linear function, explain why the ratio of the rise and run is always the same.

29. ▰▰ **GRIDDED RESPONSE** Lionel charted the growth rate of his new puppy for several weeks. He measured the weight of the puppy in pounds and plotted the values on the graph below.

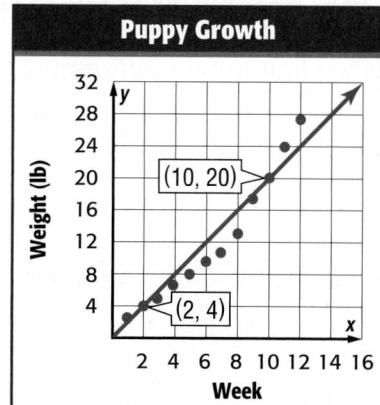

He drew a line passing through points (2, 4) and (10, 20) on the graph to estimate the puppy's weight during any week. What is the slope of the line he drew?

30. Line *AB* represents a steep hill.

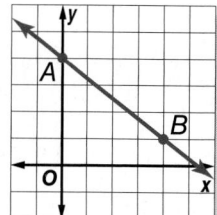

What is the slope of the hill?

A. $-\dfrac{4}{3}$ **C.** $\dfrac{3}{4}$

B. $-\dfrac{3}{4}$ **D.** $\dfrac{4}{3}$

31. Line *AB* goes through points $A(-4, -3)$ and $B(-2, 0)$. What is the slope of the line?

F. $-\dfrac{3}{2}$ **H.** $\dfrac{2}{3}$

G. $-\dfrac{2}{3}$ **I.** $\dfrac{3}{2}$

More About **Graphs** ....

Qualitative graphs are graphs that are used to represent situations that do not necessarily have numerical values. For example, Graph A could describe a car that is accelerating at a constant rate. Graph B could describe a car that is traveling at a constant rate for a certain amount of time and then accelerates at a constant rate.

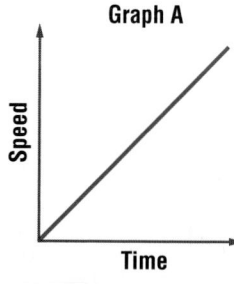

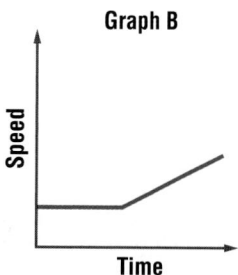

32. Describe a situation that could be represented by Graph C.

33. Draw a qualitative graph that could be used to represent the following situation: Ava takes a ski lift to the top of a mountain and then skis down.

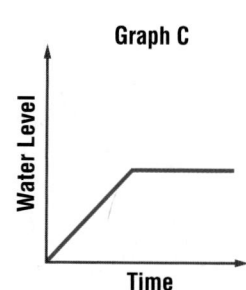

Explore

Proportional and Nonproportional Relationships

Main Idea

Compare and contrast proportional and nonproportional linear relationships.

 8.EE.5

In this lab, you will use models to explore two different relationships.

ACTIVITIES

STEP 1 Using centimeter cubes, build the two patterns shown.

Pattern	A	B
Figures		
Figure Number	0 1 2 3	0 1 2 3

STEP 2 Let x represent the figure number and y represent the number of cubes in each tower. Copy and complete the table for each pattern. Then graph and label each data set on separate coordinate planes.

x	Function Rule	y
0		
1		
2		
3		
4		
x		

Analyze the Results

1. Compare and contrast the models of patterns A and B.
2. Compare and contrast the function rules for patterns A and B.
3. Compare and contrast the graphs of patterns A and B from Step 2.
4. Which pattern represents a proportional relationship? Explain. How can you tell this from the function rule shown in the table? from the graph?

Main Idea

Use direct variation to solve problems.

Vocabulary

direct variation
constant of variation

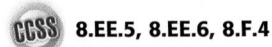

CCSS 8.EE.5, 8.EE.6, 8.F.4

Direct Variation

COMPUTERS The graph shows the output of a color printer.

1. What is the constant rate of change, or slope, of the line?

2. Is the total number of pages printed always proportional to the printing time? Explain.

3. Compare the constant rate of change to the constant ratio.

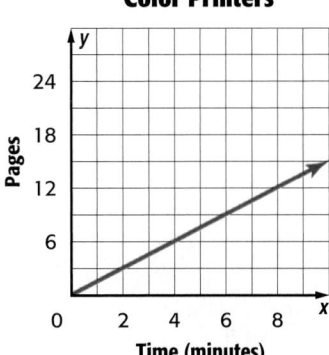

Color Printers

In the example above, the number of minutes and the number of pages printed both vary. But the ratio of pages printed to minutes, 1.5 pages per minute, remains constant.

When the ratio of two variable quantities is constant, their relationship is called a **direct variation**. The constant ratio is called the **constant of variation**.

REAL-WORLD EXAMPLE Find a Constant Ratio

1 **FUNDRAISER** The amount of money Robin has raised for a bike-a-thon is shown in the graph. Determine the amount that Robin raises for each mile she rides.

Since the graph of the data forms a line, the rate of change is constant. Use the graph to find the constant ratio.

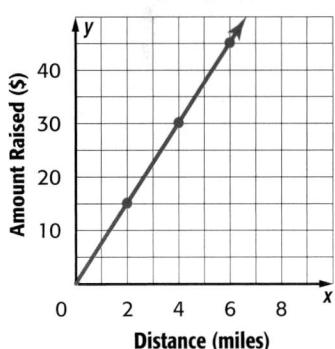

Money Raised

$$\frac{\text{amount raised}}{\text{distance}} \rightarrow \frac{15}{2} \text{ or } \frac{7.5}{1} \quad \frac{30}{4} \text{ or } \frac{7.5}{1} \quad \frac{45}{6} \text{ or } \frac{7.5}{1}$$

Robin raises $7.50 for each mile she rides.

 CHECK Your Progress

a. **SKYDIVING** Two minutes after a skydiver opens his parachute, he has descended 1,900 feet. After 5 minutes, he has descended 4,750 feet. If the distance varies directly with the time, at what rate is the skydiver descending?

In a direct variation equation, the constant rate of change, or slope, is assigned a special variable, k.

 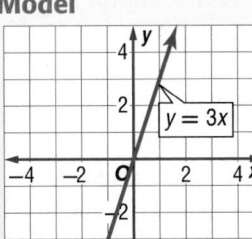

REAL-WORLD EXAMPLE

Solve a Direct Variation

2 **PETS** Refer to the information at the left. Assume that the age of a dog varies directly as its equivalent age in human years. What is the human-year age of a dog that is 6 years old?

Write an equation of direct variation. Let x represent the dog's actual age and let y represent the human-equivalent age.

$y = kx$ Direct variation
$21 = k(3)$ $y = 21, x = 3$
$7 = k$ Simplify.
$y = 7x$ Replace k with 7.

Use the equation to find y when $x = 6$.

$y = 7x$
$y = 7(6)$ $x = 6$
$y = 42$ Multiply.

A dog that is 6 years old is 42 years old in human years.

 CHECK Your Progress

b. SHOPPING A grocery store sells 6 oranges for $2. How much would it cost to buy 10 oranges? Round to the nearest cent if necessary.

In a direct variation, the constant of variation k is a constant rate of change. When the x-value changes by an amount a, then the y-value will change by the corresponding amount ka. In the previous example, when a dog's age x increased by 1 year, its human year equivalent y increased by 7 years.

Real-World Link · · · ·

Most pets age at a different rate than their human companions. For example, a 3-year-old dog is often considered to be 21 in human years.

Not all relationships with a constant rate of change are proportional. Likewise, not all linear functions are direct variations.

EXAMPLES Identify Direct Variation

Determine whether each linear function is a direct variation. If so, state the constant of variation.

Miles, x	25	50	75	100
Gallons, y	1	2	3	4

Compare the ratios to check for a common ratio.

$$\frac{\text{gallons}}{\text{miles}} \rightarrow \quad \frac{1}{25} \qquad \frac{2}{50} \text{ or } \frac{1}{25} \qquad \frac{3}{75} \text{ or } \frac{1}{25} \qquad \frac{4}{100} \text{ or } \frac{1}{25}$$

Since the ratios are the same, the function is a direct variation. The constant of variation is $\frac{1}{25}$.

4

Hours, x	2	4	6	8
Earnings, y	36	52	68	84

$$\frac{\text{earnings}}{\text{hours}} \rightarrow \quad \frac{36}{2} \text{ or } \frac{18}{1} \qquad \frac{52}{4} \text{ or } \frac{13}{1} \qquad \frac{68}{6} \text{ or } \frac{11.33}{1} \qquad \frac{84}{8} \text{ or } \frac{10.50}{1}$$

The ratios are not the same. The function is not a direct variation.

CHECK Your Progress

c.

Days, x	5	10	15	20
Height, y	12.5	25	37.5	50

d.

Time, x	4	6	8	10
Distance, y	12	16	20	24

Study Tip

Direct Variations Notice that the graph of a direct variation, which is a proportional linear relationship, is a line that passes through the origin.

Key Concept Linear Functions

Proportional

Table

x	−2	−1	1	2
y	−4	−2	2	4
$\frac{y}{x}$	2	2	2	2

Graph

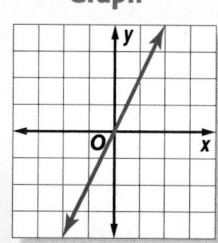

Equation

$y = 2x$

Nonproportional

Table

x	−2	−1	1	2
y	−5	−3	1	3
$\frac{y}{x}$	$\frac{5}{2}$	3	1	$\frac{3}{2}$

Graph

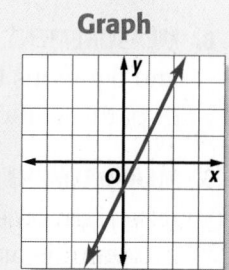

Equation

$y = 2x - 1$

Example 1

1. MANUFACTURING The number of computers built varies directly with the number of hours the production line operates. What is the ratio of computers built to hours of production?

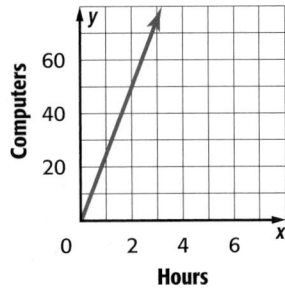

Example 2

2. TRANSPORTATION A charter bus travels 210 miles in $3\frac{1}{2}$ hours. Assuming that the distance traveled is directly proportional to the time traveled, how far will the bus travel in 6 hours?

Examples 3 and 4

3. Determine whether the linear function is a direct variation. If so, state the constant of variation.

Hours, x	2	3	4	5
Miles, y	116	174	232	290

Practice and Problem Solving

● = Step-by-Step Solutions begin on page R1.
Extra Practice begins on page EP2.

Example 1

4. GARDENING Katrina planted ornamental grass seeds. After the grass breaks the soil surface, its height varies directly with the number of days. What is the rate of growth?

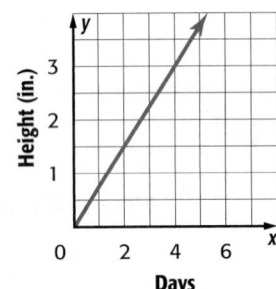

5 JOBS The amount Dusty earns varies directly with the number of newspapers he delivers. How much does Dusty earn for each newspaper delivery?

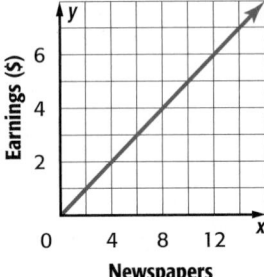

Example 2

6. SUBMARINES Ten minutes after a submarine is launched from a research ship, it is 25 meters below the surface. After 30 minutes, the submarine has descended 75 meters. At what rate is the submarine diving?

7. MOVIES The Stratton family rented 3 DVDs for $10.47. The next weekend, they rented 5 DVDs for $17.45. What is the rental fee for a DVD?

8. MEASUREMENT Hector used 3 gallons of paint to cover 1,050 square feet and 5 gallons to paint an additional 1,750 square feet. How many gallons of paint would he need to cover 2,800 square feet?

9. MEASUREMENT The weight of an object on Mars varies directly with its weight on Earth. An object that weighs 70 pounds on Mars weighs 210 pounds on Earth. If an object weighs 160 pounds on Earth, how much would it weigh on Mars?

10. ELECTRONICS The height of a wide-screen television screen varies directly with its width. A manufacturer makes a television screen that is 60 centimeters wide and 33.75 centimeters high. Find the height of a television screen that is 90 centimeters wide.

11. BAKING A cake recipe requires $2\frac{3}{4}$ cups of flour for 12 servings. How much flour is required to make a cake that serves 30?

Examples 3 and 4 **Determine whether each linear function is a direct variation. If so, state the constant of variation.**

12.

Pictures, x	5	6	7	8
Profit, y	20	24	28	32

13.

Minutes, x	200	400	600	800
Cost, y	65	115	165	215

14.

Age, x	10	11	12	13
Grade, y	5	6	7	8

15.

Price, x	10	15	20	25
Tax, y	0.70	1.05	1.40	1.75

ALGEBRA If y varies directly with x, write an equation for the direct variation. Then find each value.

16. If $y = -12$ when $x = 9$, find y when $x = -4$.

17. Find y when $x = 10$ if $y = 8$ when $x = 20$.

18. If $y = -6$ when $x = -14$, what is the value of x when $y = -4$?

19. Find x when $y = 25$, if $y = 7$ when $x = 8$.

20. Find y when $x = 5$, if $y = 12.6$ when $x = 14$.

21 MEASUREMENT The number of centimeters in a measure varies directly with the number of inches. Find the measure of an object in centimeters if it is 50 inches long.

Inches, x	6	9	12	15
Centimeters, y	15.24	22.86	30.48	38.10

22. MATH IN THE MEDIA Find examples of linear graphs in a newspaper or magazine, on television, or on the Internet. Determine if the graph is a direct variation. Then find the slope of the line and interpret its meaning.

Real-World Link · · · ·
The aspect ratio of a television screen is the ratio of its width to its height. Standard screens have an aspect ratio of 4:3 while wide-screen televisions have an aspect ratio of 16:9.

H.O.T. Problems

23. OPEN ENDED Identify values for x and y in a direct variation relationship where $y = 9$ when $x = 16$.

24. CHALLENGE The amount of stain needed to cover a wood surface is directly proportional to the area of the surface. If 3 pints are required to cover a square deck with a side of 7 feet, how many pints of stain are needed to paint a square deck with a side of 10 feet 6 inches?

25. WRITE MATH Write a direct variation equation. Then triple the x-value and explain how to find the corresponding change in the y-value.

26. Students in a science class recorded lengths of a stretched spring, as shown in the table below.

Length of Stretched Spring	
Distance Stretched, x (centimeters)	Mass, y (grams)
0	0
2	12
5	30
9	54
12	72

Which equation best represents the relationship between the distance stretched x and the mass of an object on the spring y?

A. $y = -6x$ C. $y = -\dfrac{x}{6}$

B. $y = 6x$ D. $y = \dfrac{x}{6}$

27. **THINK SOLVE EXPLAIN** **SHORT RESPONSE** Nicole read 24 pages during a 30-minute independent reading period. At this rate, how many pages would she read in 45 minutes?

28. To make fruit punch, Kelli adds 8 ounces of pineapple juice for every 12 ounces of orange juice. Suppose she uses 32 ounces of orange juice. Which proportion can she use to find x, the number of ounces of pineapple juice needed to make the punch?

F. $\dfrac{8}{12} = \dfrac{32}{x}$

G. $\dfrac{8}{x} = \dfrac{32}{12}$

H. $\dfrac{8}{12} = \dfrac{x}{32}$

I. $\dfrac{x}{12} = \dfrac{8}{32}$

Find the slope of each line. (Lesson 1C)

29.

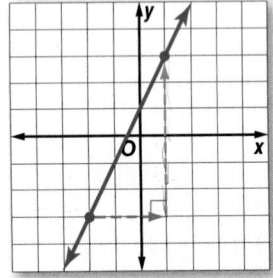

30.

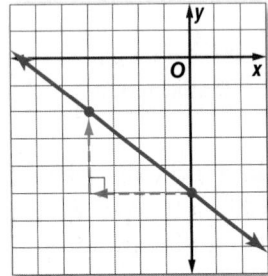

31.

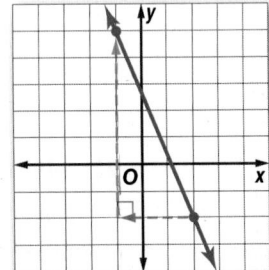

32. **SALES** The cost for printing calendars by Calendars Galore is shown in the graph. Find the constant rate of change. Then determine whether a proportional relationship exists between the printing cost and the number of calendars printed. Explain your reasoning.

(Lesson 1A)

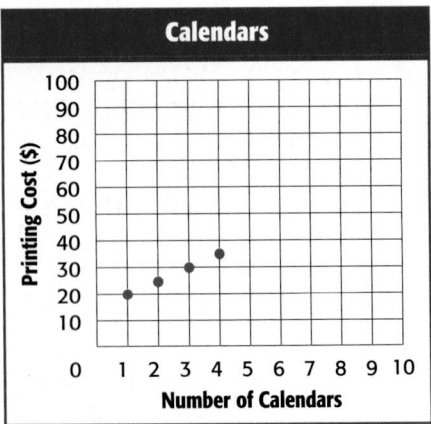

Main Idea
Graph linear equations using the slope and y-intercept.

 Vocabulary
slope-intercept form
y-intercept
boundary
half-plane

 Get ConnectED

 CCSS 8.EE.5, 8.EE.6, 8.F.3, 8.F.4

Slope-Intercept Form

 **GASOLINE** The graph represents the cost of gasoline at \$3.50 per gallon.

1. Write an equation that represents the cost of gasoline at \$3.50 per gallon and a drink that costs \$2.

2. Graph the equation from Exercise 1.

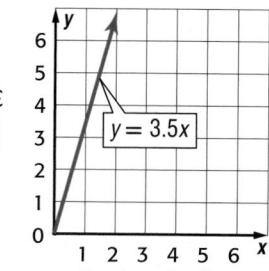

Gasoline

$y = 3.5x$

Proportional linear functions can be written in the form $y = kx$, where k is the constant of variation, or slope of the line.

Nonproportional linear functions can be written in the form $y = mx + b$. This is called the **slope-intercept form**. When an equation is written in this form, m is the slope and b is the y-intercept. The **y-intercept** of a line is the y-coordinate of the point where the line crosses the y-axis.

 EXAMPLES Find Slopes and y-intercepts

State the slope and the y-intercept of the graph of each equation.

① $y = \frac{2}{3}x - 4$

$y = \frac{2}{3}x + (-4)$ Write the equation in the form $y = mx + b$.

$y = mx + b$ $m = \frac{2}{3}, b = -4$

The slope of the graph is $\frac{2}{3}$, and the y-intercept is -4.

② $x + y = 6$

$x + y = 6$ Write the original equation.

$x - x + y = 6 - x$ Subtract x from each side.

$y = 6 - x$ Simplify.

$y = -1x + 6$ Write the equation in the form $y = mx + b$. Recall that $-x$ means $-1x$.

$y = mx + b$ $m = -1, b = 6$

The slope of the graph is -1 and the y-intercept is 6.

CHECK Your Progress

a. $y = -5x + 3$ **b.** $y = \frac{1}{4}x - 6$ **c.** $y - x = 5$

③ Write an equation of a line in slope-intercept form with a slope of −3 and a y-intercept of −4.

$y = mx + b$ Slope-intercept form

$y = -3x + (-4)$ Replace m with −3 and b with −4.

$y = -3x - 4$ Simplify.

④ Write an equation in slope-intercept form for the graph shown.

The y-intercept is 4. From (0, 4), you move down 1 unit and right 2 units to another point on the line.

So, the slope is $-\dfrac{1}{2}$.

$y = mx + b$ Slope-intercept form

$y = -\dfrac{1}{2}x + 4$ Replace m with $-\dfrac{1}{2}$ and b with 4.

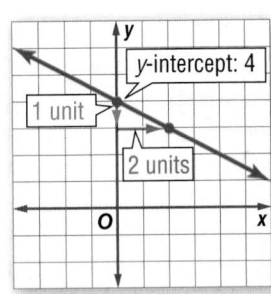

✔️ **CHECK Your Progress**

d. Write an equation in slope-intercept form for the graph shown.

e. Write an equation of a line in slope-intercept form with a slope of $\dfrac{3}{4}$ and a y-intercept of −3.

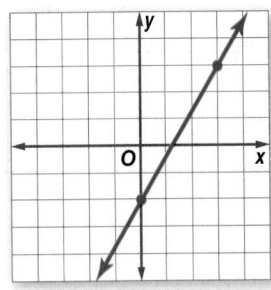

✏️ **EXAMPLE** Graph Using Slope-Intercept Form

⑤ Graph $y = -\dfrac{3}{2}x - 1$ using the slope and y-intercept.

Step 1 Find the slope and y-intercept.

$y = -\dfrac{3}{2}x - 1$ slope $= -\dfrac{3}{2}$, y-intercept $= -1$

Step 2 Graph the y-intercept −1.

Step 3 Write the slope $-\dfrac{3}{2}$ as $\dfrac{-3}{2}$. Use it to locate a second point on the line.

$m = \dfrac{-3}{2}$ ← change in y: down 3 units
            ← change in x: right 2 units

Step 4 Draw a line through the two points.

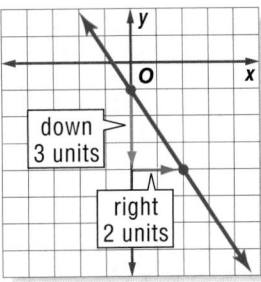

> **Study Tip**
>
> **Check for Accuracy** To check your graph, substitute the x- and y-values of another point on your graph into the equation. For Example 5, test the point (2, −4).
>
> $y = -\dfrac{3}{2}x - 1$
>
> $-4 = -\dfrac{3}{2}(2) - 1$
>
> $-4 = -3 - 1$
>
> $-4 = -4$ ✓

✔️ **CHECK Your Progress**

Graph each equation using the slope and y-intercept.

f. $y = x + 3$ **g.** $y = \dfrac{1}{2}x - 1$ **h.** $y = -\dfrac{4}{3}x + 2$

EXAMPLES Graph an Equation to Solve Problems

ACTIVITIES Student Council is selling spirit T-shirts during spirit week. It costs $20 for the design and $5 to print each shirt. The cost y to print x shirts is given by $y = 5x + 20$.

6 Graph the equation to find the number of shirts that can be printed for $50.

$y = 5x + 20$ slope = 5, y-intercept = 20

Plot the point (0, 20). Locate another point up 5 and right 1. Draw the line. The x-coordinate is 6 when the y-coordinate is 50, so the number of T-shirts is 6.

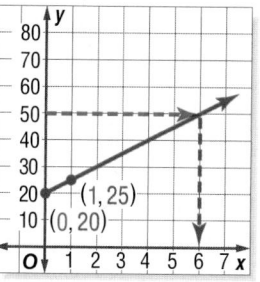

7 Interpret the slope and the y-intercept.

The slope 5 represents the cost in dollars per T-shirt. The y-intercept 20 is the one-time charge in dollars for the design.

CHECK Your Progress

TRANSPORTATION A taxi fare y can be determined by the equation $y = 0.50x + 3.50$, where x is the number of miles traveled.

i. Graph the equation to find the cost of traveling 8 miles.

j. Interpret the slope and the y-intercept.

Real-World Link
Shirt designs can be created on a computer, then sent to a company to screen print the shirts. Each color requires a separate screen for the ink to pass through.

✓ CHECK Your Understanding

Examples 1 and 2   **1.** State the slope and the y-intercept for the graph of $2x + y = 3$.

Examples 3 and 4   **2.** Write an equation in slope intercept form for the graph shown.

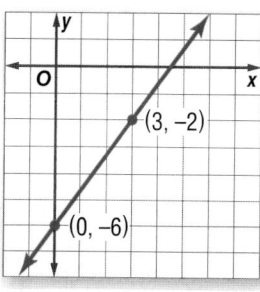

Write an equation of a line in slope-intercept form with the given slope and y-intercept.

3. slope: 4, y-intercept: 2

4. slope: $-\frac{1}{4}$, y-intercept: 5

Example 5   **5** Graph $y = \frac{1}{3}x - 2$ using the slope and the y-intercept.

Examples 6 and 7   **6. SCHOOL** Liam is reading a 254-page book for school. He can read 40 pages in one hour. The equation for the number of pages he has left to read is $y = 254 - 40x$, where x is the number of hours he reads.

a. Graph the equation to find how many pages Liam has left after 3 hours.

b. Interpret what the slope and the y-intercept represent.

Practice and Problem Solving

= **Step-by-Step Solutions** begin on page R1.
Extra Practice begins on page EP2.

Examples 1 and 2   State the slope and the *y*-intercept for the graph of each equation.

7. $y = 3x + 4$

8. $y = -5x + 2$

9. $y = \frac{1}{2}x - 6$

10. $y = -\frac{3}{7}x - \frac{1}{7}$

11. $y - 2x = 8$

12. $3x + y = -4$

Examples 3 and 4   Write an equation of a line in slope-intercept form with the given slope and *y*-intercept.

13. slope: $-\frac{3}{4}$, *y*-intercept: -2

14. slope: $\frac{5}{6}$, *y*-intercept: 8

15. slope: $-\frac{2}{3}$, *y*-intercept: -10

16. slope: $-\frac{1}{4}$, *y*-intercept: 5

Write an equation in slope-intercept form for each graph shown.

17.
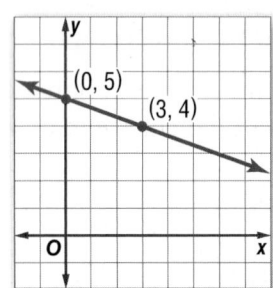

18.
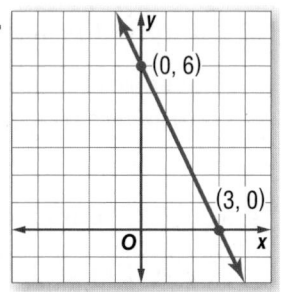

19.
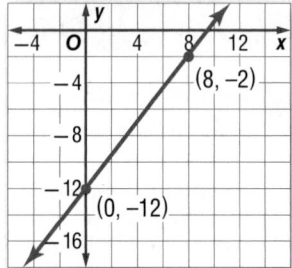

Example 5   Graph each equation using the slope and the *y*-intercept.

20. $y = \frac{1}{3}x - 5$

21. $y = -x + \frac{3}{2}$

22. $y = -\frac{4}{3}x + 1$

23. $y = \frac{3}{2}x - 4$

24. $y + 2x = -3.5$

25. $1.5 = y - 3x$

Examples 6 and 7   **26. TRAVEL** The Viera family is traveling from Philadelphia to Orlando for vacation. The equation $y = 1,000 - 65x$ represents the distance remaining in their trip after *x* hours.

  a. Graph the equation to find the distance remaining after 6 hours.

  b. Interpret the slope and the *y*-intercept.

27. BOATING The Lakeside Marina charges a $35 rental fee for a boat in addition to charging $15 an hour for usage. The total cost *y* of renting a boat for *x* hours can be represented by the equation $y = 15x + 35$.

  a. Graph the equation to find the total cost for a 3-hour rental.

  b. Interpret the slope and the *y*-intercept.

28. GEOMETRY Use the pair of angles at the right.

  a. Write the equation in slope-intercept form.

  b. Graph the equation.

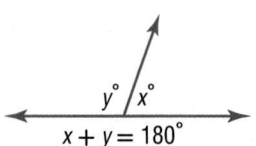

29 **INSECTS** The equation $y = 15x + 37$ can be used to approximate the temperature *y* in degrees Fahrenheit based on the number of chirps *x* a cricket makes in 15 seconds. Graph the equation to estimate the number of chirps a cricket will make in 15 seconds if the temperature is 80°F.

30. ⟳ **MULTIPLE REPRESENTATIONS** Jacquie has 20 postcards in her postcard collection. She decides that from now on, every time she goes on vacation she will buy 8 postcards to add to her collection.

   a. **ALGEBRA** Write an equation to represent the number of postcards y she will have after any number of vacations x.

   b. **GRAPHS** Graph the equation from part **a** on a coordinate plane.

   c. **NUMBERS** What are the slope and y-intercept of the line?

   d. **WORDS** Explain why the line "slopes up" by 8 for each vacation.

   e. **WORDS** Why does the line cross the y-axis at 20?

31. The table shows points that lie on a line.

   a. Find the slope and y-intercept of the line.

   b. Describe how the slope and y-intercept appear on the graph of the line.

   c. Use the slope and y-intercept to find the equation of the line in slope-intercept form.

x	0	1	2	3
y	1	5	9	13

32. **GRAPHIC NOVEL** Refer to the graphic novel frame below for Exercises a–b.

   a. Write an equation in slope-intercept form for the total cost of any number of tickets at 7 tickets for $5.

   b. Write an equation in slope-intercept form for the total cost of a wristband for all you can ride.

H.O.T. Problems

33. **OPEN ENDED** Draw the graph of a line that has a y-intercept but no x-intercept. What is the slope of the line?

34. **REASONING** What is the slope and y-intercept of a vertical line?

35. 📝 **WRITE MATH** Write a real-world problem that involves a linear relationship. Describe how the slope and y-intercept would appear in these three representations of the problem: table, equation, and graph.

36. Which **best** represents the graph of $y = 3x + 4$?

A.
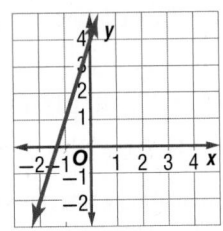

B.
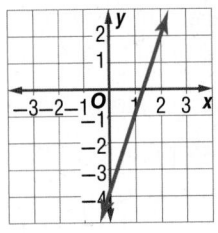

C.
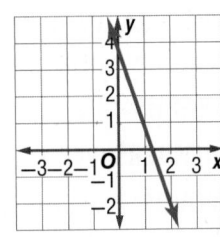

D.
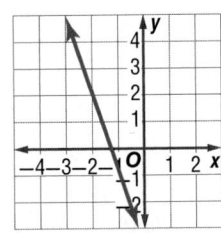

More About **Graphing on the Coordinate Plane**

Like a linear equation in two variables, the solution set of an inequality in two variables can be graphed on a coordinate plane. An equation defines the **boundary** or edge and divides the coordinate plane into two **half-planes**.

✏️ **EXAMPLE**

Graph $y \leq 2x + 3$.

Step 1 Graph $y = 2x + 3$. It is the boundary for the inequality. Since y is less than *or equal* to $2x + 3$, the boundary will be a solid line.

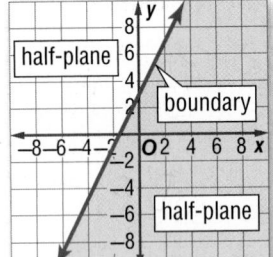

Step 2 Select a point in one of the half-planes and test it in the original inequality. Use $(0, 0)$.

$$y \leq 2x + 3$$

$$0 \leq 2(0) + 3$$

$$0 \leq 3 \checkmark$$

The statement $0 \leq 3$ is true, so the section of the graph that contains $(0, 0)$ is also part of the solution. To show this, shade that section.

In the Example, the boundary was a solid line because y is less than or equal to $2x + 3$. When the inequality symbol does not contain an "=", a dashed line on a coordinate plane indicates that the boundary is *not* part of the solution set.

Graph each inequality on a coordinate plane.

37. $y \geq x - 2$

38. $y \leq x + 3$

39. $y > 3x + 1$

40. $y < 2x - 2$

41. $y \leq \frac{2}{3}x$

42. $y > \frac{1}{4}x$

Main Idea

Graph a function using the x- and y-intercepts.

 Vocabulary

standard form
x-intercept

 Get Connect **ED**

 **8.EE.5**

Graph Functions Using Intercepts

MOVIES Mrs. Hodges spent $80 on movie tickets and drinks for her son and his friends. The total cost of x movie tickets and y drinks is represented by the function $8x + 4y = 80$.

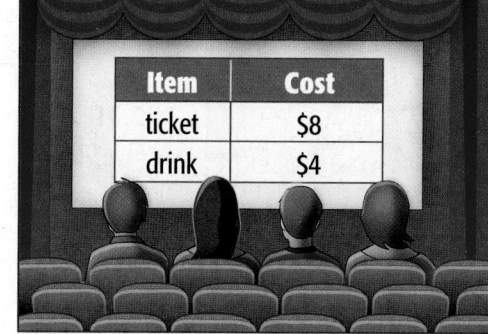

Item	Cost
ticket	$8
drink	$4

1. Describe a method you could use to graph the function. Then graph the function.

2. What do the points (0, 20) and (10, 0) represent?

The equation $8x + 4y = 80$ is written in standard form. **Standard form** is when an equation is written in the form $Ax + By = C$.

The **x-intercept** of a function is the x-coordinate of the point where the graph crosses the x-axis. Since any linear function can be graphed using two points, you can use the x- and y-intercepts to graph a function.

EXAMPLES **Graph a Function Using Intercepts**

1 State the x- and y-intercepts of $2x + y = 4$. Graph the function.

Step 1 Find the x-intercept.

To find the x-intercept, let $y = 0$.

$2x + y = 4$ Write the equation.
$2x + (0) = 4$ Replace y with 0.
$2x = 4$ Simplify.
$x = 2$ Divide each side by 2.

The x-intercept is 2.

Step 2 Find the y-intercept.

To find the y-intercept, let $x = 0$.

$2x + y = 4$ Write the equation.
$2(0) + y = 4$ Replace x with 0.
$y = 4$ Simplify.

The y-intercept is 4.

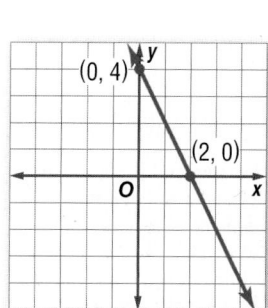

Step 3 Graph the points (2, 0) and (0, 4) on a coordinate plane. Then connect the points.

2 State the *x*- and *y*-intercepts of $-1.5x + y = -9$. Graph the function.

Step 1 Find the *x*- and *y*-intercepts.

x-intercept	*y*-intercept
$-1.5x + y = -9$	$-1.5x + y = -9$
$-1.5x + (0) = -9$	$-1.5(0) + y = -9$
$-1.5x = -9$	$y = -9$
$x = 6$	The *y*-intercept is -9.

The *x*-intercept is 6.

Step 2 Graph the points $(6, 0)$ and $(0, -9)$ on a coordinate plane. Then connect the points.

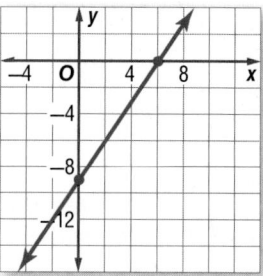

☑ **CHECK Your Progress**

a. $3x + 2y = 6$ **b.** $-5x - 2y = 10$ **c.** $\frac{1}{3}x + y = 5$

📝 **REAL-WORLD EXAMPLES**   Using Intercepts

YEARBOOKS Mauldin Middle School wants to make $4,740 from yearbooks. Print yearbooks *x* cost $60 and digital yearbooks *y* cost $15. This can be represented by the function $60x + 15y = 4,740$.

3 Use the *x*- and *y*-intercepts to graph the function.

x-intercept	*y*-intercept
$60x + 15y = 4,740$	$60x + 15y = 4,740$
$60x + 15(0) = 4,740$	$60(0) + 15y = 4,740$
$60x = 4,740$	$15y = 4,740$
$x = 79$	$y = 316$

4 Interpret the *x*- and *y*-intercepts.
The *x*-intercept is at the point $(79, 0)$. This means they can sell 79 print yearbooks and 0 digital yearbooks to earn $4,740.

The *y*-intercept is at the point $(0, 316)$. This means they can sell 0 print yearbooks and 316 digital yearbooks to earn $4,740.

Yearbook Sales

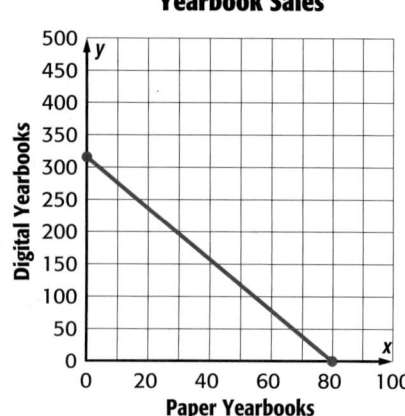

☑ **CHECK Your Progress**

d. LUNCH Mr. Davies spent $230 on lunch for his class. Sandwiches *x* cost $6 and drinks *y* cost $2. This can be represented by the function $6x + 2y = 230$. Graph the function. Then interpret the *x*- and *y*-intercepts.

Examples 1 and 2 State the *x*- and *y*-intercepts of each function. Then graph the function.

1. $5x + 4y = 20$

2. $6x - 2y = 18$

3. $-3x + 6y = 12$

4. $-4x - 6y = 24$

5. $\frac{3}{2}x + 10y = 15$

6. $\frac{3}{4}x - \frac{3}{5}y = 12$

Examples 3 and 4 **7. DISPLAYS** A store sells juice boxes in packages of 6 boxes and 8 boxes. They have 288 total juice boxes. This is represented by the function $6x + 8y = 288$. Graph the function. Then interpret the *x*- and *y*-intercepts.

Practice and Problem Solving

● = Step-by-Step Solutions begin on page R1.
Extra Practice begins on page EP2.

Examples 1 and 2 State the *x*- and *y*-intercepts of each function. Then graph the function.

8. $8x + 4y = 28$

9 $12x + 9y = 15$

10. $5x - 3y = 30$

11. $16x - 12y = -8$

12. $-6x - 8y = 20$

13. $-9x + 15y = -45$

14. $-12x - 8y = -4$

15. $20x - 6y = 4$

16. $-10x + 15y = -5$

17. $-\frac{2}{3}x - \frac{3}{4}y = 12$

18. $\frac{5}{6}x + \frac{1}{2}y = 5$

19. $-\frac{1}{4}x - \frac{2}{5}y = -3$

Examples 3 and 4 **20. SHOPPING** The table shows the cost for a clothing store to buy jeans and khakis. The total cost for Saturday's shipment, $1,800, is represented by the function $15x + 20y = 1,800$. Graph the function. Then interpret the *x*- and *y*-intercepts.

	Jeans	Khakis
Cost per Pair ($)	15	20
Amount Shipped	*x*	*y*

21. ZOOS The total number of legs, 1,500, on four-legged and two-legged animals in a zoo can be represented by the function $4x + 2y = 1,500$. Graph the function. Then interpret the *x*- and *y*-intercepts.

22. AMUSEMENT PARKS The table shows the group rate for admission tickets for adults and children to an amusement park.

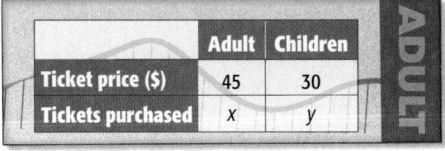

	Adult	Children
Ticket price ($)	45	30
Tickets purchased	*x*	*y*

a. The total cost of a group's tickets is $1,350. Write a function to represent the number of adults' and children's tickets purchased.

b. What are the *x*- and *y*-intercepts and what do they represent?

c. Graph the function. Use the graph to find the number of children's tickets purchased if 20 adult tickets were purchased.

23 **MEASUREMENT** The perimeter of a rectangle that is *x* units wide and *y* units long is 24 centimeters.

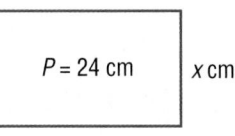

a. Write an equation for the perimeter.

b. Find the *x*- and *y*-intercepts. Does either intercept make sense as a solution for this situation? Explain.

24. FIND THE ERROR Carmen is finding the x-intercept of the function $3x - 4y = 12$. Find her mistake and correct it.

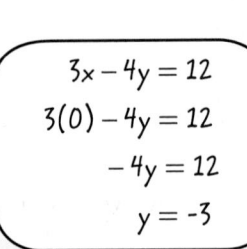

$$3x - 4y = 12$$
$$3(0) - 4y = 12$$
$$-4y = 12$$
$$y = -3$$

25. ✏️ **WRITE MATH** Describe two different methods for graphing a function. Which method do you prefer to use? Why?

Test Practice

26. The function $12x - 10y = 600$ represents the total amount Student Council spent on supplies for a school fundraiser. What is the x-intercept of the function?

  A. -60 **C.** 50

  B. -50 **D.** 60

27. The y-intercept of which of the following functions is 5?

  F. $4x - 5y = 30$

  G. $4x - 6y = 30$

  H. $4x + 5y = 30$

  I. $4x + 6y = 30$

28. Which of the following is the graph of the function $2x + 3y = 6$?

A. 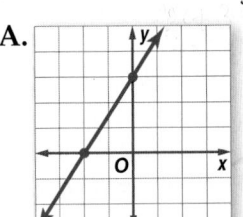   **C.**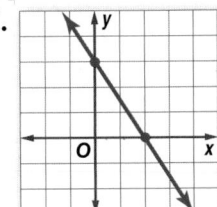

B. 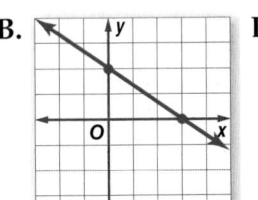   **D.**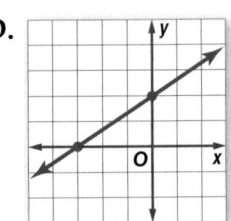

Spiral Review

State the slope and y-intercept for the graph of each equation. (Lesson 2A)

29. $y = 3x + 4$ **30.** $y = -\frac{1}{2}x - 5$ **31.** $2x - 3y = 3$

32. TRAVEL One and a half hours after leaving its station, a train has traveled 202.5 miles. At this rate, how far will the train travel after 5 hours? (Lesson 1E)

33. MONEY The table shows the amount of money Annie earned working at her job. The points in the table lie on a line. (Lesson 1C)

  a. What is the slope of the line?

  b. What does the slope of the line represent?

Hours, x	Money Earned ($), y
2	13.00
3	19.50
4	26.00
5	32.50

Main Idea

Use technology to investigate situations to determine if they display linear behavior.

*Get Connect*ED

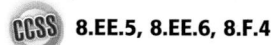 8.EE.5, 8.EE.6, 8.F.4

Graphing Technology:
Model Linear Behavior

In this activity, you will examine a situation using a data collection device and a graphing calculator to determine if this situation displays linear behavior.

ACTIVITY

1 **STEP 1** Connect a motion detector to your calculator. Start the data collection program by pressing [APPS] (CBL/CBR) [ENTER], and then select Ranger, Applications, Meters, Dist Match.

STEP 2 Place the detector on a desk or table so that it can read the motion of a walker.

STEP 3 Mark the floor at a distance of 1 and 6 meters from the detector. Have a partner stand at the 1-meter mark.

STEP 4 When you press the button to begin collecting data, have your partner begin to walk away from the detector at a slow but steady pace.

STEP 5 Stop collecting data when your partner passes the 6-meter mark.

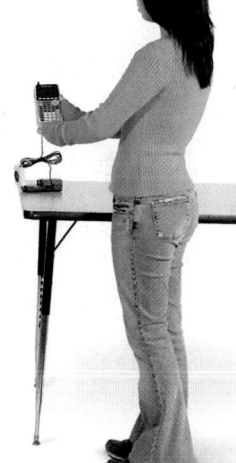

STEP 6 Press [ENTER] to display a graph of the data. The x-values represent equal intervals of time in seconds. The y-values represent the distances from the detector in meters.

Analyze the Results

1. Describe the DISTANCE graph of the data. Does the relationship between time and distance appear to be linear? Explain.

2. Use the [TRACE] feature on your calculator to find the y-intercept on the graph. Interpret its meaning.

3. Press [STAT] 1 and record the time data from [L1] and the distance data from [L2] in a table like the one shown. Then use these data to calculate the rate of change $\frac{distance}{time}$ for several pairs of points. What do you notice?

List L1	List L2

4. Does your answer to Exercise 3 support your conclusion about the graph in Exercise 1? Explain.

5. **MAKE A PREDICTION** Predict how your graph and answers to Exercise 3 would change if the person in the activity were to

   a. move at a steady but *quick* pace *away* from the detector.

   b. move at a steady but *slow* pace *toward* the detector.

6. **COLLECT THE DATA** Repeat the activity and answer Exercises 1 through 3 again for each of the situations described in Exercise 5.

7. **MAKE A CONJECTURE** How could you change the situation to be one that does not display linear behavior?

8. **COLLECT THE DATA** Repeat the activity and answer Exercises 1 through 3 again for the situation you described in Exercise 7.

ACTIVITY

Families of graphs are graphs that are related in some manner. In this activity, you will study families of linear graphs.

2 **STEP 1** Enter each of the following equations: $y = -2x + 4$, $y = -2x + 1$, and $y = -2x - 3$.

STEP 2 Press [ZOOM] 6 to graph the equations.

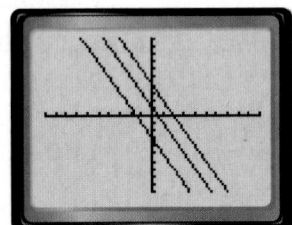

Analyze the Results

9. Compare the three equations and their graphs.

10. **MAKE A CONJECTURE** Consider equations of the form $y = ax + b$, where the value of a is constant but the value of b varies. What do you think is true for the graphs of these equations?

11. Use your calculator to graph $y = 2x + 3$, $y = -x + 3$, and $y = -3x + 3$. Compare the three equations and their graphs.

12. **MAKE A CONJECTURE** Consider equations of the form $y = ax + b$, where the value of a varies but the value of b remains constant. What do you think is true for the graphs of these equations?

Mid-Chapter Check

1. **MONEY** Is the relationship between the amount of money owed on a CD player and the number of payments linear? If so, find the constant rate of change. (Lesson 1A)

Amount Owed ($)	180	150	120	90
Number of Payments	1	2	3	4

2. **DISTANCE** Is the relationship between the time and the distance traveled linear? If so, find the constant rate of change. (Lesson 1A)

Distance (ft)	0	13	26	39
Time (s)	0	2	4	6

Find the slope of the line that passes through each pair of points. (Lesson 1C)

3. $A(2, 5)$, $B(3, 1)$

4. $C(-1, 2)$, $D(-5, 2)$

5. $E(5, 2)$, $F(2, -3)$

6. $G(4, 3)$, $H(-2, -6)$

7. **MULTIPLE CHOICE** Which graph has a negative slope? (Lesson 1C)

A.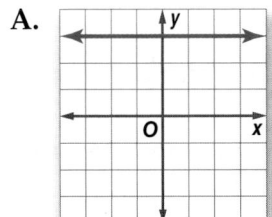

C.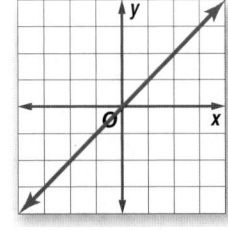

B.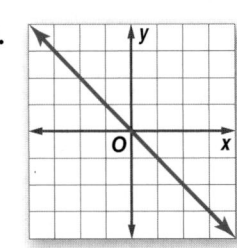

D.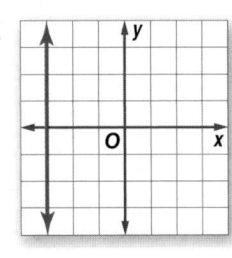

8. **BAKING** Ernesto baked 3 cakes in $2\frac{1}{2}$ hours. Assuming that the number of cakes baked is directly proportional to the number of hours, how many cakes can he bake in $7\frac{1}{2}$ hours? (Lesson 1E)

Write each equation in slope-intercept form. (Lesson 2A)

9. $2x + 3y = 12$

10. $-4x - 6y = 16$

11. $-5x + 2y = 15$

12. $3x - 4y = 12$

13. **MONEY** The total money y Aaron earned mowing lawns is shown by the equation $y = 15x + 25$. What does the slope represent? (Lesson 2A)

14. **MULTIPLE CHOICE** Which equation is the slope-intercept form for the line below? (Lesson 2A)

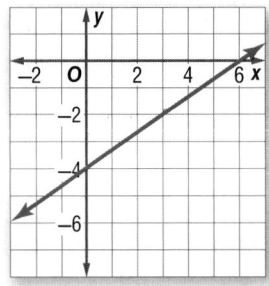

F. $y = -4x + 6$ H. $y = \frac{3}{2}x - 4$

G. $y = \frac{2}{3}x - 4$ I. $y = 6x - 4$

State the x- and y-intercepts of each function. Then graph the function. (Lesson 2B)

15. $2x + 3y = 18$

16. $-4x - 5y = 10$

17. $-6x - 4y = 12$

18. $5x + 2y = 15$

Problem-Solving Investigation

Main Idea Guess, check, and revise to solve problems.

🚶 📝 **P.S.I. TEAM ✛**

Guess, Check, and Revise

ADRIENNE: My class is going to the museum to see an art exhibit. Student admission is $2 and adult admission is $4. We spent $66 on 30 tickets.

YOUR MISSION: Guess, check, and revise to find the number of students and adults going to the museum.

Understand	The student cost is $2 and the adult cost is $4. There are 30 people on the trip.
Plan	Make a guess and check to see if it is correct. If it is not, then revise.
Solve	Find the combination that gives a total of $66. In the list, s is the number of students and a is the number of adults on the trip.

s	a	$2s + 4a$	Check
26	4	$2(26) + 4(4) = 68$	too high
29	1	$2(29) + 4(1) = 62$	too low
28	2	$2(28) + 4(2) = 64$	still too low
27	3	$2(27) + 4(3) = 66$	correct

So, 27 students and 3 adults are going to the museum.

Check	$27 + 3 = 30$ and $2(27) + 4(3) = 66$; the guess is correct. ✔

Analyze the Strategy

1. For the problem above, 23 students and 5 adults would also spend $66 to get into the museum. Explain why this could not be the correct solution.

2. 📝 **WRITE MATH** Write a problem that could be solved by *guess, check, and revise.* Then write the steps you would take to find the solution.

Mixed Problem Solving

- Guess, check, and revise.
- Draw a diagram.
- Make a table.
- Choose an operation.

Use the *guess, check, and revise* strategy to solve Exercises 3–5.

3. NUMBER THEORY A number is squared and the result is 576. Find the number.

4. COINS Gerardo has $2.50 in quarters, dimes, and nickels. If he has 18 coins, how many of each coin does he have?

5 SHOPPING Shyla was buying gifts for each of her 8 cousins. She bought everyone either a ring for $6 or a toy for $7. If she spent a total of $53, how many of each did she buy?

Use any strategy to solve Exercises 6–13.

6. MEASUREMENT The length ℓ of the rectangle below is longer than its width w. List the possible whole number dimensions for the rectangle and identify the dimensions that give the greatest perimeter.

$$A = 36 \text{ in}^2 \quad w$$
$$\ell$$

7. NUMBERS Name three numbers that have a sum of 23 if the greatest number is 9 more than the least number.

8. NEWSPAPERS The list at the right shows the number of letters in the first 20 words of an article on the front page of a newspaper. What number of letters occurs most often?

The Daily Reader

3	5	7	4
4	7	6	4
5	5	3	6
8	4	5	5
6	5	7	7

9. MOVING Marcel is moving and wants to put all of his 20 CD cases in one box. Give two possible dimensions for a box that would hold the cases with no space leftover.

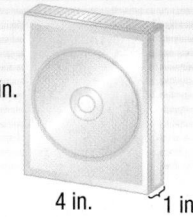

5 in.
4 in.   1 in.

10. HOBBIES Baseball cards come in packages of 8 and 12. Brighton bought some of each type for a total of 72 baseball cards. How many of each package did he buy?

11. SIBLINGS Three siblings have a combined age of 108 years. The oldest is 8 years older than the youngest. What are the ages of the siblings?

12. ANALYZE TABLES One hundred twenty students at Blendon Middle School could sign up to hear three different speakers for career day. Seventy students heard the nurse speak, 37 heard the firefighter, and 63 heard the Webmaster. Some students heard more than one speaker. The results are shown below.

Number of Students	Speaker
15	all three
20	nurse and firefighter
30	Webmaster and nurse
12	firefighter only

a. How many students signed up only for Webmaster?

b. How many students did *not* sign up for nurse?

13. NUMBER SENSE Find the product of

$$1 - \frac{1}{2}, 1 - \frac{1}{2}, 1 - \frac{1}{3}, 1 - \frac{1}{4}, \ldots, 1 - \frac{1}{48},$$
$$1 - \frac{1}{49}, \text{ and } 1 - \frac{1}{50}.$$

Explore

Main Idea

Find one solution for a set of two equations.

Get ConnectED

CCSS   **8.EE.8, 8.EE.8a**

Graphing Technology:
Systems of Equations

INTERNET SHOPPING Web site A charges $3 plus $1 per pound to ship an item. Web site B charges $1 plus $2 per pound to ship the same item. For an object that weighs x pounds, the charges for Web site A are represented by $y = x + 3$. The charges for Web site B are represented by $y = 2x + 1$.

📟 ACTIVITY

Use a graphing calculator to model a function table for $y = x + 3$ and $y = 2x + 1$. Then use the table to find the total cost to ship objects that weigh 0, 1, 2, or 3 pounds.

STEP 1 Press Y= to access the function list. Then enter each equation.

STEP 2 Set up the function table. Press 2nd [TBLSET] to display the table setup screen. Press ▼ ▼ ▶ ENTER to highlight Indpnt: Ask. Then press ▼ ENTER to highlight Depend: Auto.

STEP 3 Access the table by pressing 2nd [TABLE]. Now key in your input values, pressing ENTER after each one.

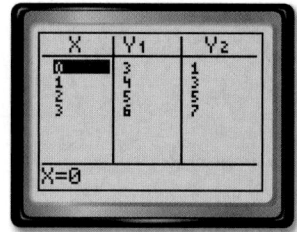

Analyze the Results

1. For what number of pounds are the shipping charges the same?

2. For what number of pounds are the charges for Web site A less than those for Web site B? greater than the ones for Web site B?

3. Press GRAPH to graph both equations.

4. At what point do the two lines intersect? What does this ordered pair represent?

Main Idea

Solve systems of equations by graphing.

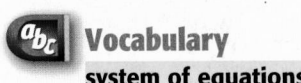 **Vocabulary**

system of equations

 **Get Connect ED**

 **8.EE.8, 8.EE.8a**

Solve Systems of Equations by Graphing

ACTIVITIES A campground offers canoe and bicycle rentals as shown.

	Deposit ($)	Cost per Hour ($)
Canoe	20.00	5.50
Bicycle	15.00	7.00

1. Write an equation to represent the total cost y of renting a canoe for any number of hours x.

2. Write an equation to represent the total cost y of renting a bicycle for any number of hours x.

3. Which item would cost less to rent for 5 hours?

Together, the equations you wrote above are called a **system of equations**. There are two equations and two different unknowns.

The ordered pair for the point of intersection of the graphs is the solution of the system.

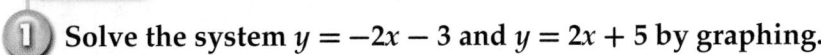 **EXAMPLE** One Solution

① **Solve the system $y = -2x - 3$ and $y = 2x + 5$ by graphing.**

Graph each equation on the same coordinate plane.

The graphs appear to intersect at $(-2, 1)$.

Check this estimate by replacing x with -2 and y with 1.

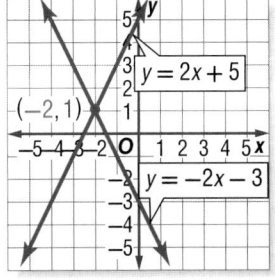

Check

$$y = -2x - 3 \qquad y = 2x + 5$$
$$1 \overset{?}{=} -2(-2) - 3 \qquad 1 \overset{?}{=} 2(-2) + 5$$
$$1 = 1 \checkmark \qquad 1 = 1 \checkmark$$

The solution of the system is $(-2, 1)$.

Study Tip

Graphing To graph each equation, you can use the slope and y-intercept or use the x- and y-intercepts.

 CHECK Your Progress

a. Solve the system $y = x - 1$ and $y = 2x - 2$ by graphing.

MOTORSPORTS Gregory's Motorsports has motorcycles (two wheels) and ATVs (four wheels) in stock. The store has a total of 45 vehicles, that, together, have 130 wheels.

2 **Write a system of equations that represents the situation.**

Let y represent the motorcycles and x represent the ATVs.

$y + x = 45$ The number of motorcycles and ATVs is 45.
$2y + 4x = 130$ The number of wheels equals 130.

Study Tip

Systems of Equations When using a table to solve a system of equations, use the guess, check, and revise strategy to find the solution.

3 **Solve the system of equations. Interpret the solution.**

Write each equation in slope-intercept form.

$$x + y = 45 \qquad\qquad 2y + 4x = 130$$
$$y = -x + 45 \qquad\qquad 2y = -4x + 130$$
$$\qquad\qquad\qquad y = -2x + 65$$

Choose values for x that could satisfy the equations.

x	$y = -x + 45$	y
10	$-(10) + 45$	35
15	$-(15) + 45$	30
20	$-(20) + 45$	25
25	$-(25) + 45$	20

x	$y = -2x + 65$	y
10	$-2(10) + 65$	45
15	$-2(15) + 65$	35
20	$-2(20) + 65$	25
25	$-2(25) + 65$	15

Both equations have the same value when $x = 20$ and $y = 25$.

You can also graph both equations on the same coordinate plane. The equations intersect at (20, 25).

The solution is (20, 25). This means that the store has 20 ATVs and 25 motorcycles.

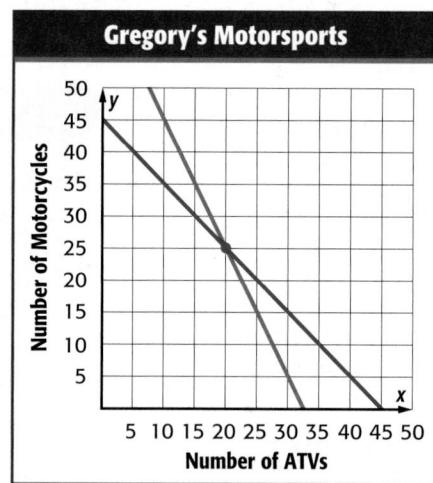

Gregory's Motorsports

CHECK Your Progress

b. CRAFTS Creative Crafts gives scrapbooking lessons x for $15 per hour plus a $10 supply charge. Scrapbooks Incorporated gives lessons for $20 per hour with no additional charges. Write and solve a system of equations to find the total cost y for each craft store. Interpret the solution.

4 Solve the system $y = 2x + 1$ and $y = 2x - 3$ by graphing.

Graph each equation on the same coordinate plane.

The graphs appear to be parallel lines. Since there is no coordinate point that is a solution of both equations, there is no solution for this system of equations.

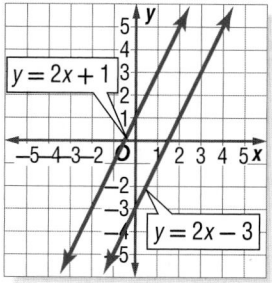

Study Tip

Slopes and Intercepts
When a linear system of equations has:

• different slopes and y-intercepts, there is one and only one solution.

• the same slope and different y-intercepts, there is no solution.

• the same slope and the same y-intercept, there is an infinite number of solutions.

5 Solve the system $y = 2x + 1$ and $y - 3 = 2x - 2$ by graphing.

Write $y - 3 = 2x - 2$ in slope-intercept form.

$$y - 3 = 2x - 2 \qquad \text{Write the equation.}$$
$$y - 3 + 3 = 2x - 2 + 3 \qquad \text{Add 3 to each side.}$$
$$y = 2x + 1 \qquad \text{Simplify.}$$

Both equations are the same. Graph the equation.

Any ordered pair on the graph will satisfy both equations. So, there are infinitely many solutions of the system.

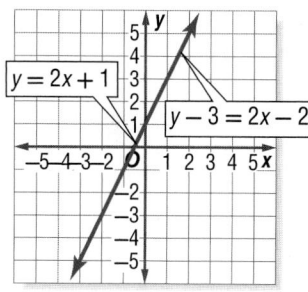

CHECK Your Progress

Solve each system of equations by graphing.

c. $y = -4x - 1$
   $y = -4x + 2$

d. $y - x = 1$
   $y = x - 2 + 3$

The graph of a system of equations indicates the number of solutions.

• If the lines intersect, there is one solution.

• If the lines are parallel, there is no solution.

• If the lines are the same, there are infinitely many solutions.

✓ **CHECK Your Understanding**

Examples 1, 4, and 5 Solve each system of equations by graphing.

1. $y = x + 3$
   $y = -2x - 3$

2. $y = 3x$
   $y - 4 = 3x$

3 $y - 6 = 2x$
   $y = 2(x + 1) + 4$

Examples 2 and 3 **4. AGE** The sum of Sally's age plus twice Tomas' age is 12. The difference of Sally's age and Tomas' age is 3. Write and solve a system of equations to find their ages. Interpret the solution.

Practice and Problem Solving

 = **Step-by-Step Solutions** begin on page R1.
Extra Practice begins on page EP2.

Examples 1, 4, and 5 **Solve each system of equations by graphing.**

5. $y = x$
$y = 2x - 4$

6. $y = -\frac{1}{2}x + 5$
$y = 3x - 2$

7. $y = 4x - 15$
$y - 4x = 16$

8. $y - 2x = 4$
$y = 2x$

9. $x + y = 7$
$3x - y = 5$

10. $x + y = -3$
$y = x - 2x - 3$

11. $y - 4x = 8$
$y = 2(2x + 4)$

12. $x + y = 3$
$y = -3(2x - 1)$

13. $-x + y = -2$
$y = 2$

Examples 2 and 3 **Write a system of equations that represents each situation. Use a table or graph to solve. Interpret the solution.**

14. PETS A pet store currently has a total of 45 cats and dogs. There are 7 more cats than dogs. Find the number of cats and dogs in the store.

15 JEWELRY Janelle is ordering materials for beaded bracelets. Babs' Beads charges $0.12 per bead with a shipping cost of $3.25. Jewels by Jo charges $0.25 per bead with no shipping costs. When would the costs be the same?

16. FINANCIAL LITERACY Mrs. Wilkins is having the baseball team's names put on the back of their jerseys. Company A charges $60 plus $5 per jersey. Company B charges a flat rate of $125.

a. There are 15 players on the baseball team. Which company offers the better deal?

b. For what number of jerseys is the cost the same for both companies?

c. Does the company you chose for part **a** always offer a better deal? Explain.

17 GRAPHIC NOVEL Refer to the graphic novel frame below for Exercises a–b.

a. Refer to Lesson 2A, Exercise 32. Graph each equation on the same coordinate plane.

b. How many rides must each person ride for the wristband to be the better deal?

18. CHALLENGE One equation in a system of equations is $y = 2x + 1$.

    a. Write a second equation so that the system has $(1, 3)$ as its only solution.

    b. Write an equation so that the system has no solution.

    c. Write an equation so that the system has infinitely many solutions.

19. **WRITE MATH** Write a real-world problem that could be represented by a system of equations. What does the point of intersection represent?

Test Practice

20. Katia baked 36 cookies. There are 8 more chocolate chip cookies than peanut butter. Which system can be used to find the number of each type of cookie?

    A. $c + p = 36$
         $p = c + 8$

    B. $c + p = 36$
         $c = p + 8$

    C. $c + p = 8$
         $p = c + 36$

    D. $c + p = 8$
         $c = p + 36$

21. Two equations in a system are shown in the graph. Which of the following statements is true?

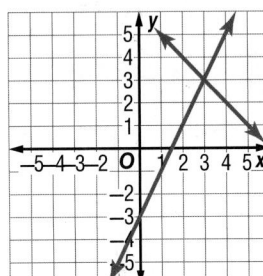

    F. The solution of the system is $(0, -3)$.

    G. The solution of the system is $(3, 3)$.

    H. The system has no solution.

    I. The system has infinitely many solutions.

Spiral Review

22. SOUVENIRS Green Gables Gift Shop sells regular postcards in packages of 5 and large postcards in packages of 3. If Román bought 16 postcards, how many packages of each did he buy? (Lesson 3A)

State the x- and y-intercepts of each function. Then graph the function. (Lesson 2B)

23. $2x + 3y = 18$
        24. $5x - 2y = 20$
        25. $-4x - 6y = 24$

Find the slope of each line. (Lesson 1C)

26.

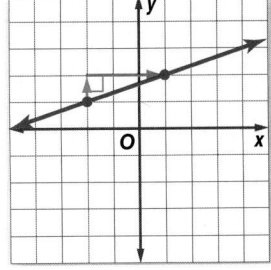

27.

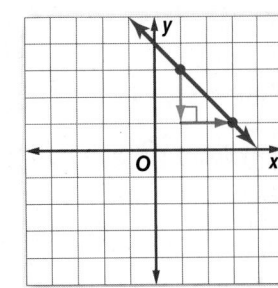

28.

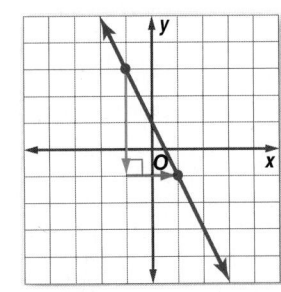

Solve Systems of Equations by Substitution

EXPLORE Mary Anne sold 20 necklaces and bracelets at the craft fair. She sold 3 times as many necklaces as bracelets. This situation can be represented by the bar diagram below.

1. Use the bar diagram to find how many bracelets and necklaces Mary Anne sold.

2. Write a system of equations to represent the situation.

3. How could you use the system of equations to solve the problem?

In the previous lesson, you learned to solve a system of equations by graphing. **Substitution** is an algebraic model that can be used to find the exact solution of a system of equations.

EXAMPLE **Solve a System by Substitution**

1 **Solve the system of equations by substitution.**
$$y = x - 5$$
$$y = 2x$$

Since y is equal to $2x$, you can replace y with $2x$ in the first equation.

$$
\begin{array}{ll}
y = \quad x - 5 & \text{Write the equation.} \\
2x = \quad x - 5 & \text{Replace } y \text{ with } 2x. \\
\underline{-x = -x} & \text{Subtract } x \text{ from each side.} \\
x = -5 & \text{Simplify.}
\end{array}
$$

Since $x = -5$ and $y = 2x$, then $y = -10$ when $x = -5$. The solution of this system of equations is $(-5, -10)$. Check the solution by graphing.

CHECK Your Progress

Solve each system of equations by substitution.

a. $y = x + 4$
$\quad\; y = 2$

b. $y = x - 6$
$\quad\; y = 3x$

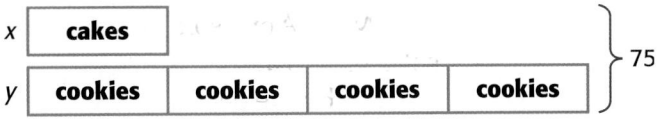

 **FUNDRAISING** A total of 75 cookies and cakes were donated for a bake sale to raise money for the football team. There were four times as many cookies donated as cakes.

② **Write a system of equations to represent this situation.**

Draw a bar diagram. Then write the system.

x	cakes			
y	cookies	cookies	cookies	cookies

} 75

$y = 4x$ There were 4 times as many cookies donated as cakes.

$x + y = 75$ The total number of cakes and cookies is 75.

③ **Solve the system by substitution. Interpret the solution.**

Since y is equal to $4x$, you can replace y with $4x$.

$x + y = 75$ Write the equation.

$x + 4x = 75$ Replace y with $4x$.

$5x = 75$ Simplify.

$\dfrac{5x}{5} = \dfrac{75}{5}$ Divide each side by 5.

$x = 15$ Simplify.

Since $x = 15$ and $y = 4x$, then $y = 60$ when $x = 15$. The solution is (15, 60). This means that 15 cakes and 60 cookies were donated.

Check   Check the solution by graphing. The graphs of the functions intersect at the point (15, 60). ✓

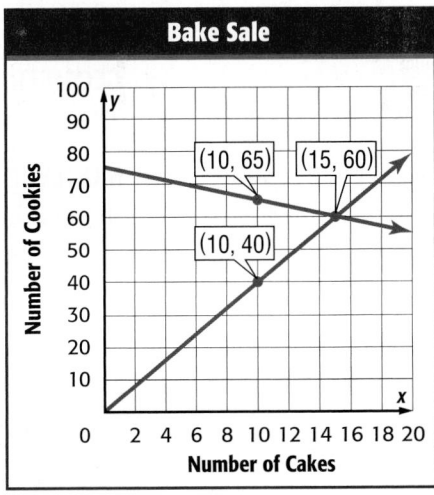

Bake Sale

 CHECK Your Progress

FOOD Mr. Thomas cooked 45 hamburgers and hot dogs at a cookout. He cooked twice as many hot dogs as hamburgers.

c. Write a system of equations to represent this situation.

d. Solve the system by substitution. Interpret the solution.

Example 1 Solve each system of equations by substitution.

1. $y = x + 7$
   $y = 4$

2. $y = x - 3$
   $y = 6$

3. $y = x + 5$
   $y = 3x$

4. $y = x - 9$
   $y = -4x$

Examples 2 and 3

5. **MOVIES** Seven adults and children went to the movies. The number of adults was one more than the number of children. Write a system of equations that represents the number of adults and children. Solve the system by substitution. Interpret the solution.

Practice and Problem Solving

● = Step-by-Step Solutions begin on page R1.
Extra Practice begins on page EP2.

Example 1 Solve each system of equations by substitution.

6. $y = x + 5$
   $y = 6$

7. $y = x + 12$
   $y = -18$

8. $y = x - 10$
   $y = -12$

9   $y = -x - 14$
   $y = 15$

10. $y = x + 15$
    $y = 2x$

11. $y = -x + 22$
    $y = 5x$

12. $y = x - 4$
    $y = -\frac{1}{2}x$

13. $y = -x - 12$
    $y = -3x$

Examples 2 and 3 Write and solve a system of equations that represents each situation. Use a bar diagram if needed. Interpret the solution.

14. **SHOPPING** Elaine bought a total of 15 shirts and pairs of pants. She bought 7 more shirts than pants. How many of each did she buy?

15. **GAMES** Together, Preston and Horatio have 49 video games. Horatio has 11 more games than Preston. How many games does each person have?

16. **GEOMETRY** The length of the rectangle is 3 meters more than the width. The perimeter is 26 meters. What are the dimensions of the rectangle?

17 **MONEY** Brad has nickels and dimes in his pocket. He has three more dimes than nickels. The coins have a total value of 90¢. Find how many of each coin he has.

18. **MULTIPLE REPRESENTATIONS** The table shows the rates at which Ajay and Tory are biking along the same trail.

Person	Rate (m/min)
Ajay	200
Tory	250

    a. ALGEBRA Suppose Ajay began the trail 300 meters ahead of Tory. Write an equation to represent the distance y each person will travel after any number of minutes x.

    b. WORDS Which person was farther along the trail after 5 minutes?

    c. TABLES Make a table showing the distance each person has traveled along the trail after biking for 1 to 10 minutes.

    d. WORDS If the rates continue, will Tory catch up to Ajay? Explain.

    e. GRAPHS Describe what the graph of the rates would look like.

19. CHALLENGE What is the solution of the system of equations $y = -1$ and $x = 7$?

20. **WRITE MATH** Give an example of three different systems of equations. One system should have no solution, the second system should have one solution, and the third system should have an infinite number of solutions. Justify your response.

Test Practice

21. Jay and Leanne are playing a game. Leanne has three times as many points as Jay. Jay has 20 fewer points than Leanne. Which system of equations can be used to find each player's points?

A. $j = 3l$
$j = l + 20$

B. $j = 3l$
$j = l - 20$

C. $l = 3j$
$l = j + 20$

D. $l = 3j$
$l = j - 20$

22. A veterinarian examined twice as many cats y as dogs x. She examined a total of 30 cats and dogs. Which system of equations represents this situation?

F. $x = 2y$
$y = 2x$

G. $y = 2x$
$x - y = 30$

H. $y = 2x$
$y = x + 30$

I. $2x = y$
$x + y = 30$

More About Systems of Equations

You can also solve systems of equations that are written in function notation. Recall that function notation is a way to name a function that is defined by an equation. For example, the equation $y = 7x - 1$ is written in function notation as $f(x) = 7x - 1$.

EXAMPLE

Solve $f(x) = g(x)$ if $f(x) = 2x + 5$ and $g(x) = x - 2$.

Since $f(x) = g(x)$, $2x + 5 = x - 2$. Solve for x.

$$
\begin{array}{ll}
2x + 5 = x - 2 & f(x) = g(x) \\
\underline{-x = -x} & \text{Subtraction Property of Equality} \\
x + 5 = -2 & \text{Simplify.} \\
\underline{-5 = -5} & \text{Subtraction Property of Equality} \\
x = -7 & \text{Simplify.}
\end{array}
$$

So, $f(-7) = 2(-7) + 5$ or -9.   Check this solution by finding $g(-7)$

The solution of this system of equations is $(-7, -9)$.

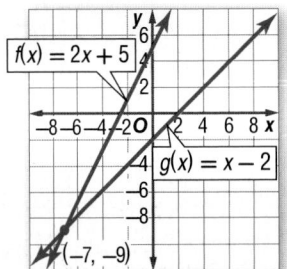

Solve $f(x) = g(x)$ for each set of functions.

23. $f(x) = 3x + 1$
$g(x) = x - 5$

24. $f(x) = 4x - 6$
$g(x) = x + 9$

25. $f(x) = 5x + 15$
$g(x) = 2x + 3$

Mastering the Music

Do you love listening to music? Are you interested in the technical aspects of music-making? If so, a career mastering CDs might be something to think about! A mastering engineer produces CD masters and is responsible for making songs sound better, having the proper spacing between songs, removing extra noises, and assuring all the songs on a CD have consistent levels of tone and balance. Having a great-sounding master helps increase radio airplay and sales of CDs for recording artists.

21st Century Careers

Are you interested in a career as a mastering engineer? Take some of the following courses in high school.

- Algebra
- Music Appreciation
- Recording Techniques
- Sound Engineering and Sound

Get ConnectED

Engineering Hits	
Number of Songs	Cost ($)
1	100
2	160
3	210
4	250

Dynamic Mastering	
Number of Songs	Cost ($)
2	120
4	240
6	360
8	480

Mastering Mix	
Number of Songs	Cost ($)
1	125
3	275
5	425
7	575

Real-World Math

Use the information in the tables to solve each problem.

1. At Engineering Hits, is the relationship between the number of songs and the cost linear? Explain your reasoning.

2. Is there a proportional linear relationship between number of songs and cost at Dynamic Mastering? Explain your reasoning.

3. Find the slope of the function represented in the Mastering Mix table. What does the slope represent?

4. Is the linear function represented in the Mastering Mix table a direct variation? Explain.

5. Write a direct variation equation to represent number of songs x and cost y at Dynamic Mastering. How much does it cost to master 11 songs?

6. For 4 or more songs at Engineering Hits, the cost varies directly as the number of songs. How much does it cost to master 6 songs?

Chapter Study Guide and Review

Be sure the following Key Concepts are noted in your Foldable.

Key Concepts

Constant Rate of Change (Lesson 1)
- Two quantities a and b have a proportional linear relationship if they have a constant ratio and a constant rate of change.

Slope (Lesson 1)
- The slope m of a line passing through points (x_1, y_1) and (x_2, y_2) is the ratio of the difference in the y-coordinates to the corresponding difference in the x-coordinates.

Direct Variation (Lesson 1)
- A direct variation is a relationship in which the ratio of y to x is a constant k.

Slope-Intercept Form (Lesson 2)
- An equation written in slope-intercept form is written as $y = mx + b$, where m is the slope and b is the y-intercept.

Systems of Equations (Lesson 3)
- Two equations together are called a system of equations.

Key Vocabulary

constant of variation
constant rate of change
direct variation
linear relationship
qualitative graph
rise
run
slope
slope-intercept form
standard form
substitution
system of equations
x-intercept
y-intercept

Vocabulary Check

Choose the correct term to complete each sentence.

1. The (x-intercept, y-intercept) has the coordinates $(0, b)$.

2. The slope formula is $m = \left(\dfrac{y_2 - y_1}{x_2 - x_1}, \dfrac{x_2 - x_1}{y_2 - y_1} \right)$.

3. The (rise, run) is the vertical change between two points on a line.

4. In the equation $y = 3x + 5$, the slope is $(3, 5)$.

5. An equation like $3x + 4y = 12$ is written in (standard, slope-intercept) form.

6. The x-intercept of the equation $2x + 3y = 6$ is $(2, 3)$.

7. A direct variation is when the ratio of two variable quantities is (constant, proportional).

Multi-Part Lesson Review

Lesson 1   Slope

Constant Rate of Change (Lesson 1A)

Determine whether the relationship between the two quantities described in each table is linear. If so, find the constant rate of change. If not, explain your reasoning.

8.

Hours	Rainfall (in.)
1	2
2	4
3	7
4	9

9.

Minutes	Cost ($)
1	7
2	14
3	21
4	28

EXAMPLE 1   The distance traveled on a car trip is shown. Is the relationship between the distance traveled and number of hours spent in the car linear? If so, find the constant rate of change. If not, explain.

Time (h)	Distance (mi)	
2	120	+120
4	240	+120
6	360	+120
8	480	

(+2 between each Time value)

As the hours increase by two, the distance increases by 120. Since the rate of change is constant, it is linear. So, the constant rate of change is $\frac{120}{2}$ or 60 miles per hour.

Slope (Lesson 1C)

Find the slope of each line that passes through each pair of points.

10. $G(6, 2)$, $H(1, 5)$ 11. $U(-4, 2)$, $R(2, -3)$

12. **SLIDES** Find the slope of a slide that descends 8 feet for every horizontal change of 14 feet.

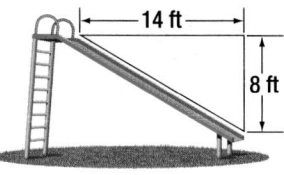
←——— 14 ft ———→
8 ft

EXAMPLE 2   Find the slope of the line that passes through $A(-3, 2)$ and $B(5, -1)$.

$m = \dfrac{y_2 - y_1}{x_2 - x_1}$ Definition of slope

$m = \dfrac{-1 - 2}{5 - (-3)}$ $(x_1, y_1) = (-3, 2), (x_2, y_2) = (5, -1)$

$m = \dfrac{-3}{8}$ or $\dfrac{3}{-8}$ Simplify.

Direct Variation (Lesson 1E)

13. **MONEY** Josiah spent $15.60 on 3 comic books. The next time, he spent $10.40 on 2 comic books. What is the cost for each comic book?

14. **FRUIT** The cost of peaches varies directly with the number of pounds bought. If 3 pounds of peaches cost $4.50, find the cost of 5.5 pounds.

EXAMPLE 3   Mrs. Dimas paid $6.48 for 8 apples. The next weekend, she paid $9.72 for 12 apples. What is the cost of each apple?

$\dfrac{\$6.48}{8 \text{ apples}}$ or $\dfrac{\$0.81}{1 \text{ apple}}$ $\dfrac{\$9.72}{12 \text{ apples}}$ or $\dfrac{\$0.81}{1 \text{ apple}}$

So, each apple costs $0.81.

Lesson 2 **Intercepts**

Slope-Intercept Form (Lesson 2A)

State the slope and y-intercept for the graph of each equation.

15. $y = 2x + 5$

16. $y = -\frac{1}{5}x + 6$

17. $y - 4x = 7$

18. $3x + y = -2$

19. Write an equation in slope-intercept form for the graph shown.

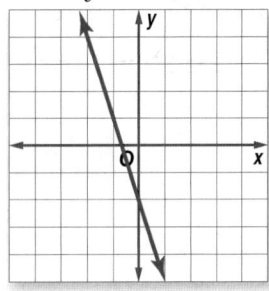

20. MONEY Felix has $100 in his savings account. He plans to add $25 each week. The equation for the amount of money y Felix has in his savings account is $y = 100 + 25x$, where x is the number of weeks. Graph the equation.

EXAMPLE 4 State the slope and y-intercept for the graph of $y = -\frac{1}{2}x + 3$.

$y = mx + b$

$y = -\frac{1}{2}x + 3$

The slope of the graph is $-\frac{1}{2}$ and the y-intercept is 3.

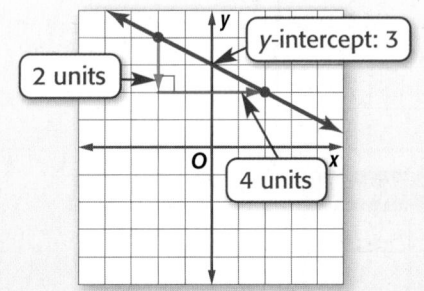

Graph Functions Using Intercepts (Lesson 2B)

State the x- and y-intercepts of each function. Then graph the function.

21. $3x + 6y = 24$

22. $2x - 5y = 20$

23. GOLF Golf balls are sold in boxes of 12 and 15. Coach Taylor bought a total of 180 golf balls for the golf team. This is represented by the function $12x + 15y = 180$.

  a. Graph the function.

  b. Interpret the x- and y-intercepts.

24. FOOD Hot dog buns are sold in packages of 8 and 12. Taro bought 192 total buns for a party. This is represented by the function $8x + 12y = 192$.

  a. Graph the function.

  b. Interpret the x- and y-intercepts.

EXAMPLE 5 State the x- and y-intercepts of $3x + 2y = 12$. Then graph the function.

x-intercept	y-intercept
To find the x-intercept, let $y = 0$.	To find the y-intercept, let $x = 0$.
$3x + 2y = 12$	$3x + 2y = 12$
$3x + 2(0) = 12$	$3(0) + 2y = 12$
$3x = 12$	$2y = 12$
$x = 4$	$y = 6$

The x-intercept is 4.   The y-intercept is 6.

Graph the points $(4, 0)$ and $(0, 6)$ on a coordinate plane. Then connect the points.

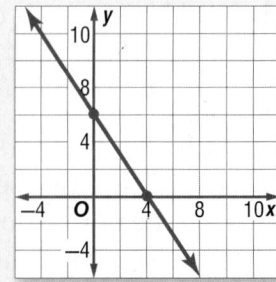

Guess, Check, and Revise (Lesson 3A)

25. FOOD A store sells apples in 2-pound bags and oranges in 5-pound bags. How many bags of each should you buy if you need exactly 11 pounds of apples and oranges?

26. BONES Each hand in the human body has 27 bones. There are 6 more bones in the fingers than in the wrist. There are 3 fewer bones in the palm than in the wrist. How many bones are in each part of the hand?

EXAMPLE 6 The product of two consecutive even whole numbers is 1,088. What are the whole numbers?

Numbers	Product	Result
24 × 26	624	too low
30 × 32	960	too low
34 × 36	1,224	too high
32 × 34	1,088	correct ✓

The numbers are 32 and 34.

Solve Systems of Equations by Graphing (Lesson 3C)

Solve each system of equations by graphing.

27. $y = 3x + 4$
$y = -2x - 6$

28. $y = -\frac{2}{3}x + 2$
$y = \frac{1}{6}x - 3$

29. FOOD Twenty-five teenagers were surveyed about food. Five more preferred pizza than steak. Write and solve a system of equations to find how many preferred steak and how many preferred pizza.

EXAMPLE 7 Solve the system $y + x = 20$ and $y = x + 8$ by graphing.

Graph each equation on the same coordinate plane.

Since the graphs intersect at (6, 14), the solution of the system is (6, 14).

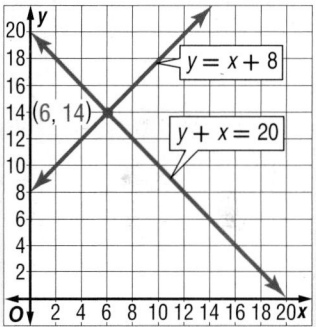

Solve Systems of Equations by Substitution (Lesson 3D)

Solve each system of equations by substitution.

30. $y = x + 5$
$y = 7$

31. $y = x - 4$
$y = -2$

32. $y = x + 3$
$y = 2x$

33. $y = x - 6$
$y = 4x$

34. SOUVENIRS Lena bought a total of 20 postcards. She bought 6 more large postcards than small. Write a system of equations that represents the postcards Lena purchased. Solve the system by substitution. Interpret the solution.

EXAMPLE 8 Solve the system of equations by substitution.

$y = x + 6$
$y = 2x$

Replace y with $2x$ in the first equation.

$y = x + 6$	Write the equation.
$2x = x + 6$	Replace y with $2x$.
$2x - x = x - x + 6$	Subtract x from each side.
$x = 6$	Simplify.

Since $x = 6$ and $y = 2x$, then $y = 12$ when $x = 6$. The solution of this system of equations is (6, 12).

1. MEASUREMENT Is the relationship between weight and number of months linear? If so, find the constant rate of change. If not, explain your reasoning.

Number of Months	Weight (lb)
4	14
6	18
8	20
10	22

Find the slope of the line that passes through each pair of points.

2. $R(-2, 5)$, $S(-2, 1)$

3. $T(2, -1)$, $U(5, -3)$

4. $V(3, -6)$, $W(4, -8)$

5. SPORTS Find the slope of the ramp below.

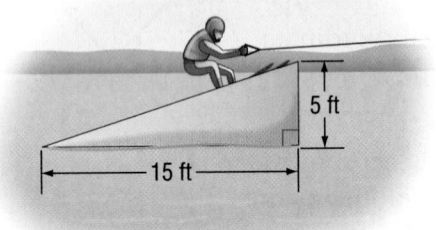

5 ft

15 ft

6. JOBS The amount Jerri earns working varies directly with the time she works. If she earns $187.50 after working 25 hours, how much will she earn working 30 hours?

7. MULTIPLE CHOICE Rico planted 18 flowers in 30 minutes. At the same rate, how many flowers would he plant in 55 minutes?

A. 30 flowers   **C.** 36 flowers

B. 33 flowers   **D.** 38 flowers

State the x- and y-intercepts of each function. Then graph the function.

8. $-3x - 6y = 12$ **9.** $-2x + 6y = 15$

10. MONEY The equation that represents the amount of money y Simon owes his grandfather after x months is $y = 120 - 15x$.

  a. Graph the equation to find how much Simon will owe after 6 months.

  b. Interpret the slope and y-intercept.

Find the slope of each line.

11. 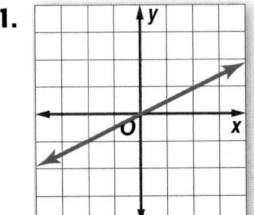   **12.**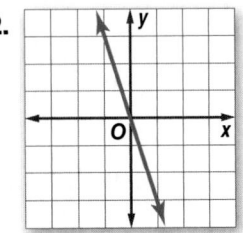

13. MUSICAL Joseph sold tickets to the school musical. He had 12 bills worth $175 for the tickets sold. If all the money was in $5 bills, $10 bills, and $20 bills, how many of each bill did he have?

14. MONEY Robert has 26 coins that are all nickels and dimes. The value of the coins is $1.85. Write and solve a system of equations that represents this situation.

15. **THINK SOLVE EXPLAIN** **EXTENDED RESPONSE** Mr. Jacobsen wants to hire a magician for his son's birthday party. The table shows the rates of two different magicians.

Magician	Travel Fee	Charge per Hour
Marlon	$30	$15
Gregory	$50	$10

Part A Define a variable. Then write a system of equations to represent the situation.

Part B Solve the system of equations. Interpret the solution.

When Will I Use This?

Mandar and Roberto in

Ramping Up

I can't believe my Dad is letting us build a bike ramp!

Yeah! We have plenty of boards. Let's go take a look at the plans.

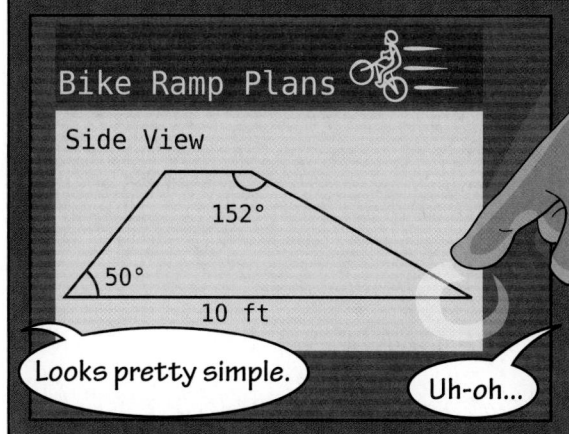

Bike Ramp Plans

Side View

152°

50°

10 ft

Looks pretty simple.

Uh-oh...

That angle isn't given.

We can't build it without that angle!

If only there was a way to figure it out.

Your Turn!
You will solve this problem in the chapter.

Are You Ready for the Chapter?

You have two options for checking prerequisite skills for this chapter.

Text Option Take the Quick Check below. Refer to the Quick Review for help.

QUICK Check

Solve each equation.

1. $49 + b + 45 = 180$
2. $t + 98 + 55 = 180$
3. $15 + 67 + k = 180$
4. $90 + 62 + 120 + a = 360$
5. $108 + m + 54 + 78 = 360$
6. $50 + 105 + c + 105 = 360$

7. **TICKETS** The table shows the number of tickets Dalila sold for the school play during the beginning of the week.

Day	Tickets
Monday	65
Tuesday	40
Wednesday	135

If she sold 360 tickets altogether, how many did she sell the rest of the week?

Evaluate each expression.

8. $(3 - 2)180$
9. $(7 - 2)180$
10. $(9 - 2)180$
11. $(11 - 2)180$
12. $(18 - 2)180$
13. $(12 - 2)180$
14. $(20 - 2)180$
15. $(25 - 2)180$

16. **NUMBER SENSE** Find the product of the difference of 5 and 2 and 180.

QUICK Review

EXAMPLE 1

Solve $82 + g + 41 = 180$.

$$82 + g + 41 = 180 \quad \text{Write the equation.}$$
$$123 + g = 180 \quad \text{Add 82 and 41.}$$
$$\underline{-123 \qquad = -123} \quad \text{Subtraction Property}$$
$$g = 57 \qquad \text{of Equality}$$

EXAMPLE 2

Solve $97 + a + 116 + 108 = 360$.

$$97 + a + 116 + 108 = 360 \quad \text{Write the equation.}$$
$$321 + a = 360 \quad \text{Add.}$$
$$\underline{-321 \qquad = -321} \quad \text{Subtraction}$$
$$a = 39 \qquad \text{Property of Equality}$$

EXAMPLE 3

Evaluate $(8 - 2)180$.

$$(8 - 2)180 = (6)180 \quad \text{Subtract 2 from 8.}$$
$$= 1{,}080 \quad \text{Multiply.}$$

 **Online Option** Take the Online Readiness Quiz.

Explore   Angle Measure

Main Idea
Measure and draw angles.

 Vocabulary
degree

 Get ConnectED

Angles are measured in units called **degrees**. If a circle is divided into 360 equal-size parts, each part has a measure of 1 degree (1°). You can use a tool called a protractor to measure an angle.

ACTIVITY   Measure an Angle

Use a protractor to measure ∠FGH.

STEP 1   Place the center point of the protractor's base on vertex *G*. Align the straight side with side $\overline{GH}$ so that the marker for 0° is on the ray.

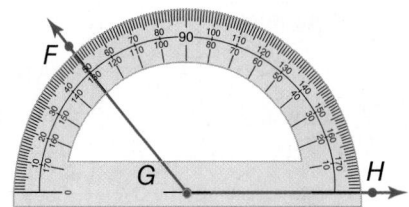

STEP 2   Use the scale that begins with 0° at $\overline{GH}$. Read where the other side of the angle, $\overline{GF}$, crosses the scale.

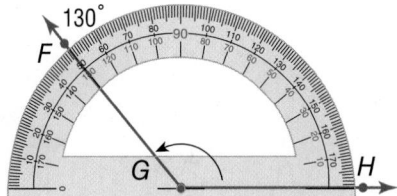

The measure of angle *FGH* is 130°. Using symbols, *m∠FGH* = 130°.

Practice and Apply

Use a protractor to find the measure of each angle.

1. ∠XZY
2. ∠SZT
3. ∠SZY
4. ∠UZX
5. ∠TZW
6. ∠UZV

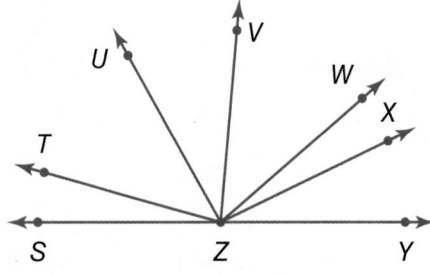

Use a protractor to draw an angle having each measurement.

7. 110°
8. 85°
9. 90°
10. 155°
11. 140°
12. 117°

Main Idea

Classify angles and identify vertical and adjacent angles.

 Vocabulary

congruent
vertex
acute angle
right angle
obtuse angle
straight angle
vertical angles
adjacent angles

 Get Connect**ED**

Classify Angles

ROLLER COASTERS The angles of descent of a roller coaster are shown.

1. The roller coaster at the right shows two angles of descent. Draw an angle between 44° and 70°.

2. Some roller coasters have an angle of descent that is 90°, known as a vertical angle of descent. Draw a vertical angle of descent.

An angle is formed by two rays that share a common endpoint and can be named in several ways.

Angles are classified according to their measure. Two angles that have the same measure are said to be **congruent**.

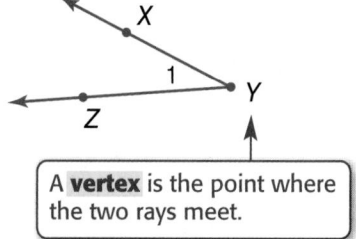

A **vertex** is the point where the two rays meet.

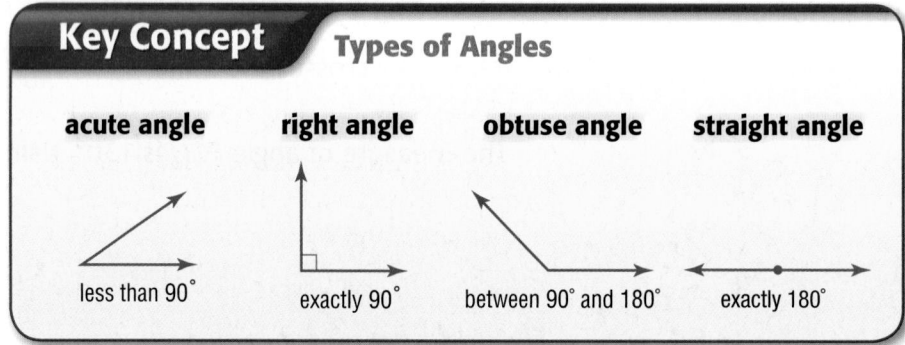

Key Concept — Types of Angles

acute angle	right angle	obtuse angle	straight angle
less than 90°	exactly 90°	between 90° and 180°	exactly 180°

Read Math

Symbols The symbol for angle is ∠.

 **EXAMPLE** Name Angles

① **Name the angle shown above the Key Concept box. Then classify it as *acute*, *right*, *obtuse*, or *straight*.**

• Use the vertex as the middle letter and a point from each side, ∠XYZ or ∠ZYX.

• Use the vertex only, ∠Y.

• Use a number, ∠1.

Since the angle is less than 90°, it is an acute angle.

CHECK Your Progress

Name each angle in four ways. Then classify each angle as *acute, right, obtuse,* or *straight.*

a.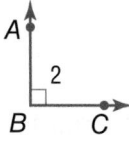

b.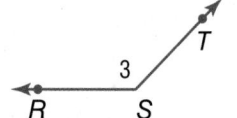

c.

Key Concept — Pairs of Angles

Words	Models	Symbols
Two angles are **vertical** if they are opposite angles formed by the intersection of two lines. Vertical angles are congruent.	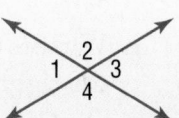 ∠1 and ∠3, ∠2 and ∠4	∠1 ≅ ∠3 ∠2 ≅ ∠4
Two angles are **adjacent** if they share a common vertex, a common side, and do not overlap.	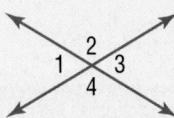	Adjacent angle pairs are ∠1 and ∠2, ∠2 and ∠3, ∠3 and ∠4, and ∠4 and ∠1.

REAL-WORLD EXAMPLE

2 **INTERSECTIONS** Identify a pair of vertical angles and adjacent angles in the diagram at the right. Justify your response.

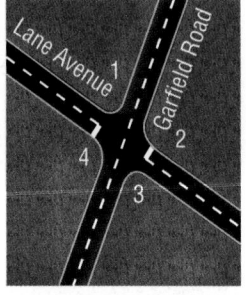

Since ∠2 and ∠4 are opposite angles formed by the intersection of two lines, they are vertical angles.

Since ∠1 and ∠2 share a common side, they are adjacent angles.

CHECK Your Progress

Refer to the diagram at the right. Identify each of the following. Justify your response.

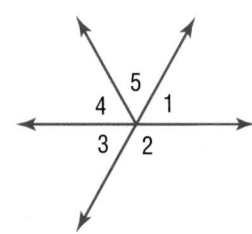

d. a pair of vertical angles

e. a pair of adjacent angles

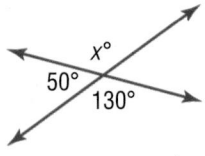

EXAMPLE Find a Missing Angle Measure

 What is the value of x in the figure at the right?

The angle labeled $x°$ and the angle labeled $130°$ are vertical angles.

Since vertical angles are congruent, the value of x is 130.

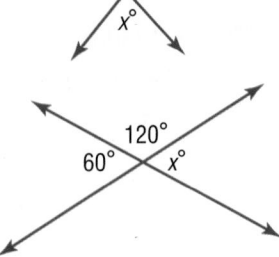

CHECK Your Progress

f. What is the value of x in the figure at the right?

g. What is the value of x in the figure at the right?

CHECK Your Understanding

Example 1 Name each angle in four ways. Then classify the angle as *acute, right, obtuse,* or *straight.*

 1.

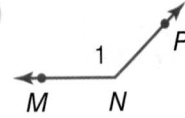

2.

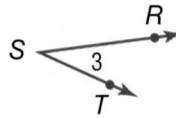

3.

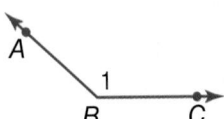

4.

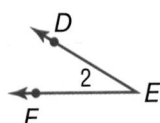

Example 2 **5. RAILROADS** Identify a pair of vertical angles and adjacent angles on the railroad crossing sign. Justify your response.

Example 3 Find the value of x in each figure.

6.

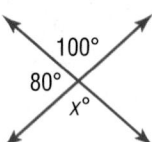

7.

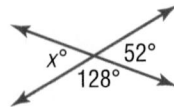

8.

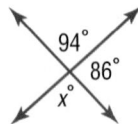

9.

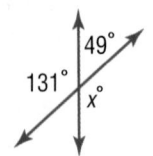

Practice and Problem Solving

 = **Step-by-Step Solutions** begin on page R1.
Extra Practice begins on page EP2.

Example 1 **Name each angle in four ways. Then classify the angle as** *acute, right,* *obtuse,* **or** *straight.*

10.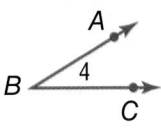

11.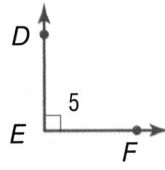

12.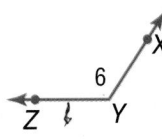

13.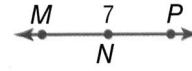

14.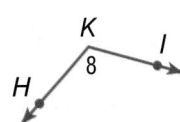

15.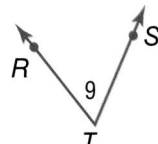

Example 2 **Refer to the diagram at the right. Identify each**
angle pair as *adjacent, vertical,* **or** *neither.*

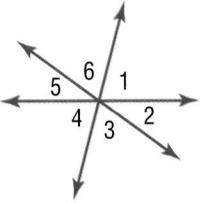

16. ∠2 and ∠5   **17.** ∠4 and ∠6   **18.** ∠3 and ∠4

19. ∠5 and ∠6   **20.** ∠1 and ∠3   **21.** ∠1 and ∠4

22. GEOGRAPHY The corner where the states of Utah,
Arizona, New Mexico, and Colorado meet is called
the Four Corners.

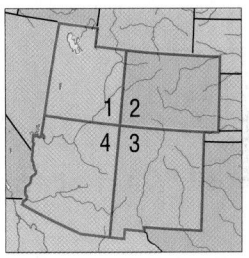

   a. Identify a pair of vertical angles. Justify your
   response.

   b. Identify a pair of adjacent angles. Justify your
   response.

Example 3 **Refer to the figure at the right to determine the**
measure of each given angle.

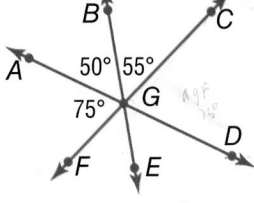

   23. ∠CGD   **24.** ∠DGE

   25. ∠FGE   **26.** ∠FGD

   27. ∠EGC   **28.** ∠BGD

29. ARCHITECTURE The John Hancock Center in
Chicago is shown at the right. Classify each
pair of angles.

   a. ∠1 and ∠2   **b.** ∠2 and ∠4

   c. ∠3 and ∠4   **d.** ∠1 and ∠3

   e. If the measure of ∠2 is 66°, what are the
   measures of the other angles?

Real-World Link · · · ·

Chicago is home to
three of the 20 tallest
buildings in the world:
the Willis Tower, Aon
Center, and the John
Hancock Center.

30. ALGEBRA Angles *ABC* and *DBE* are vertical
angles. If the measure of ∠*ABC* is 40°, what
is the measure of ∠*ABD*?

31. OPEN ENDED Draw examples of angles that represent real-world objects. Be sure to include at least three of the following angles: acute, right, obtuse, straight, vertical, and adjacent. Verify by measuring the angles.

32. REASONING Explain how you can use a protractor to measure the angle shown. Find the measure of the angle.

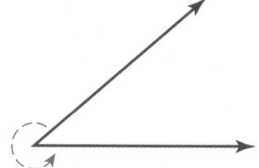

CHALLENGE Determine whether each statement is *true* or *false*. If the statement is true, provide a diagram to support it. If the statement is false, explain why.

33. A pair of obtuse angles can also be vertical angles.

34. A pair of straight angles can also be adjacent angles.

35.  **WRITE MATH** Describe the differences between vertical and adjacent angles.

Test Practice

36. Which statement is true?

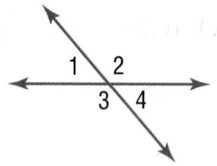

  A. ∠1 and ∠4 are adjacent angles.

  B. ∠2 and ∠3 are vertical angles.

  C. ∠3 and ∠4 are vertical angles.

  D. ∠2 and ∠3 are adjacent angles.

37. Which word best describes the angle marked in the figure?

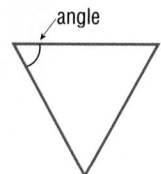

  F. acute

  G. obtuse

  H. right

  I. straight

38. 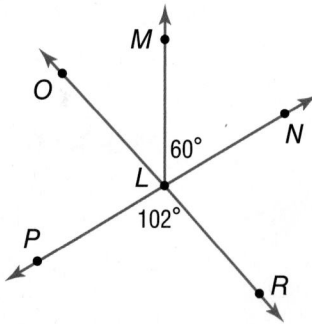 **GRIDDED RESPONSE** Refer to the figure in Exercise 36. Suppose the measure of ∠1 is 40°. What is the measure, in degrees, of ∠4?

39. **THINK SOLVE EXPLAIN** **EXTENDED RESPONSE** In the figure below, the measure of ∠NLM is 60° and the measure of ∠PLR is 102°.

Part A Write and solve an equation to find the measure of ∠NLR.

Part B Classify ∠OLM by its measure. Explain.

Part C Is ∠OLP the same as ∠OPL? Explain.

Main Idea

Identify complementary and supplementary angles and find missing angle measures.

Vocabulary

complementary angles
supplementary angles
paragraph proof

Get ConnectED

Complementary and Supplementary Angles

 Explore Refer to ∠A shown at the right.

1. Classify it as *acute, right, obtuse,* or *straight*.

2. Copy the angle onto a piece of paper. Then draw a ray that separates the angle into two congruent angles. Label these angles ∠1 and ∠2.

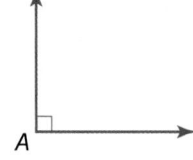

3. What is $m\angle 1$ and $m\angle 2$?

4. What is the sum of $m\angle 1$ and $m\angle 2$?

5. Copy the original angle onto a piece of paper. Then draw a ray that separates the angle into two noncongruent angles. Label these angles ∠3 and ∠4.

6. What is true about the sum of $m\angle 3$ and $m\angle 4$? Verify by measuring the angles.

7. Complete Exercises 1–6 for ∠B shown below.

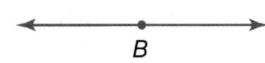

A special relationship exists between two angles with a sum of 90°. A special relationship also exists between two angles with a sum of 180°.

Read Math

Symbols The symbol $m\angle 1$ means *the measure of angle 1.*

Key Concept — Pairs of Angles

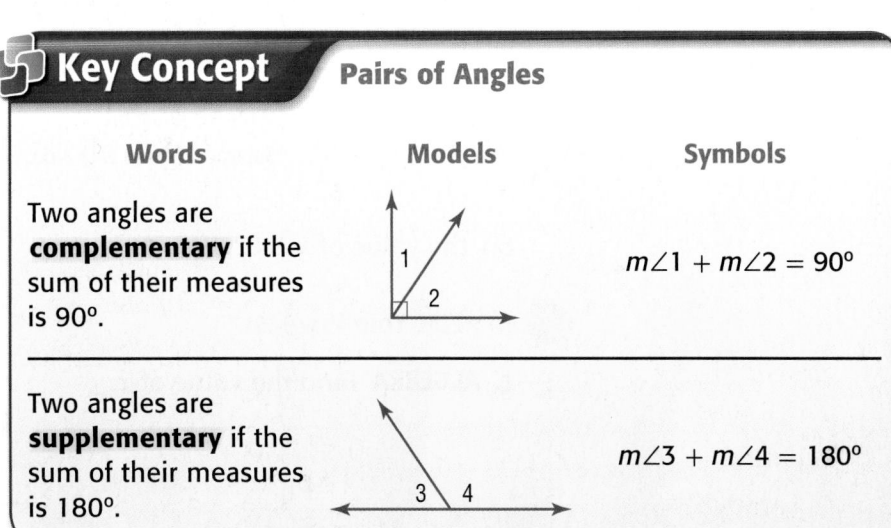

Words	Models	Symbols
Two angles are **complementary** if the sum of their measures is 90°.		$m\angle 1 + m\angle 2 = 90°$
Two angles are **supplementary** if the sum of their measures is 180°.		$m\angle 3 + m\angle 4 = 180°$

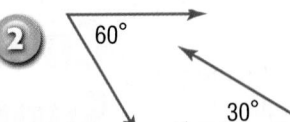 **EXAMPLES** Identify Angles

Identify each pair of angles as *complementary*, *supplementary*, or *neither*.

1

∠1 and ∠2 form a straight angle. So, the angles are supplementary.

2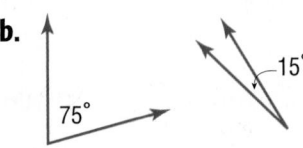

60° + 30° = 90°
The angles are complementary.

CHECK Your Progress

a.

85° | 90°

b.

75° ... 15°

You can use angle relationships to find missing measures.

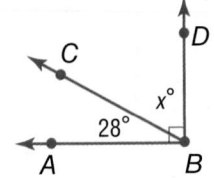 **EXAMPLE** Find a Missing Angle Measure

3 **ALGEBRA** Find the value of *x*.

Since the two angles form a right angle, they are complementary.

Words	The sum of the measures of ∠ABC and ∠CBD	is	90°.
Variable	Let *x* represent the measure of ∠CBD.		
Equation	28 + *x*	=	90

$$28 + x = 90 \quad \text{Write the equation.}$$
$$\underline{-28 \quad\quad = -28} \quad \text{Subtract 28 from each side.}$$
$$x = 62$$

So, the value of *x* is 62.

CHECK Your Progress

c. ALGEBRA Find the value of *x*.

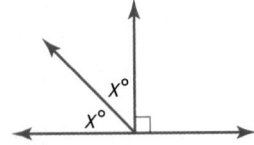

Examples 1 and 2 Identify each pair of angles as *complementary, supplementary,* or *neither.*

1.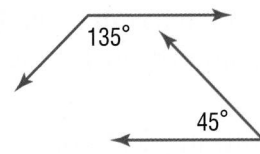

2.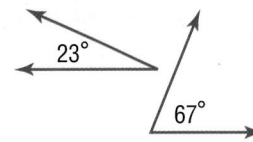

Example 3 **3. ALGEBRA** Find the value of x.

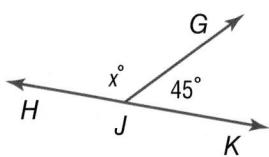

Practice and Problem Solving

⬤ = Step-by-Step Solutions begin on page R1.
Extra Practice begins on page EP2.

Examples 1 and 2 Identify each pair of angles as *complementary, supplementary,* or *neither.*

4.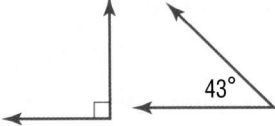

5.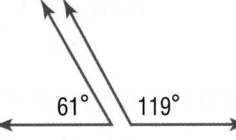

6.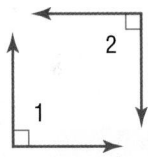

7.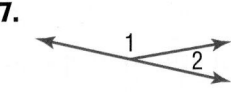

8.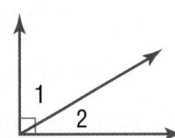

9.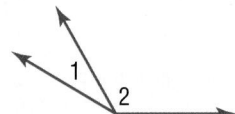

Example 3 **10. ALGEBRA** If $\angle A$ and $\angle B$ are complementary and the measure of $\angle B$ is 67°, what is the measure of $\angle A$?

11 ALGEBRA What is the measure of $\angle J$ if $\angle J$ and $\angle K$ are supplementary and the measure of $\angle K$ is 115°?

12. SCHOOL SUPPLIES What is the measure of the angle given by the opening of the scissors?

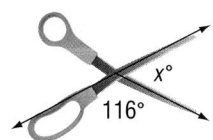

13 SKATEBOARDING A skateboard ramp forms a 43° angle as shown. Find the measure of the unknown angle.

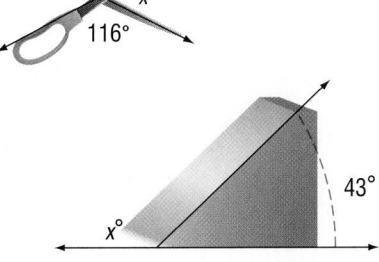

Use the figure at the right to name the following.

14. a pair of supplementary angles

15. a pair of complementary angles

16. a pair of vertical angles

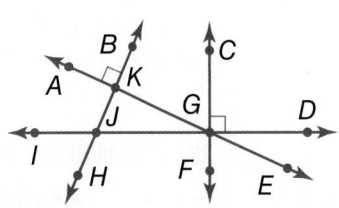

17. GEOMETRY Use the figure at the right.

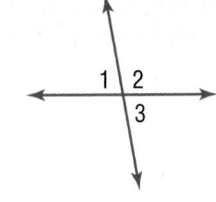

  a. Are ∠1 and ∠2 vertical angles, adjacent angles, or neither? ∠2 and ∠3? ∠1 and ∠3?

  b. Write an equation representing the sum of $m\angle1$ and $m\angle2$. Then write an equation representing the sum of $m\angle2$ and $m\angle3$.

  c. Solve the equations you wrote in part **b** for $m\angle1$ and $m\angle3$, respectively. What do you notice?

  d. MAKE A CONJECTURE Use your answer from part **c** to make a conjecture as to the relationship between vertical angles.

Determine whether each statement is *always*, *sometimes*, or *never* true. Explain your reasoning.

18. Vertical angles have the same angle measure.

19. Two right angles are complementary.

20. Two obtuse angles are supplementary.

21 Two vertical angles are complementary.

22. 🔄 MULTIPLE REPRESENTATIONS Line *a* passes through (1, 4) and (−4, −1). Line *b* passes through (−3, 4) and (2, −1).

QUICK Review

Slope-Intercept Form

Equations in the form $y = mx + b$ are in slope-intercept form.

  a. GRAPHS Graph each line on the same coordinate plane. Describe the lines.

  b. ALGEBRA Write an equation in slope-intercept form for each line.

  c. NUMBERS What is the slope of each line?

  d. GRAPHS Repeat parts **a–c** with line *c* that passes through (−2, 3) and (−5, −2) and line *d* that passes through (−5, −2) and (0, −5).

  e. WORDS Perpendicular lines meet at right angles. What do you notice about the slopes of perpendicular lines?

H.O.T. Problems

23. REASONING When a basketball hits a hard, level surface, it bounces off at the same angle at which it hits. Use the figure to find the angle at which the ball hit the floor.

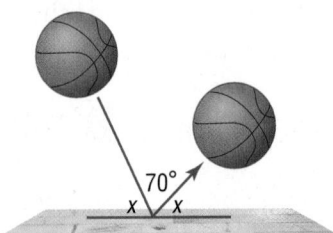

24. CHALLENGE Angles *E* and *F* are complementary. If $m\angle E = x - 10$ and $m\angle F = x + 2$, find the measure of each angle.

25. ✍️ WRITE MATH How are vertical, adjacent, complementary, and supplementary angles related?

26. Which angle pairs are NOT supplementary?

A.

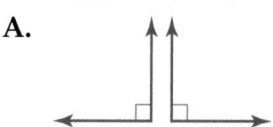

B.

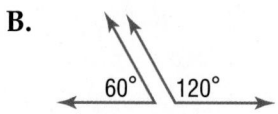

60° 120°

C.

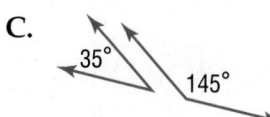

35°

145°

D.

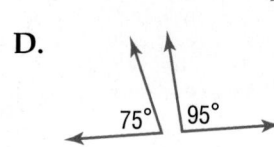

75° 95°

27. ▰▰▱ **GRIDDED RESPONSE** The angle at which the light ray hits the water is equal to the angle at which the light ray is reflected from the water. What is the measure of the angle in degrees at which the light ray is reflected from the water?

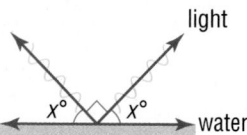

light

x°   x°   water

28. In the figure below, $m\angle YXZ = 35°$ and $m\angle WXV = 40°$. What is $m\angle ZXW$?

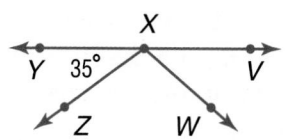

X

Y   35°   V

Z   W

F. 180°   **H.** 75°

G. 105°   **I.** 15°

More About   Angle Relationships

A **paragraph proof** consists of a paragraph explaining why a conjecture is true, the information that is given, and the statement to be proven.

 **EXAMPLE**

Refer to the figure at the right. Write a paragraph proof to show that $\angle 1 \cong \angle 3$.

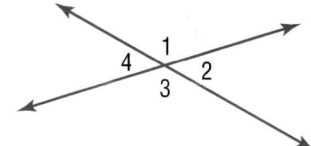

4 1
  2
3

GIVEN: two intersecting lines

PROVE: $\angle 1 \cong \angle 3$

PROOF: Since $\angle 1$ and $\angle 2$ form a straight line, $\angle 1$ is supplementary to $\angle 2$. Since $\angle 3$ and $\angle 2$ form a straight line, $\angle 3$ is supplementary to $\angle 2$. Since $\angle 1$ and $\angle 3$ are supplementary to the same third angle, they must have the same measure. Therefore, $\angle 1 \cong \angle 3$.

29. Refer to the figure at the right. $\angle 1$ is congruent to $\angle 2$, and $\angle 1$ and $\angle 2$ are supplementary. Write a paragraph proof to show that $\angle 1$ is a right angle and $\angle 2$ is a right angle.

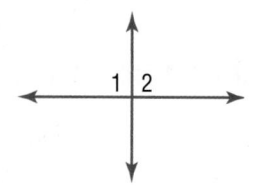

1 2

Problem-Solving Investigation

Main Idea Solve problems by using logical reasoning.

🏃 ✐ P.S.I. TERM ➕

Use Logical Reasoning

CHRIS: I know that an acute angle measures less than 90°.

YOUR MISSION: Given a right triangle, use logical reasoning to make a conjecture about the sum of the measures of the two acute angles of any right triangle.

Understand	Investigate the angle measures of right triangles to see whether there is a pattern.
Plan	Draw several right triangles, measure each angle, and look for a pattern.
Solve	

It appears that the sum of the measures of the acute angles of a right triangle is 90°. So, the acute angles are complementary.

Check	You can try several more examples to see whether your conjecture appears to be true. But at this point, it is just a conjecture, not an actual proof.

Analyze the Strategy

1. *Inductive reasoning* is the process of making a conjecture after observing several examples. Determine whether Chris used inductive reasoning. Explain.

2. 📝 **WRITE MATH** Write about a situation in which you use inductive reasoning.

Mixed Problem Solving

= **Step-by-Step Solutions** begin on page R1.
Extra Practice begins on page EP2.

- Use logical reasoning.
- Look for a pattern.
- Determine reasonable answers.
- Choose an operation.

Use *logical reasoning* to solve Exercises 3–5.

3. GEOMETRY Draw several rectangles and their diagonals. Measure the lengths of their diagonals. What seems to be true about the measures of the lengths of the diagonals of a rectangle?

4. JOBS Anne, Ichiro, Karina, and Amal all have summer jobs. One mows lawns, one lifeguards at the city pool, one delivers newspapers, and one babysits. From the clues below, list each person and their job.

- Amal does not wear a swimming suit for his job.
- Karina's wage depends on the number of children she watches.
- Ichiro lives next door to the person who delivers newspapers.
- Anne is an excellent swimmer.

5. NUMBER SENSE Write each fraction in the table as a decimal. Then use logical reasoning to write the decimal equivalents for the fractions $\frac{3}{11}$, $\frac{6}{11}$, and $\frac{9}{11}$.

Fraction	Decimal
$\frac{1}{11}$	
$\frac{4}{11}$	
$\frac{8}{11}$	

Use any strategy to solve Exercises 6–11.

6. GEOMETRY Right triangles are arranged according to the pattern below. If each triangle has an area of 12 square inches, find the area of the pattern formed by the fifth figure.

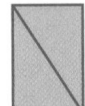 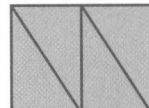 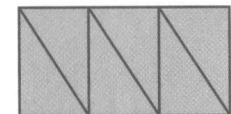

7. PHONES Paulo and Clarissa are each looking over their phone bill and at the number of hours they spent on the phone. Paulo said, "If I was on the phone twice as long, I would have been on the phone as long as you." Clarissa replied, "If I was on the phone twice as long, I would have been on the phone four times as long as you." How much time could each person have spent on the phone?

8. BALLET A group of ballet dancers are practicing a formation on stage. The first row has one dancer and each additional row has two more dancers than the previous row. If there are 25 dancers, how many rows can they form?

9. SHOPPING Lakisha needs $8\frac{1}{4}$ pounds of ground beef for a party. She can buy ground beef in 1-pound, $2\frac{1}{2}$-pound, or 3-pound packages. If Lakisha wants to spend the least amount of money, which packages should she buy and how many?

10. MEASUREMENT The circumference of Earth around the equator is 24,901.55 miles. The circumference through the North and South Poles is 24,859.82 miles. How much greater is the circumference of Earth around the equator than through the poles?

11. BIRDS The arctic tern has the longest migration of any bird. Each year, it flies over 21,750 miles. If the average lifespan of an arctic tern is 20 years, how many miles, on average, will it have flown in the course of its life?

Parallel Lines

Main Idea

Examine angle relationships formed by parallel lines and a transversal.

CCSS 8.G.5, 8.G.6

Certain pairs of lines have special relationships.

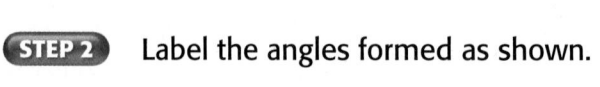

STEP 1 Draw two horizontal lines on notebook paper and a line that intersects both of those lines as shown.

STEP 2 Label the angles formed as shown.

STEP 3 Use a protractor and angle relationships you have previously learned to find the measure of each numbered angle and record it in a table.

Angle	Measure
1	
2	
3	

STEP 4 Color the angles that have the same measure.

Analyze the Results

1. What is the relationship between the two horizontal lines?

2. Congruent angles are angles that have the same measure. Describe the position of the congruent angles.

3. What do you notice about the measures of angles that are side by side?

MAKE A CONJECTURE If the measure of ∠1 in the figure at the right is 40°, determine the measure of each given angle without using a protractor. Then check your conjecture by measuring with a protractor.

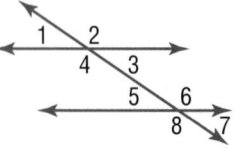

4. ∠2
5. ∠3
6. ∠4
7. ∠5
8. ∠6
9. ∠7
10. ∠8

Main Idea

Identify relationships of angles formed by two parallel lines cut by a transversal.

Vocabulary

perpendicular lines
parallel lines
transversal
interior angles
exterior angles
alternate interior angles
alternate exterior angles
corresponding angles

 8.G.5

Lines

RAILROADS In the United States, the standard distance between rails on a railroad track is 4 feet 8.5 inches. The diagram at the right shows a road crossing over railroad tracks.

1. Measure angles 1 and 2. Record the measures.

2. Make a conjecture about the measure of angle 3. Then measure the angle to verify your conjecture.

Perpendicular lines are lines that intersect at right angles.
Parallel lines are two lines in a plane that never intersect or cross.

A red right angle symbol indicates that lines *m* and *n* are perpendicular. →

Red arrowheads indicate that lines *p* and *q* are parallel.

$m \perp n$ $p \parallel q$

A line that intersects two or more lines is called a **transversal**. If the two lines are cut by a transversal, then these special angles are formed.

Key Concept   Transversals and Angles

When a transversal intersects two parallel lines, eight angles are formed.

Interior angles lie inside the parallel lines. $\angle 3, \angle 4, \angle 5, \angle 6$

Exterior angles lie outside the parallel lines. $\angle 1, \angle 2, \angle 7, \angle 8$

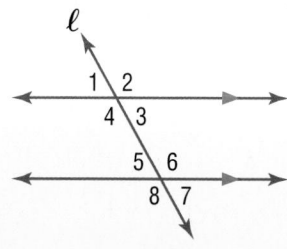

Alternate interior angles are interior angles that lie on opposite sides of the transversal. **Examples:** $\angle 4 \cong \angle 6, \angle 3 \cong \angle 5$

Alternate exterior angles are exterior angles that lie on opposite sides of the transversal. **Examples:** $\angle 1 \cong \angle 7, \angle 2 \cong \angle 8$

Corresponding angles are those angles that are in the same position on the two lines in relation to the transversal.
Examples: $\angle 1 \cong \angle 5, \angle 2 \cong \angle 6, \angle 4 \cong \angle 8, \angle 3 \cong \angle 7$

 **Read Math**

Parallel and Perpendicular Lines
Read $m \perp n$ as *m is perpendicular to n*. Read $p \parallel q$ as *p is parallel to q*.

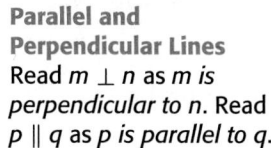

Classify Relationships

Use the figure to classify each pair of angles as *alternate interior*, *alternate exterior*, or *corresponding*.

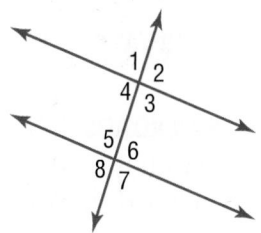

1 **∠1 and ∠7**

∠1 and ∠7 are exterior angles that lie on opposite sides of the transversal. They are alternate exterior angles.

2 **∠2 and ∠6**

∠2 and ∠6 are in the same position on the two lines. They are corresponding angles.

CHECK Your Progress

a. Refer to the figure above. Classify the relationship between ∠4 and ∠6. Explain.

REAL-WORLD EXAMPLE **Find an Angle Measure**

3 **BOOKCASES** A furniture designer built the bookcase shown. Line *a* is parallel to line *b*. If $m\angle 2 = 105°$, find $m\angle 6$ and $m\angle 3$. Justify your answer.

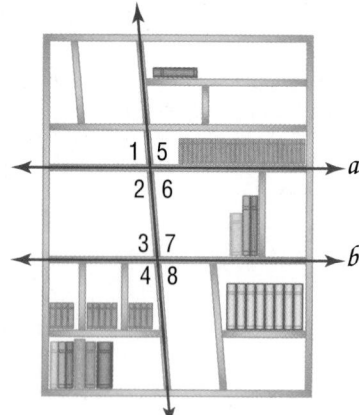

Since ∠2 and ∠6 are supplementary, the sum of their measures is 180°.
$m\angle 6 = 180° - 105°$ or 75°

Since ∠6 and ∠3 are interior angles that lie on opposite sides of the transversal, they are alternate interior angles. Alternate interior angles are congruent.
$m\angle 3 = 75°$

CHECK Your Progress

b. Refer to the situation above. Find $m\angle 4$. Justify your answer.

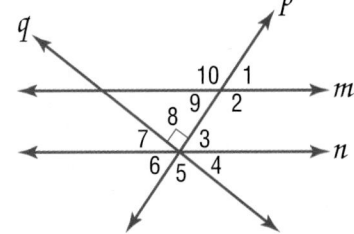

✎ EXAMPLE Find the Missing Measures

4 In the figure, line *m* is parallel to line *n*, and line *q* is perpendicular to line *p*. The measure of ∠1 is 40°. What is the measure of ∠7?

Since ∠1 and ∠6 are alternate exterior angles, $m\angle 6 = 40°$.

Since ∠6, ∠7, and ∠8 form a straight line, the sum of their measures is 180°.

$$40 + 90 + m\angle 7 = 180$$

So, $m\angle 7$ is 50°.

✓ CHECK Your Progress

c. Refer to the figure above. Find the measure of ∠4.

✓ CHECK Your Understanding

Examples 1 and 2 Classify each pair of angles as *alternate interior, alternate exterior,* or *corresponding.*

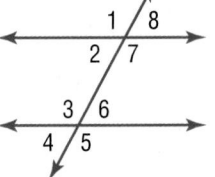

   1. ∠4 and ∠8   **2.** ∠5 and ∠7

   3. ∠3 and ∠7   **4.** ∠6 and ∠8

Example 3 **STAIRS** Refer to the porch stairs shown. Line *m* is parallel to line *n* and *m*∠7 is 35°. Find each given angle measure. Justify your answer.

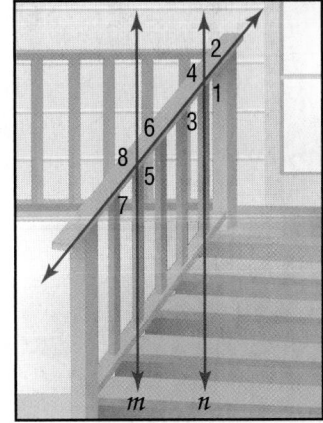

   5 *m*∠1

   6. *m*∠2

   7. *m*∠3

   8. *m*∠6

Example 4 Refer to the figure at the right. Line *a* is parallel to line *b* and *m*∠2 is 135°. Find each given angle measure. Justify your answer.

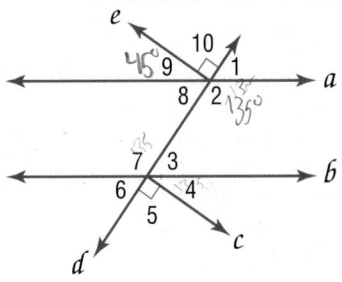

   9. *m*∠9

   10. *m*∠7

   11. *m*∠3

   12. *m*∠4

Practice and Problem Solving

= Step-by-Step Solutions begin on page R1.
Extra Practice begins on page EP2.

Examples 1 and 2

Classify each pair of angles as *alternate interior, alternate exterior,* or *corresponding.*

13. ∠2 and ∠4   **14.** ∠3 and ∠6

15. ∠1 and ∠3   **16.** ∠2 and ∠7

17. ∠1 and ∠8   **18.** ∠4 and ∠5

Example 3

19. **ART** In the quilt design on the barn below, line *a* is parallel to line *b*. If $m\angle 1 = 120°$, find $m\angle 2$ and $m\angle 3$. Justify your answers.

20. **FLAGS** In the flag below, line *a* is parallel to line *b*. If $m\angle 1 = 150°$, find $m\angle 4$ and $m\angle 7$. Justify your answers.

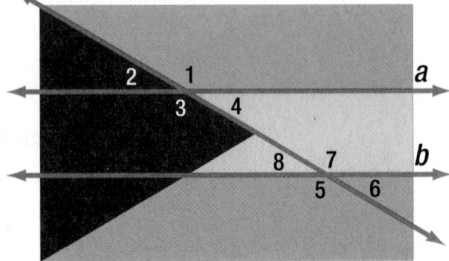

Example 4

Refer to the figure at the right. Line *s* is parallel to line *t*, $m\angle 2$ is 110° and $m\angle 11$ is 137°. Find each given angle measure. Justify your answer.

21. $m\angle 7$ 70°   **22.** $m\angle 6$ 110°

23. $m\angle 8$ 110°   **24.** $m\angle 13$ 110°

25. $m\angle 3$ **26.** $m\angle 4$

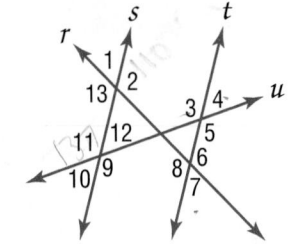

27. **CITY PLANNING** Refer to the street map of Washington, D.C. If K Street and Constitution Avenue are parallel and $m\angle 1 = 22°$, classify the relationship between ∠1 and ∠2. Then find $m\angle 2$. Explain your reasoning.

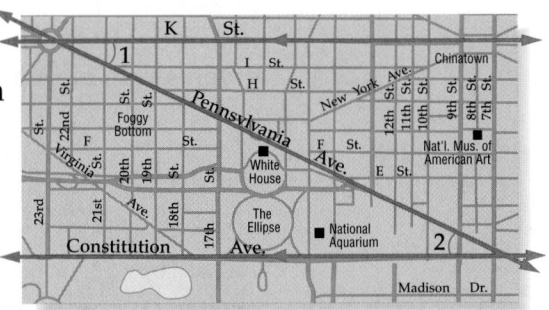

28. **ARCHITECTURE** The Leaning Tower of Pisa located in the town of Pisa, Italy, is one of the most famous architectural wonders in the world. Refer to the image at the left. If $m\angle 1 = 84.5°$, classify the relationship between ∠1 and ∠3. Then find $m\angle 2$. Explain your reasoning.

Real-World Link · · · ·

Since construction began on the Leaning Tower of Pisa in 1173, engineers have tried to stop the progression of its lean.

29. **ALGEBRA** The parallel lines at the right are cut by a transversal. Find the value of *x*.

a. Angles 1 and 2 are corresponding angles, $m\angle 1 = 45°$, and $m\angle 2 = (x + 25)°$.

b. Angles 3 and 4 are alternate interior angles, $m\angle 3 = 2x°$, and $m\angle 4 = 80°$.

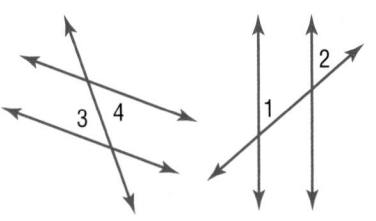

30. Use the figure at the right. List all pairs of each type of angle below.

a. vertical

b. complementary

c. supplementary

d. corresponding

e. alternate interior

f. alternate exterior

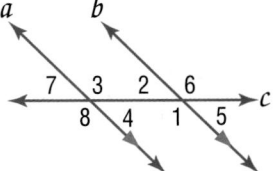

31. RESEARCH Use the Internet to find a map of your city. Identify any vertical, complementary, supplementary, corresponding, and right angles that are formed by the roads.

32. GEOMETRY Describe a method you could use to find the value of x in the figure at the right without using a protractor.

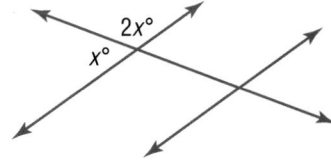

33 GRAPHIC NOVEL Refer to the graphic novel frame below for Exercises a–b.

a. Describe a method you could use to find the missing angle.

b. Use your method from part **a** to find the measure of the missing angle.

H.O.T. Problems

34. OPEN ENDED Draw a pair of parallel lines cut by a transversal. Estimate the measure of one angle and label it. Without using a protractor, label all the other angles with their approximate measure.

35. CHALLENGE In the figure at the right, quadrilateral *ABCD* is a parallelogram. Side *CD* has been extended to include point *E*. Make a conjecture about the relationship of ∠*DAB* and ∠*ADC*. Justify your reasoning.

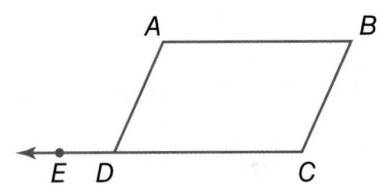

36. ✏️ **WRITE MATH** If two parallel lines are cut by a transversal, what relationship exists between interior angles that are on the same side of the transversal?

37. Lines *a* and *b* are parallel in the figure below. Find the value of *x*.

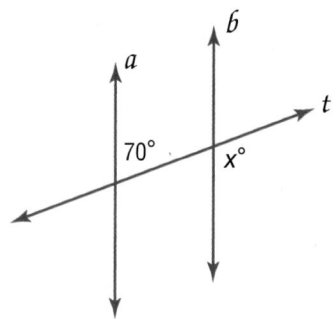

    A. 70   **C.** 100

    B. 80   **D.** 110

38. In the figure below, line *x* is parallel to line *y* and line *z* is perpendicular to $\overrightarrow{AB}$. The measure of ∠1 is 50°. What is the measure of ∠2?

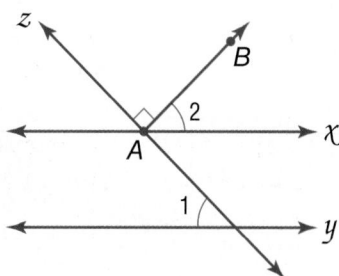

    F. 40°   **H.** 90°

    G. 50°   **I.** 130°

39. Which of the following statements is NOT true concerning ∠A, ∠B, and ∠C labeled on the glass pyramid structure at the Louvre in Paris, France?

    A. ∠B and ∠C are obtuse angles.

    B. ∠A and ∠C are vertical angles.

    C. ∠A and ∠B are alternate interior angles.

    D. ∠A and ∠C are congruent.

40. Which of the following is true about the angles of parallel lines cut by a transversal?

    F. Vertical angles are supplementary.

    G. Alternate exterior angles are supplementary.

    H. Alternate interior angles are complementary.

    I. Corresponding angles are congruent.

Spiral Review

41. Find the value of *x* in the figure at the right. (Lesson 1C)

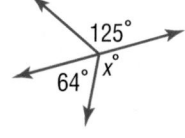

42. SALES A card store sells cards for $1.55 each. If you buy 3, you get 1 free. How many cards did Blanca get if she spent $18.60 before tax? Use the *logical reasoning* strategy. (Lesson 1D)

Refer to the diagram at the right. Identify each pair of angles as *adjacent*, *vertical*, or *neither*. (Lesson 1B)

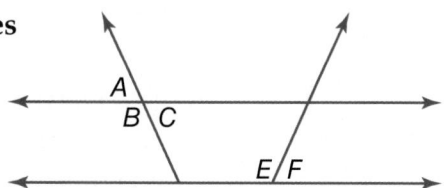

43. ∠A and ∠B   **44.** ∠A and ∠C

45. ∠C and ∠E   **46.** ∠E and ∠F

Name each angle in four ways. Then classify each angle as _acute, right, obtuse,_ or _straight._ (Lesson 1B)

1.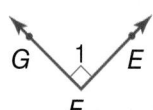

2.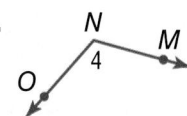

Find the value of x in each figure. (Lessons 1B and 1C)

3.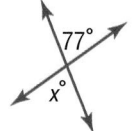

4.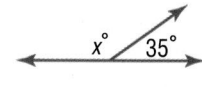

5.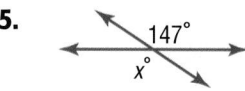

6.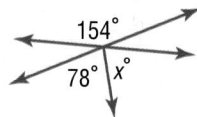

Refer to the figure below. (Lessons 1B and 1C)

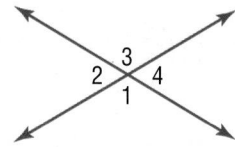

7. Identify a pair of vertical angles.

8. Identify a pair of adjacent angles.

9. Identify all pairs of supplementary angles.

10. Are there any pairs of complementary angles? Explain.

11. MULTIPLE CHOICE Which angle is complementary to $\angle CBD$? (Lesson 1C)

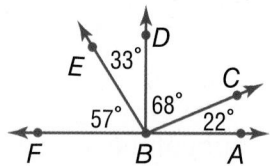

   A. $\angle ABC$ **C.** $\angle DBE$

   B. $\angle FBC$ **D.** $\angle EBF$

Classify each pair of angles as _complementary, supplementary,_ or _neither._ (Lesson 1C)

12.   **13.**

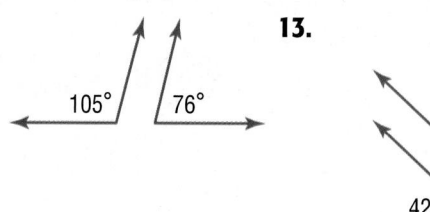

14. Consider the following pattern.

$$1^2 = 1$$
$$11^2 = 121$$
$$111^2 = 12,321$$

Use logical reasoning to find the next equation. Explain your reasoning. (Lesson 1D)

Refer to the figure below. Classify each pair of angles as _alternate interior, alternate exterior,_ or _corresponding._ (Lesson 2B)

15. $\angle 7$ and $\angle 1$

16. $\angle 2$ and $\angle 6$

17. $\angle 6$ and $\angle 4$

18. $\angle 2$ and $\angle 8$

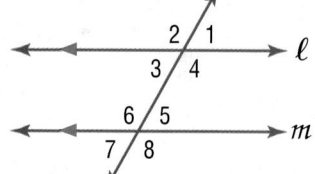

19. MULTIPLE CHOICE In the figure below, line a is parallel to line b.

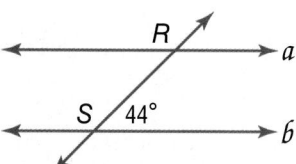

How is the relationship between $\angle R$ and $\angle S$ classified? (Lesson 2B)

   F. alternate interior angles

   G. alternate exterior angles

   H. corresponding angles

   I. vertical angles

Triangles

Main Idea

Explore the relationship among the angles of a triangle.

Vocabulary
proof

CCSS 8.G.5

Triangle means *three angles*. In this lab, you will explore how the three angles of a triangle are related.

ACTIVITY   Angles in a Triangle

① **STEP 1**   Draw a triangle similar to the one shown below on notebook or construction paper.

STEP 2   Label the corners 1, 2, and 3. Then tear off each corner.

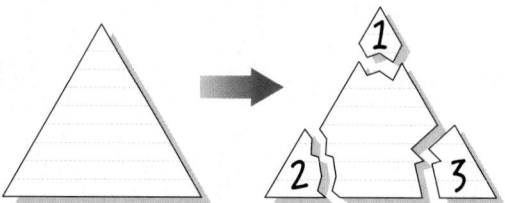

STEP 3   Rearrange the torn pieces so that the corners all meet at one point as shown.

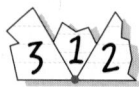

STEP 4   Repeat steps 1 and 2 with two differently shaped triangles.

Analyze the Results

1. What does each torn corner represent?

2. The point where these three corners meet is the vertex of another angle as shown. Classify this angle as *acute, right, obtuse,* or *straight*. Explain.

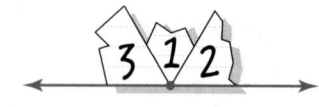

3. **MAKE A CONJECTURE** What is the sum of the measures of angles 1, 2, and 3 for each of your triangles? Verify your conjecture by measuring each angle using a protractor. Then find the sum of these measures for each triangle.

Activity 1 worked for specific triangles. How do we know it works for all triangles? This activity uses rules we already know to find a relationship that is true for any triangle. This is called a proof. A **proof** is a logical argument in which each statement that is made is supported by a statement that is accepted as true.

ACTIVITY Angles in a Triangle

2 **STEP 1** Draw a pair of parallel lines.

STEP 2 Draw a transversal as shown. Label ∠1 and ∠2.

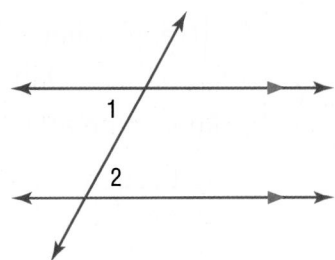

STEP 3 Draw a second transversal as shown. Label ∠3 and ∠4. Label the triangle formed by these lines *ABC.*

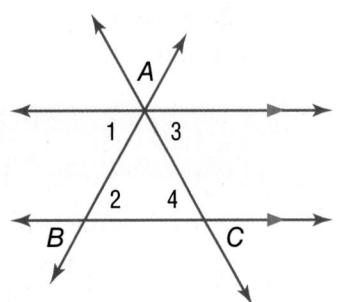

Analyze the Results

4. Classify the relationship between ∠1 and ∠2. What is true about this pair of angles?

5. Classify the relationship between ∠3 and ∠4. What is true about this pair of angles?

6. What type of angle is formed by ∠1, ∠3, and ∠*BAC*? What is the sum of the measures of ∠1, ∠3, and ∠*BAC*?

7. What can you conclude about the sum of the measures of the angles in △*ABC*? Explain your reasoning.

8. **MAKE A CONJECTURE** Based on this activity, what is the sum of the measures of the angles of any triangle?

Main Idea

Find missing angle measures in triangles.

Vocabulary

triangle
acute triangle
right triangle
obtuse triangle
scalene triangle
isosceles triangle
equilateral triangle

 Get ConnectED

Triangles

  **Explore**

Step 1 Draw the triangle shown at the right on dot paper. Then cut it out.

Step 2 Measure each angle of the triangle and label each angle with its measure.

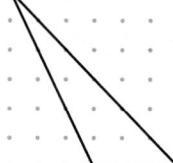

The triangle shown has two acute angles. Since the third angle is obtuse, the triangle is an *obtuse* triangle.

1. Repeat the activity with nine other triangles.

2. Sort your triangles into three groups based on the third angle measures. Name the groups *acute, right,* and *obtuse.*

A **triangle** is formed by three line segments that intersect only at their endpoints. A point where the segments intersect is a vertex.

 Key Concept   **Angles of a Triangle**

Words	The sum of the measures of the angles of a triangle is 180°.	Model

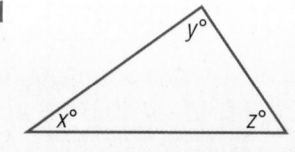

Symbols $x + y + z = 180°$

REAL-WORLD EXAMPLE   **Find Angle Measures**

 FLAGS Find the value of x in the Antigua and Barbuda flag.

$$x + 55 + 90 = \quad 180 \quad \text{Write the equation.}$$
$$x + 145 = \quad 180 \quad \text{Simplify.}$$
$$\underline{-145 = -145} \quad \text{Subtract.}$$
$$x = \quad 35 \quad \text{Simplify.}$$

The value of x is 35.

 CHECK Your Progress

a. In $\triangle XYZ$, if $m\angle X = 72°$ and $m\angle Y = 74°$, what is $m\angle Z$?

Segments $\overline{AB}$ is read as segment AB. So the sides of the triangle below are $\overline{AB}$, $\overline{AC}$, and $\overline{BC}$.

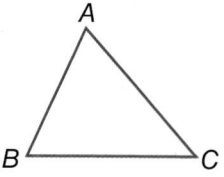

EXAMPLE Use Ratios to Find Angle Measures

2 The measures of the angles of $\triangle ABC$ are in the ratio 1:4:5. What are the measures of the angles?

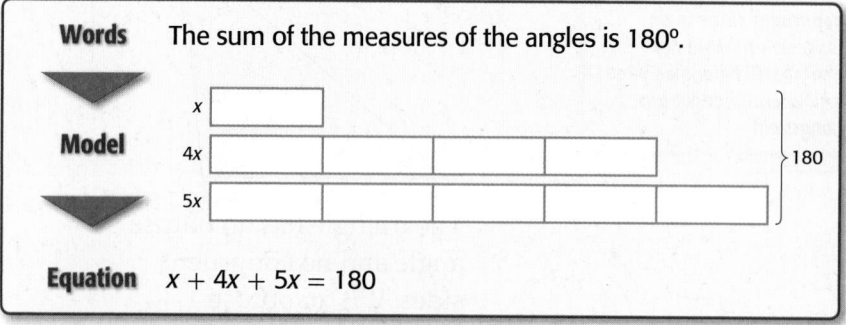

| Words | The sum of the measures of the angles is 180°. |

Model

x

4x

5x

180

Equation $x + 4x + 5x = 180$

$$x + 4x + 5x = 180 \quad \text{Write the equation.}$$
$$10x = 180 \quad \text{Combine like terms.}$$
$$x = 18 \quad \text{Simplify.}$$

Since $x = 18$, $4x = 4(18)$ or 72, and $5x = 5(18)$ or 90.
The measures of the angles are 18°, 72°, and 90°.

CHECK Your Progress

b. The measures of the angles of $\triangle LMN$ are in the ratio 2:4:6. What are the measures of the angles?

All triangles have at least two acute angles. Triangles can be classified by the angle measure of its third angle and by their sides. Congruent sides are sides that have the same length.

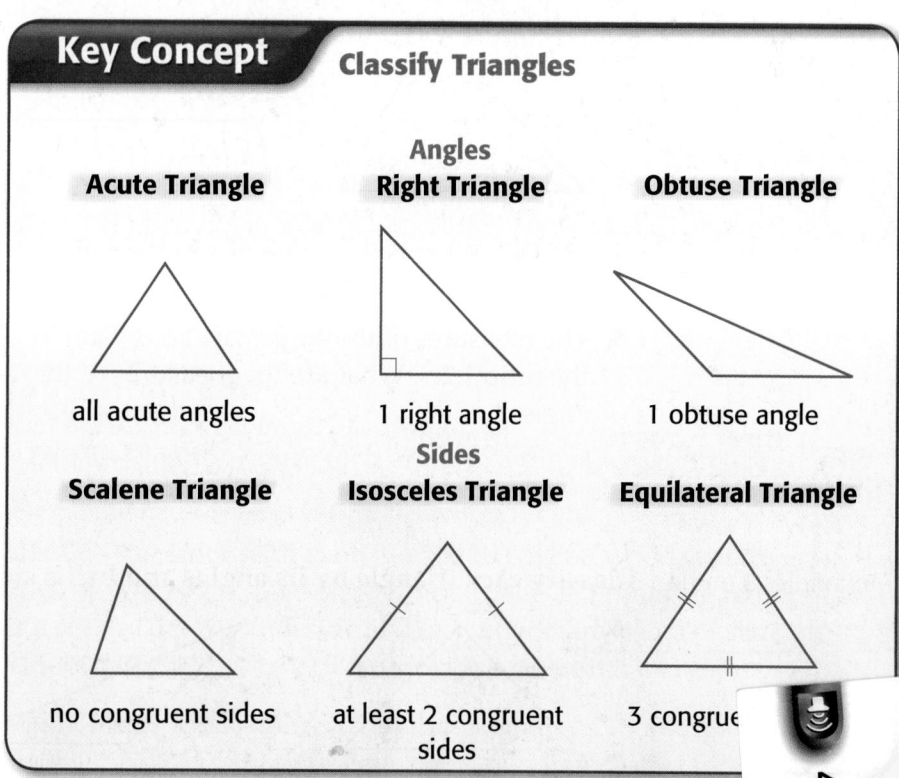

Key Concept Classify Triangles

Angles

Acute Triangle   **Right Triangle**   **Obtuse Triangle**

all acute angles   1 right angle   1 obtuse angle

Sides

Scalene Triangle   **Isosceles Triangle**   **Equilateral Triangle**

no congruent sides   at least 2 congruent   3 congrue
                          sides

 **EXAMPLES** Classify Triangles

Study Tip

Congruent Angles
The angles opposite the congruent sides in an isosceles triangle are congruent. All angles in an equilateral triangle are congruent.

Classify each triangle by its angles and by its sides.

③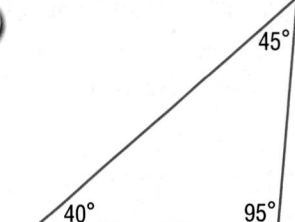

④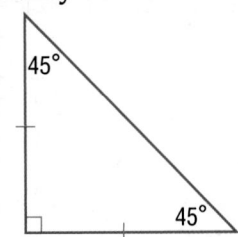

The triangle has an obtuse angle and no congruent sides. It is an obtuse scalene triangle.

The triangle has a right angle and two congruent sides. It is a right isosceles triangle.

 CHECK Your Progress

c.
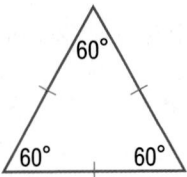

d.
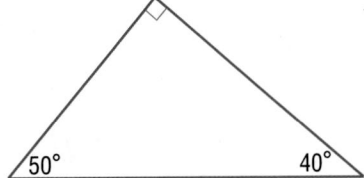

✓ **CHECK Your Understanding**

Example 1 | **Find the value of x in each triangle.**

①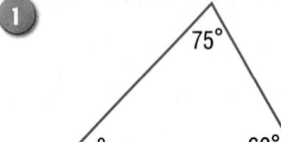

2.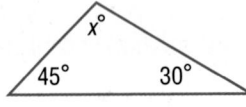

3.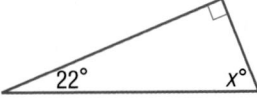

4. SAILING What is the value of x in the sail of the sailboat at the right?

Example 2 | **5.** The measures of the angles of △*LMN* are in the ratio 1:2:5. What are the measures of the angles?

Examples 3 and 4 | **Classify each triangle by its angles and by its sides.**

6.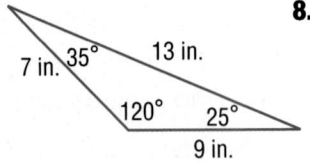

7.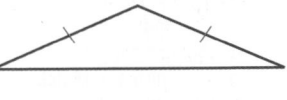

8.

Two- and Three-Dimensional Geometry

Practice and Problem Solving

● = **Step-by-Step Solutions** begin on page R1.
Extra Practice begins on page EP2.

Example 1 **Find the value of x in each triangle with the given angle measures.**

9.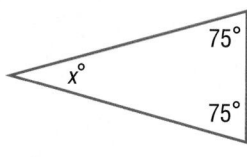

10.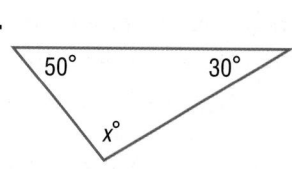

11.

12. $70°, 60°, x°$

13. $x°, 60°, 25°$

14. $x°, 35°, 25°$

15. **SKYSCRAPERS** The diagram below shows the view of the top of Fountain Place in Dallas. What is the value of x?

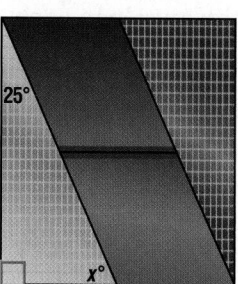

16. **PARKS** An A-frame picnic shelter at George Rogers Clark Historic Park in Ohio is shown below. What is the value of x?

Example 2

17. The measures of the angles of $\triangle RST$ are in the ratio 2:4:9. What are the measures of the angles?

18. The measures of the angles of $\triangle DEF$ are in the ratio 2:4:4. What are the measures of the angles?

19. The measures of the angles of $\triangle XYZ$ are in the ratio 3:3:6. What are the measures of the angles?

20. The measures of the angles of $\triangle XYZ$ are in the ratio 4:5:6. What are the measures of the angles?

Examples 3 and 4 **Classify each triangle with the given angle and side measures.**

21.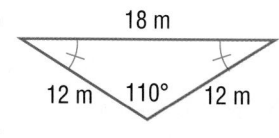

22.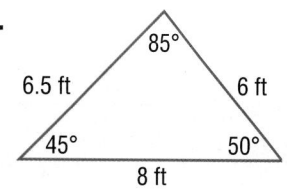

23.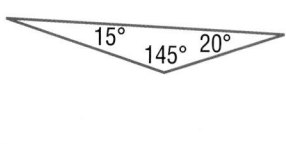

24. angles: 100°, 45°, 35°; sides: 9 in., 11 in., 13 in.

25. angles: 90°, 45°, 45°; sides: 8 cm, 8 cm, 9 cm

26. angles: 70°, 30°, 80°; sides: 5 m, 6 m, 7 m

27 What is the measure of the third angle of a triangle if one angle measures 25° and the second angle measures 50°?

28. What is the third angle measure of a right triangle if one of the angle measures is 32°?

29. ALGEBRA In △*ABC* the measure of angle *A* is 2*x* + 3, the measure of angle *B* is 4*x* + 2, and the measure of angle *C* is 2*x* − 1. What are the measures of the angles?

Determine whether each triangle described below can be drawn. If the triangle can be drawn, draw it. If *not*, explain why.

30. acute scalene   **31.** obtuse equilateral   **32.** right isosceles

ALGEBRA The measure of the sides of a triangle are given. Classify each triangle by its sides.

33. *m, m, m*   **34.** 3*n*, 2*n*, 3*n*   **35.** $2a, a, 2\frac{1}{2}a$

36. Apply what you know about angles and lines to find the values of *x* and *y* in the figure at the right.

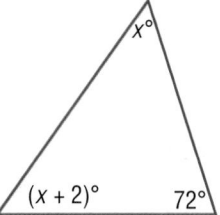

ALGEBRA Find the measures of the angles in each triangle.

37 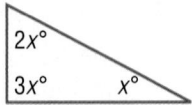   **38.** 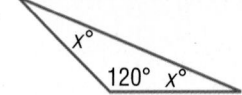   **39.**

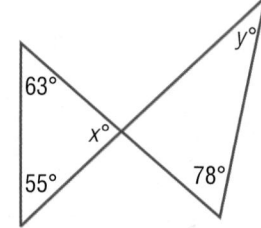

40. MATH IN THE MEDIA Find examples of triangles in a magazine or on the Internet. Classify each triangle by its angles and by its sides.

H.O.T. Problems

41. CHALLENGE Find the values of *x* and *y* in the figure at the right.

42. OPEN ENDED Draw an obtuse scalene triangle using a ruler and protractor. Label each side and angle with its measure.

43. FIND THE ERROR Alma is finding the measures of the angles in a triangle that have the ratio 1:3:5. Find her mistake and correct it.

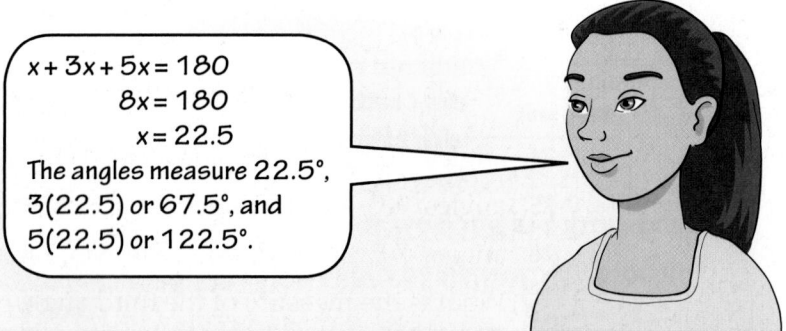

x + 3x + 5x = 180
8x = 180
x = 22.5
The angles measure 22.5°,
3(22.5) or 67.5°, and
5(22.5) or 122.5°.

44. ✏️ WRITE MATH Explain why a triangle must always have at least two acute angles. Include drawings in your explanation.

45. When viewed from the front, the base of an upright fan has a triangular face with the angle measures shown. What is the value of x?

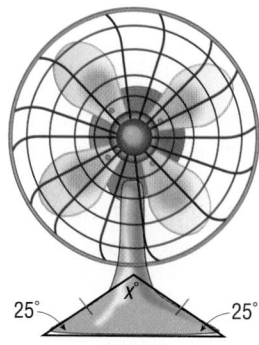

25°   x   25°

A. 40   C. 105

B. 100   D. 130

46. What is always true about the relationship between the measures of two acute angles of any right triangle?

F. They are equivalent.

G. They are complementary.

H. They are supplementary.

I. They are scalene.

47. ☐ **SHORT RESPONSE** Triangle ABC is isosceles. The measure of ∠B is 48° and the measures of ∠A and ∠C are equal. What is the measure of ∠A?

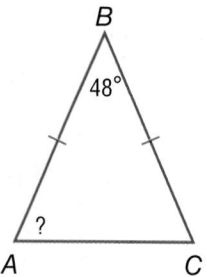

B
48°

?
A   C

48. A triangle has angles measuring 25° and 60°. What is the measure of the triangle's third angle?

A. 15°

B. 85°

C. 95°

D. 115°

49. CITY SERVICES The street maintenance vehicles for the city of Centerburg cannot safely make turns less than 70°. Should the proposed site of the new maintenance garage at the northeast corner of Park and Main be approved? Explain. (Lesson 2B)

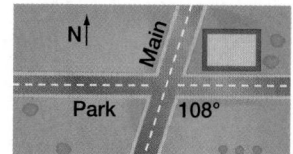

50. JUICE Nate has a large container of pineapple juice, an empty 5-pint container, and an empty 4-pint container. Explain how Nate can use only these containers to measure 2 pints of juice. (Lesson 1D)

51. ALGEBRA ∠A and ∠B are complementary, and the measure of ∠A is 39°. What is the measure of ∠B? (Lesson 1C)

Name each angle in four ways. Then classify the angle as *acute, right, obtuse,* **or** *straight.* (Lesson 1B)

52.

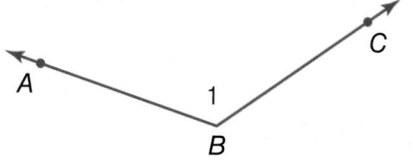

A   1
B

C

53.

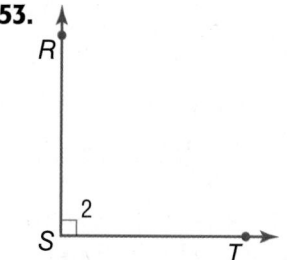

R

2
S   T

54.

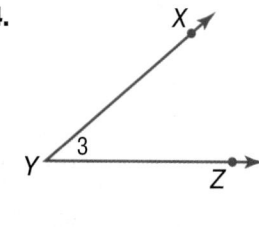

X

3
Y   Z

Explore Quadrilaterals

Main Idea

Investigate the properties of special quadrilaterals.

Get ConnectED

Four-sided figures are called *quadrilaterals*. In this lab, you will explore the properties of different types of quadrilaterals.

ACTIVITY

STEP 1   Draw the quadrilaterals shown on grid paper.

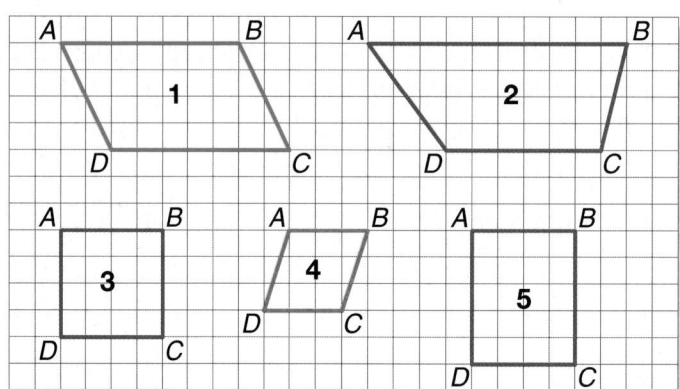

STEP 2   Use a ruler and a protractor to measure the sides and angles of each quadrilateral. Record your results in a table.

Analyze the Results

1. Describe any similarities or patterns in the angle measurements.

2. Describe any similarities or patterns in the side measurements.

3. Cut out the quadrilaterals you drew in the activity. Then sort them into categories according to their similarities and differences. Arrange and record your categories in a two-circled Venn diagram.

4. Create another Venn diagram illustrating a different way of categorizing these quadrilaterals.

5. **WRITE MATH** Did you find shapes that did not fit a category? Where did you place these shapes? Did any shapes have properties allowing them to belong to more than one category? Could you arrange these quadrilaterals into a three-circled Venn diagram? If so, how?

Main Idea

Identify and classify quadrilaterals.

Vocabulary

quadrilateral
trapezoid
parallelogram
rhombus

Quadrilaterals

VIDEO GAMES The general shape of a video game controller is shown.

1. Describe the angles inside the four-sided figure.

2. Which sides of the figure appear to be parallel?

3. Which sides appear to be congruent?

A **quadrilateral** is a closed figure with four sides and four angles. Quadrilaterals are named based on their sides and angles. The diagram shows how quadrilaterals are related. Notice how it goes from the most general to the most specific.

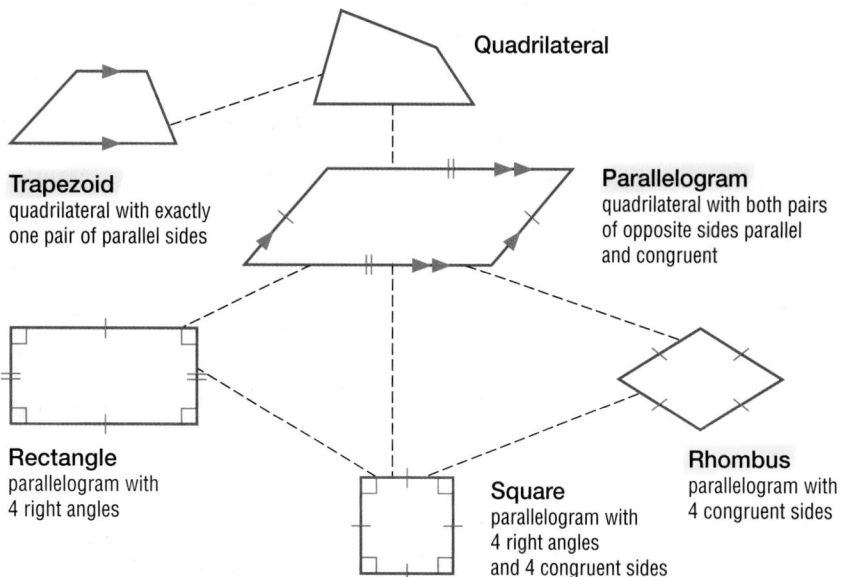

Quadrilateral

Trapezoid
quadrilateral with exactly one pair of parallel sides

Parallelogram
quadrilateral with both pairs of opposite sides parallel and congruent

Rectangle
parallelogram with 4 right angles

Square
parallelogram with 4 right angles and 4 congruent sides

Rhombus
parallelogram with 4 congruent sides

The name that *best* describes a quadrilateral is the one that is most specific.

- If a quadrilateral has all the properties of a parallelogram and a rhombus, then the *best* description of the quadrilateral is a rhombus.

- If a quadrilateral has all the properties of a parallelogram, rhombus, rectangle, and square, then the *best* description of the quadrilateral is a square.

✏ EXAMPLES Draw and Classify Quadrilaterals

Draw a quadrilateral that satisfies each set of conditions. Then classify each quadrilateral with the name that best describes it.

1 a parallellogram with four right angles and four congruent sides

Draw one right angle. The two segments should be congruent.

Draw a second right angle that shares one of the congruent segments. The third segment drawn should be congruent to the first two segments drawn.

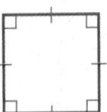

Connect the fourth side of the quadrilateral. All four angles should be right angles, and all four sides should be congruent.

The figure is a square.

Study Tip

Check for Reasonableness
Use a ruler and a protractor to measure the sides and angles to verify that your drawing satisfies the given conditions.

2 a quadrilateral with opposite sides parallel

Draw two parallel sides of equal length. Connect the endpoints of these two sides so that two new parallel sides are drawn.

The figure is a parallelogram.

✅ CHECK Your Progress

a. Draw a quadrilateral with exactly one pair of parallel sides. Then classify it.

All quadrilaterals can be separated into two triangles, A and B. Since the sum of the angle measures of each triangle is 180°, the sum of the angle measures of the quadrilateral is 2 · 180, or 360°.

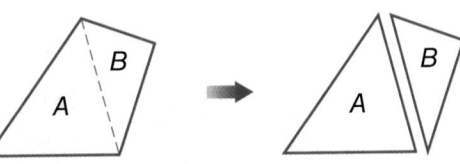

🧩 Key Concept   Angles of a Quadrilateral

Words	The sum of the measures of the angles of a quadrilateral is 360°.
Model	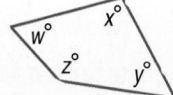
Algebra	$w + x + y + z = 360$

 **EXAMPLE** **Find a Missing Measure**

3 **ALGEBRA** Find the value of x in the quadrilateral shown.

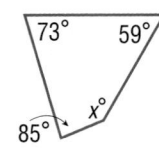

Write and solve an equation.

Words	The sum of the measures is 360°.
Variable	Let x represent the missing measure.
Equation	$85 + 73 + 59 + x = 360$

$$85 + 73 + 59 + x = 360 \quad \text{Write the equation.}$$

$$217 + x = 360 \quad \text{Simplify.}$$

$$\underline{-217 \qquad = -217} \quad \text{Subtraction Property of Equality}$$

$$x = 143$$

So, the value of x is 143.

Study Tip

Check for Reasonableness Since $\angle x$ is an obtuse angle, $m\angle x$ should be between 90° and 180°. Since $90° < 143° < 180°$, the answer is reasonable.

 CHECK Your Progress

b. ALGEBRA Find the value of x in the quadrilateral shown.

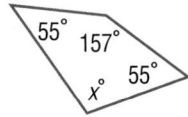

✓ **CHECK Your Understanding**

Examples 1 and 2 **Classify each quadrilateral with the name that best describes it.**

1.

2.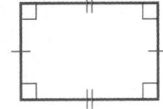

3.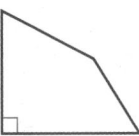

4. BOATS The photo shows a sailboat called a schooner. What type of quadrilateral does the indicated sail best represent?

Example 3 **5** **ALGEBRA** In quadrilateral $DEFG$, $m\angle D = 57°$, $m\angle E = 78°$, and $m\angle G = 105°$. What is $m\angle F$?

ALGEBRA **Find the missing angle measure in each quadrilateral.**

6.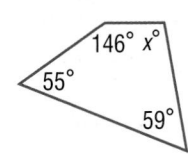

7.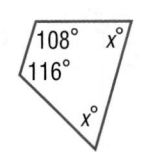

8.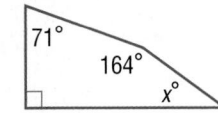

Practice and Problem Solving

 = **Step-by-Step Solutions** begin on page R1.
Extra Practice begins on page EP2.

Examples 1 and 2 Classify each quadrilateral with the name that best describes it.

9.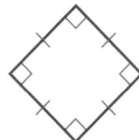

10.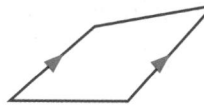

11.

12.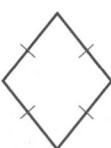

13.

14.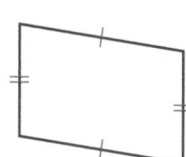

Example 3 **ALGEBRA** Find the missing angle measure in each quadrilateral.

15.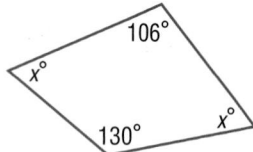

16.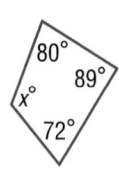

17.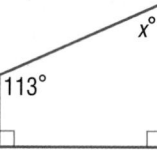

18.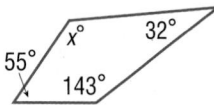

19.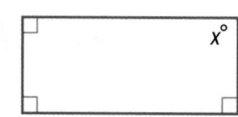

20.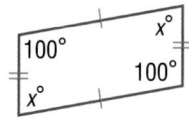

21. ALGEBRA Find $m\angle B$ in quadrilateral $ABCD$ if $m\angle A = 87°$, $m\angle C = 135°$, and $m\angle D = 22°$.

22. ALGEBRA What is $m\angle X$ in quadrilateral $WXYZ$ if $m\angle W = 45°$, $m\angle Y = 128°$, and $\angle Z$ is a right angle?

23. ART Identify the shapes of the tiles used in the stained glass window at the right. Use the name that best describes the tiles.

24. MEASUREMENT Find each of the missing angle measures a, b, c, and d in the figure at the right. Justify your answers.

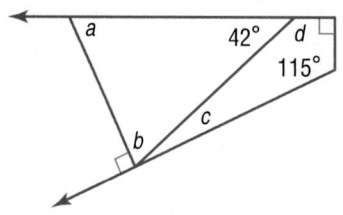

ALGEBRA Find the value of x in each quadrilateral.

25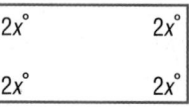

26.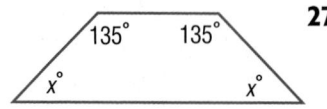

27.

28. GRAPHIC NOVEL Refer to the graphic novel frame below for Exercises a–b.

a. Classify the quadrilateral made by the boards for the ramp. Explain your reasoning.

b. Find the measures of the two missing angles using properties of quadrilaterals and parallel lines.

DRAWING QUADRILATERALS Determine whether each figure described below can be drawn. If the figure can be drawn, draw it. If not, explain why.

29 a trapezoid with three right angles

30. a trapezoid with two congruent sides

31. a quadrilateral that is both a rhombus and a rectangle

H.O.T. Problems

32. CHALLENGE The table gives the properties of several parallelograms. Property A states that both pairs of opposite sides are parallel and congruent.

Parallelogram	Properties
1	A, C
2	A, B, C
3	A, B

a. If property C states that all four sides are congruent, classify parallelograms 1–3. Justify your response.

b. If parallelogram 3 is a rectangle, describe Property B. Justify your response.

REASONING Determine whether each statement is *sometimes, always,* or *never* true. Explain your reasoning.

33. A quadrilateral is a trapezoid.

34. A trapezoid is a parallelogram.

35. A square is a rectangle.

36. A rhombus is a square.

37. ✍ **WRITE MATH** The diagonals of a rectangle are congruent, and the diagonals of a rhombus are perpendicular. Based on this information, what can you conclude about the diagonals of a square? of a parallelogram? Explain your reasoning.

38. Identify the name that does NOT describe the quadrilateral shown.

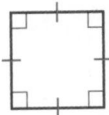

- **A.** square
- **B.** rectangle
- **C.** rhombus
- **D.** trapezoid

39. Which statement is always true about a rhombus?

- **F.** It has 4 right angles.
- **G.** The sum of the measures of the angles is 180°.
- **H.** It has exactly one pair of parallel sides.
- **I.** It has 4 congruent sides.

40. The figure represents a top view of a garden. What is the value of *x*?

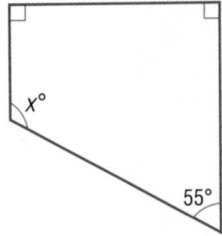

- **A.** 45
- **B.** 55
- **C.** 90
- **D.** 125

41. In the kite shown below, ∠*X* is congruent to ∠*Z*. What is the measure of ∠*WXY*?

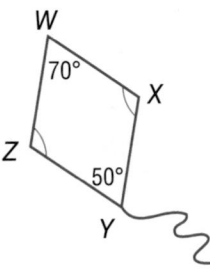

- **F.** 240°
- **G.** 130°
- **H.** 120°
- **I.** 110°

Spiral Review

Classify each triangle by its angles and by its sides. (Lesson 3B)

42.

43.

44.

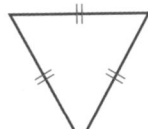

In the figure at the right, line *a* is parallel to line *b* and line *t* is a transversal. If *m*∠3 = 115°, find the measure of each angle. Justify your answers. (Lesson 2B)

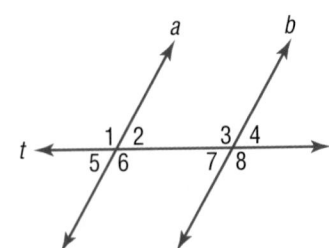

45. ∠6

46. ∠5

47. ∠1

48. ∠2

49. LUNCH James, Morgan, Kelley, and Nina are each eating a different lunch. One is eating tacos, one a pizza, one a sandwich, and one a bowl of soup. Kelley needs a spoon to eat her lunch. Morgan does not like tacos and is eating sliced bread with her lunch. Nina is sitting next to the person eating pizza. Who is eating tacos? Use the *logical reasoning* strategy. (Lesson 1D)

Main Idea
Find the sum of the angle measures of a polygon and the measure of an interior angle of a regular polygon.

 Vocabulary
polygon
interior angle
equiangular
regular polygon

 Get Connect ED

Polygons and Angles

 GEOGRAPHY Study the shapes of the different states below.

 Colorado  Missouri  Utah  New Mexico  Georgia

1. Sort the states into two different groups based on their shapes. Explain your reasoning.

2. Use a map of the United States to find other states that would fit into each group.

A **polygon** is a simple closed figure formed by three or more line segments. The segments intersect only at their endpoints.

Polygons			Not Polygons		

Polygons can be classified by the number of sides they have.

Polygon	pentagon	hexagon	heptagon	octagon	nonagon	decagon
Number of Sides	5	6	7	8	9	10

 EXAMPLE Classify Polygons

① **Determine whether the figure is a polygon. If it is, classify the polygon. If it is not a polygon, explain why.**

The figure has 7 sides that intersect at their endpoints only. It is a heptagon.

 CHECK Your Progress

a.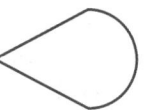

b.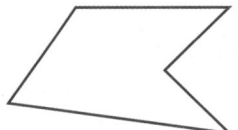

You can use the sum of the angle measures of a triangle to find the sum of the interior angle measures of various polygons. An **interior angle** is an angle that lies inside a polygon.

Number of Sides	Sketch of Figure	Number of Triangles	Sum of Angle Measures
3		1	$1(180°) = 180°$
4		2	$2(180°) = 360°$
5		3	$3(180°) = 540°$
6		4	$4(180°) = 720°$

Key Concept — Interior Angle Sum of a Polygon

Words   The sum of the measures of the interior angles of a polygon is $(n − 2)180$, where n represents the number of sides.

Symbols $S = (n − 2)180$

 EXAMPLE   Sum of Interior Angle Measures

2 Find the sum of the measures of the interior angles of a decagon.

$S = (n − 2)180$ Write an equation.

$S = (10 − 2)180$ A decagon has 10 sides. Replace n with 10.

$S = (8)180$ or $1{,}440$ Simplify.

The sum of the measures of the interior angles of a decagon is $1{,}440°$.

CHECK Your Progress

Find the sum of the interior angle measures of each polygon.

c. hexagon   **d.** octagon   **e.** 15-gon

A polygon that is equilateral (all sides congruent) and **equiangular** (all angles congruent) is called a **regular polygon**. Since all the angles of a regular polygon are congruent, their measures are equal.

3 **NATURE** Each chamber of a bee honeycomb is a regular hexagon. Find the measure of an interior angle of a regular hexagon.

Step 1 Find the sum of the measures of the angles.

$S = (n - 2)180$ Write an equation.

$S = (6 - 2)180$ Replace n with 6.

$S = (4)180$ or 720   Simplify.

The sum of the measures of the interior angles is 720°.

Step 2 Divide 720 by 6, the number of interior angles, to find the measure of one interior angle. So, the measure of one interior angle of a regular hexagon is 720° ÷ 6 or 120°.

 **CHECK Your Progress**

Find the measure of one interior angle in each regular polygon. Round to the nearest tenth if necessary.

f. octagon   **g.** heptagon   **h.** 20-gon

✓ CHECK Your Understanding

Example 1   Determine whether each figure is a polygon. If it is, classify the polygon. If it is not a polygon, explain why.

1. 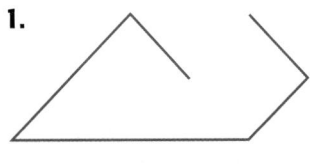   **2.**    **3.**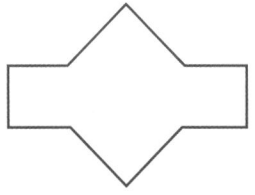

Example 2   Find the sum of the interior angle measures of each polygon.

     4. quadrilateral   **5** nonagon   **6.** 12-gon

Example 3   **7. QUILTING** The quilt pattern shown is made of repeating equilateral triangles. What is the measure of one interior angle of an equilateral triangle?

Practice and Problem Solving

 = **Step-by-Step Solutions** begin on page R1.
Extra Practice begins on page EP2.

Example 1 Determine whether each figure is a polygon. If it is, classify the polygon. If it is not a polygon, explain why.

8.

9.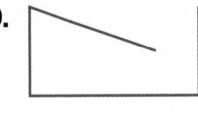

10.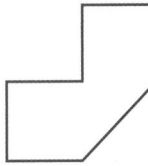

11.

12.

13.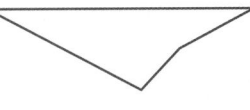

Example 2 Find the sum of the interior angle measures of each polygon.

14. pentagon   15. heptagon   16. 11-gon

17. 14-gon   18. 19-gon   19. 24-gon

Example 3 20. **ART** The sculpture at the right consists of repeating regular pentagons and hexagons. Find the measure of one interior angle of a pentagon.

Find the measure of one interior angle in each regular polygon. Round to the nearest tenth if necessary.

21. nonagon   22. decagon   ㉓ 13-gon   24. 16-gon

ART A tessellation is a repetitive pattern of polygons that fit together without overlapping and without gaps between them. For each tessellation, find the measure of each angle at the circled vertex. Then find the sum of the angles.

25.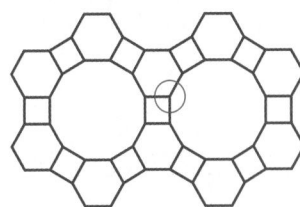

26.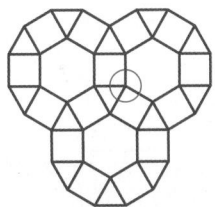

㉗ **ARCHITECTURE** The surface of the dome of Spaceship Earth in Orlando consists of repeating equilateral triangles as shown. Find the measure of each angle in each outlined triangle. Then make a conjecture about the interior angle measures in equilateral triangles of different sizes.

28. CHALLENGE How many sides does a regular polygon have if the measure of an interior angle is 160°? Justify your answer.

29. REASONING Draw three nonregular hexagons.

　　a. MAKE A CONJECTURE What is the sum of the interior angles of nonregular hexagons? Explain.

　　b. Use a protractor to find the measures of the interior angles of each figure. What is the sum of the interior angles?

30. ✏ **WRITE MATH** Explain the relationship between the number of sides of a regular polygon and the measure of each interior angle.

☑ Test Practice

31. A stained glass window is in the shape of a regular hexagon.

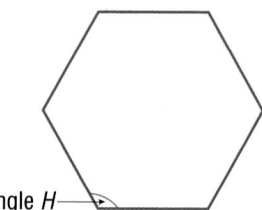

Angle H

What is the measure, in degrees, of ∠H in the window?

A. 1,080 　　　**C.** 180

B. 720 　　　　**D.** 120

32. ▦ **GRIDDED RESPONSE** What is the sum of the measures, in degrees, of the interior angles of an octagon?

33. After the first two folds of an origami paper design, the paper is shaped like a square with two isosceles triangles removed from two adjacent corners.

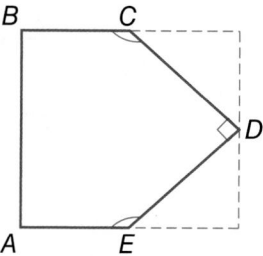

If angle AED is congruent to angle BCD, what is the measure of angle AED?

F. 45° 　　　**H.** 135°

G. 90° 　　　**I.** 160°

34. ALGEBRA In quadrilateral *ABCD*, *m*∠*A* = 65°, *m*∠*B* = 124°, and *m*∠*C* = 57°. What is *m*∠*D*? (Lesson 3D)

35. The measures of the sides of a triangle are 4 inches, 4 inches, and 7 inches. Classify the triangle by its sides. (Lesson 3B)

36. In the figure, line *m* is parallel to line *n*. Name four pairs of corresponding angles. (Lesson 2B)

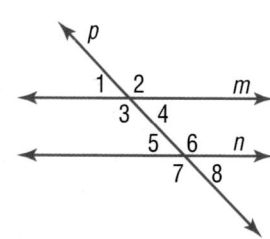

Intersection of Geometric Figures

Main Idea

Identify the intersection of two geometric figures in the plane.

When two lines intersect, there are three possibilities. The lines can intersect at zero points, one point, or infinitely many points. If the two lines intersect at infinitely many points, they are the same line.

Zero points	One point	Infinitely many points

ACTIVITY Intersection of a Line and a Circle

① **STEP 1** On a piece of paper, draw a circle.

STEP 2 Somewhere on the same paper, draw a line.

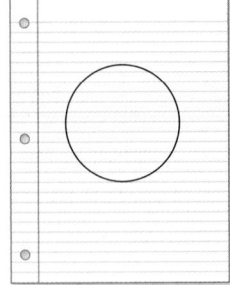

Analyze the Results

1. Describe the point(s) of intersection, if any, between the circle and the line.

2. Draw several arrangements of a line and a circle. What are the different possibilities for the intersection of a circle and a line?

As with other figures, two circles can intersect in several different ways.

Zero points	One point	Two points

2 **STEP 1** Fold a piece of paper into fourths by first folding it horizontally, then folding it vertically.

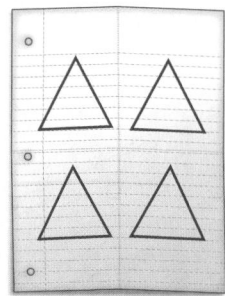

STEP 2 Open the paper and draw a triangle in each of the four sections as shown.

STEP 3 Draw a line in each of the four sections so that it intersects the triangle at zero points, one point, two points, and infinitely many points.

Analyze the Results

3. Describe the location of the line when it intersects the triangle at infinitely many points.

4. Compare the placements of the lines when the intersections result in one point and two points.

Practice and Apply

5. **GEOMETRY** Describe the possible points of intersection between a circle and a square. Justify your answer by drawing the possibilities.

6. **GEOMETRY** Draw three triangles that intersect at only one point.

7. **GEOMETRY** Draw two hexagons that intersect at two points.

8. **GEOMETRY** A line, a circle, and a rhombus intersect. Draw a sketch of what this might look like. At how many points do all three figures intersect?

ART **Refer to the art design shown.**

9. At how many points does the circle intersect the triangle?

10. At how many points do the two rectangles intersect each other?

11. At how many points does rectangle *A* intersect the triangle?

12. Name two figures that do not intersect each other.

13. Name three figures that all intersect at the same point.

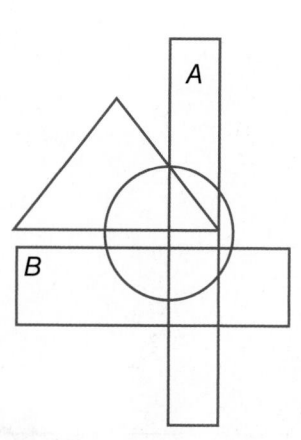

ANIMATION IS A BLAST!

Do you have creativity, a love of animation, and the motivation to work very hard? Then you should think about a career in animation. Animators create realistic movement in digital environments. They use the latest software and apply basic principles of animation to create characters that have both movement and weight. As an animator, you could also specialize in a particular area by becoming a story artist, a layout or background artist, a texture and lighting specialist, or a production designer.

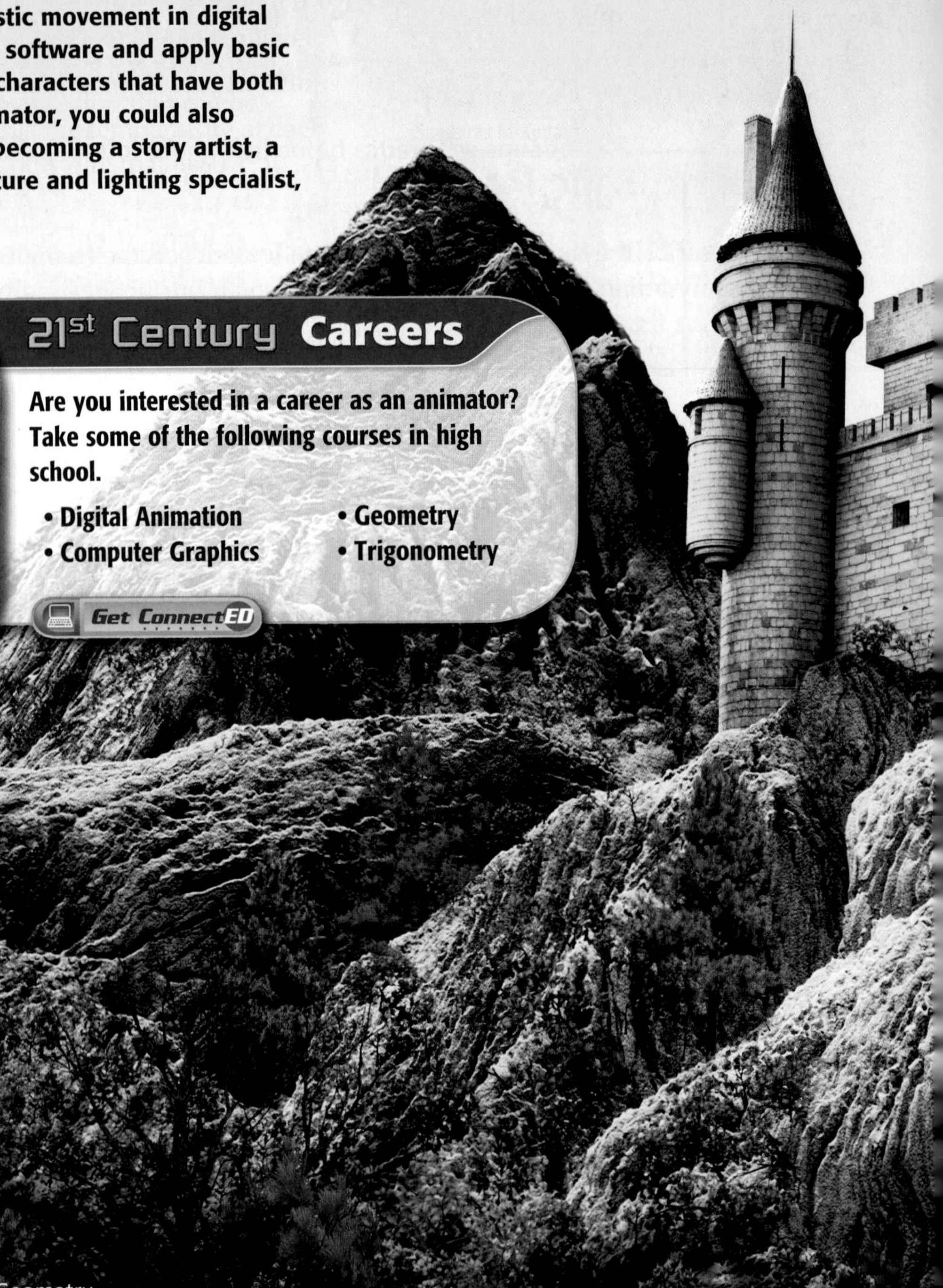

21st Century Careers

Are you interested in a career as an animator? Take some of the following courses in high school.

- Digital Animation
- Computer Graphics
- Geometry
- Trigonometry

Get ConnectED

Real-World Math

Use the figures above to solve each problem.

1. Classify triangle *ADE* by its angles and by its sides.

2. If *m∠ADE* is 65°, what is *m∠DAE*?

3. Classify quadrilateral *ABCD*.

4. What is *m∠DAB*?

5. Classify triangle *XYZ* by its angles and sides.

6. What is *m∠XZY*?

Explore

Three-Dimensional Figures

Main Idea

Draw three-dimensional figures.

Get ConnectED

A *three-dimensional figure* has length, width, and height. You can draw three-dimensional figures using different points of view, called perspectives. For example, the perspective of the photograph of the spiral staircase is a bottom view.

ACTIVITY

1. Use cubes to build the figure that has the top, side, and front views shown. Then draw your model on isometric dot paper.

top

side

front

STEP 1 Build the base using the top view. The base is a 2 by 3 rectangle.

STEP 2 Complete the figure using the side view. The first and second rows are 1 unit high. The third row is 2 units high.

STEP 3 Check the figure using the front view. The overall width is 2 units. The overall height is 2 units.

STEP 4 Draw your model on isometric dot paper. Label the front and the side of your figure.

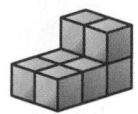

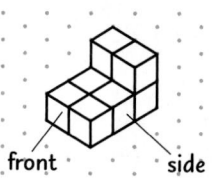
front   side

Analyze the Results

1. Which view, *top*, *side*, or *front*, shows that a three-dimensional figure has multiple heights?

2. Build your own figure using up to 20 cubes and draw it on isometric dot paper. Then draw the figure's top, side, and front views. Explain your reasoning.

Another way to draw three-dimensional figures made of stacked cubes is by using a *top-count view.* Imagine looking down on the cubes from above. The numbers in the squares identify the number of cubes in each stack.

ACTIVITY

2 The top-count view of a three-dimensional figure is shown. Use cubes to build the figure. Then draw the figure on isometric dot paper.

STEP 1 The greatest number on the top-count view is 3. Therefore, the height of the solid is 3 units, and it has three layers. Build the first layer.

STEP 2 Build the second and third layers, adding the appropriate number of cubes.

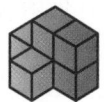 

STEP 3 Draw the figure on isometric dot paper. Label the front and the side of the figure.

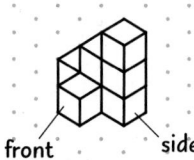

Analyze the Results

3. Which view, a top-side-front view or a top-count view, makes it easier to build a three-dimensional figure? Explain your reasoning.

Practice and Apply

The top, side, and front views of three-dimensional figures are shown. Use cubes to build each figure. Then draw your model on isometric dot paper, labeling its front and side.

4.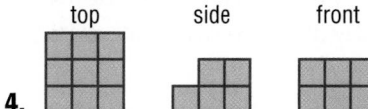

5.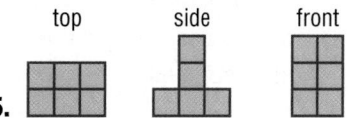

The top-count views of three-dimensional figures are shown. Use cubes to build each figure. Then draw each figure on isometric dot paper, labeling its front and side.

6.

7.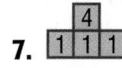

8.

9.

Main Idea

Identify and draw three-dimensional figures.

 Vocabulary

coplanar
parallel
solid
polyhedron
edge
face
vertex
diagonal
prism
base
pyramid
cylinder
cone
cross section

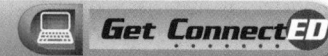

Properties of Three-Dimensional Figures

MONUMENTS A two-dimensional figure, like a rectangle, has two dimensions: length and width. A three-dimensional figure, like a building, has three dimensions: length, width, and height.

1. Name the two-dimensional shapes that make up the sides of the Washington Monument.

2. If you observed the building from directly above, what two-dimensional figure would you see?

3. How are two- and three-dimensional figures related?

The figure at the right shows rectangle *ABCD*. Lines *AB* and *DC* are **coplanar** because they lie in the same plane. They are also **parallel** because they will never intersect, no matter how far they are extended.

Just as two lines in a plane can intersect or be parallel, there are different ways that planes may be related in space.

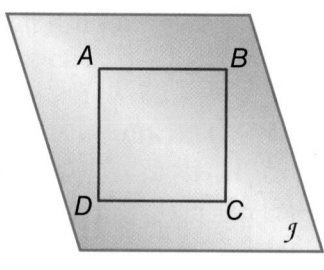

Intersect in a Line   Intersect at a Point   No Intersection

These are called *parallel planes.*

Intersecting planes can also form three-dimensional figures or **solids**. A **polyhedron** is a solid with flat surfaces that are polygons. Some terms associated with three-dimensional figures are *edge*, *face*, *vertex*, and *diagonal*.

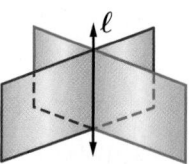
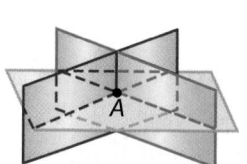
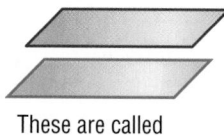

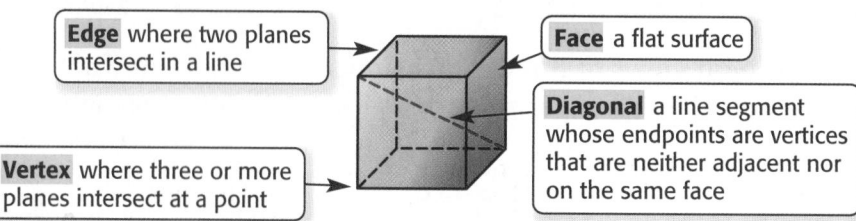

Edge where two planes intersect in a line

Face a flat surface

Vertex where three or more planes intersect at a point

Diagonal a line segment whose endpoints are vertices that are neither adjacent nor on the same face

QUICK Review

Polygons

Sides	Name
5	pentagon
6	hexagon
7	heptagon
8	octagon
9	nonagon
10	decagon

A **prism** is a polyhedron with two parallel, congruent faces called **bases**. A **pyramid** is a polyhedron with one base that is a polygon and faces that are triangles. Prisms and pyramids are named by the shape of their bases.

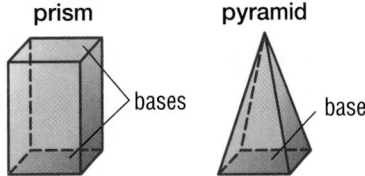

There are also solids that are not polyhedrons. A **cylinder** is a three-dimensional figure with congruent, parallel bases that are circles connected with a curved side. A **cone** has one circular base and a vertex connected by a curved side.

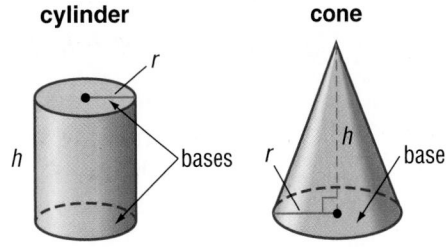

Study Tip

Common Error In the drawing of a rectangular prism, the bases do not have to be on the top and bottom. Any two parallel rectangles are bases. In a triangular pyramid, any face is a base.

 EXAMPLES **Identify Solids**

Identify the figure. Then name the bases, faces, edges, and vertices.

1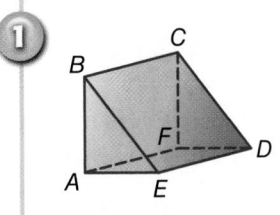

The figure has two parallel congruent bases that are triangles, so it is a triangular prism.
bases ABE, FCD
faces $ABE, FCD, BCDE, FAED, ABCF$
edges $\overline{AB}, \overline{BE}, \overline{EA}, \overline{FC}, \overline{CD}, \overline{DF}, \overline{BC}, \overline{ED}, \overline{AF}$
vertices A, B, C, D, E, F

2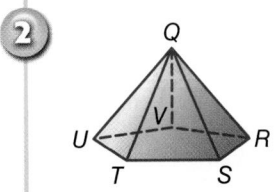

The figure has one base that is a pentagon, so it is a pentagonal pyramid.
base $RSTUV$
faces $RSTUV, QVR, QRS, QST, QTU, QUV$
edges $\overline{QR}, \overline{QS}, \overline{QT}, \overline{QU}, \overline{QV}, \overline{VR}, \overline{RS}, \overline{ST}, \overline{TU}, \overline{UV}$
vertices Q, R, S, T, U, V

 CHECK Your Progress

a. 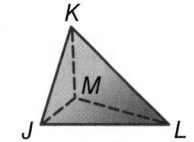   **b.** 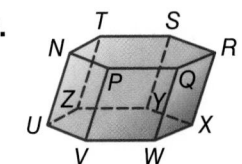   **c.**

You can use three-dimensional drawings of objects to describe how different parts of the objects are related in space.

Analyze Drawings

3 FURNITURE The photo shows a garden bench. Draw and label the top, front, and side views of the bench.

Top   Front   Side

CHECK Your Progress

d. TOOLBOX Draw and label the top, front, and side views of the toolbox shown.

A well-landscaped lawn and garden can increase the value of a home up to 15%.

The intersection of a solid and a plane is called a **cross section** of the solid.

Identify Cross Sections

4 Describe the shape resulting from a vertical, angled, and horizontal cross section of a cylinder.

Vertical Slice   Angled Slice   Horizontal Slice

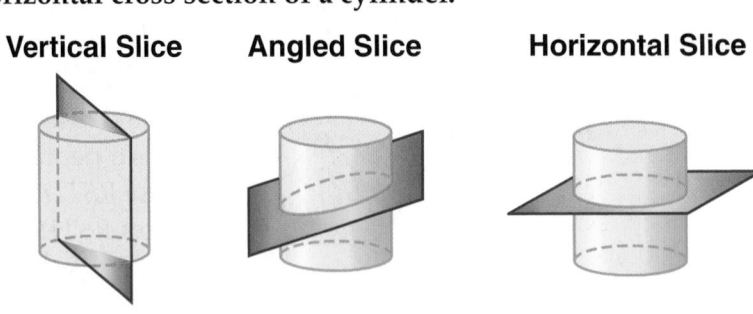

The cross section is a rectangle.

The cross section is an oval.

The cross section is a circle.

CHECK Your Progress

e. Describe the shape resulting from a vertical, angled, and horizontal cross section of a square pyramid.

Examples 1 and 2 **Identify each figure. Then name the bases, faces, edges, and vertices.**

1.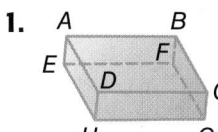

2.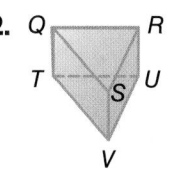

3.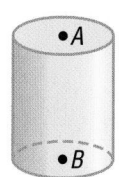

Example 3 4. **AQUARIUMS** Draw and label the top, front, and side views of the aquarium shown.

Example 4 **Describe the shape resulting from each cross section.**

5.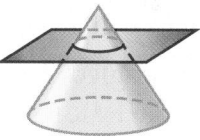

6.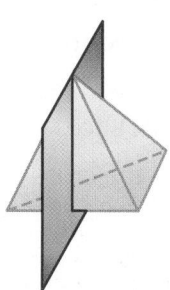

7.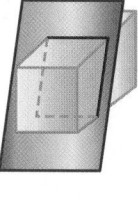

Practice and Problem Solving

● = Step-by-Step Solutions begin on page R1.
Extra Practice begins on page EP2.

Examples 1 and 2 **Identify each figure. Then name the bases, faces, edges and vertices.**

8.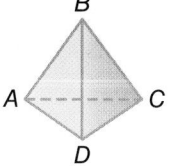

9.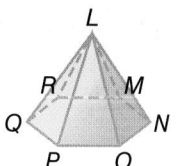

10.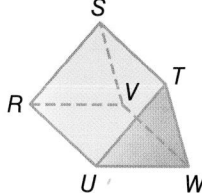

Example 3 11. **BUILDINGS** Draw and label the top, front, and side views of the building shown.

12. **TENT** Draw and label the top, front, and side views of the tent.

Example 4 Describe the shape resulting from each cross section.

13. 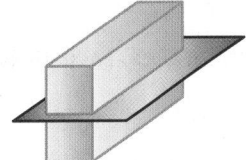 **14.** 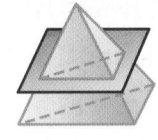 **15.**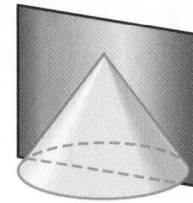

16. 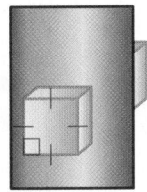 **17** 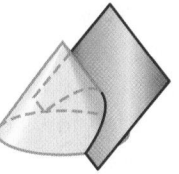 **18.**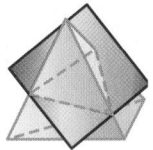

19. State whether the following conjecture is *true* or *false*. If *false*, provide a counterexample.

Two planes in three-dimensional space can intersect at one point.

20. SPORTS A standard basketball is shaped like a *sphere*.

   a. Draw a basketball with a vertical, angled, and horizontal slice.

   b. Describe the cross section made by each slice.

H.O.T. Problems

21. OPEN ENDED Draw the cross sections of a polyhedron, cylinder, or cone. Exchange papers with another student. Identify the three-dimensional figures represented by the cross sections.

22. FIND THE ERROR Brian is identifying the figure below. Find his mistake and correct it.

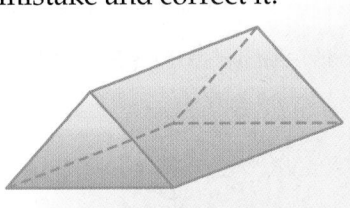

The figure has a triangular base. It is a triangular pyramid.

CHALLENGE Determine whether each statement is *always*, *sometimes*, or *never* true. Explain your reasoning.

23. A prism has 2 bases and 4 faces.   **24.** A pyramid has parallel faces.

25. WRITE MATH Explain whether a top-front-side view diagram *always* provides enough information to draw a figure. If not, provide a counterexample.

26. Benita received the gift box shown.

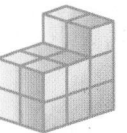

Which drawing **best** represents the top view of the gift box?

A.

B.

C.

D.

27. Which of the following is NOT an example of a polyhedron?

    F. cylinder

    G. rectangular prism

    H. octagonal pyramid

    I. triangular prism

28. Which of the following represents a side view of the figure below?

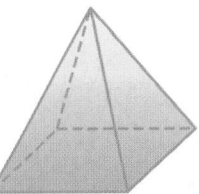

A.   **C.**

B.   **D.**

29. The figure below is a square pyramid.

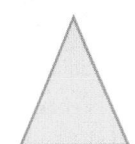

Which of the following is NOT a cross section from the square pyramid?

F.   **H.**

G.   **I.**

 **Spiral Review**

Find the measure of one interior angle in each regular polygon. Round to the nearest tenth if necessary. (Lesson 3E)

30. triangle   **31.** pentagon   **32.** heptagon   **33.** nonagon

Classify each quadrilateral with the name that best describes it. (Lesson 3D)

34. 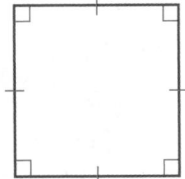   **35.**    **36.**

Polyhedrons

Main Idea

Identify the relationship between the faces, vertices, and edges of a polyhedron.

Get ConnectED

Leonhard Euler (1707–1783) was a Swiss mathematician and physicist. He made many discoveries in mathematics, including a relationship among the faces, vertices, and edges of any polyhedra.

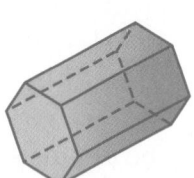   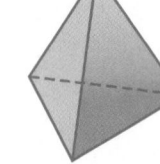   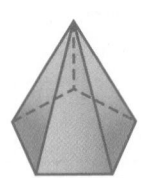   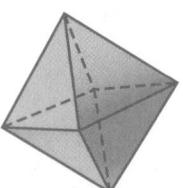

hexagonal prism   **tetrahedron**   **pentagonal pyramid**   **octahedron**

ACTIVITY   Faces, Vertices, and Edges

Refer to Lesson 4B and the figures shown above. Complete the table below that gives the number of faces, vertices, and edges for some common polyhedra.

Polyhedron	Number of Faces, f	Number of Vertices, v	Number of Edges, e	Sum of Faces and Vertices, $f + v$
Cube				
Triangular Prism				
Hexagonal Prism				
Tetrahedron				
Square Pyramid				
Pentagonal Pyramid				
Octahedron				

Analyze the Results

1. Describe the relationship between the number of edges e and the sum of the number faces and vertices $f + v$.

2. Write an algebraic rule that represents the relationship you found in the Activity between the number of edges of a polyhedron and the sum of the number of its faces and vertices.

3. Using your rule, find the number of edges of a polyhedron that has 11 faces and 8 vertices.

4. **GEOMETRY** A soccer ball is a polyhedron that has 12 pentagonal faces, 20 hexagonal faces, and 90 edges. How many vertices are in a soccer ball?

Chapter Study Guide and Review

Be sure the following Key Concepts are noted in your Foldable.

Angle Measure

Key Concepts

Angle Measure (Lesson 1)

- Two angles are adjacent if they have the same vertex, share a common side, and do not overlap.
- Two angles are vertical if they are opposite angles formed by the intersection of two lines.
- Two angles are complementary if the sum of their measures is 90°.
- Two angles are supplementary if the sum of their measures is 180°.

Lines (Lesson 2)

- Lines that intersect at right angles are perpendicular.
- Two lines in the same plane that never intersect are parallel.
- When two parallel lines are cut by a transversal, congruent angle pairs are formed: alternate interior angles, alternate exterior angles, and corresponding angles.

Angle Relationships in Polygons (Lesson 3)

- The sum of the angle measures in a triangle is 180°.
- The sum of the angle measures in a quadrilateral is 360°.
- The sum of the measures of the interior angles of a polygon is $(n - 2)180$.

Key Vocabulary

acute angle	perpendicular lines
adjacent angles	polygon
alternate exterior angles	polyhedron
alternate interior angles	prism
complementary angles	pyramid
corresponding angles	regular polygon
cross section	right angle
edge	solid
exterior angles	straight angle
face	supplementary angles
interior angle	transversal
interior angles	vertex
obtuse angle	vertical angles
parallel lines	

Vocabulary Check

State whether each statement is *true* or *false*. If *false*, replace the underlined word or number to make a true sentence.

1. $m\angle 1$ is read as the <u>measure</u> of angle 1.

2. A polygon with angles that are all congruent is said to be <u>equilateral</u>.

3. Two angles with measures adding to 180° are called <u>complementary</u> angles.

4. A <u>hexagon</u> is a polygon with 6 sides.

5. An angle with a measure less than 90° is called a <u>right angle</u>.

6. The <u>vertex</u> is where the sides of an angle meet.

7. A <u>trapezoid</u> has both pairs of opposite sides parallel.

8. All triangles have at least two <u>obtuse</u> angles.

Multi-Part Lesson Review

Lesson 1 **Angle Measure**

Classify Angles (Lesson 1B)

Refer to the figure at the right. Identify each pair of angles. Justify your answer.

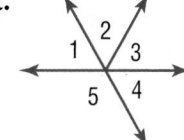

9. a pair of vertical angles

10. a pair of adjacent angles

11. Find the value of x in the figure at the right.

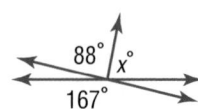

EXAMPLE 1 Refer to the figure below. Identify a pair of vertical angles.

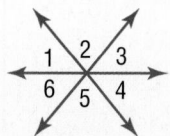

$\angle 1$ and $\angle 4$ are opposite angles formed by the intersection of two lines.

$\angle 1$ and $\angle 4$ are vertical angles.

Complementary and Supplementary Angles (Lesson 1C)

Classify each pair of angles as *complementary, supplementary,* or *neither.*

12. 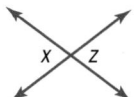   13.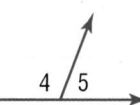

Find the value of x in each figure.

14. 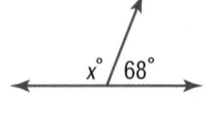   15.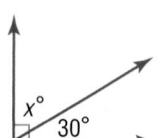

EXAMPLE 2 Find the value of x.

$$x + 27 = 90$$
$$\underline{-27 = -27}$$
$$x = 63$$

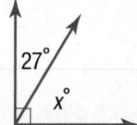

PSI: Use Logical Reasoning (Lesson 1D)

Solve each problem using logical reasoning.

16. **GEOMETRY** Draw several squares and connect the opposite vertices. Then measure the four angles that are formed by the intersecting diagonals on each square. What seems to be true about the diagonals of a square?

17. **LINES** Hazen, Riley, and Alyca are standing in a line. If there is only one person in front of Hazen and Riley is not at the end, in what order are they standing?

EXAMPLE 3 Use logical reasoning to find the next number.

3, 5, 8, 12, 17, …

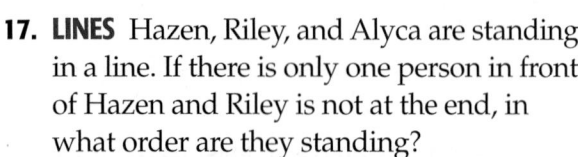

Since the numbers increase by 2, 3, 4, and 5, the next number will increase by 6. The next number is 23.

Lesson 2   Lines

Lines (Lesson 2B)

Refer to the figure below. Classify each pair of angles as *alternate interior*, *alternate exterior*, or *corresponding*.

18. ∠8 and ∠6

19. ∠1 and ∠5

20. ∠4 and ∠2

21. ∠3 and ∠7

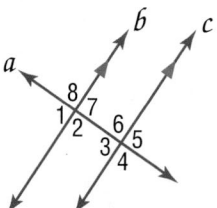

22. **ARCHITECTURE**
Parallel lines are cut by a transversal. If $m\angle 1 = 86°$, find $m\angle 2$ and $m\angle 3$.

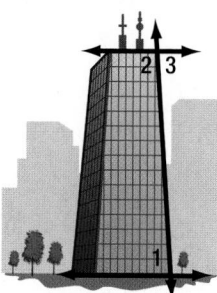

EXAMPLE 4   Classify ∠3 and ∠5 as *alternate interior*, *alternate exterior*, or *corresponding*. If $m\angle 3 = 65°$, find $m\angle 5$.

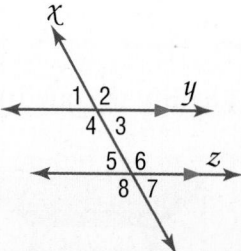

Since ∠3 and ∠5 are interior angles that lie on opposite sides of the transversal, they are alternate interior angles. Since ∠3 and ∠5 are alternate interior angles, they are congruent. So, $m\angle 5 = 65°$.

Lesson 3   Angle Relationships in Polygons

Triangles (Lesson 3B)

Find the value of x in each triangle.

23.
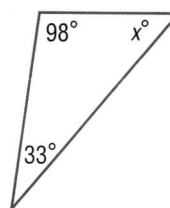

24.
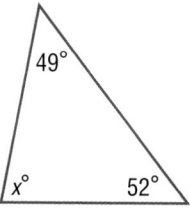

Classify each triangle by its angles and by its sides.

25.
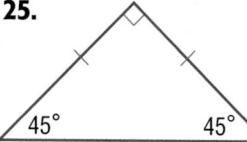

26.
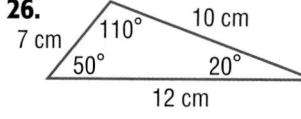

EXAMPLE 5   Find the value of x in the triangle shown.

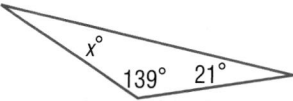

The sum of the angle measures in a triangle is 180°. So, $x + 139 + 21 = 180$.

$$x + 139 + 21 = 180 \quad \text{Write the equation.}$$
$$x + 160 = 180 \quad \text{Simplify.}$$
$$\underline{-160 = -160} \quad \text{Subtract.}$$
$$x = 20$$

Lesson 3 ### Angle Relationships in Polygons (continued)

Quadrilaterals (Lesson 3D)

Classify each quadrilateral with the name that best describes it.

27.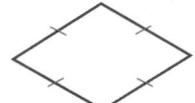

28.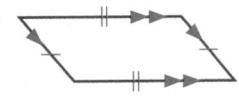

29. TABLES Identify the quadrilateral outlined.

Find the value of x in each quadrilateral.

30.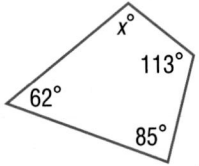

31.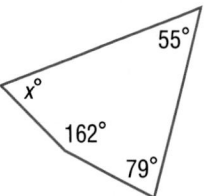

EXAMPLE 6 Classify the quadrilateral shown below.

The quadrilateral has exactly one pair of parallel sides, so it is a trapezoid.

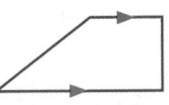

EXAMPLE 7 Find the value of x in the quadrilateral shown.

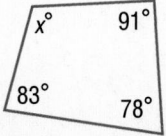

The sum of the angle measures in a quadrilateral is 360°.

$$x + 91 + 78 + 83 = 360$$
$$x + 252 = 360 \quad \text{Simplify.}$$
$$\underline{- 252 = -252} \quad \text{Subtract.}$$
$$x = 108$$

Polygons and Angles (Lesson 3E)

32. Determine whether the figure at the right is a polygon. If it is, classify the polygon. If it is not a polygon, explain why.

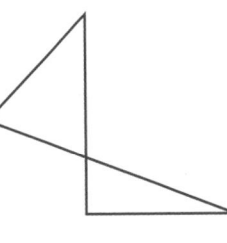

Find the sum of the measures of the interior angles of each polygon.

33. decagon

34. 32-gon

Find the measure of one interior angle in each regular polygon. Round to the nearest tenth if necessary.

35. heptagon

36. pentagon

37. RUGS Find the measure of one interior angle of a rug shaped like a regular octagon.

EXAMPLE 8 Find the measure of one interior angle of a regular hexagon.

Find the sum of the measures of the angles.

$S = (n - 2)180$ Write an equation.
$S = (6 - 2)180$ Replace n with 6.
$S = (4)180$ Subtract.
$S = 720$ Multiply.

The sum of the measures of the interior angles is 720°.

Divide 720° by 6, the number of interior angles. So, the measure of one interior angle of a regular hexagon is 720° ÷ 6 or 120°.

Properties of Three-Dimensional Figures (Lesson 4B)

Identify each figure. Then name the bases, faces, edges, and vertices.

38.

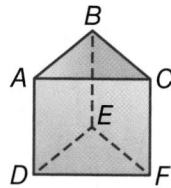

39.

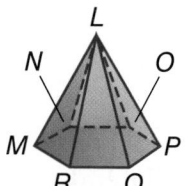

40. Describe the shape resulting from a vertical, angled, and horizontal cross section of a triangular pyramid.

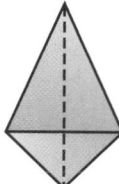

41. Describe the shape resulting from a vertical cross section of a cone.

42. Describe the shape resulting from an angled cross section of a cube.

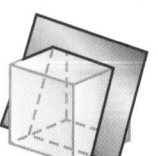

EXAMPLE 9 Name the bases, faces, edges, and vertices of the rectangular prism.

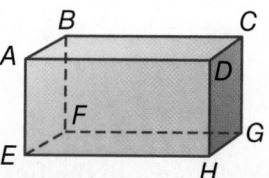

bases ABCD and EFGH, ABFE and DCGH, ADHE and BCGF

faces ABCD, EFGH, ABFE, DCGH, ADHE, BCGF

edges $\overline{AB}, \overline{BC}, \overline{CD}, \overline{AD}, \overline{EF}, \overline{FG}, \overline{GH}, \overline{EH}, \overline{AE}, \overline{BF}, \overline{CG}, \overline{DH}$

vertices A, B, C, D, E, F, G, H

EXAMPLE 10 Describe the shape resulting from a vertical, angled, and horizontal cross section of a triangular prism.

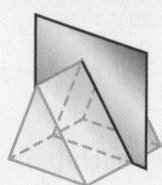

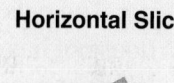

Vertical Slice	Angled Slice	Horizontal Slice
The cross section is a triangle.	The cross section is a triangle.	The cross section is a rectangle.

Name each angle in four ways. Then classify each angle as *acute*, *right*, *obtuse*, or *straight*.

1.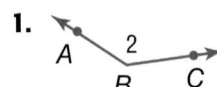

2.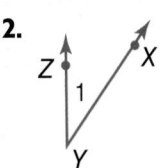

3. Find the value of x in the figure.

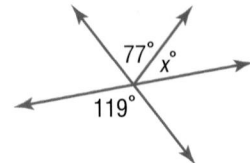

Classify each pair of angles as *complementary*, *supplementary*, or *neither*.

4.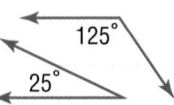

5.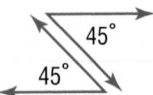

6. **PETS** Henry, Marcie, and Yulia each have a pet. The pets are a dog, cat, and a bird. Yulia does not have a cat and Henry has a dog. What type of pet does each person have?

Refer to the figure below. Classify each pair of angles as *alternate interior*, *alternate exterior*, or *corresponding*.

7. $\angle 1$ and $\angle 5$

8. $\angle 2$ and $\angle 7$

9. $\angle 4$ and $\angle 8$

10. $\angle 3$ and $\angle 6$

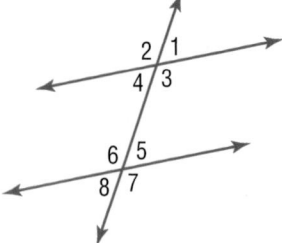

11. **MULTIPLE CHOICE** In triangle ABC, $m\angle A = 62°$ and $m\angle C = 44°$. What is $m\angle B$?

   A. 90°   C. 64°

   B. 74°   D. 42°

12. **SPORTS** Classify the triangle shown by its angles.

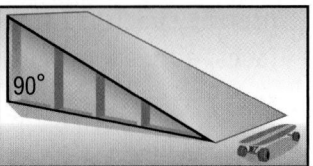

ALGEBRA Find the value of x in each quadrilateral.

13.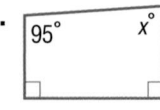

14.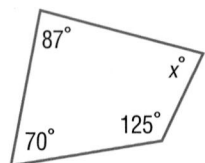

ALGEBRA Find the sum of the measures of the interior angles of each regular polygon. Then, find the measures of one interior angle.

15. octagon

16. 15-gon

17. **GEOMETRY** Identify the figure. Then name the bases, faces, edges, and vertices.

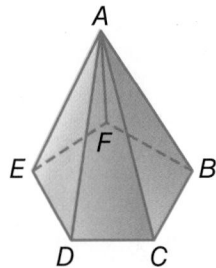

18. **THINK SOLVE EXPLAIN** **EXTENDED RESPONSE** The quilt pattern is made using triangles and quadrilaterals.

Part A Classify the triangles and quadrilaterals in the pattern. Use the names that **best** describe the figures.

Part B Without measuring, determine the measures of the angles of the triangles. Explain.

Preparing for Standardized Tests

⚞✎ Gridded Response: Decimals

When a gridded-response answer is a decimal, include the decimal point in its own answer box at the top of the grid. Then fill in the decimal point bubble below it.

TEST EXAMPLE

In the quadrilateral shown, ∠RQT is congruent to ∠STQ. What is the measure, in degrees, of ∠TQR?

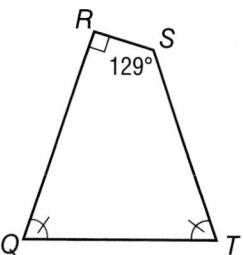

$$90 + 129 + x + x = 360 \quad \text{Write the equation.}$$

$$219 + 2x = 360 \quad \text{Simplify.}$$

$$\underline{-219 \qquad = -219} \quad \text{Subtract 219 from each side.}$$

$$\frac{2x}{2} = \frac{141}{2} \quad \text{Divide each side by 2.}$$

$$x = 70.5 \quad \text{The measure of } \angle TQR \text{ is 70.5 degrees. Grid in 70.5.}$$

Correct

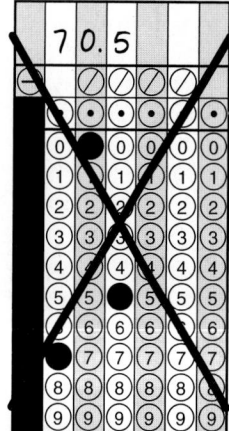

or

NOT Correct

Place the decimal point in its own answer box.

⚞✎ Work on It

In triangle *JKL*, the measure of angle *J* is 25.8 degrees and angle *K* is a right angle. What is the measure, in degrees, of angle *L*? Fill in your answer on an answer grid.

Test Hint

With a decimal like 0.5, you do not need to include the 0 on the answer grid. So, 0.5 could be correctly gridded as 0.5 or as .5.

✓ Test Practice

Read each question. Then fill in the correct answer on the answer document provided by your teacher or on a sheet of paper.

1. Which of the following two angles are complementary?

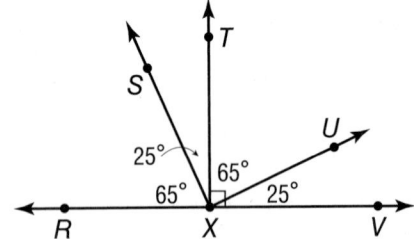

A. ∠RXS and ∠TXU

B. ∠SXT and ∠TXU

C. ∠RXS and ∠SXV

D. ∠SXR and ∠SXV

2. The diameter of a red blood cell is about 0.00074 centimeter. Which expression represents this number in scientific notation?

F. 7.4×10^4

G. 7.4×10^3

H. 7.4×10^{-3}

I. 7.4×10^{-4}

3. ≡≡✎ **GRIDDED RESPONSE** Greta packs tomatoes in boxes that weigh 1.4 kilograms when empty. The average tomato weighs 0.2 kilogram, and the total weight of a box filled with tomatoes is 11 kilograms. How many tomatoes are packed in each box?

4. What is the solution of the inequality below?

$$4n - 8 \leq 40$$

A. $n \leq 8$

B. $n \leq 12$

C. $n \geq 8$

D. $n \geq 12$

5. In the figure below, line x is parallel to line y.

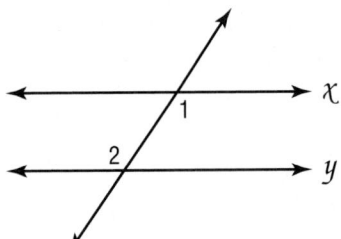

What type of angles are ∠1 and ∠2?

F. vertical angles

G. alternate interior angles

H. alternate exterior angles

I. corresponding angles

6. Which point on the number line best represents $\sqrt{8}$?

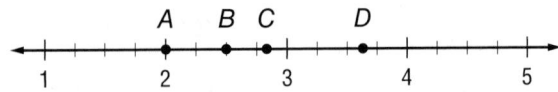

A. point A

B. point B

C. point C

D. point D

7. THINK SOLVE EXPLAIN **SHORT RESPONSE** In triangle ABC, $m\angle A = 55°$ and $m\angle B = 35°$. Classify the triangle by its angles.

8. A stained glass window is in the shape of a regular decagon. What is the measure of one interior angle of the decagon?

F. 1,800°

G. 1,440°

H. 180°

I. 144°

9. THINK SOLVE EXPLAIN **SHORT RESPONSE** The square root of 250 is between which two whole numbers?

10. What is the measure of ∠1 in the figure?

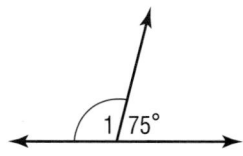

A. 15°   **C.** 100°

B. 25°   **D.** 105°

11. The table shows the atomic weights of certain elements.

Element	Atomic Weight (amu)
Argon	39.948
Lead	207.2
Mercury	200.59
Oxygen	15.9994
Titanium	47.867
Zinc	65.38

Which element has an atomic weight that is exactly 160.642 less than the atomic weight of mercury?

F. argon   **H.** oxygen

G. titanium   **I.** zinc

12. Jesse purchased a new digital camera for $499 and a printer for $299 including tax. He plans to pay the total amount in 6 equal monthly payments. What is a reasonable estimate of the amount he will pay each month?

A. $66.50   **C.** $155.00

B. $133.00   **D.** $165.00

13. Which of the following is a correct drawing of a quadrilateral with all sides congruent and with four right angles?

F.

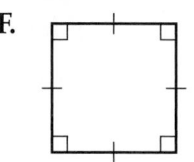

G.

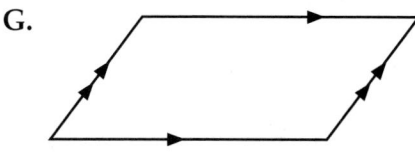

H.

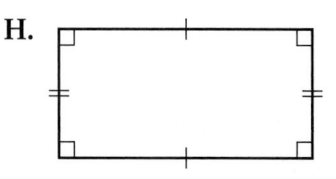

I.

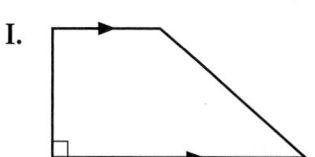

14. **EXTENDED RESPONSE** Use triangle XYZ to answer the following questions.

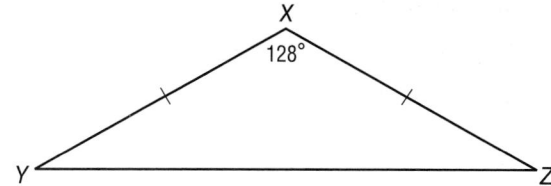

Part A   Classify angle X.

Part B   Classify angle Y.

Part C   Classify the triangle by its sides and by its angles.

Part D   If ∠Y is congruent to ∠Z, find the measure of ∠Z. Explain.

NEED EXTRA HELP?														
If You Missed Question...	1	2	3	4	5	6	7	8	9	10	11	12	13	14
Go to Chapter-Lesson...	7-1C	2-2B	3-2B	3-4A	7-2B	2-3C	7-3B	7-3E	2-3C	7-1C	1-1B	1-1D	7-3D	7-3B

Triangles and Transformations

connectED.mcgraw-hill.com

 Investigate

 Animations

 Vocabulary

 Multilingual eGlossary

 Learn

 Personal Tutor

 Virtual Manipulatives

 Graphing Calculator

 Audio

 Foldables

 Practice

 Self-Check Practice

Worksheets

Assessment

The ☆BIG Idea

How does the transformation of a figure on the coordinate plane affect the congruency, orientation, location, and symmetry of an image?

 FOLDABLES® Study Organizer

Make this Foldable to help you organize your notes.

Triangles and Transformations

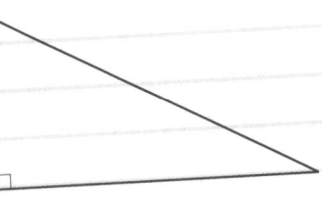

Review Vocabulary

right triangle   triángulo rectángulo a triangle having one right angle

Key Vocabulary

English	Español
hypotenuse	hipotenusa
indirect measurement	medición indirecta
legs	catetos
Pythagorean Theorem	Teorema de Pitágoras

When Will I Use This?

Your Turn! You will solve this problem in the chapter.

Are You Ready for the Chapter?

You have two options for checking prerequisite skills for this chapter.

Text Option Take the Quick Check below. Refer to the Quick Review for help.

QUICK Check

Evaluate each expression.

1. $2^2 + 4^2$ **2.** $3^2 + 3^2$

3. $5^2 + 8^2$ **4.** $3^2 + 4^2$

5. $12^2 - 7^2$ **6.** $13^2 - 5^2$

7. $10^2 - 6^2$ **8.** $5^2 - 4^2$

9. AGES Find the sum of the squares of Patty's age and Warren's age if Patty is 13 years old and Warren is 15 years old.

QUICK Review

EXAMPLE 1

Evaluate $5^2 + 9^2$.

$$5^2 + 9^2 = (5)(5) + (9)(9) \quad \text{Definition of powers}$$
$$= 25 + 81 \quad \text{Multiply.}$$
$$= 106 \quad \text{Simplify.}$$

EXAMPLE 2

Evaluate $15^2 - 12^2$.

$$15^2 - 12^2$$
$$= (15)(15) - (12)(12) \quad \text{Defintion of powers}$$
$$= 225 - 144 \text{ or } 81 \quad \text{Multiply. Then simplify.}$$

Graph and label each point on a coordinate plane.

10. $A(2, 4)$ **11.** $B(-1, -3)$

12. $C(0, -4)$ **13.** $D(3, -2)$

14. $E(-4, -3)$ **15.** $F(3, 0)$

16. $G(-2, 4)$ **17.** $H(2, -1)$

18. MAPPING Blake's house is located at point $(-3, -4)$ on the coordinate plane below. Describe the location of his house in respect to the playground.

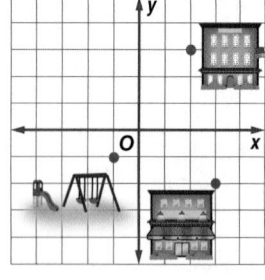

EXAMPLE 3

Graph $X(-3, -2)$, $Y(-1, 2)$, and $Z(2, 3)$ on a coordinate plane.

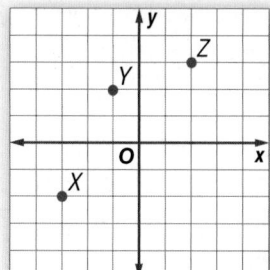

Start at the origin. The first number in each ordered pair is the *x*-coordinate. The second number in each ordered pair is the *y*-coordinate.

Online Option Take the Online Readiness Quiz.

Studying Math

Use a Web

A *web* can help you understand how math concepts are related to each other. To make a web, write the major topic in the center of a piece of paper. Then, draw "arms" from the center for as many categories as you need.

Try using a web to help you understand how math concepts are linked together.

Here is a partial web for the major topic of *triangles*.

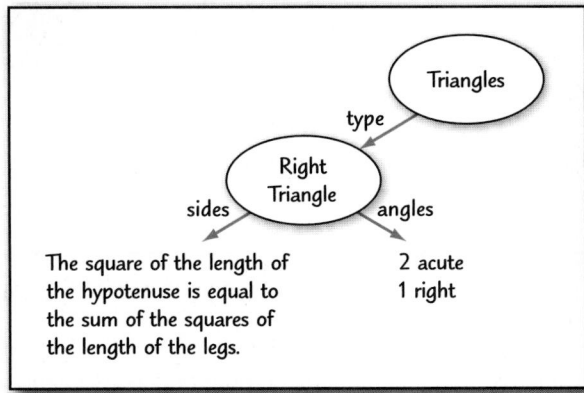

Practice

1. Continue the web above by adding arms for *scalene* and *obtuse triangles*.

2. You know that the set of rational numbers contains fractions, percents, terminating and repeating decimals, and integers. You also know that there are proper and improper fractions, and that integers are whole numbers and their opposites. Complete the web below for *rational numbers*.

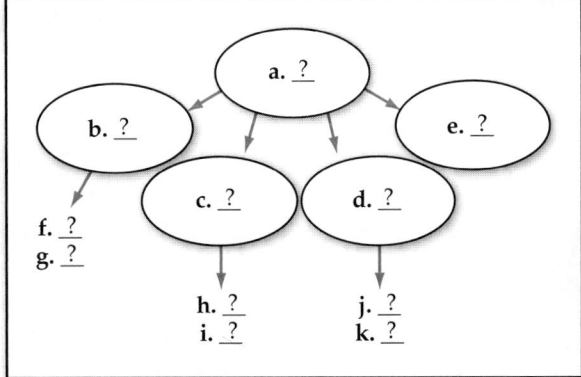

Problem-Solving Investigation

Main Idea Solve problems by drawing a diagram.

 **P.S.I. TEAM +**

Draw a Diagram

STELLA: The Service Club is planting flowers every two feet around the perimeter of the student patio. The patio is shaped like a right triangle with sides that measure 12 feet, 16 feet, and 20 feet.

YOUR MISSION: Draw a diagram to find how many flowers the Service Club needs to buy.

Understand	Flowers will be planted every two feet around the perimeter of the triangular patio that measures 12, 16, and 20 feet per side.
Plan	Draw a diagram showing the patio and the number of flowers needed.
Solve	The patio measures 12 feet, 16 feet, and 20 feet on each side. Since the flowers are planted every two feet, the Service Club will need to buy 24 flowers.
Check	Since flowers are placed every two feet, divide each length by 2. Then add. $12 \div 2 = 6$, $16 \div 2 = 8$, $20 \div 2 = 10$. Since $6 + 8 + 10 = 24$, the answer is correct. ✔

Analyze the Strategy

1. Describe another way to find the number of flowers the Service Club needs to buy.

2. **WRITE MATH** Write a problem that is more easily solved by drawing a diagram. Then draw a diagram and solve the problem.

Mixed Problem Solving

 = **Step-by-Step Solutions** begin on page R1.
Extra Practice begins on page EP2.

- Guess, check and revise.
- Look for a pattern.
- Determine reasonable answers.
- Choose an operation.

Use the _draw a diagram_ strategy to solve Exercises 3–5.

3 **TRAVEL** Mrs. Rogers is driving to the lake. After 45 miles, she is $\frac{5}{6}$ of the way there. How many more miles does she need to travel to reach the lake?

4. WATER A 500-gallon water tank is being filled. Eighty gallons are in the tank after 6 minutes. How many minutes will it take to fill the tank?

5. GEOMETRY A stock clerk is piling baseballs in the shape of a square-based pyramid as shown. If the pyramid is to have five layers, how many baseballs will he need?

Use any strategy to solve Exercises 6–13.

6. AGES Jake, Sanchez, Hao, Mike, and Toshio are friends. Jake is not the youngest. Mike is younger than Jake but older than Hao. Hao is older than Sanchez and Toshio. Sanchez is not the youngest. Write the boys' names in order from youngest to oldest.

7. FLIGHTS A DC-11 jumbo jet carries 345 passengers with 38 in first class and the rest in coach. For a day flight, a first-class ticket from Los Angeles to Chicago costs $650, and a coach ticket costs $230. What will be the ticket sales if the flight is full?

8. SCRAPBOOKS A scrapbook page measures 12 inches long by 12 inches wide. How many 3-inch by 5-inch horizontal photographs can be placed on the page if $\frac{1}{2}$ inch is placed between each photo and at least 1 inch is left as a margin on all four sides?

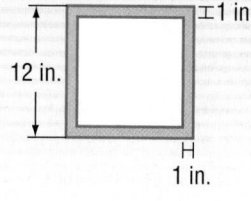

9. PARKS The Morris family is driving to an amusement park. They drive 162 miles on the highway and 63 miles through towns. If the car can go 21 miles on 1 gallon of gas in town and 36 miles on 1 gallon of gas on the highway, how many gallons of gas will the Morris family use?

10. DESSERTS The table below shows the type of desserts the guests at Nikita's birthday party chose.

Dessert	Number of People
Cake	12
Ice cream	8
Both	5

How many total people had dessert?

11. MONEY Junco has 9 coins in her pocket that total $1. The coins are nickels, dimes, and quarters. She has 2 more nickels than quarters. How many of each coin does she have?

12. MEASUREMENT It takes 20 minutes to cut a log into 5 equal-size pieces. How long will it take to cut a similar log into 3 equal-size pieces?

13. GEOMETRY The sides of a triangle are in the ratio 2:3:4. If the perimeter of the triangle is 81 feet, find the length of each side.

Main Idea

Identify similar polygons and find missing measures of similar polygons.

 Vocabulary

similar polygons
corresponding parts
scale factor

 Get ConnectED

 8.G.4

Similar Polygons

Explore Follow the steps below to discover how the triangles are related.

Step 1 Copy both triangles onto tracing paper.

Step 2 Measure and record the sides of each triangle.

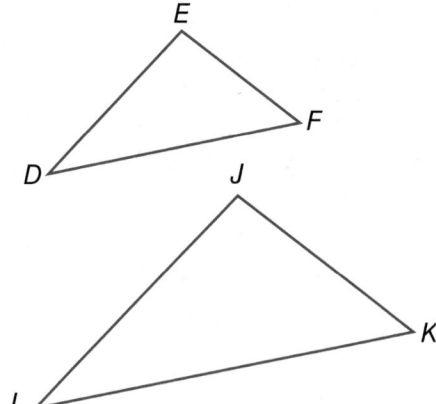

1. Name all pairs of angles in the same relative position. Cut out the triangles. Compare the angles by matching them up. What do you notice about the measure of these angles?

2. Express the ratios $\frac{DF}{LK}$, $\frac{EF}{JK}$, and $\frac{DE}{LJ}$ as decimals to the nearest tenth.

3. What do you notice about the ratios of the sides of these triangles?

Polygons that have the same shape are called **similar polygons**. In the Key Concept box, triangle ABC is similar to triangle XYZ. This is written as $\triangle ABC \sim \triangle XYZ$. The parts of similar figures that "match" are called **corresponding parts**.

Key Concept   **Similar Polygons**

Words   Two polygons are similar if
• their corresponding angles are congruent and
• the measures of their corresponding sides are proportional.

Model

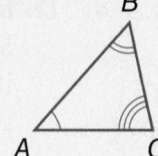 

$\triangle ABC \sim \triangle XYZ$

Symbols $\angle A \cong \angle X$, $\angle B \cong \angle Y$, $\angle C \cong \angle Z$, and $\frac{AB}{XY} = \frac{BC}{YZ} = \frac{AC}{XZ}$

Read Math

Congruence The symbol $\cong$ is read *is congruent to*. Arcs are used to show congruent angles.

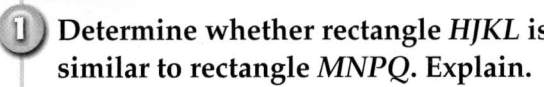

EXAMPLE Identify Similar Polygons

1) **Determine whether rectangle HJKL is similar to rectangle MNPQ. Explain.**

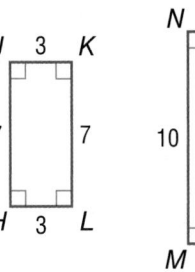
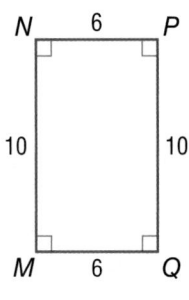

First, check to see if corresponding angles are congruent.

Since the two polygons are rectangles, all of their angles are right angles. Therefore, all corresponding angles are congruent.

Next, check to see if corresponding sides are proportional.

$$\frac{HJ}{MN} = \frac{7}{10} \qquad \frac{JK}{NP} = \frac{3}{6} \text{ or } \frac{1}{2} \qquad \frac{KL}{PQ} = \frac{7}{10} \qquad \frac{LH}{QM} = \frac{3}{6} \text{ or } \frac{1}{2}$$

Since $\frac{7}{10}$ and $\frac{1}{2}$ are not equivalent, the rectangles are *not* similar.

 CHECK Your Progress

a. Determine whether $\triangle ABC$ is similar to $\triangle XYZ$. Explain.

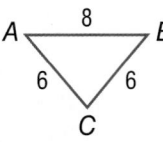
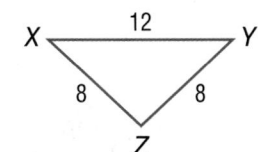

Scale factor is the ratio of the lengths of two corresponding sides of two similar polygons. You can use the scale factor of similar figures to find missing measures.

EXAMPLE Find Missing Measures

2) **GEOMETRY** Given that polygon $WXYZ \sim$ polygon $ABCD$, find the missing measure.

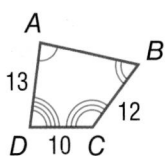
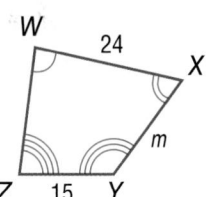

Find the scale factor from polygon $WXYZ$ to polygon $ABCD$.

scale factor: $\frac{YZ}{CD} = \frac{15}{10}$ or $\frac{3}{2}$

So, a length on polygon $WXYZ$ is $\frac{3}{2}$ times as long as the corresponding length on polygon $ABCD$. Let m represent the measure of $\overline{XY}$.

$m = \frac{3}{2}(12)$ Write the equation.

$m = 18$ Multiply.

 CHECK Your Progress

Find each missing measure above.

b. WZ **c.** AB

Square A ~ square B with a scale factor of 3:2. Notice that the ratio of their perimeters is 12:8 or 3:2.

3 m

Square A

2 m

Square B

Square	Perimeter
A	12 m
B	8 m

Key Concept   Ratios of Similar Figures

Words   If figure *B* is similar to figure *A* by a scale factor, then the perimeter *P* of figure *B* is equal to the perimeter *P* of figure *A* times the scale factor.

Symbols   *P* of figure *B* =
P of figure *A* • scale factor

Model

a

Figure *A*

b

Figure *B*

REAL-WORLD EXAMPLE   Find Perimeter

3 GARDENS A landscape architect is examining the scale drawing shown for a garden. In the scale drawing, the perimeter of the garden is 64 inches. The actual length of $\overline{AB}$ is 18 feet. What is the perimeter of the actual garden?

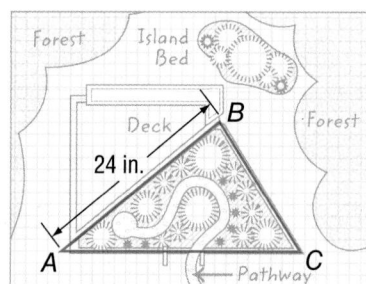

Step 1   The actual length is proportional to the length in the drawing with a ratio of $\frac{18 \text{ ft}}{24 \text{ in.}}$. Find the scale factor.

$$\frac{18 \text{ ft}}{24 \text{ in.}} = \frac{216 \text{ in.}}{24 \text{ in.}} \text{ or } \frac{9}{1}$$ Convert feet to inches and divide out units.

Step 2   Find the perimeter of the actual garden.

perimeter of garden = perimeter of drawing • scale factor

$P = 64 \cdot 9 \text{ or } 576$ Substitute. Then simplify.

The perimeter of the actual garden is 576 inches or 48 feet.

CHECK Your Progress

d. CRAFTS Two quilting squares are shown. The scale factor is 2:3. What is the perimeter of square *TUVW*?

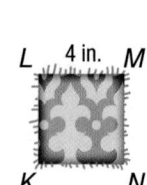

4 in.

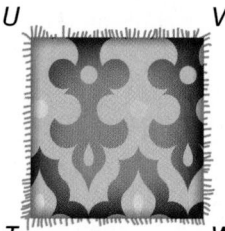

Example 1 Determine whether each pair of polygons is similar. Explain.

1.
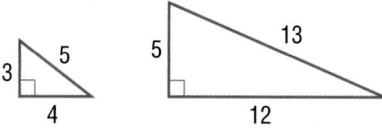

2.
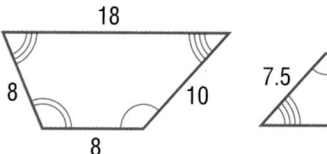

Example 2 **3.** In the figure at the right, △FGH ~ △KLJ. Find each missing side measure.

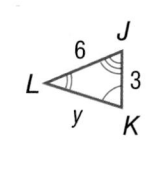

Example 3 **4. PARKS** The city of Brice is planning to build a skate park. An architect designed the area shown at the right. In the plan, the perimeter of the park is 80 inches. If the actual length of $\overline{WX}$ is 50 feet, what will be the perimeter of the actual skate park?

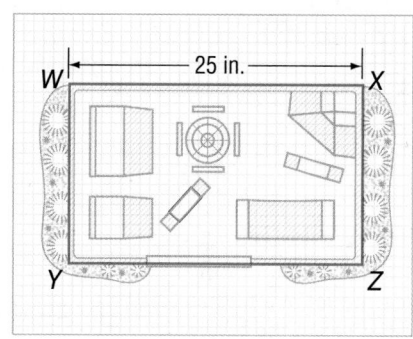

Practice and Problem Solving

● = Step-by-Step Solutions begin on page R1.
Extra Practice begins on page EP2.

Example 1 Determine whether each pair of polygons is similar. Explain.

5.
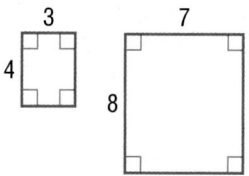

6.
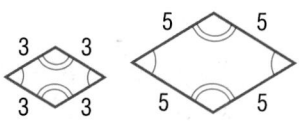

7.
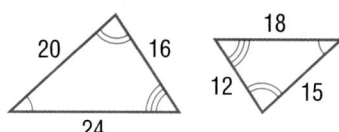

8.
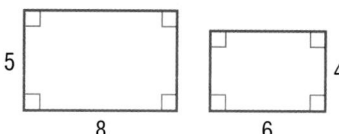

Example 2 Each pair of polygons is similar. Find each missing side measure.

9
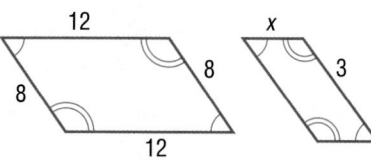

10.
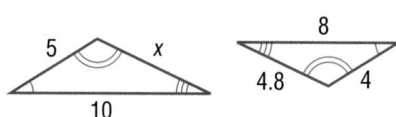

11.
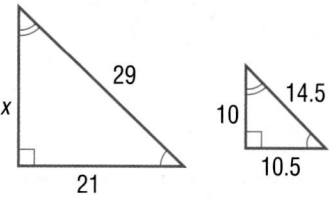

12.
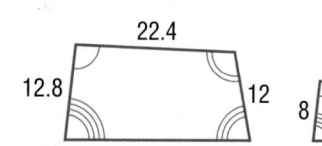

Example 3 **13** **FENCING** Mrs. Henderson wants to build a fence around the rectangular garden in her backyard. In the scale drawing, the perimeter of the garden is 14 inches. If the actual length of $\overline{AB}$ is 20 feet, how many feet of fencing will she need?

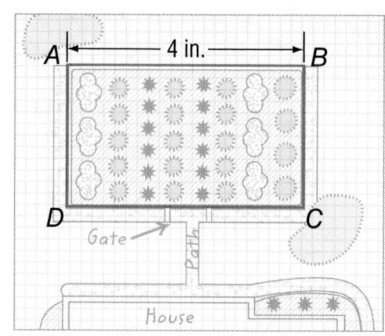

14. **ART** Isaiah is making a mosaic using different pieces of tile. The tiles shown at the right are similar. If the perimeter of the larger tile is 23 centimeters, what is the perimeter of the smaller tile?

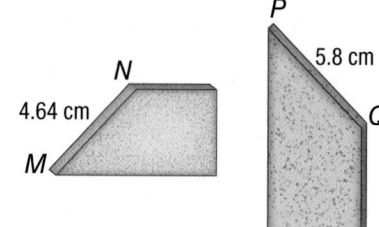

15. **GEOMETRY** The figures below are similar.

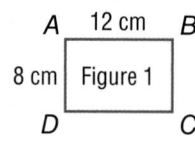 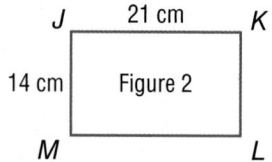

a. Find the area of both figures.

b. Compare the scale factor of the side lengths and the ratio of the areas.

16. **LIFE SCIENCE** Refer to the information at the left. If one of the bones of the model is 8.25 centimeters long, how long is the actual bone in a human ear?

H.O.T. Problems

17. **FIND THE ERROR** Roberto is finding the missing measure in the similar figures below. Find his mistake and correct it.

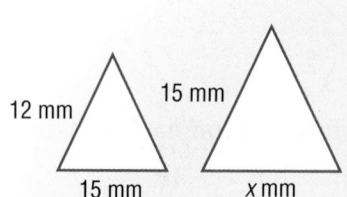 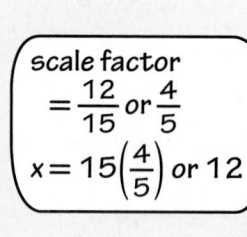

scale factor
$= \frac{12}{15}$ or $\frac{4}{5}$

$x = 15\left(\frac{4}{5}\right)$ or 12

18. **CHALLENGE** Suppose two rectangles are similar with a scale factor of 2. What is the ratio of their areas? Explain.

WRITE MATH Determine whether each statement is *always, sometimes,* or *never* true. Explain your reasoning.

19. Any two rectangles are similar.   20. Any two squares are similar.

21. Triangle *FGH* is similar to triangle *RST*.

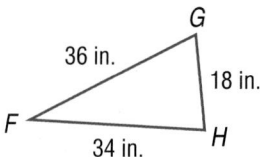
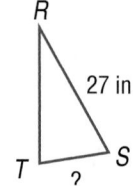

What is the length of $\overline{TS}$?

A. $13\frac{1}{2}$ inches   **C.** 24 inches

B. $22\frac{2}{3}$ inches   **D.** $25\frac{1}{2}$ inches

22. Quadrilateral *RSTU* is similar to quadrilateral *WXYZ*. Which of the following statements is NOT always true?

F. $\angle RST \cong \angle WXY$

G. $\frac{ST}{XY} = \frac{TU}{YZ}$

H. $\angle TUR \cong \angle YZW$

I. $\overline{RS} = \overline{WX}$

23. Quadrilateral *ABCD* is similar to quadrilateral *WXYZ*.

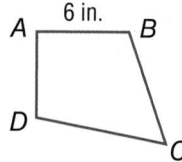
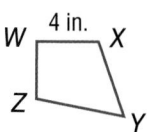

If the perimeter of quadrilateral *ABCD* is 54 inches, what is the perimeter of quadrilateral *WXYZ*?

A. 13.5 inches   **C.** 27 inches

B. 24 inches   **D.** 36 inches

24. $\triangle DEF \sim \triangle GHI$. What is the value of $\overline{GH}$ if $\overline{EF}$ is 6 meters, $\overline{DE}$ is 9 meters, and $\overline{HI}$ is 10 meters?

F. 5.4 m

G. 9 m

H. 15 m

I. 19.4 m

More About **Similar Triangles** . . . .

If two triangles are similar, the ratio of their areas is equal to the *square* of their scale factor. The similar triangles below have a scale factor of $\frac{2}{3}$.

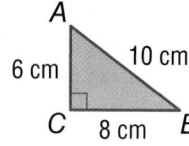
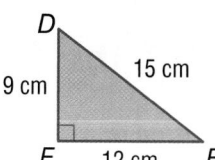

area of triangle *ABC* = 24 cm² 　　area of triangle *DEF* = 54 cm²

$$\frac{\text{area of triangle } ABC}{\text{area of triangle } DEF} = \frac{24}{54} = \frac{4}{9} \text{ or } \left(\frac{2}{3}\right)^2$$

Triangle *LMN* is similar to triangle *RST*. What is the area of triangle *RST*?

25.
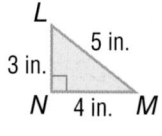
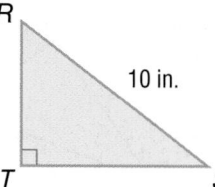

26.
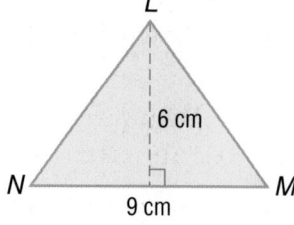
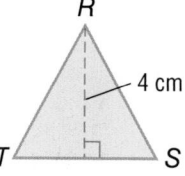

Similar Triangles

Main Idea

Investigate parallel lines and similar triangles.

CCSS 8.G.5

If two triangles are similar, then their corresponding angles are congruent. In addition, if two angles of one triangle are congruent to two angles of another triangle, then the triangles are similar.

ACTIVITY **Similar Triangles**

STEP 1 Draw a pair of parallel lines.

STEP 2 Draw two transversals as shown. Label ∠1, ∠2, ∠3, and ∠4. Label the triangles *RST* and *USV*.

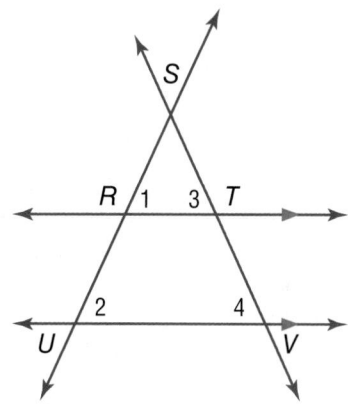

Analyze the Results

1. What type of angles are ∠1 and ∠2? What is true about this pair of angles?

2. What type of angles are ∠3 and ∠4? What is true about this pair of angles?

3. What can you conclude about △*RST* and △*USV*? Explain your reasoning.

4. Determine whether △*ABC* is similar to △*EDC* in the figure shown. Justify your answer.

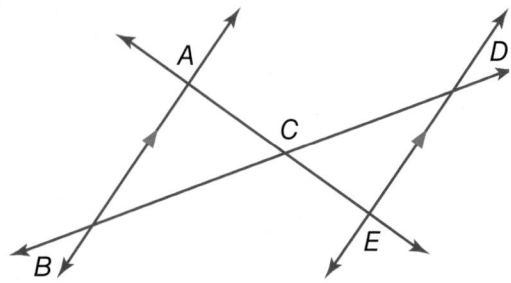

Main Idea

Solve problems involving similar triangles.

 **Vocabulary**

indirect measurement

Get ConnectED

Indirect Measurement

HISTORY Thales is known as the first Greek scientist, engineer, and mathematician. Legend says that he was the first to determine the height of the pyramids in Egypt by examining the shadows made by the Sun. He considered three points: the top of the objects, the lengths of the shadows, and the bases.

1. What appears to be true about the corresponding angles in the two triangles?

2. If the corresponding sides are proportional, what could you conclude about the triangles?

Indirect measurement allows you to use properties of similar polygons to find distances or lengths that are difficult to measure directly. The type of indirect measurement Thales used is called *shadow reckoning*. He measured his height and the length of his shadow then compared it with the length of the shadow cast by the pyramid.

$$\frac{\text{Thales' shadow}}{\text{pyramid's shadow}} = \frac{\text{Thales' height}}{\text{pyramid's height}}$$

 **EXAMPLE** Use Shadow Reckoning

1 **CITY PROPERTY** A fire hydrant 2.5 feet high casts a 5-foot shadow. How tall is a street light that casts a 26-foot shadow at the same time? Let h represent the height of the street light.

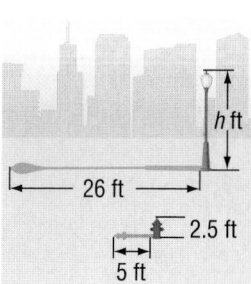

Shadow	Height

hydrant →
street light → $\dfrac{5}{26} = \dfrac{2.5}{h}$ ← hydrant
← street light

$5h = 26 \cdot 2.5$ Find the cross products.

$5h = 65$ Multiply.

$\dfrac{5h}{5} = \dfrac{65}{5}$ Divide each side by 5.

$h = 13$

The street light is 13 feet tall.

CHECK Your Progress

a. STREETS At the same time a 2-meter street sign casts a 3-meter shadow, a nearby telephone pole casts a 12.3-meter shadow. How tall is the telephone pole?

You can also use similar triangles that do not involve shadows to find missing measurements.

🏃 EXAMPLE Use Indirect Measurement

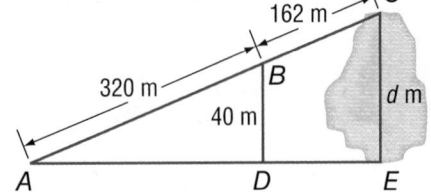

② LAKES In the figure at the right, triangle *DBA* is similar to triangle *ECA*. Ramon wants to know the distance across the lake.

$$\dfrac{AB}{AC} = \dfrac{BD}{CE}$$ $\overline{AB}$ corresponds to $\overline{AC}$ and $\overline{BD}$ corresponds to $\overline{CE}$.

$$\dfrac{320}{482} = \dfrac{40}{d}$$ Replace *AB* with 320, *AC* with 482, and *BD* with 40.

$$320d = 40 \cdot 482$$ Find the cross products.

$$\dfrac{320d}{320} = \dfrac{19{,}280}{320}$$ Multiply. Then divide each side by 320.

$$d = 60.25$$

The distance across the lake is 60.25 meters.

CHECK Your Progress

b. STREETS Find the length of Kentucky Lane.

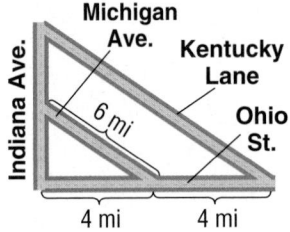

✓ CHECK Your Understanding

Examples 1 and 2 In Exercises 1 and 2, the triangles are similar.

① TREES How tall is the tree?

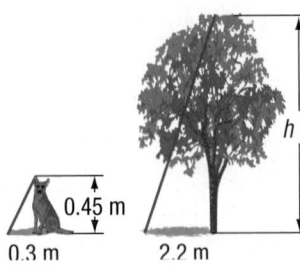

2. WALKING Find the distance from the house to the street light.

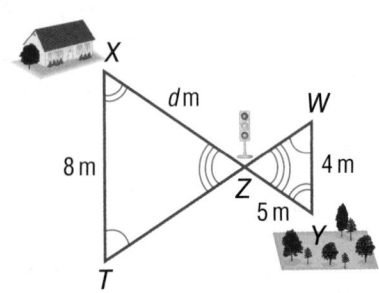

Practice and Problem Solving

⬤ = **Step-by-Step Solutions** begin on page R1.
Extra Practice begins on page EP2.

Example 1   **In Exercises 3–8, the triangles are similar.**

3. BUILDING How tall is the building?   **4. FLAGS** How tall is the taller flagpole?

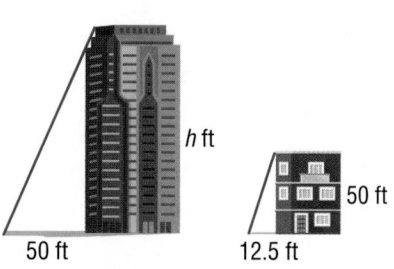

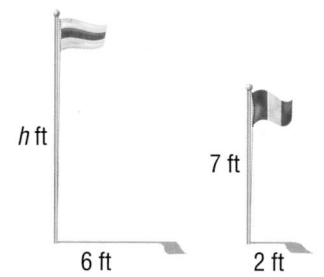

Example 2   **⑤ PARKS** How far is it from the log ride to the pirate ship?

6. CREEKS About how long is the log that goes across the creeks?

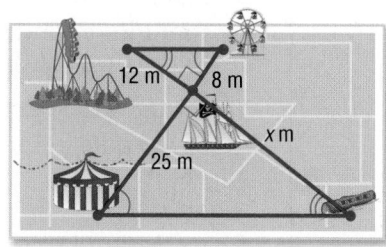

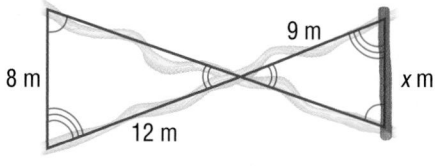

7. CONSTRUCTION Find the height of the brace.

8. LAKES How deep is the water 62 meters from the shore?

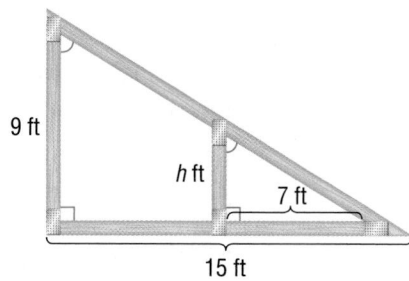

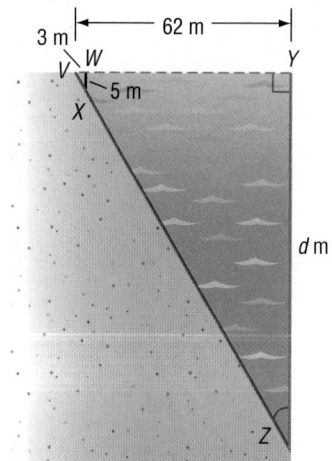

For Exercises 9 and 10, draw a diagram.

⑨ FERRIS WHEELS The Giant Wheel at Cedar Point in Ohio is one of the tallest Ferris wheels in the country at 136 feet tall. If the Giant Wheel casts a 34-foot shadow, write and solve a proportion to find the height of a nearby man who casts a $1\frac{1}{2}$-foot shadow.

10. HEIGHT A 78-inch-tall man casts a shadow that is 54 inches long. At the same time, a nearby building casts a 48-foot-long shadow. About how tall is the building? Write and solve a proportion.

11. **OPEN ENDED** Describe a situation that requires indirect measurement. Explain how to solve the problem.

12. **CHALLENGE** You cut a square hole $\frac{1}{4}$-inch wide in a piece of cardboard. With the cardboard 30 inches from your face, the moon fits exactly into the square hole. If the moon is about 240,000 miles from Earth, estimate the moon's diameter.

13. ✏️ **WRITE MATH** What measures must be known in order to calculate the height of tall objects using shadow reckoning?

Test Practice

14. Mila must determine the height of the statue to make a scale drawing of it.

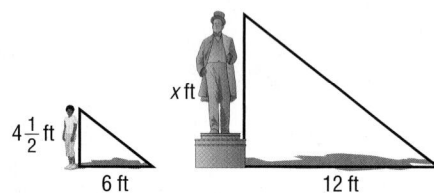

Mila is $4\frac{1}{2}$ feet tall, and her shadow is 6 feet long. At the same time, the statue's shadow is 12 feet long. What is the height of the statue?

A. $8\frac{1}{4}$ ft

B. 9 ft

C. $13\frac{1}{2}$ ft

D. 24 ft

15. As shown below, Lenno used similiar triangles to find the height of a telephone pole. When he stood 7 feet from a mirror laying on the ground, he could see the top of the pole in the mirror.

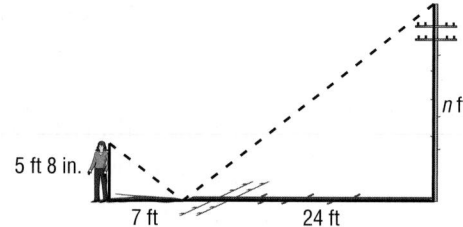

Which is closest to the height of the telephone pole ?

F. 50 ft

G. 40 ft

H. 20 ft

I. 10 ft

Spiral Review

Determine whether each pair of polygons is similar. Explain. (Lesson 1B)

16.

17.

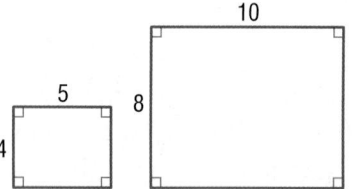

18. **GIFTS** Tammy wants to buy a card and a balloon for her mother's birthday. She is deciding among 5 different cards and 4 different balloons. If she buys only one card and one balloon, how many different combinations can be purchased? Use the *draw a diagram* strategy. (Lesson 1A)

Main Idea

Use the tangent ratio to find missing measures of right triangles.

Vocabulary

trigonometry
trigonometric ratio
tangent ratio

**Get ConnectED**

The Tangent Ratio

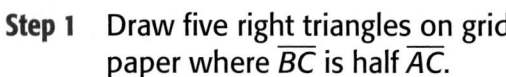**Explore** The triangle at the right is a right triangle where $\overline{BC}$ is half $\overline{AC}$.

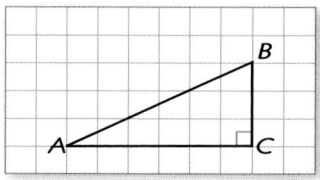

Step 1 Draw five right triangles on grid paper where $\overline{BC}$ is half $\overline{AC}$.

Step 2 Measure $\angle A$ in each triangle.

1. Are each of the triangles from Step 1 similar? Explain.

2. **MAKE A CONJECTURE** Draw a right triangle where $m\angle A \approx 27°$. Predict the relationship between the opposite and adjacent sides of the angle. Test your conjecture by measuring $\overline{AC}$ and $\overline{BC}$.

3. Draw four right triangles where $\angle A = 45°$ and $\angle C = 90°$. Find the ratio $\frac{BC}{AC}$ for each triangle. What can you conclude about the lengths of the legs in this type of triangle?

Trigonometry is the study of the properties of triangles. A **trigonometric ratio** is a ratio of the lengths of two sides of a right triangle. One type of trigonometric ratio is the tangent ratio.

The **tangent ratio** compares the measure of the leg opposite an angle with the measure of the leg adjacent to that angle. The symbol for the tangent of angle A is tan A.

Key Concept **Tangent Ratio**

Words If $\angle A$ is an acute angle of a right triangle,

$$\tan \angle A = \frac{\text{measure of leg opposite } \angle A}{\text{measure of leg adjacent to } \angle A}.$$

Symbols $\tan A = \dfrac{a}{b}$

Model

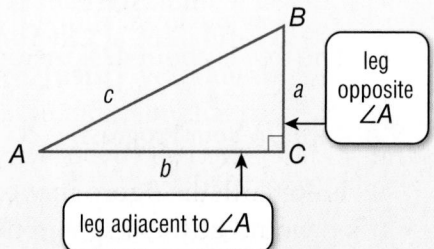

leg opposite $\angle A$

leg adjacent to $\angle A$

EXAMPLE Find the Tangent

1 Find tan A in the triangle at the right. Round to the nearest hundredth. Explain its meaning.

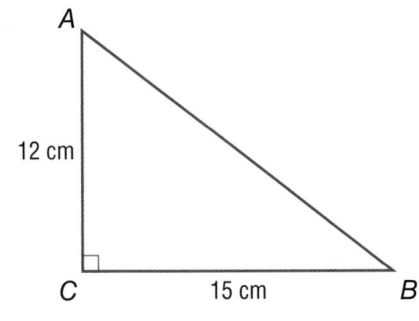

$\tan A = \dfrac{\text{measure of leg opposite } \angle A}{\text{measure of leg adjacent to } \angle A}$

$\tan A = \dfrac{BC}{AC}$

$= \dfrac{15}{12}$ or 1.25

The side opposite $\angle A$ is 1.25 times as long as the side adjacent to $\angle A$.

CHECK Your Progress

a. Find tan B in triangle ABC above. Explain its meaning.

Study Tip

Calculators Make sure that your graphing calculator is in "degree" mode when finding the tangent of an angle measured in degrees. Press the MODE button and check that "degree" is highlighted instead of "radian."

You can find the tangent of a given angle using a calculator. Press the TAN button followed by the measure of the angle.

EXAMPLE Use Tangent to Find Missing Lengths

2 **TREES** Mr. Mullins is looking at the top of a tree in his backyard at a 22° angle. If he is sitting 10 meters from the base of the tree, how tall is the tree? Round to the nearest hundredth.

$\tan A = \dfrac{\text{opposite leg}}{\text{adjacent leg}}$	Write the tangent ratio.
$\tan 22° = \dfrac{x}{10}$	Substitution
$10(\tan 22°) = 10 \cdot \dfrac{x}{10}$	Multiply each side by 10.
10 ☒ TAN 22 ENTER 4.040262258	Use a calculator.
$x = 4.04$	Simplify.

The tree is about 4.04 meters tall.

CHECK Your Progress

b. Refer to the figure above. Suppose $\angle A$ measured 15°. What would be the height of the tree? Round to the nearest hundredth.

Example 1   Find the tangent of each acute angle. Round to the nearest hundredth. Explain its meaning.

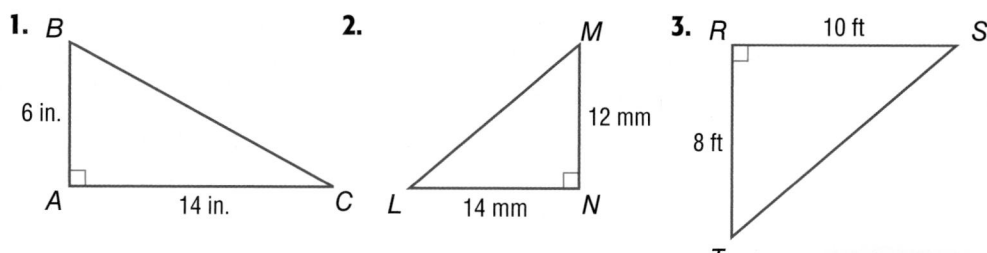

1.
B
6 in.
A   14 in.   *C*

2.
M
12 mm
L   14 mm   *N*

3.
R   10 ft   *S*
8 ft
T

Example 2   **4. BALLOONING** A hot air balloon is directly above a tree. An observer is standing 500 feet from the tree. He is looking at the balloon at a 24° angle. If the observer is 6 feet tall, how far above the ground is the top of the balloon? Round to the nearest hundredth.

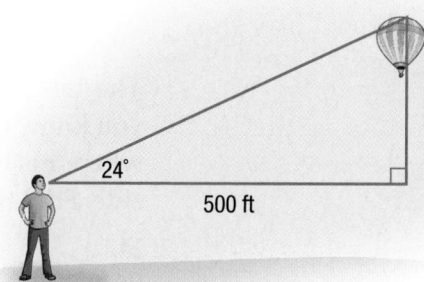

24°   500 ft

Practice and Problem Solving

● = Step-by-Step Solutions begin on page R1.
Extra Practice begins on page EP2.

Example 1   Find the tangent of each acute angle. Explain its meaning.

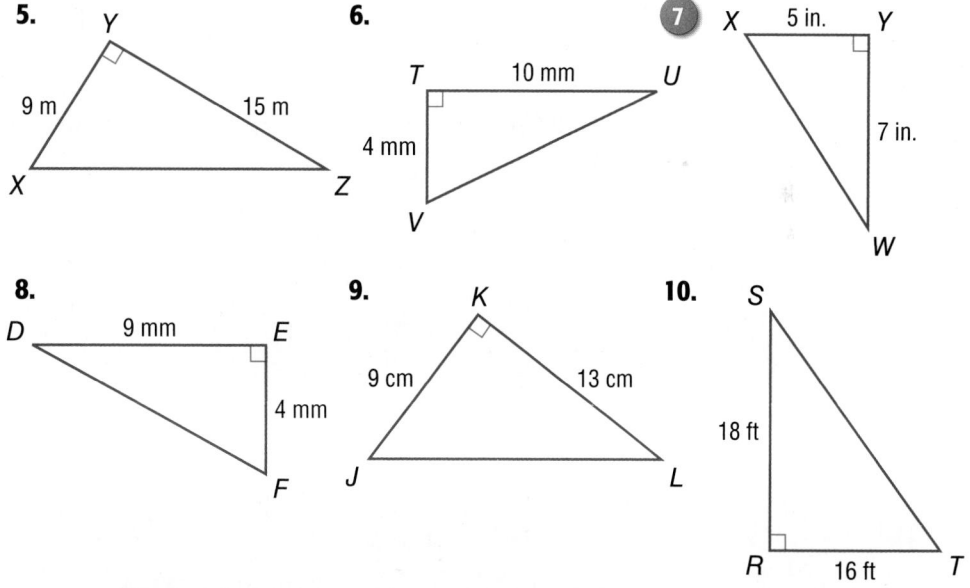

5.
Y
9 m   15 m
X   *Z*

6.
T   10 mm   *U*
4 mm
V

7   *X*   5 in.   *Y*
7 in.
W

8.
D   9 mm   *E*
4 mm
F

9.
K
9 cm   13 cm
J   *L*

10.
S
18 ft
R   16 ft   *T*

Example 2   **11. PLANES** An airplane is flying at an altitude of 10,000 feet. The pilot wants to increase the altitude of the plane at an angle of 1.5° over the next 100 miles. What will be the plane's change in altitude? Round to the nearest hundredth. (*Hint:* 1 mi = 5,280 ft)

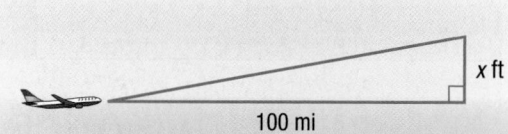

x ft
100 mi

12. SIGHTSEEING In order to see past a cliff from an observation tower, you must look out at a 15° angle to see over the edge of the cliff. How far from the edge of the cliff is the tower? Round to the nearest hundredth.

13. WHEELCHAIRS The horizontal distance from the end of a wheelchair ramp to the base of a building is 32 feet. If the angle of inclination for the ramp is 3.6°, how tall is the ramp? Round to the nearest hundredth.

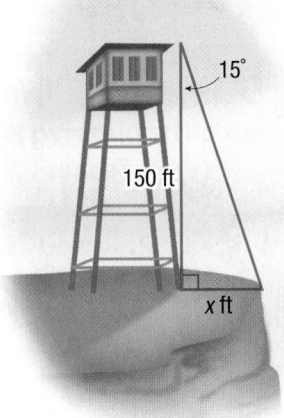

The [tan⁻¹] feature on a calculator allows you to find an angle measure when you know the lengths of the sides of the triangle. Press $\boxed{\text{2nd}}$ [tan⁻¹] $x \div y$, where x is the length of the opposite side and y is the length of the adjacent side. Find the measure of each acute angle. Round to the nearest degree.

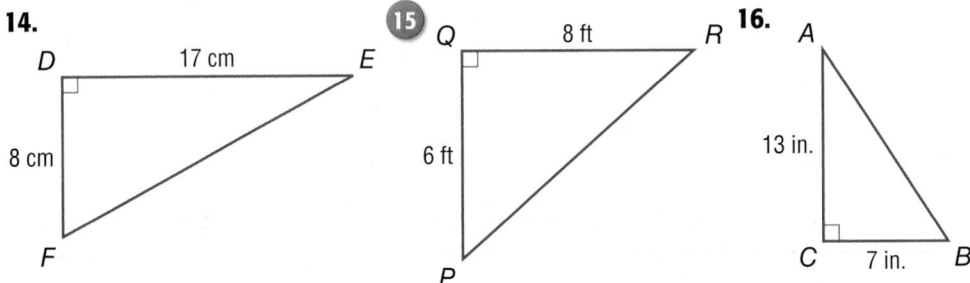

14.

D —— 17 cm —— E

8 cm

F

15.

Q —— 8 ft —— R

6 ft

P

16.

A

13 in.

C —— 7 in. —— B

17. GRAPHIC NOVEL Refer to the graphic novel frame below for Exercises a–b.

a. Suppose the brochure says that the parasailer will be 300 feet behind the boat. Draw a diagram to represent the boat, the rope, and the parasailer.

b. The angle of ascent from the boat to the parasailer is 35°. Find the height of the parasailer above the water. Round to the nearest hundredth.

18. **OPEN ENDED** The smallest angle of a right triangle measures 30°. Give the possible lengths of the sides opposite and adjacent to the angle.

19. **CHALLENGE** Find the value of x in the figure at the right.

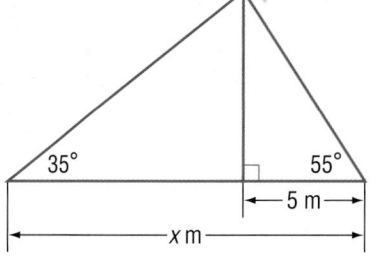

20. **WRITE MATH** If the tangent ratio of an angle is 1, what is true about the measures of the legs of the triangle? Explain.

Test Practice

21. What is the tangent of angle A?

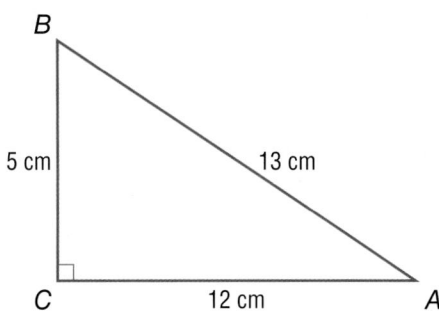

A. $\frac{12}{13}$ C. $\frac{12}{5}$

B. $\frac{5}{12}$ D. $\frac{13}{12}$

22. Kamal is 5 feet 5 inches tall. If the Sun's rays hit Kamal at a 52° angle, what will be the approximate length of Kamal's shadow?

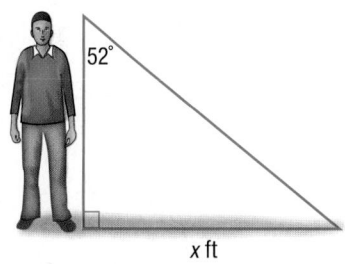

F. 5 ft H. 8 ft

G. 7 ft I. 9 ft

23. **FLAGPOLE** A 10-foot tall flagpole casts a 4-foot shadow. At the same time, a nearby tree casts a 25-foot shadow. What is the height of the tree? (Lesson 1D)

24. **PHOTOGRAPHY** Eva wants to enlarge the picture at the right and frame it. The scale factor from the original picture to the enlarged picture is to be $\frac{5}{2}$. Find the dimensions of the enlarged picture. (Lesson 1B)

 6 in.

4 in.

25. **ROCK CLIMBING** Shannon is climbing up a wall. Every 5 minutes she climbs 6 feet but then loses her footing, slips back 1 foot, and then rests for 1 minute. Where will she be after 36 minutes? Use the *draw a diagram* strategy. (Lesson 1A)

Mid-Chapter Check

1. **SEATING** The school theater is arranged in sections so that each row has the same number of seats. Spencer is seated in the 5th row from the front and the 3rd row from the back. His seat is 6th from the left and 2nd from the right. Draw a diagram to determine the number of seats in Spencer's section of the theater. (Lesson 1A)

Determine whether each pair of polygons below is similar. Explain. (Lesson 1B)

2.
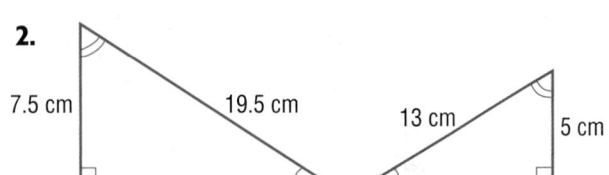

3.
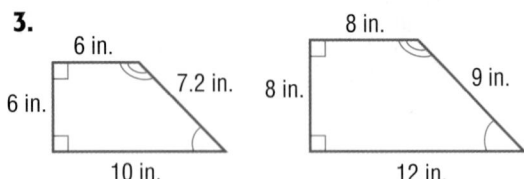

4. **MULTIPLE CHOICE** The triangles below are similar.

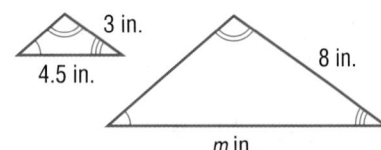

What is the missing measure? (Lesson 1B)

A. 1.69 in.   C. 11 in.

B. 9.5 in.   D. 12 in.

5. **QUILTING** Ariana is making a quilt. On the pattern, the quilt is 6 inches wide and has a perimeter of 33 inches. If the actual quilt is 4 feet wide, what is its perimeter? (Lesson 1B)

6. **SURVEYING** In the figure below, the triangles are similar. How wide is the river? (Lesson 1D)

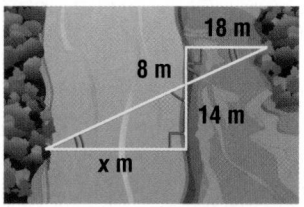

7. **MAILBOXES** At the same time a mailbox casts an 18-inch shadow, a nearby tree casts a 234-inch shadow. If the mailbox is 4 feet tall, how tall is the tree? (Lesson 1D)

8. **MULTIPLE CHOICE** What is tan Z in the triangle below? (Lesson 1E)

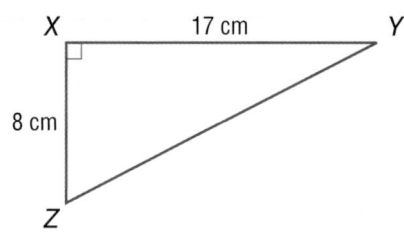

F. $\frac{8}{17}$ H. 8

G. $2\frac{1}{8}$ I. 17

Find the tangent of each acute angle. Round to the nearest hundredth. Explain its meaning. (Lesson 1E)

9.   10.
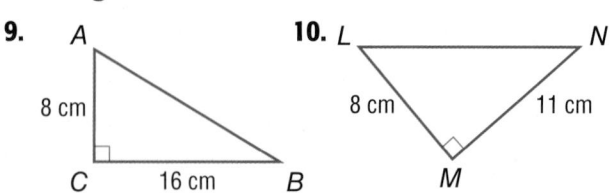

11.   12.
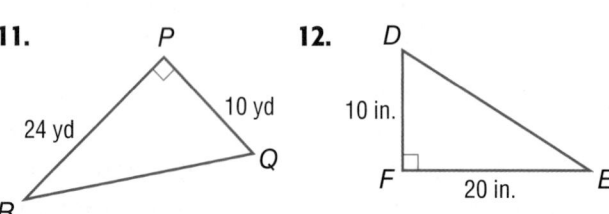

Explore

Main Idea

Find the relationship among the sides of a right triangle.

Right Triangle Relationships

You can use centimeter grid paper to find the area of squares and triangles. In this lab, you will investigate the relationship among the sides of a right triangle.

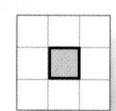 Area = 1 cm²

 Area = $\frac{1}{2}$ cm²

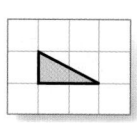 Area = 1 cm²

ACTIVITY

STEP 1 Draw each figure on centimeter grid paper. In each figure, the sides of three squares form a right triangle.

Triangle 1
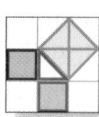

Triangle 2
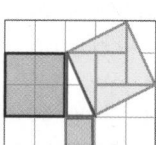

Triangle 3
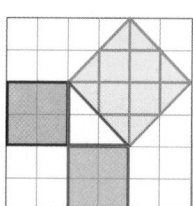

STEP 2 Find the area of each square that is attached to the triangle. Record this information in a table.

Analyze the Results

1. What relationship exists among the areas of the three squares bordering each triangle?

2. On centimeter grid paper, draw a right triangle with the two shorter sides 3 centimeters and 4 centimeters long. If squares were attached to each side of this triangle, what would be the area of each square? Use a ruler to measure the length of the third side of the triangle.

3. **MAKE A CONJECTURE** Determine the length of the longest side of a right triangle if the lengths of the two shorter sides are 6 centimeters and 8 centimeters long.

Main Idea

Use the Pythagorean Theorem.

Vocabulary

legs
hypotenuse
Pythagorean Theorem
converse

Get ConnectED

CCSS 8.EE.2, 8.G.7

The Pythagorean Theorem

Explore When viewed from the side, the shape of some wooden skateboarding ramps is a right triangle. The dimensions of four possible ramps of this type are given in the table. Copy this table.

Ramp Design	Height, H (ft)	Base, B (ft)
A	3	4
B	6	8
C	5	12
D	7	24

Step 1 Draw a side-view model of each ramp on grid paper, letting the width of one grid equal 1 foot.

Step 2 Cut out each ramp and use your grid paper to find the length of the ramp, which is the longest side of your model. Write these measures in a new column labeled *length, L (ft)*.

Step 3 Finally, add a column labeled $H^2 + B^2$. Calculate each of these values and place them in your table.

1. What is the relationship between the values in the $H^2 + B^2$ column and the values in the L column?

2. How could you use a value in the $H^2 + B^2$ column to find a corresponding value in the L column?

Recall that a right triangle is a triangle with one right angle.

The **legs** are the sides that form the right angle.

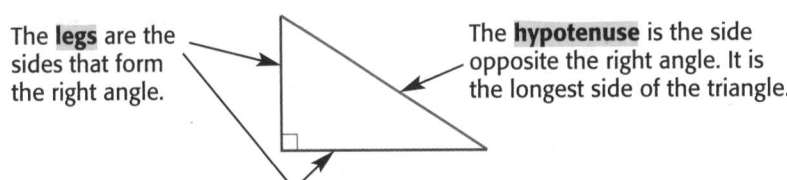

The **hypotenuse** is the side opposite the right angle. It is the longest side of the triangle.

The **Pythagorean Theorem** describes the relationship between the lengths of the legs and the hypotenuse for *any* right triangle.

Key Concept Pythagorean Theorem

Words In a right triangle, the sum of the squares of the lengths of the legs is equal to the square of the length of the hypotenuse.

Model

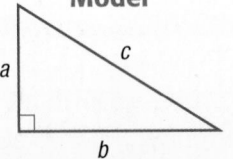

Symbols $a^2 + b^2 = c^2$

You can use the Pythagorean Theorem to find the length of a side of a right triangle when you know the other two sides.

EXAMPLES Find a Missing Length

Write an equation you could use to find the length of the missing side of each right triangle. Then find the missing length. Round to the nearest tenth if necessary.

Read Math

Right Angle The symbol ⌐ indicates an angle with a measure of 90°.

(1)

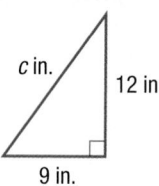

c in. 12 in.

9 in.

$a^2 + b^2 = c^2$	Pythagorean Theorem
$9^2 + 12^2 = c^2$	Replace a with 9 and b with 12.
$81 + 144 = c^2$	Evaluate 9^2 and 12^2.
$225 = c^2$	Add 81 and 144.
$\pm\sqrt{225} = c$	Definition of square root
$c = 15$ or -15	Simplify.

The equation has two solutions, 15 and −15. However, the length of a side must be positive. So, the hypotenuse is 15 inches long.

(2)

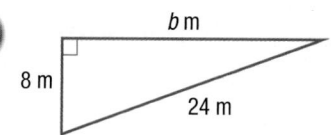

b m

8 m

24 m

$a^2 + b^2 = c^2$	Pythagorean Theorem
$8^2 + b^2 = 24^2$	Replace a with 8 and c with 24.
$64 + b^2 = 576$	Evaluate 8^2 and 24^2.
$64 - 64 + b^2 = 576 - 64$	Subtract 64 from each side.
$b^2 = 512$	Simplify.
$b = \pm\sqrt{512}$	Definition of square root
$b \approx 22.6$ or -22.6	Use a calculator.

The length of side b is about 22.6 meters.

Study Tip

Check for Reasonableness The hypotenuse is always the longest side in a right triangle. Since 22.6 is less than 24, the answer is reasonable.

CHECK Your Progress

a.

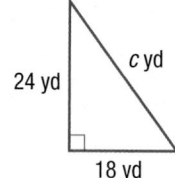

c yd

24 yd

18 yd

b.   3 mi

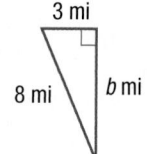

8 mi b mi

c.

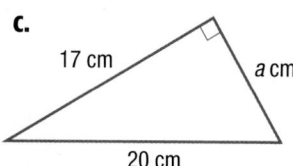

17 cm a cm

20 cm

If you reverse the parts of the Pythagorean Theorem, you have formed its **converse**. The converse of the Pythagorean Theorem is also true.

Key Concept | **Converse of Pythagorean Theorem**

If the sides of a triangle have lengths a, b, and c units such that $a^2 + b^2 = c^2$, then the triangle is a right triangle.

 EXAMPLE **Identify a Right Triangle**

3 The measures of three sides of a triangle are 5 inches, 12 inches, and 13 inches. Determine whether the triangle is a right triangle.

$a^2 + b^2 = c^2$	Pythagorean Theorem
$5^2 + 12^2 \overset{?}{=} 13^2$	$a = 5, b = 12, c = 13$
$25 + 144 \overset{?}{=} 169$	Evaluate 5^2, 12^2, and 13^2.
$169 = 169$ ✓	Simplify.

The triangle is a right triangle.

CHECK Your Progress

Determine whether each triangle with sides of given lengths is a right triangle. Justify your answer.

d. 36 mi, 48 mi, 60 mi   **e.** 4 ft, 7 ft, 5 ft

✓ **CHECK Your Understanding**

Examples 1 and 2 Write an equation you could use to find the length of the missing side of each right triangle. Then find the missing length. Round to the nearest tenth if necessary.

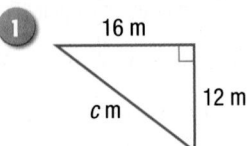

1 16 m, c m, 12 m

2. c mm, 100 mm, 200 mm

3. The hypotenuse of a right triangle is 12 inches, and one of its legs is 7 inches. Find the length of the other leg. Round to the nearest tenth if necessary.

Example 3 Determine whether each triangle with sides of given lengths is a right triangle. Justify your answer.

✗ **4.** 5 in., 10 in., 12 in.   **5.** 9 m, 40 m, 41 m

Practice and Problem Solving

● = **Step-by-Step Solutions** begin on page R1.
Extra Practice begins on page EP2.

Examples 1 and 2 Write an equation you could use to find the length of the missing side of each right triangle. Then find the missing length. Round to the nearest tenth if necessary.

6.

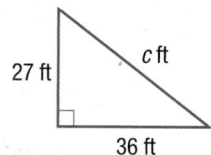

27 ft, c ft, 36 ft

7.

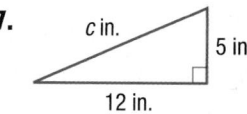

c in., 5 in., 12 in.

8.

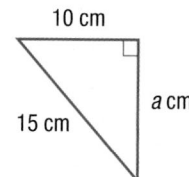

10 cm, a cm, 15 cm

9.

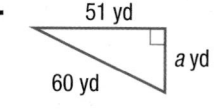

51 yd, a yd, 60 yd

10.

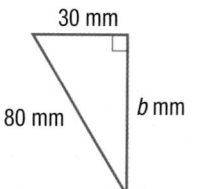

30 mm, b mm, 80 mm

11.

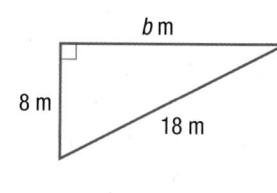

b m, 8 m, 18 m

Example 3 Determine whether each triangle with sides of given lengths is a right triangle. Justify your answer.

12. 28 yd, 195 yd, 197 yd

13. 30 cm, 122 cm, 125 cm

14. 24 m, 143 m, 145 m

15. 135 in., 140 in., 175 in.

16. 56 ft, 65 ft, 16 ft

17. 44 cm, 70 cm, 55 cm

18. POSTAGE An envelope is classified as a *large* envelope if the length exceeds 11.5 inches. Is the envelope below a large envelope? Explain.

19 **GEOGRAPHY** Calculate the length of the diagonal of the state of Wyoming.

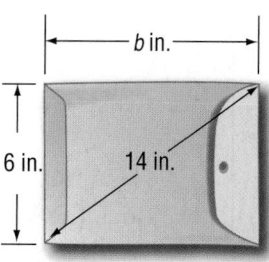

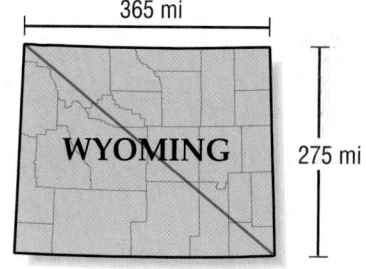

Write an equation you could use to find the length of the missing side of each right triangle. Then find the missing length. Round to the nearest tenth if necessary.

20. b, 99 mm; c, 101 mm

21 a, 48 yd; b, 55 yd

22. a, 17 ft; c, 20 ft

23. a, 23 in.; b, 18 in.

24. b, 4.5 m; c, 9.4 m

25. b, 5.1 m; c, 12.3 m

26. TRAVEL The Research Triangle in North Carolina is formed by Raleigh, Durham, and Chapel Hill. Is this triangle a right triangle? Explain.

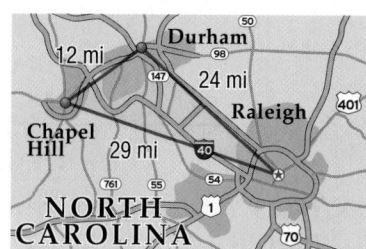

27. OPEN ENDED State three measures that could be the side measures of a right triangle. Justify your answer.

28. FIND THE ERROR Danielle is writing an equation to find the length of the third side of the right triangle. Find her mistake and correct it.

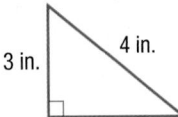
3 in.   4 in.

$a^2 = 3^2 + 4^2$

29. CHALLENGE The whole numbers 3, 4, and 5 are called Pythagorean triples because they satisfy the Pythagorean Theorem. Find three other sets of Pythagorean triples.

30. ✍ **WRITE MATH** Explain why you can use any two sides of a right triangle to find the third side.

✔ Test Practice

31. What is the perimeter of triangle *ABC*?

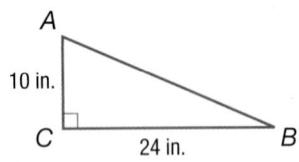
A
10 in.
C   24 in.   B

A. 26 in.   **C.** 60 in.

B. 34 in.   **D.** 68 in.

32. [THINK SOLVE EXPLAIN] **SHORT RESPONSE** The base of a ten-foot ladder stands six feet from a house.

10 ft
6 ft

How many feet up the side of the house does the ladder reach?

Spiral Review

Find the tangent of each acute angle. Explain its meaning. (Lesson 1E)

33.

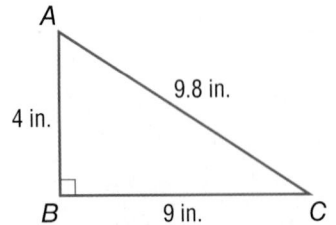
A
9.8 in.
4 in.
B   9 in.   C

34.

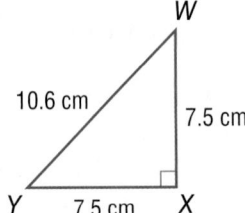
W
10.6 cm
7.5 cm
Y   7.5 cm   X

35. TREES A tree casts a shadow of 24 feet while Rishi casts a shadow of 8 feet. If Rishi is 6 feet tall, how tall is the tree? (Lesson 1D)

Main Idea

Solve problems using the Pythagorean Theorem.

CCSS 8.EE.2, 8.G.7

Use the Pythagorean Theorem

PARASAILING In parasailing, a towrope is used to attach a parasailer to a boat.

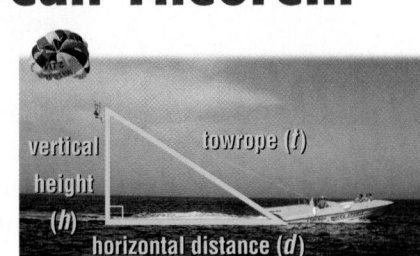

1. What type of triangle is formed by the horizontal distance, the vertical height, and the length of the towrope?

2. Write an equation that can be used to find the length of the towrope.

The Pythagorean Theorem can be used to solve a variety of problems.

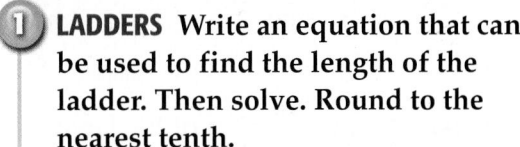

REAL-WORLD EXAMPLE **Solve a Right Triangle**

1 **LADDERS** Write an equation that can be used to find the length of the ladder. Then solve. Round to the nearest tenth.

Notice that the distance from the building, the building itself, and the ladder form a right triangle. Use the Pythagorean Theorem.

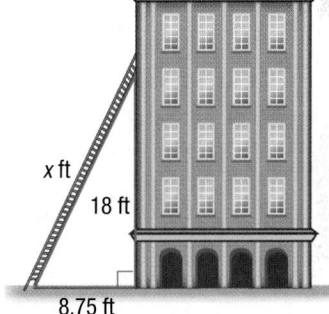

$$a^2 + b^2 = c^2 \quad \text{Pythagorean Theorem}$$

$$8.75^2 + 18^2 = c^2 \quad \text{Replace } a \text{ with 8.75 and } b \text{ with 18.}$$

$$76.5625 + 324 = c^2 \quad \text{Evaluate } 8.75^2 \text{ and } 18^2.$$

$$400.5625 = c^2 \quad \text{Add 76.5625 and 324.}$$

$$\pm\sqrt{400.5625} = c \quad \text{Definition of square root}$$

$$\pm 20.0 \approx c \quad \text{Use a calculator.}$$

Since length cannot be negative, the ladder is about 20 feet long.

 CHECK Your Progress

a. AVIATION Write an equation that can be used to find the distance between the planes. Then solve. Round to the nearest tenth.

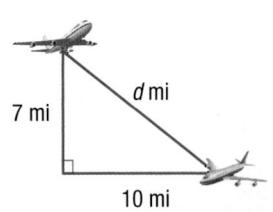

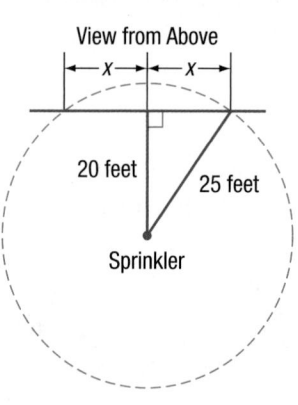

REAL-WORLD EXAMPLE

2 **LAWN CARE** A circular lawn sprinkler with a range of 25 feet is placed 20 feet from the edge of a lawn. Find the length of the section of the lawn's edge that is within the range of the sprinkler.

View from Above

20 feet

25 feet

Sprinkler

The distance of the sprinkler from the lawn's edge, the sprinkler's range, and a section of the lawn's edge all form a right triangle. The section of the lawn's edge within the range of the sprinkler is twice the section forming the right triangle.

Use the Pythagorean Theorem.

$$a^2 + b^2 = c^2 \qquad \text{Pythagorean Theorem}$$

$$20^2 + x^2 = 25^2 \qquad a = 20, b = x, \text{ and } c = 25.$$

$$400 + x^2 = 625 \qquad \text{Evaluate } 20^2 \text{ and } 25^2.$$

$$400 - 400 + x^2 = 625 - 400 \qquad \text{Subtract 400 from each side.}$$

$$x^2 = 225 \qquad \text{Simplify.}$$

$$x = \pm\sqrt{225} \qquad \text{Definition of square root}$$

$$x = 15 \text{ or } -15 \qquad \text{Simplify.}$$

The length of the section of the lawn's edge within the sprinkler's range is $x + x$ or $15 + 15 = 30$ feet.

CHECK Your Progress

b. STAIRS Mr. Parsons wants to build a new banister for the staircase shown. If the *rise* of the stairs of a building is 5 feet and the *run* is 12 feet, what will be the length of the new banister?

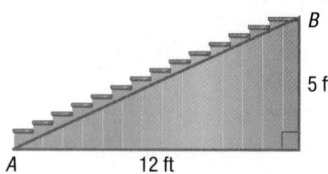

B

5 ft

A   12 ft

c. TENNIS In a singles tennis match, a serve is from one corner to inside the service zone as shown. If a serve is the entire length of the diagonal, how wide is the singles tennis court?

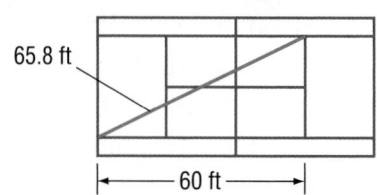

65.8 ft

60 ft

Example 1   Write an equation that can be used to answer the question. Then solve. Round to the nearest tenth if necessary.

1. What is the height of the tent?

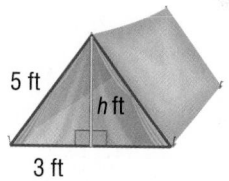

5 ft
h ft
3 ft

2. How high is the wheelchair ramp?

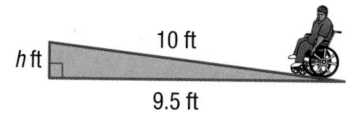

h ft   10 ft
9.5 ft

3. GEOMETRY An *isosceles* right triangle is a right triangle in which both legs are equal in length. If one leg of an isosceles triangle is 4 inches long, what is the length of the hypotenuse?

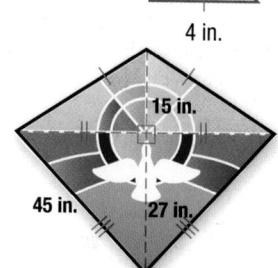

4 in.

Example 2   **4. WINDOWS** Shanise designed a stained glass window in the shape of a kite. What is the perimeter of the window?

15 in.

45 in.   27 in.

Practice and Problem Solving

● = Step-by-Step Solutions begin on page R1.
Extra Practice begins on page EP2.

Example 1   Write an equation that can be used to answer the question. Then solve. Round to the nearest tenth if necessary.

5 How far up the tree is the cat?

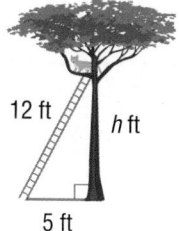

12 ft   *h* ft

5 ft

6. How deep is the water?

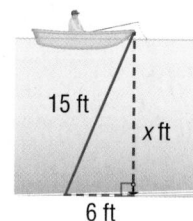

15 ft   *x* ft

6 ft

7. How far away is the bird?

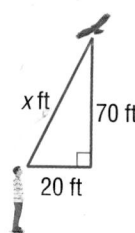

x ft   70 ft

20 ft

Example 2   **8. CAMP** Refer to the map of the Woodlands Camp at the right. Round your answers to the nearest tenth.

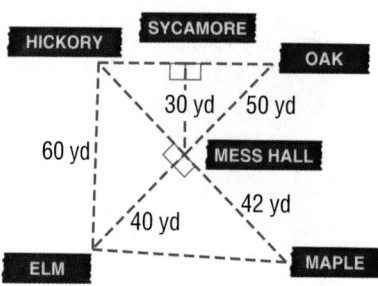

HICKORY   SYCAMORE   OAK
30 yd   50 yd
60 yd   MESS HALL
42 yd
40 yd
ELM   MAPLE

a. How far is it from Sycamore cabin to Oak cabin?

b. A camper in Hickory cabin wants to visit a friend in Elm cabin. How much farther is it if she walks to the Mess Hall first?

c. A group of campers walk from Elm cabin to Maple cabin, then to the Mess Hall. How far do they walk?

9. DISTANCE Larry wants to go from his house to his grandmother's house. How much distance is saved if he takes Main Street instead of Market and Exchange?

10. GEOGRAPHY Suppose Greenville, Rock Hill, and Columbia form a right triangle. What is the distance from Columbia to Greenville?

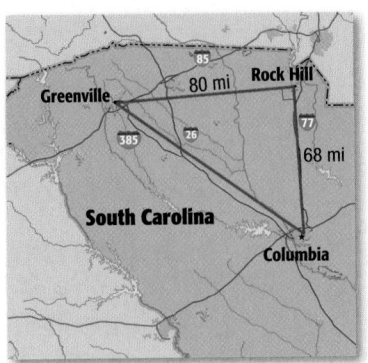

11. GIFTS Rodrigo is buying a $5\frac{1}{2}$-foot-long fishing rod for his father for his birthday. He wants to put it in a box so that his dad will not be able to guess what is in the box. The box he wants to use is 4 feet long and 4 feet wide. Will the pole fit in the box? Justify your reasoning.

12. GEOMETRY Find the length of the hypotenuse *AB*. The length of segment *AD* is congruent to the length of segment *DE*. Round your answer to the nearest tenth.

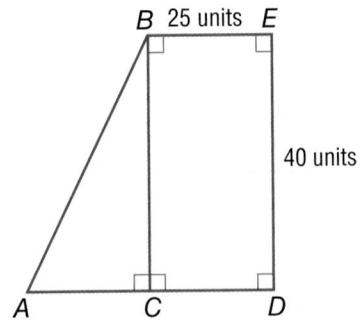

13. GRAPHIC NOVEL Refer to the graphic novel frame below for Exercises a–b.

a. What do you notice about the triangle formed by the rope, the distance from the boat, and the distance above the water?

b. The brochure says that the rope is 550 feet long and the parasailer is 450 feet behind the boat. Find the parasailer's height above the water.

·14. ENTERTAINMENT Connor loves to watch movies in the widescreen format on his television. He wants to buy a new television with a screen that is at least 25 inches by 13.6 inches. What diagonal-size television meets Connor's requirements?

Find the missing measure in each figure below. Round to the nearest tenth if necessary.

15

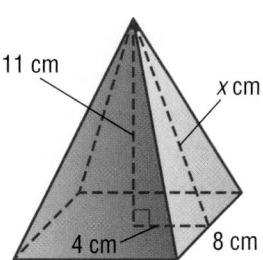

11 cm
x cm
4 cm
8 cm

16.

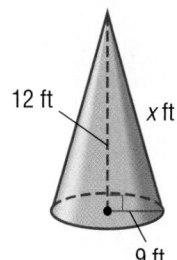

12 ft
x ft
9 ft

17.

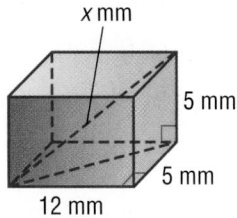

x mm
5 mm
5 mm
12 mm

18.

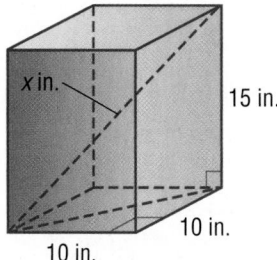

x in.
15 in.
10 in.
10 in.

19. MATH IN THE MEDIA Find examples of right triangles in a magazine or on the Internet. Estimate appropriate measures for the sides of the right triangles. Then check for reasonableness using the Pythagorean Theorem.

H.O.T. Problems

20. OPEN ENDED Write a problem that can be solved by using the Pythagorean Theorem. Then explain how to solve the problem.

21. Which One Doesn't Belong? Each set of numbers represents the side measures of a triangle. Identify the set that does not belong with the other three. Explain your reasoning.

| 3–4–5 | 12–35–37 | 3–5–7 | 6–8–10 |

22. CHALLENGE Suppose a ladder 20 feet long is placed against a vertical wall 20 feet high. How far would the top of the ladder move down the wall by pulling out the bottom of the ladder 5 feet? Explain your reasoning.

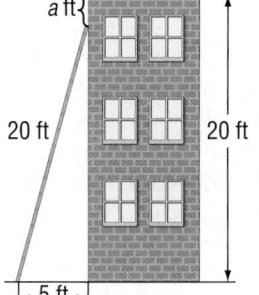

a ft
20 ft
20 ft
←5 ft→

23. [E] WRITE MATH The length of the hypotenuse of an isosceles right triangle is $\sqrt{288}$ units. Explain how to find the length of a leg.

24. Ms. Jimenez designed a rectangular garden. She plans to build a walkway through the garden as shown.

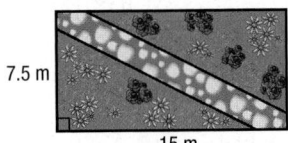

7.5 m

15 m

Which measure is closest to the length of the walkway?

A. 8 m

B. 11 m

C. 17 m

D. 23 m

25. Brayden is building the model bridge shown below.

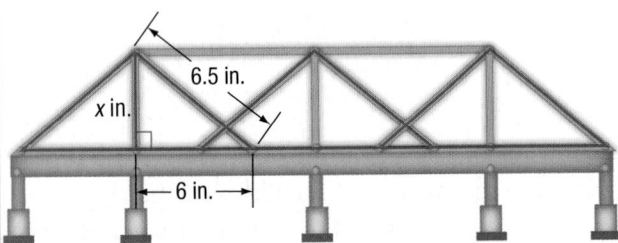

6.5 in.

x in.

6 in.

How long must he cut the piece of wood for one of the vertical support beams?

F. 8.8 in.

G. 3 in.

H. 2.5 in.

I. 0.5 in.

26. Anita is decorating for a party. She attached a streamer to the top of a wall and to the floor 4 feet away from the wall. Assuming the wall is perpendicular to the ground, what is the approximate height of the wall?

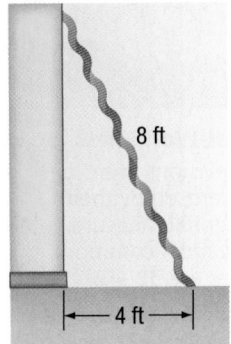

8 ft

4 ft

A. 4 ft

B. 7 ft

C. 9 ft

D. 12 ft

27. **THINK SOLVE EXPLAIN** **EXTENDED RESPONSE** The triangle shown below is a right triangle.

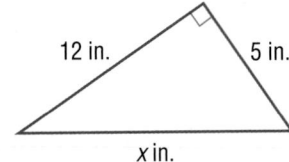

12 in.   5 in.

x in.

Part A Write an equation that can be used to find the length of the missing side of the triangle.

Part B Solve the equation in Part A. What is the length of the missing side?

Part C Give possible dimensions for another right triangle that is similar to the triangle shown.

Spiral Review

28. **GEOMETRY** Determine whether a triangle with sides 20 inches, 48 inches, and 52 inches long is a right triangle. Justify your answer. (Lesson 2B)

29. What is the height of the triangle shown at the right? Round to the nearest tenth. (Lesson 1E)

30. **MONUMENTS** The Washington Monument is the world's tallest *obelisk*, which is a pillar that tapers towards a pyramidal top. It stands a little over 555 feet tall. If the monument casts a shadow that is 190 feet long at the same time a nearby man casts a shadow that is 2 feet long, how tall is the man? Round to the nearest tenth. (Lesson 1D)

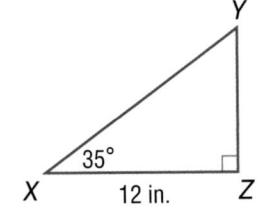

Y

35°

X   12 in.   Z

Main Idea

Find the distance between two points on the coordinate plane.

Vocabulary

Distance Formula

8.G.8

Distance on the Coordinate Plane

MOUNTAIN BIKING Evan was biking on a trail. A map of the trail is shown at the right. His brother timed his ride from point A to point B.

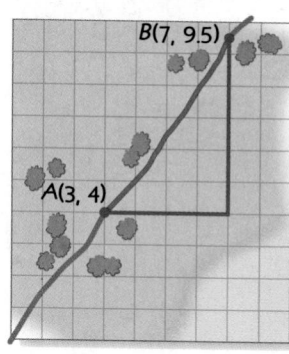

1. What does each colored line on the graph represent?

2. What type of triangle is formed by the lines?

3. What are the lengths of the two blue lines?

Recall that a coordinate plane is formed by the intersection of a vertical and horizontal number line at their zero points. The coordinate plane is separated into quadrants as shown.

You can use the Pythagorean Theorem to find the distance between two points on the coordinate plane.

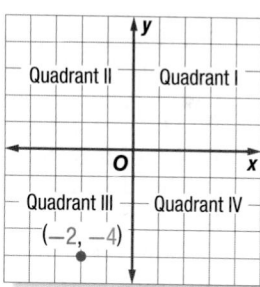

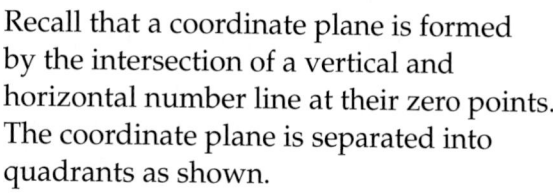 **EXAMPLE** **Find Distance on the Coordinate Plane**

① Graph the ordered pairs (3, 0) and (7, −5). Then find the distance c between the two points.

$a^2 + b^2 = c^2$ Pythagorean Theorem

$4^2 + 5^2 = c^2$ Replace a with 4 and b with 5.

$41 = c^2$ $4^2 + 5^2 = 16 + 25$ or 41

$\pm\sqrt{41} = \sqrt{c^2}$ Definition of square root

$\pm 6.4 \approx c$ Use a calculator.

The points are about 6.4 units apart.

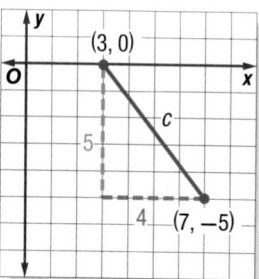

Study Tip

Distance To find the distance between two points on the coordinate plane, graph the points then draw a right triangle with c as the hypotenuse.

 CHECK Your Progress

Graph each pair of ordered pairs. Then find the distance between the points. Round to the nearest tenth.

a. (2, 0), (5, −4)   **b.** (1, 3), (−2, 4)   **c.** (−3, −4), (2, −1)

② **MAPS** On the map, each unit represents 45 miles. West Point, New York, is located at $\left(1\frac{1}{2}, 2\right)$ and Annapolis, Maryland, is located at $\left(-1\frac{1}{2}, -1\frac{1}{2}\right)$. What is the approximate distance between West Point and Annapolis?

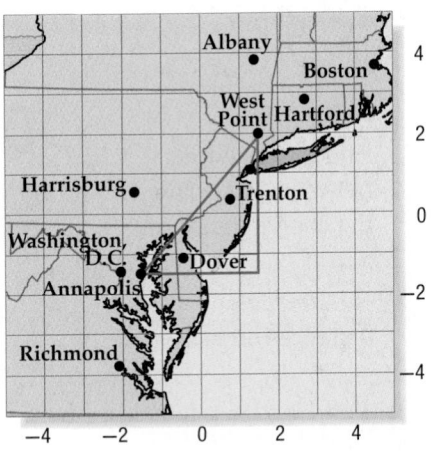

Let c represent the distance between West Point and Annapolis. Then $a = 3$ and $b = 3.5$.

$$a^2 + b^2 = c^2 \qquad \text{Pythagorean Theorem}$$
$$3^2 + 3.5^2 = c^2 \qquad \text{Replace } a \text{ with 3 and } b \text{ with 3.5.}$$
$$21.25 = c^2 \qquad 3^2 + 3.5^2 = 9 + 12.25 \text{ or } 21.25$$
$$\pm\sqrt{21.25} = \sqrt{c^2} \qquad \text{Definition of square root}$$
$$\pm 4.6 \approx c \qquad \text{Use a calculator.}$$

Since each map unit equals 45 miles, the distance between the cities is 4.6 · 45 or about 207 miles.

Real-World Link · · · ·

The United States Military Academy, also known as West Point, graduates more than 900 officers each year. The same is true for the United States Naval Academy, which is located in Annapolis.

✓ **CHECK Your Progress**

d. SPORTS Cromwell Field is located at $\left(2\frac{1}{2}, 3\frac{1}{2}\right)$ and Dedeaux Field at $\left(1\frac{1}{2}, 4\frac{1}{2}\right)$ on a map. Graph these points. If each map unit is 0.1 mile, about how far apart are the fields?

You can also use the **Distance Formula** to find the distance between two points on the coordinate plane. The Distance Formula is based on the Pythagorean Theorem as shown below.

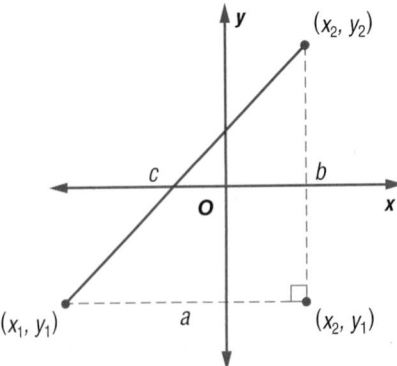

$$c^2 = a^2 + b^2 \qquad \text{Pythagorean Theorem}$$
$$c^2 = (x_2 - x_1)^2 + (y_2 - y_1)^2 \qquad \text{Substitute. The length of side } a \text{ is } (x_2 - x_1), \text{ and the length of side } b \text{ is } (y_2 - y_1).$$
$$c = \sqrt{(x_2 - x_1)^2 + (y_2 - y_1)^2} \qquad \text{Definition of square root}$$

Study Tip

Distance Formula Because distance is always positive, you can use either endpoint as (x_1, y_1). The result will be the same.

EXAMPLES   The Distance Formula

Use the Distance Formula to find the distance between each pair of points. Round to the nearest tenth if necessary.

3 $A(-3, 4)$, $B(2, -1)$

$d = \sqrt{(x_2 - x_1)^2 + (y_2 - y_1)^2}$	Distance Formula
$AB = \sqrt{[2 - (-3)]^2 + (-1 - 4)^2}$	$(x_1, y_1) = (-3, 4)$, $(x_2, y_2) = (2, -1)$
$AB = \sqrt{5^2 + (-5)^2}$	Simplify.
$AB = \sqrt{25 + 25}$	Evaluate 5^2 and $(-5)^2$.
$AB = \sqrt{50}$	Add 25 and 25.
$AB \approx \pm 7.1$	Use a calculator.

So, the distance between points A and B is about 7.1 units.

Check Use the Pythagorean Theorem.

$a^2 + b^2 = c^2$	Pythagorean Theorem
$5^2 + 5^2 = c^2$	Replace a with 5 and b with 5.
$50 = c^2$	$5^2 + 5^2 = 25 + 25$ or 50
$\pm\sqrt{50} = c$	Definition of square root
$\pm 7.1 \approx c$	$7.1 = 7.1$ The answer is correct. ✓

4 $X(5, -4)$, $Y(-3, -2)$

$d = \sqrt{(x_2 - x_1)^2 + (y_2 - y_1)^2}$	Distance Formula
$XY = \sqrt{(-3 - 5)^2 + [-2 - (-4)]^2}$	$(x_1, y_1) = (5, -4)$, $(x_2, y_2) = (-3, -2)$
$XY = \sqrt{(-8)^2 + 2^2}$	Simplify.
$XY = \sqrt{64 + 4}$	Evaluate $(-8)^2$ and 2^2.
$XY = \sqrt{68}$	Add 64 and 4.
$XY \approx \pm 8.2$	Simplify.

So, the distance between points X and Y is about 8.2 units.

CHECK Your Progress

e. $C(3, 4)$, $D(-2, 0)$ **f.** $S(-1, -3)$, $T(3, -5)$

Example 1 Graph each pair of ordered pairs. Then find the distance between the points. Round to the nearest tenth if necessary.

1. $(1, 5), (3, 1)$ **2.** $(-1, 0), (2, 7)$ **3.** $(-5.5, -2), (2.5, 3)$

4. BOATING Two boats leave an island at the same time. One of the boats travels 12 miles east and then 16 miles north. The second boat travels 24 miles south and then 18 miles west. Use the Pythagorean Theorem to find the distance between the boats.

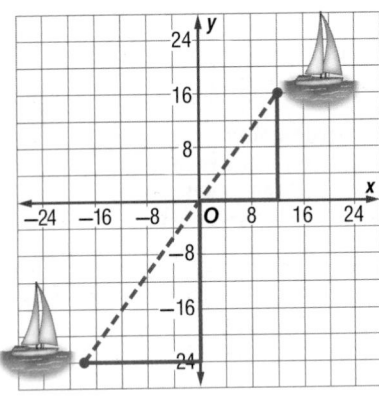

Example 2 **5. PARKS** On a park map, the ranger station is located at $(2.5, 3.5)$ and the nature center is located at $(0.5, 4)$. Each unit in the map is equal to 0.5 mile. Graph the ordered pairs. What is the approximate distance between the ranger station and the nature center?

Examples 3 and 4 Use the Distance Formula to find the distance between each pair of points. Round to the nearest tenth if necessary.

6. $A(-4, 2), B(1, 3)$ **7.** $M(0, 5), N(-3, -2)$ **8.** $R(6, 10), S(-7, -3)$

Practice and Problem Solving

● = Step-by-Step Solutions begin on page R1.
Extra Practice begins on page EP2.

Example 1 Graph each pair of ordered pairs. Then find the distance between the points. Round to the nearest tenth if necessary.

9. $(4, 5), (2, 2)$ **10.** $(6, 2), (1, 0)$ **11.** $(-3, 4), (1, 3)$

12. $(-5, 1), (2, 4)$ **13.** $(2.5, -1), (-3.5, -5)$ **14.** $(4, -2.3), (-1, -6.3)$

Example 2 **15 NAVIGATION** A ferry sets sail from an island located at $(4, 12)$ on the map at the right. Its destination is Ferry Landing B at $(6, 2)$. How far will the ferry travel if each unit on the grid is 0.5 mile?

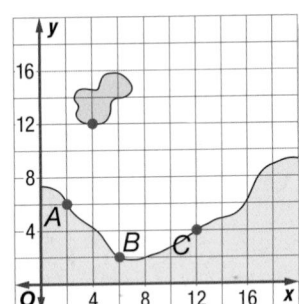

16. GEOGRAPHY On a map of Minnesota, Mankato is located at $(3, 2.5)$, and Duluth is located at $(8.5, 14.5)$. Each unit on the map equals 16.5 miles. Graph the ordered pairs. What is the approximate distance between the cities?

Use the Distance Formula to find the distance between each pair of points. Round to the nearest tenth if necessary.

17. $C(-5, -3), D(-4, -2)$ **18.** $W(1, 7), X(-2, -4)$ **19.** $Y(3.5, 1), Z(-4, 2.5)$

20. $G(-6.25, 5), H(-3.75, 2)$ **21.** $K(8\frac{1}{2}, 12), L(-6\frac{3}{4}, 7\frac{1}{2})$ **22.** $P(-9\frac{1}{4}, -7\frac{1}{2}), Q(-4, 5)$

Find the perimeter and area of each rectangle. Round to the nearest tenth if necessary.

23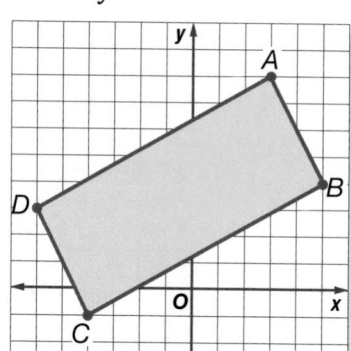

24.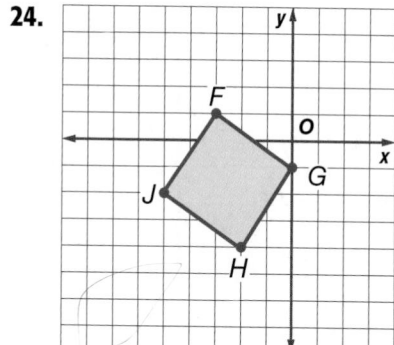

25. TRAVEL Chicago, Illinois, has a longitude of 88°W and a latitude of 42°N. Indianapolis, Indiana, is located at 86°W and 40°N. At this longitude/latitude, each degree is about 53 miles. Find the distance between Chicago and Indianapolis.

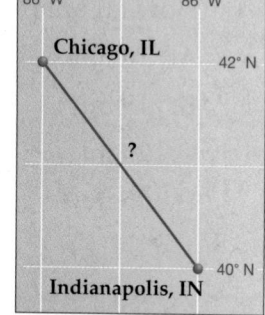

26. GEOMETRY One point is located at $(-5, 4)$, and another point is located at $(-8, -2)$. Find the distance between the points.

27. GEOMETRY Refer to the figure at the right.

   a. Verify the Pythagorean Theorem using the square built on each side of $\triangle ABC$. Explain your reasoning.

   b. Verify the length of AC using the Distance Formula.

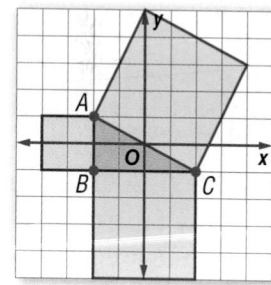

H.O.T. Problems

28. REASONING Layla needs to find the distance between the points $A(-2.4, 3.7)$ and $B(4.6, -1.3)$. Explain a method she could use to find the length. Then find the length.

29. CHALLENGE Apply what you have learned about distance on the coordinate plane to determine possible coordinates of the endpoints of a line segment that is neither horizontal nor vertical and has a length of 5 units.

30. ✏️ **WRITE MATH** In your own words, explain how to find the length of a non-vertical and a non-horizontal segment whose endpoints are (x_1, y_1) and (x_2, y_2).

31. Two cars leave a house in Richmond, Virginia. One of the cars travels 8 miles north and then 6 miles east. The second car travels 12 miles south and then 9 miles west.

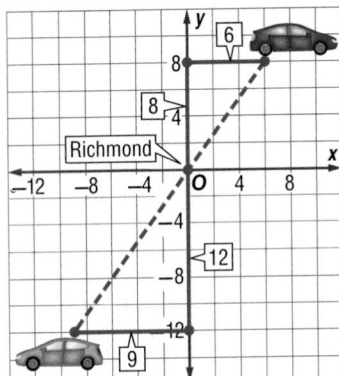

How far apart are the cars?

A. 10 mi

B. 15 mi

C. 20 mi

D. 25 mi

32. **EXTENDED RESPONSE** Mr. Brady designed a treasure hunt for his students. The hunt begins at the flagpole. The first clue is hidden 5 units north of the flagpole. The second clue is located 6 units east of the flagpole.

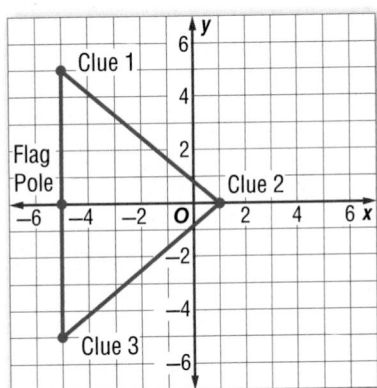

Part A Clue 2 says that Clue 3 is located 5 units south of the flagpole. What are the coordinates of Clue 3?

Part B To the nearest tenth, calculate the shortest distance from Clue 2 to Clue 3.

33. CHESS A knight moves two spaces over then one space up or down. About how far from its starting position is a knight that makes two moves, both of which are 2 spaces right and 1 space up? (Lesson 2C)

GEOMETRY Find the missing side of each right triangle. Round to the nearest tenth if necessary. (Lesson 2B)

34.

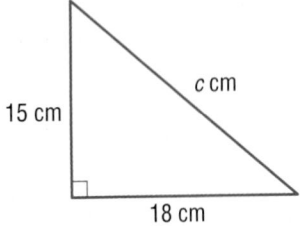

35.

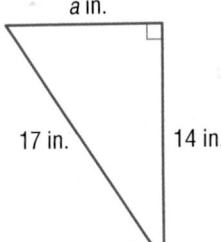

36.

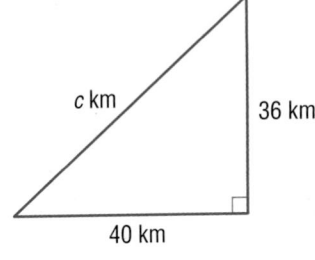

37. HOUSES A house is 25 feet tall, and a tree next to it is 40 feet tall. The tree casts a shadow that is 60 feet long. How long is the house's shadow at the same time? (Lesson 1D)

Main Idea

Graph and analyze slope triangles.

Get ConnectED

CCSS 8.EE.6, 8.F.4

Slope Triangles

Refer to the graph at the right. Triangle ABC is formed by the rise, run, and section of the line between points A and B.

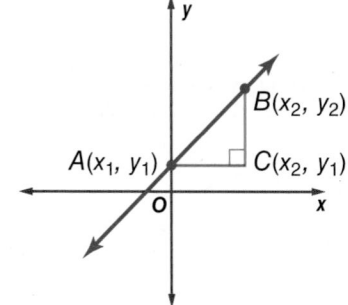

ACTIVITY

STEP 1 Graph $y = x + 1$ on a sheet of grid paper. Make the graph as large as possible.

STEP 2 Select two points on the line and create a right triangle.

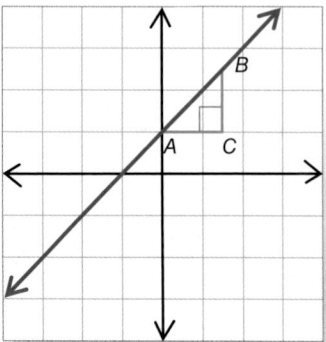

Analyze the Results

1. What is the slope of the line?

2. Select two different points on the line. Make another right triangle from these two points. Is the slope of this triangle the same as the first? Explain.

3. What do you notice about the two triangles?

4. The triangles in the activity above are called *slope triangles*. Write a definition for the term slope triangle.

5. **MAKE A CONJECTURE** What do you think is true of any two slope triangles formed on the same line? Explain.

Main Idea

Use special right triangles to solve problems.

 8.G.7

Special Right Triangles

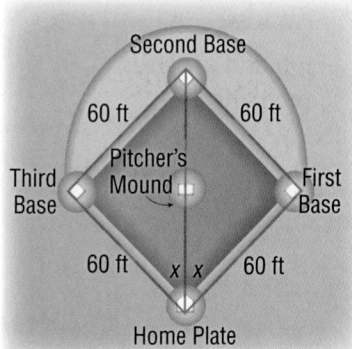

SOFTBALL A softball field is in the shape of a square with each base and home plate 60 feet apart. The diagonal from home plate to second base goes through the pitcher's mound.

1. What is the angle measure of each angle x at home plate if the diagonal divides it in half?

2. What is the approximate distance from home plate to second base?

In the triangles below, the corresponding angles have the same measure and the corresponding sides are proportional with a scale factor of 2. This suggests that all 45°-45°-90° triangles are similar.

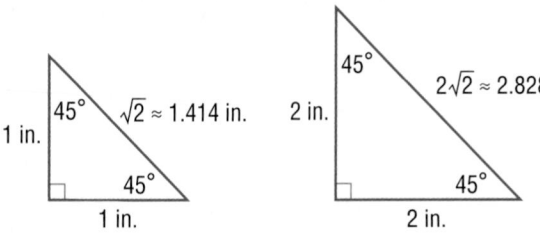

EXAMPLE   **45°-45°-90° Triangles**

① Triangle *ABC* and triangle *XYZ* are 45°-45°-90° triangles. Find the length of the hypotenuse in △*XYZ*.

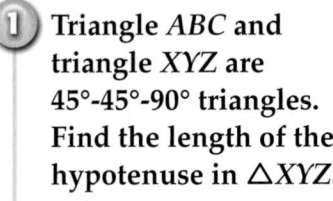

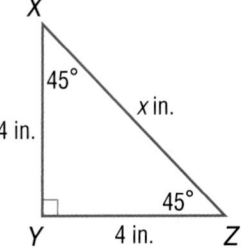

The scale factor from

△*ABC* to △*XYZ* is $\frac{4}{1}$ or 4. Use the scale factor to find the hypotenuse.

$x = 4 \cdot \sqrt{2}$ Multiply the length of $\overline{AC}$ by the scale factor, 4.

$\quad = 4\sqrt{2}$

So, the hypotenuse of △*XYZ* measures $4\sqrt{2}$ inches.

 CHECK Your Progress

a. Triangle *DEF* is a 45°-45°-90° triangle. If the legs measure 3 meters, what is the length of the hypotenuse?

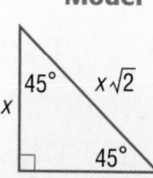

Key Concept — 45°-45°-90° Triangles

Words	In a 45°-45°-90° triangle, the length of the hypotenuse is $\sqrt{2}$ times the length of a leg.	Model

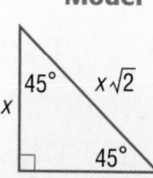

Symbols   hypotenuse = leg · $\sqrt{2}$

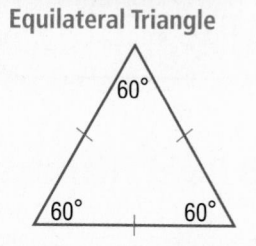

QUICK Review

Equilateral Triangle

Study the following equilateral triangle. The altitude of an equilateral triangle divides the triangles into two 30°-60°-90° triangles.

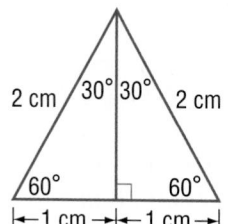 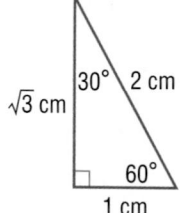

This example shows that the hypotenuse of a 30°-60°-90° triangle is twice as long as the shorter leg of the same triangle. Using the Pythagorean Theorem, the length of the third side is $\sqrt{3}$ centimeters long.

Just as 45°-45°-90° triangles are similar, 30°-60°-90° triangles are similar.

Study Tip

Exact Answers When you use a calculator to find a square root of a non-perfect square, you are finding an approximate answer. The exact answer is when you leave the solution in radical form.

EXAMPLE — Find Missing Measures of a 30°-60°-90° Triangle

2 Triangle *ABC* and triangle *DEF* are 30°-60°-90° triangles. Find the exact length of the missing measures.

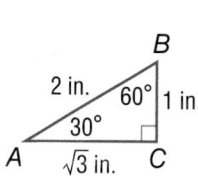 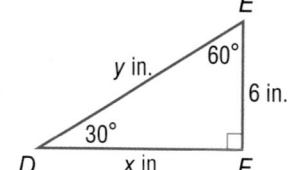

The scale factor from △*ABC* to △*DEF* is 6. Use the scale factor to find the missing measures.

$y = 6 \cdot 2$ or 12   Multiply the length of $\overline{AB}$ by the scale factor.

So, $\overline{DE}$ is 12 inches.

$x = 6 \cdot \sqrt{3}$ or $6\sqrt{3}$ Multiply the length of $\overline{AC}$ by the scale factor.

So, $\overline{DF}$ is $6\sqrt{3}$ inches.

CHECK Your Progress

b. Find the exact length of the missing measures in the 30°-60°-90° triangle.

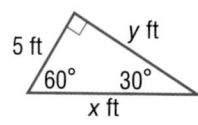

Key Concept — 30°-60°-90° Triangles

Words

In a 30°-60°-90° triangle, the length of the hypotenuse is twice the length of the shorter leg, and the length of the longer leg is $\sqrt{3}$ times the length of the shorter leg.

Model

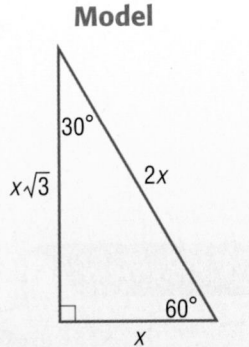

Symbols

hypotenuse = 2 · shorter leg
longer leg = $\sqrt{3}$ · shorter leg

REAL-WORLD EXAMPLE — Use Special Right Triangles

3 TOWERS The radio tower shown is 100 feet tall. It is supported by two cables that run from the top of the tower to points on the ground 100 feet from the base of the tower. What is the length of one of the cables? Round to the nearest tenth.

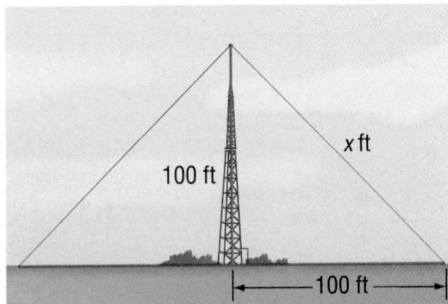

The triangle shown is a 45°-45°-90° triangle. The hypotenuse of a 45°-45°-90° triangle is $\sqrt{2}$ times the length of a leg.

$$\text{hypotenuse} = \text{leg} \cdot \sqrt{2} \quad \text{Relationship between sides}$$
$$x = 100 \cdot \sqrt{2} \quad \text{Substitute.}$$

This is a real-world situation. Use a calculator to find the decimal approximation of the length.

100 [×] [2nd] [$\sqrt{\ }$] 2 [ENTER] 141.4213562

$100 \cdot \sqrt{2} \approx 141.4$ Round to the nearest tenth.

The cable is about 141.4 feet long.

CHECK Your Progress

c. RAMPS The skateboard ramp shown is a 30°-60°-90° triangle. The height of the ramp is 24 feet. Find the length of the ramp.

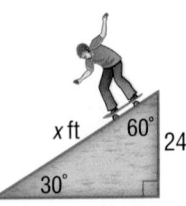

d. CONSTRUCTION A 15-foot square backdrop for the school play is supported by a diagonal brace. Find the length of the diagonal brace. Round to the nearest tenth.

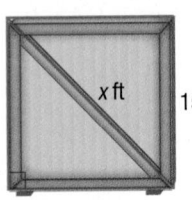

Examples 1 and 2 **Find each missing measure.**

1.
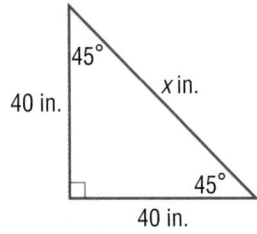

2.
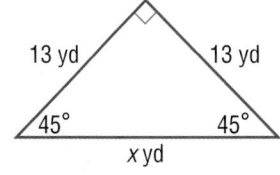

3.
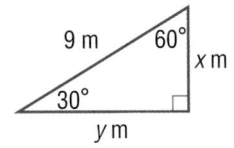

4.
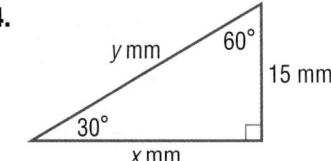

5. In a 45°-45°-90° triangle, a leg is 30 centimeters long. Find the exact length of the hypotenuse.

6. The shorter leg of a 30°-60°-90° triangle is 17 millimeters long. Find the exact length of the longer leg and the length of the hypotenuse.

Example 3 7. **BILLIARDS** A billiards table measures 4 feet wide. Find each missing measure. Round to the nearest tenth.

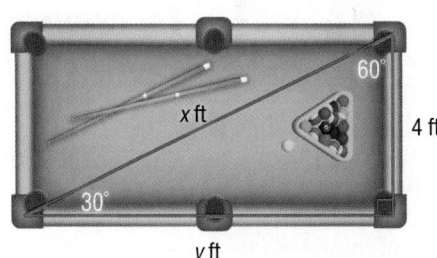

Practice and Problem Solving

● = Step-by-Step Solutions begin on page R1.
Extra Practice begins on page EP2.

Examples 1 and 2 **Find each missing measure.**

8.
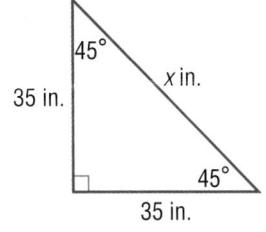

9.
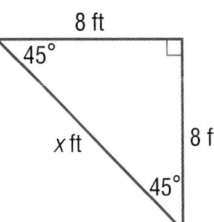

10.
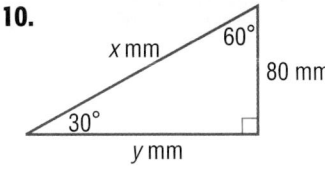

11.
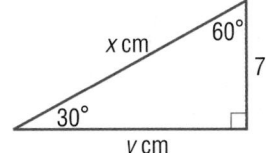

12.
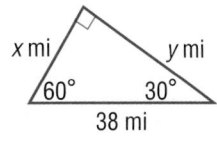

13.
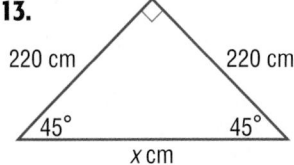

14. In a 45°-45°-90° triangle, a leg is 8.5 yards long. Find the exact length of the hypotenuse.

15. In a 30°-60°-90° triangle, the hypotenuse is 68 centimeters long. Find the exact lengths of the legs.

Example 3 **16. TOWN SQUARES** The town square of Richville is shown below. Each block is 500 feet long. What is the approximate distance x between the post office and the library?

17. GARDENING The triangular garden shown below is in the shape of a 30°-60°-90° triangle. What is the length of the hypotenuse?

18. GEOMETRY Find the perimeter of the triangle shown to the nearest tenth of a meter.

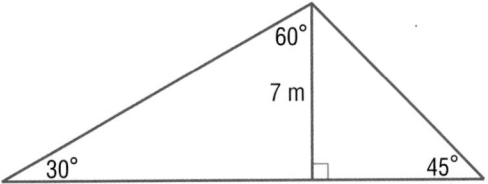

19 GYMNASTICS The mat that gymnasts perform their floor routines on is square with an area of 1,600 square feet. Find the exact length of the diagonal of the floor mat.

Find each missing measure.

20.

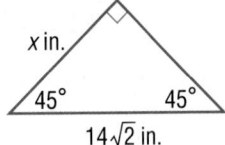

21.

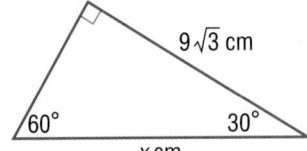

22. GEOMETRY The regular hexagon shown is made up of six equilateral triangles. The *apothem* of a regular polygon is the distance along a segment from the center of the polygon perpendicular to a side. If the apothem of the regular hexagon is $7\sqrt{3}$ millimeters, find the perimeter of the hexagon.

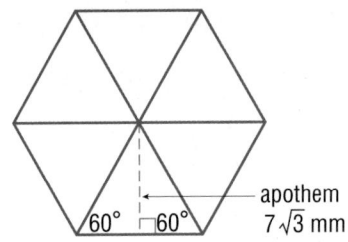

23. GEOMETRY Find the length of x in the cone at the right.

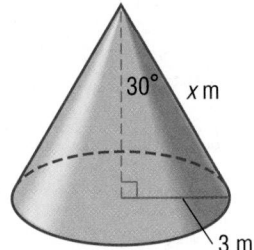

24. CONSTRUCTION A conveyor belt hauls packages of shingles to a roof that is under construction. The base of the conveyor belt makes a 30° angle with the ground, and the belt is 24 feet long. How far from the house is the base of the conveyor belt? Round to the nearest tenth.

In Example 1, notice that the vertices of the image can also be found by adding 2 to the x-coordinates and –5 to the y-coordinates.

Key Concept — Translations in the Coordinate Plane

Words When a figure is translated, the x-coordinate of the preimage changes by the value of the horizontal translation *a*. The y-coordinate of the preimage changes by the vertical translation *b*.

Model

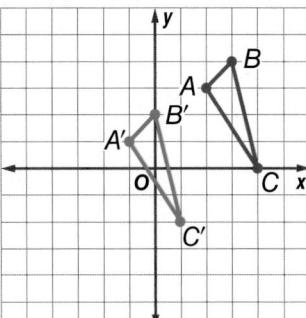

Symbols $(x, y) \rightarrow (x + a, y + b)$

In a translation, every point is moved the same distance in the same direction. So, a translation can be described using an ordered pair.

✎ REAL-WORLD EXAMPLE

2 **COMPUTER GRAPHICS** A computer image is being translated to create the illusion of movement. Use translation notation to describe the translation from point *A* to point *B*.

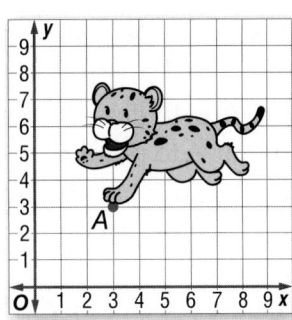 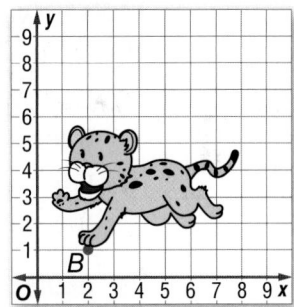

Point *A* is located at (3, 3). Point *B* is located at (2, 1).

$$(x, y) \rightarrow (x + a, y + b)$$
$$(3, 3) \rightarrow (3 + a, 3 + b) \rightarrow (2, 1)$$

$3 + a = 2$	$3 + b = 1$
$a = -1$	$b = -2$

So, the translation is $(x - 1, y - 2)$, 1 unit to the left and 2 units down.

✓ CHECK Your Progress

c. COMPUTER GRAPHICS Refer to the figure in Example 2. If point *A* was at (1, 5), use translation notation to describe the translation from point *A* to point *B*.

Example 1 Graph △XYZ with vertices X(−4, −4), Y(−3, −1), and Z(2, −2). Then graph the image of △XYZ after each translation, and write the coordinates of its vertices.

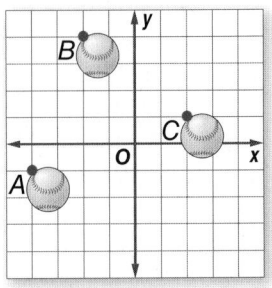

    1. 3 units right and 4 units up

    2. 2 units left and 3 units down

Example 2 **3. ANIMATION** The baseball at the right was filmed using stop-motion animation so it appears to be thrown in the air. Use translation notation to describe the translation from point A to point B.

Practice and Problem Solving

= **Step-by-Step Solutions** begin on page R1.
Extra Practice begins on page EP2.

Example 1 Graph each figure with the given vertices. Then graph the image of the figure after the indicated translation, and write the coordinates of its vertices.

    4. △ABC with vertices A(1, 2), B(3, 1), and C(3, 4) translated 2 units left and 1 unit up

    5. rectangle JKLM with vertices J(−3, 2), K(3, 5), L(4, 3), and M(−2, 0) translated 1 unit right and 4 units down

Example 2 **DIGITAL GAMES** Use the image of the race car at the right.

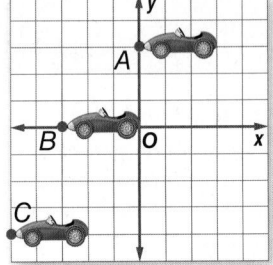

    6. Use translation notation to describe the translation from point A to point B.

    7. Use translation notation to describe the translation from point B to point C.

    8. GEOMETRY Triangle ABC has vertices A(4, 3), B(−7, 0), and C(6, 5). When translated, A′ has coordinates (−1, 3). Find the coordinates of B′ and C′. Then describe the translation of triangle ABC.

    9 **GEOMETRY** Quadrilateral KLMN has vertices K(−2, −2), L(1, 1), M(0, 4), and N(−3, 5). It is first translated by (x + 2, y − 1) and then translated by (x − 3, y + 4). When a figure is translated twice, a double prime symbol is used. Find the coordinates of K″L″M″N″ after both translations.

    10. SCIENCE A diagram of a DNA double helix is shown below. Look for a pattern. Copy the double helix and indicate where this pattern repeats or is translated. Find how many translations of the original pattern are shown in the diagram.

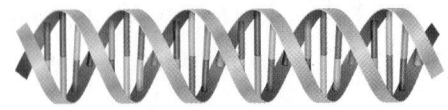

11. REASONING A figure is translated by $(x - 5, y + 7)$. Then the result is translated by $(x + 5, y - 7)$. Without graphing, what is the final position of the figure? Explain your reasoning.

12. CHALLENGE What are the coordinates of the point (x, y) after being translated m units left and n units up?

13. ✏️ **WRITE MATH** Write a real-world problem in which you would need to translate a figure. Then solve your problem.

✓ Test Practice

14. If $\triangle PQR$ is translated 4 units right and 3 units up, what are the coordinates of R'?

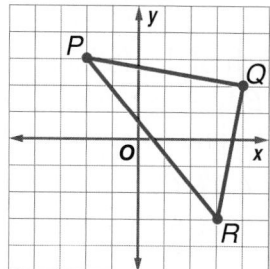

A. $(-1, -6)$

B. $(7, 0)$

C. $(-1, 0)$

D. $(7, -6)$

15. What are the coordinates of C' of trapezoid $ABCD$ after a translation 3 units right and 7 units down?

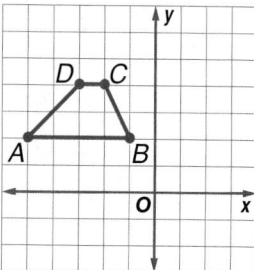

F. $(1, 3)$

G. $(5, 7)$

H. $(1, -3)$

I. $(-9, 1)$

Spiral Review

Find the exact length of each missing side. (Lesson 2F)

16.
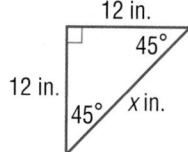

17.
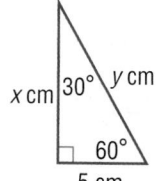

18.
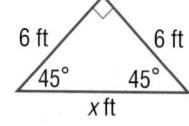

Graph each set of ordered pairs. Then find the distance between the points. Round to the nearest tenth. (Lesson 2D)

19. $(1, 4), (6, -3)$

20. $(-1, 5), (3, -2)$

21. $(-5, -2), (-1, 0)$

Reflections

 Explore

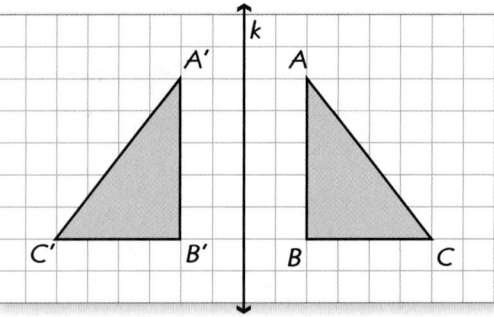

Step 1 Draw △*ABC*, △*A′B′C′*, and line *k* on grid paper as shown.

Step 2 Measure the distance from point *A* to line *k* and from point *A′* to line *k*.

Step 3 Measure the distance from each of the other four vertices to line *k*.

1. Compare the distance between each of the following: point *A* to line *k* and point *A′* to line *k*, point *B* to line *k* and point *B′* to line *k*, and point *C* to line *k* and point *C′* to line *k*.

2. Compare and contrast △*ABC* and △*A′B′C′*.

A **reflection** is a mirror image of the original figure. It is the result of a transformation of a figure over a line called a **line of reflection**. In a reflection, each point of the preimage and its image are the same distance from the line of reflection. So, in a reflection, the image is congruent to the preimage.

EXAMPLE **Reflect a Figure Over the *x*-axis**

1 **Triangle *ABC* has vertices *A*(5, 2), *B*(1, 3) , and *C*(−1, 1). Graph the figure and its reflected image over the *x*-axis. Then find the coordinates of the vertices of the reflected image.**

The *x*-axis is the line of reflection. So, plot each vertex of *A′B′C′* the same distance from the *x*-axis as its corresponding vertex on *ABC*.

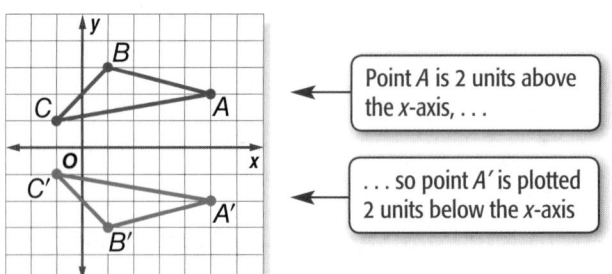

Point *A* is 2 units above the *x*-axis, . . .

. . . so point *A′* is plotted 2 units below the *x*-axis

The coordinates are *A′*(5, −2), *B′*(1, −3), and *C′*(−1, −1).

CHECK Your Progress

a. Rectangle *GHIJ* has vertices *G*(3, −4), *H*(3, −1), *I*(−2, −1), and *J*(−2, −4). Graph the figure and its image after a reflection over the *x*-axis. Then find the coordinates of the reflected image.

Study Tip

Reflections Notice that in a reflection, the orientation of a preimage and its image are reversed.

EXAMPLE Reflect a Figure Over the *y*-axis

2 **Quadrilateral *KLMN* has vertices *K*(2, 3), *L*(5, 1), *M*(4, −2), and *N*(1, −1). Graph the figure and its reflection over the *y*-axis. Then find the coordinates of the vertices of the reflected image.**

The *y*-axis is the line of reflection. So, plot each vertex of *K'L'M'N'* the same distance from the *y*-axis as its corresponding vertex on *KLMN*.

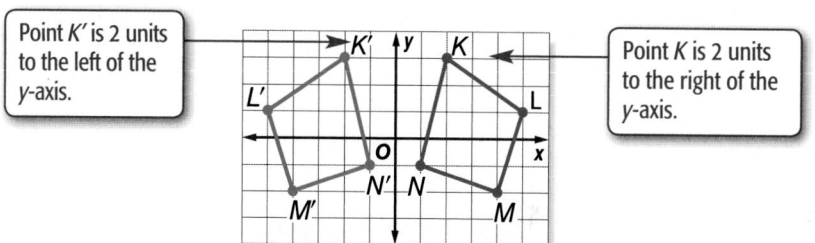

Point *K'* is 2 units to the left of the *y*-axis.

Point *K* is 2 units to the right of the *y*-axis.

The coordinates are *K'*(−2, 3), *L'*(−5, 1), *M'*(−4, −2), and *N'*(−1, −1).

CHECK Your Progress

b. Triangle *PQR* has vertices *P*(1, 5), *Q*(3, 7), and *R*(5, −1). Graph the figure and its reflection over the *y*-axis. Then find the coordinates of the reflected image.

Key Concept   Reflections in the Coordinate Plane

	Over the *x*-axis	Over the *y*-axis
Words	To reflect a figure over the *x*-axis, multiply the *y*-coordinates by −1.	To reflect a figure over the *y*-axis, multiply the *x*-coordinates by −1.
Symbols	(*x*, *y*) → (*x*, −*y*)	(*x*, *y*) → (−*x*, *y*)
Models	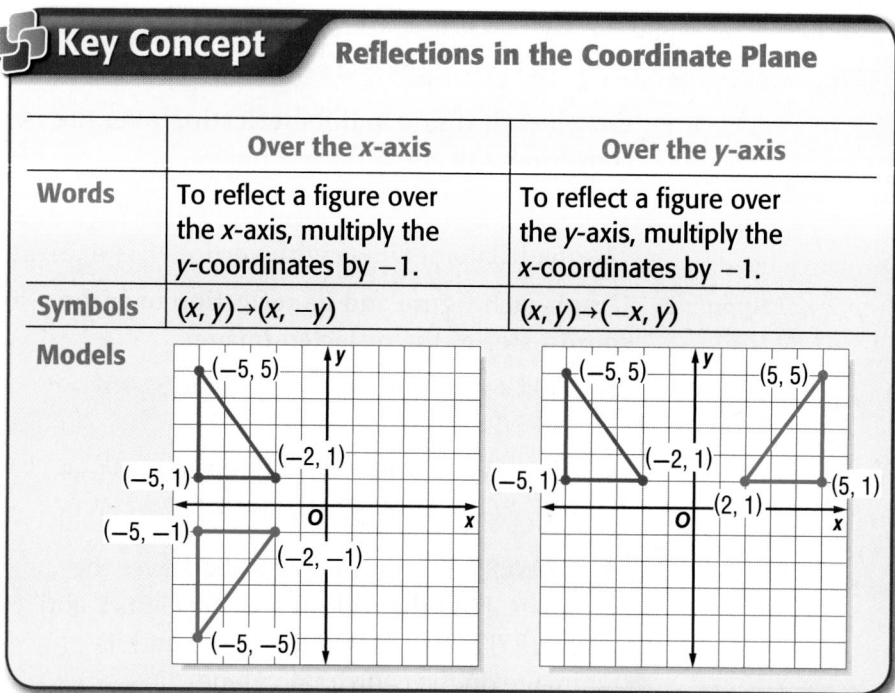	

REAL-WORLD EXAMPLE Use a Reflection

3 **ART** The figure below is reflected over the y-axis. Find the coordinates of point A' and point B'. Then sketch the figure and its image on the coordinate plane.

Point A is located at $(1, 4)$. Point B is located at $(2, 1)$. Since the figure is being reflected over the y-axis, multiply the x-coordinates by -1.

$$A(1, 4) \rightarrow A'(-1, 4)$$
$$B(2, 1) \rightarrow B'(-2, 1)$$

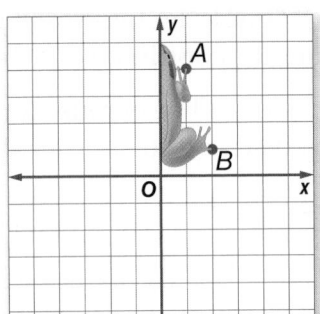 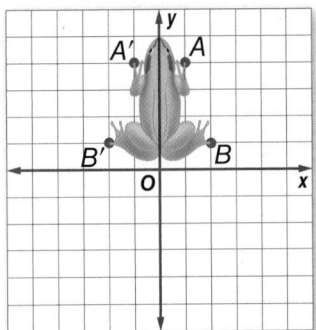

CHECK Your Progress

c. ART The figure at the right is reflected over the x-axis. Find the coordinates of point A' and point B'. Then sketch the figure and its image on the coordinate plane.

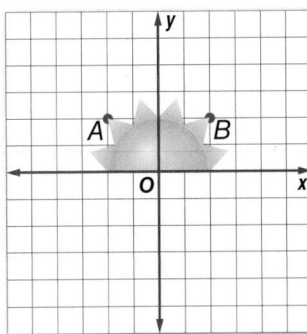

CHECK Your Understanding

Example 1 Graph each figure and its reflection over the x-axis. Then find the coordinates of the reflected image.

1. $\triangle ABC$ with vertices $A(5, 8)$, $B(1, 2)$, and $C(6, 4)$

2. quadrilateral $DEFG$ with vertices $D(-4, 6)$, $E(-2, -3)$, $F(2, 2)$, and $G(4, 9)$

Example 2 Graph each figure and its reflection over the y-axis. Then find the coordinates of the reflected image.

3. $\triangle QRS$ with vertices $Q(2, -5)$, $R(5, -5)$, and $S(2, 3)$

4. parallelogram $WXYZ$ with vertices $W(-4, -2)$, $X(-4, 3)$, $Y(-2, 4)$, and $Z(-2, -1)$

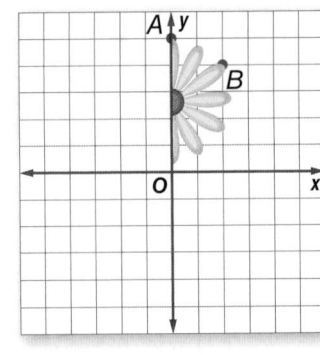

Example 3 **5. FLOWERS** The figure is reflected over the y-axis. Find the coordinates of point A' and point B'. Then sketch the figure and its image on the coordinate plane.

Practice and Problem Solving

● = **Step-by-Step Solutions** begin on page R1.
Extra Practice begins on page EP2.

Example 1 **Graph each figure and its reflection over the *x*-axis. Then find the coordinates of the reflected image.**

6. *TUV* with vertices *T*(−6, −1), *U*(−2, −3), and *V*(5, −4)

7. *MNP* with vertices *M*(2, 1), *N*(−3, 1), and *P*(−1, 4)

8. square *ABCD* with vertices *A*(2, 4), *B*(−2, 4), *C*(−2, 8), and *D*(2, 8)

9 *WXYZ* with vertices *W*(−1, −1), *X*(4, 1), *Y*(4, 5), and *Z*(1, 7)

Example 2 **Graph each figure and its reflection over the *y*-axis. Then find the coordinates of the reflected image.**

10. △*RST* with vertices *R*(−5, 3), *S*(−4, −2) , and *T*(−2, 3)

11. △*GHJ* with vertices *G*(4, 2), *H*(3, −4), and *J*(1, 1)

12. parallelogram *HIJK* with vertices *H*(−1, 3), *I*(−1, −1), *J*(2, −2), and *K*(2, 2)

13. quadrilateral *DEFG* with vertices *D*(1, 0), *E*(1, −5), *F*(4, −1), and *G*(3, 2)

Example 3 **14. ANIMALS** The figure below is reflected over the *y*-axis. Find the coordinates of point *A′* and point *B′*. Then sketch the figure and its image on the coordinate plane.

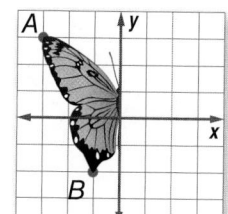

15. QUILTS The figure below is reflected over the *x*-axis. Find the coordinates of point *A′* and point *B′*. Then sketch the figure and its image on the coordinate plane.

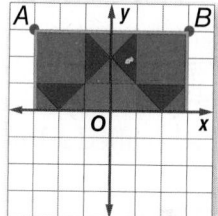

The coordinates of a point and its image after a reflection are given. Describe the reflection as over the *x*-axis or *y*-axis.

16. *A*(−3, 5) → *A′*(3, 5)

17 *M*(3, 3) → *M′*(3, −3)

18. *X*(−1, −4) → *X′*(−1, 4)

19. *W*(−4, 0) → *W′*(4, 0)

H.O.T. Problems

20. OPEN ENDED Make a tessellation using a combination of translations and reflections of polygons. Explain your method.

21. CHALLENGE Triangle *JKL* has vertices *J*(−7, 4), *K*(7, 1), and *L*(2, −2). Without graphing, find the new coordinates of the vertices of the triangle after a reflection first over the *x*-axis and then over the *y*-axis.

22. WRITE MATH Draw a figure on a coordinate plane and its reflection over the *y*-axis. Explain how the *x*- and *y*-coordinates of the reflected figure relate to the *x*- and *y*-coordinates of the original figure. Then repeat, this time reflecting the figure over the *x*-axis.

23. The figure shown was transformed from Quadrant II to Quadrant III.

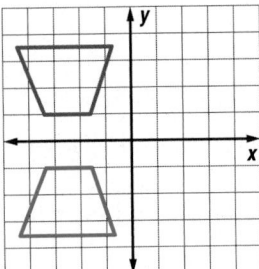

This transformation best represents which of the following?

A. translation 2 units up

B. translation 2 units down

C. reflection over the *x*-axis

D. reflection over the *y*-axis

24. If *ABCD* is reflected over the *x*-axis and translated 5 units to the right, which is the resulting image of point *B*?

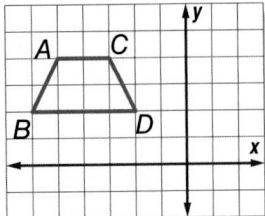

F. $(-1, -2)$

G. $(-11, 2)$

H. $(-1, 2)$

I. $(11, 2)$

More About Reflections

Figures that match exactly when folded in half have **line symmetry**. The figures at the right have line symmetry. Each fold line is called a **line of symmetry**.

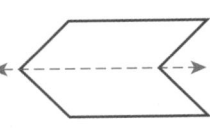
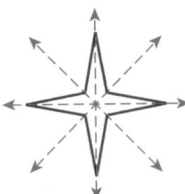

EXAMPLES

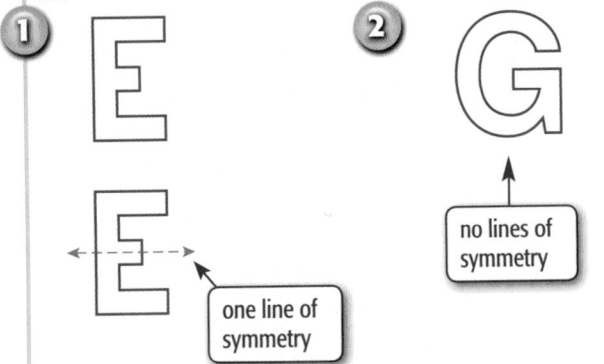

① one line of symmetry

② no lines of symmetry

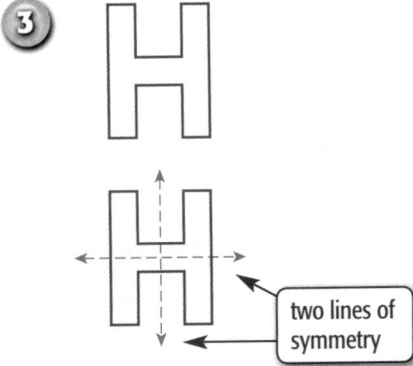

③ two lines of symmetry

Determine whether each figure has line symmetry. If so, copy the figure and draw all lines of symmetry.

25.

26.

27.

Explore Rotational Symmetry

Main Idea

Identify rotational symmetry.

 Vocabulary

rotational symmetry
angle of rotation

 Get ConnectED

 8.G.1, 8.G.1a, 8.G.1b, 8.G.1c, 8.G.3

LOGOS Many products have logos so people can easily identify the products. A figure has **rotational symmetry** if it can be rotated or turned less than 360° about its center so that the figure looks exactly as it does in its original position. Does the first aid logo have rotational symmetry?

First aid box

ACTIVITY

STEP 1 Copy the outline of the first aid logo onto both a piece of tracing paper and a transparency. Label one vertex *A*.

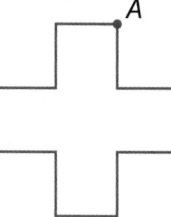

STEP 2 Place the transparency over the outline on your tracing paper. Put your pencil point at the center of the cross to hold the transparency in place. Turn the transparency from its original position so that the two figures match.

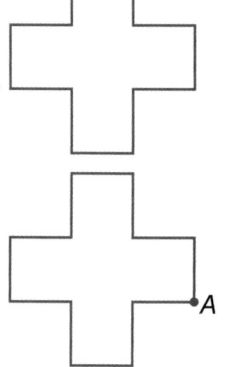

STEP 3 Continue turning the transparency until the logo is back to its original position.

Analyze the Results

1. Does the figure have rotational symmetry? Explain.

2. The degree measure of the angle through which the figure is rotated is called the **angle of rotation**. Find the first angle of rotation by dividing 360° by the total number of times the figures matched.

3. List the other angles of rotation by adding the measure of the first angle of rotation to the previous angle measure. Stop when you reach 360°.

Determine whether the figure has rotational symmetry. Write *yes* or *no*. If *yes*, name its angle(s) of rotation.

4.

5.

6.

Main Idea

Graph rotations on the coordinate plane.

Vocabulary

rotation
center of rotation

Get ConnectED

CCSS 8.G.1, 8.G.1a, 8.G.1b, 8.G.1c, 8.G.3

Rotations

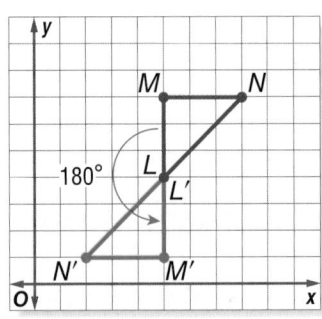

PRIZES Lijuan is spinning the prize wheel shown.

1. If the section labeled 10 makes three complete turns clockwise, how many degrees will it have traveled?

2. Does the center of the wheel change if the wheel is spun counterclockwise as opposed to clockwise?

3. Does the distance from the center to the edge change as it spins?

A **rotation** is a transformation in which a figure is rotated, or turned, about a fixed point. The **center of rotation** is the fixed point. A rotation does not change the size or shape of the figure. So, the preimage and the image are congruent.

 EXAMPLE **Rotate a Figure About a Point**

1 **Triangle *LMN* has vertices *L*(5, 4), *M*(5, 7), and *N*(8, 7). Graph the figure and its image after a counterclockwise rotation of 180° about vertex *L*. Then give the coordinates of the vertices for △*L'M'N'*.**

Step 1 Graph the original triangle.

Step 2 Graph the rotated image. Use a protractor to measure an angle of 180° with *M* as one point on the ray and *L* as the vertex. Mark off a point the same distance as *ML*. Label this point *M'* as shown.

Step 3 Repeat Step 2 for point *N*. Since *L* is the point at which △*LMN* is rotated, *L'* will be in the same position as *L*.

So, the coordinates of the vertices of △*L'M'N'* are *L'*(5, 4), *M'*(5, 1), and *N'*(2, 1).

CHECK Your Progress

a. Rectangle *ABCD* has vertices *A*(−7, 4), *B*(−7, 1), *C*(−2, 1), and *D*(−2, 4). Graph the figure and its image after a clockwise rotation of 90° about vertex *C*. Then give the coordinates of the vertices for rectangle *A'B'C'D'*.

Figures can also be rotated about the origin. The rotations shown below are clockwise rotations about the origin.

Key Concept — Rotations in the Coordinate Plane

Words A rotation is a transformation around a fixed point. Each point of the original figure and its image are the same distance from the center of rotation.

Models

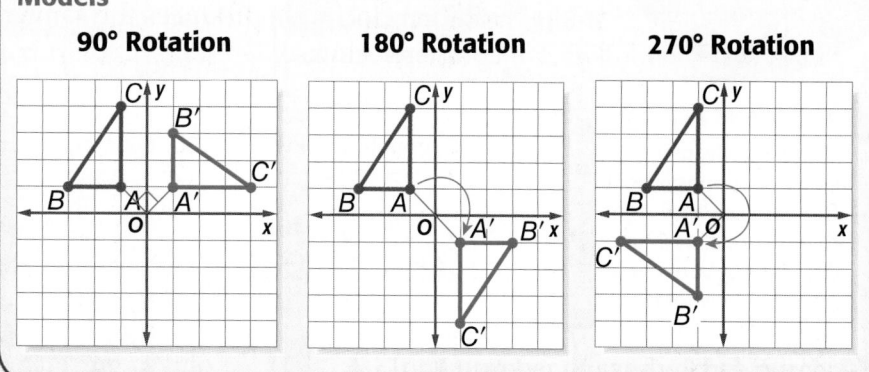

90° Rotation   180° Rotation   270° Rotation

EXAMPLE Rotate a Figure About the Origin

2 Triangle *DEF* has vertices *D*(−4, 4), *E*(−1, 2), and *F*(−3, 1). Graph the figure and its image after a clockwise rotation of 90° about the origin. Then give the coordinates of the vertices for △*D'E'F'*.

Step 1 Graph △*DEF* on a coordinate plane.

Step 2 Sketch segment $\overline{EO}$ connecting point *E* to the origin. Sketch another segment, $\overline{E'O}$, so that the angle between point *E*, *O*, and *E'* measures 90° and the segment is congruent to $\overline{EO}$.

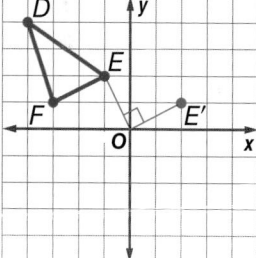

Step 3 Repeat Step 2 for points *D* and *F*. Then connect the vertices to form △*D'E'F'*.

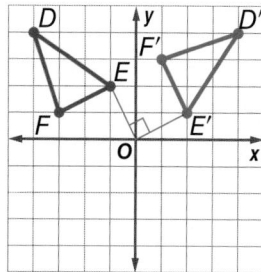

So, the coordinates of the vertices of △*D'E'F'* are *D'*(4, 4), *E'*(2, 1), and *F'*(1, 3).

CHECK Your Progress

b. Quadrilateral *MNPQ* has vertices *M*(2, 7), *N*(6, 6), *P*(6, 3), and *Q*(2, 3). Graph the figure and its image after a counterclockwise rotation of 270° about the origin. Then give the coordinates of the vertices for quadrilateral *M'N'P'Q'*.

REAL-WORLD EXAMPLE

3 **FOLK ART** Copy and complete the barn sign shown so that the completed figure has rotational symmetry with 90°, 180°, and 270° as its angles of rotation.

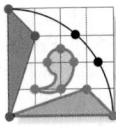

Use the procedure described in Example 2 and the points indicated to rotate the figure 90°, 180°, and 270° counterclockwise. A 90° rotation clockwise produces the same rotation as a 270° rotation counterclockwise.

90° counterclockwise

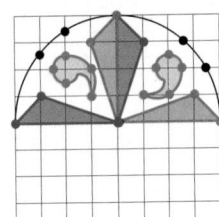

180° counterclockwise

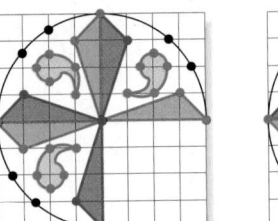

90° clockwise

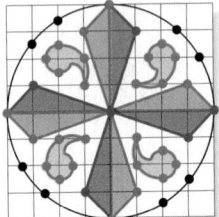

 CHECK Your Progress

c. SYMBOLS Copy and complete the symbol for recycling shown so that the completed figure has rotational symmetry with 120° and 240° as its angles of rotation.

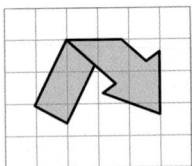

CHECK Your Understanding

Examples 1 and 2   Graph △XYZ and its image after each rotation. Then give the coordinates of the vertices for △X′Y′Z′.

1. 270° counterclockwise about vertex X

2. 180° clockwise about the origin

Example 3   **3. ARCHITECTURE** Copy and complete the window so that the completed figure has rotational symmetry with 45°, 90°, 135°, 180°, 225°, 270°, and 315° as its angles of rotation.

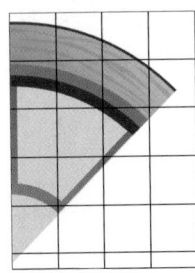

Practice and Problem Solving

= **Step-by-Step Solutions** begin on page R1.
Extra Practice begins on page EP2.

Example 1 **Graph quadrilateral *ABCD* and its image after each rotation. Then give the coordinates of the vertices for quadrilateral *A'B'C'D'*.**

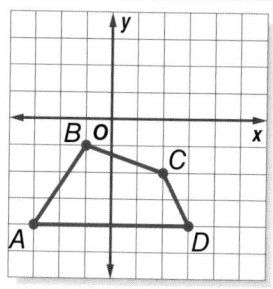

  4. 90° clockwise about vertex *A*

  5. 180° counterclockwise about vertex *D*

Example 2 **6.** Quadrilateral *EFGH* has vertices *E*(1, −1), *F*(3, −5), *G*(7, −5), and *H*(6, −1). Graph the figure and its rotated image after a counterclockwise rotation of 90° about the origin. Then give the coordinates of the vertices for quadrilateral *E'F'G'H'*.

  7 Triangle *RST* has vertices *R*(−7, 8), *S*(−7, 2), and *T*(−2, 2). Graph the figure and its rotated image after a clockwise rotation of 180° about the origin. Then give the coordinates of the vertices for triangle *R'S'T'*.

Example 3 **8. CARS** A partial hubcap is shown. Copy and complete the figure so that the completed hubcap has rotational symmetry of 90°, 180°, and 270°.

9. FOOD A piece of pizza is shown. Copy and complete the figure so that the entire pizza has rotational symmetry of 60°, 120°, 180°, 240°, and 300°. How many slices are needed to complete the pizza?

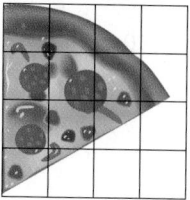

10. Identify each transformation as a *translation*, *reflection*, or *rotation*.

  a. **b.** **c.**

  11 STATES Which capital letters in VIRGINIA produce the same letter after being rotated 180°?

  12. MATH IN THE MEDIA Find an example of a non-circular item in a magazine or on the Internet that has rotational symmetry. Then name its angle(s) of rotation.

13. OPEN ENDED Draw a figure on the coordinate plane. Then rotate the figure 270° counterclockwise about one of its vertices.

14. CHALLENGE Triangle *ABC* has vertices *A*(0, 4), *B*(0, −2), and *C*(2, 0). The triangle is reflected over the *x*-axis. Then the image is rotated 180° counterclockwise about the origin. What are the coordinates of the final image?

15. CHALLENGE Triangle *QRS* is translated 7 units right. Then its image is rotated 90° clockwise about the origin. The vertices of triangle *Q″R″S″* are *Q″*(6, −1), *R″*(0, −1), and *S″*(0, −7). Find the coordinates of △*QRS*.

16. WRITE MATH A figure is rotated 360° clockwise about the origin. Describe the placement of the image in relation to the original figure.

✔ Test Practice

17. Square *JKLM* is rotated about the origin. Which of the following describes the rotation?

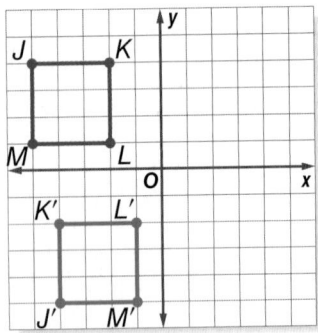

   A. 90° clockwise

   B. 90° counterclockwise

   C. 180° clockwise

   D. 270° counterclockwise

18. **EXTENDED RESPONSE** Triangle *ABC* is shown.

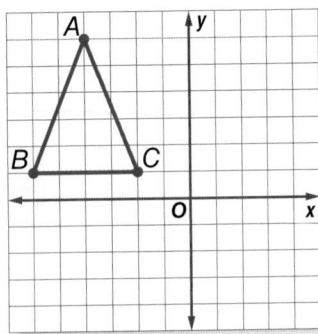

Part A Graph the figure after a clockwise rotation of 180° about the origin.

Part B List the coordinates of the vertices for triangle *A′B′C′*.

19. Use the graph of △*ABC* shown at the right.

   a. What are the coordinates of △*A′B′C′* when △*ABC* is reflected over the *x*-axis? (Lesson 3B)

   b. Graph △*ABC* and its image after it is translated 2 units right and 1 unit up. (Lesson 3A)

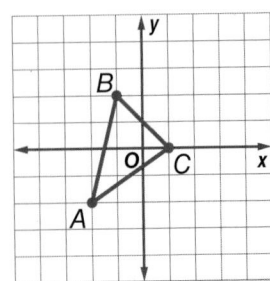

Main Idea
Use the scale factor to graph dilations on the coordinate plane.

 **Vocabulary**
dilation

 Get Connect**ED**

 **8.G.3**

Dilations

TECHNOLOGY Necie wants to insert the photograph shown on her blog. The current size of the photo is 480 pixels by 640 pixels.

1. Suppose she wants to reduce the art to 120 pixels by 160 pixels. Compare and contrast the original art and the reduction.

2. What is the scale factor from the original to the reduction?

A **dilation** is a transformation that enlarges or reduces a figure by a scale factor. Since the figure is enlarged or reduced by a scale factor, the preimage and the image are similar figures.

Key Concept **Dilations in the Coordinate Plane**

Words When the center of dilation in the coordinate plane is the origin, each coordinate of the preimage is multiplied by the scale factor k to find the coordinates of the image.

Model

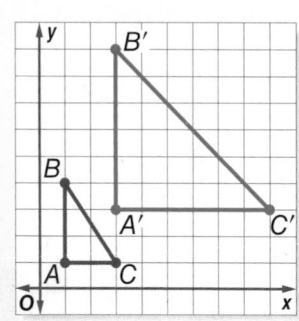

dilation with scale factor = 3

Symbols $(x, y) \rightarrow (kx, ky)$

EXAMPLE **Enlargement in the Coordinate Plane**

① A triangle has vertices $A(0, 0)$, $B(8, 0)$, and $C(3, -2)$. Find the coordinates of the triangle after a dilation with a scale factor of 4.

The dilation is $(x, y) \rightarrow (4x, 4y)$. Multiply the coordinates of each vertex by 4.

$A(0, 0) \rightarrow (4 \cdot 0,\ 4 \cdot 0) \rightarrow (0, 0)$
$B(8, 0) \rightarrow (4 \cdot 8,\ 4 \cdot 0) \rightarrow (32, 0)$
$C(3, -2) \rightarrow [4 \cdot 3,\ 4 \cdot (-2)] \rightarrow (12, -8)$

So, the coordinates after the dilation are $A'(0, 0), B'(32, 0)$, and $C'(12, -8)$.

 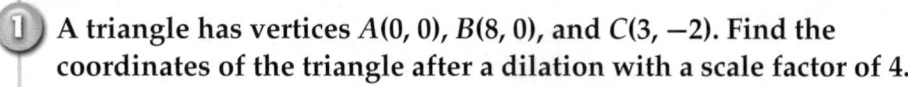

CHECK Your Progress

a. A figure has vertices $W(-2, 4)$, $X(1, 4)$, $Y(-3, -1)$, and $Z(3, -1)$. Find the coordinates of the figure after a dilation with a scale factor of 2.

EXAMPLE Reduction in the Coordinate Plane

② A figure has vertices $F(0, 6)$, $G(1, 4)$, $H(-2, 4)$, and $I(-3, 5)$. Graph the figure and the image of the figure after a dilation with a scale factor of $\frac{1}{2}$.

The dilation is $(x, y) \rightarrow \left(\frac{1}{2}x, \frac{1}{2}y\right)$. Multiply the coordinates of each vertex by $\frac{1}{2}$.

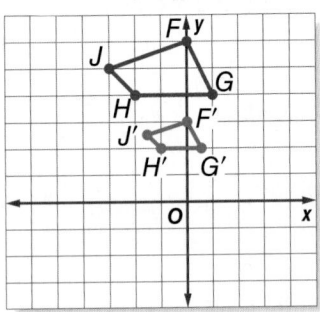

$$F(0, 6) \rightarrow \left(\frac{1}{2} \cdot 0, \frac{1}{2} \cdot 6\right) \rightarrow (0, 3)$$

$$G(1, 4) \rightarrow \left(\frac{1}{2} \cdot 1, \frac{1}{2} \cdot 4\right) \rightarrow \left(\frac{1}{2}, 2\right)$$

$$H(-2, 4) \rightarrow \left(\frac{1}{2} \cdot (-2), \frac{1}{2} \cdot 4\right) \rightarrow (-1, 2)$$

$$I(-3, 5) \rightarrow \left(\frac{1}{2} \cdot (-3), \frac{1}{2} \cdot 5\right) \rightarrow \left(-1\frac{1}{2}, 2\frac{1}{2}\right)$$

CHECK Your Progress

b. A triangle has vertices $R(-3, 6)$, $S(0, 6)$, and $T(-3, -6)$. Graph the figure and the image of the figure after a dilation with a scale factor of $\frac{2}{3}$.

REAL-WORLD EXAMPLE Find a Scale Factor

③ **EARTH SCIENCE** Through a microscope, the image of a grain of sand with a 0.25-millimeter diameter appears to have a diameter of 11.25 millimeters. What is the scale factor of the dilation?

Write a ratio comparing the diameters of the two images.

$$\frac{\text{diameter in dilation}}{\text{diameter in original}} = \frac{11.25}{0.25}$$

$$= 45$$

So, the scale factor of the dilation is 45.

CHECK Your Progress

c. **PHOTOGRAPHY** Lucas wants to enlarge a 3- by 5-inch photo to a $7\frac{1}{2}$ by $12\frac{1}{2}$-inch photo. What is the scale factor of the dilation?

Examples 1 and 2 Find the coordinates of the vertices of each figure after a dilation with the given scale factor k. Then graph the original image and the dilation.

    1. $A(3, 5)$, $B(0, 4)$, $C(-2, -2)$; $k = 2$

    2. $J(0, -4)$, $K(0, 6)$, $L(4, 4)$, $M(4, 2)$; $k = \frac{1}{4}$

Example 3  **TECHNOLOGY** Mrs. Bowen's homeroom is creating a Web page for their school's Intranet site. They need to reduce a scanned photograph so it is 720 pixels by 320 pixels. If the scanned photograph is 1,080 pixels by 480 pixels, what is the scale factor of the dilation?

Practice and Problem Solving

> ● = **Step-by-Step Solutions** begin on page R1.
> **Extra Practice** begins on page EP2.

Examples 1 and 2 Find the coordinates of the vertices of each figure after a dilation with the given scale factor k. Then graph the original image and the dilation.

    4. $C(1, 4)$, $A(2, 2)$, $T(5, 5)$; $k = 2$ **5.** $V(-3, 4)$, $X(-2, 0)$, $W(1, 2)$; $k = 3$

    6. $k = \frac{3}{4}$ **7.** $k = \frac{2}{5}$

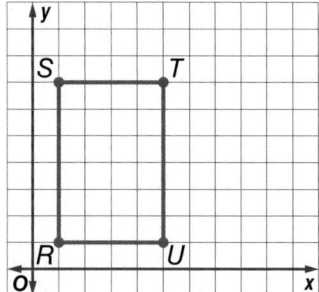   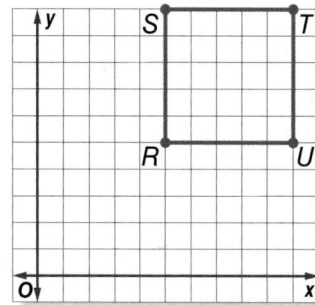

Example 3 **8. GRAPHIC DESIGN** A graphic designer created a logo on $8\frac{1}{2}$- by 11-inch paper. In order to be placed on a business card, the logo needs to be $1\frac{7}{10}$ inches by $2\frac{1}{5}$ inches. What is the scale factor of the dilation?

9. GAMES Darian wants to build a regulation-size pool table that is 9 feet in length. The plans he ordered are 18 by 36 inches. What is the scale factor of the dilation he must use to build the regulation pool table?

10. TECHNOLOGY Carmen wants to project a 3-inch square photograph onto a wall to create a 96-inch square image. If the projector makes the dimensions of the image twice as large for each yard that it is moved away from the wall, how far away should Carmen place the projector?

11 GEOMETRY A triangle has vertices $A(-2, 3)$, $B(0, 0)$, and $C(1, 1)$.

    a. Find the coordinates of the triangle if it is reflected over the x-axis, then dilated by a scale factor of 3.

    b. Find the coordinates if the original triangle is dilated by a scale factor of 3, then reflected over the x-axis.

    c. Are the two transformations commutative? Explain.

Real-World Link · · · ·
Common display resolutions for video projectors are in the ratio 4:3 or 4:2.25.

12. REASONING A figure has a vertex at the point $(-4, -6)$. The figure is dilated with the center at the origin with a scale factor of 5. The resulting image is then dilated with a scale factor of $\frac{3}{5}$.

    a. What are the coordinates of the vertex in the final image?

    b. How do they compare with those of the original image?

    c. Can you predict the scale factor of a compound dilation? Explain.

13. CHALLENGE The coordinates of two rectangles are shown in the table. Is rectangle $WXYZ$ a dilation of rectangle $ABCD$? Explain.

14. ✍ **WRITE MATH** Explain how you can determine if a dilation is a reduction or an enlargement based on the scale factor.

WXYZ		ABCD	
W	(a, b)	A	$(4a, 2b)$
X	(a, c)	B	$(4a, 2c)$
Y	(d, b)	C	$(4d, 2b)$
Z	(d, c)	D	$(4d, 2c)$

✔ Test Practice

15. A rectangle has vertices $Q(-5, -6)$, $P(-5, 1)$, $R(3, -6)$, and $S(3, 1)$. Which of the following figures is the image after a dilation?

    A. $Q'(-5, 6)$, $P'(-5, -1)$, $R'(3, 6)$, $S'(3, -1)$

    B. $Q'(5, -6)$, $P'(5, 1)$, $R'(-3, -6)$, $S'(-3, 1)$

    C. $Q'(-10, -12)$, $P'(-10, 2)$, $R'(6, -12)$, $S'(6, 2)$

    D. $Q'(-6, -5)$, $P'(1, -5)$, $R'(-6, 3)$, $S'(1, 3)$

16. ✍ **GRIDDED RESPONSE** What is the scale factor of the dilation?

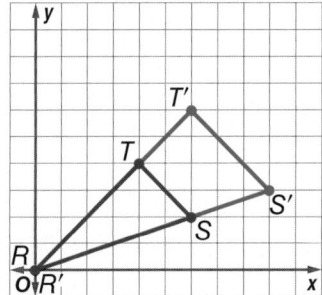

Spiral Review

17. ALPHABET Which capital letters, if any, in the printed alphabet have rotational symmetry? Identify the angle of rotation for each. (Lesson 3D)

18. Graph polygon $ABCDE$ with vertices $A(-5, -3)$, $B(-2, 1)$, $C(-3, 4)$, $D(-1, 2)$, and $E(-1, -3)$. Then graph the image of the figure after a reflection over the y-axis and write the coordinates of its vertices. (Lesson 3B)

19. GEOMETRY Find the missing measure in the right triangle shown. Round to the nearest tenth if necessary. (Lesson 2B)

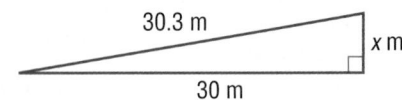

Extend

Composition of Transformations

Main Idea

Draw compositions of translations, reflections, and rotations.

 Vocabulary

composition of transformations

 Get ConnectED

 8.G.2, 8.G.4

GRAPHIC ARTS Graphic artists often use several transformations to create designs. When a transformation is applied to a figure and then another transformation is applied to the image, the result is called a **composition of transformations**.

ACTIVITY

1 **Create a composition of transformations.**

STEP 1 Fold a piece of paper vertically into three equal sections.

STEP 2 Draw an arrow in the first section.

STEP 3 Draw the reflection of the arrow over the fold in the second section.

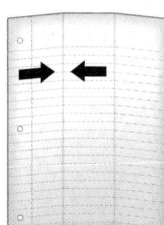

STEP 4 Draw a reflection of the 2nd arrow over the fold in the third section.

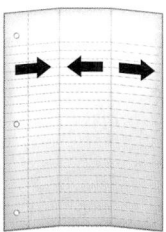

STEP 5 Repeat Steps 1 through 4 with another figure.

Analyze the Results

1. How are the original figure and the final figure related?

2. Would the final image be the same if the second reflection was over a horizontal line? Explain.

ACTIVITY

2 Use transformations to create a border.

> **STEP 1** Choose an image and draw it on a coordinate plane.
>
> **STEP 2** Translate the image. In this activity, the image is translated 2 units to the right.

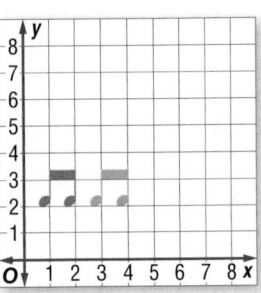

> **STEP 3** Reflect the image across a horizontal line. In this activity, the image is reflected across the line $y = 2$.
>
> **STEP 4** Repeat the process to create your border.

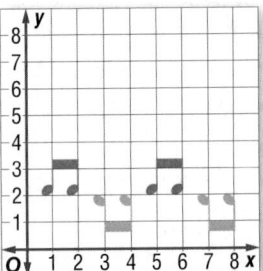

Analyze the Results

3. How are the original figure and the final figure related?
4. Would the border appear different if the reflection was done before the translation?

Practice and Apply

Describe the transformations combined to create the outlined patterns shown.

5.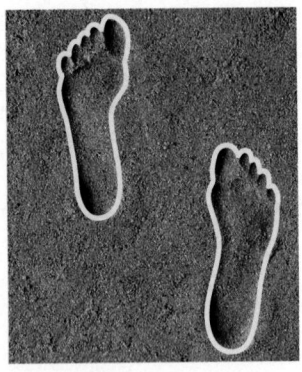

6.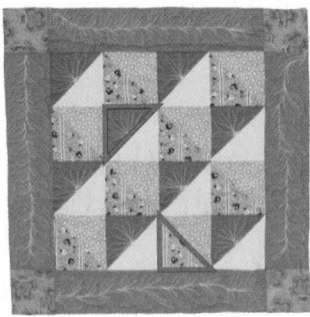

7.

Tessellations can be a result of compositions of transformations. They are described by counting the number of sides of the polygons that form a vertex. Begin with the smallest number and move around the vertex, point A, clockwise. The tessellation shown is a 3, 4, 6, 4 tessellation and is created by translations, rotations, and reflections. Find the transformations used in the following tessellations.

8. 3, 3, 3, 3, 6

9. 3, 6, 3, 6

10. 3, 12, 12

11. 3, 3, 4, 3, 4

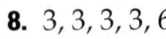

Chapter Study Guide and Review

Be sure the following Key Concepts are noted in your Foldable.

Key Concepts

Indirect Measurement (Lesson 1)
- Use properties of similar polygons to find distances or lengths that are difficult to measure.

Pythagorean Theorem (Lesson 2)
- In a right triangle, the square of the length of the hypotenuse is equal to the sum of the squares of the lengths of the legs.

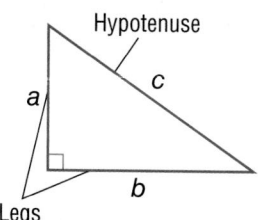

Transformations (Lesson 3)
- A translation slides a figure from one position to another without turning it.
- A reflection is a mirror image of the original figure.
- A rotation is a transformation in which a figure is rotated about a fixed point.
- A dilation enlarges or reduces a figure by a scale factor.

Key Vocabulary

converse
corresponding parts
dilation
Distance Formula
hypotenuse
indirect measurement
legs
Pythagorean Theorem
reflection

rotation
scale factor
similar polygons
tangent ratio
transformation
translation
trigonometric ratio
trigonometry

Vocabulary Check

Choose the correct term to complete each sentence.

1. If the measures of the sides of a triangle are 6 inches, 8 inches, and 10 inches, then the triangle (is, is not) a right triangle.

2. A (dilation, reflection) is a transformation in which the image and the preimage are congruent.

3. The hypotenuse is the (shortest, longest) side of a right triangle.

4. The Pythagorean Theorem states that the sum of the squares of the lengths of the (legs, angles) of a right triangle equals the square of the length of the hypotenuse.

5. (Indirect measurement, Scale factor) is a technique that uses similar triangles to find missing lengths.

Multi-Part Lesson Review

Lesson 1 Similar Triangles

PSI: Draw a Diagram (Lesson 1A)

Solve. Use the *draw a diagram* strategy.

6. **CONCERTS** Nina, Tyrese, Leslie, and Ethan are going to a rock concert. In how many different ways can they enter the concert?

7. **PHYSICAL SCIENCE** A tennis ball is dropped from 12 feet above the ground. It hits the ground and bounces up half as high as it fell. This is true for each successive bounce. What height does the ball reach on the fourth bounce?

EXAMPLE 1 A photographer is taking a class picture. She places 8 students in the first row. Each additional row has 4 more students in it. If there is a total of 80 students, how many rows will there be?

Draw a diagram with 8 Xs in row one and then add 4 to each new row.

XXXXXXXX	1st row
XXXXXXXXXXXX	2nd row

There are a total of 5 rows.

Similar Polygons (Lesson 1B)

8. In the figure, the polygons are similar. Find the missing measure.

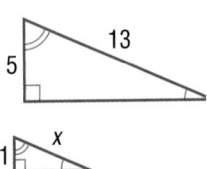

9. **MEASUREMENT** If square D has a perimeter of 49 feet and square F has a perimeter of 64 feet, what is the scale factor from square D to square F?

EXAMPLE 2 Rectangle $GHJK$ is similar to rectangle $PQSR$. Find the value of x.

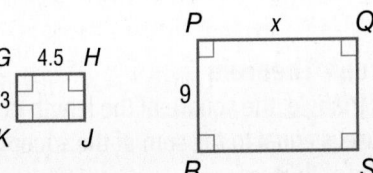

The scale factor is $\dfrac{PR}{GK}$, which is $\dfrac{9}{3}$ or $\dfrac{3}{1}$.

$x = 3(4.5)$ or 13.5 Write the equation. Then multiply.

Indirect Measurement (Lesson 1D)

10. **TREES** A 36-foot tree casts a 9-foot shadow at the same time a building casts a 15-foot shadow. How tall is the building?

11. **BUILDINGS** A building casts an 18.5-foot shadow. How tall is the building if a 10-foot tall sculpture nearby casts a 7-foot shadow?

EXAMPLE 3 A flagpole casts a shadow that is 6 meters long. A boy casts a shadow that is 1.2 meters long. If the flagpole is 7.5 meters tall, how tall is the boy?

flagpole ⟶ $\dfrac{6}{1.2} = \dfrac{7.5}{x}$ ⟵ flagpole
boy ⟶　　　　　　 ⟵ boy

$$1.2 \cdot 7.5 = 6 \cdot x$$
$$9 = 6x$$
$$1.5 = x$$

The boy is 1.5 meters tall.

Similar Triangles (continued)

The Tangent Ratio (Lesson 1E)

Find the tangent of each acute angle. Round to the nearest hundredth.

12.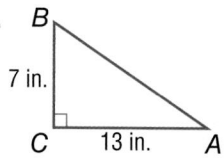

13.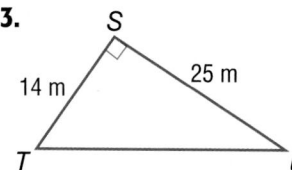

14. ROADS The end of an exit ramp from a highway is 17.5 feet above the highway. The angle of inclination of the ramp is 3°. How long is the base of the ramp? Round to the nearest tenth.

EXAMPLE 4 Find tan A. Round to the nearest hundredth.

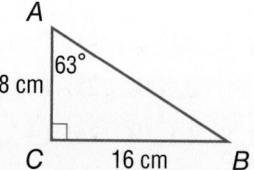

$$\tan A = \frac{BC}{AC}$$

$$= \frac{16}{8} \text{ or } 2$$

The Pythagorean Theorem

The Pythagorean Theorem (Lesson 2B)

Write an equation you could use to find the length of the missing side of each right triangle. Then find the missing length. Round to the nearest tenth if necessary.

15. a, 5 in.; c, 6 in.   **16.** a, 6 cm; b, 7 cm

17. GEOMETRY Lolita drew a right triangle where the hypotenuse was 17 inches and one of the legs was 8 inches. What was the length of the third side?

EXAMPLE 5 Write an equation you could use to find the length of the hypotenuse of the triangle. Then find the missing length.

$$a^2 + b^2 = c^2$$
$$3^2 + 5^2 = c^2$$
$$9 + 25 = c^2$$
$$34 = c^2$$
$$\pm\sqrt{34} = c$$
$$\pm 5.8 \approx c$$

The hypotenuse is about 5.8 meters long.

Use the Pythagorean Theorem (Lesson 2C)

Write an equation that can be used to answer the question. Then solve. Round to the nearest tenth if necessary.

18. How high does the ladder reach?

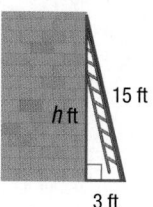

19. How wide is the kite?

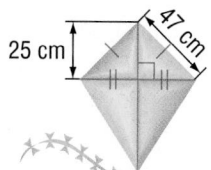

EXAMPLE 6 Write an equation that can be used to find the height of the pole h. Then solve.

$$3.5^2 + h^2 = 13^2$$
$$12.25 + h^2 = 169$$
$$h^2 = 156.75$$
$$h = \pm\sqrt{156.75}$$
$$h \approx \pm 12.5$$

The pole is about 12.5 meters high.

The Pythagorean Theorem (continued)

Distance on the Coordinate Plane (Lesson 2D)

Graph each pair of ordered pairs. Then find the distance between the points. Round to the nearest tenth if necessary.

20. $(0, -3)$, $(5, 5)$ 21. $(-1, 2)$, $(4, 8)$

22. $(-2, 1.5)$, $(2, 3.6)$ 23. $(-6, 2)$, $(-4, 5)$

24. $(3, 4.2)$, $(-2.1, 0)$ 25. $(-1, 3)$, $(2, 4)$

26. **GEOMETRY** The coordinates of points R and S are $(4, 3)$ and $(1, 6)$. What is the distance between the points? Round to the nearest tenth if necessary.

EXAMPLE 7 Graph the ordered pairs $(2, 3)$ and $(-1, 1)$. Then find the distance between the points.

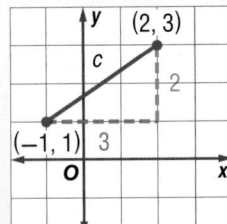

$$a^2 + b^2 = c^2$$
$$3^2 + 2^2 = c^2$$
$$9 + 4 = c^2$$
$$13 = c^2$$
$$\pm\sqrt{13} = c$$
$$\pm 3.6 \approx c$$

The distance is about 3.6 units.

Special Right Triangles (Lesson 2F)

Find each missing measure.

27.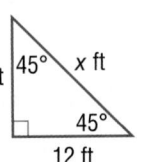

28.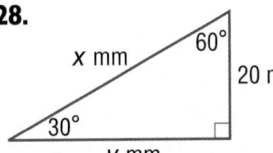

29. The shorter leg of a 30°-60°-90° triangle is 12 inches long. Find the exact lengths of the longer leg and the hypotenuse.

EXAMPLE 8 Triangle *DEF* is a 45°-45°-90° triangle. Find the length of the hypotenuse.

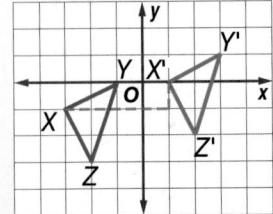

Since this is a 45°-45°-90° triangle, multiply the length of $\overline{EF}$ by $\sqrt{2}$.

$x = 5\sqrt{2}$ $\overline{EF} = 5$

So, the hypotenuse of $\triangle DEF$ is $5\sqrt{2}$ inches.

Transformations

Translations (Lesson 3A)

Graph $\triangle ABC$ with vertices $A(2, 2)$, $B(3, 5)$, and $C(5, 3)$. Then graph the image of $\triangle ABC$ after each translation and write the coordinates of its vertices.

30. 1 unit right and 4 units down

31. 2 units left and 3 units up

32. **CHESS** A knight moves two spaces up and one space to the left. If the starting point is represented by $P(1, 4)$, what are the coordinates of its stopping point?

EXAMPLE 9 Graph $\triangle XYZ$ with vertices $X(-3, -1)$, $Y(-1, 0)$, and $Z(-2, -3)$ and its image after a translation 4 units right and 1 unit up.

The coordinates of the image are $X'(1, 0)$, $Y'(3, 1)$, and $Z'(2, -2)$.

Reflections (Lesson 3B)

Graph parallelogram *QRST* and its reflection over the given axis. Then find the coordinates of the reflected image.

33. *x*-axis

34. *y*-axis

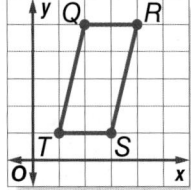

EXAMPLE 10 Graph △*FGH* with vertices *F*(1, −1), *G*(3, 1), and *H*(2, −3) and its image after a reflection over the *y*-axis.

The coordinates of the vertices of the image are *F*′(−1, −1), *G*′(−3, 1), and *H*′(−2, −3).

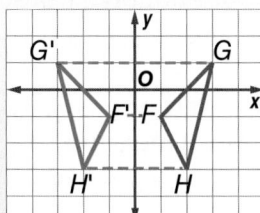

Rotations (Lesson 3D)

Rectangle *MNPQ* has vertices *M*(3, 8), *N*(10, 8), *P*(10, 2), and *Q*(3, 2). Graph the figure and its image after each rotation. Then give the coordinates of the vertices for rectangle *M*′*N*′*P*′*Q*′.

35. 270° clockwise about the origin

36. 180° counterclockwise about vertex *Q*

EXAMPLE 11 Graph △*PQR* with vertices *P*(1, 4), *Q*(3, 1), and *R*(1, 1) and its image after a counterclockwise rotation of 90° about the origin. Then give the coordinates of the vertices for △*P*′*Q*′*R*′.

The coordinates of the vertices of the image are *P*′(−4, 1), *Q*′(−1, 3), and *R*′(−1, 1).

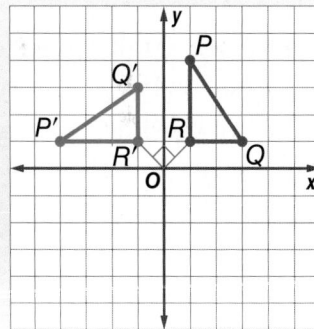

Dilations (Lesson 3E)

Find the coordinates of the vertices of each figure after a dilation with the given scale factor *k*.

37. *A*(−2, −3), *B*(−2, 3), *C*(3, −3); *k* = 3

38. *L*(−4, −1), *M*(4, −1), *N*(4, 6), *P*(−6, −6); $k = \frac{1}{2}$

EXAMPLE 12 A triangle has vertices *X*(4, 1), *Y*(4, 7), and *Z*(10, 1). Find the coordinates of the triangle after a dilation with a scale factor of 2.

Multiply the coordinates of each vertex by 2.

X(4, 1) → *X*(2 · 4, 2 · 1) → *X*′(8, 2)

Y(4, 7) → *Y*(2 · 4, 2 · 7) → *Y*′(8, 14)

Z(10, 1) → *Z*(2 · 10, 2 · 1) → *Z*′(20, 2)

1. SCHOOL Of the 30 students in a science class, 19 like to do chemistry labs, 15 prefer physical science labs, and 7 like to do both. How many students like chemistry labs but not physical science labs? Use the *draw a diagram* strategy.

Each pair of polygons is similar. Find each missing measure.

2.

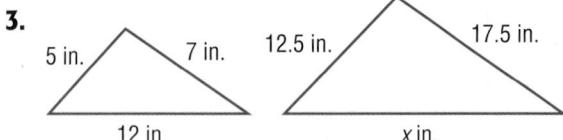

6 cm 4 cm
12 cm x cm

3.

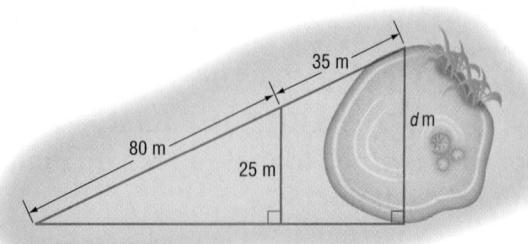

5 in. 7 in. 12.5 in. 17.5 in.
12 in. x in.

4. MULTIPLE CHOICE A child $4\frac{1}{2}$ feet tall casts a 6-foot shadow. A nearby statue casts a 12-foot shadow. How tall is the statue?

A. $8\frac{1}{4}$ ft   **C.** $13\frac{1}{2}$ ft

B. 9 ft   **D.** 24 ft

5. PONDS In the figure, the triangles are similar. What is the width of the pond? Round to the nearest tenth.

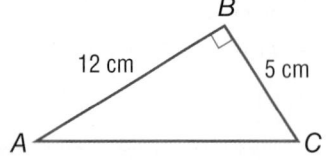

35 m
d m
80 m
25 m

6. Find tan A in the triangle at the right.

B
12 cm 5 cm
A C

Determine whether each triangle with sides of given lengths is a right triangle. Justify your answer.

7. 12 in., 20 in., 24 in.   **8.** 34 cm, 30 cm, 16 cm

9. 15 ft, 25 ft, 20 ft   **10.** 27 yd, 14 yd, 35 yd

Graph each pair of ordered pairs. Then find the distance between points. Round to the nearest tenth if necessary.

11. $(-2, -2)$, $(5, 6)$ **12.** $(-0.5, 0.25)$, $(0.25, -0.75)$

Find each missing measure.

13.

9.5 in.
45°
9.5 in.
45° x in.

14.

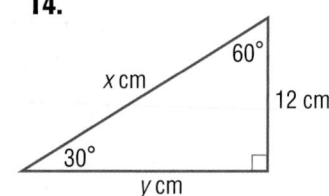

60°
x cm 12 cm
30°
y cm

Polygon *HJKL* has vertices $H(-6, 2)$, $J(4, 4)$, $K(7, -2)$, and $L(-2, -4)$. Graph the figure and its image after each transformation. Then give the coordinates of the vertices for polygon *H′J′K′L′*.

15. translation 3 units left and 2 units up

16. dilation with a scale factor of $\frac{1}{2}$

17. EXTENDED RESPONSE A survey team calculated the distance across a river from point A to point B.

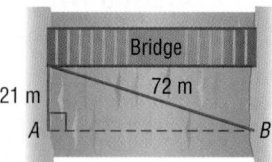

Bridge
21 m 72 m
A B

Part A Write an equation that can be used to find the distance across the water.

Part B Use the equation from Part **A** to find the width of the river at this point. Round to the nearest tenth.

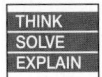

Preparing for Standardized Tests

THINK SOLVE EXPLAIN

Short-Response: Showing Work

Short-response questions require that you show your work. You may be able to get partial credit even if your answer is not entirely correct.

TEST EXAMPLE

A skateboard park has two connecting ramps as shown in the diagram.

Part A *What is the height of the ramps?*

Height _____3 meters_____

Part B *A skateboarder goes from point A to point B to point D. What is the total length of the skateboarder's path?*

$$a^2 + b^2 = c^2$$
$$(3)^2 + (7)^2 = c^2$$
$$9 + 49 = c^2$$
$$58 = c^2$$
$$\pm\sqrt{58} = c$$
$$\pm 7.6 \approx c$$

Do not forget to write the formula and show your work.

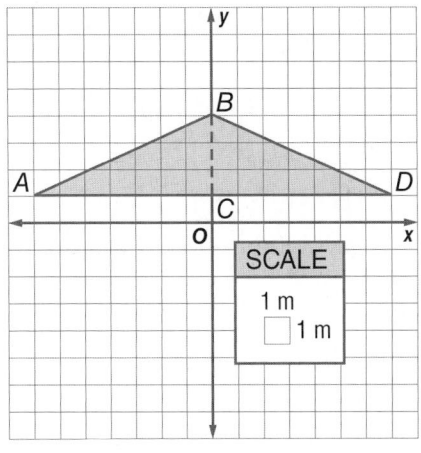

Since length cannot be negative, the distance from A to B is about 7.6 meters. Since the triangles are the same, the distance from B to D is also 7.6 meters. So, the path is 7.6 × 2 or about 15.2 meters long.

Length of Path _____about 15.2 meters_____

Work on It

The ramps at the skateboard park above are changed so that point B is at (0, 6), A is at (−8, 1), and D is at (8, 1). C is unchanged.

Part A What is the height of the ramps?

Height _____

Part B A skateboarder starts at point B and skates down to point D. How far does the skateboarder skate?

Length of Path _____

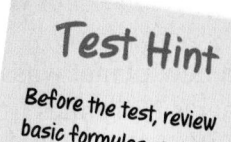

Test Hint

Before the test, review basic formulas such as the Pythagorean Theorem. If you know a few important formulas and their common uses, you will be more prepared.

Read each question. Then fill in the correct answer on the answer document provided by your teacher or on a sheet of paper.

1. Justin is flying a kite as shown below.

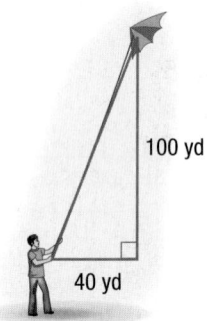

100 yd

40 yd

Which of the following is closest to the length of the string?

A. 70 yd

B. 92 yd

C. 108 yd

D. 146 yd

2. ≡≡⟋ **GRIDDED RESPONSE** The triangles in the figure below are similar.

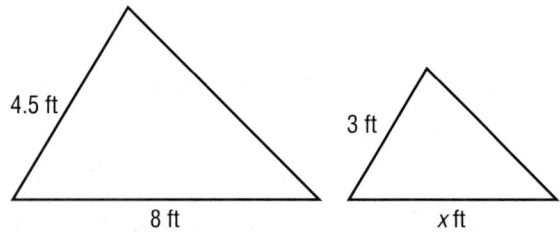

4.5 ft

8 ft

3 ft

x ft

What is the value of x to the nearest tenth?

3. In 2003, a new planet was discovered. This new planet is 10^{10} miles from the Sun. Which of the following represents this number in standard form?

F. 10,000,000,000 mi

G. 10,000,000 mi

H. 10,000 mi

I. 100 mi

4. Which of the following is closest to the perimeter of the triangle below?

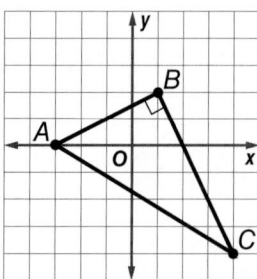

A. 17 units

C. 21 units

B. 19 units

D. 23 units

5. [THINK SOLVE EXPLAIN] **SHORT RESPONSE** Shanelle purchased a new computer for $1,099 and a computer desk for $699 including tax. She plans to pay the total amount in 24 equal monthly payments. What will be the amount of her monthly payments?

6. The area of a square is 20 square inches. Which **best** represents the length of a side of the square?

F. 4.5 inches

H. 10 inches

G. 5 inches

I. 11 inches

7. The proposed location of a new water tower intersects a section of an existing service road. What is x, the inside length of the section of road that is intersected by the water tower?

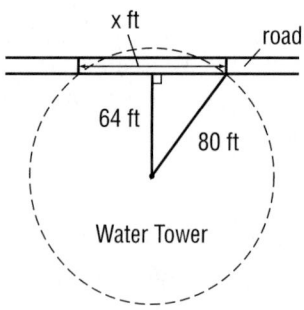

x ft

road

64 ft

80 ft

Water Tower

A. 36 ft

C. 96 ft

B. 48 ft

D. 112 ft

8. Which irrational number is closest to the number 5?

F. $\sqrt{30}$ H. $\sqrt{20}$

G. $\sqrt{27}$ I. $\sqrt{18}$

9. **SHORT RESPONSE** Use scientific notation to find the product of 25,000,000 and 160,000.

10. Which of the following is the graph of $y = \frac{2}{3}x + 2$?

A.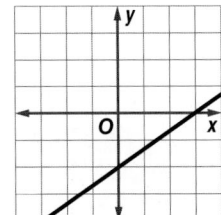

C.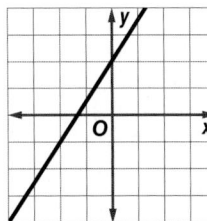

B.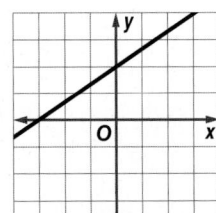

D.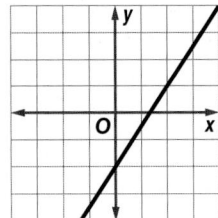

11. What is the height in feet of the skateboard ramp shown below?

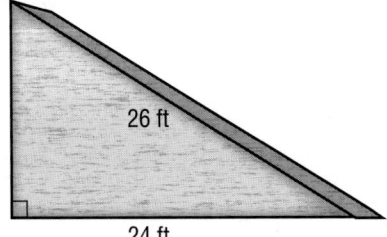

F. 10 ft   H. 25 ft

G. 22 ft   I. 34 ft

12. **GRIDDED RESPONSE** Niles casts a shadow that is 8 feet long at the same time his dog casts a shadow that is 30 inches long. If Niles is $5\frac{1}{2}$ feet tall, how many inches tall is his dog? Round to the nearest tenth.

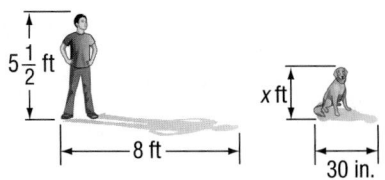

13. **EXTENDED RESPONSE** Refer to the park map below.

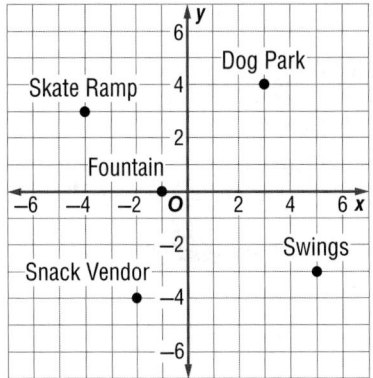

Part A   At what point is the snack vendor located?

Part B   At what point is the dog park?

Part C   What is the distance between the dog park and the snack vendor?

Part D   Each unit on the map represents 15 meters. What is the approximate distance between the snack vendor and the dog park?

NEED EXTRA HELP?													
If You Missed Question...	1	2	3	4	5	6	7	8	9	10	11	12	13
Go to Chapter-Lesson...	8-2C	8-1B	2-1A	8-2D	1-1D	2-3C	8-2C	2-3C	2-2C	5-3C	8-2C	8-1D	8-2D

Units of Measure

connectED.mcgraw-hill.com

 Investigate

Animations

 Vocabulary

 Multilingual eGlossary

 Learn

Personal Tutor

 Virtual Manipulatives

 Graphing Calculator

 Audio

 Foldables

 Practice

 Self-Check Practice

 Worksheets

 Assessment

The ☆BIG Idea

How are unit ratios used to convert measurements between different measurement systems?

FOLDABLES®
Study Organizer

Make this Foldable to help you organize your notes.

Units of Measure

Review Vocabulary

inch pulgada customary unit of length, there are 12 inches in one foot

$\frac{1}{2}$ in. $\frac{1}{4}$ in.

1 in.

0 in.   1   2   3

Key Vocabulary

English	Español
Celsius	Celsius
derived unit	unidad derivada
Fahrenheit	Fahrenheit
literal equation	ecuación literal

When Will I Use This?

Are You Ready for the Chapter?

You have two options for checking prerequisite skills for this chapter.

Text Option Take the Quick Check below. Refer to the Quick Review for help.

QUICK Check

Solve each equation.

1. $19 = 2a - 7$
2. $35 = 7b - 28$
3. $-5c + 17 = 22$
4. $32 - 6d = 50$
5. $\frac{e}{3} - 6 = -7$
6. $\frac{f}{-4} - 12 = -10$
7. $\frac{g - 6}{2} = 4$
8. $\frac{h + 1}{5} = -1$
9. **SPORTS** In football, a touchdown is worth 6 points, an extra point is worth one point, and a field goal is worth 3 points. A team scored a total of 30 points. Use the equation $6t + 12 = 30$ to find the number of touchdowns t a team scored.

Find each product or quotient.

10. $8 \cdot 2.54$
11. $9 \cdot 0.305$
12. $16 \cdot 29.57$
13. $4 \cdot 3.78$
14. $35 \cdot \frac{5}{9}$
15. $42 \cdot \frac{5}{9}$
16. $\frac{9}{5} \cdot 16$
17. $\frac{9}{5} \cdot 32$
18. **MEASUREMENT** What is the area of the rectangle below?

1.2 cm

1.5 cm

QUICK Review

EXAMPLE 1

Solve $3a + 4 = 13$.

$$3a + 4 = 13 \quad \text{Write the equation.}$$
$$\underline{ - 4 = -4} \quad \text{Subtraction Property of Equality}$$
$$3a = 9$$
$$\frac{3a}{3} = \frac{9}{3} \quad \text{Division Property of Equality}$$
$$a = 3 \quad \text{Simplify.}$$

EXAMPLE 2

Find $6 \cdot 2.54$.

$$\begin{array}{r} {}^{3}\,{}^{2} \\ 2.54 \\ \times\ 6 \\ \hline 15.24 \end{array}$$ two decimal places

Count two decimal places from the right.

EXAMPLE 3

Find the area of a rectangle with a length of 9.8 inches and a width of 4.6 inches.

$$A = \ell w \quad \text{Area of a rectangle}$$
$$A = (9.8)(4.6) \quad \text{Replace } \ell \text{ with 9.8 and } w \text{ with 4.6.}$$
$$A = 45.08 \quad \text{Multiply.}$$

The area of the rectangle is 45.08 square inches.

Online Option Take the Online Readiness Quiz.

Main Idea

Solve literal equations for an indicated variable.

 Vocabulary

literal equation

 Get Connect**ED**

Literal Equations

BALLOONING A hot air balloon travels 13.5 miles at a rate of about 9 miles per hour.

1. Use the equation $d = rt$ where d is the distance traveled, r is the rate, and t is the time in hours to find the total time of the balloon ride.

2. Suppose the trip lasted for $2\frac{1}{4}$ hours. How far would the balloon travel at the same rate?

A **literal equation** is an equation or formula that has more than one variable. Some examples of literal equations are shown.

You can apply the properties of equality to solve for any of the variables.

Type	Equation
Perimeter of a Rectangle	$P = 2(\ell + w)$
Area of a Rectangle	$A = \ell w$
Area of a Triangle	$A = \frac{1}{2}bh$
Distance, Rate, and Time	$d = rt$
Circumference of a Circle	$C = 2\pi r$

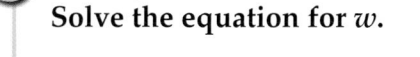 **EXAMPLE** Literal Equations

1. The equation $A = \ell w$ can be used to find the area of a rectangle. Solve the equation for w.

$A = \ell w$ Write the equation.

$\dfrac{A}{\ell} = \dfrac{\ell w}{\ell}$ Division Property of Equality

$\dfrac{A}{\ell} = w$ Simplify.

So, $w = \dfrac{A}{\ell}$.

 CHECK Your Progress

Solve each equation for the indicated variable.

a. $C = 2\pi r$, for r **b.** $d = rt$, for r

REAL-WORLD EXAMPLE **Find a Missing Measure**

Study Tip

Order of Operations In the expression $2(\ell + w)$, you start with w, add ℓ, and then multiply by 2. To solve for ℓ, divide by 2 and subtract w.

2 **CRAFTS** Molly is making a picture frame for her friends. The frame will have a lace border. She has 32 inches of lace for the border. What is the greatest width of the frame she can make?

9 in.

Use the equation $P = 2(\ell + w)$.

Step 1 Solve the equation for w.

$$P = 2(\ell + w) \qquad \text{Write the equation.}$$

$$\frac{P}{2} = \frac{2(\ell + w)}{2} \qquad \text{Division Property of Equality}$$

$$\frac{P}{2} = \ell + w \qquad \text{Simplify.}$$

$$\frac{P}{2} - \ell = \ell - \ell + w \qquad \text{Subtraction Property of Equality}$$

$$\frac{P}{2} - \ell = w \qquad \text{Simplify.}$$

Step 2 Find the width.

$$\frac{P}{2} - \ell = w \qquad \text{Write the equation.}$$

$$\frac{32}{2} - 9 = w \qquad \text{Replace } P \text{ with 32 and } \ell \text{ with 9.}$$

$$16 - 9 = w \qquad \text{Divide.}$$

$$7 = w \qquad \text{Subtract.}$$

The greatest width of the picture frame she can make is 7 inches.

 CHECK Your Progress

c. PATIOS The area A of a circular patio is 154 square feet. Solve the equation $A = \pi r^2$ for r. Then find the radius r of the patio to the nearest tenth.

✓ **CHECK Your Understanding**

Example 1 Solve each equation for the indicated variable.

1. $V = \ell wh$, for w **2.** $A = \ell w$, for ℓ **3** $I = prt$, for t

Example 2 **4. GEOMETRY** A parallelogram has an area A of 240 square centimeters.

a. Solve the equation $A = bh$ for b.

b. The height h of the parallelogram is 10 centimeters. Find the length of the base b.

Practice and Problem Solving

⬤ = Step-by-Step Solutions begin on page R1.
Extra Practice begins on page EP2.

Example 1 Solve each equation for the indicated variable.

5. $A = s^2$, for s **6.** $A = \frac{1}{2}bh$, for h **7.** $D = \frac{m}{v}$, for v

8. $P = 2(\ell + w)$, for ℓ **9.** $S = 6s^2$, for s **10.** $A = \pi r^2$, for r

Example 2   **11** **FINANCIAL LITERACY** The equation $I = prt$ is used to determine the simple interest I earned on money invested p at a certain rate r for a certain period of time t.

     a. Solve the equation for r.

     b. Find the interest rate r if \$2,500 was invested for 2 years and earned \$362.50 in interest.

12. **ALGEBRA** The equation for a line in slope-intercept form is $y = mx + b$.

     a. Solve the equation for m.

     b. A line goes through (2, 5), and the y-intercept b is 4. Find the slope m of the line.

13 **PARTY** Meredith wants to buy cone shaped bags to fill with candy for a party.

     a. The equation $V = \frac{1}{3}\pi r^2 h$ can be used to find the volume V of a cone with radius r and height h. Solve the equation for r.

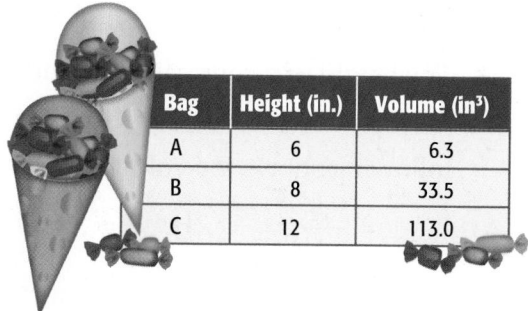

Bag	Height (in.)	Volume (in³)
A	6	6.3
B	8	33.5
C	12	113.0

     b. Meredith wants to choose the widest bag. Which bag should she choose? Explain your reasoning.

H.O.T. Problems

14. **FIND THE ERROR** Dion is solving the equation $V = \pi r^2 h$ for r. Find his mistake and correct it.

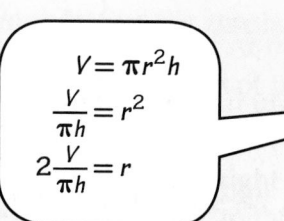

$$V = \pi r^2 h$$
$$\frac{V}{\pi h} = r^2$$
$$2\frac{V}{\pi h} = r$$

15. **CHALLENGE** The surface area $S.A.$ of a rectangular prism is found using the equation $S.A. = 2\ell w + 2wh + 2\ell h$, where ℓ is the length, h is the height, and w is the width. Solve the equation for ℓ.

16. **📝 WRITE MATH** Explain how to solve an equation for one of the variables.

17. The equation $A = \frac{1}{2}h(b_1 + b_2)$ is used to find the area of a trapezoid. Which of the following is equivalent to $A = \frac{1}{2}h(b_1 + b_2)$?

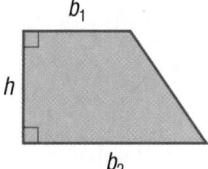

A. $2Ah - b_2 = b_1$

B. $\frac{A}{2} - (b_1 + b_2) = h$

C. $\frac{2A}{b_1 + b_2} = h$

D. $\frac{Ah}{2} - b_2 = b_1$

18. The height of a cone can be found using the equation $h = \frac{3V}{\pi r^2}$. Which of the following shows the volume of a cone?

F. $V = \frac{3h}{\pi r^2}$ **H.** $V = 3\pi r^2 h$

G. $V = \frac{h}{3\pi r^2}$ **I.** $V = \frac{1}{3}\pi r^2 h$

19. The equation $I = prt$ is used to find the simple interest I earned on money invested p at a certain rate r over a period of time t in years. Which of the following shows the equation solved for p?

A. $\frac{I}{rt} = p$

B. $\frac{rt}{I} = p$

C. $Irt = p$

D. $\frac{It}{r} = p$

20. **EXTENDED RESPONSE** The mailing tube below has a volume V of 170 cubic inches. Its height h is 24 inches.

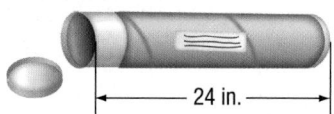

Part A Solve the equation $V = \pi r^2 h$ for r, the radius of the tube.

Part B Find the radius of the tube.

 **Literal Equations**

The standard form of a linear equation is a literal equation that, when solved for y, will be written in slope-intercept form.

EXAMPLE

Write the equation $-4x + 2y = 10$ in slope-intercept form.

$-4x + 2y = 10$	Write the equation.
$2y = 4x + 10$	Add 4x to each side. Simplify.
$y = 2x + 5$	Divide each side by 2. Simplify.

So, $-4x + 2y = 10$ written in slope-intercept form is $y = 2x + 5$.

Write each equation in slope-intercept form.

21. $-3x + 4y = 16$ **22.** $2x + 2y = -9$ **23.** $5x - y = 4$

Main Idea

Convert temperatures between the Fahrenheit and Celsius scales.

 Vocabulary

degree
Celsius (°C)
Fahrenheit (°F)
Kelvin (K)

 Get Connect ED

Convert Temperatures

Explore Use a thermometer that has both a Fahrenheit (°F) and Celsius (°C) scale to measure the temperature of the items listed.

Item	Temperature	
	(°F)	**(°C)**
Temperature of glass of ice water		
Temperature of cold water from faucet		
Temperature of hot water from faucet		

1. Copy the table and record your findings.

Use your findings to predict the temperature of each item.

2. cold glass of milk   **3.** hot cup of coffee   **4.** frozen dessert

Temperature is the measure of hotness or coldness of an object or environment. It is measured as **degrees** on a temperature scale.

In the Celsius temperature scale, temperature is measured in degrees **Celsius (°C)**. Water freezes at 0°C and boils at 100°C.

In the Fahrenheit temperature scale, temperature is measured in degrees **Fahrenheit (°F)**. Water freezes at 32°F and boils at 212°F.

The thermometers at the right show common temperatures in degrees Celsius and degrees Fahrenheit.

In the United States, the Fahrenheit temperature scale is used. Most other countries use the Celsius temperature scale.

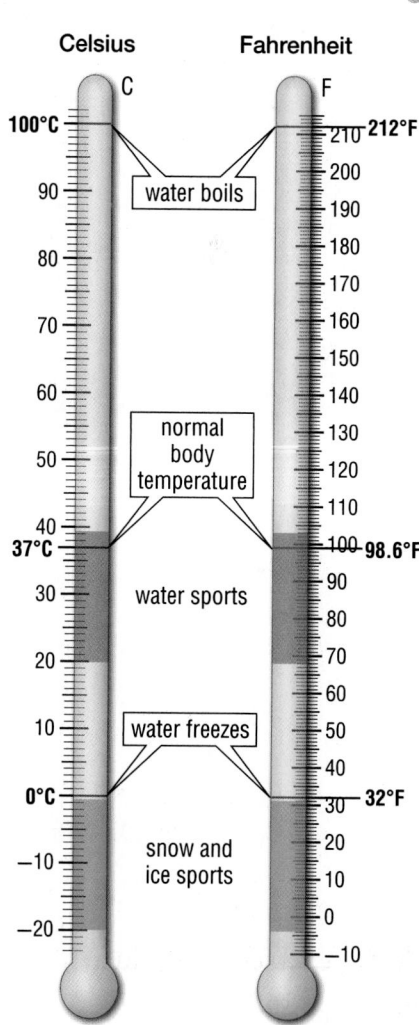

You can use the equation $F = \frac{9}{5}C + 32$ where F is the temperature in degrees Fahrenheit and C is the temperature in degrees Celsius, to *convert* or change temperatures from degrees Celsius to degrees Fahrenheit.

✏️ EXAMPLES Convert Temperatures

Complete each conversion. Round to the nearest hundredth if necessary.

1 **35°C = ■°F**

$F = \frac{9}{5}C + 32$ Write the equation.

$F = \frac{9}{5}(35) + 32$ Replace C with 35.

$F = 63 + 32$ Multiply.

$F = 95$ Simplify.

So, 35°C = 95°F.

QUICK Review

Multiplying Fractions

$$\frac{9}{5} \cdot \frac{35}{1} = \frac{9}{\cancel{5}} \cdot \frac{\cancel{35}^{7}}{1}$$
$$= 9 \cdot 7$$
$$= 63$$

2 **−28°C = ■°F**

$F = \frac{9}{5}C + 32$ Write the equation.

$F = \frac{9}{5}(-28) + 32$ Replace C with −28.

$F = -50.4 + 32$ Multiply.

$F = -18.4$ Simplify.

So, −28°C = −18.4°F.

✔️ CHECK Your Progress

a. 15°C = ■°F   **b.** −2°C = ■°F

✏️ REAL-WORLD EXAMPLES Convert Temperatures

3 **Solve the equation $F = \frac{9}{5}C + 32$ for C to convert temperatures from degrees Fahrenheit to degrees Celsius.**

$F = \frac{9}{5}C + 32$ Write the equation.

$F - 32 = \frac{9}{5}C + 32 - 32$ Subtract 32 from each side.

$F - 32 = \frac{9}{5}C$ Simplify.

$\frac{5}{9}(F - 32) = \frac{9}{5}C \cdot \frac{5}{9}$ Multiply each side by $\frac{5}{9}$.

$\frac{5}{9}(F - 32) = C$ Simplify.

So, to convert from degrees Fahrenheit to degrees Celsius, use the equation $C = \frac{5}{9}(F - 32)$.

④ ANIMALS The average body temperature of a horse is 100° Fahrenheit. Write this temperature in degrees Celsius.

Method 1   Use the equation $F = \frac{9}{5}C + 32$.

$F = \frac{9}{5}C + 32$	Write the equation.
$100 = \frac{9}{5}C + 32$	Replace *F* with 100.
$\underline{-32 = \qquad -32}$	Subtract 32 from each side.
$68 = \frac{9}{5}C$	Simplify.
$\frac{5}{9} \cdot 68 = \frac{5}{9} \cdot \frac{9}{5}C$	Multiply each side by $\frac{5}{9}$.
$37.78 \approx C$	Simplify.

 Real-World Link
The average lifespan for a horse is about 20–25 years. However, the oldest horse on record reached the age of 62.

Method 2   Use the equation $C = \frac{5}{9}(F - 32)$.

$C = \frac{5}{9}(F - 32)$	Write the equation.
$C = \frac{5}{9}(100 - 32)$	Replace *F* with 100.
$C = \frac{5}{9}(68)$	Subtract.
$C \approx 37.78$	Simplify.

The average body temperature of a horse is about 37.78°C.

CHOOSE Your Method

c. SCIENCE The melting point of the element lead is 621°F. Write this temperature in degrees Celsius.

✓ CHECK Your Understanding

Examples 1–4   Complete each conversion. Round to the nearest hundredth if necessary.

1. 10°C = ▨ °F   **2.** 22°C = ▨ °F   **3.** −6°C = ▨ °F

4. 78°F = ▨ °C   **5.** 30°F = ▨ °C   **6.** −10°F = ▨ °C

⑦ WEATHER Refer to the information at the right. Write both temperatures in degrees Celsius.

8. WATER The freezing point of salt water is about −21.1°C. About what temperature is this in degrees Fahrenheit?

High Temperature	Low Temperature
79°F	**68°F**

Practice and Problem Solving

 = Step-by-Step Solutions begin on page R1.
Extra Practice begins on page EP2.

Examples 1–4   **Complete each conversion. Round to the nearest hundredth if necessary.**

9. $45°C = \blacksquare °F$ **10.** $12°C = \blacksquare °F$ **11.** $113°F = \blacksquare °C$

12. $64°F = \blacksquare °C$ **13** $0°C = \blacksquare °F$ **14.** $-5°C = \blacksquare °F$

15. $-15°F = \blacksquare °C$ **16.** $-36°F = \blacksquare °C$ **17.** $16.5°C = \blacksquare °F$

18. WEATHER The average high temperatures during July for several European cities are listed in the table. Convert each temperature to degrees Fahrenheit.

City	Temperature (°C)
Athens, Greece	32
Berlin, Germany	23
Dublin, Ireland	19
Paris, France	24
Rome, Italy	31

19 TECHNOLOGY A computer's fan turns on when the hard drive reaches a temperature of 40°C. What is this temperature in degrees Fahrenheit?

Replace each ● with <, >, or = to make a true statement.

20. $15°C ● 59°F$ **21.** $-30°C ● 20°F$ **22.** $28°F ● -2.5°C$

23. 🔁 **MULTIPLE REPRESENTATIONS** The equation $F = \frac{9}{5}C + 32$ is used to convert temperatures in degrees Celsius to degrees Fahrenheit.

   a. GRAPHS Graph the equation on a coordinate grid.

   b. WORDS Interpret the y-intercept, x-intercept, and slope of the line.

   c. NUMBERS Use the graph to predict what temperature in degrees Celsius is equal to 25°F.

24. GRAPHIC NOVEL Refer to the graphic novel frame below for Exercises a–b.

   a. Contrast 28°F with 28°C to explain why Sarah thought she should pack for cold weather.

   b. Find the average June temperature for Madrid in degrees Fahrenheit.

25. **REASONING** The hottest temperature ever recorded on Earth was 136° in Azizia, Libya, in 1922. Is this temperature given in degrees Fahrenheit or degrees Celsius? Explain your reasoning.

26. **CHALLENGE** Is there a temperature at which the number in degrees Celsius is the same as the number in degrees Fahrenheit? If so, find it. If not, explain why not.

27. **WRITE MATH** Is 15° Fahrenheit or 15° Celsius a more reasonable temperature for a day of ice skating? Explain your reasoning.

Test Practice

28. Which of the following is a reasonable temperature for the activity shown in the illustration below?

    A. 25°C

    B. 35°F

    C. 80°C

    D. 150°F

29. Which of the following shows the correct steps to convert 75°F to degrees Celsius?

    F. Add 32 to 75. Then multiply by 5 and divide by 9.

    G. Multiply 75 by 9. Then divide by 5 and add 32.

    H. Subtract 32 from 75. Then multiply by 5 and divide by 9.

    I. Multiply 75 by 9. Then divide by 5 and subtract 32.

30. **THINK SOLVE EXPLAIN** **SHORT RESPONSE** Convert 20°C to degrees Fahrenheit. Show all of your work.

More About Temperature

Another temperature scale is the **kelvin (K)** scale. *Absolute zero*, or −273.15°C, is represented by 0 K. Temperatures on this scale are represented by *kelvins* (K), not degrees.

To convert from °C to kelvins, use the formula K = °C + 273.15.

Convert each temperature from °C to kelvins.

31. the boiling point of water   **32.** 27°C

33. −5°C   **34.** −15°C

Convert each temperature from °F to kelvins. Round to the nearest hundredth if necessary.

35. 34°F   **36.** 92°F

37. −3°F   **38.** −7°F

Problem-Solving Investigation

Main Idea Solve problems by determining reasonable answers.

Determine Reasonable Answers

VIOLETA: The instructions for an experiment in science say to heat a solution to about 125°C. The thermometer I am using is only in degrees Fahrenheit.

YOUR MISSION: Is 250°F a reasonable estimate for the temperature of the solution?

Understand	The solution needs to be heated to about 125°C. Violeta's thermometer measures temperatures in degrees Fahrenheit.
Plan	Use mental math to determine a reasonable answer.
Solve	The formula to convert degrees Celsius to degrees Fahrenheit is $F = \frac{9}{5}C + 32$. Think $\frac{9}{5}$ is close to $\frac{10}{5}$ or 2, 125°C is close to 120°C, and 32 is close to 30. $2 \times 120 + 30 = 240 + 30$ or 270 125°C ≈ 270°F Since 270°F ≈ 250°F, 250°F is a reasonable estimate.
Check	Convert 125°C to °F. $F = \frac{9}{5}(125) + 32 = 225 + 32$ or 257. Since 257 is close to 250, the answer is reasonable. ✔

Analyze the Strategy

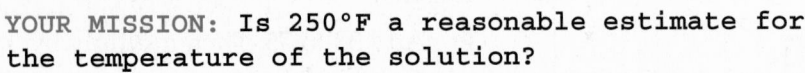

1. Explain why determining a reasonable answer was an appropriate strategy for solving the above problem.

2. **✍ WRITE MATH** Explain why mental math skills are important when finding reasonable answers.

Mixed Problem Solving

- Determine reasonable answers.
- Work backward.
- Look for a pattern.
- Choose an operation.

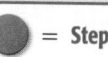 = Step-by-Step Solutions begin on page R1.
Extra Practice begins on page EP2.

Determine reasonable answers to solve Exercises 3–5.

3. MONEY Suki wants to buy an MP3 player that costs $129. She found it on sale for 75% of the original price. Would the sale price be about $30, $60, or $90? Explain.

4. CLOTHES Geraldo received a $100 gift card from a local store. He found a shirt for $24 and a pair of pants for $43.99. Does he have enough money to buy a pair of shoes for $35.99? Explain.

5. MEASUREMENT One pound is approximately equal to 0.454 kilogram. A shipping company charges $1.25 per pound to ship an item. If a package weighs 5 kilograms, will it cost $9, $12, or $15 to ship the package? Explain.

Use any strategy to solve Exercises 6–13.

6. TRAVEL Amit is saving money for a camping trip. He needs $54 for food and $320 for equipment. He has saved $150. If he can save $40 each week, in how many weeks will he have enough money?

7. NUMBER THEORY Study the pattern.

$$1 \times 1 = 1$$
$$11 \times 11 = 121$$
$$111 \times 111 = 12,321$$
$$1111 \times 1111 = 1,234,321$$

Without doing the multiplication, find $1,111,111 \times 1,111,111$.

8. JEWELRY Bethany is making a necklace with a pattern of blue, green, and white beads as shown below. What percent of the necklace will be white beads?

9. PARTY PLANNING Emmett is preparing for a surprise party. He bought invitations which cost $\frac{1}{4}$ of the money he had. Then he bought decorations, which cost $\frac{1}{2}$ of what remained. He had $15 left for a cake. How much money did he have at the beginning of the shopping trip?

10. PETS In a recent survey, 44% of students at Davison High School own a cat. If there are 1,532 students in the school, is 600, 675, or 715 a reasonable estimate for the number of students who own a cat? Explain.

11. POPULATION About 12.25% of the people in the U.S. live in California. If the U.S. population is about 297,000,000, estimate the population of California.

12. TRUCKS The five most popular colors for an SUV/Truck are listed in the table shown. If a plant manufactured 1,500 trucks one month, how many were *not* white?

Color	Percent Manufactured
White	26
Other	21
Silver	16
Gray	13
Black	13
Red	11

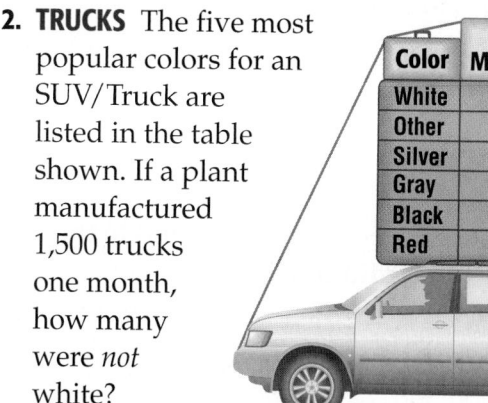

13. MEASUREMENT The entrance of a new convention center will need 1.8×10^5 square feet of ceramic tile. The tiles measure 1 foot by 2 feet and are sold in boxes of 48. How many boxes of tiles are needed to tile the entrance?

Solve each equation for the indicated variable. (Lesson 1A)

1. $C = \pi d$, for d

2. $S = 4\pi r^2$, for r

3. $A = \frac{1}{2}bh$, for b

4. $V = \frac{1}{3}Bh$, for B

5. **FOOD** A box of cereal has a volume V of 306 cubic inches. (Lesson 1A)

   a. Solve the equation $V = \ell wh$ for w.

   b. The height h of the box is 12 inches and the length ℓ of the box is 8.5 inches. Find the width w of the box.

6. **MULTIPLE CHOICE** The perimeter of a rectangle is found using the equation $P = 2(\ell + w)$. Which of the following shows the equation correctly solved for ℓ? (Lesson 1A)

   A. $\ell = \dfrac{P}{2w}$

   B. $\ell = 2P - w$

   C. $\ell = \dfrac{P}{2} - w$

   D. $\ell = \dfrac{P - w}{2}$

Complete each conversion. Round to the nearest hundredth if necessary. (Lesson 1B)

7. $-18°C = \blacksquare °F$ 8. $28°C = \blacksquare °F$

9. $12°F = \blacksquare °C$ 10. $58°F = \blacksquare °C$

11. $-20.9°C = \blacksquare °F$ 12. $105.7°F = \blacksquare °C$

13. **WEATHER** The record high temperature in Florida is 109 degrees Fahrenheit. It was set in Monticello in 1931. Convert the temperature to degrees Celsius. Round to the nearest hundredth of a degree. (Lesson 1B)

14. **SCIENCE** The boiling point of some common solutions are shown in the table. Find each boiling point in degrees Fahrenheit. Round to the nearest hundredth if necessary. (Lesson 1B)

Element	Boiling Point (°C)
Ether	35
Petroleum	210
Toluene	110.6
Water	100

15. **MULTIPLE CHOICE** Which equation is used to convert from degrees Celsius to degrees Fahrenheit? (Lesson 1B)

   F. $F = \dfrac{5}{9}(C - 32)$

   G. $F = \dfrac{5}{9}(C + 32)$

   H. $F = \dfrac{9}{5}C - 32$

   I. $F = \dfrac{9}{5}C + 32$

16. **FOOD** The table below shows the amount the Schaffer family spent on groceries each week last month.

Week	Total Spent ($)
1	121.59
2	168.54
3	98.67
4	141.78

If their grocery budget is $500 per month, did they stay within their budget? Explain. (Lesson 1C)

17. **DONATIONS** The Student Council raised $1,550 by selling tickets to a teachers versus students basketball game. If they donate 40% of the money to charity, is it reasonable to expect that they will donate $800? Explain. (Lesson 1C)

Main Idea

Convert customary and metric units of length, weight or mass, capacity, and time.

 Vocabulary
unit ratio

 *Get Connect*ED

Convert Length, Weight/Mass, Capacity, and Time

TRACK AND FIELD Jesse Owens set a record of 9.4 seconds for the 100-yard dash at the Big Ten track meet in Ann Arbor, Michigan, on May 25, 1935. The next year at the 1936 Olympic Games in Berlin, Germany, he matched the world record of 10.3 seconds in the 100-meter race.

1. A *yard* is a unit of length in the customary system. Name another unit of length in the customary system.

2. A *meter* is a unit of length in the metric system. Name another unit of length in the metric system.

3. Which is longer: one yard or one meter?

The relationships among the most commonly used customary and metric units of measure are shown in the table.

Key Concept	Measurement Conversions
Customary Units	**Metric Units**
Length	
1 foot (ft) = 12 inches (in.)	1 meter (m) = 1,000 millimeters (mm)
1 yard (yd) = 3 feet	1 meter = 100 centimeters (cm)
1 mile (mi) = 5,280 feet	1 kilometer (km) = 1,000 meters
Weight/Mass	
1 pound (lb) = 16 ounces (oz)	1 gram (g) = 1,000 milligrams (mg)
1 ton (T) = 2,000 pounds	1 kilogram (kg) = 1,000 grams
Capacity	
1 cup (c) = 8 fluid ounces (fl oz)	1 liter (L) = 1,000 milliliters (mL)
1 pint (pt) = 2 cups	1 kiloliter (kL) = 1,000 liters
1 quart (qt) = 2 pints	
1 gallon (gal) = 4 quarts	
Time	
1 minute (min) = 60 seconds (s)	1 week (wk) = 7 days
1 hour (h) = 60 minutes	1 year (yr) = 365 days
1 day (d) = 24 hours	

Each of the relationships in the table can be written as a unit ratio. A **unit ratio** is a ratio in which the denominator is 1 unit.

$$\frac{3\ \text{ft}}{1\ \text{yd}} \qquad \frac{2{,}000\ \text{lb}}{1\ \text{T}} \qquad \frac{1{,}000\ \text{m}}{1\ \text{km}} \qquad \frac{24\ \text{h}}{1\ \text{d}}$$

The value of each ratio is 1. To convert from larger units to smaller units, multiply by the appropriate unit ratio. To convert from smaller units to larger units, multiply by the reciprocal of the appropriate unit ratio.

REAL-WORLD EXAMPLE   Convert Measures

1 **BANNERS** Carleta needs 450 centimeters of material to make a banner for a parade. How many meters of material does she need?

$450\ \text{cm} = 450\ \text{cm} \cdot \dfrac{1\ \text{m}}{100\ \text{cm}}$ — Since 1 meter = 100 centimeters, multiply by $\frac{1\ \text{m}}{100\ \text{cm}}$.

$= 450\ \cancel{\text{cm}} \cdot \dfrac{1\ \text{m}}{100\ \cancel{\text{cm}}}$ — Divide out common units, leaving the desired unit, meter.

$= \dfrac{450}{100}\ \text{m or } 4.5\ \text{m}$ — Multiply.

Carleta needs 4.5 meters of material.

CHECK Your Progress

Complete each conversion.

a. 27 yd = ■ ft   **b.** 5 km = ■ m   **c.** 4,000 g = ■ kg

Real-World Link

The Rose Parade is one of the longest parades in the United States. It follows a 5.5-mile-long route and lasts about 2.5 hours.

You can also use dimensional analysis to convert *between* measurement systems. Recall that dimensional analysis is the process of including units of measurement when you compute.

Key Concept   Measurement Conversions

Customary to Metric	Metric to Customary
Length	
1 in. ≈ 2.54 cm	1 cm ≈ 0.394 in.
1 ft ≈ 0.305 m	1 m ≈ 3.279 ft
1 yd ≈ 0.914 m	1 m ≈ 1.094 yd
1 mi ≈ 1.609 km	1 km ≈ 0.621 mi
Capacity	
1 fl oz ≈ 29.574 mL	1 mL ≈ 0.034 fl oz
1 pt ≈ 0.473 L	1 L ≈ 2.114 pt
1 qt ≈ 0.946 L	1 L ≈ 1.057 qt
1 gal ≈ 3.785 L	1 L ≈ 0.264 gal
Weight/Mass	
1 oz ≈ 28.35 g	1 g ≈ 0.035 oz
1 lb ≈ 0.454 kg	1 kg ≈ 2.203 lb

EXAMPLES Convert Between Systems

2 Complete each conversion. Round to the nearest hundredth.
9 cm ≈ ▓ in.

Method 1

Use 1 in. ≈ 2.54 cm.

$9 \text{ cm} \approx 9 \text{ cm} \cdot \dfrac{1 \text{ in.}}{2.54 \text{ cm}}$ Since 1 in. ≈ 2.54 cm, multiply by $\frac{1 \text{ in.}}{2.54 \text{ cm}}$.

$\approx 9 \text{ c\cancel{m}} \cdot \dfrac{1 \text{ in.}}{2.54 \text{ c\cancel{m}}}$ Divide out common units, leaving the desired unit, inch.

$\approx \dfrac{9 \text{ in.}}{2.54}$ or 3.54 in.   Multiply.

Method 2

Use 1 cm ≈ 0.394 in.

$9 \text{ cm} \approx 9 \text{ cm} \cdot \dfrac{0.394 \text{ in.}}{1 \text{ cm}}$ Since 1 cm ≈ 0.394 in., multiply by $\frac{0.394 \text{ in.}}{1 \text{ cm}}$.

$\approx 9 \text{ c\cancel{m}} \cdot \dfrac{0.394 \text{ in.}}{1 \text{ c\cancel{m}}}$ Divide out common units, leaving the desired unit, inch.

$\approx 9 \cdot 0.394 \text{ in.}$ or 3.54 in.   Multiply.

So, 9 centimeters is approximately 3.54 inches.

3 12 lb ≈ ▓ kg

Method 1

Use 1 kg ≈ 2.203 lb

$12 \text{ lb} \approx 12 \text{ lb} \cdot \dfrac{1 \text{kg}}{2.203 \text{ lb}}$ Since 1 kg ≈ 2.203 lb, multiply by $\frac{1 \text{ kg}}{2.203 \text{ lb}}$.

$\approx 12 \text{ \cancel{lb}} \cdot \dfrac{1 \text{ kg}}{2.203 \text{ \cancel{lb}}}$ Divide out common units, leaving the desired unit, kilogram.

$\approx \dfrac{12 \text{ kg}}{2.203}$ or 5.45 kg   Divide.

Method 2

Use 1 lb ≈ 0.454 kg.

$12 \text{ lb} \approx 12 \text{ lb} \cdot \dfrac{0.454 \text{ kg}}{1 \text{ lb}}$ Since 1 lb ≈ 0.454 kg, multiply by $\frac{0.454 \text{ kg}}{1 \text{ lb}}$.

$\approx 12 \text{ \cancel{lb}} \cdot \dfrac{0.454 \text{ kg}}{1 \text{ \cancel{lb}}}$ Divide out common units, leaving the desired unit, kilogram.

$\approx 12 \cdot 0.454 \text{ kg}$ or 5.45 kg   Multiply.

So, 12 pounds is approximately 5.45 kilograms.

CHOOSE Your Method

d. 6 oz = ▓ g   **e.** 5 km = ▓ mi

f. 6 yd = ▓ m   **g.** 2 L = ▓ qt

Example 1   **Complete.**

1. 5 lb = ■ oz
2. $8\frac{2}{3}$ yd = ■ ft
3. 630 min = ■ h
4. 1.6 yr = ■ d
5. 686 cm = ■ m
6. 65 L = ■ mL

7. **FISH** The average weight of bass in a certain pond is 40 ounces. How many pounds does a bass weigh?

Examples 2 and 3   **Complete each conversion. Round to the nearest hundredth if necessary.**

8. 6 in. ≈ ■ cm
9. 1.6 cm ≈ ■ in.
10. 4 qt ≈ ■ L
11. 50 mL ≈ ■ fl oz
12. 17 mi ≈ ■ km
13. 19 kg ≈ ■ lb

14. **ANIMALS** A snake on display at the zoo is 54 centimeters long. How many inches long is the snake?

Practice and Problem Solving

● = **Step-by-Step Solutions** begin on page R1.
Extra Practice begins on page EP2.

Example 1   **Complete.**

15. 2 mi = ■ ft
16. 5 gal = ■ qt
17. 8.25 kg = ■ g
18. 9.25 L = ■ mL
19. 20 wk = ■ d
20. $1\frac{1}{4}$ h = ■ min
21. 72 in. = ■ ft
22. 104 oz = ■ lb
23. 4,570 mm = ■ m
24. 2,500 g = ■ kg
25. 120 min = ■ h
26. 49 d = ■ wk

27. How many pounds are in 76 ounces?

28. Convert 11,400 milligrams to grams.

Examples 2 and 3   **Complete each conversion. Round to the nearest hundredth if necessary.**

29. 3 in. ≈ ■ cm
30. 10 in. ≈ ■ cm
31. 17 cm ≈ ■ in.
32. 10.2 cm ≈ ■ in.
33. 4 L ≈ ■ qt
34. 20 mL ≈ ■ fl oz
35. 4,000 lb ≈ ■ kg
36. 42.5 kg ≈ ■ lb
37. 5 gal ≈ ■ L
38. 17 m ≈ ■ yd
39. 3.4 qt ≈ ■ mL
40. 4.25 kg ≈ ■ lb

41. How many kilograms are in 1.5 tons?

42. Convert 8,200 milligrams to ounces.

43. **WORLD RECORDS** The table below shows different world records.

Record	Length
Longest distance jumped on a pogo stick	37.18 km
Greatest distance walked with a milk bottle balanced on the head	130.3 km
Longest lawn mower ride	23,487.5 km
Longest ears on dog	34.9 cm

a. How many yards long was the distance jumped on a pogo stick?

b. How many inches long are the longest ears on a dog?

c. How many meters long was the longest lawn mower ride?

Replace each ● with <, >, or = to make a true statement.

44. 5 gal ● 18 L   **45.** 10 oz ● 0.3 kg   **46.** 8 mi ● 12.6 km

47. 46 fl oz ● 1.2 L   **48.** 3.5 lb ● 1,600 g   **49.** 2 T ● 1,816 kg

50. CAPACITY A standard coffee cup holds one cup of liquid.

    a. How many liters does one coffee cup hold?

    b. A water tower shaped like a coffee cup is the world's largest coffee cup. It has a capacity of 150,000 gallons of water. How many times more liters does the tower hold than a regular coffee cup?

Order each set of measurements from least to greatest.

51 10 ft, 3.1 m, 3 yd, 300 cm   **52.** 1,500 mL, 50 fl oz, 6 c, 1.55 L

53. 150 lb, 67.9 kg, 2,350 oz, 67,000 g   **54.** 27 m, 2,850 cm, 1,100 in., 90 ft

55. GRAPHIC NOVEL Refer to the graphic novel frame below for Exercises a–b.

    a. The airline says that Sarah's luggage cannot be more than 30 kilograms. Is her luggage under this limit? Explain.

    b. While in Spain, Sarah purchased souvenirs that have a mass of 13 kilograms. If she packs all of the souvenirs in her luggage, will it still be under the limit? Explain.

H.O.T. Problems

56. OPEN ENDED Use the Internet or another source to find a recipe in which the ingredients are given in grams or milliliters. Convert each measurement to customary units.

57. CHALLENGE Using the information at the beginning of the lesson, explain how you can compare the 100-yard dash and the 100-meter dash. Compare Owens's records in the two events.

58. WRITE MATH Compare and contrast the metric system and the customary system. In which system is it easier to convert? Explain.

59. THINK SOLVE EXPLAIN **EXTENDED RESPONSE** The table shows the lengths of the longest rivers in the world.

River	Length (mi)
Nile	4,145
Amazon	4,000
Mississippi-Missouri	3,740
Yangtze	3,720

Part A About how many kilometers long is the Nile River?

Part B How many yards longer is the Nile River than the Amazon River?

60. A piece of poster board measures 22 inches by 28 inches. Which of the following metric approximations is the same?

A. 5 cm by 7 cm

B. 0.5 m by 0.7 m

C. 50 mm by 70 mm

D. 5 m by 7 m

61. How many millimeters are in 5 meters?

F. 0.5

G. 50

H. 500

I. 5,000

Spiral Review

62. LIFE EXPECTANCY The average life expectancy in the United States is about 77 years of age. In 1901, the average life expectancy was about 63% of this number. Would 30, 48, or 60 years of age be a reasonable life expectancy for the year 1901? Explain. (Lesson 1C)

Complete each conversion. Round to the nearest hundredth if necessary.
(Lesson 1B)

63. $20°C = $ ■ $°F$

64. $48°C = $ ■ $°F$

65. $-5°C = $ ■ $°F$

66. $81°F = $ ■ $°C$

67. $-8°F = $ ■ $°C$

68. $46°F = $ ■ $°C$

69. CANDLES Adrienne wants to buy candles like the ones shown to put on a shelf in her room. (Lesson 1A)

a. The equation $V = \frac{1}{3}s^2h$ can be used to find the volume of a square pyramid with side length s and height h. Solve the equation for h.

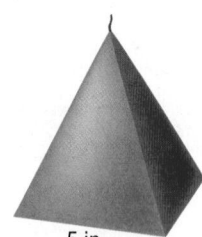

2 in.

5 in.

$V = 63$ in³ $V = 62.5$ in³

b. The equation $V = \pi r^2h$ can be used to find the volume of a cylinder with radius r and height h. Solve the equation for h.

c. If the shelf in Adrienne's room is 6 inches tall, which candle should she buy: the pyramid or the cylinder? Explain.

Main Idea
Determine the accuracy and precision of a set of data.

 Vocabulary
accuracy
precision

Accuracy and Precision

All measurements taken in the real world are approximations. Accuracy and precision are two terms used to describe measurements. **Accuracy** is how close a measurement is to some accepted, true value. An accurate measurement is considered correct. **Precision** is the ability of a measurement to be consistently reproduced. Precise measurements are close to each other.

The targets shown display different degrees of accuracy and precision.

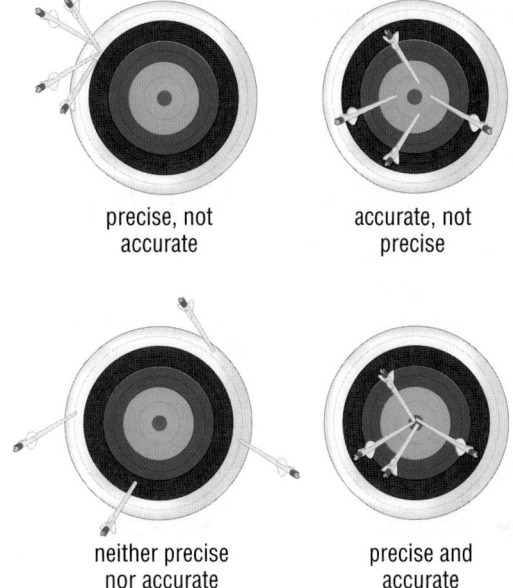

precise, not accurate

accurate, not precise

neither precise nor accurate

precise and accurate

ACTIVITY

1 Work with a partner to simulate the accuracy and precision of ten shots on goal in soccer. Cut a piece of notebook paper in half to represent the goal.

> **STEP 1** Place the goal on the floor. Have one person stand 3 feet away from the goal and toss 10 counters toward it to represent 10 shots on goal.

> **STEP 2** The other person should draw a diagram that shows where each counter lands.

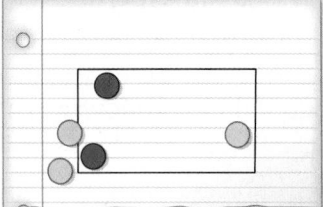

> **STEP 3** Switch roles and repeat the activity.

Analyze the Results

1. Categorize each partner's goal as precise, accurate, neither, or both.

2. Compare your results with your classmates' results. Find the percentage of results that are accurate, precise, neither, or both.

3. **WRITE MATH** During soccer practice, one player always hits the left crossbar of the goal. Does this represent accuracy, precision, neither, or both? Explain.

ACTIVITY

2. Work with a partner. Each person should fold a sheet of paper in half lengthwise. Then fold it in half along the width. Carefully tear the paper along the fold lines. You should each have 4 rectangles that are approximately the same size.

STEP 1 Without measuring, each student should draw a line segment estimated to be 5.0 centimeters long. Place the paper aside, facedown. Repeat the process 3 times.

STEP 2 With a centimeter ruler, measure and label each line segment you drew. Record your results in a table like the one shown.

Student 1	Measured Length	Line 1	Line 2	Line 3	Line 4
	5 centimeters				
Student 2	Measured Length	Line 1	Line 2	Line 3	Line 4
	5 centimeters				

Analyze the Results

4. Which of the two sets of estimates is more accurate? more precise? Explain.

5. The percent of error can be found by using the expression $\frac{|5.0 - \text{measurement}|}{5.0} \times 100$. Find the percent of error for each line. What does the percent of error tell you about the accuracy and precision of your lines?

6. Design an experiment to determine the difference between accuracy and precision. Your experiment should contain the following components.

Component	Description
Purpose	What do you want to find out?
Hypothesis	What do you think will happen?
Materials	What materials will you need to conduct the experiment?
Procedure	What steps will be taken during the experiment?
Data	How will you collect and record the data?
Conclusion	What conclusion can you draw based on your hypothesis?

Main Idea
Convert rates using dimensional analysis.

Vocabulary
unit rate
derived unit

Get ConnectED

Convert Rates

TENNIS Andy Roddick set a world record for the fastest tennis serve at 153 miles per hour.

1. How many feet are in one mile?

2. How many minutes are in one hour?

3. Write a ratio that compares one mile to the number of feet in one mile.

4. Write a ratio that compares one hour to the number of minutes in one hour.

A rate is a ratio that compares two quantities with different types of units such as $5 for 2 pounds or 130 miles in 2 hours. When a rate is simplified so it has a denominator of 1, it is called a **unit rate**. A **derived unit** is a unit that is derived from a measurement system base unit, such as length, mass, or time. Examples of derived units include square feet, cubic meter, and miles per hour.

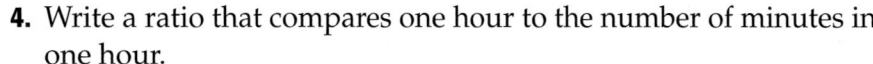
REAL-WORLD EXAMPLE **Use Dimensional Analysis**

1 **SPEED** The speed limit on a certain highway is 70 miles per hour. Convert 70 miles per hour to miles per minute.

Since 1 hour = 60 minutes, use $\frac{1\,h}{60\,min}$.

$$\frac{70\,mi}{1\,h} = \frac{70\,mi}{1\,h} \cdot \frac{1\,h}{60\,min} \qquad \text{Multiply by } \tfrac{1\,h}{60\,min}.$$

$$= \frac{70\,mi}{1\,\cancel{h}} \cdot \frac{1\,\cancel{h}}{60\,min} \qquad \text{Divide out common units.}$$

$$= \frac{70\,mi}{60\,min} \qquad \text{Multiply.}$$

$$\approx \frac{1.17\,mi}{1\,min} \qquad \begin{array}{l}\text{Divide both the numerator and denominator}\\ \text{by 60 to get a denominator of 1 minute.}\end{array}$$

So, 70 miles per hour is about 1.17 miles per minute.

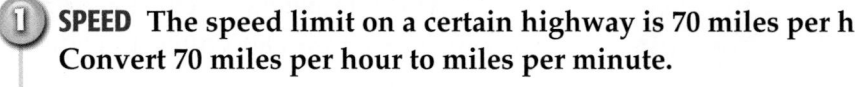
CHECK Your Progress

a. **MONEY** Mr. Roswell earns a salary of $65,000 per year. How much is this per week?

2 **TENNIS** Refer to the information at the beginning of the lesson. Convert 153 miles per hour to feet per second.

Use 1 mile = 5,280 feet and 1 hour = 3,600 seconds.

$$\frac{153 \text{ mi}}{1 \text{ h}} = \frac{153 \text{ mi}}{1 \text{ h}} \cdot \frac{5,280 \text{ ft}}{1 \text{ mi}} \cdot \frac{1 \text{ h}}{3,600 \text{ s}}$$ Multiply by $\frac{5,280 \text{ ft}}{1 \text{ mi}}$ and $\frac{1 \text{ h}}{3,600 \text{ s}}$.

$$= \frac{153 \text{ mi}}{1 \text{ h}} \cdot \frac{\overset{22}{5,280} \text{ ft}}{1 \text{ mi}} \cdot \frac{1 \text{ h}}{\underset{15}{3,600} \text{ s}}$$ Divide out common units.

$$= 153 \cdot 22 \text{ ft} \cdot \frac{1}{15 \text{ s}}$$ Simplify.

$$= \frac{224.4 \text{ ft}}{1 \text{ s}}$$ Multiply.

So, 153 miles per hour is equivalent to 224.4 feet per second.

✓ **CHECK Your Progress**

b. POOLS Water is being drained from a swimming pool at a rate of 150 gallons per hour. How many quarts is this per minute?

Real-World Link · · · ·
Sloths are found in Central and South America. They live in trees and can sleep hanging upside down from a tree for up to 18 hours.

REAL-WORLD EXAMPLE   **Convert Rates Between Systems**

3 **ANIMALS** A sloth traveled 3.8 kilometers in 2 hours. How fast is this in feet per second?

To convert kilometers to feet, use conversion factors relating kilometers to miles and miles to feet.

To convert hours to seconds, use conversion factors relating hours to minutes and minutes to seconds.

$$\frac{3.8 \text{ km}}{2 \text{ h}} \cdot \frac{1 \text{ mi}}{1.609 \text{ km}} \cdot \frac{5,280 \text{ ft}}{1 \text{ mi}} \cdot \frac{1 \text{ h}}{60 \text{ min}} \cdot \frac{1 \text{ min}}{60 \text{ s}}$$

$$= \frac{3.8 \text{ km}}{2 \text{ h}} \cdot \frac{1 \text{ mi}}{1.609 \text{ km}} \cdot \frac{5,280 \text{ ft}}{1 \text{ mi}} \cdot \frac{1 \text{ h}}{60 \text{ min}} \cdot \frac{1 \text{ min}}{60 \text{ s}}$$ Divide out common units.

$$= \frac{20,064 \text{ ft}}{11,584.8 \text{ s}}$$ Multiply.

$$\approx \frac{1.73 \text{ ft}}{1 \text{ s}}$$ Divide.

The sloth's top speed is about 1.73 feet per second.

✓ **CHECK Your Progress**

c. CARS A vehicle can travel 33 kilometers on 3 liters of gasoline. How many miles per gallon is this?

Examples 1–3   **Complete each conversion. Round to the nearest hundredth if necessary.**

1. 50 mi/h ≈ ■ ft/s   **2.** 30 lb/day ≈ ■ oz/h

3. 15 km/min ≈ ■ mi/h   **4.** 350 cm/s ≈ ■ in./min

5 **FOOD** In a recent competition, Joey Chestnut ate 7.5 pounds of chicken wings in 12 minutes. How many ounces is this per second?

6. **SWIMMING** In the 2008 Olympics, Michael Phelps won the 200-meter butterfly with a time of 112.03 seconds. How many miles is this per hour?

Practice and Problem Solving

= **Step-by-Step Solutions** begin on page R1.
Extra Practice begins on page EP2.

Examples 1–3   **Complete each conversion. Round to the nearest hundredth if necessary.**

7. 70 mi/h ≈ ■ ft/s   **8.** 1,500 mg/day ≈ ■ g/wk

9. 225 L/h ≈ ■ mL/s   **10.** 16 fl oz/h ≈ ■ mL/min

11. 52 mi/h ≈ ■ km/min   **12.** 15 gal/h ≈ ■ L/min

13. 175 kg/day ≈ ■ oz/min   **14.** 200 m/min ≈ ■ ft/s

15. **SPACE** The moon revolves around Earth at a rate of 1.03 kilometers per second. How many miles is this per hour?

16. **INSECTS** A certain type of cockroach traveled 14.7 feet in 3 seconds. How many miles is this per hour?

17 **FOOD** The average American eats about 3 pounds of peanut butter annually. How many milligrams of peanut butter is this per day?

18. **ROLLER COASTERS** The table lists the fastest roller coasters on three continents.

a. Convert each speed to feet per second.

b. Order the roller coasters from greatest to least speeds.

Fastest Roller Coasters		
Continent	**Name**	**Speed**
Asia	Dodonpa	172 km/h
Europe	Ring Racer	217 km/h
North America	Kingda Ka	128 mi/h

19. Order the following rates from least to greatest: 25 oz/min, 95 lb/h, 40 kg/h

20. **PHYSICS** To find the velocity v of an object, divide the distance d the object travels by the time t it takes the object to travel that distance.

a. Find the velocity in meters per second of a car that travels 20 miles in 17 minutes.

b. Find the velocity in miles per hour of a thrown ball that travels 18.45 meters in 0.4 second.

21. **MATH IN THE MEDIA** In a newspaper or a magazine, find a speed that is written in the customary system. Then use dimensional analysis to convert the rate to appropriate units in the metric system.

Real-World Link · · · ·
The Kingda Ka roller coaster accelerates to its top speed in 3.5 seconds. This is faster than any other coaster in the United States.

22. Which One Doesn't Belong? Select the rate that does not have the same value as the other three. Explain your reasoning.

| 65 mi/h | 95.34 ft/s | 112.63 km/h | 29.05 m/s |

23. CHALLENGE The top speed of a roller coaster is 76 miles per hour. If the coaster traveled 136 meters in 4.3 seconds, is it traveling at top speed? Explain.

24. WRITE MATH Compare and contrast the conversion factors $\dfrac{1\text{ ft}}{0.305\text{ m}}$ and $\dfrac{0.305\text{ m}}{1\text{ ft}}$ when converting 40 meters per minute to feet per second.

Test Practice

25. Which of the following is equivalent to 120 kilometers per hour?

A. 2 km/s

B. 2 km/min

C. 12 km/min

D. 720 km/s

26. A runner in a marathon averaged 5.78 miles per hour for the race. Which of the following is NOT equivalent to 5.78 miles per hour?

F. 2.58 m/s

G. 2.83 yd/s

H. 9.3 km/h

I. 15.5 m/min

27. THINK SOLVE EXPLAIN EXTENDED RESPONSE The gas efficiencies for three different cars are shown in the table below.

Car	Gas Efficiency
A	28 mi/gal
B	8.5 km/L
C	14.4 m/mL

Part A Convert the efficiency for Cars B and C to miles per gallon.

Part B Order the cars from least to greatest gas efficiency.

More About Unit Rates

Suppose you were shopping for shampoo and the store offered two different sizes at two different prices. Which one is the better buy? You can use unit rates to compare the cost per unit to determine the better buy.

28. Find the unit rate, or $\dfrac{\text{cost}}{\text{ounce}}$, for each bottle of shampoo. Which is the better buy? Explain your reasoning.

Clean & Fresh Shampoo $3.60 16 oz

Clean & Fresh Shampoo $5.25 24 oz

29. SHOPPING A 17-ounce box of cereal costs $4.89. A 21-ounce box of cereal costs $5.69. Which is the better buy? Explain your reasoning.

30. COMIC BOOKS An online comic book store is offering 3 comic books for $7.20 or 2 comic books for $5.60. Which is the better buy? Explain your reasoning.

Main Idea

Convert units of area and volume between customary and metric systems.

 **Get ConnectED**

Convert Units of Area and Volume

Explore Mr. Johanssen is building a play set in his backyard. The play set will cover an area that is 3 yards long and 2 yards wide.

2 yd

3 yd

1. What is the area of the rectangle?

2. What is the length and width of the rectangle in feet?

3. Copy the rectangle onto centimeter grid paper. Draw lines in blue marking each yard. Then draw lines in red marking each foot.

4. What is the area of the rectangle in square feet? How does this compare to the area in square yards?

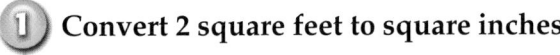

Some units of area in the customary system are square inch (in^2), square foot (ft^2), square yard (yd^2), and square mile (mi^2). Some units of area in the metric system are square centimeter (cm^2) and square meter (m^2).

Just as you used unit ratios to convert units of length, you can use unit ratios when you convert units of area.

EXAMPLE Convert Units of Area

1 **Convert 2 square feet to square inches.**

$$2 \text{ ft}^2 = 2 \times \cancel{ft} \times \cancel{ft} \times \frac{12 \text{ in.}}{1 \cancel{ft}} \times \frac{12 \text{ in.}}{1 \cancel{ft}} \quad \text{Multiply by } \frac{12 \text{ in.}}{1 \text{ ft}}.$$

$$= 2 \times 12 \text{ in.} \times 12 \text{ in.}$$

$$= 288 \text{ in}^2$$

So, $2 \text{ ft}^2 = 288 \text{ in}^2$.

 CHECK Your Progress

Complete each conversion.

a. $1.5 \text{ ft}^2 = \blacksquare \text{ in}^2$

b. $45 \text{ ft}^2 = \blacksquare \text{ yd}^2$

c. $24 \text{ cm}^2 = \blacksquare \text{ m}^2$

d. $3.2 \text{ km}^2 = \blacksquare \text{ m}^2$

Some units of volume in the customary system are cubic inch (in³), cubic foot (ft³), cubic yard (yd³), and cubic mile (mi³). Some units of volume in the metric system are cubic centimeter (cm³), and cubic meter (m³).

Study Tip

Look Back You can review conversion factors in Lesson 2A.

REAL-WORLD EXAMPLE **Convert Units of Volume**

2 **BUILDING** How many cubic yards of concrete will a builder need for a rectangular driveway that has a volume of 132 cubic feet?

$$132 \text{ ft}^3 = 132 \times \text{ft} \times \text{ft} \times \text{ft} \times \frac{1 \text{ yd}}{3 \text{ ft}} \times \frac{1 \text{ yd}}{3 \text{ ft}} \times \frac{1 \text{ yd}}{3 \text{ ft}} \quad \text{Multiply by } \frac{1 \text{ yd}}{3 \text{ ft}}.$$

$$= \frac{132 \text{ yd}^3}{27}$$

$$\approx 4.89 \text{ yd}^3$$

The builder needs about 4.89 cubic yards of concrete.

CHECK Your Progress

e. How many cubic meters of concrete are needed for a sidewalk that has a volume of 280,000 cubic centimeters?

f. A homeowner needs 150 cubic feet of mulch. Mulch is sold by the cubic yard. How many cubic yards does she need to buy?

You can also use conversion factors to convert area and volume *between* the customary and metric systems.

EXAMPLES **Convert Between Systems**

Complete each conversion. Round to the nearest hundredth if necessary.

3 **12 cm² ≈ ▪ in²**

$$12 \text{ cm}^2 \approx 12 \times \text{cm} \times \text{cm} \times \frac{1 \text{ in.}}{2.54 \text{ cm}} \times \frac{1 \text{ in.}}{2.54 \text{ cm}} \quad \text{Multiply by } \frac{1 \text{ in.}}{2.54 \text{ cm}}.$$

$$\approx \frac{12 \text{ in}^2}{6.45} \qquad\qquad \text{Multiply.}$$

$$\approx 1.86 \text{ in}^2 \qquad\qquad \text{Simplify.}$$

4 **47 ft³ ≈ ▪ m³**

$$47 \text{ ft}^3 \approx 47 \times \text{ft} \times \text{ft} \times \text{ft} \times \frac{0.305 \text{ m}}{1 \text{ ft}} \times \frac{0.305 \text{ m}}{1 \text{ ft}} \times \frac{0.305 \text{ m}}{1 \text{ ft}} \quad \begin{array}{l}\text{Multiply by}\\ \frac{0.305 \text{ m}}{1 \text{ ft}}.\end{array}$$

$$\approx 1.33 \text{ m}^3$$

CHECK Your Progress

g. $25 \text{ mi}^2 \approx \blacksquare \text{ km}^2$ **h.** $23 \text{ in}^3 \approx \blacksquare \text{ cm}^3$

i. $750 \text{ ft}^2 \approx \blacksquare \text{ m}^2$ **j.** $212 \text{ km}^3 \approx \blacksquare \text{ mi}^3$

The metric system is unique because it connects volume and capacity. One liter is equal to 1,000 cubic centimeters (1L = 1,000 cc).

 REAL-WORLD EXAMPLE **Convert Units of Volume**

5 FISH How many liters of water are needed to completely fill the aquarium shown?

Step 1 Find the volume.

$V = \ell wh$

$V = 50 \text{ cm} \cdot 25 \text{ cm} \cdot 30 \text{ cm}$

$V = 37{,}500 \text{ cm}^3$

30 cm

25 cm

50 cm

Step 2 Find the capacity.

$$\frac{37{,}500 \text{ cm}^3}{1} \times \frac{1 \text{ L}}{1{,}000 \text{ cm}^3} = \frac{\overset{37.5}{\cancel{37{,}500} \text{ cm}^3}}{1} \times \frac{1 \text{ L}}{\underset{1}{\cancel{1{,}000} \text{ cm}^3}}$$

$$= 37.5 \text{ L}$$

So, 37.5 liters of water are needed to fill the aquarium.

CHECK Your Progress

k. POOLS How many liters of water are needed to fill an inflatable swimming pool that is 160 centimeters long, 160 centimeters wide, and 35 centimeters tall?

CHECK Your Understanding

Examples 1 and 2   **Complete each conversion.**

1. $3 \text{ ft}^2 = \blacksquare \text{ in}^2$

2. $10.8 \text{ cm}^2 = \blacksquare \text{ mm}^2$

3. $4.3 \text{ yd}^3 = \blacksquare \text{ ft}^3$

4. $2{,}400 \text{ cm}^3 = \blacksquare \text{ m}^3$

5 REMODELING Suppose you have a room that has an area of 270 square feet. How many square yards of carpet would cover this room?

Examples 3 and 4   **Complete each conversion. Round to the nearest hundredth if necessary.**

6. $10 \text{ ft}^2 \approx \blacksquare \text{ m}^2$

7. $144 \text{ in}^2 \approx \blacksquare \text{ cm}^2$

8. $25 \text{ m}^3 \approx \blacksquare \text{ yd}^3$

9. $130 \text{ cm}^3 \approx \blacksquare \text{ in}^3$

Example 5   **10. COOKING** A casserole dish is 33 centimeters long, 23 centimeters wide, and 5 centimeters deep. How many liters can the casserole dish hold?

Practice and Problem Solving

= Step-by-Step Solutions begin on page R1.
Extra Practice begins on page EP2.

Examples 1 and 2 **Complete each conversion.**

11. $1.6 \text{ yd}^2 = \blacksquare \text{ ft}^2$ 12. $10.4 \text{ ft}^2 = \blacksquare \text{ in}^2$

13. $2.8 \text{ m}^2 = \blacksquare \text{ cm}^2$ 14. $4,654 \text{ cm}^2 = \blacksquare \text{ m}^2$

15. **BIOLOGY** The total surface area of the average adult's skin is about 21.5 square feet. Convert this measurement to square inches.

16. **BALLOONS** A standard hot air balloon holds about 2,000 cubic meters of hot air. How many cubic centimeters is this?

17. **LANDSCAPING** A landscape architect is designing the outside of a new restaurant. She needs 5 cubic yards of stone to cover a certain area. Will 100 cubic feet of stone be enough? If not, how many cubic feet are needed?

Examples 3 and 4 **Complete each conversion. Round to the nearest hundredth if necessary.**

18. $25 \text{ m}^2 \approx \blacksquare \text{ yd}^2$ 19. $240 \text{ in}^2 \approx \blacksquare \text{ cm}^2$

20. $2 \text{ mi}^3 \approx \blacksquare \text{ km}^3$ 21. $10 \text{ ft}^3 \approx \blacksquare \text{ m}^3$

Example 5 22. **JUICE** A carton of juice is 20 centimeters long, 12 centimeters wide, and 30 centimeters tall. How many liters of juice can the carton hold?

23. **MEDICINE** One dosage of a common vaccine is 0.5 milliliter of medicine. How many cubic centimeters is this?

24. **SCIENCE** *Density* is the ratio of an object's mass to its volume. The table shows the mass and volume of elements.

Element	Mass	Volume
Sodium	0.1 kg	102,999 mm³
Mercury	1 kg	73,820 mm³
Platinum	5 kg	233,100 mm³

    a. Find the density of each element in kilograms per cubic millimeter.

    b. Convert each ratio from part **a** to grams per cubic centimeter.

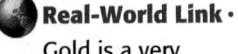
Real-World Link
Gold is a very dense material. Pure gold has a density of 19.29 grams per cubic centimeter. $10 million worth of gold bars will fit within one cubic foot.

25. **PAINT** One gallon of paint can cover 400 square feet of wall. How many square meters will one gallon of paint cover?

26. **MEASUREMENT** Refer to the information at the left. To the nearest hundredth, find the mass in grams of a gold bar that is 0.75 inch by 1 inch by 0.75 inch.

27. **MICROWAVES** The inside of a microwave oven has a volume of 1.2 cubic feet and measures 18 inches wide and 10 inches long. Using the formula $V = \ell wh$, find the depth of the microwave to the nearest tenth.

Complete each conversion. Round to the nearest hundredth if necessary.

28. $150 \text{ g/cm}^3 \approx \blacksquare \text{ oz/in}^3$ 29. $1,600 \text{ fl oz/ft}^3 \approx \blacksquare \text{ L/m}^3$

30. $675 \text{ mL/m}^3 \approx \blacksquare \text{ fl oz/in}^3$ 31. $0.1 \text{ g/cm}^3 \approx \blacksquare \text{ oz/yd}^3$

32. FIND THE ERROR Hana is converting 48 square meters to square feet. Find her mistake and correct it.

$$48\ m^2 = 48 \times m \times m \times \frac{0.305\ m}{1\ ft} \times \frac{0.305\ m}{1\ ft}$$
$$\approx 4.47\ ft^2$$

33. CHALLENGE A hectare is a metric unit of area approximately equal to 10,000 square meters or 2.47 acres. The base of the Great Pyramid of Khufu in Egypt is a square with sides that measure 230 meters. About how many acres does the base cover?

34. WRITE MATH Compare and contrast converting units of area and volume.

Test Practice

35. Mr. Meade is tiling his kitchen floor. The area to be covered measures 5 yards by 3 yards. He is using 6-inch square tiles to cover the floor. How many tiles are needed to cover the floor?

   A. 4   **C.** 540

   B. 36   **D.** 3,240

36. Approximately how many cubic feet are there in six cubic meters?

   F. 5.89   **H.** 41.31

   G. 29.31   **I.** 211.47

37. GRIDDED RESPONSE A cereal box has a volume of 176 cubic inches. How many cubic centimeters of cereal are in the box if the box is completely full?

38. MEASUREMENT The speed limit on a Canadian highway is 100 kilometers per hour. Approximately how fast can you drive on this highway in miles per hour? (Lesson 2B)

Complete. (Lesson 2A)

39. 22 ft = ■ yd   **40.** 4 lb = ■ oz   **41.** 600 g = ■ mg

Keeping Cars *Cruising*

Do you know a lot about cars and have strong analytical skills? If so, a career in automotive technology might be a perfect fit for you. Automotive service technicians inspect, maintain, and repair vehicles that run on gasoline, electricity, or alternative fuels. They must be able to work with electronic components and use electronic diagnostic equipment and digital manuals. Automotive technicians must also have good reading, mathematics, and computer skills.

21st Century Careers

Are you interested in a career in automotive technology? Take some of these courses in high school.

- Algebra
- Automotive Electronics
- Computer-Controlled Fuel Systems
- Intro to Automotive Technology
- Physics

 Get ConnectED

National Hot Rod Association Cars		
Drag Racing Car	**Fact**	**Top Speed (mph)**
Top Fuel	It is 25 feet long.	335
Funny Car	It covers the quarter-mile in 4.6 seconds.	330
Pro Stock	• It must weigh a mimimum of 2,350 pounds, including the driver. • The gasoline in its fuel system flows at $7\frac{1}{2}$ gallons per minute.	205

Real-World Math

For each problem, use the information in the table. Round to the nearest tenth.

1. Find the length of a Top Fuel drag racing car in meters.

2. How many meters does a Funny Car travel in 4.6 seconds?

3. What is the minimum weight of a Pro Stock drag racing car and driver in kilograms?

4. What is the top speed of the Funny Car in kilometers per minute?

5. How fast is the Top Fuel car in meters per second?

6. What is the rate of the gasoline flow for the Pro Stock car in liters per second?

FOLDABLES Study Organizer

Be sure the following Key Concepts are noted in your Foldable.

Units of Measure

Key Concepts

Literal Equations (Lesson 1)
• A literal equation is an equation or formula that has more than one variable.

• Literal equations can be solved for any variable using the Properties of Equality.

Convert Temperatures (Lesson 1)
• Use the equations below to convert between temperature scales.

$$C = \frac{5}{9}(F - 32)$$

$$F = \frac{9}{5}C + 32$$

Convert Units of Measure (Lesson 2)
• You can multiply by a unit ratio to convert from larger units to smaller units.

• To convert from smaller units to larger units, multiply by the reciprocal of the appropriate unit ratio.

• Use dimensional analysis to convert rates.

Key Vocabulary

accuracy

Celsius (°C)

degree

derived unit

Fahrenheit (°F)

literal equation

precision

unit rate

unit ratio

Vocabulary Check

Choose the correct term or numbers to complete each sentence.

1. A (unit rate, derived unit) is a unit that comes from a measurement system base unit.

2. The appropriate temperature for a warm summer day is (75° Fahrenheit, 75° Celsius).

3. A (literal equation, derived unit) is a formula with more than one variable.

4. When converting measurements, multiply by the (unit ratio, reciprocal).

5. The equation $A = \ell w$ is an example of a (literal equation, unit rate).

Multi-Part Lesson Review

Lesson 1 **Literal Equations**

Literal Equations (Lesson 1A)

Solve each equation for the indicated variable.

6. $a^2 + b^2 = c^2$, for b **7.** $D = \frac{m}{v}$, for m

8. $d = rt$, for t **9.** $A = \frac{1}{2}h(b_1 + b_2)$, for b_1

10. MONEY A quarter has a circumference C of about 76.8 millimeters.

   a. Solve the equation $C = 2\pi r$ for r.

   b. What is the radius r of a quarter? Round to the nearest tenth.

11. STORAGE A cylindrical storage container has a volume V of 4.7 cubic feet.

   a. Solve the equation $V = \pi r^2 h$ for r.

   b. The height h of the container is 1.5 feet. Find the radius r of the container to the nearest tenth.

EXAMPLE 1 Solve the equation $V = \frac{1}{3}\pi r^2 h$ for h.

$V = \frac{1}{3}\pi r^2 h$	Write the equation.
$V \cdot 3 = \frac{1}{3}\pi r^2 h \cdot 3$	Multiply each side by 3.
$3V = \pi r^2 h$	Simplify.
$\frac{3V}{\pi r^2} = \frac{\pi r^2 h}{\pi r^2}$	Divide each side by πr^2.
$\frac{3V}{\pi r^2} = h$	Simplify.

So, $h = \frac{3V}{\pi r^2}$.

Convert Temperatures (Lesson 1B)

Complete each conversion. Round to the nearest hundredth if necessary.

12. $25°C = \blacksquare \,°F$ **13.** $-15°C = \blacksquare \,°F$

14. $65°F = \blacksquare \,°C$ **15.** $20°F = \blacksquare \,°C$

16. $42.5°C = \blacksquare \,°F$ **17.** $101.8°F = \blacksquare \,°C$

18. SPACE The high daytime surface temperatures for some of the planets in our solar system are shown in the table. Convert each temperature to degrees Fahrenheit.

Planet	Temperature (°C)
Earth	58
Jupiter	−153
Mars	20
Mercury	427
Venus	480

EXAMPLE 2 Convert 30°C to degrees Fahrenheit. Round to the nearest hundredth if necessary.

$F = \frac{9}{5}C + 32$	Write the equation.
$F = \frac{9}{5}(30) + 32$	Replace C with 30.
$F = 54 + 32$	Multiply.
$F = 86$	Simplify.

EXAMPLE 3 Convert −8°F to degrees Celsius. Round to the nearest hundredth if necessary.

$C = \frac{5}{9}(F - 32)$	Write the equation.
$C = \frac{5}{9}(-8 - 32)$	Replace F with −8.
$C = \frac{5}{9}(-40)$	Subtract.
$C \approx -22.22$	Multiply.

Lesson 1 ## Literal Equations (continued)

PSI: Determine Reasonable Answers (Lesson 1C)

Determine a reasonable answer.

19. **ECOLOGY** In a survey of 1,413 consumers, 6% said they would be willing to pay more for recycled products in order to protect the environment. Is 8.4, 84, or 841 a reasonable estimate for the number of consumers willing to pay more? Explain.

20. **PIZZA** Twelve friends share three large pizzas. If they split the cost evenly among themselves, and each pizza cost $11.95, will each person pay about $2, $3, or $4? Explain.

21. **TEMPERATURE** A chemical solution needs to be heated to about 160°C. Is 300°F, 330°F, or 375°F a more reasonable estimate for the temperature of the solution in degrees Fahrenheit? Explain.

EXAMPLE 4 Philip's flight departed at 9:10 A.M. and arrived at 3:15 P.M., Eastern Standard Time. While in flight, Philip checked his watch and estimated that he had completed about 63% of the trip. Is 11 A.M., 12 P.M., or 1 P.M. a reasonable estimate for the time that Philip checked his watch?

The total duration of the trip is 365 minutes, or 6 hours and 5 minutes. One half, or 50%, of the trip would be 3 hours and $2\frac{1}{2}$ minutes after departure, or about 12:12 P.M. Since 63% is greater than 50%, 1 P.M. is the only reasonable answer.

Lesson 2 ## Convert Units of Measure

Convert Length, Weight/Mass, Capacity, and Time (Lesson 2A)

Complete each conversion. Round to the nearest hundredth if necessary.

22. 200 oz = ■ lb
23. 3.5 mi = ■ ft
24. 1,200 mg = ■ g
25. 26 m = ■ mm
26. 5 in. ≈ ■ cm
27. 25 km ≈ ■ mi
28. 12 L ≈ ■ gal
29. 30 oz ≈ ■ g
30. 120 m ≈ ■ ft
31. 20 pt ≈ ■ L

32. Which is greater: a 10-pound weight or a 5-kilogram weight?

33. Which is greater: a 5,000-meter race or a 4-mile race?

34. **ANIMALS** The average weight of a Clydesdale horse is 1,900 pounds. About how many kilograms does a Clydesdale weigh?

EXAMPLE 5 Which has a greater capacity: a bottle containing 32 fluid ounces of spring water or a bottle containing 1 liter of spring water?

Use the unit ratios $\frac{29.574 \text{ mL}}{1 \text{ fl oz}}$ and $\frac{1 \text{L}}{1,000 \text{ mL}}$.

$32 \text{ fl oz} \approx 32 \text{ fl oz} \cdot \frac{29.574 \text{ mL}}{1 \text{ fl oz}} \cdot \frac{1 \text{L}}{1,000 \text{ mL}}$

$\approx 32 \cdot \frac{29.574 \text{ L}}{1,000}$ or 0.95 L

The 1-liter bottle contains more water.

Convert Rates (Lesson 2C)

Complete each conversion. Round to the nearest hundredth if necessary.

35. 1,500 km/day ≈ ▩ ft/s

36. 730 m/h ≈ ▩ in./min

37. 65 mi/h ≈ ▩ m/min

38. 8 gal/day ≈ ▩ mL/s

39. **TRACK** In the 2008 Olympics, Usain Bolt set a world record in the 100-meter dash with a time of 9.69 seconds. How many miles is this per hour?

40. **SPACE** Halley's Comet travels at a rate of about 70 kilometers per second. How many miles per hour is this?

EXAMPLE 6 Convert 70 miles per hour to feet per second.

Use 1 mile = 5,280 feet and 1 hour = 3,600 seconds.

$$\frac{70 \text{ mi}}{1 \text{ h}} = \frac{70 \text{ mi}}{1 \text{ h}} \cdot \frac{5,280 \text{ ft}}{1 \text{ mi}} \cdot \frac{1 \text{ h}}{3,600 \text{ s}}$$

$$= \frac{70 \text{ mi}}{1 \text{ h}} \cdot \frac{5,280 \text{ ft}}{1 \text{ mi}} \cdot \frac{1 \text{ h}}{3,600 \text{ s}}$$

$$= 70 \cdot 5,280 \text{ ft} \cdot \frac{1}{3,600 \text{ s}}$$

$$\approx \frac{369,600 \text{ ft}}{3,600 \text{ s}} \text{ or } \frac{102.67 \text{ ft}}{1 \text{ s}}$$

So, 70 miles per hour is equivalent to 102.67 feet per second.

Convert Units of Area and Volume (Lesson 2D)

Complete each conversion. Round to the nearest hundredth if necessary.

41. 120 yd^3 = ▩ ft^3

42. 10 cm^2 = ▩ mm^2

43. 16 m^2 ≈ ▩ ft^2

44. 4 yd^2 ≈ ▩ m^2

45. 120 cm^3 ≈ ▩ in^3

46. 1.34 m^3 ≈ ▩ yd^3

47. **PARKING** The area of a parking lot is 375,000 square feet. How many square meters is the parking lot?

48. **STORAGE** The total capacity of a certain storage unit is about 23 cubic meters. How many cubic feet is the storage unit?

EXAMPLE 7 Convert 15 square centimeters to square inches.

Use the unit ratio $\frac{1 \text{ in.}}{2.54 \text{ cm}}$.

$$15 \text{ cm}^2 \approx 15 \text{ cm}^2 \cdot \frac{1 \text{ in.}}{2.54 \text{ cm}} \cdot \frac{1 \text{ in.}}{2.54 \text{ cm}}$$

$$\approx \frac{15 \text{ in}^2}{2.54 \cdot 2.54} \text{ or } 2.33 \text{ in}^2$$

So, 15 cm^2 ≈ 2.33 in^2.

Solve each equation for the indicated variable.

1. $A = \pi r^2$, for r **2.** $a^2 + b^2 = c^2$, for a

3. $V = \pi r^2 h$, for h **4.** $V = \frac{1}{3}\pi r^2 h$, for r

5. GEOMETRY A triangle has an area A of 24 square meters.

    a. Solve the equation $A = \frac{1}{2}bh$ for b.

    b. The height h of the triangle is 6 meters. Find the length of the base b.

Complete each conversion. Round to the nearest hundredth if necessary.

6. $40°C = \blacksquare$ °F   **7.** $-10°C = \blacksquare$ °F

8. $84°F = \blacksquare$ °C   **9.** $38°F = \blacksquare$ °C

10. $15.5°C = \blacksquare$ °F   **11.** $72.6°F = \blacksquare$ °C

12. ANIMALS The normal body temperature for several animals is shown below.

Animal	Temperature (°C)
Camel	39.5
Chicken	42.0
Cow	38.5
Squirrel	36.2

    a. Convert each temperature to degrees Fahrenheit.

    b. Order the animals from least to greatest body temperature.

13. MULTIPLE CHOICE Which of the following is a more reasonable temperature for a snowy day?

    A. 35°C   **C.** 75°C

    B. 35°F   **D.** 75°F

14. FINANCIAL LITERACY Sandra estimated that about 35% of her $420 paycheck was deducted for taxes and insurance. Did about $100, $150, or $200 get deducted from her pay?

Complete each conversion. Round to the nearest hundredth if necessary.

15. $18 \text{ pt} \approx \blacksquare$ L   **16.** $15 \text{ m} \approx \blacksquare$ yd

17. $25 \text{ kg} \approx \blacksquare$ lb   **18.** $250 \text{ ft} \approx \blacksquare$ m

19. RECORDS The world's largest baseball bat is 120 feet long. About how many meters long is the baseball bat?

Complete each conversion. Round to the nearest hundredth if necessary.

20. $50 \text{ gal/h} \approx \blacksquare$ L/min

21. $45 \text{ yd/s} \approx \blacksquare$ km/h

22. CARS A certain car travels an average of 18 miles per gallon of gasoline. How many kilometers can it travel per one liter of gasoline?

Complete each conversion. Round to the nearest hundredth if necessary.

23. $15 \text{ ft}^2 = \blacksquare$ yd²   **24.** $108 \text{ ft}^2 \approx \blacksquare$ m²

25. $150 \text{ cm}^3 = \blacksquare$ mm³   **26.** $45 \text{ in}^3 \approx \blacksquare$ cm³

27. THINK SOLVE EXPLAIN **EXTENDED RESPONSE** The table shows the heights of different mountain peaks in Colorado.

Mountain Peak	Height (ft)
Mt. Elbert	14,440
La Plata Peak	14,336
Pike's Peak	14,115

    Part A Convert each height to meters.

    Part B How many meters taller is Mt. Elbert than Pike's Peak?

    Part C The tallest mountain in the world is Mt. Everest in Nepal. It has a height of 8,850 meters. How does the height of Mt. Everest compare to the height of Mt. Elbert?

Preparing for Standardized Tests

 Extended-Response: Reference Sheet

Basic formulas and measurement conversions may be provided on a reference sheet in the test booklet. Quickly review this reference sheet before the test so that you are familiar with the information.

TEST EXAMPLE

The wood of an African blackwood tree is very heavy. It weighs an average of 1,200 ounces per cubic foot.

Part A   **What is the weight of the wood in pounds per cubic foot? Explain your reasoning.**

To convert 1,200 ounces per cubic foot to pounds per cubic foot, multiply $\dfrac{1{,}200 \text{ oz}}{ft^3}$ by $\dfrac{1 \text{ lb}}{16 \text{ oz}}$.

$$\frac{1{,}200 \; \cancel{oz}}{ft^3} \times \frac{1 \text{ lb}}{16 \; \cancel{oz}} = 75 \text{ lb/}ft^3$$

Part B   **What is the weight of the wood in kilograms per cubic meter? Use the conversions at the right if necessary.**

1 ft ≈ 30.48 cm
1 ft^3 ≈ 0.028 m^3
1 lb ≈ 0.454 kg

First, I multiply $\dfrac{75 \text{ lb}}{ft^3}$ by the ratio $\dfrac{1 \, ft^3}{0.028 \, m^3}$ so that the ft^3 units cancel out.

$$\frac{75 \text{ lb}}{\cancel{ft^3}} \times \frac{1 \; \cancel{ft^3}}{0.028 \, m^3} \approx 2678.57 \text{ lb/}m^3$$

Then, I multiply by the ratio $\dfrac{0.454 \text{ kg}}{1 \text{ lb}}$ so that the lb units cancel out.

$$\frac{2678.57 \; \cancel{lb}}{m^3} \times \frac{0.454 \text{ kg}}{1 \; \cancel{lb}} = 1216.07 \text{ kg/}m^3$$

 Work on It

A roller coaster is traveling 28 meters per second. Use the conversion factor below.

$$1 \text{ km} \approx 0.62 \text{ mi}$$

Part A   How many kilometers per second is the roller coaster traveling?

Part B   What is the speed of the roller coaster in miles per hour?

Test Hint

If you find that you cannot answer every part of an open-ended question, do as much as you can. You may earn partial credit.

Read each question. Then fill in the correct answer on the answer document provided by your teacher or on a sheet of paper.

1. A flag is being made that has an area of six square feet. Approximately how many square meters of fabric is this?
(1 ft ≈ 0.305 m)

A. 0.56 m²

B. 1.83 m²

C. 19.67 m²

D. 64.50 m²

2. A park is shaped like a rectangle with the dimensions shown below. Which of the following is closest to the length of a diagonal of the park?

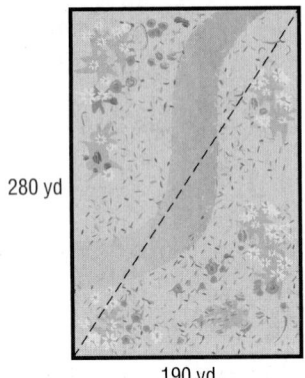

280 yd

190 yd

F. 165 yd

G. 290 yd

H. 340 yd

I. 405 yd

3. ✎ **GRIDDED RESPONSE** Two siblings agreed to split the cost of a television and a DVD player evenly. They spent a total of $335.00 on the television and $95 on the DVD player. How much, in dollars, did each sibling pay?

4. Marin spent $1\frac{1}{2}$ hours practicing the piano last week. How many seconds are in $1\frac{1}{2}$ hours?

A. 90 seconds

B. 540 seconds

C. 3,600 seconds

D. 5,400 seconds

5. THINK SOLVE EXPLAIN **SHORT RESPONSE** The table below shows the total cost of DVD rentals at Movies Plus.

Number of DVDs	Total Cost ($)
1	4.95
2	6.95
3	8.95
4	10.95
5	12.95

Write an expression that can be used to find the total cost of any number of DVD rentals.

6. What is the approximate distance between point *A* and point *B*?

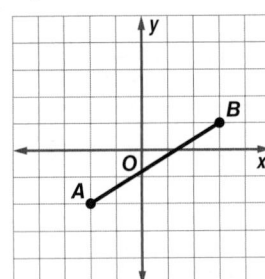

F. 4 units

G. 5.8 units

H. 6.2 units

I. 34 units

7. Which number line shows the solution set of the inequality below?

$$-3n + 6 \le -18$$

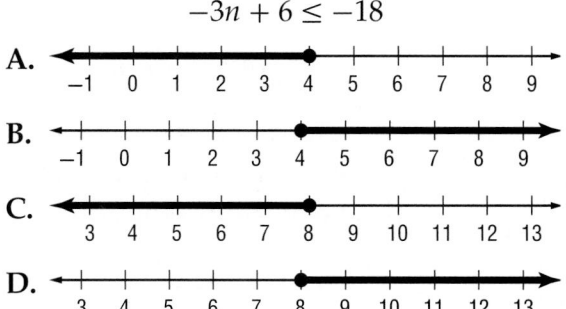

A.

B.

C.

D.

8. Which of the following is an irrational number?

F. -2

G. $-\dfrac{7}{8}$

H. $\sqrt{3}$

I. $\sqrt{4}$

9. **GRIDDED RESPONSE** What is the slope of the line below?

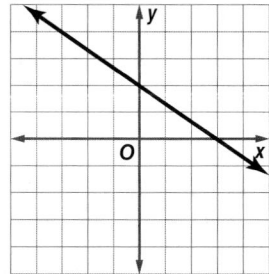

10. **THINK SOLVE EXPLAIN** **SHORT RESPONSE** T.J. is $1\frac{1}{2}$ meters tall. About how tall is he in feet and inches?

11. The circumference of Earth around the equator is 2.49×10^4 miles. Which of the following represents the circumference in standard notation?

A. 249 mi **C.** 24,900 mi

B. 2,490 mi **D.** 249,000 mi

12. **THINK SOLVE EXPLAIN** **EXTENDED RESPONSE** The volume of the ice cream cone below is found using the equation $V = \frac{1}{3}\pi r^2 h$, where r is the radius and h is the height of the cone.

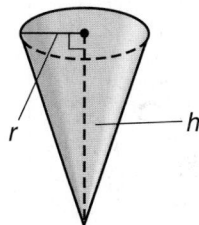

Part A Solve the equation $V = \frac{1}{3}\pi r^2 h$ for r.

Part B The volume of the ice cream cone is about 5 cubic inches. If the height of the cone is 5 inches, what is the radius to the nearest tenth?

Part C What is the volume of the cone in cubic centimeters?

NEED EXTRA HELP?												
If You Missed Question...	1	2	3	4	5	6	7	8	9	10	11	12
Go to Chapter-Lesson...	9-2D	8-2C	1-1D	9-2A	5-2B	8-2D	3-4A	2-3D	6-1C	9-2A	2-2B	9-1A

Data Analysis and Statistics

connectED.mcgraw-hill.com

The ☆BIG Idea

How can data displays help to analyze and make conjectures about data?

FOLDABLES®
Study Organizer

Make this Foldable to help you organize your notes.

1 Analyze Data
2 Box-and-Whisker Plots
3 Scatter Plots
4 Vocabulary

Review Vocabulary

Area of Great Lakes

circle graph
gráfica circular
a graph used to compare parts of a data set to the whole set of data

Lake Ontario 8%
Lake Erie 11%
Lake Superior 33%
Lake Huron 24%
Lake Michigan 24%

Key Vocabulary

English	Español
box-and-whisker plot	diagrama de caja y patillas
line of best fit	recta de mejor ajuste
measures of variation	medidas de variación
scatter plot	diagrama de dispersión

When Will I Use This?

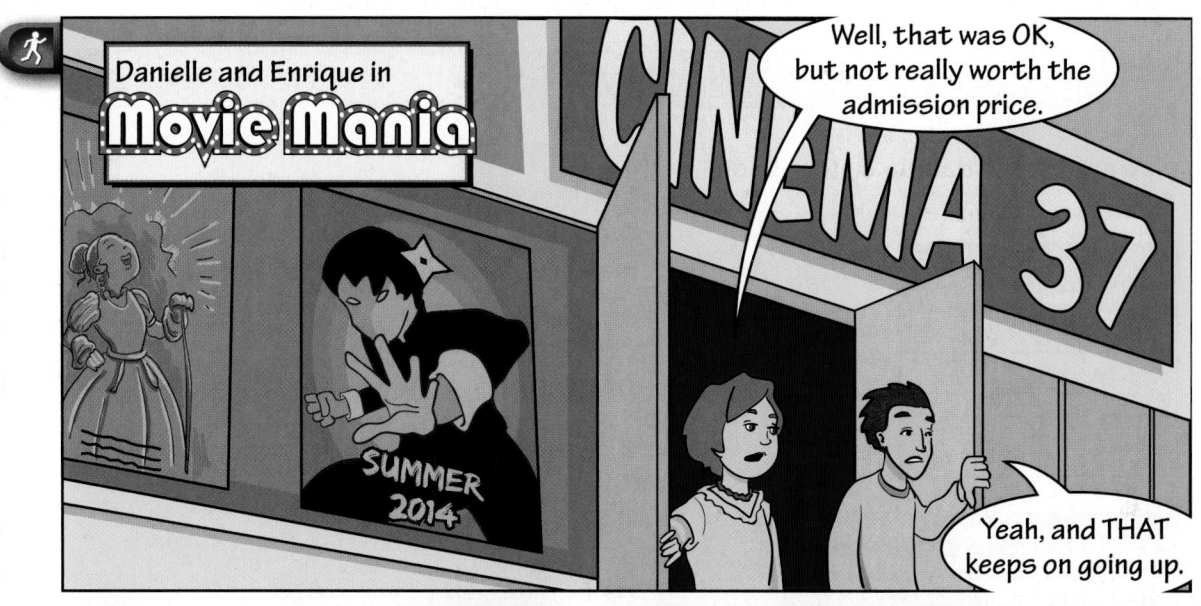

Are You Ready for the Chapter?

You have two options for checking prerequisite skills for this chapter.

Text Option Take the Quick Check below. Refer to the Quick Review for help.

QUICK Check

Display each set of data in a stem-and-leaf plot.

1.

Life Spans of Certain Mammals (years)	
20	35
11	9
12	25
10	24

2.

Points Scored by the Winning Team in the Super Bowl					
35	33	16	23	16	24
14	24	16	21	32	27
35	31	27	26	27	38
38	46	39	42	20	55
20	37	52	30	49	27
35	31	34	23	34	20
48	32	24	21	29	17

Find the mean (average) for each data set. Round to the nearest tenth if necessary.

3. 14, 17, 20, 16, 13

4. 52, 36, 17, 41, 18, 29, 28, 32

5. 18, 20, 16, 15, 17, 20, 15, 11

6. **BASEBALL** In 12 games last season, the school baseball team scored 5, 11, 2, 0, 4, 8, 9, 6, 7, 4, 1, and 2 runs. What was the average number of runs scored per game?

QUICK Review

EXAMPLE 1

AGES The ages of people performing in a play are 12, 18, 24, 49, 38, 29, 27, 18, 7, 54, 33, 38, 30, 32, and 24. Display the data in a stem-and-leaf plot.

STEP 1 Find the least and the greatest number. Then identify the greatest place value digit in each number. The least value, 7, has 0 in the tens place. The greatest value, 54, has 5 in the tens place.

STEP 2 Draw a vertical line and write the stems from 0 to 5 to the left of the line. Then write the leaves in ascending order to the right.

Ages

Stem	Leaf
0	7
1	2 8 8
2	4 4 7 9
3	0 2 3 8 8
4	9
5	4

$1|2 = 12$ years

EXAMPLE 2

FOOTBALL On five plays, the Hawks completed passes for 15, 3, 8, 4, and 5 yards. What was the average number of yards per completed pass?

Find the sum of the numbers. Then divide by how many numbers are in the set.

$$\frac{15 + 3 + 8 + 4 + 5}{5} = \frac{35}{5}$$
$$= 7$$

The Hawks gained an average of 7 yards per completed pass.

 **Online Option** Take the Online Readiness Quiz.

Main Idea
Find the mean, median, and mode of a set of data.

 Vocabulary
measures of central tendency
mean
median
mode

Measures of Central Tendency

OLYMPIC MEDALS Use the table to answer each question.

1. What number(s) appear the most in the bronze category?

2. What is the *average* number of medals won by the United States in the bronze category?

3. Place the numbers in the bronze category in order from least to greatest. What is the middle number?

United States' Summer Olympics Medals 1976–2008			
Year	Gold	Silver	Bronze
1976	34	35	25
1980	0	0	0
1984	83	61	30
1988	36	31	27
1992	37	34	37
1996	44	32	25
2000	40	24	33
2004	35	39	29
2008	36	38	36

Measures of central tendency are numbers that describe the center of a data set. They are the mean, median, and mode.

Key Concept	Measures of Central Tendency
mean	sum of the data divided by the number of items in the set; commonly called the *average*
median	middle number of the data ordered from least to greatest, or the mean of the middle two numbers
mode	number or numbers that occur most often

 EXAMPLE Find Measures of Central Tendency

1 The ages, in years, of the people seated at a table are 22, 18, 24, 32, 24, and 18. Find the mean, median, and mode of the data set.

Mean $\dfrac{22 + 18 + 24 + 32 + 24 + 18}{6} = \dfrac{138}{6}$ or 23 years old

Median 18, 18, $\underbrace{22,\ 24,}$ 24, 32 Arrange in order from least to greatest.

$\dfrac{22 + 24}{2} = 23$ years old

Mode The data set has two modes, 18 and 24 years old.

 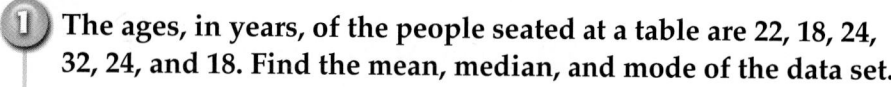

a. Refer to the table on the previous page. Find the mean, median, and mode for the gold medals won.

Sometimes one or two measures of central tendency are more representative of the data than the other measure(s). Different circumstances determine which measure of central tendency is most appropriate to describe a data set.

 REAL-WORLD EXAMPLE

2 **INSECTS** Select the appropriate measure of central tendency to describe the data in the table. Justify your reasoning.

Find the mean, median, and mode of the data.

Most Common Insects

Species	Number of Known Species (thousands)
Beetles	400
Butterflies and moths	165
Ants, bees, and wasps	140
True flies	120
Bugs	90
Caddisflies	10

Mean

$$\frac{400 + 165 + 140 + 120 + 90 + 10}{6}$$

$$= \frac{925}{6} \approx 154.2$$

The mean is about 154.2 thousand.

Median Arrange the numbers from least to greatest.
10, 90, 120, 140, 165, 400

The median is $\frac{120 + 140}{2}$ or 130 thousand.

Mode Since each number only occurs once, there is no mode.

The mean, 154.2 thousand, is greater than more than half the number of known species. So, the mean is not the appropriate measure of central tendency. There is no mode.

Since half of the data values are above and below the median, the median is the appropriate measure of central tendency.

 CHECK Your Progress

b. **COMPUTERS** Select the appropriate measure of central tendency to describe the data in the table. Justify your reasoning.

Computer Model	L100	L150	NX250	NX300	PC150	PC250
Hard Drive (gigabytes)	640	250	500	640	720	640

Example 1   **1. DRIVING** The people in Mr. Kendrick's office drive 10, 3, 17, 1, 8, 6, 12, and 15 miles to work. Find the mean, median, and mode of the data.

Example 2   **2. TEACHERS** The table at the right shows the number of years that educators at South Middle School have been teaching.

    a. Find the mean, median, and mode of the data.

    b. Select the appropriate measure of central tendency to describe the data. Justify your reasoning.

Number of Years Teaching at South Middle School	
Ms. Malan	27
Mr. Sliger	11
Mrs. Lindley	9
Ms. Nolasco	6
Mr. Wyatt	5
Mrs. Clarke	3

Practice and Problem Solving

  ● = **Step-by-Step Solutions** begin on page R1.
     Extra Practice begins on page EP2.

Example 1   **Find the mean, median, and mode of each data set. Round to the nearest tenth if necessary.**

    3. points scored by each of five basketball players: 9, 8, 15, 8, 20

    4. prices, in dollars, of shoes: 78, 80, 75, 73, 84, 81, 84, 79

    5 Find the mean, median, and mode of the data in the stem-and-leaf plot.

    6. There are 18, 22, 19, 24, and 20 students in Mr. Jacobs' math classes. Find the mean, median, and mode of the data.

Tallest Buildings in Dallas, Texas

Stem	Leaf
2	7 9 9
3	0 1 1 1 3 3 4 4 6 6 7
4	0 2 2 5 9
5	0 0 0 0 2 5 6 8
6	0
7	2

2 | 7 = 27

Example 2   **Select the appropriate measure of central tendency to describe the data in each table. Justify your reasoning.**

7.

Fastest Roller Coasters	
Coaster	Speed (mph)
Dodonpa	107
Kingda Ka	128
Millennium Force	93
Phantom's Revenge	82
Steel Dragon 2000	95
Superman: The Escape	100
Top Thrill Dragster	120
Tower of Terror	100

8.

Known Moons of Planets	
Planet	Number of Moons
Mercury	0
Venus	0
Earth	1
Mars	2
Jupiter	63
Saturn	34
Uranus	27
Neptune	13

9 **FIELD TRIP** If Gregory earns an 85% average on five tests in Spanish, he can attend the class trip to the Hispanic Cultural Museum. His current test scores are 94%, 82%, 78%, and 80%. Find the minimum test score Gregory needs to earn on the fifth test in order to attend the class trip.

···**10.** **BIRDS** The table lists the incubation periods for pet birds.

    a. Find the mean, median, and mode of the incubation periods of all the birds.

    b. Select the appropriate measure of central tendency to describe the data. Justify your reasoning.

    c. Using the measures of central tendency of the parrots and of the cockatoos, determine which species, parrot or cockatoo, seems to have the greater incubation period. Justify your reasoning.

Number of Days of Incubation Periods for Pet Birds	
Australian King Parrot	20
Glossy Cockatoo	30
Major Mitchell's Cockatoo	26
Princess Parrot	21
Red-Tailed Cockatoo	30
Red-Winged Parrot	21
Regent Parrot	21
Superb Parrot	20
White-Tailed Cockatoo	29
Yellow-Tailed Cockatoo	29

Real-World Link ····
The incubation period is timed from when the hen starts sitting on the eggs until the chick hatches.

H.O.T. Problems

11. **OPEN ENDED** Construct a data set that has a mode of 10 and a median of 7.

12. **FIND THE ERROR** Jasmine is finding the median of 62, 64, 63, 60, 65, 65, and 70. Find her mistake and correct it.

> 62, 64, 63, <u>60</u>, 65, 65, 70
> The median is 60.

13. **REASONING** Determine whether the following statement is *always*, *sometimes*, or *never* true. Explain your reasoning.
    All measures of central tendency must be members of the set of data.

14. **CHALLENGE** Give a counterexample to show that the following statement is false.
    The median is always representative of the data.

15. **WRITE MATH** Find data from a newspaper or magazine. Write a problem that asks for the measures of central tendency. Tell which measure is most representative of the data.

16. The speeds, in miles per hour, of several cars on a busy street are shown.

> 42, 38, 44, 35, 50, 38

Which measure of central tendency would make the speeds appear the fastest?

A. mode   C. mean

B. median   D. all of the above

17. The stem-and-leaf plot shows the costs, in dollars, of shirts at a boutique. What is the median of the data?

Shirt Costs ($)

Stem	Leaf
0	9
1	2 4 5 8
2	1 7
3	7

1 | 2 = $12

F. $15.00   H. $18.00

G. $16.50   I. $19.10

More About  **Measures of Central Tendency**

Measures of central tendency for the same data set can vary greatly. Sometimes choosing one measure of central tendency may be misleading.

EXAMPLE **Misleading Statistics**

JOBS A company claims its average employee salary is $37,778 per year. Use the table to determine if the claim is valid or misleading.

The claim uses the word "average." So find the mean, median, and mode of the data.

Salaries of Quick-Go Employees		
Job	Number of Employees	Salary
Scheduler	5	$22,000
Driver	10	$27,000
Manager	2	$50,000
President	1	$200,000

Mean: $\dfrac{5 \times \$22{,}000 + 10 \times \$27{,}000 + 2 \times \$50{,}000 + \$200{,}000}{18}$ or about $37,778.

Median: $27,000

Mode: $27,000

In this company, most of the employees make much less than $37,778. The claim is misleading.

Determine if the following claims are valid or misleading. Explain.

18. A car company claims the average miles per gallon (mpg) for a certain car is 39. They tested five cars with the following results: 40 mpg, 41 mpg, 35 mpg, 37 mpg, 42 mpg.

19. A teacher claims the average test grade in one class was 85%. The test scores were 92%, 78%, 85%, 65%, 85%, 42%, 77%, 71%, 80%, and 85%.

Extend

Graphing Technology:
Mean and Median

Main Idea

Use the mean and median to describe and compare data sets.

MONEY The students at Wedgewood Middle School are raising money for charity. The table shows the amount of money that was raised by each homeroom.

Money Raised ($)							
Seventh Grade				**Eighth Grade**			
145	98	75	120	150	132	165	95
105	130	190	85	110	105	90	130
127	115	90	110	115	120	108	100

Use a graphing calculator to find the mean and median of each data set.

 **ACTIVITY**

STEP 1 Clear the existing data by pressing STAT ENTER ▲ CLEAR ENTER. Then enter the data for the seventh grade in L1 and the data for eighth grade in L2.

STEP 2 Display a list of statistics for the data in L1 by pressing STAT ▶ ENTER 2nd [L1].

> The first value, $\bar{x}$, is the mean. →

```
1-Var Stats
 x̄=115.8333333
 Σx=1390
 Σx²=171458
 Sx=30.82157843
 σx=29.50941469
↓n=12
```

The mean amount of money for the seventh grade is $115.83.

Use the down arrow key to locate Med. The median amount of money for the seventh grade is $112.50.

Analyze the Results

1. Repeat Step 2 to find the mean and median for the eighth grade. Remember to press [L2] when displaying the statistics.

2. How do the mean and median for each set compare? Which measure would you use to show that the eighth grade raised more money?

3. Research and collect two data sets that can be compared using the mean and median.
   a. Choose a measure that best describes each data set.
   b. How do the measures compare?
   c. Describe how the measure you chose might be useful to someone looking at your data.

Main Idea

Determine and describe how changes in data values impact measures of central tendency.

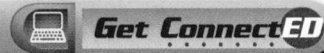

Changes in Data

FOOD DRIVE Miss Roberts divided her class into groups for a canned food drive. The group with the highest average of canned goods per person wins a pizza party. The total donations for the winning group are shown in the table at the right.

Group 3	
Student	**Cans Donated**
Jairo	13
Cesar	10
Aisha	11
Bailey	44
Ling	12

1. Find the mean of the data.

2. Is the mean representative of the entire group? Explain your reasoning.

3. **MAKE A CONJECTURE** If Bailey's donations are removed from the total, would the mean increase or decrease? Explain.

Measures of central tendency may change if data are added or removed from a set.

REAL-WORLD EXAMPLE Describe Changes in Data

1. **SCHOOL** Julio has the following scores on five quizzes: 90, 85, 90, 60, and 100. Describe how the mean, median, and mode will change if his teacher drops his lowest score.

Calculate each measure with and without the lowest test score.

	including the lowest score	without the lowest score
Mean	$\dfrac{90 + 85 + 90 + 60 + 100}{5} = 85$	$\dfrac{90 + 85 + 90 + 100}{4} = 91.25$
Median	60, 85, ⑨⓪ 90, 100	85, 90, 90, 100 ⑨⓪
Mode	90	90

If Julio's lowest quiz score is dropped, his mean quiz score will increase. His median score and mode score stay the same.

CHECK Your Progress

a. **MONEY** Darci deposited $35, $10, $25, and $50 into her savings account last month. Describe how the mean, median, and mode will change if she deposits $44 this week.

2 **TICKETS** The stem-and-leaf plot shows the number of tickets sold for different showings of the same movie over one weekend. Which measure of central tendency will change the most if the greatest number of tickets is eliminated?

Movie Tickets Sold	
Stem	**Leaf**
3	0 4 6 6 7 8 8
4	2 2 2 4 5 6 8
5	2 4
6	7 7
7	2
8	
9	
10	9 3|4 = 34 tickets

Calculate each measure with and without the greatest number of tickets.

	with the greatest number	without the greatest number
Mean	$\dfrac{30 + 34 + \ldots + 109}{20} \approx 49.0$	$\dfrac{30 + 34 + \ldots + 72}{19} \approx 45.8$
Median	43	42
Mode	42	42

If the greatest number of tickets is eliminated, the mean changes the most.

✔️ **CHECK Your Progress**

b. BOWLING Horace's bowling scores are shown in the table. Which measure of central tendency will change the most if his lowest score is removed?

Bowling Scores			
164	128	151	138
158	162	130	162
109	134	157	137

☑️ **CHECK Your Understanding**

Example 1 **1** **STUDY TIME** Brianna studied 1 hour, 3 hours, 2 hours, and 2 hours over four days. Describe how the mean, median, and mode will change if she would have studied 2 hours instead of 1 hour on one of the days.

Example 2 **2. SURVEYS** A restaurant conducted a survey asking its customers to rate the new menu using a scale of 1 to 20. The results of the survey are shown in the line plot below. Which measure of central tendency will change the most if the highest survey score is removed?

Restaurant Survey Results

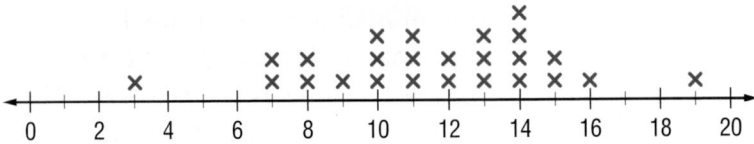

= **Step-by-Step Solutions** begin on page R1.
Extra Practice begins on page EP2.

Example 1   **3. BASEBALL** The table gives the seating capacity of several baseball parks. Describe how the mean, median, and mode will change if the data for Yankee Stadium are not included.

Seating Capacity of Baseball Parks	
Comerica Park	40,120
Kauffman Stadium	40,793
Progressive Field	43,405
Tropicana Field	43,772
U.S. Cellular Field	40,615
Yankee Stadium	52,325

4. RUNNING Natalie runs 4 miles on Mondays, 3.5 miles on Wednesdays, and 4.5 miles on Fridays. Describe how her weekly mean, median, and mode will change if Natalie chooses to add a 3.5-mile jog on Sundays.

Example 2   **5 CAMERAS** A store wants to advertise an average price for their digital cameras. The prices of their cameras are shown in the table. If the cameras with the greatest and least price are excluded, which measure of central tendency will change the most?

Prices of Digital Cameras					
100	180	250	200	130	180
250	280	90	300	300	750
150	130	200	180	100	350

6. SALES The Liberty Middle School student council is selling school T-shirts. Monday through Thursday, they sold 34, 52, 42, and 40 T-shirts. If they sell 65 T-shirts on Friday, which measure of central tendency for the week will change the most from Thursday to Friday?

Describe how the mean is affected if the indicated value is removed from the data set.

7. number of words in a magazine article: 100, 118, 115, 97, 40, 100

8. length in inches of spools of ribbon: 60, 48, 36, 144, 72

9. cost in dollars for shirts: 25, 22, 18, 16, 60, 30, 25

10. students in a classroom: 40, 42, 35, 41, 39, 36, 38, 16, 36, 35, 38, 41

11 TESTS Mr. Mitchell's science test has eight essay questions. Students must answer all of the questions, but he will drop the two lowest essay scores before calculating a final score. Liliana's essay scores were 8, 6, 5, 8, 9, 10, 5, and 7 points. What will be the difference between the mean, median, and mode of her original grades and the mean, median, and mode of her grades after Mr. Mitchell drops the two lowest essay scores?

H.O.T. Problems

12. **OPEN ENDED** Create a data set that contains from 8 to 12 values such that the mean is greater than the median but when the largest value is removed, the median is greater than the mean.

13. **REASONING** Which measure of central tendency is most likely to be affected by the removal of a data value that is much larger or much smaller than the rest of the data? Explain your reasoning.

CHALLENGE Each table shows five test scores for a student. Use the information provided to find possible test scores for the sixth test score x.

14. The 6th test score changed the median by 0.5 point and changed the mean by 2 points.

Test	Score
1	63
2	75
3	80
4	78
5	79
6	x

15. The 6th test score changed the median by 5 points and changed the mean by 1 point.

Test	Score
1	80
2	80
3	90
4	92
5	78
6	x

16. **WRITE MATH** Describe a real-world set of data where one measure of central tendency is more appropriate to use to describe the data. Explain.

Test Practice

17. Brooke grew pumpkins to enter into the competition at the fair. Five of her pumpkins will be weighed.

If the weight of the largest pumpkin is included in the data, which measure would be **most** affected?

A. mean

B. median

C. mode

D. all of the above

18. Isaac earned the amounts of money shown below mowing lawns.

$25, $20, $30, $25

If he earns another $30, which of the following statements would be true?

F. The mode would not be affected.

G. The mean would decrease.

H. The median would decrease.

I. The mean will increase.

19. Anthony received the scores below on his last four tests.

85, 91, 94, 96

If he receives a 92 on the next test, which of the following would increase?

A. mean

B. median

C. mode

D. all of the above

Main Idea

Find the measures of variation of a set of data.

Vocabulary

measures of variation
range
quartiles
lower quartile
upper quartile
interquartile range
outlier

Get Connect**ED**

Measures of Variation

 **TEEN SPENDING** The average amount of money teens spend each week is given in the table.

1. Find the median of the data.

2. Organize the data into two groups: the top half and the bottom half. How many data values are in each group?

3. What is the median of each group?

4. Find the difference between the two numbers from Exercise 3.

Top Ten Countries Average Weekly Teen Spending	
Norway	$49.70
Sweden	$41.70
Brazil	$41.30
Argentina	$40.50
Hong Kong	$38.00
United States	$37.60
Denmark	$37.40
Singapore	$34.10
Greece	$32.90
France	$31.30

Measures of variation are used to describe the distribution of the data. One measure of variation is the range. The **range** of a data set is the difference between the greatest number (maximum) and least number (minimum) in the set.

 EXAMPLE Find Range

① **Find the range of the data in the stem-and-leaf plot.**

Ages

Stem	Leaf
0	3 3 3 4 4 5 5 5 6 6 6 7 7 7 8 8 8 9 9 9
1	0 0 0 0 1 1 1 1 2 2 3 4 4 7 8 8
2	5 8 8 9
3	0 0 2 3 5 5 6 8 8 9
4	0 2 5 5 5
5	0
6	3
7	0 5
8	0 *1	2 = 12 years*

The greatest value is 80 years, and the least value is 3 years. So, the range is 80 − 3, or 77 years.

CHECK Your Progress

a. Find the range of the data at the beginning of the lesson about weekly teen spending.

b. The weights of several puppies are 11.5, 10.4, 9.8, 12.1, 13.9, and 10.7 pounds. Find the range of the data.

Quartiles are values that divide a set of data into four equal parts. Recall that the median separates a set of data into two equal parts.

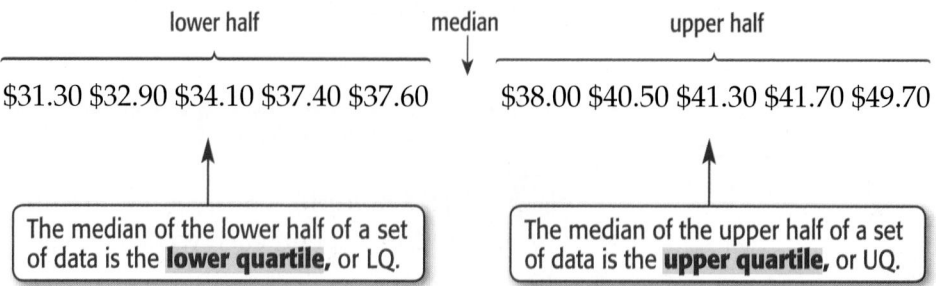

lower half	median	upper half
$31.30 $32.90 $34.10 $37.40 $37.60		$38.00 $40.50 $41.30 $41.70 $49.70

The median of the lower half of a set of data is the **lower quartile**, or LQ.

The median of the upper half of a set of data is the **upper quartile**, or UQ.

So, one half of the data lie between the lower quartile and the upper quartile. Another measure of variation is the **interquartile range**.

Key Concept | Interquartile Range

The interquartile range is the range of the middle half of the data. It is the difference between the upper quartile and the lower quartile.

EXAMPLE **Find Measures of Variation**

2 **MOVIES** Find the measures of variation for the data in the table.

Range   204 − 20 or 184 films

Median, Upper Quartile, and Lower Quartile

Order the numbers from least to greatest.

lower quartile   median   upper quartile

20   24   41   85   85   123   139   204

$\frac{24 + 41}{2} = 32.5$ $\frac{85 + 85}{2} = 85$ $\frac{123 + 139}{2} = 131$

Themes of Sports Films	
Sport	**Films**
Auto Racing	85
Baseball	85
Basketball	41
Boxing	204
Football	123
Golf	24
Horse Racing	139
Wrestling	20

The median is 85, the lower quartile is 32.5, and the upper quartile is 131.

Interquartile Range   upper quartile − lower quartile

$$= 131 − 32.5 \text{ or } 98.5$$

Study Tip

Value of Interquartile Range A small interquartile range means that the data in the middle of the set are close together. A large interquartile range means that the data in the middle are spread out.

CHECK Your Progress

c. ENTERTAINMENT Determine the measures of variation for the data in the table.

DVD Prices at Various Stores ($)			
14.95	19.99	24.99	17.99
14.99	14.95	23.49	15.89
15.99	21.95	17.99	15.99

Data that are more than 1.5 times the value of the interquartile range beyond either quartile are called outliers. An **outlier** is a data value that is either much *larger* or much *smaller* than the median.

 EXAMPLE Find Outliers

3 **WINDS** Find any outliers for the data in the table.

Find the interquartile range.
$$12.4 - 9.0 = 3.4$$

Multiply the interquartile range by 1.5.
$$3.4 \times 1.5 = 5.1$$

Now subtract 5.1 from the lower quartile and add 5.1 to the upper quartile.
$$9.0 - 5.1 = 3.9 \qquad 12.4 + 5.1 = 17.5$$

Average Speeds of Winds	
Station	**Speed (mph)**
Mt. Washington, NH	35.1
Boston, MA	12.4
Buffalo, NY	11.8
Detroit, MI	10.2
Lexington, KY	9.1
Pittsburgh, PA	9.0
Phoenix, AZ	6.2

← upper quartile
← median
← lower quartile

The only outlier is 35.1 because it is greater than 17.5.

CHECK Your Progress

d. MOVIES Find any outliers for the data about themes of sports films in Example 2.

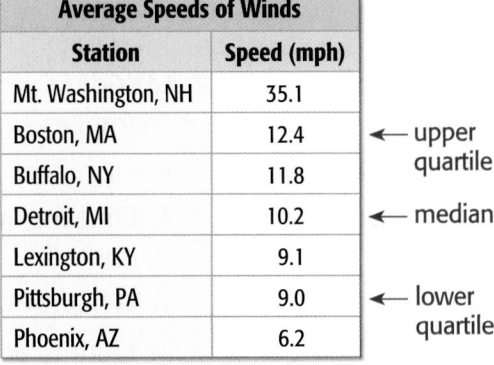 **REAL-WORLD EXAMPLE** Describe Data

4 **SLEEP** Use the measures of variation to describe the data in the table.

Find the measures of variation.

The range is $19.9 - 1.9$, or 18.

The median is 11.25.

The upper quartile is 17.05.

The lower quartile is 4.55.

Number of Hours of Sleep for Selected Animals	
Brown Bat	19.9
Giant Armadillo	18.1
Infant Human	16.0
Cat	12.1
Bottle-Nosed Dolphin	10.4
Gray Seal	6.2
Horse	2.9
Giraffe	1.9

The interquartile range is $17.05 - 4.55$, or 12.5.

Fifty percent of the animals got more than 11.25 hours of sleep. Twenty-five percent of the animals get at or above 17.05 hours of sleep.

Real-World Link · · ·
The brain of a dolphin appears to sleep one hemisphere at a time.

 CHECK Your Progress

e. CYCLING Use the measures of variation to describe the data in the table at the right.

Number of Tour de France Wins	
France	36
Belgium	18
United States	11
Italy	9
Spain	9

Examples 1–3 **1. TESTS** Find the measures of variation and any outliers for the class test scores shown.

Test Scores	
Stem	**Leaf**
7	2 2 3 8 8
8	0 0 1 3 6 7
9	0 1 1 2 3

$7 \,|\, 2 = 72\%$

Examples 1–4 **2. LANGUAGE** The table shows data about languages spoken in the home.

    a. Determine the range of the data.

    b. Find the median and the upper and lower quartiles.

    c. What is the interquartile range of the data?

    d. Identify any outliers.

    e. Use the measures of variation to describe the data in the table.

U.S. Non-English Language Spoken at Home	
Language	**Speakers (millions)**
Spanish	28.1
Chinese	2.0
French	1.6
German	1.4
Tagalog	1.2
Vietnamese	1.0
Italian	1.0
Korean	0.9

Practice and Problem Solving

 = **Step-by-Step Solutions** begin on page R1.
Extra Practice begins on page EP2.

Examples 1–4 **Find the measures of variation and any outliers for each set of data. Then use the measures of variation to describe the data.**

3.

Annual Production of Maple Syrup (gallons)	
Vermont	430,000
Maine	265,000
New York	210,000
Wisconsin	76,000
Michigan	59,000

4.

Calories Burned per Minute of Exercise	
Jogging (6 mph)	8
Jumping Rope	7
Basketball	7
Soccer	6
Bicycling (9.4 mph)	5
Downhill Skiing	5
Walking (4 mph)	4

5

Number of Species in the Animal Kingdom	
Arthropods	1,100,000
Fish	24,500
Birds	9,000
Mammals	9,000
Reptiles	8,000
Amphibians	5,000

6.

Number of U.S. Shuttle Launches 1981–2005	
1981–1985	23
1986–1990	15
1991–1995	35
1996–2000	28
2001–2005	13

7. GOLF Brandon's final score in several golf tournaments relative to par were $-1, -2, 4, -6, 3, -1,$ and -3. Rashan's scores were $-5, 5, 0, 4, -1, -4,$ and -3. Find the measures of variation for both persons' scores. Then describe any similarities or differences in the measures of variation.

8. RIVERS The table shows lengths of rivers in two continents.

    a. Which continent has a greater range of length of rivers?

    b. Find the measures of variation for each continent.

    c. Compare the modes and the interquartile ranges.

    d. Select the appropriate measure of central tendency or range to describe the lengths of rivers for each continent. Justify your response.

    e. Describe the lengths of rivers of Africa and South America, using both the measures of central tendency and variation.

Length (miles) of Principal Rivers			
Africa		South America	
4,160	700	4,000	1,300
2,900	660	2,485	1,100
2,590	500	2,100	1,000
1,700		2,013	1,000
1,300		1,988	1,000
1,100		1,750	956
1,100		1,677	910
1,020		1,600	808
1,000		1,584	400
1,000		1,400	150

9. EARTHQUAKES The line plot shows the magnitudes of U.S. earthquakes.

    a. Find the range, mean, median, mode, upper and lower quartiles, and the interquartile range.

    b. Identify any outliers.

    c. Use the measures of variation to describe the data in the line plot.

Magnitude of Earthquakes in the Central U.S., September 2005

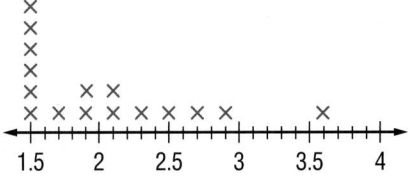

10. BRIDGES The table lists the ten longest suspension bridges in the world.

    a. Find the length of the Golden Gate Bridge if the median is 4,000 feet.

    b. Find the length of the Akashi Kaikyo Bridge if the range is 2,566 feet.

    c. The 11th longest suspension bridge in the world is the Tagus River Bridge in Portugal, with a length of 3,323 feet. Describe how the measures of variation are affected if this data value is included.

10 Longest Suspension Bridges in the World		
Bridge	Country	Length (ft)
Akashi Kaikyo	Japan	y
Great Belt Link	Denmark	5,328
Humber River	England	4,626
Verrazano Narrows	United States	4,260
Golden Gate	United States	x
Mackinac Straits	United States	3,800
Minami Bian-Seto	Japan	3,668
Second Bosphorous	Turkey	3,576
First Bosphorous	Turkey	3,524
George Washington	United States	3,500

11. MATH IN THE MEDIA Find a set of data that contains an outlier in a newspaper or magazine, on television, or on the Internet. Describe the changes to the different measures of variation if the outlier is removed. Are the measures of variation misleading with the outlier? without? Explain.

12. **OPEN ENDED** Create a list of data with at least eight numbers that has an interquartile range of 20 and one outlier.

13. **CHALLENGE** Create two different sets of data that have the same range but different interquartile ranges. Then create two different sets of data that have the same median and same quartiles, but different ranges.

14. **REASONING** Determine whether the following statement is *always*, *sometimes*, or *never* true. Justify your answer.

      The range is affected by outliers.

15. **REASONING** *True* or *false*. Of the mean, median, or mode, the mean is the one most affected by an outlier. Justify your reasoning. If false, give a counterexample.

16.  **WRITE MATH** Explain why the interquartile range is not affected by very high or low values in the data.

Test Practice

17. Which of the following statements is never true concerning the measures of variation of a set of data?

    A. Half of the data lie between the lower quartile and the upper quartile.

    B. Three fourths of the data lie above the lower quartile.

    C. The median, the lower quartile, and the upper quartile separate the data into three equal parts.

    D. 50% of the data lie below the median.

18. The numbers of Grand Slam singles titles won by twelve tennis players are shown below.

    | 14, 8, 7, 6, 5, 5, 10, 11, 8, 8, 6, 7 |

    Which of the following statements is NOT supported by these data?

    F. Half of the titles won are below 7.5 and half are above 7.5.

    G. The spread of the data is 9 titles.

    H. An outlier of the data is 11 titles.

    I. About one fourth of the titles won are at or above 9 titles.

19. **AGES** The ages of people working in an office are shown in the stem-and-leaf plot. Find the mean, median, and mode of the data. (Lesson 1A)

20. **GOLF** In a golf league, the golfers are allowed to drop their highest score before calculating their average score. Seki has scores of 103, 98, 125, 96, 100, 95, and 98. Which measure of central tendency will be affected the most by dropping the highest score? Explain. (Lesson 1C)

Ages of Office Workers

Stem	Leaf
2	3 5 8 8
3	1 2 3 3 6 9
4	2 5 7
5	1 3

2|3 = 23 years

Main Idea

Display and interpret data in a box-and-whisker plot.

 **Vocabulary**

box-and-whisker plot

Get ConnectED

Box-and-Whisker Plots

ELEVATION The table gives the elevation of several cities.

1. What is the least value in the data?
2. What is the lower quartile of the data?
3. What is the median of the data?
4. What is the upper quartile of the data?
5. What is the greatest value in the data?
6. Name any outliers.

Elevation of Selected United States Cities	
City	Elevation (feet)
Boston, MA	330
Mobile, AL	209
Baltimore, MD	193
Richmond, VA	164
Hartford, CT	162
New York, NY	158
Anchorage, AK	130
Atlantic City, NJ	114
Wilmington, DE	92

A **box-and-whisker plot** uses a number line to show the distribution of a set of data by using five values. The *box* is drawn around the middle half of the data, and the *whiskers* extend from each quartile to the extreme data points that are not outliers. A vertical line is drawn through the box at the median.

EXAMPLE **Construct a Box-and-Whisker Plot**

1. **ELEVATION Use the data in the table above to construct a box-and-whisker plot.**

   Step 1 Draw a number line that includes the least and greatest numbers in the data.

   Step 2 Mark the extremes, the median, and the upper and lower quartile above the number line. Check for outliers. If an outlier exists, mark the greatest value that is not an outlier. Use an asterisk (*) to indicate an outlier. It is not connected to a whisker.

   Step 3 Draw the box and the whiskers.

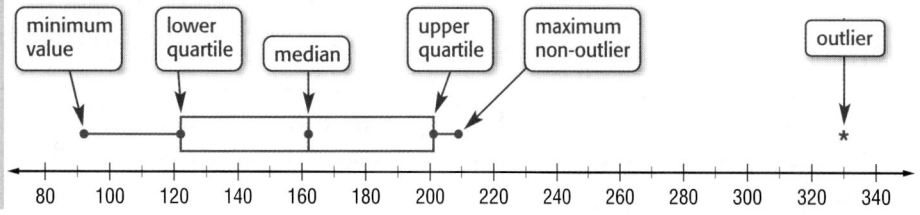

CHECK Your Progress

Construct a box-and-whisker plot for each data set.

a. Prices, in dollars, of admission to a hockey game:
42, 38, 42, 45, 43, 65, 55, 50, 34, 36, 40, 35

b. Low temperatures for various cities:
52, 58, 67, 63, 47, 44, 52, 28, 49, 65, 52, 59

Box-and-whisker plots separate data into four parts. Although the parts usually differ in length, each part contains one fourth of the data.

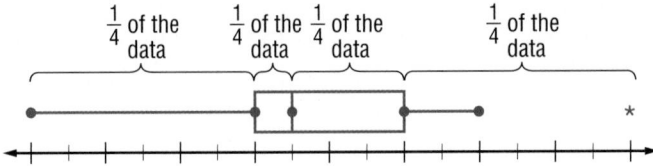

A long whisker or box indicates that the data have a greater range. A short whisker or box indicates the data have a lesser range.

EXAMPLE Interpret Data

2 **SOFTBALL** What does the length of the box-and-whisker plot tell you about the data?

Home Runs Hit in a Softball Season

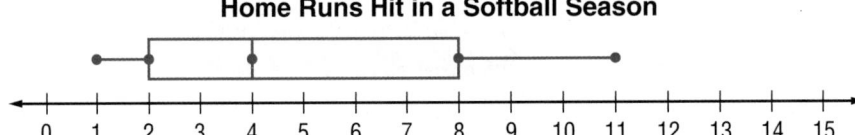

The data between the median and the upper quartile are more spread out than the data between the median and the lower quartile. The whisker at the right is longer than the whisker at the left, so the data above the upper quartile are more spread out than the data below the lower quartile.

CHECK Your Progress

c. WORK Compare data between the median and the upper quartile and the data between the median and the lower quartile.

**Average Daily Commute Time (minutes)
to Work for Selected U.S. States**

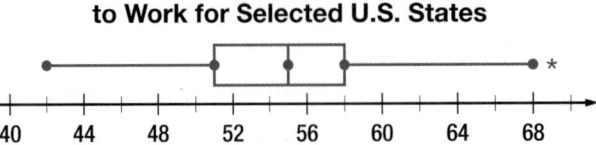

Example 1 Construct a box-and-whisker plot for each data set.

1. hours per month volunteering at the community center: 38, 43, 36, 37, 32, 37, 29, 51

2. points earned on a test: 100, 70, 70, 90, 50, 90, 50, 90, 100, 50, 90, 100, 90, 50, 25, 80

Example 2 **3. FISH** Refer to the box-and-whisker plot below.

Number of Fish in Various Ponds

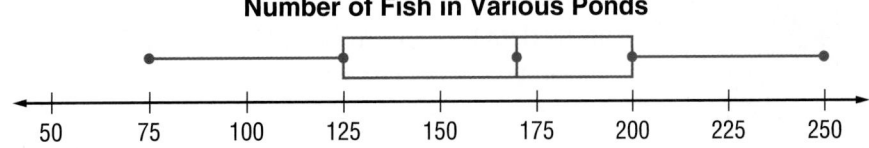

a. What is the interquartile range of the data?

b. Three fourths of the ponds have at least how many fish?

Practice and Problem Solving

> ● = Step-by-Step Solutions begin on page R1.
> Extra Practice begins on page EP2.

Example 1 Construct a box-and-whisker plot for each data set.

4. pages read: 49, 45, 55, 32, 28, 53, 26, 38, 35, 35, 51

5. miles traveled: 77, 85, 72, 76, 95, 90, 73, 82, 82, 80, 73

6. student enrollment: 540, 460, 520, 350, 500, 480, 475, 525, 450, 515

7. prices of bicycles: 225, 245, 220, 270, 350, 280, 230, 240, 225, 270

Example 2 **8. HISTORY** The box-and-whisker plot shows the populations of the 13 original states.

Population of Thirteen Original States, 1790 (thousands)

a. Approximately what percent of the states had populations greater than 100,000?

b. How does the length of the whisker after the upper quartile represent the data?

9 **ZOOS** The box-and-whisker plot below shows the areas of the largest zoos in the United States.

Areas (acres) of the Ten Largest Zoos in the United States

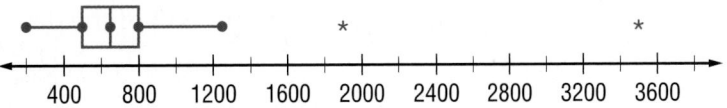

a. How many outliers are in the data?

b. Describe the distribution of the data. What can you say about the areas of the major zoos in the U.S.?

10. PARKS Refer to the table at the right.

    a. Construct a box-and-whisker plot for the data set. Then determine in which interval the data are the most spread out.

    b. Describe how the box-and-whisker plot would change if the data for California and Florida were not included.

State and National Parkland of Selected States	
State	**Total Acres per 10 Square Miles of Land**
California	616.6
Florida	611.2
Arizona	412.8
Michigan	176.6
North Carolina	172.8
Minnesota	79.5
Texas	72.7
Ohio	58.3
Georgia	25.1

11. FOOD The cost of a hot dog at the different NFL stadiums is shown in the table.

    a. Construct a box-and-whisker plot for the data set.

    b. Which has a greater range, the prices above the median or the prices below the median?

Hot Dog Prices ($)			
3.50	3.50	4.75	4.75
4.50	4.75	5.00	5.25
3.00	3.75	3.00	3.00
5.00	4.00	4.00	5.00
4.25	3.50	3.00	4.00
5.00	3.50	4.00	3.50
5.00	4.50	4.00	5.75
4.25	3.50	4.00	3.50

12. TRACK Marcia's times in seconds for the 100-meter dash this year are 13.14, 13.07, 12.94, 12.99, 13.04, 13.12, 12.97, 13.06, 12.91, and 12.98.

    a. Construct a box-and-whisker plot of the data.

    b. The box-and-whisker plot below shows Valerie's times for the 100-meter dash. Compare the times of Marcia and Valerie. Who had the faster times this year?

Valerie's Time (s)

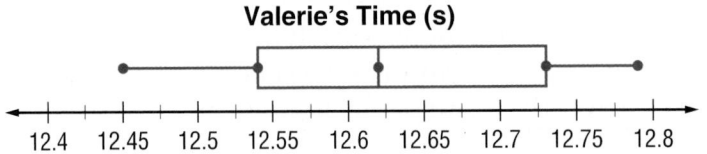

13. LAKES The table shows the acres covered by some lakes.

    a. Construct a box-and-whisker plot for the data set.

    b. How does the lower half of the data compare to the upper half of the data?

Lake	Acres
Crooked Lake	5,538
Cypress Lake	4,097
Deerpoint Lake	5,000
Lake Minnehaha	2,261
Lake Monroe	9,406
Lake Rosalie	4,597
Lake Washington	4,362
Lake Weir	5,685
Levy Lake	4,556
Santa Fe Lake	4,721

14. **OPEN ENDED** Create a data set that could be represented by the box-and-whisker plot at the right.

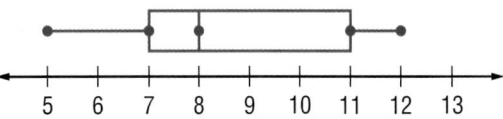

15. **REASONING** The lower quartile, median, and upper quartile of a data set are x, y, and 70, respectively. If a box-and-whisker plot were to be made from these data, give possible values for x and y according to each of the following conditions.

   a. The median separates the box into two quartiles, each with the same range.

   b. The box between the median and the upper quartile is twice as long as the box between the median and the lower quartile.

16. **WRITE MATH** Explain the advantage of using a box-and-whisker plot to display data.

Test Practice

17. Which box-and-whisker plot represents the data set 18, 22, 31, 25, 30, 19, 26, 24, and 35?

   A.

   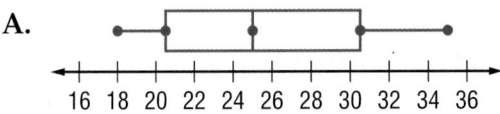

   B.

   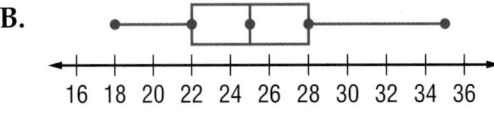

   C.

   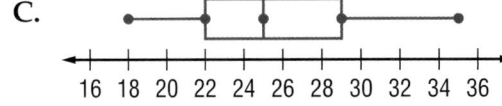

   D.

   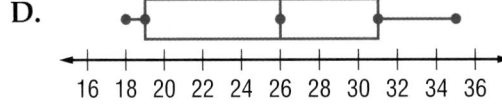

18. Which of the following statements is NOT true about the box-and-whisker plot below?

   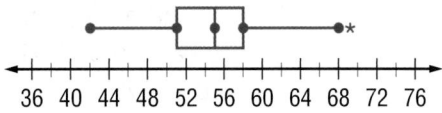

   F. The value 69 is an outlier.

   G. Half of the data is above 55.

   H. $\frac{1}{4}$ of the data is in the interval 58–69.

   I. There are more data values in the interval 42–51 than there are in the interval 55–58.

Spiral Review

19. The costs, in dollars, of videogames are 73, 52, 31, 54, 46, 28, 47, 49, and 58. Find the measures of variation and any outliers for the data. (Lesson 2A)

20. **BASKETBALL** The number of baskets Arnaldo made in each game this season are shown in the table at the right. Describe how the mean, median, and mode would change if he had made 12 baskets instead of 2 baskets in one game. (Lesson 1C)

Baskets Made		
14	12	9
7	12	10
14	7	8
12	2	10

Main Idea

Compare data in box-and-whisker plots.

 **Vocabulary**

double box-and-whisker plot

Double Box-and-Whisker Plots

BASKETBALL The points scored in a recent season by the top nine players on two different basketball teams are shown below.

Team	Points Scored								
Florida	552	522	375	327	305	293	155	144	109
Northwestern	402	348	334	180	159	113	92	70	69

1. Find the median of each set of data.

2. How do the medians compare?

3. MAKE A CONJECTURE Describe a method you could use to easily describe both sets of data.

You can use a double box-and-whisker plot to compare two data sets. A **double box-and-whisker plot** consists of two box-and-whisker plots graphed on the same number line. The two graphs can line up vertically or horizontally. Because of the way the graphs line up, a double box-and-whisker plot can also be called a *parallel box plot*.

 REAL-WORLD EXAMPLES **Compare Data**

1 **BASKETBALL** Use the data in the table above to construct a double box-and-whisker plot.

Step 1 Find the measures of variation for each team. There are no outliers.

	Florida	**Northwestern**
Least value	109	69
Lower quartile	$\frac{155 + 144}{2} = 149.5$	$\frac{92 + 70}{2} = 81$
Median	305	159
Upper quartile	$\frac{522 + 375}{2} = 448.5$	$\frac{348 + 334}{2} = 341$
Greatest value	552	402

Step 2 Construct the box-and-whisker plots. Draw a number line that includes the least and greatest numbers for both data sets.

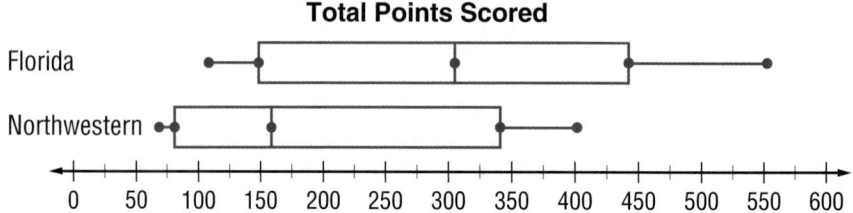

Total Points Scored

2 Refer to the box-and-whisker plots in Example 1. Compare the points scored by the top scorers of each team.

In general, Florida's players scored more points than Northwestern. The median number of points for Florida is close to the upper quartile for Northwestern. This means that 50% of Florida's players scored more points than 75% of Northwestern's players.

✓ **CHECK Your Progress**

SHOPPING The stem-and-leaf plot shows the number of customers who shopped at two different stores last month.

Store A	Stem	Store B
2 2 4 4 4	3	9
1 3 5 8 8 9	4	2 4 4
0 0 2 2 3 5 7 8 9 9	5	1 5
2 3 3 8	6	3 3 4 4 8
	7	1 5 5 8
2 4 4 5	8	1 2 2 4
2	9	0 2 3 3 5 5 7
	10	0 2 2 3

2 | 3 = 32 customers   3 | 9 = 39 customers

a. Construct a double box-and-whisker plot for the data.

b. Use the plots to compare the customers of Store A and Store B.

✓ **CHECK Your Understanding**

Examples 1 and 2   **1** **POPULATION** The table shows the populations of the states in two regions of the United States.

a. Construct a double box-and-whisker plot for the data.

b. Compare the populations of the two regions.

Regional Population (millions)						
West			**Southeast**			
36.5	0.9	2.6	4.6	9.4	2.9	6.0
4.8	2.5	6.4	2.8	4.2	8.9	7.6
1.5	3.7	0.5	18.1	4.3	4.3	1.8

Example 2   **2. HEIGHT** Refer to the double box-and-whisker plot below that shows the height of girls and boys in a class.

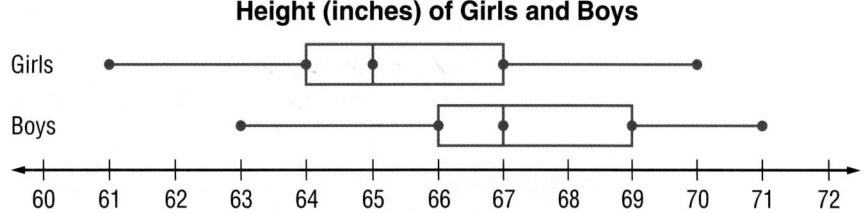

Height (inches) of Girls and Boys

a. Compare the heights of the girls and the boys.

b. What percent of the girls and what percent of the boys are 67 inches or shorter?

Practice and Problem Solving

= **Step-by-Step Solutions** begin on page R1.
Extra Practice begins on page EP2.

Examples 1 and 2

3. BRIDGES The table shows the lengths of the longest bridges in the U.S. and Europe.

Longest Bridges (thousand meters)			
United States		**Europe**	
38.4	36.7	17.2	11.7
29.3	24.1	7.8	6.8
17.7	12.9	6.6	6.1
11.3	10.9	5.1	5.0
8.9	8.9	4.3	3.9

   a. Construct a double box-and-whisker plot for the data.

   b. Compare the bridge lengths in the U.S. to the European bridges.

4. FOOTBALL The stem-and-leaf plot shows the total points scored in two bowl games.

   a. Construct a double box-and-whisker plot for the data.

   b. Compare the total points scored in the bowl games.

Total Points Scored 1989–2008

Rose Bowl	Stem	Orange Bowl
	1	5 9
6 7	2	2 6 7
6 7 7 7	3	0 4 7
2 8 8	4	1 1 1 5 9
0 1 8 8	5	5 7 9
6 9 9	6	2 9
3 5 9	7	4 9
0	8	

6 | 2 = 26 points   *1 | 5 = 15 points*

Example 2

5 ROLLER COASTERS Refer to the box-and-whisker plots below.

Speed (miles per hour) of Roller Coasters

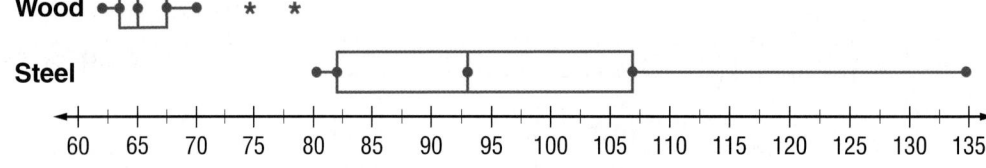

   a. What percent of wood roller coasters travel at least 67 miles per hour?

   b. What percent of steel roller coasters travel at least 82 miles per hour?

   c. In general, do steel roller coasters travel faster or slower than wood roller coasters? Justify your reasoning.

6. MULTIPLE REPRESENTATIONS Refer to the box-and-whisker plots below.

Average Number of Sunny Days Per Year

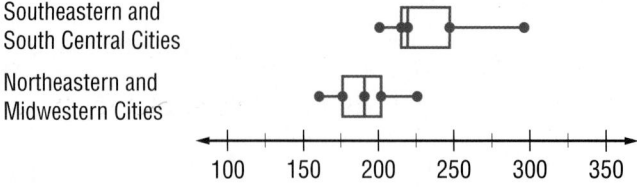

   a. NUMERICAL What percent of the data for the Southeastern and South Central cities is above the lower quartile for the Northeastern and Midwestern cities?

   b. ANALYTICAL Compare the average sunny days for the two groups of cities.

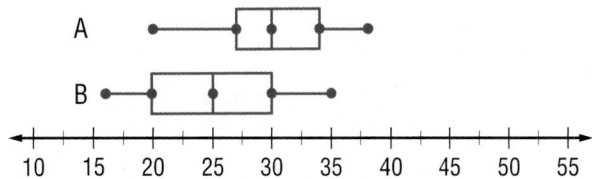

7. **OPEN ENDED** Create two data sets that contain 15 to 20 data values that could have been used to make the double box-and-whisker plot below.

8. **WRITE MATH** Explain how using a double box-and-whisker plot is useful when comparing data.

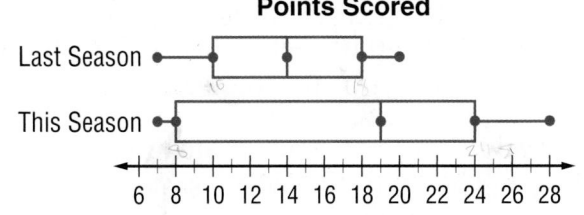
Test Practice

9. Mr. Kendrick used box-and-whisker plots to compare the points scored per game by the football team this season to last season.

Points Scored

Last Season

This Season

6 8 10 12 14 16 18 20 22 24 26 28

How many points higher was the median number of points scored this season than the median number of points scored last season?

A. 0 points   C. 5 points

B. 2 points   D. 8 points

10. Refer to the box-and-whisker plot in Exercise 9. How many points higher was the interquartile range for this season than the interquartile range for the last season?

F. 4 points   H. 8 points

G. 6 points   I. 10 points

11. **SHORT RESPONSE** The double box-and-whisker plot below shows the cost of different items at two clothing stores.

Clothing Prices ($)

Store A

Store B

10 15 20 25 30 35 40 45 50 55 60

At which store do 75% of the items cost more than $18?

12. The ages of people at a movie are 22, 25, 36, 42, 33, 76, 45, 53, 44, 36, 37, and 29 years. Draw a box-and-whisker plot for the data.
(Lesson 2B)

13. **GEOGRAPHY** The table at the right shows the lengths of major U.S. rivers. (Lesson 2A)

a. Determine the measures of variation for the data.

b. Determine any outliers of the data.

c. Use the measures of variation to describe the data.

River	Length (mi)
Arkansas	1,459
Colorado	1,450
Columbia	1,243
Mississippi	2,348
Ohio	981
Rio Grande	1,900

Main Idea

Use a graphing calculator to make box-and-whisker plots.

Graphing Technology:
Box-and-Whisker Plots

TESTS The grades on Miss Romero's last math test are shown in the table. Use a graphing calculator to construct a box-and-whisker plot of the data.

Miss Romero's Math Test Scores							
78	94	85	92	72	56	89	92
90	84	98	82	75	100	94	87
92	85	94	70	78	95	70	80

ACTIVITY

STEP 1 Clear any existing lists. Then enter the data into L1. Input each number and press ENTER.

STEP 2 Press 2nd [STAT PLOT] ENTER to choose the first plot. Highlight On, the modified box-and-whisker plot for the type, L1 for the Xlist, and 1 as the frequency.

STEP 3 Press WINDOW and choose an appropriate range for the x values. The window 50 to 110 with a scale of 4 includes all of these data.

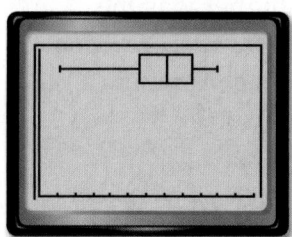

STEP 4 Press GRAPH. Press TRACE and the arrow keys to determine the five key data points of your graph.

Analyze the Results

1. What are the values of the five key data points of the graph? What do these values represent?

2. What percent of the test scores are below 78?

3. What percent of the test scores are above the median? What percent of the test scores are below the median?

4. What percent of the scores are between 56 and 86?

5. Suppose you earned a grade of 80. Describe what percent of students scored higher and what percent scored lower than you.

Mid-Chapter Check

Find the mean, median, and mode of each data set. Round to the nearest tenth if necessary. (Lesson 1A)

1. sales tax, as a percent, for several counties: 4.5, 6, 5.75, 5, 6.25, 5.5

2. length, in seconds, of several commercials: 35, 41, 17, 22, 25, 33, 17

3. ages of people in a pottery class: 16, 25, 42, 58, 24, 29, 36, 32, 38

4. **MULTIPLE CHOICE** The table gives the scores on a recent history test. Which measure of central tendency would make the scores appear highest? (Lesson 1A)

History Test Scores									
77	82	65	92	77	87	100	83	77	78
45	73	67	87	82	59	75	77	68	85
82	75	87	52	87	79	85	82	87	85

A. mean   C. median

B. mode   D. all of the above

5. **SCHOOL** Melanie has the following scores on six tests: 92, 83, 95, 71, 98, and 96. Describe how the mean, median, and mode will change if her teacher drops her lowest test score. (Lesson 1C)

6. **WEATHER** Find the measures of variation and any outliers for the data below. Then use the measures of variation to describe the data. (Lesson 2A)

Average Monthly High Temperatures (°F) for Tucson, Arizona		
66	82	99
70	90	99
97	66	94
85	74	74

Construct a box-and-whisker plot for each data set. (Lesson 2B)

7. pages in magazines: 42, 38, 42, 45, 43, 80, 55, 50, 34, 36, 40, 35

8. costs, in dollars, of video games: 52, 58, 67, 63, 47, 44, 52, 15, 49, 65, 52, 59

9. **MULTIPLE CHOICE** Mrs. Yearling used box-and-whisker plots to compare the money raised by the seventh- and eighth-grade students at Washington Middle School.

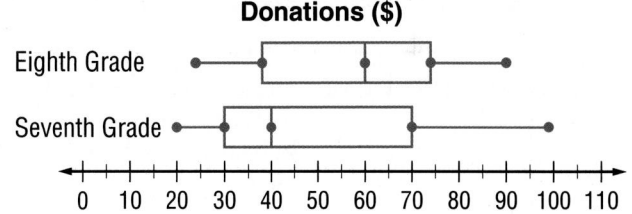

Donations ($)

How many dollars higher was the median amount of money raised by the eighth graders than the median amount raised by the seventh graders? (Lesson 2C)

F. $2   H. $10

G. $5   I. $20

10. **SCHOOL** The table shows the number of students with a "B" average at two elementary schools by grade. (Lesson 2C)

Grade	Jackson	Jefferson
1st	27	13
2nd	22	28
3rd	17	36
4th	12	9
5th	33	27
6th	26	30

a. Construct a double box-and-whisker plot for the data.

b. Compare the students with "B" averages at Jackson Elementary versus the students at Jefferson Elementary.

Problem-Solving Investigation

Main Idea Solve problems by using a graph.

🚶 ✏️ 📷   **P.S.I. TEAM +**

Use a Graph

AURELIO: I ranked ten Web sites from 1 to 10 with
a ranking of 1 being the most popular. Then, I
created a graph showing the download times of
these Web sites.

YOUR MISSION: Use a graph to find out if the most
popular Web site has the fastest download time.

Understand	You want to know whether the most popular Web site ranked by Aurelio has the fastest download time.	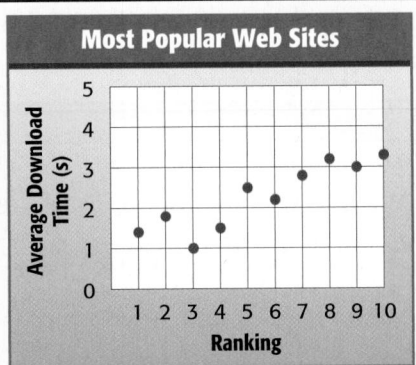
Plan	Make a graph and study the data.	
Solve	The graph shows that, in general, the more popular Web sites are faster than the less popular Web sites. However, the fastest Web site, represented by (3, 1), is not the most popular Web site. So, there are some exceptions to this pattern.	
Check	Look at the graph. Two Web sites have a higher rating than the fastest Web site.	

Analyze the Strategy

1. Explain what the ordered pair (1, 1.4) represents in terms of the question posed.

2. Find a graph in a newspaper, in a magazine, or on the Internet. Write a sentence describing any patterns in the data.

= **Step-by-Step Solutions** begin on page R1.
Extra Practice begins on page EP2.

- Use a graph.
- Look for a pattern.
- Use logical reasoning.
- Choose an operation.

Use a graph to solve Exercises 3 and 4.

3. ALGEBRA The blue line shows the weekly cost of a car rental at Company A. The green line shows the weekly cost of a car rental at Company B. If you wish to rent a car for one week and drive 60 miles, which company charges the lesser amount?

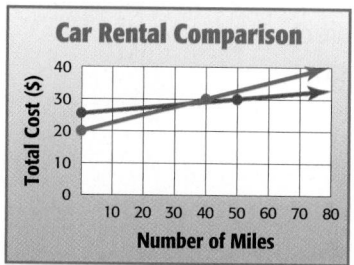

4. JOBS A popular Web site listed the results of a survey about the top 10 most glamorous jobs. Make a graph of the data. Does the most glamorous job have the highest median salary?

Job	Votes (%)	Median Salary ($)
Airline pilot	7.2	105,000
Commercial real estate developer	3.8	184,000
Event planner	4.5	52,000
Fashion designer	24.8	42,000
Interior designer	5.5	48,000
Investment banker	5.9	234,000
Meteorologist	3.8	85,000
Photographer	7.2	50,000
Public relations specialist	4.8	78,000
Surgeon	8.3	270,000

Use any strategy to solve Exercises 5–9.

5. NUMBERS Find the next two numbers in the sequence 4, 0, −4, −8, … .

6. CLUBS The table below shows the math club membership from 2006 to 2011.

Math Club Membership	
Year	Number of Students
2006	20
2007	21
2008	30
2009	34
2010	38
2011	45

a. Make a graph of the data.

b. Describe how the number of math club memberships changed from 2006 to 2011.

c. What is a reasonable prediction for the membership in 2012 if this membership trend continues?

7. SCHOOL COLORS The graph shows the results of a favorite color survey. To the nearest percent, what percent of the students chose purple and orange?

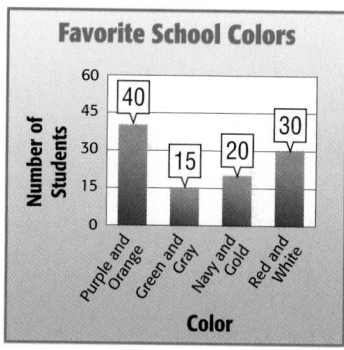

8. MEASUREMENT You need 2 cups of water to mix paint. You have an empty 3-cup container and an empty 5-cup container that do not have any markings on them. Explain how you can use these containers to measure the 2 cups of water you need.

9. STATISTICS The results of a survey showed that 34% of eighth graders wanted to take an extra language class. The school's policy states that there must be at least 32 students interested in the class. If 105 eighth graders were surveyed, is this enough students for an extra language class?

Explore Scatter Plots

Main Idea

Use a scatter plot to investigate the relationship between two sets of data.

Get ConnectED

CCSS 8.SP.1

CIRCLES Delmar and Lucira are trying to determine whether or not a relationship exists between the distance around a circle, the *circumference,* and the distance across a circle, the *diameter.* They found different circular objects in their classroom to measure.

ACTIVITY

STEP 1 Measure and record the distance *d* across the circular part of an object, such as a can, through its center.

STEP 2 Place the object on a piece of paper. Mark the point where the object touches the paper on both the object and on the paper.

STEP 3 Carefully roll the object so that it makes one complete rotation. Then mark the paper again.

STEP 4 Finally, measure the distance *C* between the marks.

STEP 5 Repeat with four more circular objects. Write the measurements of each as ordered pairs (*d, C*). Graph the ordered pairs on a coordinate plane.

Analyze the Results

1. Is there a noticeable trend in the data? If so, describe the trend.

2. How does the distance *C* between the marks compare to the distance *d* through the center?

3. Use your graph from Step 5 to estimate the circumference *C* of a circle with a diameter *d* of 10 inches.

Main Idea
Construct and make conjectures about scatter plots.

 **Vocabulary**
scatter plot

 **Get ConnectED**

 **8.SP.1**

Scatter Plots

Explore Measure a partner's height in inches. Then ask your partner to stand with his or her arms extended parallel to the floor. Measure the distance from the end of the longest finger on one hand to the longest finger on the other hand. Write these measures as the ordered pair (height, arm span) on the board.

1. Graph each of the ordered pairs listed on the board.

2. Examine the graph. Do you think there is a relationship between height and arm span? Explain.

A **scatter plot** shows the relationship between a data set with two variables graphed as ordered pairs on a coordinate plane. For example, the data set year and number of visitors can be displayed as a scatter plot.

EXAMPLE Construct a Scatter Plot

1. **TELEVISION** Construct a scatter plot of the number of viewers who watched new seasons of a certain television show.

Let the horizontal axis, or x-axis, represent the number of seasons. Let the vertical axis, or y-axis, represent the number of viewers. Then graph the ordered pairs (season, viewers).

Television Ratings	
Season	**Viewers (millions)**
1	31.7
2	26.3
3	25.0
4	24.7
5	22.6
6	22.1

Television Ratings

CHECK Your Progress

a. **ALLIGATORS** Construct a scatter plot of the weight of an alligator at various times after hatching.

Weeks	0	9	18	27	34	43	49
Weight (pounds)	6	8.6	10	13.6	15	17.2	19.8

Scatter plots often show a pattern, trend, or relationship between the variables.

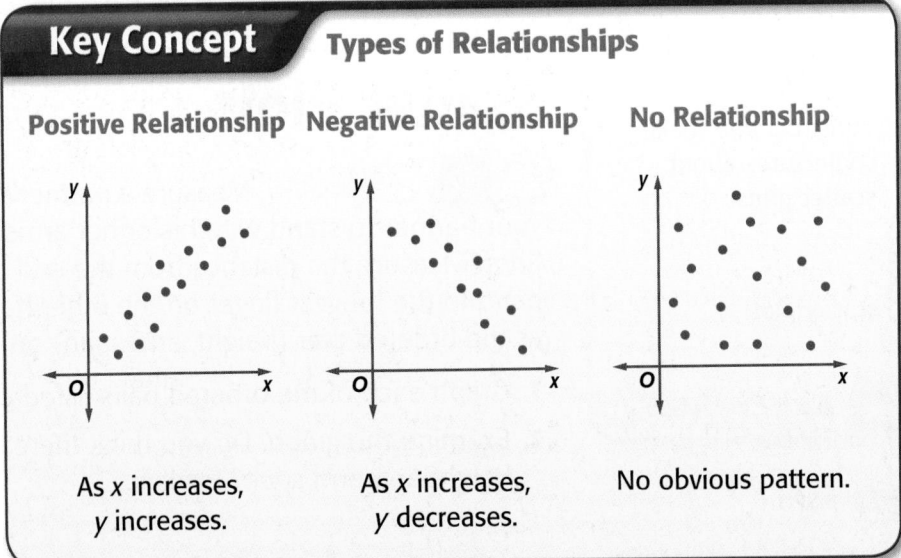

Key Concept   **Types of Relationships**

Positive Relationship   **Negative Relationship**   **No Relationship**

As *x* increases, *y* increases.

As *x* increases, *y* decreases.

No obvious pattern.

Real-World Link · · · ·
The first portable MP3 player was released in 1998. It had a storage capacity of 16 megabytes.

REAL-WORLD EXAMPLE   **Interpret Scatter Plots**

2 **MP3 PLAYERS** Explain whether the scatter plot of the data for the amount of memory in an MP3 player and the cost shows a *positive*, *negative*, or *no* relationship.

As the amount of memory increases, the cost increases. Therefore, the scatter plot shows a positive relationship.

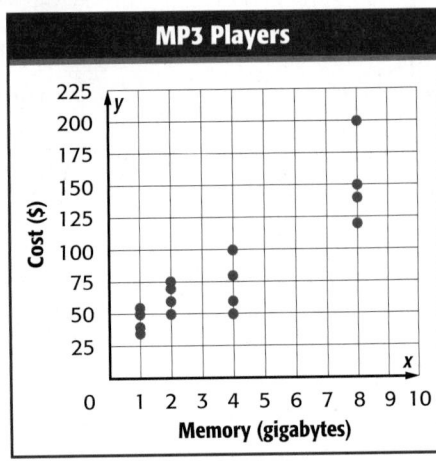

CHECK Your Progress

b. Explain whether the scatter plot of the time elapsed and temperature of water shows a *positive*, *negative*, or *no* relationship.

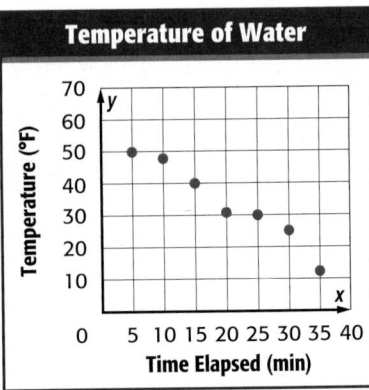

ICE CREAM The table shows how many pints of ice cream were sold at an ice cream shop during two weeks in May.

Day	1	2	3	4	5	6	7	8	9	10	11	12	13	14
Pints Sold	24	10	34	30	22	15	19	32	20	35	23	29	29	13

Real-World Link · · · ·

In 2006, the production of ice cream and other frozen desserts was about 1.55 billion gallons.

3 **Construct a scatter plot of the data.**

Let the horizontal axis represent the day and let the vertical axis represent the pints sold.

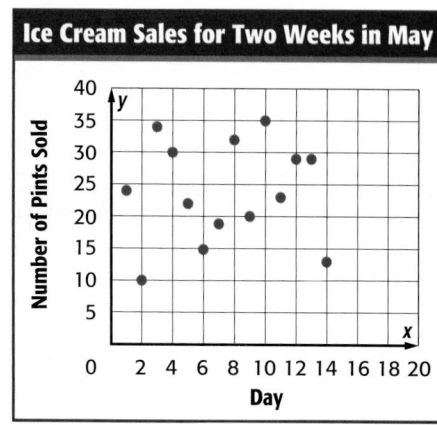

4 **Explain whether the scatter plot shows a *positive*, *negative*, or *no* relationship.**

The day does not affect the pints sold. So, the scatter plot shows no relationship.

5 **If a relationship exists, make a conjecture about the number of pints that will be sold on the 20th day.**

Since the scatter plot shows no relationship between the data, it is not possible to predict how many pints will be sold on the 20th day.

CHECK Your Progress

SWIMMING The table shows the winning times for the men's Olympic 100-meter freestyle swim.

Year	1956	1960	1964	1968	1972	1976	1980
Winning Time (s)	55.4	55.2	53.4	52.2	51.22	49.99	50.4
Year	1984	1988	1992	1996	2000	2004	2008
Winning Time (s)	49.8	48.63	49.02	48.74	48.3	48.17	47.21

c. Construct a scatter plot of the data.

d. Explain whether the scatter plot shows a *positive*, *negative*, or *no* relationship.

e. If a relationship exists, make a conjecture about the winning time in the 2016 Olympics.

Examples 1–5 **1** **MANUFACTURING** The table shows the number of units produced in a certain number of hours at a manufacturing plant.

Time (h)	8	19	16	40	34	8	40	19	34
Units Produced	20	41	28	60	49	28	63	40	58

a. Construct a scatter plot of the data.

b. Explain whether the scatter plot shows a *positive, negative,* or *no* relationship.

c. Make a conjecture about the number of units produced in 50 hours.

Practice and Problem Solving

> ● = **Step-by-Step Solutions** begin on page R1.
> **Extra Practice** begins on page EP2.

Example 1 **Construct a scatter plot for each data set.**

2.

Month	1	2	3	4	5	6	7	8	9	10
Number of Visitors	208	245	423	432	412	626	647	620	402	356

3.

Year	1	2	3	4	5	6	7	8
Number of Books	27	38	24	47	58	65	63	68

Example 2 **Explain whether the scatter plot of the data for each of the following shows a *positive, negative,* or *no* relationship.**

4.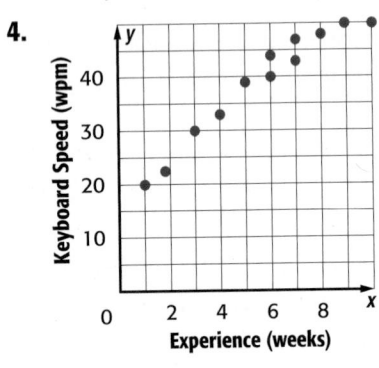

5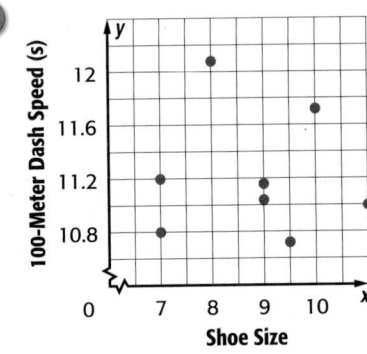

Examples 3–5 **6. SCHOOL** The table shows the amount of time different students studied for a test and their test scores.

Time (min)	10	15	20	25	30	35	40	45
Test Score	65	68	67	78	79	85	89	92

a. Construct a scatter plot of the data.

b. Explain whether the scatter plot shows a *positive, negative,* or *no* relationship.

c. If a relationship exists, make a conjecture about the test score for a student who studied for 60 minutes.

7 E-MAIL The table shows the number of junk E-mails Petra received over the last 10 days.

Day	1	2	3	4	5	6	7	8	9	10
Number of E-Mails	10	12	15	10	11	8	20	10	10	9

a. Construct a scatter plot of the data.

b. Explain whether the scatter plot shows a *positive*, *negative*, or *no* relationship.

c. If a relationship exists, make a conjecture about the number of junk E-mails on Day 15.

8. GRAPHIC NOVEL Refer to the graphic novel frame below for Exercises a–b.

It sure is getting expensive to go to the movies.

Year	Average Price ($)
1996	4.42
1997	4.59
1998	4.69
1999	5.08
2000	5.39
2001	5.66
2002	5.81
2003	6.03
2004	6.21
2005	6.41
2006	6.55
2007	6.88

a. Construct a scatter plot of the data. The values for the horizontal axis should be years since 1995.

b. Do the data represent a *positive*, *negative*, or *no* relationship? Explain.

9. MATH IN THE MEDIA Find some sports data in a newspaper or on the Internet. Construct a scatter plot of the data and explain if the scatter plot shows a *positive*, *negative*, or *no* relationship.

H.O.T. Problems

10. NUMBER SENSE Suppose a scatter plot shows that as the values of *x* decrease, the values of *y* decrease. Does the scatter plot show a *positive*, *negative*, or *no* relationship?

11. CHALLENGE Determine whether the following statement is *always*, *sometimes*, or *never* true. Justify your answer.

> *A scatter plot that shows a positive relationship suggests that the relationship is proportional.*

12. WRITE MATH What are the inferences that can be drawn from sets of data points having a positive relationship and a negative relationship?

13. A car owner tracked the value of a car using a scatter plot. Which description **best** represents the relationship of the car's value?

Lifetime Value of Car

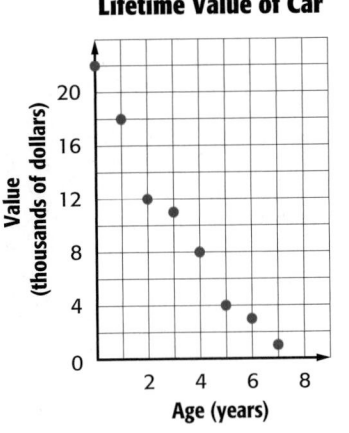

A. negative relationship

B. no relationship

C. positive relationship

D. cannot be determined

 **Scatter Plots** ....

It is possible to measure the strength of a relationship in a scatter plot using a numerical value. One value is the Quadrant Count Ratio, or QCR. A QCR of 1 or −1 indicates a very strong relationship, while a QCR of 0 indicates no relationship.

EXAMPLE Using the Quadrant Count Ratio

Find and interpret the quadrant count ratio for the data from Example 2 in this lesson.

MP3 Players

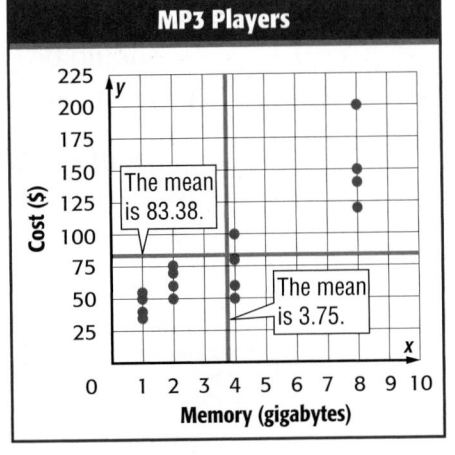

Step 1 Draw a vertical line at the mean value for memory and a horizontal line at the mean value for cost to create four quadrants.

Step 2 Count the number n of data values in each quadrant. Use the formula below.

$$QCR = \frac{[n(Q1) + n(Q3)] - [n(Q2) + n(Q4)]}{\text{total number of values } n}$$

$$= \frac{[5 + 8] - [0 + 3]}{16}$$

$$= 0.625$$

The value of the QCR suggests an above average positive relationship between the amount of memory in an MP3 player and the cost of the MP3 player.

Refer to the following Exercises from the lesson. Find and interpret the QCR for each.

14. Exercise 3   15. Exercise 6   16. Exercise 7

Explore Lines of Best Fit

Main Idea

Use models to predict.

CCSS 8.SP.2

SWIMMING Sashi and Stacie found the data below showing the winning times in the Olympics for the women's 100-meter freestyle swim. They want to predict the winning time in the 2024 Olympics.

100-Meter Freestyle Winning Times							
Year	1956	1960	1964	1968	1972	1976	1980
Winning Time (s)	62.0	61.2	59.5	60.0	58.59	55.65	54.79
Year	1984	1988	1992	1996	2000	2004	2008
Winning Time (s)	55.92	54.93	54.65	54.5	53.83	53.84	53.12

ACTIVITY

STEP 1
Construct a scatter plot for the data.

STEP 2
Use a piece of uncooked spaghetti to make a line that goes through most of the data points.

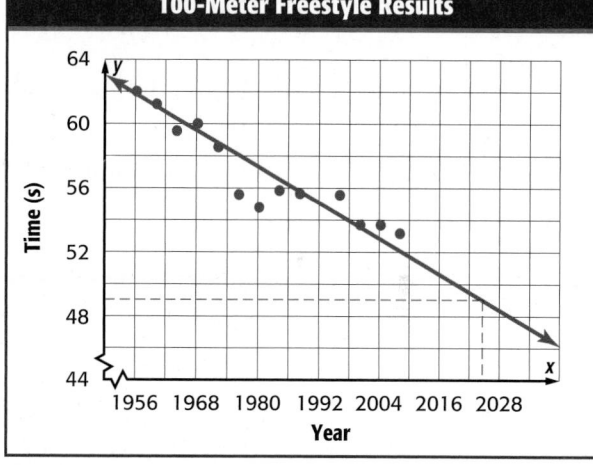

STEP 3
Look at the point where the spaghetti has an *x*-value of 2024. When the *x*-value is 2024, the *y*-value is about 49.

A prediction for the winning time in 2024 is 49 seconds.

Analyze the Results

1. Is it possible to use this method to make a prediction from any scatter plot? Explain.

2. Refer to the scatter plot above. Is this method always valid when making a prediction? Explain.

3. Research and collect a set of data that has a positive or negative relationship. Display your data in a scatter plot. Use the method in the activity to make a prediction about the data.

Main Idea
Draw lines of best fit and use them to make predictions about data.

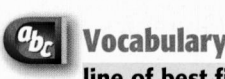

Vocabulary
line of best fit

Get Connect**ED**

8.SP.2, 8.SP.3

Lines of Best Fit

COOKIES The table shows the average annual cost of one pound of chocolate chip cookies.

Years Since 1995	Average Cost ($)	Years Since 1995	Average Cost ($)
0	2.47	7	2.59
1	2.58	8	2.81
2	2.61	9	2.65
3	2.54	10	2.67
4	2.60	11	2.88
5	2.59	12	2.70
6	2.44	13	2.85

1. What year corresponds to 0 years since 1995? 8 years since 1995?

2. If the data were displayed in a scatter plot, would the scatter plot show a *positive*, *negative*, or *no* relationship?

3. Would a more reasonable prediction for the cost of cookies in 2010 be $2.65 or $3.00? Explain.

When data are collected, the points graphed usually do not form a straight line, but may approximate a linear relationship. A **line of best fit** is a line that is very close to most of the data points.

REAL-WORLD EXAMPLES Make Conjectures

COOKIES Refer to the information in the table above.

1 Construct a scatter plot using the data. Then draw a line that seems to best represent the data.

Graph each of the data points. Draw a line that fits the data.

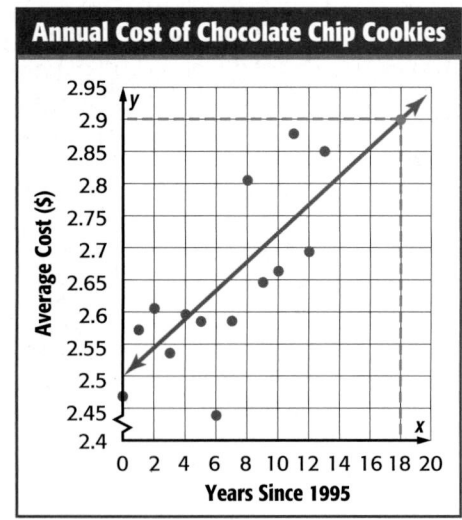
Annual Cost of Chocolate Chip Cookies

2 Use the line of best fit to make a conjecture about the cost of cookies in 2013.

Extend the line so that you can estimate the y-value for an x-value of $2013 - 1995$ or 18. The y-value for 18 is about $2.90. We can predict that in 2018, a pound of chocolate chip cookies will cost $2.90.

CELL PHONES The scatter plot shows the number of cellular service subscribers in the U.S.

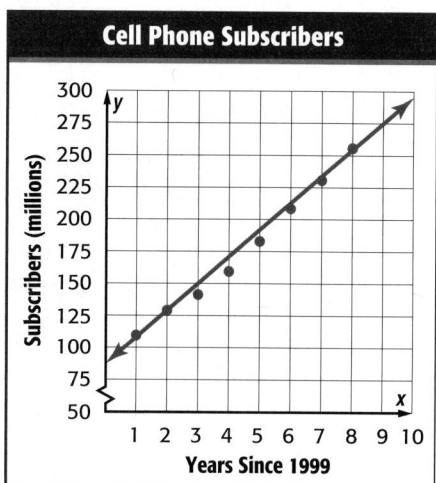

Cell Phone Subscribers

3) Write an equation in slope-intercept form for the line of best fit that is drawn.

Choose any two points on the line. They may or may not be data points. The line passes through points (3, 150) and (9, 275). Use these points to find the slope, or rate of change, of the line.

$$m = \frac{y_2 - y_1}{x_2 - x_1} \qquad \text{Definition of slope}$$

$$m = \frac{275 - 150}{9 - 3} \qquad (x_1, y_1) = (3, 150) \text{ and } (x_2, y_2) = (9, 275)$$

$$m = \frac{125}{6} \text{ or about } 20.83 \qquad \text{The slope is } 20.83.$$

The y-intercept is 85 because the line of fit crosses the y-axis at about the point (0, 85).

$$y = mx + b \qquad \text{Slope-intercept form}$$
$$y = 20.83x + 85 \qquad \text{Replace } m \text{ with } 20.83 \text{ and } b \text{ with } 85.$$

The equation for the line of best fit is $y = 20.83x + 85$.

4) Use the equation to make a conjecture about the number of cellular subscribers in 2015.

The year 2015 is 16 years after 1999.

$$y = 20.83x + 85 \qquad \text{Equation for the line of best fit}$$
$$y = 20.83(16) + 85 \qquad \text{Replace } x \text{ with } 16.$$
$$y = 418.28 \qquad \text{Simplify.}$$

In 2015, there will be about 418.28 million cellular subscribers.

 CHECK Your Progress

EDUCATION The scatter plot shows the graduation rate of high school students.

a. Write an equation in slope-intercept form for the line of best fit that is drawn.

b. Use the equation to make a conjecture about the graduation rate in 2020.

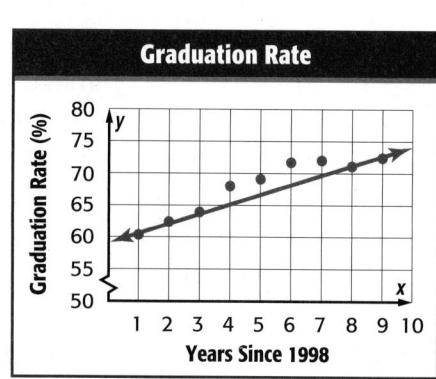

Graduation Rate

Slope

Find the slope of the line that passes through A (1, 4) and B (3, 7).

$$m = \frac{y_2 - y_1}{x_2 - x_1}$$
$$= \frac{(7 - 4)}{(3 - 1)}$$
$$= \frac{3}{2}$$

Study Tip

Estimation Drawing a line of best fit using the method in this lesson is an estimation. Therefore, it is possible to draw different lines to approximate the same data.

Examples 1 and 2

1. LIFE EXPECTANCY The table shows the life expectancy, in years, for people born in certain years.

Years Since 1900	0	10	20	30	40	50	60	70	80	90	100
Life Expectancy	47.3	50.0	54.1	59.7	62.9	68.2	69.7	70.8	73.7	75.4	77.1

a. Construct a scatter plot of the data. Then draw a line that best represents the data.

b. Use the line of best fit to make a conjecture about the life expectancy for a person born in 2020.

Examples 3 and 4

2. TRAVEL The scatter plot shows the amount of fuel remaining in a car after traveling a certain distance.

a. Write an equation in slope-intercept form for the line of best fit that is drawn.

b. Use the equation to make a conjecture about the gallons of fuel remaining after 225 miles.

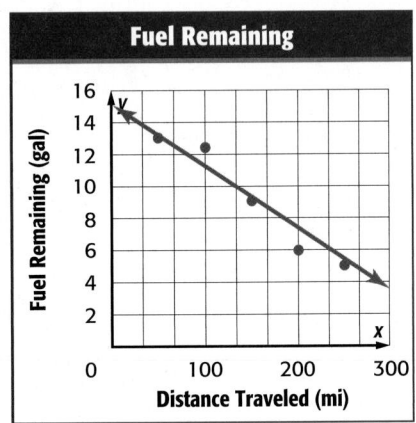

Practice and Problem Solving

● = **Step-by-Step Solutions** begin on page R1.
Extra Practice begins on page EP2.

Examples 1 and 2

3 BUSINESS The results of a survey about women's shoe sizes and heights are shown.

a. Construct a scatter plot of the data. Then draw a line that best represents the data.

b. Use the line of best fit to make a conjecture about the height of a female who wears a size 5 shoe.

Height (inches) and Shoe Size			
Shoe Size	Height	Shoe Size	Height
8	66	$6\frac{1}{2}$	65
8	65	9	68
$7\frac{1}{2}$	65	$6\frac{1}{2}$	62
7	62	7	64
7	61	$5\frac{1}{2}$	62
10	70	5	60
7	62	9	67
9	65	6	59
9	65	$7\frac{1}{2}$	63
9	68	$9\frac{1}{2}$	66

4. NUTRITION The table shows fat and Calories for fast food sandwiches.

Fat (grams)	21	10	14	21	30	34	32	37	27	26	18	7
Calories	490	280	330	430	530	590	540	590	550	470	450	340

a. Construct a scatter plot of the data. Then draw a line that best represents the data.

b. Use the line of best fit to make a conjecture about the number of grams of fat in a sandwich with 350 Calories.

5 SPORTS The scatter plot shows the number of girls who participate in ice hockey.

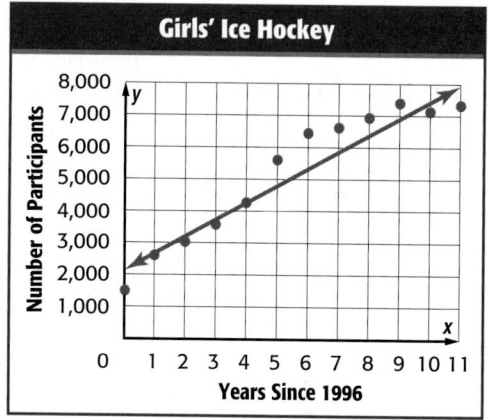

a. Write an equation in slope-intercept form for the line of best fit that is drawn.

b. Use the equation to make a conjecture about the number of girls that will participate in ice hockey in 2020.

6. EXERCISE The table shows the number of Calories burned when walking laps around a track.

a. Construct a scatter plot of the data. Then draw a line that best represents the data.

b. Write an equation in slope-intercept form for the line of best fit.

c. Use the equation to make a conjecture about the number of Calories burned if someone walks 15 laps.

Laps Completed	Calories Burned
1	30
2	70
3	80
4	112
5	150
6	170
7	225

7. GRAPHIC NOVEL Refer to the graphic novel frame below for Exercises a–b.

a. Draw a line that best represents the data in your scatter plot.

b. Write an equation in slope-intercept form for the line of best fit. Make a conjecture about the cost of a movie ticket in 2020.

8. **OPEN ENDED** Use a newspaper or the Internet to find a scatter plot that consists of at least seven data points. Draw a line of best fit and write an equation for the line.

9. **✍ WRITE MATH** Why do we estimate a line of best fit for a scatter plot?

☑ Test Practice

10. **THINK SOLVE EXPLAIN** **EXTENDED RESPONSE** The table shows the wind chill temperatures for different wind speeds when the outside temperature is 30°F.

Wind Chill Temperatures at 30°F			
Wind Speed (mph)	Temperature (°F)	Wind Speed (mph)	Temperature (°F)
5	25	25	16
10	21	30	15
15	19	35	14
20	17	40	13

Part A Construct a scatter plot of the data. Then draw a line of best fit.

Part B Write an equation for the line of best fit.

Part C Use the equation to make a conjecture about the wind chill with a 60 mile per hour wind.

11. The scatter plot shows the average prices for a ticket to an NFL game.

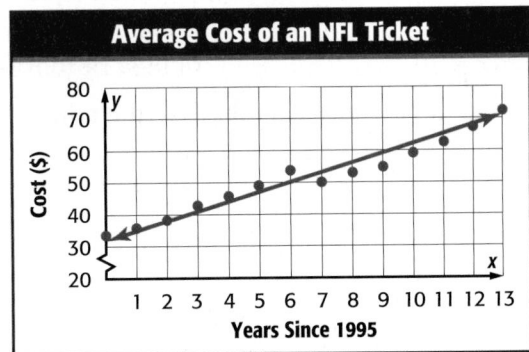

Which of the following is the **most** reasonable equation for the line of best fit?

A. $y = \frac{1}{3}x + 34$ **C.** $y = 3x + 34$

B. $y = -\frac{1}{3}x + 34$ **D.** $y = -3x + 34$

 **Spiral Review**

12. **STATISTICS** Determine whether a scatter plot of a student's age and how many siblings he or she has might show a *positive*, *negative*, or *no* relationship. (Lesson 3C)

13. **CITIES** The table shows the populations of the largest U.S. cities. (Lesson 3A)

   a. Construct a graph of the data.

   b. Describe how the population of San Diego changed from 1950 to 2000.

   c. Which city had the greatest percent increase from 1950 to 2000?

Population of Largest U.S. Cities (millions)		
City	1950	2000
New York, NY	7.89	8.01
Los Angeles, CA	1.97	3.69
Chicago, IL	3.62	2.9
Houston, TX	0.6	1.95
Philadelphia, PA	2.07	1.52
Phoenix, AZ	0.11	1.32
San Diego, CA	0.33	1.22

Extend

Main Idea

Create scatter plots and calculate lines of best fit using technology.

 8.SP.2, 8.SP.3

Graphing Technology:
Scatter Plots

LEISURE Construct a scatter plot to determine if a relationship exists between the weekly number of hours spent watching television and the weekly number of hours spent exercising.

Weekly Television (h)	17	20	11	10	15	38	5	25
Weekly Exercise (h)	5	4.5	7.5	8	6.5	1	7.5	3
Weekly Television (h)	25	32	5	17	40	28	20	30
Weekly Exercise (h)	2.5	3.5	6	7	0.5	5	4	1.5

ACTIVITY

1 **STEP 1**   Clear the existing data by pressing $\boxed{\text{STAT}}$ $\boxed{\text{ENTER}}$ $\boxed{\blacktriangle}$ $\boxed{\text{CLEAR}}$ $\boxed{\text{ENTER}}$

STEP 2   Next, enter the data. Input the number of weekly hours spent watching television in L1 and press $\boxed{\text{ENTER}}$. Then enter the weekly hours spent exercising in L2.

STEP 3   Turn on the statistical plot by pressing $\boxed{\text{2nd}}$ [STAT PLOT] $\boxed{\text{ENTER}}$ $\boxed{\text{ENTER}}$. Select the scatter plot and confirm L1 as the Xlist, L2 as the Ylist, and the square as the mark.

STEP 4   Graph the data by pressing $\boxed{\text{ZOOM}}$ 9. Use the Trace feature and the left and right arrow keys to move from one point to another.

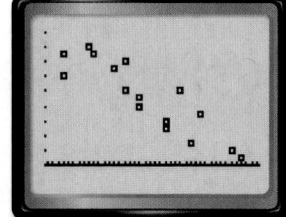

So, as the number of weekly hours spent watching television increases, the number of weekly hours spent exercising decreases. Therefore, the scatter plot shows a negative relationship.

Practice and Apply

1. **WEATHER** Use a graphing calculator to make a scatter plot of the following weather data. Store the data in L3 and L4 and use Plot 2 to create the graph. Then determine whether the data have a *positive*, *negative*, or *no* relationship. Explain your reasoning.

Average Monthly Temperature (°F)	77	42	45	55	57	63	76	65
Average Monthly Rainfall (in.)	6.0	4.8	7.0	3.2	6.8	4.8	5.7	7.2

Average Monthly Temperature (°F)	67	73	51	81	84	86	64	43
Average Monthly Rainfall (in.)	2.6	5.5	5.9	6.3	7.9	4.2	6.3	4.5

ACTIVITY

2 **LEISURE** Find and graph a line of best fit for the data in Activity 1.

STEP 1 Access the CALC menu by pressing [STAT] [▶].

STEP 2 Select 4 to find a line of best fit in the form $y = ax + b$. Press [2nd] [L1] [,] [2nd] [L2] [ENTER] to find a line of best fit for the data in lists L1 and L2.

STEP 3 Graph the line of best fit in Y1 by pressing [Y=] and then [VARS] 5 to access the Statistics… menu. Use the [▶] and [ENTER] keys to select EQ and then press 1 to select RegEQ, the line of best fit equation. Finally, press [GRAPH].

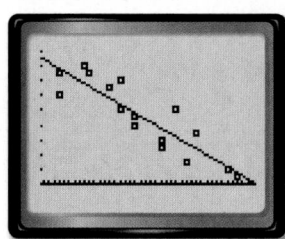

Analyze the Results

2. **MAKE A PREDICTION** Use the [TRACE] feature to predict the average number of hours of exercise someone who watches 35 hours of television would get.

3. **COLLECT THE DATA** Collect a set of data that can be represented in a scatter plot. Use a graphing calculator to determine whether the data have a *positive*, *negative*, or *no* relationship and whether there appears to be any outlying values. Then use the calculator to find a line of best fit and to make a prediction.

Main Idea

Select an appropriate display for a set of data.

Select an Appropriate Display

WEATHER Mr. Watkin's class charted the high temperatures in various cities. The graphs show four ways they displayed the data.

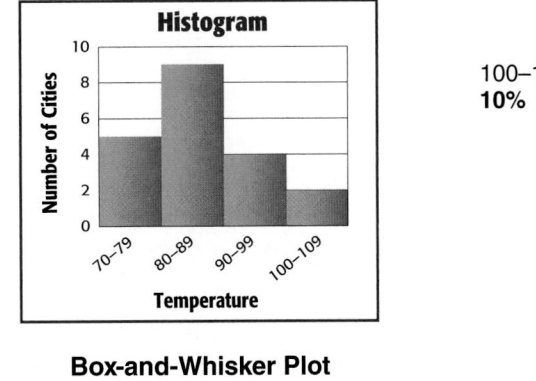

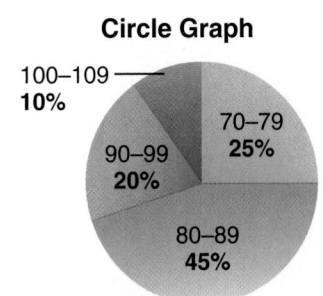

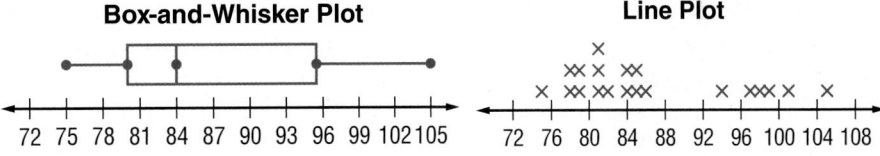

1. Which display(s) show how many cities had a temperature of exactly 79° F? between 100° and 109°?

To decide what type of display to use, analyze the data and determine what the display should show.

Key Concept | **Statistical Displays**

Type of Display	Best Used to...
Bar Graph	show the number of items in specific categories
Box-and-Whisker Plot	show measures of variation for a set of data
Circle Graph	compare parts of the data to the whole
Double Bar Graph	compare two sets of categorical data
Double Line Graph	compare change over a period of time for two data sets
Histogram	show frequency of data divided into equal intervals
Line Graph	show change over a period of time
Line Plot	show how many times each number occurs
Scatter Plot	show relationship between a data set with two variables
Stem-and-Leaf Plot	list all individual numerical data in condensed form

Select an Appropriate Display

1 **SCHEDULES** Select an appropriate display to show the parts of a day taken up by different activities. Justify your reasoning.

Since the display will show the parts of a whole, a circle graph would be an appropriate display to represent these data.

✓ CHECK Your Progress

Select an appropriate display. Justify your reasoning.

a. the population of the United States arranged by age intervals

b. the spread of the average top speeds of 100 cars

🌐 Real-World Link · · · · ·

In 2008, U.S. box office sales were $9.6 billion.

EXAMPLE **Construct an Appropriate Display**

2 **MOVIES** Select an appropriate type of display for the data below to show the relationship between year and U.S. movie admissions. Justify your reasoning. Then construct the display.

Annual Movie Admissions (billions)								
Year	2001	2002	2003	2004	2005	2006	2007	2008
Admissions	1.44	1.60	1.52	1.48	1.38	1.39	1.40	1.34

A scatter plot shows the relationship between two variables. So, a scatter plot would be an appropriate display to show the relationship between year and U.S. movie admissions.

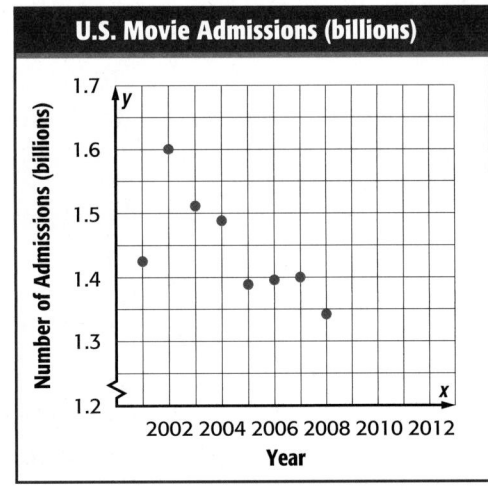

✓ CHECK Your Progress

c. **OCEANS** The table lists the areas in square miles of five oceans. Select an appropriate type of display to compare the areas of the oceans. Then construct the display.

Ocean Areas	
Ocean	Area (sq. mi)
Arctic	5,427,000
Atlantic	29,637,900
Indian	26,469,900
Pacific	60,060,700
Southern	7,848,300

Example 1 Select an appropriate display for each situation. Justify your reasoning.

1 the number of students ordering yearbooks by grade

2. the sales of a particular brand of shoes compared to the total

Example 2 **3. SCHOOL** Select an appropriate type of display for showing how the data varies. Justify your reasoning. Then construct the display.

Test Scores, Period 4														
98	77	89	63	71	79	81	96	81	85	81	92	77	68	72
74	85	72	85	92	91	73	85	77	78	67	91	88	74	88

Practice and Problem Solving

● = **Step-by-Step Solutions** begin on page R1.
Extra Practice begins on page EP2.

Example 1 Select an appropriate display for each situation. Justify your reasoning.

4. the number of cell phone subscribers for the past 5 years

5. point totals for the top 10 NASCAR drivers

6. the portion of a family's budget assigned to each category

7. the median of the exam scores for one class

8. gas mileage for 2012 cars

9 number of Americans who speak Spanish, French, and/or German

Example 2 Select an appropriate display for each situation. Justify your reasoning. Then construct the display.

10.

Favorite Sports for Girls (ages 6–17, in millions)	
Bicycling	10.1
Walking/Hiking	9.0
Bowling	8.9
Volleyball	7.6
Basketball	6.2
Soccer	6.2
In-Line Skating	5.5

11.

Average Height of Females	
Age (years)	Height (inches)
10	56.4
11	59.6
12	61.4
13	62.6
14	63.7
15	63.8

12. ANIMALS Refer to the table at the right. Construct an appropriate display of the data.

13. MUSIC A survey asked teens what they liked most about a song. Of those who responded, 59 said the music only, 41 said the lyrics only, 18 said they liked both equally, and 5 said they did not like either. Construct an appropriate display of this data.

Federally Endangered Animals, U.S.	
Type	Number of Species
Birds	75
Fish	74
Invertebrates	204
Mammals	69
Reptiles and Amphibians	26

14. **MUSIC** Refer to the displays below. Select which display is most appropriate to answer each question. Justify your reasoning. Then answer the question.

8th Grade Music Preference

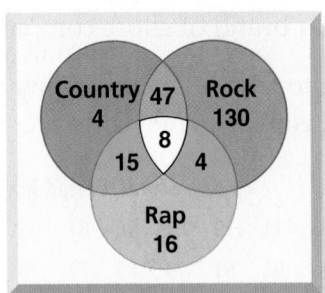

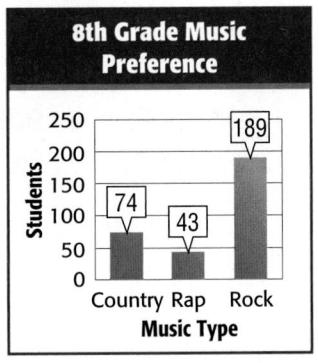

a. How many students like only country music?

b. How many students like rock music?

15. **COLLECT THE DATA** Conduct a survey of your classmates about sports using data that can be presented in a Venn diagram. Then draw the Venn diagram.

16. **CELL PHONES** Refer to the plot at the right. Construct an appropriate display to represent this data in each of the following questions. Then answer the questions.

Number of Text Messages Received on Saturday

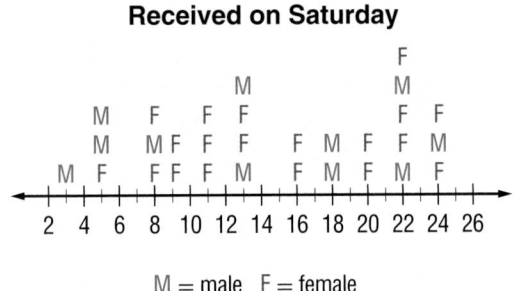

a. Compare the median for the number of text messages received by males and females.

b. What percent of people are female and received more than 10 text messages that day?

17. **MATH IN THE MEDIA** Find a set of data in a newspaper or magazine, on television, or on the Internet. Select an appropriate display for the data, and construct the display.

H.O.T. Problems

CHALLENGE State whether the following statements are *always*, *sometimes*, or *never* true. Justify your response.

18. A circle graph can be used to display data from a histogram.

19. A line graph can be used to display data from a Venn diagram.

20. A box-and-whisker plot can be used to display data from a line plot.

21. **WRITE MATH** Compare and contrast bar graphs and histograms. Explain when it is appropriate to use a histogram rather than a bar graph.

22. THINK SOLVE EXPLAIN **SHORT RESPONSE** Roger polled 24 classmates to find out the average number of hours each student spends online each week. Which of the following displays would be most appropriate to show the individual student responses? Explain your reasoning.

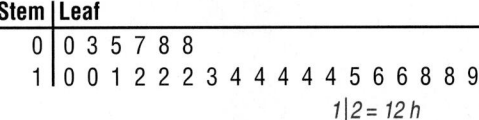

Number of Hours Spent Online Each Week

```
Stem | Leaf
   0 | 0 3 5 7 8 8
   1 | 0 0 1 2 2 2 3 4 4 4 4 4 5 6 6 8 8 9
                          1|2 = 12 h
```

Graph A

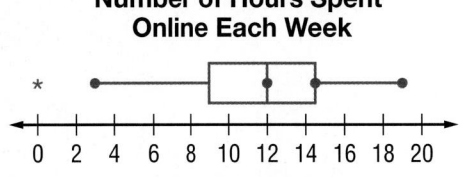

Number of Hours Spent Online Each Week

Graph B

 More About **Statistical Graphs**

Two graphs that represent the same data may look quite different. If different vertical scales are used, each graph will give a different visual impression.

EXAMPLES **Misleading Graphs**

JOBS The graphs show the unemployment rate in the U.S. over ten years.

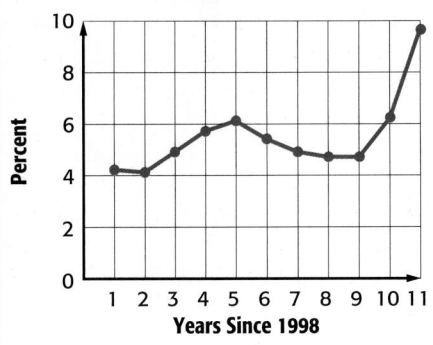

U.S. Unemployment Rate Graph A

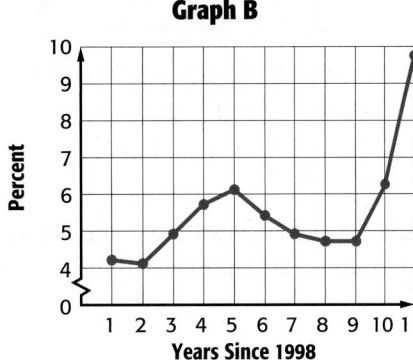

U.S. Unemployment Rate Graph B

1 **Why do the graphs look different?**

The vertical scales differ.

2 **Which graph appears to show that unemployment rates decreased sharply from 2003 to 2006? Explain your reasoning.**

Graph B; The vertical scale on Graph A is 2 units while the vertical scale on Graph B is 1 unit. Any changes will be more noticeable on Graph B.

23. Construct a graph for Exercise 10 to give the impression that in-line skating appears to be as popular as bowling.

24. Construct a graph for Exercise 11 to give the impression that the average height of females between 10 and 15 years of age appears to be almost constant.

Promoting the Games

Are you creative and competitive? Would you enjoy a job working in the sports business? If so, you should consider a career in sports marketing. Sports marketers use statistics to develop plans to promote sporting events, such as state athletic games. They also work for professional and college sports teams, Olympic athletes, and sporting event venues. Their job is to develop merchandise and plan events that promote an athlete's or team's popularity, thereby increasing sales.

21st Century Careers

Are you interested in a career as a sports marketer? Take some of the following courses in high school.

- Calculus for Business
- Principles of Marketing
- Sport, Recreation, and Entertainment Essentials
- Statistical Methods

 Get ConnectED

Girls' 50-Yard Freestyle Results (s)

25.32	28.23	29.10	30.78
26.74	28.27	29.11	30.87
26.79	28.67	29.87	31.41
27.20	28.75	29.95	31.54
27.39	28.80	29.99	31.89
27.46	28.84	30.11	33.16
27.75	29.09	30.76	33.42

Real-World Math

Use the information in the table to solve each problem.

1. What is the range of the times?

2. Find the median, the upper and lower quartiles, and the interquartile range of the data.

3. Identify any outliers.

4. Use measures of variation to describe the data.

5. Make a box-and-whisker plot of the data. Use a number line from 21 to 37.

6. What percent of the girls had times less than 29.095 seconds?

FOLDABLES®
Study Organizer

Be sure the following Key Concepts are noted in your Foldable.

Key Concepts

Measures of Central Tendency (Lesson 1)
- The mean of a set of data is the sum of the data divided by the number of items in the data set.
- The median of a set of data is the middle number of the ordered data, or the mean of the middle two numbers.
- The mode or modes of a set of data is the number or numbers that occur most often.

Measures of Variation (Lesson 2)
- The range of a set of data is the difference between the greatest and the least numbers in the set.
- The interquartile range is the range of the middle half of the data. It is the difference between the upper quartile and the lower quartile.
- Box-and-whisker plots use a number line to show the distribution of a set of data.

Scatter Plots (Lesson 3)
- In a positive relationship, *y* increases as *x* increases.
- In a negative relationship, *y* decreases as *x* increases.
- If there is no relationship, no obvious pattern exists between *x* and *y*.

Key Vocabulary

box-and-whisker plot

double box-and-whisker plot

interquartile range

line of best fit

lower quartile

mean

measures of central tendency

measures of variation

median

mode

outlier

quartiles

range

scatter plot

upper quartile

Vocabulary Check

State whether each sentence is *true* or *false*. If *false*, replace the underlined word or number to make a true sentence.

1. A <u>function table</u> is a graph that shows the relationship between a set of data with two variables.

2. A(n) <u>variation</u> is a piece of data that is more than 1.5 times the value of the interquartile range beyond the quartiles.

3. The range is one of the <u>measures of central tendency</u>.

4. The <u>mean</u> is the sum of the data divided by the number of pieces of data.

5. A line of best fit is a line that is very close to most of the data points in a <u>scatter plot</u>.

6. The <u>mode</u> is the middle number of a set of ordered data.

7. The <u>range</u> is the difference between the greatest and least values in a set of data.

8. The <u>median</u> is a data value that is quite separated from the rest of the data.

Multi-Part Lesson Review

Lesson 1 **Analyze Data**

Measures of Central Tendency (Lesson 1A)

9. The height of various slides is 20, 19, 15, 15, 18, 15, and 3 feet. Find the mean, median, and mode of the data.

10. **FOOD DRIVE** Miss Hollern's homeroom collected 18, 22, 34, 17, and 5 cans of food each day last week. Select the appropriate measure of central tendency to describe the data. Justify your answer.

EXAMPLE 1 The numbers of grams of fat in various candy bars are 9, 8, 9, 8, 13, 9, and 24. Find the mean, median, and mode. Round to the nearest tenth if necessary.

Mean $\dfrac{8+8+9+9+9+13+24}{7}$ or 11.4 g

Median 8, 8, 9, ⑨, 9, 13, 24

Mode 9 grams occurs most frequently.

Changes in Data (Lesson 1C)

11. **PLANTS** The heights in inches of plants are listed below. Describe how the mean, median, and mode will change if the plant listed as 42 inches tall should have been listed as 24 inches tall.

    1, 1, 2, 4, 4, 5, 6, 7, 7, 8, 9, 10, 11, 12, 12, 12, 13, 14, 17, 18, 18, 19, 21, 23, 42

12. **SALES** A booster club sold 61, 71, 68, and 70 magnets. If they sell 98 magnets next week, which measure of central tendency will change the most?

EXAMPLE 2 Pilan has the following scores on five tests: 82, 73, 60, 90, and 95. Which measure will change the most if his teacher drops his lowest score?

	with lowest score	without lowest score
Mean	80	85
Median	82	86
Mode	none	none

The mean will change the most.

Lesson 2 **Box-and-Whisker Plots**

Measures of Variation (Lesson 2A)

Find the measures of variation and any outliers for each set of data.

13. miles from school to home: 12, 2, 3, 2, 3, 3, 4, 5, 4, 6, 1

14. hours spent listening to music: 7, 5, 7, 3, 7, 8, 9, 8

EXAMPLE 3 The hours spent studying by different students are 10, 9, 2, 9, 3, 9, 4, 5, 6, 9, and 9 hours. Find the measures of variation.

Range 10 − 2 or 8
Median 2, 3, 4, 5, 6, ⑨, 9, 9, 9, 9, 10
Lower quartile 2, 3, ④, 5, 6
Upper quartile 9, 9, ⑨, 9, 10
Interquartile range 9 − 4 or 5

Lesson 2 **Box-and-Whisker Plots** (continued)

Box-and-Whisker Plots (Lesson 2B)

15. Mr. Hamre spent 7, 2, 7, 8, 8, 9, 7, and 5 hours working last week. Construct a box-and-whisker plot for the data.

16. PETS The numbers of pets various students have are 3, 6, 2, 3, 1, 3, 6, 4, 5, 4, and 2. Construct a box-and-whisker plot for the data. What do the lengths of the parts of the plot tell you?

EXAMPLE 4 The duration of various plane flights are 9, 2, 3, 9, 5, 6, 7, 9, 4, 10, and 9 hours. Construct a box-and-whisker plot for the data set.

Lengths of Plane Flights (hours)

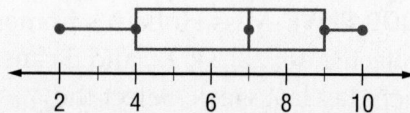

Double Box-and-Whisker Plots (Lesson 2C)

17. SPORTS The times for the members of two track teams in the 400-meter dash are shown.

400-Meter Dash Times

Team A	Stem	Team B
1 4 4 8 8 9	6	1 2 3 8
0 1 2 3	7	4 5 6 8
	8	0 2

$1\,|\,6 = 61\,s$ $7\,|\,4 = 74\,s$

a. Construct a double box-and-whisker plot for the data.

b. Compare the times of Team A versus Team B.

EXAMPLE 5 Compare the average monthly temperatures for the two cities in the plots below.

Average Monthly Temperature (°F)

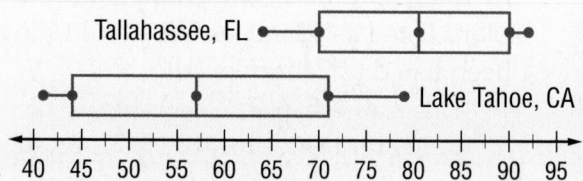

More than 50% of the temperatures in Tallahassee are greater than the temperatures in Lake Tahoe.

Lesson 3 **Scatter Plots**

PSI: Use a Graph (Lesson 3A)

18. BASKETBALL The table shows the average number of points per game for a basketball team for the last several seasons. Use a graph to make a prediction about the team's average number of points next season.

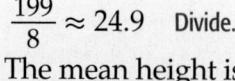

Season	1	2	3	4	5	6
Points	62	65	64	68	71	74

EXAMPLE 6 The graph shows the heights of maple trees. Find the mean height of the trees. Round to the nearest tenth.

$21 + 24 + 26 + 18 +$
$29 + 30 + 23 + 28 =$
199 Add the heights.

$\dfrac{199}{8} \approx 24.9$ Divide.

The mean height is about 24.9 feet.

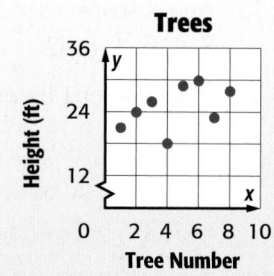

Scatter Plots (continued)

Scatter Plots (Lesson 3C)

19. SHOPPING Construct a scatter plot of the number of customers to visit a store.

Week	1	2	3	4	5	6
Customers	542	601	589	610	648	670
Week	7	8	9	10	11	12
Customers	631	620	723	754	885	910

20. Determine whether a scatter plot of the day of the week and temperature might show a *positive*, *negative*, or *no* relationship.

EXAMPLE 7 Determine whether the graph below shows a *positive*, *negative*, or *no* relationship.

Since there is no obvious pattern, there is no relationship.

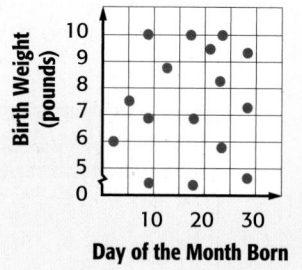

Lines of Best Fit (Lesson 3E)

21. ATTENDANCE The number of students who attended a volleyball game are shown.

Volleyball Game	1	2	3	4	5	6	7
Number of Students	28	30	37	35	36	39	40

a. Construct a scatter plot using the data. Draw a line that seems to best represent the data.

b. Write an equation in slope-intercept form for the line of best fit.

c. Use the equation to make a conjecture about the number of students who will attend Game 12.

EXAMPLE 8 The scatter plot shows the enrollment for a school district. Use the line of best fit to make a conjecture about enrollment in 2016.

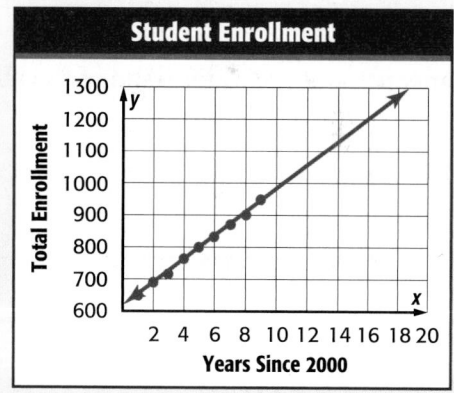

In 2016, the school enrollment will be about 1,200 students.

Select an Appropriate Display (Lesson 3G)

22. TEST SCORES Select an appropriate display to show student scores on an English test in numerical order.

23. CHORES Is a circle graph appropriate to represent how many hours in a day are spent on chores? Justify your answer.

EXAMPLE 9 Select an appropriate display for the number of hockey players compared to the total number of athletes.

An appropriate display would be a circle graph because you are comparing a part to the whole.

1. MUSIC The table shows the number of days the members of the eighth-grade band practiced their instruments last week. Find the measures of central tendency for the data. Round to the nearest tenth if necessary.

Number of Days Students Practiced Musical Instruments Last Week								
7	5	7	6	5	7	6	3	6
4	2	0	6	4	5	2	5	7
4	0	7	5	3	7	1	3	0

2. MULTIPLE CHOICE Chantal has the following scores on four quizzes.

$$70, 85, 85, 90$$

If her teacher drops her lowest score, which of the following would increase?

A. mode

C. median

B. mean

D. none of the above

3. PICNICS The ages of the people at a picnic are listed.

$$75, 36, 25, 26, 19, 32, 35,$$
$$38, 16, 23, 22, 40, 17$$

a. Find the mean, median, and mode of the data.

b. Select the appropriate measure of central tendency or range to describe the data. Justify your reasoning.

c. Find the measures of variation for the data.

d. Identify any outliers.

e. Construct a box-and-whisker plot for the data.

f. When one more person joined the picnic, the mean age was 30. How old was the person who joined the picnic?

4. SCORES The double box-and-whisker plot shows the scores of two different classes on a biology test.

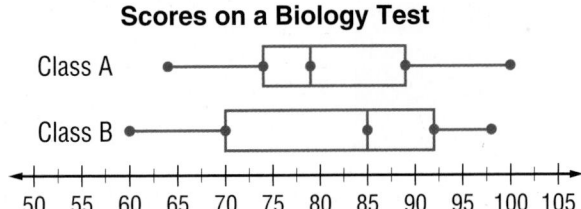

Scores on a Biology Test

How does the median score for Class A compare with the median score for Class B?

5. BASEBALL The graph shows the number of hits by each player on a baseball team. How many hits did player 7 have?

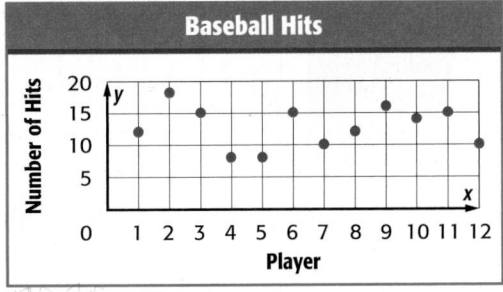

Baseball Hits

6. Determine whether a scatter plot of the number of Calories burned and length of time exercising might show a *positive*, *negative*, or *no* relationship.

7. THINK SOLVE EXPLAIN EXTENDED RESPONSE The table shows the distance a car travels.

Distance (mi)	50	100	150	200	250
Gas (gal)	2	6	8	15	18

Part A Construct a scatter plot for the data and draw a line of best fit.

Part B Write an equation for the line of best fit.

Part C Use your equation to estimate the gas needed to travel 375 miles.

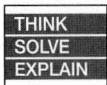

Short Response: Drawing Graphs

When drawing statistical graphs for short-response questions, be sure to include all necessary labels and information in order to receive full credit for that part of the question.

TEST EXAMPLE

The ages and heights of nine students are shown below.

Ages and Heights of Students								
Age	11	12	12	13	13	13	14	14	15
Height (in.)	58	59	61	63	62	60	64	65	65

Use the information in the table to construct a scatter plot of the data.

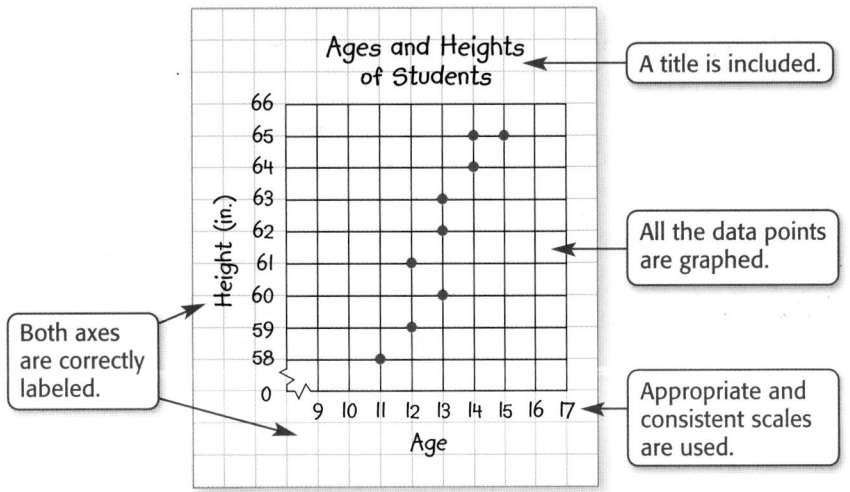

Work on It

Refer to the data in the table above.

Construct a box-and-whisker plot to represent the heights of the students.

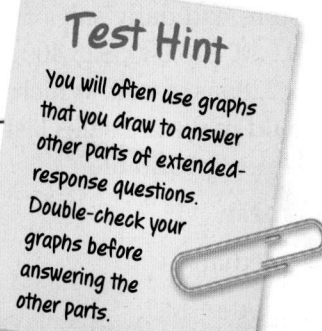

Test Hint

You will often use graphs that you draw to answer other parts of extended-response questions. Double-check your graphs before answering the other parts.

Read each question. Then fill in the correct answer on the answer document provided by your teacher or on a sheet of paper.

1. The scatter plot below shows the cost of computer repairs in relation to the number of hours the repair takes.

Cost of Computer Repairs

(scatter plot: y-axis "Total Cost ($)" labeled 5, 10, 15, 20, 25, 30, 35, 40, 45, 50, 55; x-axis "Number of Hours" labeled 1 2 3 4 5 6 7)

Based on the information in the scatter plot, which statement is a valid conclusion?

A. As the length of time increases, the cost of the repair increases.

B. As the length of time increases, the cost of the repair stays the same.

C. As the length of time decreases, the cost of the repair increases.

D. As the length of time increases, the cost of the repair decreases.

2. A store had daily sales of $15,696, $23,400, $19,080, $18,000, $23,400, $17,604, and $15,228 last week. Which data measure would make the sales last week appear the most profitable?

F. mean

G. median

H. mode

I. range

3. 📝 **GRIDDED RESPONSE** Erin jogged along the track around the outer edge of a park. She ran two miles along the one edge and then 3 miles along the other edge. She then cut across the park as shown by the dotted line. To the nearest tenth, how many miles was Erin's shortcut across the park?

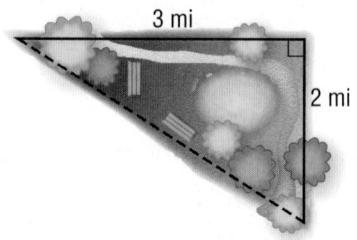

4. Rakim's French test scores are shown below.

| 86, 84, 80, 65, 90, 75, 88 |

Which measure of data would change the most if his lowest quiz score was dropped?

A. mean

B. median

C. mode

D. none of the above

5. 📝 THINK SOLVE EXPLAIN **SHORT RESPONSE** Laurie wants to buy a new sweater that costs $45. The sweater is on sale for 25% off and 6.75% sales tax will be applied to the purchase. What will be the total cost of the sweater?

6. The moon is about 3.84×10^5 kilometers from Earth. Which of the following represents this number in standard notation?

F. 38,400,000 kilometers

G. 3,840,000 kilometers

H. 384,000 kilometers

I. 38,400 kilometers

7. An ice cream store surveyed 100 of its customers about their favorite flavor. The results are shown in the table. If the store uses only these data to order ice cream, what conclusion can be drawn from the data?

Flavor	Frequency
Chocolate chip	40
Vanilla	15
Cookie dough	20
Chocolate	15
Other	10

A. More than half of each order should be chocolate chip and cookie dough ice cream.

B. Half of the order should be vanilla and chocolate ice cream.

C. Only chocolate, cookie dough, and vanilla ice cream should be ordered.

D. About one third of the order should be vanilla and chocolate chip ice cream.

8. THINK SOLVE EXPLAIN **SHORT RESPONSE** The graph of rectangle *LMNP* is shown below.

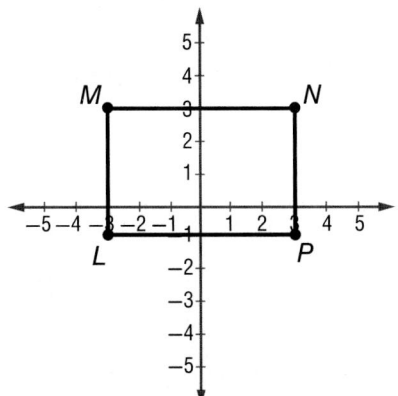

What is the area of rectangle *LMNP*?

9. Alisha's average math test score was 82. Which of the following students has the same average math test score as Alisha?

F. Jenny earned 492 points on 6 tests.

G. Frankie earned 352 points on 4 tests.

H. Benicio earned 468 points on 6 tests.

I. Dontonio earned 344 points on 4 tests.

10. THINK SOLVE EXPLAIN **EXTENDED RESPONSE** The table shows the amount of time different students studied for an exam and the scores they received.

Student	Study Time	Test Score
Patrick	30 min	75
LaDonne	50 min	89
Marlena	1 hr 10 min	93
Jason	25 min	72
Joaquim	1 hour	91
Carla	45 min	83
Heather	1 hr 15 min	90

Part A Why is a scatter plot a good representation of the data?

Part B Graph the data. Do the data represent a *positive*, *negative*, or *no* relationship?

Part C Draw a line that seems to best represent the data. Then write an equation in slope-intercept form for the line of best fit.

Part D Use the equation to make a conjecture about the test score of a student that studied for 1 hour 30 minutes.

NEED EXTRA HELP?										
If You Missed Question...	1	2	3	4	5	6	7	8	9	10
Go to Chapter-Lesson...	10-3C	10-1A	8-2C	10-1C	1-3A	2-2B	5-2B	8-2D	10-1A	10-3C

Probability and Combinations

Investigate

 Animations

 Vocabulary

 Multilingual eGlossary

Learn

 Personal Tutor

 Virtual Manipulatives

 Graphing Calculator

 Audio

 Foldables

Practice

 Self-Check Practice

 Worksheets

 Assessment

The ☆BIG Idea

How are theoretical and experimental probability and sampling used to make predictions about events?

FOLDABLES Study Organizer

Make this Foldable to help you organize your notes.

Review Vocabulary

survey encuesta a question or a set of questions designed to collect data about a specific group of people, or population

What Type of Sport Do You Like to Watch?
Football
Basketball

Key Vocabulary

English	Español
dependent events	eventos dependientes
independent events	eventos independientes
outcome	resultado
simulation	simulacro

When Will I Use This?

You have two options for checking prerequisite skills for this chapter.

Text Option Take the Quick Check below. Refer to the Quick Review for help.

QUICK Check

Write each fraction in simplest form.

1. $\dfrac{48}{72}$ 2. $\dfrac{35}{60}$ 3. $\dfrac{21}{99}$

4. $\dfrac{21}{63}$ 5. $\dfrac{75}{120}$ 6. $\dfrac{24}{44}$

7. **TRAVEL** On a trip to California, Dustin drove 4 hours out of 18 hours. Write this portion of time spent driving as a fraction in simplest form.

Solve each problem.

8. Find 35% of 90.

9. Find 42% of 340.

10. What is 60% of 220?

11. What is 5% of 72?

12. 75 is what percent of 375?

13. What percent of 260 is 117?

14. **SURVEY** Anna surveyed 144 students. Of them, 82% said that pizza is their favorite lunch. How many students surveyed said their favorite lunch is pizza?

15. **SPORTS** During a recent football practice, Will attempted 25 field goals. He made 15 of them. Find the percent of field goals Will made.

QUICK Review

EXAMPLE 1

Write $\dfrac{45}{51}$ in simplest form.

$\div 3$

$\dfrac{45}{51} = \dfrac{15}{17}$ Divide the numerator and denominator by their GCF, 3.

$\div 3$

EXAMPLE 2

Find 20% of 170.

$\dfrac{a}{b} = \dfrac{p}{100}$ Use the percent proportion.

$\dfrac{a}{170} = \dfrac{20}{100}$ Replace b with 170 and p with 20.

$a \cdot 100 = 170 \cdot 20$ Find the cross products.

$100a = 3{,}400$ Multiply.

$\dfrac{100a}{100} = \dfrac{3{,}400}{100}$ Divide each side by 100.

$a = 34$

34 is 20% of 170.

Online Option Take the Online Readiness Quiz.

Main Idea

Count outcomes by using a tree diagram or the Fundamental Counting Principle.

 Vocabulary

outcome
event
sample space
tree diagram
Fundamental Counting Principle
random
probability
odds in favor
odds against

 *Get Connect*ED

Count Outcomes

TRACK At an awards banquet for the school track team, each person can select a meal from the options shown.

1. How many different appetizers are available? main dishes? desserts?

2. Make a list showing all of the different meals you could have at the banquet.

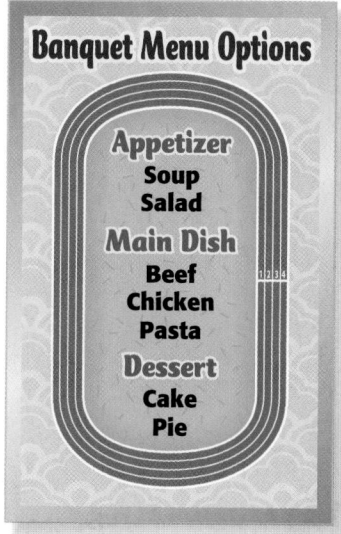

Banquet Menu Options

Appetizer
Soup
Salad

Main Dish
Beef
Chicken
Pasta

Dessert
Cake
Pie

An **outcome** is any one of the possible results of an action. One outcome is choosing salad, pasta, and pie. An **event** is an outcome or a collection of outcomes.

An organized list of outcomes, called a **sample space**, can help you determine the total number of possible outcomes for an event. One type of organized list is a **tree diagram**.

 EXAMPLE Use a Tree Diagram

1. **BANQUET** Draw a tree diagram to determine the number of different meals described above.

Appetizer	Main Dish	Dessert	Outcome
	beef	cake	soup, beef, cake
		pie	soup, beef, pie
soup	chicken	cake	soup, chicken, cake
		pie	soup, chicken, pie
	pasta	cake	soup, pasta, cake
		pie	soup, pasta, pie
	beef	cake	salad, beef, cake
		pie	salad, beef, pie
salad	chicken	cake	salad, chicken, cake
		pie	salad, chicken, pie
	pasta	cake	salad, pasta, cake
		pie	salad, pasta, pie

There are 12 different meal choices for the banquet.

 CHECK Your Progress

a. A dime and a penny are tossed. Draw a tree diagram to determine the number of outcomes.

You can also find the total number of outcomes by multiplying. This principle is known as the **Fundamental Counting Principle**.

> **Key Concept**   **Fundamental Counting Principle**
>
> If event *M* has *m* possible outcomes and event *N* has *n* possible outcomes, then event *M* followed by event *N* has $m \cdot n$ possible outcomes.

 REAL-WORLD EXAMPLE

2   **BASEBALL** Use the information at the left to determine how many different infield teams are possible when one player is selected for each position.

There are 14 choices for each position.

$$14 \ \times \ 14 \ \times \ 14 \ \times \ 14 \ \times \ 14 \ = \ 537{,}824$$

There are 537,824 possible infield teams.

 CHECK Your Progress

b. DINING A restaurant offers a choice of 3 types of pasta with 5 types of sauce. Each pasta entrée comes with or without a meatball. How many different entrées are available?

Outcomes occur at **random** if each outcome is equally likely to occur. In this situation, the **probability** of an event is the ratio of the number of outcomes in that event to the total number of outcomes.

 REAL-WORLD EXAMPLE   **Find Probability**

3   **CLASSES** Julio can take six classes each day: math, science, social studies, English, Spanish, and music. What is the probability his first three classes are science, music, and math, in that order?

First, find the number of possible outcomes. There are 6 choices for the first class, 5 choices for the second class, and 4 choices for the third class.

$$6 \ \times \ 5 \ \times \ 4 \ = \ 120$$

There are 120 possible outcomes. There is one way Julio's classes can be science, music, and math, in that order.

$P(\text{science, music, math}) = \dfrac{1}{120}$ or 0.8%   There is 1 order out of 120.

 CHECK Your Progress

c. Two number cubes are rolled. What is the probability that the sum of the numbers on the cubes is 12?

QUICK Review

Probability
The probability that an event will happen can be any number from 0 to 1, inclusive. Probability can be written as a fraction, decimal, or percent.

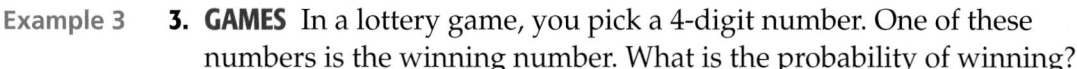
Example 1 **1.** The number cube shown is tossed twice. Draw a tree diagram to determine the number of possible outcomes.

Example 2 **2.** **CODES** At Wyler Middle School, a student's identification code is the first two letters of the student's last name followed by the last four numbers of his or her Social Security number. How many different student identification codes are possible?

Example 3 **3.** **GAMES** In a lottery game, you pick a 4-digit number. One of these numbers is the winning number. What is the probability of winning?

Practice and Problem Solving

● = Step-by-Step Solutions begin on page R1.
Extra Practice begins on page EP2.

Example 1 **Draw a tree diagram to determine the number of possible outcomes.**

 4. A penny, a nickel, and a dime are tossed.

 5. A number cube is rolled and a penny is tossed.

 6. A white or red ball cap comes in small, medium, large, or extra large.

 7. The Sweet Treats Shoppe offers single-scoop ice cream in chocolate, vanilla, or strawberry, and two types of cones, regular or sugar.

Example 2 **Use the Fundamental Counting Principle to find the number of possible outcomes.**

 8. The day of the week is picked at random and a number cube is rolled.

 9. A number cube is rolled 3 times.

 10. There are 5 true-false questions on a history quiz.

 11 There are 4 choices for each of 5 multiple-choice questions on a science quiz.

 12. **BAGELS** At Bonnie's Bagels, you can choose from five different types of bagels, four different spreads, and four different toppings. How many different bagel combinations are possible?

 13. **VEHICLES** A state's license plates are issued with 2 letters, followed by 2 numbers and a letter. How many different license plates could the state issue?

Example 3 **14.** **FLOWERS** Tatiana and her brother are each looking for flowers to give their aunt for her birthday. Tatiana likes either red roses or yellow tulips. Her brother likes blue irises, yellow daisies, red tulips, or white gardenias. If they each choose a flower at random, what is the probability that they select the same color flower?

15. PHONE NUMBERS The first three digits of a phone number are based on the area of the state you live in. The next four digits are random numbers. What is the probability of the last four digits being the current year?

16. ELECTRONICS The table shows various options for a digital music player.

a. How many different players are available, based on storage capacity and color?

b. If an FM radio tuner is also available as an option, how many players are available?

Storage Capacity	Colors	
256 megabytes	blue	purple
512 megabytes	red	pink
1 gigabyte	green	silver
2.5 gigabytes	white	black

17 One marble is drawn from each bag. Use a tree diagram to answer each question.

a. What is the probability that at least one marble will be blue?

b. What is the probability that at least one marble will be yellow?

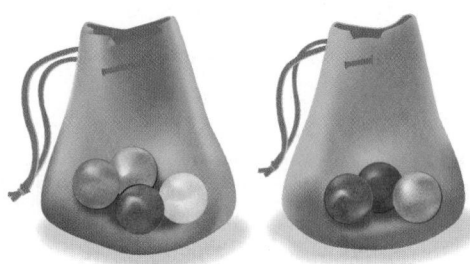

18. LUNCHES Parent volunteers made lunches for an 8th-grade field trip. Each lunch had a peanut butter and jelly or a deli-meat sandwich; a bag of potato chips or pretzels; an apple, an orange, or a banana; and juice, water, or soda. One of each possible lunch combination was made.

a. How many different lunch combinations were made?

b. How many of these combinations contained an apple?

c. If the lunches are handed out randomly, what is the probability that a student receives a lunch containing a banana?

d. What is the probability of a student receiving a lunch with potato chips and soda?

e. Suppose 4 types of meat were used for the deli-meat sandwiches. What is the probability that a student receives one specific type of sandwich?

H.O.T. Problems

19. OPEN ENDED Give an example of a situation that has 15 possible outcomes.

20. NUMBER SENSE A pizza shop has regular, hand-tossed, and thin crusts; two different cheeses; and four toppings. Without calculating the number of possible outcomes, how many more pizzas can they make if they add a deep-dish crust to their menu?

21. CHALLENGE Write an algebraic expression to find the number of possible outcomes if a number cube is rolled x times.

22. WRITE MATH Describe a possible advantage for using a tree diagram rather than the Fundamental Counting Principle.

Real-World Link · · · ·
One of the first personal computers had a memory of 16 kilobytes. Today, the average personal computer has a memory of 4 gigabytes or 250,000 times more memory.

23. The Nunez family purchased a travel package for their upcoming vacation. They can choose from three rental cars, four hotels, and two theme parks. How many travel packages are possible for their trip?

   A. 9   **C.** 24

   B. 12   **D.** 48

24. A quarter is tossed and the number cube shown is rolled. What is the probability that the coin lands on heads and a 4 is rolled, in that order?

   F. $\frac{1}{12}$ **H.** $\frac{2}{3}$

   G. $\frac{1}{6}$ **I.** $\frac{1}{2}$

25. The Yogurt Spot advertises that there are 1,512 ways to enjoy a one-topping sundae. They offer six flavors of frozen yogurt, six different sizes, and toppings. How many toppings do they offer?

   A. 1,500 toppings   **C.** 36 toppings

   B. 42 toppings   **D.** 12 toppings

26. THINK/SOLVE/EXPLAIN **SHORT RESPONSE** A school cafeteria offers sandwiches with three types of meat and two types of bread, as shown in the table. How many possible sandwich combinations are available?

Bread	Meat
Wheat	Ham
Rye	Turkey
	Beef

More About Outcomes

The **odds in favor** of an event occurring is the ratio that compares the number of ways the event *can* occur to the number of ways the event *cannot* occur. The **odds against** an event occurring is the ratio that compares the number of ways the event *cannot* occur to the number of ways the event *can* occur.

✎ 📷 **EXAMPLES**

1. **A number cube is rolled. Find the odds in favor of rolling a 6.**

   the number of ways the event *can* occur ⟶ 1:5 ⟵ the number of ways the event *cannot* occur

   So, the odds in favor of rolling a 6 is 1:5.

2. **A number cube is rolled. Find the odds against rolling a 3 or a 4.**

   the number of ways the event *cannot* occur ⟶ 4:2 ⟵ the number of ways the event *can* occur

   So, the odds against rolling a 3 or a 4 is 4:2 or 2:1.

The spinner shown at the right is spun. Find each odd.

27. the odds in favor of landing on an even number

28. the odds against landing on a 7 or an 8

29. the odds against landing on a multiple of 3

Explore Permutations

Main Idea

Find permutations.

GAMES Four friends are choosing their game pieces one at a time from among a yellow piece, a green piece, a red piece, and a blue piece. In how many ways can they choose the first two game pieces?

ACTIVITY

What do you need to find? the different arrangements of 2 of the 4 game pieces

STEP 1 Write yellow, green, red, and blue on index cards.

| yellow | green | red | blue |

STEP 2 Find each arrangement of the first 2 game pieces. Record your results in a table like the one shown.

Trial	Arrangement
1	yellow, green
2	
3	

STEP 3 Rearrange the cards and record each arrangement until there are no more new arrangements.

Analyze the Results

1. How many different arrangements did you make for the first two pieces?

2. How many different game pieces could you pick for the first place?

3. Once you picked the first-place game piece, how many game pieces could you pick for the second place?

4. Use the Fundamental Counting Principle to determine the number of arrangements for first and second places.

5. How do the numbers in Exercises 1 and 4 compare?

6. **MAKE A CONJECTURE** Suppose there were five game pieces. In how many ways can they choose the first two game pieces?

Main Idea

Find the number of permutations of objects.

Vocabulary

permutation

Get Connect**ED**

Permutations

LUNCH Christina, Mulan, and Sophia are standing in line for lunch.

1. Draw a tree diagram to show all of the possible arrangements of the three students.

2. How many different ways can the three students line up?

3. Describe another method for finding the number of arrangements.

When deciding the number of ways the students can line up, order is important. An arrangement or listing in which order is important is called a **permutation**.

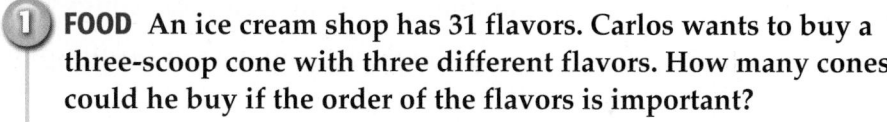

EXAMPLE **Find a Permutation**

1 **FOOD** An ice cream shop has 31 flavors. Carlos wants to buy a three-scoop cone with three different flavors. How many cones could he buy if the order of the flavors is important?

There are 31 choices for the first scoop, 30 choices for the second scoop, and 29 choices for the third scoop. Use the Fundamental Counting Principle.

31 × 30 × 29 = 26,970

Carlos could buy 26,970 different cones.

CHECK Your Progress

a. In a race with 7 runners, in how many ways can the runners end up in first, second, and third place?

The symbol $P(31, 3)$ represents the number of permutations of 31 things taken 3 at a time.

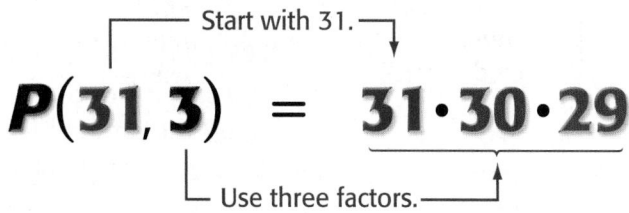

Start with 31.

$$P(31, 3) = 31 \cdot 30 \cdot 29$$

Use three factors.

 **EXAMPLES** Use Permutation Notation

Find each value.

② $P(8, 3)$

$P(8, 3) = 8 \cdot 7 \cdot 6$ or 336   8 things taken 3 at a time

③ $P(6, 6)$

$P(6, 6) = 6 \cdot 5 \cdot 4 \cdot 3 \cdot 2 \cdot 1$ or 720   6 things taken 6 at a time

 CHECK Your Progress

b. $P(12, 2)$ **c.** $P(4, 4)$ **d.** $P(10, 5)$

Permutations can be used when finding probabilities of real-world situations.

REAL-WORLD EXAMPLE Find Probability

④ **MUSIC** Ashley's CD player has a setting that allows the songs on a CD to play in a random order. She puts in a CD that contains 10 songs. What is the probability that the CD player will randomly play the first three songs in order?

First find the permutation of ten things taken three at a time or $P(10, 3)$.

10 songs	→ →	Choose 3

$P(10, 3) = 10 \cdot 9 \cdot 8$ ← 10 choices for the 1st song
9 choices for the 2nd song
8 choices for the 3rd song

$= 720$

So, there are 720 different ways to play the first 3 songs. Since you want the first three songs in order, there is only 1 of the 720 ways to do this.

$$P(\text{playing the first three songs in order}) = \frac{1 \text{ success}}{720 \text{ outcomes}}$$

So, the probability that the first 3 songs will play in order is $\frac{1}{720}$ or about 0.1%.

 CHECK Your Progress

e. NUMBERS Consider all of the four-digit numbers that can be formed using the digits 1, 2, 3, and 4, where no digit is used twice. Find the probability that one of these numbers picked at random is between 1,000 and 2,000.

Example 1   **1. SPORTS** In how many ways can the Wildcats' coach pick the first 3 batters out of the 9 players on the baseball team?

Examples 2 and 3   **Find each value.**

2. $P(5, 3)$ **③** $P(7, 4)$ **4.** $P(12, 5)$ **5.** $P(8, 8)$

Example 4   **6. PHONE NUMBERS** Ivy knows that the last four digits of her friend's phone number are 0, 3, 5, and 6, but she cannot remember the exact order. She knows that the first digit is 5. What is the probability that Ivy randomly selects the digits in the correct order?

Practice and Problem Solving

 = **Step-by-Step Solutions** begin on page R1.
Extra Practice begins on page EP2.

Example 1   **7. CODES** A security system has a keypad with 10 digits. How many four-digit codes are available if no digit is repeated?

8. ENTERTAINMENT Of the 10 games at the theater's arcade, Tyrone plans to play 3 different games. In how many orders can he play the 3 games?

9. MUSIC A disc jockey has 12 different songs he plans to play in the next hour. In how many ways can he pick the next 3 songs?

10. CONSTRUCTION A contractor can build 11 different model homes. She only has 4 lots. In how many ways can she put a different house on each lot?

Examples 2 and 3   **Find each value.**

11. $P(6, 3)$ **12.** $P(9, 2)$ **13.** $P(5, 5)$ **14.** $P(7, 7)$

15. $P(14, 5)$ **16.** $P(12, 4)$ **17.** $P(25, 4)$ **18.** $P(100, 3)$

Example 4   **LETTERS** Each arrangement of the letters in the word *quilt* is placed on a piece of paper. One paper is selected at random. Find each probability.

19. P(arrangement begins with q)   **20.** P(arrangement ends with lt)

㉑ GAMES Tanisha and Eric are playing Tic-Tac-Toe. Each player takes turns placing an X or an O in any of the nine locations that are empty. In how many different ways can the first 3 moves of the game occur?

22. SOCCER The teams of the Eastern Conference of Major League Soccer are listed at the right. If there are no ties for placement in the conference, in how many ways can the teams finish the season from first to last place?

MLS Eastern Conference
Chicago Fire
Columbus Crew
D.C. United
Kansas City Wizards
New England Revolution
New York Red Bulls
Toronto FC

23. **FIND THE ERROR** Sara is evaluating $P(7, 3)$. Find her mistake and correct it.

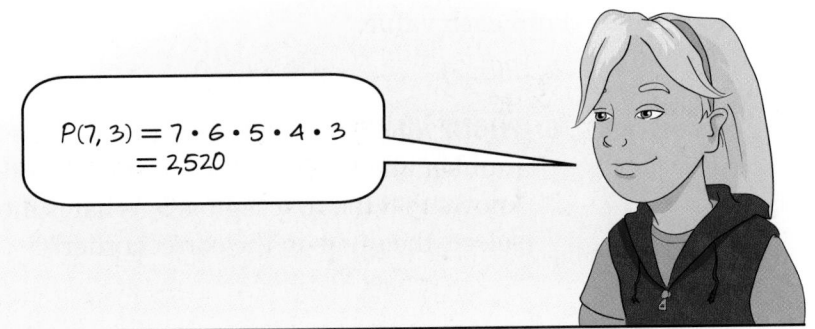

$$P(7, 3) = 7 \cdot 6 \cdot 5 \cdot 4 \cdot 3$$
$$= 2,520$$

24. **REASONING** If $P(9, 9) = 362,880$, what is $P(10, 10)$? Explain your reasoning.

25. **CHALLENGE** Compare $P(n, n)$ and $P(n, n - 1)$, where n is any whole number greater than one. Explain your reasoning.

26. **✍ WRITE MATH** Write a real-world problem for which you would use a permutation to solve. Then solve your problem.

✔ Test Practice

27. How many seven-digit phone numbers are available if a digit can be used only once and the first digit cannot be 0?

   A. 5,040

   B. 544,320

   C. 604,800

   D. 10,000,000

28. The table shows the classes Miguel can choose from for the first three periods of the day.

Classes
Math
Science
Social Studies
Language Arts

   In how many ways can Miguel arrange his first three classes?

   F. 3

   G. 9

   H. 12

   I. 24

29. The school talent show is featuring 13 acts. In how many ways can the talent show coordinator order the first 5 acts?

   A. 6,227,020,800

   B. 371,293

   C. 154,440

   D. 1,287

30. The schools listed below are finalists in a science competition. In how many ways can they finish in first, second, and third place?

Finalists
Chester Middle School
Glenwood Middle School
Lincoln Middle School
River Valley Middle School
South Middle School

   F. 15

   G. 20

   H. 60

   I. 120

Main Idea

Find the number of combinations of objects.

Vocabulary

combination

Get ConnectED

Combinations

ICE CREAM The Inside Scoop Ice Cream Shop offers the six flavors shown. Maria wants scoops of two different flavors on her cone.

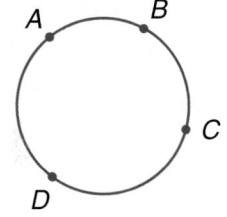

vanilla
chocolate
strawberry
rocky road
butter pecan
peach

1. Make a list of all the different cones Maria can buy.

2. How many different ice cream cones were on the list?

3. Is the number of ice cream cones equal to $P(6, 2)$? Explain.

In this case, the order of the scoops of ice cream is not important. An arrangement or listing where order is not important is called a **combination**.

 **EXAMPLE** **Find a Combination**

1. **GEOMETRY** Four points are located on a circle. How many line segments can be drawn with these points as endpoints?

In drawing the line segments, the order of the endpoints is not important. This arrangement is a combination.

First list all possible permutations of endpoints A, B, C, and D taken two at a time. Then cross off arrangements that are the same as another one.

$$\overline{AB} \quad \overline{AC} \quad \overline{AD} \quad \overline{BA} \quad \overline{BC} \quad \overline{BD}$$
$$\overline{CA} \quad \overline{CB} \quad \overline{CD} \quad \overline{DA} \quad \overline{DB} \quad \overline{DC}$$

> $\overline{AB}$ and $\overline{BA}$ are the same in this case, so cross off one of them.

There are only six *different* arrangements. So, there are six ways to draw the line segment. Check by drawing the segments.

 **CHECK Your Progress**

a. **HANDSHAKES** If there are 8 people in a room, how many handshakes will occur if each person shakes hands with every other person?

The symbol $C(4, 3)$ represents the number of combinations of 4 things taken 3 at a time.

$$C(4, 3) = \frac{P(4, 3)}{3 \cdot 2 \cdot 1}$$

the number of combinations of 4 things taken 3 at a time

the number of ways 3 things can be arranged

the number of permutations of 4 things taken 3 at a time

EXAMPLES Combinations and Permutations

MUSIC The makeup of a symphony is shown in the table at the right.

Symphony Makeup	
Instrument	**Number**
Strings	45
Woodwinds	8
Brass	8
Percussion	3
Harps	2

2 **A group of 3 musicians from the strings section will talk to students at Madison Middle School. Is this a *combination* or a *permutation*?**

This is a combination because the order is not important.

3 **How many possible groups could talk to the students?**

$$C(45, 3) = \frac{45 \cdot 44 \cdot 43}{3 \cdot 2 \cdot 1} \qquad \text{45 string musicians taken 3 at a time}$$

$$= \frac{\overset{15}{45} \cdot \overset{22}{44} \cdot 43}{\underset{1}{\cancel{3}} \cdot \underset{1}{\cancel{2}} \cdot 1} \text{ or } 14{,}190 \qquad \text{Divide out common factors.}$$

There are 14,190 different groups that could talk to the students.

4 **One member from the strings section will talk to students at Brown Middle School, another to students at Oak Avenue Middle School, and another to students at Jefferson Junior High. Is this a *combination* or a *permutation*?**

Since it makes a difference which member goes to which school, order is important. This is a permutation.

5 **Refer to Example 4. In how many possible ways can the strings members talk to the students?**

$$P(45, 3) = 45 \cdot 44 \cdot 43 \text{ or } 85{,}140 \qquad \text{Definition of } P(45, 3)$$

There are 85,140 ways for the members to talk to the students.

CHECK Your Progress

b. **FOOD** At a restaurant, customers can select three burger toppings: catsup, mustard, pickles, lettuce, onions, or cheese. Suppose the layering of the toppings is *not* important. Does the possible number of burgers with three toppings represent a *combination* or *permutation*? How many three-topping burgers are possible?

Example 1 **1** **GEOMETRY** Eight points are located on a circle. How many line segments can be drawn with these points as endpoints?

Examples 2–5 **Determine whether each situation is a *permutation* or a *combination*. Then find the number of possible outcomes.**

2. writing a four-digit number using the numbers 0 through 9 with no digit used more than once

3. choosing 3 shirts to pack for vacation from a choice of 7 shirts

Practice and Problem Solving

= Step-by-Step Solutions begin on page R1.
Extra Practice begins on page EP2.

Example 1 4. How many three-topping pizzas can be ordered from the list of toppings at the right?

5. How many different starting squads of 6 players can be picked from 10 volleyball players?

Pizza Toppings		
anchovies	sausage	onions
bacon	green peppers	black olives
ham	hot peppers	green olives
pepperoni	mushrooms	pineapple

Examples 2–5 **Determine whether each situation is a *permutation* or a *combination*. Then find the number of possible outcomes.**

6. choosing a committee of 5 from the 36 members of a class

7. choosing 2 co-captains of the basketball team with 14 players

8. choosing the placement of 9 model cars in a line from a selection of 12 models

9 choosing 4 out of 12 photographs to display at specific locations

10. **GRAPHIC NOVEL** Refer to the graphic novel frame below for Exercises a–b.

a. The four candidates will each be giving a speech. Is the number of ways in which the candidates can give their speeches a *permutation* or a *combination*? Explain.

b. In how many possible ways can the candidates give their speeches?

11. **OPEN ENDED** Describe a situation that could be represented by $C(15, 5)$.

12. **Which One Doesn't Belong?** Identify the situation that does not belong with the other three. Explain your reasoning.

| choosing 3 toppings for the pizzas to be served at the party | choosing 3 members for the decorating committee | choosing 3 people to chair 3 different committees | choosing 3 desserts to serve at the party |

13. **CHALLENGE** Is the value of $P(x, y)$ *sometimes*, *always*, or *never* greater than the value of $C(x, y)$? Explain. Assume x and y are positive integers and $x \geq y$.

14. **WRITE MATH** Give an example of a situation in which you would use a combination. Then, change the situation so that you need to use a permutation. Explain the difference between the situations.

Test Practice

15. Which situation is represented by $C(8, 3)$?

   A. the number of arrangements of 8 people in a line

   B. the number of ways to pick 3 out of 8 vegetables to add to a salad

   C. the number of ways to pick 3 out of 8 students to be the first, second, and third contestant in a spelling bee

   D. the number of ways 8 people can sit in a row of 3 chairs

16. The enrollment for Centerville Middle School is shown. How many different four-person committees could be formed from the students in the 8th grade?

Centerville Middle School		
Class	**Boys**	**Girls**
6th grade	42	47
7th grade	55	49
8th grade	49	53

   F. 211,876　　H. 4,249,575

   G. 292,825　　I. 149,059,680

 **Spiral Review**

Find each value. (Lesson 1C)

17. $P(7, 2)$

18. $P(15, 4)$

19. $P(20, 5)$

20. $P(7, 7)$

21. **SCHOOL** At the school cafeteria, students can choose from 4 entrees and 3 beverages. How many different lunches of one entree and one beverage can be purchased at the cafeteria? (Lesson 1A)

Menu

Entrees	Beverages
macaroni and cheese	milk
chicken nuggets	fruit juice
pizza	lemonade
turkey sandwich	

Main Idea

Find the probability of independent and dependent events.

 Vocabulary

compound event
independent events
dependent events
disjoint events

Get ConnectED

Compound Events

 MONEY Reginald has 3 state quarters, one from Colorado, one from Montana, and one from Washington.

1. If Reginald picks one quarter without looking, what is the probability it is from Colorado?

2. Suppose he tosses the coin. What is the probability it lands heads up?

3. Make a tree diagram to find the probability of choosing a Colorado quarter that lands heads up.

4. How are the answers to Exercises 1, 2, and 3 related?

In the situation above, choosing a quarter and tossing heads is a compound event. A **compound event** consists of two or more simple events. Since choosing a quarter does not affect tossing heads, the two events are called **independent events**. The outcome of one event does not affect the outcome of the other event.

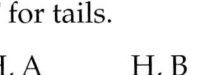 **EXAMPLES** Independent Events

1 **A coin is tossed, and the spinner shown is spun. Find the probability of tossing heads and spinning a consonant.**

List the sample space. Use H for heads and T for tails.

H, A   H, B   H, C

T, A   T, B   T, C

P(H and a consonant) = $\dfrac{\text{number of times heads and a consonant occurs}}{\text{number of possible outcomes}}$

P(H and a consonant) = $\dfrac{2}{6}$ or $\dfrac{1}{3}$

So, the probability is $\dfrac{1}{3}$ or about 33%.

 CHECK Your Progress

A number cube is rolled, and the spinner in Example 1 is spun. Find each probability.

a. P(4 and a consonant)   **b.** P(odd and B)

The probability in Example 1 can also be found by multiplying the probabilities of each event. $P(\text{H}) = \frac{1}{2}$ and $P(\text{consonant}) = \frac{2}{3}$. So, $P(\text{H and C}) = \frac{1}{2} \cdot \frac{2}{3}$ or $\frac{1}{3}$.

Key Concept — Probability of Independent Events

Words   The probability of two independent events can be found by multiplying the probability of the first event by the probability of the second event.

Symbols $P(A \text{ and } B) = P(A) \cdot P(B)$

 REAL-WORLD EXAMPLE

2 **SHOPPING** Brianna is buying a new outfit. She is choosing among 2 red, 1 purple, 3 pink, or 4 yellow tops. For pants, she is choosing between jeans or capris. If Brianna chooses a top and pants at random, find the probability that she chooses a yellow top and jeans.

$P(\text{yellow top and jeans})$

$\quad = P(\text{yellow top}) \cdot P(\text{jeans})$

$\quad = \dfrac{4}{10} \cdot \dfrac{1}{2}$ 4 out of 10 tops are yellow.
   1 out of 2 pants are jeans.

$\quad = \dfrac{\overset{2}{\cancel{4}}}{10} \cdot \dfrac{1}{\underset{1}{\cancel{2}}} = \dfrac{2}{10}$ or $\dfrac{1}{5}$ Simplify.

So, the probability is $\frac{1}{5}$ or 20%.

Study Tip

Reasonable Answer
You can check your answer in Example 2 by listing the sample space or by making a tree diagram.

 CHECK Your Progress

c. SHOPPING If khakis and shorts are added to Brianna's pant choices, find the probability that she chooses a red top and capris.

If the outcome of one event affects the outcome of a second event, the events are called **dependent events**. Just as in independent events, the probabilities of dependent events can be found by multiplying the probabilities of each event. However, now the probability of the second event depends on the fact that the first event has already occurred.

Key Concept — Probability of Dependent Events

Words   If two events, A and B, are dependent, then the probability of both events occurring is the product of the probability of A and the probability of B after A occurs.

Symbols $P(A \text{ and } B) = P(A) \cdot P(B \text{ following } A)$

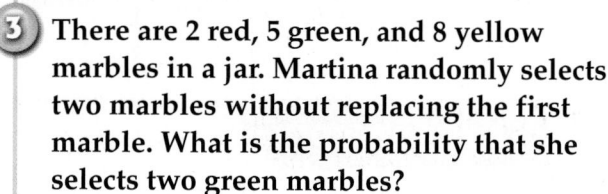 **EXAMPLE** Dependent Events

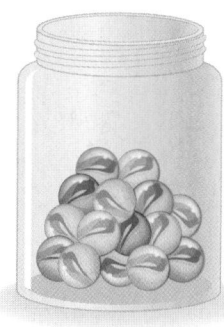

3 There are 2 red, 5 green, and 8 yellow marbles in a jar. Martina randomly selects two marbles without replacing the first marble. What is the probability that she selects two green marbles?

Since the first marble is not replaced, the first event affects the second event. These are dependent events.

$P(\text{first marble is green}) = \dfrac{5}{15}$ ← number of green marbles
← total number of marbles

$P(\text{second marble is green}) = \dfrac{4}{14}$ ← number of green marbles left
← total number of marbles left

$P(\text{two green marbles}) = \dfrac{\overset{1}{\cancel{5}}}{\underset{3}{\cancel{15}}} \cdot \dfrac{\overset{2}{\cancel{4}}}{\underset{7}{\cancel{14}}} \text{ or } \dfrac{2}{21}$

So, the probability of selecting two green marbles is $\dfrac{2}{21}$, or about 9.5%.

 CHECK Your Progress

d. There are 4 blueberry, 6 raisin, and 2 plain bagels in a bag. Javier randomly selects two bagels without replacing the first bagel. Find the probability that he selects a raisin bagel and then a plain bagel.

When you toss a coin, either heads *or* tails will turn up. Tossing heads and tossing tails are examples of **disjoint events**, or events that cannot happen at the same time. Disjoint events are also called *mutually exclusive events*.

Study Tip

Disjoint Events
When finding the probabilities of disjoint events, the word *or* is usually used.

EXAMPLE Disjoint Events

4 A number cube is rolled. What is the probability of rolling an odd number or a 6?

These are disjoint events since it is impossible to roll an odd number and a 6 at the same time.

$P(\text{odd number or 6}) = \dfrac{4}{6}$ ← There are four favorable outcomes: 1, 3, 5, or 6.
← There are 6 total possible outcomes.

So, the probability of rolling an odd number or a 6 is $\dfrac{4}{6}$, or $\dfrac{2}{3}$.

CHECK Your Progress

e. Twenty-six cards are labeled, each with a letter of the alphabet, and placed in a box. A single card is randomly selected. What is the probability that the card selected will be labeled with the letter M or the letter T?

Notice that the probability in Example 4 can also be found by adding the probabilities of each event. $P(\text{odd number}) = \frac{3}{6}$ and $P(6) = \frac{1}{6}$. So, $P(\text{odd number or 6}) = \frac{3}{6} + \frac{1}{6}$ or $\frac{4}{6}$.

Key Concept   **Probability of Disjoint Events**

Words   If two events, *A* and *B*, are disjoint, then the probability that either *A* or *B* occurs is the sum of their probabilities.

Symbols $P(A \text{ or } B) = P(A) + P(B)$

✓ CHECK Your Understanding

Example 1   **A number cube is rolled, and the spinner is spun. Find each probability.**

 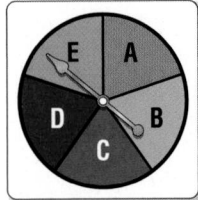

1. $P(5 \text{ and } E)$

2. $P(2 \text{ and a vowel})$

3. $P(3 \text{ and a consonant})$

4. $P(\text{factor of 6 and } D)$

Example 2   ⑤ **CLOTHES** Loretta has 2 pairs of black pants, 3 pairs of blue pants, and 1 pair of tan pants. She also has 4 white and 2 red shirts. If Loretta chooses a pair of pants and a shirt at random, what is the probability that she will choose a pair of black pants and a white shirt?

6. **MARBLES** A bag contains 12 marbles. Four are red, 3 are white, and 5 are blue. A marble is randomly selected, its color recorded, and then the marble is returned to the bag. A second marble is randomly selected. Find the probability that both marbles selected are blue.

Example 3   7. The digits 0–9 are each written on a slip of paper and placed in a hat. Two slips of paper are randomly selected, without replacing the first. What is the probability that the number 0 is drawn first and then a 7 is drawn?

Example 4   **A number cube is rolled. Find each probability.**

8. $P(4 \text{ or } 5)$

9. $P(3 \text{ or even number})$

10. $P(1 \text{ or a multiple of } 2)$

11. $P(6 \text{ or a number less than } 3)$

Practice and Problem Solving

● = **Step-by-Step Solutions** begin on page R1.
Extra Practice begins on page EP2.

Example 1 **A coin is tossed, and a number cube is rolled. Find each probability.**

12. P(heads and 1)

13. P(tails and a multiple of 3)

A set of five cards is labeled 1–5. A second set of ten cards contains the following colors: 2 red, 3 purple, and 5 green. One card from each set is selected. Find each probability.

14. P(5 and green)

15 P(odd and red)

16. P(prime and purple)

17. P(even and yellow)

Example 2 **18. MUSIC** Denzel is listening to a CD that contains 12 songs. If he presses the random button on his CD player, what is the probability that the first two songs played will be the first two songs listed on the CD?

19. JUICE POPS Lakita has two boxes of juice pops with an equal number of pops in each flavor. Find the probability of randomly selecting a grape juice pop from the first box and randomly selecting a juice pop from the second box that is *not* grape.

Example 3 **20. FRUIT** Francesca randomly selects two pieces of fruit from a basket containing 8 oranges and 4 apples without replacing the first fruit. Find the probability that she selects two oranges.

21. SCHOOL The names of 24 students, of which 14 are boys and 10 are girls, in Mr. Santiago's science class are written on cards and placed in a jar. Mr. Santiago randomly selects two cards without replacing the first to determine which students will present their lab reports today. Find the probability that two girls are selected.

Example 4 **A day of the week is randomly selected. Find each probability.**

22. P(Monday or Tuesday)

23. P(a day beginning with T or Friday)

24. P(a weekday or Saturday)

25. P(Wednesday or a day with 6 letters)

A coin is tossed twice, and a letter is randomly picked from the word *event*. Find each probability.

26. P(heads, tails, *not* V)

27 P(two tails and a vowel)

28. 🧩 **MULTIPLE REPRESENTATIONS** Use the fact that the probability for a boy or a girl is each $\frac{1}{2}$.

   a. TABULAR Copy and complete the table that gives the probability that all the children in a family are boys.

   b. NUMERICAL Predict the probability that, in a family of ten children, all ten are boys.

   c. ALGEBRAIC Predict the probability that, in a family of n children, all n are boys.

Number of Children	P(all boys)
1	$\frac{1}{2}$
2	$\frac{1}{2} \cdot \frac{1}{2}$ or $\frac{1}{4}$
3	■
4	■
5	■

29. FIND THE ERROR Enrique tosses a coin twice. He wants to determine the probability of tossing two heads. Find his mistake and correct it.

$$P\,(\text{two heads}) = \frac{1}{2} + \frac{1}{2}$$
$$= 1$$

30. CHALLENGE Suppose the spinner is designed so that for each spin there is a 40% probability of spinning red and a 20% chance of spinning blue. What is the probability of spinning two reds and then one blue?

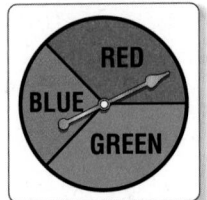

31. REASONING A shelf has books A, B, and C on it. You randomly pick two books without replacing the first book. What is the probability of picking books A and B? If you replace the first book, how does the probability of picking books A and B change?

32. WRITE MATH How are the probabilities of independent events and dependent events similar? How are they different?

Test Practice

33. A jar contains 8 white marbles, 4 green marbles, and 2 purple marbles. If Darla picks one marble from the jar without looking, what is the probability that it will be either white or purple?

A. $\frac{5}{7}$ 　　C. $\frac{2}{7}$

B. $\frac{4}{7}$ 　　D. $\frac{1}{7}$

34. What is the probability of spinning a red, the number 1, and the letter A on the three spinners below?

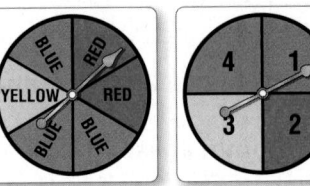  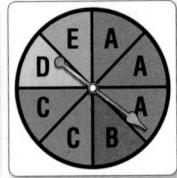

F. $\frac{1}{3}$ 　　G. $\frac{1}{32}$ 　　H. $\frac{1}{12}$ 　　I. $\frac{1}{64}$

Spiral Review

35. FOOD Derek can order a burrito with the ingredients shown in the table. How many different burritos with two fillings are possible? (Lesson 1D)

36. CHEERLEADING There are 10 girls on a cheerleading squad. In how many ways can the coach choose two girls for captain and co-captain? (Lesson 1C)

Burrito Fillings
lettuce
cheese
beans
rice

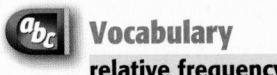

Relative Frequency

Main Idea

Use an experiment to determine probability.

Vocabulary

relative frequency

Get ConnectED

GAMES In a board game, you must roll two number cubes to determine the number of spaces you can move. If you roll doubles, or two of the same number, you double the sum of the cubes as your move.

You can conduct an experiment to find the relative frequency of rolling doubles using a pair of number cubes. **Relative frequency** is the ratio of the number of experimental successes to the total number of experimental attempts.

ACTIVITY

STEP 1 The table shows all of the possible outcomes for rolling two number cubes. What is the probability of rolling doubles?

(1,1)	(2, 1)	(3,1)	(4, 1)	(5,1)	(6, 1)
(1, 2)	(2, 2)	(3, 2)	(4, 2)	(5, 2)	(6, 2)
(1, 3)	(2, 3)	(3, 3)	(4, 3)	(5, 3)	(6, 3)
(1, 4)	(2, 4)	(3, 4)	(4, 4)	(5, 4)	(6, 4)
(1, 5)	(2, 5)	(3, 5)	(4, 5)	(5, 5)	(6, 5)
(1, 6)	(2, 6)	(3, 6)	(4, 6)	(5, 6)	(6, 6)

STEP 2 Roll two number cubes and record the outcomes in a frequency table. Repeat the experiment 50 times.

STEP 3 Find the relative frequency of rolling doubles, or the ratio $\dfrac{\text{number of times doubles were rolled}}{\text{number of rolls}}$.

Analyze the Results

1. Are the ratios in Steps 1 and 3 the same? Explain any differences.

2. If the number cubes were rolled 100 times, would you expect the results to be the same? 1,000 times? Explain.

A spinner is divided into 4 equal sections. Find the relative frequencies and probabilities of each of the following. Express your answer as a decimal.

Section	A	B	C	D
Frequency	27	40	10	23

3. A

4. B

5. C

6. D

7. **WRITE MATH** Compare the relative frequencies and the probabilities. Explain any differences.

Main Idea

Find experimental and theoretical probabilities and use them to make predictions.

 Vocabulary

experimental probability
theoretical probability
fair game

 Get ConnectED

Experimental and Theoretical Probability

 **CARDS** Juno randomly selected a card from a set of cards and replaced it. She repeated the process 60 times. The results are shown in the table.

Color	Green	Red	Yellow	Blue
Frequency	16	10	20	14

1. Find the relative frequency for choosing each color.

2. If there were 100 cards, about how many blue cards would you expect to find?

3. **MAKE A CONJECTURE** Based on Juno's results, describe the set of cards.

Probabilities that are based on the relative frequencies of the outcomes obtained by conducting an experiment are called **experimental probabilities**.

Probabilities based on known characteristics or facts are called **theoretical probabilities**. Theoretical probability tells you what *should* happen in an experiment.

EXAMPLES Theoretical and Experimental Probability

(1) **What is the theoretical probability of rolling a double 1 using two number cubes?**

The theoretical probability is $\frac{1}{6} \cdot \frac{1}{6}$ or $\frac{1}{36}$.

(2) **The graph shows the results of an experiment in which two number cubes were rolled. According to the experimental probability, is a sum of 12 likely to occur?**

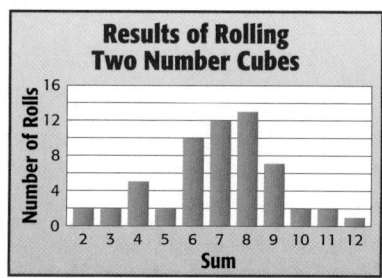

Only 1 of the 58 sums is 12. So, the experimental probability of rolling a sum of 12 is $\frac{1}{58}$. It is not likely that a sum of 12 will occur.

CHECK Your Progress

a. Refer to the graph above. According to the experimental probability, which sum is most likely to occur?

Real-World Link · · · ·
The average American cell phone user replaces his or her cell phone every 1.5 years.

 **REAL-WORLD EXAMPLE**

3 **TIME** Three hundred people were surveyed as to how they keep track of time. What is the experimental probability that a person uses his or her cell phone?

Method	Number Who Use This Method
Cell phone	185
Clock	58
Watch	57

There were 300 people surveyed and 185 use their cell phone to keep track of time. The experimental probability is $\frac{185}{300}$ or about 62%.

CHECK Your Progress

b. What is the experimental probability that a person uses his or her watch to keep track of time?

You can use past performance to predict future events.

EXAMPLE **Use Probability to Predict**

4 **MANUFACTURING** At a plant that manufactures lightbulbs, an inspector finds that the probability that a lightbulb is *not* defective is $\frac{8}{11}$. Is this probability experimental or theoretical? Explain.

This is an experimental probability since it is based on what has already happened.

If the company wants to have 10,000 non-defective lightbulbs, how many lightbulbs will they need to make?

This problem can be solved using a proportion.

> 8 out of 11 lightbulbs are not defective. → $\frac{8}{11} = \frac{10,000}{x}$ ← 10,000 out of x lightbulbs should not be defective.

Solve the proportion.

$\frac{8}{11} = \frac{10,000}{x}$ Write the proportion.

$8 \cdot x = 11 \cdot 10,000$ Find the cross products.

$8x = 110,000$ Multiply.

$\frac{8x}{8} = \frac{110,000}{8}$ Divide each side by 8.

$x = 13,750$ The company should make 13,750 lightbulbs.

CHECK Your Progress

c. **SURVEYS** In a recent survey of 150 people, 18 responded that they were left-handed. If 2,500 people are surveyed, how many would be expected to be left-handed?

Example 1

1. Use the table that shows the results of tossing three coins, one at a time, 50 times.

Result	Frequency	Result	Frequency
HHH	6	TTT	3
HHT	5	TTH	6
HTH	10	THT	5
HTT	5	THH	10

   a. What is the theoretical probability of tossing exactly two heads?

Example 2

   b. Find the experimental probability of tossing exactly two heads.

   c. Based on the experimental probability, how likely is it that a toss of three coins will have two heads? Explain.

2. Use the table at the right showing the results of a survey of shoes that students are wearing.

Shoe	Number of Students
Athletic	48
Sandals	33
Dress	28
Boots	11

Example 3

   a. What is the probability that the next student to walk by will be wearing athletic shoes?

Example 4

   b. Out of the next 75 students, how many would you expect to be wearing dress shoes?

Practice and Problem Solving

 = **Step-by-Step Solutions** begin on page R1.
Extra Practice begins on page EP2.

Examples 1, 2, and 4

3. Use the table that shows the results of spinning an equally divided 8-section spinner.

Number on Spinner	Frequency
1	8
2	5
3	9
4	4
5	10
6	6
7	5
8	3

   a. Compare the theoretical and experimental probabilities of the spinner landing on 5.

   b. Based on the experimental probability, how many times would you expect the spinner to land on 3 if the spinner is spun 200 times?

   c. Jarred predicts that the spinner will land on 4 or 8 on the next spin. Is this a reasonable prediction? Explain.

Example 3

4. **SURVEYS** In a survey, 120 out of 200 students said their favorite food was pizza. What is the experimental probability that a student chose pizza as his or her favorite food?

5. **TENNIS** During the first four weeks of tennis season, Jeanette won 24 out of 30 matches.

   a. What is the probability that she will win her next match?

   b. If she competes in 50 matches this season, how many should she be expected to win?

6. **SPORTS** In a survey of 90 students at Genoa Middle School, 42 liked to watch basketball and 24 liked to watch soccer. If there are 300 students in the middle school, how many would you expect to like to watch soccer?

7. **AWARDS** Use the table that shows how many films from each category won the Best Picture award.

Category	Number of Films
Drama	32
Historical/Epic	13
Comedy	10
Musical	9
War	6
Action	4
Western	3
Suspense	2

   a. What is the probability that the next movie to win will be a comedy?

   b. Out of 100 winners, how many would you expect to be action films?

8. **CARS** Of the last 80 cars sold at a dealership, 35 of them were SUVs. What is the experimental probability that the next car they sell will be an SUV?

9. **BASEBALL** Use the table that shows the batting results of a baseball player for a season.

   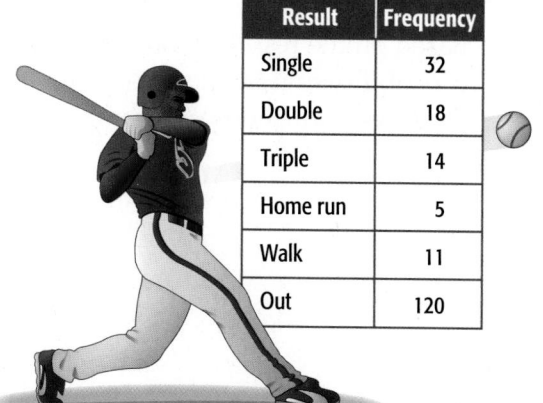

Result	Frequency
Single	32
Double	18
Triple	14
Home run	5
Walk	11
Out	120

   a. Based on the results, how likely is it that the player would be out after his next turn batting?

   b. The next time the player is at bat, how likely is it for him to hit a single or a double?

10. **FOOD** The manager of a school cafeteria asked selected students to pick their favorite menu item. The results of the survey are shown in the table. If the cafeteria serves 350 lunches, and students can choose only one lunch, how many hamburgers could the manager expect to sell?

Menu Item	Students
Hot dog	22
Hamburger	19
Pizza	30
Taco	16
Chicken strips	13

11. **MATH IN THE MEDIA** Find examples of probability in a newspaper or magazine, on television, or on the Internet. Explain whether the experimental or theoretical probability was used in the media report.

H.O.T. Problems

12. **OPEN ENDED** Two hundred fifty people are surveyed about their favorite color. Make a table of possible results if the experimental probability that the favorite color is blue is 40%.

13. **CHALLENGE** A survey found that 75 students out of 200 own a skateboard and that 280 students out of 400 own a bicycle. What is the probability that a student has both a skateboard and a bicycle?

14. **WRITE MATH** Explain why you would *not* expect the theoretical probability of an event and the experimental probability of the same event to always be the same.

15. The results of a survey about what students say is the hardest part of going to school are shown in the table.

Issue	Number of Students
Tests	72
Paying attention	38
Homework	36
Organization	32
Presentations	22

Based on the results, what is the probability that the next student surveyed will choose "organization"?

A. $\frac{8}{25}$

B. $\frac{9}{50}$

C. $\frac{4}{25}$

D. $\frac{4}{50}$

16. **GRIDDED RESPONSE** Shannon spun the spinner shown and recorded her results.

Number on Spinner	Frequency
1	20
2	10
3	2
4	40
5	8

What is the experimental probability of landing on the number five?

More About Probability

A **fair game** is a game in which each player has an equally likely chance of winning. When playing a game, it is often useful to determine the theoretical probability of each player winning to determine if a game is fair.

ACTIVITY

In a card game, five cards numbered 1–5 are shuffled and laid facedown. Player 1 chooses 2 cards and finds the sum. If the sum is even, Player 1 wins. If the sum is odd, Player 2 wins. The cards are then reshuffled and laid facedown. Then it is Player 2's turn.

Card	1	2	3	4	5
1					
2					
3					
4					
5					

STEP 1 Copy and complete the table to find all of the possible sums of two cards.

STEP 2 Play 10 rounds of the game and then record the results.

17. Find the theoretical and experimental probability of each player winning. Compare the experimental probabilities to the theoretical probabilities.

18. Based on the theoretical probabilities of each player winning, is this a fair game? Explain your reasoning.

19. Design a game that is *unfair*. Explain how the game is unfair.

Geometric Probability

Main Idea

Find the probability of an event based on area.

You can use geometric models to explore probability.

JOBS In order to determine which job you must complete on Saturday, your parents use a spinner like the one shown.

STEP 1 Construct a spinner like the one shown.

STEP 2 Spin the spinner 50 times and record your results in a table like the one shown.

Clean Your Room	Yard Work	Dishes	Babysit	Laundry

STEP 3 Use the results to find the probability of spinning each job.

Analyze the Results

1. Recall that the degree measure of a complete circle is 360. Use the ratio $\dfrac{\text{angle measure of the event}}{360°}$ to find the probability of the spinner landing on each job.

2. Compare the results from Exercise 1 and Step 3.

Exercise 2 suggests the following relationship between probability and area.

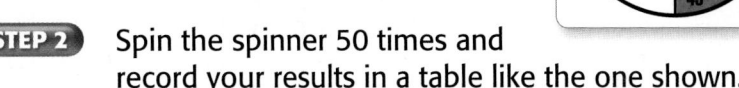

Key Concept **Probability and Area**

Words The probability of landing in a region of a target is the ratio of the area of the region to the area of the target.

Symbols $P(\text{region}) = \dfrac{\text{area of region}}{\text{area of target}}$

Practice and Apply

Find the probability that a randomly thrown dart will land in the shaded region of each dartboard. Write the probability as a fraction, decimal, and percent.

3.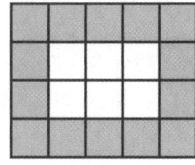

4.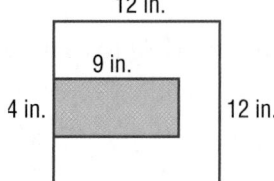

5.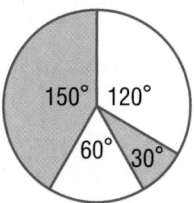

6.

7.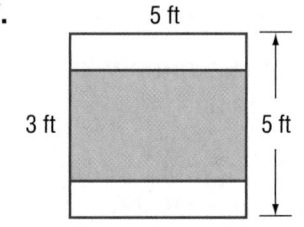

8.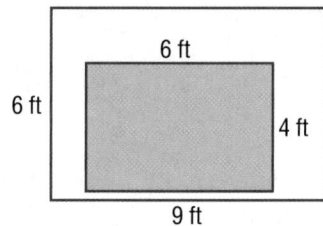

9. GOLF Two golf balls land in the rectangular region at the right. What is the probability that both of them land on the green?

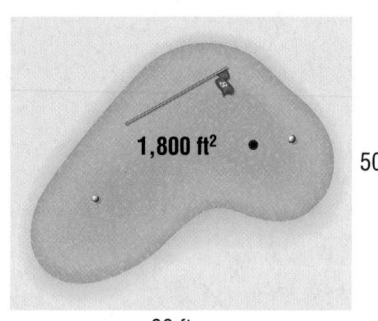

10. PATIOS A rectangular patio is 8 feet by 10 feet. At the center of the patio is a circular rug that is 7 feet in diameter. If a coin is dropped at random on the patio, what is the probability, rounded to the nearest whole percent, that the coin lands on the rug?

11. GAMES A dartboard has four circular rings surrounding a bull's-eye. The circles have radii of 1, 2, 3, 4, and 5 units. Suppose a dart is equally likely to hit any point on the board. Is the dart more likely to hit in the outermost ring or inside the region consisting of the bull's-eye and the two innermost rings?

12. Draw a dartboard in which the probability of a dart landing in the shaded area is 60%.

13. PUZZLES A tangram, a puzzle that originated in China, consists of 7 pieces that form a square, as shown. These pieces can be rearranged to form shapes of animals, people, and other objects. Suppose a counter is randomly dropped on the tangram. Find the probability that the counter will land on the small square. (*Hint:* The side of the small square is $\frac{1}{4}$ of the diagonal of the large square.)

Mid-Chapter Check

1. **BREAKFAST** Draw a tree diagram to determine the number of one-bread and one-beverage outcomes using the breakfast choices listed. (Lesson 1A)

Breakfast Choices

toast   muffin   bagel
coffee   milk   juice

2. **FASHION** Regina has three necklaces, three pairs of earrings, and two bracelets. How many combinations of the three types of jewelry are possible? (Lesson 1A)

3. **MUSIC** Five band members play the flute. In how many ways can these members be chosen for the first, second, and third chairs of the flute section? (Lesson 1C)

Find each value. (Lessons 1C and 1D)

4. $P(5, 3)$

5. $P(6, 2)$

6. $C(5, 5)$

7. $C(7, 6)$

8. **STUDENT COUNCIL** In how many ways can 2 student council members be elected from 7 candidates? (Lesson 1D)

9. **MULTIPLE CHOICE** Roman has ten cards numbered 1 to 10. What is the probability of picking two even-numbered cards one after another, if the first card picked is replaced? (Lesson 2A)

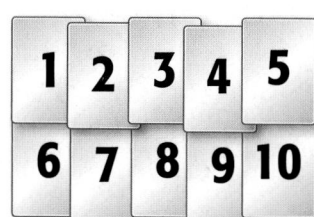

1 2 3 4 5

6 7 8 9 10

A. $\frac{1}{5}$ B. $\frac{2}{9}$ C. $\frac{1}{4}$ D. $\frac{3}{8}$

10. **MULTIPLE CHOICE** A bag contains 4 red, 20 blue, and 6 green candies. Kevin picks one at random and keeps it. Then Amy picks a candy. What is the probability that they each select a red candy? (Lesson 2A)

F. $\frac{1}{150}$ H. $\frac{2}{145}$

G. $\frac{1}{15}$ I. $\frac{1}{870}$

11. **FOOD** Two hundred twenty-five high school freshmen were asked to name their favorite hot lunch. One hundred thirty-five students named tacos as their favorite. If an additional 80 freshmen are asked, how many would be expected to choose tacos? (Lesson 2C)

12. **MUSIC** A survey asked 500 teens what formats of music they had purchased in the past two months. The table shows the results. (Lesson 2C)

Format	Number Purchased
CD	180
Download	320

a. What is the experimental probability that a teen purchased a CD in the past two months?

b. What is the experimental probability that a teen purchased a music download in the past two months?

13. **VOLLEYBALL** In her last 30 serves, Emily served the ball over the net 18 times. Based on this, how many of the next 50 serves should she expect to go over the net? (Lesson 2C)

Problem-Solving Investigation

Main Idea Solve problems by acting it out.

P.S.I. TEAM +

e-Mail: ACT IT OUT

MADISON: I have a quiz in Spanish class this Friday. I wonder if tossing a coin would be a good way to answer a 5-question true-false quiz.

YOUR MISSION: Act it out to determine if tossing a coin is a good way to answer a true-false quiz.

Understand	You know there are five true-false questions on the quiz. You can carry out an experiment to test if tossing a coin would be a good way to answer the questions and get a good grade.
Plan	Toss a coin 5 times. If the coin shows tails, the answer is T. If the coin shows heads, the answer is F. Do three trials.
Solve	Suppose the correct answers are T, F, F, T, F. Let's circle them in each trial.

Answers	T	F	F	T	F	Number Correct
Trial 1	Ⓣ	T	Ⓕ	F	T	2
Trial 2	F	Ⓕ	T	Ⓣ	Ⓕ	3
Trial 3	Ⓣ	Ⓕ	T	F	T	2

	Since the experiment produced 2–3 correct answers on a 5-question quiz, it shows that tossing a coin to answer a true-false quiz is *not* the way to get a good grade.
Check	Check by doing several more trials.

Analyze the Strategy

1. Explain an advantage of using the *act it out* strategy to solve a problem.

2. **WRITE MATH** Write a problem that could be solved by acting it out. Then solve the problem.

Mixed Problem Solving

 = **Step-by-Step Solutions** begin on page R1.
Extra Practice begins on page EP2.

- Act it out.
- Guess, check, and revise.
- Make a table.
- Choose an operation.

Use the *act it out* strategy to solve Exercises 3–5.

3. **TICKETS** Abby, Constance, Miranda, and Jasmine are standing in line to buy tickets for a concert. In how many different ways can they stand in line?

4. **CLOTHES** Julie is getting ready for school. She can choose from three pairs of jeans and five blouses. How many outfits can Julie create if all of the combinations coordinate?

5. **HOMEWORK** Bryan has homework in math, science, reading, and art. If he plans on doing math homework first, list the number of ways in which he can complete the four homework assignments.

Use any strategy to solve Exercises 6–13.

6. **GIFTS** Charlotte bought gifts for each of her 7 teachers. She bought everyone a mug for $5 or a candle for $3. She spent a total of $27. How many of each gift did she buy?

7. **ELECTIONS** About $\frac{2}{3}$ of the eighth-grade class voted for Sonya to be the class president. If there are 350 students in the eighth-grade class, how many voted for Sonya?

8. **QUIZ** A Spanish quiz has 5 questions worth 2 points each and 5 questions worth 5 points each. If Tom received 23 points on the quiz, how many of each type of question did he get correct?

9. **YEARBOOKS** Out of the 20 students working on the yearbook, 3 will be selected to sell and promote the yearbook. How many different combinations of 3 students can be chosen?

10. **BOATS** Karen would like to rent either a rowboat or a paddleboat. The prices to rent each type of boat are listed in the table below. In addition to the cost per hour, the rowboat requires a $10 deposit.

Boat	Cost per Hour
Rowboat	$5.50
Paddleboat	$10.50

For how many hours must she rent the boats for the cost to be the same for each boat?

11. **CANDY** The ratio of green jelly beans to red jelly beans is 3:4. Danielle eats four green jelly beans, and the ratio is now 1:2. How many green jelly beans did Danielle have at the beginning?

12. **TRAVEL** Mary Anne surveyed the students in her class about their favorite method of travel to a vacation destination. The responses are shown below. What percent of the students preferred traveling by car or train? Round to the nearest whole percent.

Travel Method	Frequency
Car	8
Plane	12
Train	3

13. **NUMBER SENSE** The probability of selecting a blue marble from a bag is $\frac{2}{5}$. The probability of selecting a red marble is $\frac{3}{10}$. The number of green marbles in the bag is double the number of yellow marbles. What is the least number of each color marble in the bag?

Main Idea

Perform probability simulations to model real-world situations involving uncertainty.

Vocabulary

simulation

Simulations

Explore You can act out rolling a number cube 50 times by using the random number generator on a graphing calculator. Enter 1 as the lower bound and 6 as the upper bound for 50 trials.

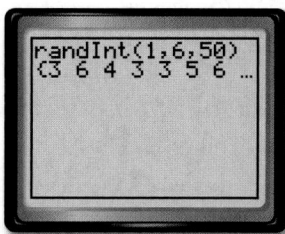

Keystrokes: MATH ◀ 5 1 , 6 , 50) ENTER

A set of 50 numbers ranging from 1 to 6 appears. Use the right arrow key to see the next number in the set. Record all 50 numbers on a separate sheet of paper.

1. Determine the experimental probability of each number generated on the graphing calculator.

2. Compare the experimental probabilities found in Exercise 1 to the theoretical probabilities of rolling an actual number cube.

A **simulation** is an experiment that is designed to act out a given situation. For example, you used a random number generator to simulate rolling a number cube. Simulations often use models to act out an event that would be impractical to perform.

EXAMPLE Describe a Simulation

1 **PRIZES** A cereal company is placing one of eight different trading cards in its boxes of cereal. If each card is equally likely to appear in a box of cereal, describe a model that could be used to simulate the cards you would find in 15 boxes of cereal.

Choose a method that has 8 possible outcomes, such as tossing 3 coins. Let each outcome represent a different card.

Toss 3 coins to simulate the cards that might be in 15 boxes of cereal. Repeat 15 times.

HHH → card 1	TTT → card 5
HHT → card 2	TTH → card 6
HTH → card 3	THT → card 7
HTT → card 4	THH → card 8

CHECK Your Progress

a. **TOYS** A restaurant is giving away 1 of 5 different toys with its children's meals. If the toys are given out randomly, describe a model that could be used to simulate which toys would be given with 6 children's meals.

2 **SCHOOL** Every student who volunteers at the concession stand during basketball games will receive a free school T-shirt. The T-shirts come in 3 different designs. Describe a model that could be used to simulate this situation. Based on your simulation, how many times must a student volunteer in order to get all 3 T-shirts?

Use a spinner divided into 3 equal sections. Assign each section one of the T-shirts. Spin the spinner until you land on each section.

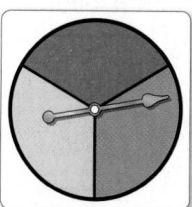 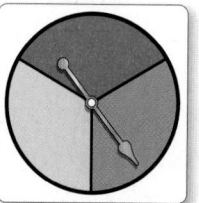 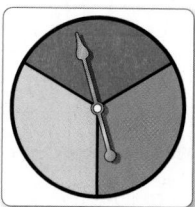

| first spin | second spin | third spin | fourth spin |

Based on this simulation, a student should volunteer 4 times in order to get all 3 T-shirts.

 CHECK Your Progress

b. CLOTHING Rodolfo must wear a dress shirt and a tie to work. Each day he picks one of his 6 ties at random. Describe a model that could be used to simulate this situation. Based on your simulation, how many days must he work in order to wear all of his ties?

Simulations can also be used to model events in which the outcomes are not equally likely.

REAL-WORLD EXAMPLE

3 **WEATHER** The weather forecast states that there is a 60% chance of rain for each of the next two days. Describe a method you could use to find the experimental probability of having rain on both of the next two days.

Place five marbles in a bag. Let 60% or $\frac{3}{5}$ of the marbles represent rain. Let 40% or $\frac{2}{5}$ of the marbles represent no rain. Randomly pick one marble to simulate the first day. Replace the marble and pick again to simulate the second day. Find the probability of rain on both days.

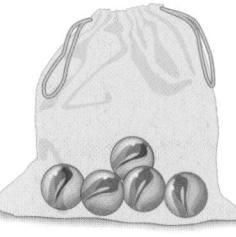

 CHECK Your Progress

c. BASKETBALL During the regular season, Jason made 80% of his free throws. Describe an experiment Jason could use to find the experimental probability of making his next two free throws.

Real-World Link · · · ·
A forecast stating that there is a 60% chance of rain means that there is a 60% chance of rain somewhere in the forecast area, not necessarily the entire area.

Examples 1 and 2 **1** **ICE CREAM** An ice cream store offers a choice of a waffle cone or sugar cone. Each cone type is equally likely to be chosen. Describe a model that could be used to simulate this situation. Based on your simulation, how many customers must order an ice cream cone in order to sell all possible combinations?

Example 3 **2. SALES** An electronics store has determined that of customers who buy a television, 45% buy a wide-screen television. Describe a model that you could use to find the experimental probability that the next three television-buying customers will buy a wide-screen television.

Practice and Problem Solving

= **Step-by-Step Solutions** begin on page R1.
Extra Practice begins on page EP2.

Example 1 **3. TESTING** The questions on a multiple-choice test each have 4 answer choices. Describe a model that you could use to simulate the outcome of guessing the correct answers to a 50-question multiple-choice test.

4. GAMES A game requires drawing balls numbered 0 through 9 for each of four digits to determine the winning number. Describe a model that could be used to simulate the selection of the number.

Example 2 **5. SNACKS** A jar of cookies contains 18 different types of cookies. Each type is equally likely to be chosen. Describe a model that could be used to simulate this situation. Based on your simulation, how many times must a cookie be chosen in order to get each type?

6. PICNICS A cooler contains 15 bottles of lemonade, 12 bottles of water, and 9 bottles of fruit punch. Each type is equally likely to be chosen. Describe a model that could be used to simulate this situation. Based on your simulation, how many times must a drink be chosen in order to get each type?

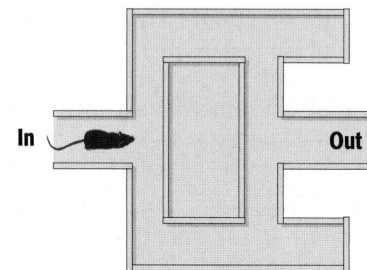

Example 3 **7. CARNIVALS** Players at a carnival game win about 30% of the time. Describe a model that could be used to find the experimental probability that the next four players will win.

8. WEATHER On average, 75% of the days in Henderson are sunny, with little or no cloud cover. Describe a model that you could use to find the experimental probability of sunny days each day for a week.

9 **SCIENCE** Suppose a mouse is placed in the maze at the right. If each decision about direction is made at random, create a simulation to determine the probability that the mouse will find its way out before coming to a dead end or going out the In opening.

10. **OPEN ENDED** Describe a situation that could be represented by a simulation. What objects could be used in this simulation?

11. **CHALLENGE** A simulation uses cards numbered 0 through 9 to generate five 2-digit numbers. A card is selected for the tens digit and not replaced. Then a card for the ones digit is drawn and not replaced. The process is repeated until all the cards are used. If the simulation is performed 10 times, about how many times could you expect a 2-digit number to begin with a 5? Explain.

12.  **WRITE MATH** Explain how using a simulation is related to experimental probability.

Test Practice

13. Marcus placed 8 blue tiles and 12 red tiles in a container. He plans to draw a tile, record its color and replace it in the container before drawing another. If he does this 50 times, how many times should he expect to draw a red tile?

   A. 8   **C.** 20

   B. 12   **D.** 30

14. Claire tosses a coin and rolls a number cube 100 times. How many times should she expect to have the coin show heads and roll a 1 or a 2?

   F. 17

   G. 33

   H. 50

   I. 66

Spiral Review

15. **GIVE-A-WAYS** A local video store has advertised that one out of every four customers will receive a free box of popcorn with their video rental. So far, 15 out of 75 customers have received popcorn. Compare the experimental and theoretical probabilities of receiving popcorn. (Lesson 2C)

16. **MONEY** Dulcinea received $25 for her birthday. She spent a total of $24.05 on a DVD. In how many different ways can she receive her change if she did not receive any pennies nor 50-cent pieces? (Lesson 3A)

17. Refer to the spinner shown. (Lesson 2A)

   a. The spinner is spun and a number cube is rolled. What is the probability of spinning an A and rolling a 2?

   b. The spinner is spun once. What is the probability of spinning an A or a B?

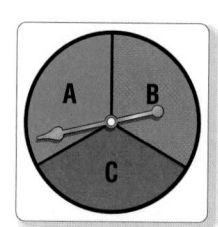

18. **PHONE NUMBERS** How many seven-digit phone numbers can be made using the numbers 0 through 9 if the first number cannot be 0? (Lesson 1A)

Extend

Graphing Technology:
Simulations

Main Idea

Use technology to perform probability simulations.

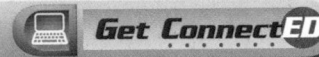

You can use a graphing calculator to simulate real-world probability experiments.

ACTIVITY

1. **TESTING** Ms. Mendez creates a test with 20 multiple-choice questions that each have 4 answer choices: A, B, C, or D. If the correct answer choices are randomized, design a probability simulation using a graphing calculator to determine the probability that 60% or more of the correct answers will be C.

Since $60\% = \frac{60}{100}$ or $\frac{3}{5}$, you want to find the probability that $\frac{3}{5}$ of 20, or 12 or more, of the 20 correct answers are C. You can use a spinner divided into 4 equal-size sections to simulate the randomizing of the correct answer choices on the test.

STEP 1 Access the probability simulator by pressing APPS ALPHA [P] and ▼ until you find Prob Sim. Then press ENTER twice.

STEP 2 Press 4 to select Spin Spinner. Notice that the spinner is already divided into 4 equal-size sections. For this simulation, let 1 represent a correct answer choice of A, 2 represent a correct answer choice of B, 3 represent a correct answer choice of C, and 4 represent a correct answer choice of D.

STEP 3 Press F3 to change the setting from 1 to 20 trials. Then press F5 to confirm this change and F2 to spin. Use the ◄ and ► keys to scroll through the results shown in the bar graph. Record the frequency of the number of times the number 3, answer choice C, was selected.

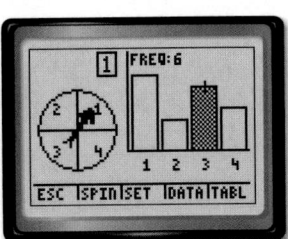

STEP 4 Return to the settings screen by pressing F3 and use the ▼, ◄, and ENTER keys to select **Yes** next to **ClearTbl.** Press F5 to confirm and F2 to spin again. Record the results as before. Repeat until a total of 20 tests have been generated.

Analyze the Results

1. What is the experimental probability of choice C being the correct answer choice for 12 or more questions?

2. Based on your experimental probability, if a student taking Ms. Mendez's test hopes to pass with a score of 60%, would you recommend guessing answer choice C for every problem? Explain.

ACTIVITY

2) **TRAVEL** The plane for a certain flight seats 72 passengers. The airline knows that customers who purchase tickets for this flight are 90% likely to check in. If the airline overbooks, selling 76 tickets, design a probability simulation to determine the probability that at most 72 passengers will check in.

You need a simulation in which one outcome is 90% likely (checking in) and another outcome is 10% likely (not checking in).

STEP 1 Access the probability simulator and select **Toss Coins.**

STEP 2 Press F3 to change the settings from 1 to 76 trials. Press F2 to select the advanced settings. Weight the coin tossed by changing the probability of the coin landing on tails to 0.9. Press F5 twice to confirm these changes.

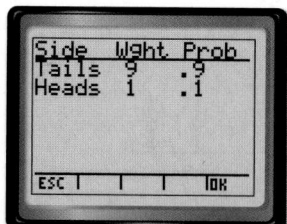

STEP 3 Press F2 to begin the coin tosses. Use the ◄ and ► keys to scroll through the results. Record the frequency of the number of times the coin landed on tails.

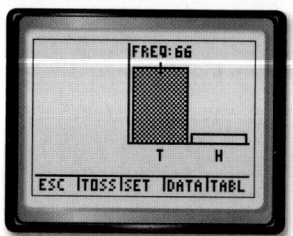

STEP 4 Repeat this simulation 20 more times.

Analyze the Results

3. What is the experimental probability that at most 72 passengers will check in for this flight?

4. **MAKE A CONJECTURE** How many tickets should be sold if the airline wants the probability of filling the flight (at least 72 check-ins) to be 95%? Use the probability simulation to check your conjecture.

Main Idea

Predict the actions of a larger group by using a sample.

 Vocabulary

sample
unbiased sample
simple random sample
stratified random sample
systematic random sample
biased sample
convenience sample
voluntary response sample

 Get ConnectED

Use Sampling to Predict

CELL PHONES A cell phone company manager wants to conduct a survey to determine what kind of ring tones people typically use.

1. Suppose she decides to survey the listeners of a rock radio station. Do you think the results would represent the entire population? Explain.

2. Suppose she decides to survey every 100th person who enters a large mall. Do you think the results would represent the entire population? Explain.

What Kind of Musical Ring Tone Do You Use?
Classical
Rock
Rap/Hip-Hop
Dance
Other

The manager of the cell phone company cannot survey everyone. A smaller group called a **sample** is chosen. A sample should be large enough to provide accurate data.

For valid results, a sample must also be chosen very carefully. An **unbiased sample** is selected so that it is representative of the entire population.

Key Concept   Unbiased Samples

Type	Description	Example
Simple Random Sample	Each item or person in the population is as likely to be chosen as any other.	Fifty phone numbers are randomly selected by a computer.
Stratified Random Sample	The population is divided into similar, nonoverlapping groups. A simple random sample is then selected from each group.	A candidate for class president picks 10 students from each homeroom to survey about school issues.
Systematic Random Sample	The items or people are selected according to a specific time or item interval.	On an assembly line, every 50th car is examined for defects.

In a **biased sample**, one or more parts of the population are favored over others. Two ways to pick a biased sample are listed below.

Key Concept   Biased Samples

Type	Description	Example
Convenience Sample	A convenience sample includes members of a population that are easily accessed.	To represent all the students attending a school, the principal surveys the students in one math class.
Voluntary Response Sample	A voluntary response sample involves only those who want to participate in the sampling.	Students who wish to express their opinion about after-school activities are asked to complete a survey.

EXAMPLES Determine Validity of Conclusions

Determine whether each conclusion is valid. Justify your answer.

1. To determine what kind of movies people like to watch, every tenth person that walks into a video rental store is surveyed. The store carries all kinds of movies. Out of 180 customers surveyed, 62 stated that they prefer action movies. The store manager concludes that about a third of all customers prefer action movies.

   The sample is an unbiased systematic random sample. The conclusion is valid.

2. A television program asks its viewers to visit a Web site to indicate their preference for two presidential candidates. 76% of the viewers who responded preferred candidate A, so the television program announced that most people prefer candidate A.

   The population is restricted to viewers who have Internet access, it is a voluntary response sample, and it is biased. The conclusion is not valid.

CHECK Your Progress

a. To determine what kind of sport junior high school students like to watch, 100 students are randomly selected from each of four junior high schools in a city. Of these, 47% like to watch football. The superintendent concludes that about half of all junior high students like to watch football.

b. To determine what people like to do in their leisure time, people at a local mall are surveyed. Of these, 82% said they like to shop. The mall manager concludes that most people like to shop during their leisure time.

3 **MASCOTS** The Student Council at a new junior high school surveyed 5 students from each of the 10 homerooms to determine what mascot students would prefer. The results of the survey are shown at the right. If there are 375 students at the school, predict how many students prefer a tiger as the school mascot.

Mascot	Number
Tornadoes	15
Tigers	28
Twins	7

The sample is an unbiased stratified random sample since students were randomly selected from each homeroom. Thus, the sample is valid.

$\frac{28}{50}$ or 56% of the students prefer a tiger. So, find 56% of 375.

$0.56 \times 375 = 210$ 56% of 375 = 0.56 × 375

So, about 210 students would prefer a tiger as the school mascot.

 CHECK Your Progress

c. AIRLINES During flight, a pilot determined that 20% of the passengers were traveling for business and 80% were traveling for pleasure. If there are 120 passengers on the next flight, how many can be expected to be traveling for pleasure?

✓ **CHECK Your Understanding**

Examples 1 and 2   **Determine whether each conclusion is valid. Justify your answer.**

1 A researcher randomly surveys ten employees from each floor of a large company to determine the number of employees who carpool to work. Of these, 31% said that they carpool. The researcher concludes that most employees do not carpool.

2. To determine the number of umbrellas the average household in the United States owns, a survey of 100 randomly selected households in Arizona is conducted. Of the households, 24 said that they own 3 or more umbrellas. The researcher concluded that 24% of the households in the United States own 3 or more umbrellas.

Example 3   **3. LUNCH** Jared randomly surveyed some students to determine their lunch habits. The results are shown in the table. If there are 268 students in the school, predict how many bring their lunch from home.

Lunch Habit	Number
Bring lunch from home	19
Buy lunch in the cafeteria	27
Other	4

Practice and Problem Solving

 = **Step-by-Step Solutions** begin on page R1.
Extra Practice begins on page EP2.

Examples 1 and 2 **Determine whether each conclusion is valid. Justify your answer.**

4. The principal of a high school randomly selects 50 students to participate in a school improvement survey. Of these, 38 said that more world language courses should be offered. As a result, the principal decides to offer two additional foreign language classes.

5 To evaluate its service, a restaurant asks its customers to call a number and complete a telephone survey. The majority of those who replied said that they prefer broccoli instead of carrots as the vegetable side dish. As a result, the restaurant decides to offer broccoli instead of carrots.

6. To determine which type of pet is preferred by most customers, the manager of a pet store surveys every 15th customer that enters the store.

7. To determine which school dance theme most students favor, 20 students from each grade level at Lakewood Middle School are surveyed. The results are shown in the table. Based on these results, the student council decides that the dance theme should be *Unforgettable*.

Theme	Number
Starry Night	20
Unforgettable	29
At the Hop	11

Example 3 **8. LAWNS** A researcher randomly surveyed 100 households in a small community to determine the number of households that use a professional lawn service. Of these, 27% of households use a professional lawn service. If there are 786 households in the community, how many can be expected to use a professional lawn service?

9. NEWS After a holiday fireworks show, a TV station surveyed 1,000 people to see if they enjoyed the show. A computer randomly generated phone numbers for the station to call. The results are shown in the table. Are the results valid? If so, of the town's 23,000 people, how many thought the fireworks show was the best one so far?

Fireworks Show	Number
Best one so far	200
Show was good	482
Did not like the show	139
Did not attend the show	179

10. PASTA A grocery store asked every 20th person who entered the store what kind of pasta they preferred. The results are shown in the table. If the store decides to restock their shelves with 450 boxes of pasta, how many boxes of lasagna should they order?

Pasta	Number
Macaroni	38
Spaghetti	56
Rigatoni	12
Lasagna	44

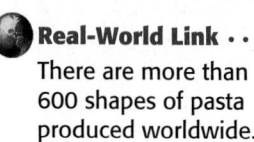

Real-World Link · · · ·
There are more than 600 shapes of pasta produced worldwide.

11. FURNITURE A furniture store manager asks the first 25 customers who enter the store if they prefer tables made of oak, cherry, or mahogany wood. Of these, 17 said they prefer cherry. If the store manager orders 80 tables in the next shipment, how many should be cherry?

12. MAGAZINES A sports magazine asks its readers to fill out a survey about favorite baseball parks. The readers can mail in the survey or go online to complete the survey. Are the results valid? If so, of the 250,000 readers, how many would choose Yankee Stadium as their favorite ballpark?

	Which of the following classic baseball stadiums is your favorite?
149	Comiskey Park
206	Fenway Park
131	Tiger Stadium
320	Wrigley Field
194	Yankee Stadium

13 HOBBIES Pedro wants to conduct a survey about the kinds of hobbies that sixth graders enjoy. Describe a valid sampling method he could use.

14. COMPARE SAMPLES Suppose you were asked to determine the approximate percent of students in your school who are left-handed without surveying every student in the school.

  a. Describe three different samples of the population that you could use to approximate the percent of students who are left-handed.

  b. Would you expect the percent of left-handed students to be the same in each of these three samples? Explain your reasoning.

  c. Describe any additional similarities and differences in your three samples.

  d. You could have surveyed every student in your school to determine the percent of students who are left-handed. Describe a situation in which it makes sense to use a sample to describe aspects of a population instead of using the entire population.

15. MATH IN THE MEDIA Find some data in a newspaper and write a real-world problem in which you would make a prediction based on samples. Identify whether your prediction is valid, based on the sample.

16. GRAPHIC NOVEL Refer to the graphic novel frame below. Brian randomly surveyed some of his classmates to find out what issues were most important to them. If there are 250 students in his class, predict the number of people that care about school lunches.

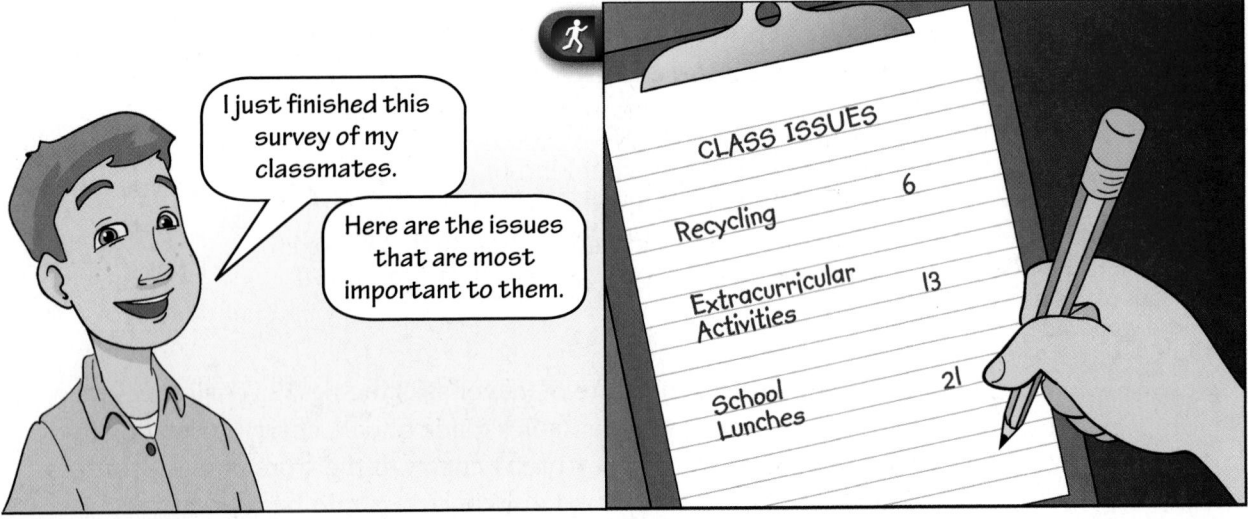

17. CHALLENGE Is it possible to create an unbiased random sample that is also a convenience sample? Explain and cite an example, if possible.

18. **WRITE MATH** Explain why the way in which a survey question is asked might influence the results that are obtained. Cite at least two examples in your explanation.

Test Practice

19. Yolanda wants to conduct a survey to determine what type of salad dressing is preferred by most students at her school. Which of the following methods is the best way for her to choose a random sample of the students at her school?

   A. Select students in her math class.

   B. Select members of the Spanish Club.

   C. Select ten students from each homeroom.

   D. Select members of the girls basketball team.

20. **SHORT RESPONSE** The manager of a zoo wanted to know which animals are most popular among visitors. She surveyed every 10th visitor to the reptile exhibit. Of these, she found that 75% like snakes. One day, there are 860 visitors to the zoo. The manager predicted about 645 of the visitors would visit the snakes. Is her claim valid? Explain.

Spiral Review

21. FUNDRAISING The Service Club is selling raffle tickets to raise money for a project. The probability of winning a prize is $\frac{1}{3}$. Design an experiment to simulate the raffle. (Lesson 3B)

22. SCHOOL Suppose the answer choices A, B, C, D, and F are written on pieces of paper and placed in a bag. Determine whether randomly selecting a piece of paper from the bag is a good way to answer a five-question multiple-choice quiz. Explain your reasoning. (Lesson 3A)

23. SCHOOL In a survey of 120 randomly selected students at Jefferson Middle School, 34% stated that science was their favorite class. Predict how many of the 858 students in the school would choose science as their favorite class. (Lesson 2C)

24. COMPUTERS Rebeka wants to buy a computer package from Computer World. The options available are shown in the table. Find the total number of different computer packages available at the store. (Lesson 1A)

Computer Memory	Printer	Router
2 gigabytes	wireless	Wireless B
3 gigabytes	all-in-one	Super C
4 gigabytes	standard inkjet laser	Wireless X

Collect Data

Main Idea

Solve a problem by collecting, organizing, displaying, and interpreting data.

MUSIC The student council wants to find out what type of music students want at this year's graduation dance. A committee is formed to collect, organize, and present their findings at the next meeting. Design and describe a way for them to accomplish this task.

To accurately draw a conclusion from data received from a sample, first decide on the best method of collecting the data. The data collection method will depend on the type of data you need to collect.

Data Collection Techniques	
survey	• Data are responses given by a sample of the population. • Used to make a general conclusion about the population.
observational study	• Data are recorded after just observing the sample. • Used to compare reactions and draw a conclusion about responses of the population.
experiment	• Data are recorded after changing the sample. • Used to make general conclusions about what will happen during an event.

ACTIVITY   **Collect and Organize Data**

1. Design and conduct a survey to find the type of music the student council should play at the dance.

   STEP 1   Make a data collection plan.
   - Be sure survey questions are worded to avoid bias. Survey questions that favor a particular answer are biased.
   - Devise a method to collect your data using an unbiased sample.

   STEP 2   Collect the data.
   - Make sure your sample size is large enough to represent the student population.
   - Record your results in a frequency table.

   STEP 3   Interpret the results.
   - Choose an appropriate type of data display and scale for your data. Then create an accurate display.
   - Write a paragraph summarizing the results. Make a recommendation to the student council about which types of music should be played at the dance.

Analyze the Results

1. Use your display to describe the distribution of your data.

2. How would you summarize the opinions of those you surveyed? Include only those statements that are clearly supported by the data.

3. Based on your analysis, what course of action would you recommend to the student council?

4. **MAKE A CONJECTURE** What factors might influence the results of your survey?

Practice and Apply

Identify each sample and determine the population from which it was selected. Then classify the type of data collection used.

5. **PETS** A pet supply company wants to know if there is a demand for sweaters for large dogs. They track the size of all the dogs that visit a dog groomer for a week.

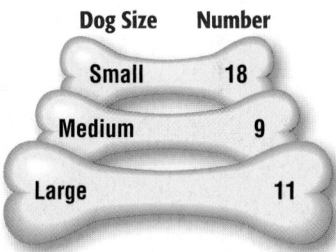

Dog Size	Number
Small	18
Medium	9
Large	11

6. **MUSIC** A radio station wants to determine what kind of music format to offer during work hours. They interview every tenth person going into an electronics store to see what kind of music they listen to.

7. **MANUFACTURING** From a batch of 10,000 lightbulbs produced, a manufacturer tests every 200th lightbulb for defects.

8. **POLITICS** To see where voters stand on a city issue, campaign workers send out questionnaires to 1,500 registered voters in a community.

9. **MULTIPLE REPRESENTATIONS** Design and conduct your own data collection project.

   a. **WRITING** Write a paragraph outlining your data collection plan. Include your topic and a question that will be answered through either observation or a survey. The topic should be meaningful to you. Describe the method you will use to gather the data. Explain why you chose that method.

   b. **ANALYTICAL** Devise a method to collect your data using an unbiased sample. Explain why you chose your sample.

   c. **CONCRETE** Conduct your observation or survey.

   d. **TABULAR** Record your results in a frequency table.

   e. **GRAPHICAL** Display your results using an appropriate graph (line, circle, histogram, and so on.).

   f. **WRITING** Write a paragraph summarizing your results. Include a description of the sample group you used, your data collection method, and any obstacles you may have encountered. Present your findings to the class.

Problem Solving in Environmental Science

In Way Too Deep!

Are you interested in the science behind natural disasters such as earthquakes, floods, landslides, and volcanic eruptions? If so, you might want to explore a career in geology. A geologist is a scientist who studies Earth, its materials, and its physical and chemical processes. Geologists use statistics, data analysis, and their knowledge of Earth's physical makeup and history to determine the probability of geologic hazards occurring.

21st Century Careers

Are you interested in a career as a geologist? Take some of the following courses in high school.

- Algebra
- Chemistry
- Computer Science
- Physics
- Statistics

 Get ConnectED

Peak Heights of the Black River, Kingstree, South Carolina

Rank of Peak	Year	Annual Maximum Peak (feet)	Recurrence Interval, R (number of years)
1	1973	19.8	116.0
2	1928	18.0	58.0
3	1945	16.1	38.7
4	1916	15.5	29.0
5	1994	15.4	23.2
6	1971	15.2	19.3
7	1925	15.2	16.6
8	1993	14.9	14.5
9	1964	14.7	12.9
10	1924	14.6	11.6
11	1893	14.5	10.6
12	1983	14.1	9.7

Real-World Math

Use the information shown below and in the table to solve each problem. Round to the nearest tenth if necessary.

Recurrence interval describes how often a river is expected to reach a certain peak. The highest recorded peak, 19.8 feet, has a recurrence interval of 116.0. This means that the river is expected to reach 19.8 feet or higher once every 116 years.

1. The probability that a river will reach a certain peak in any given year equals 1 ÷ recurrence interval, or $\frac{1}{R}$. Find the probability of the Black River reaching each height.
   a. maximum recorded peak: 19.8 feet
   b. moderate flood level: 14.1 feet

2. The Black River has major flooding when its height is above 16.0 feet.
   a. The table shows the top twelve of 115 years of data. What fraction of the 115 years had major flooding?
   b. Predict how many years in the next 80 that the Black River will have major flooding.

3. A meteorologist reports that in any given year, there is a 58% chance that the Black River will reach a height of 18 feet. Is this a valid conclusion? Explain.

Chapter Study Guide and Review

FOLDABLES
Study Organizer

Be sure the following Key Concepts are noted in your Foldable.

Tree Diagrams | Fundamental Counting Principle | Probability | Independent Events | Dependent Events | Experimental Probability | Theoretical Probability | Sampling

Key Concepts

Outcomes (Lesson 1)
- If event *M* can occur in *m* ways and is followed by event *N* that can occur in *n* ways, then the event *M* followed by the event *N* can occur in $m \cdot n$ ways.

Probability (Lesson 2)
- The probability of two independent events can be found by multiplying the probability of the first event by the probability of the second event.

$$P(A \text{ and } B) = P(A) \cdot P(B)$$

- If two events, *A* and *B*, are dependent, then the probability of both events occurring is the product of the probability of *A* and the probability of *B* after *A* occurs.

$$P(A \text{ and } B) = P(A) \cdot P(B \text{ following } A)$$

Data Collection (Lesson 3)
- An unbiased sample is representative of an entire population.
- A biased sample favors one or more parts of a population over others.

Key Vocabulary

biased sample

compound events

convenience sample

dependent events

event

experimental probability

Fundamental Counting Principle

independent events

outcome

population

probability

random

relative frequency

sample

sample space

simple random sample

simulation

stratified random sample

systematic random sample

theoretical probability

tree diagram

unbiased sample

voluntary response sample

Vocabulary Check

Choose the correct term to complete each sentence.

1. A list of all possible outcomes is called the (sample space, event).

2. The (population, probability) of an event is the ratio of the number of ways the event can occur to the total number of outcomes.

3. A (combination, compound event) consists of two or more simple events.

4. For (independent, dependent) events, the outcome of one does not affect the other.

5. (Theoretical, Experimental) probability is based on known characteristics or facts.

6. A (simple random sample, convenience sample) is a biased sample.

Multi-Part Lesson Review

Lesson 1 **Outcomes**

Count Outcomes (Lesson 1A)

A spinner with three equal sections labeled A, B, and C is spun and a number cube is rolled.

7. Draw a tree diagram to show the possible outcomes.

8. Find the probability of spinning a B and rolling a 3.

9. Find the probability of spinning a vowel and rolling an even number.

10. **BIKES** A bicycle shop has two styles of boys' bikes. They each come in one of three colors and one of two sizes. How many different bicycles are available?

EXAMPLE 1 A bakery sells 5 different flavors of cake in 3 different sizes with 1, 2, or 3 layers. How many choices does a customer have?

number of flavors	×	number of sizes	×	number of layers	=	total number of cakes
5	×	3	×	3	=	45

The customer has 45 choices.

Permutations (Lesson 1C)

Find each value.

11. $P(6, 1)$ **12.** $P(4, 4)$

13. $P(5, 3)$ **14.** $P(7, 2)$

15. $P(10, 3)$ **16.** $P(4, 1)$

17. **NUMBER THEORY** How many 3-digit whole numbers can you write using the digits 1, 2, 3, 4, 5, and 6 if no digit can be used twice?

EXAMPLE 2 Find $P(10, 4)$.

$P(10, 4)$ represents the number of permutations of 10 things taken 4 at a time.

$$P(10, 4) = 10 \cdot 9 \cdot 8 \cdot 7$$
$$= 5{,}040$$

Combinations (Lesson 1D)

Find each value.

18. $C(5, 5)$ **19.** $C(4, 3)$

20. $C(12, 2)$ **21.** $C(9, 5)$

22. $C(3, 1)$ **23.** $C(7, 2)$

24. **PETS** How many different pairs of puppies can be selected from a litter of 8 puppies?

EXAMPLE 3 Find $C(4, 2)$.

$C(4, 2)$ represents the number of combinations of 4 things taken 2 at a time.

$$C(4, 2) = \frac{4 \cdot 3}{2 \cdot 1} \qquad \text{4 things taken 2 at a time}$$

$$= \frac{\overset{2}{\cancel{4}} \cdot 3}{\underset{1}{\cancel{2}} \cdot 1} \text{ or } 6 \qquad \text{Divide out common factors.}$$

Chapter Study Guide and Review

Lesson 2 Probability

Compound Events (Lesson 2A)

A basket of candy contains 2 grape, 3 orange, and 5 cherry candies. The candy is not replaced once selected. Find each probability.

25. P(two orange)

26. P(grape then cherry)

27. P(orange then grape)

28. **TIES** Mr. Nolasco has 4 black ties, 3 gray ties, 2 maroon ties, and 1 brown tie. If he selects two ties without looking, what is the probability that he will pick two black ties?

EXAMPLE 4 A coin is tossed, and the spinner shown is spun. Find the probability of tossing tails and spinning an odd number.

$P(\text{tails}) = \frac{1}{2}$

$P(\text{even number}) = \frac{1}{2}$

$P(\text{tails and even number}) = \frac{1}{2} \cdot \frac{1}{2} \text{ or } \frac{1}{4}$

Experimental and Theoretical Probability (Lesson 2C)

29. **SPELLING** On a spelling test, Angie misspells 2 out of the first 10 words.

   a. What is the probability that she will misspell the next spelling word?

   b. If the spelling test has 25 words on it, how many words would you expect Angie to misspell?

30. A group of three coins are each tossed 20 times. The results are shown in the table.

Outcome	Frequency
0 heads, 3 tails	2
1 head, 2 tails	8
2 heads, 1 tail	6
3 heads, 0 tails	4

   a. What is the experimental probability that there will be one head and two tails?

   b. What is the experimental probability that there will be three heads and zero tails?

EXAMPLE 5 To move ahead in a game, each player spins a spinner with four sections of equal size. The results from one game are shown in the table.

Spin	Frequency
red	15
blue	7
purple	8
yellow	10

What is the experimental probability of a spin landing on purple?

Since purple happened 8 out of 40 tries, the experimental probability is $\frac{8}{40}$ or 20%.

EXAMPLE 6 If the spinner in Example 5 was spun 15 times, how many times would you expect the spinner to land on purple?

The experimental probability of the spinner landing on purple is 20%. Twenty percent of 15 is 3, so the spinner should land on purple 3 times out of 15.

694 Probability and Combinations

Multi-Part Lesson Review

Data Collection

PSI: Act It Out (Lesson 3A)

Solve. Use the *act it out* strategy.

31. **READING** In English class, each student must select 4 short stories from a list of 5 short stories to read. How many different combinations of short stories could a student read?

32. **CARPENTRY** Jaime has $14\frac{1}{4}$ feet of lumber. She uses $2\frac{7}{8}$ feet for a bookshelf. Does Jaime have enough lumber for four more identical shelves? Explain.

EXAMPLE 7 The Spirit Club is making a banner using three sheets of paper. How many different banners consisting of vertical stripes can they make using their school colors of black, orange, and white?

Use three index cards labeled black, orange, and white to model the different banners.

There are six different combinations they can make.

Simulations (Lesson 3B)

33. **PETS** A pet store gives dog biscuits to dogs at the checkout. The biscuits are given out randomly and come in green, brown, red, and yellow. Describe a model that could be used to simulate the color of biscuit the next 10 dogs would receive.

EXAMPLE 8 On average, one out of every 6 people buying a ticket for a movie is given a free pass. Describe a model to simulate whether each of the next 2 people buying tickets receives a pass.

Use a number cube where rolling a 1 represents receiving a pass and rolling a 2, 3, 4, 5, or 6 represents *not* receiving a pass.

Use Sampling to Predict (Lesson 3D)

34. **LUNCH** The school cafeteria cashier surveyed every 10th student in line to determine the students' favorite type of pizza. Of the students surveyed, 46% preferred cheese pizza. If there are 375 students in the school, predict how many students prefer cheese pizza.

EXAMPLE 9 A music store manager asks her customers to complete an online survey. Based on the results shown, the music store manager orders more country music. Is this conclusion valid?

Type of Music	Number
Country	34
Hip-Hop	26
Rock	15

This conclusion is not valid. This is a biased voluntary response sample. Not all customers participated in the survey.

Practice Chapter Test

1. **PICTURES** Students posing for their school pictures have the following options. How many different pictures can be taken?

Choices for School Pictures
5 different backgrounds
3 different poses
2 different treatments

2. **ACTIVITIES** Ms. Hawthorne randomly selects 2 students from 6 volunteers to be on the school activities committee. If Roberto and Joel volunteer, what is the probability that they will both be selected?

3. **MULTIPLE CHOICE** Mr. Delgadillo wants to know if the eighth-grade students want to take a field trip to the art museum. How should he conduct a valid survey?

   A. Ask students in the art club.

   B. Ask the parents of the students.

   C. Ask every tenth eighth-grader who enters the school.

   D. Make an announcement and ask students to come and tell him.

Two coins are tossed 20 times. No heads were tossed 4 times, one head was tossed 9 times, and 2 heads were tossed 7 times.

4. What is the experimental probability of two heads?

5. What is the experimental probability of one head?

6. Draw a tree diagram to show the outcomes of tossing two coins.

7. Compare the experimental probability with the theoretical probability of getting two heads when two coins are tossed. Explain any differences.

A jar contains 4 blue, 7 red, 6 yellow, 8 green, and 3 white tiles. Once a tile is selected, it is not replaced. Find each probability.

8. $P(2 \text{ blue})$

9. $P(\text{white then green})$

10. $P(\text{two tiles that are neither yellow nor red})$

11. **BASKETBALL** Coach Corwin has 15 players on his basketball team. He wants to try different groups of players to see how they work together. In how many ways can he pick a team of 5?

12. **VOLUNTEERING** Student Council surveyed four homerooms to find out how many hours students volunteer each year. The results are shown in the table. If there are 864 students at the school, how many can be expected to volunteer 21–40 hours?

Number of Hours	Number of Students
0–10	38
11–20	26
21–40	10
40 or more	6

13. **SOFTBALL** Miranda gets a hit 25% of the times she is at bat. Describe a model to find the experimental probability that she will get a hit at each of her next three at-bats.

14. **THINK SOLVE EXPLAIN** **EXTENDED RESPONSE** To determine the favorite flavor of ice cream, a random survey is administered at an ice cream shop. Of those surveyed, 65% responded that vanilla is their favorite flavor. It is concluded that vanilla is the favorite flavor of the shop's customers.

   Part A Is this conclusion valid? Explain.

   Part B If the conclusion is valid, how many out of 2,100 people would you expect to name vanilla as their favorite ice cream?

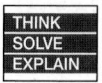 **Short Response: Make an Organized List**

Sometimes making an organized list can help you solve problems. A tree diagram is an organized list that shows all of the possible outcomes of an event.

TEST EXAMPLE

A store has two different types of backpacks: ones with wheels and ones without. Each type of backpack also comes in 5 different colors: red, blue, brown, black, and green. How many types of backpacks does the store sell? Show your work or explain in words how you found your answer.

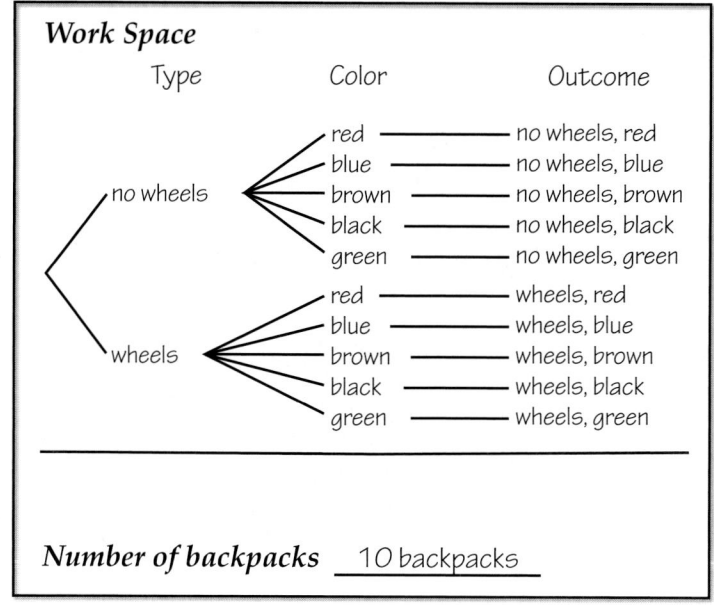

Number of backpacks ___10 backpacks___

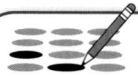 **Work on It**

A dime, a penny, a nickel, and a quarter are tossed. How many different outcomes of heads and tails are there?

Number of outcomes _____

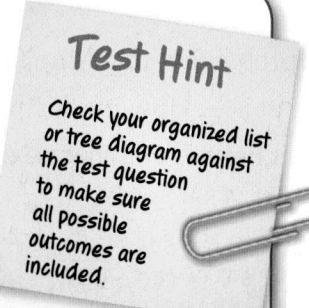

Test Hint

Check your organized list or tree diagram against the test question to make sure all possible outcomes are included.

Read each question. Then fill in the correct answer on the answer document provided by your teacher or on a sheet of paper.

1. The table below shows all of the possible outcomes of a 3-panel light switch being turned on or off.

1ˢᵗ switch	2ⁿᵈ switch	3ʳᵈ switch
ON	ON	ON
ON	ON	OFF
ON	OFF	ON
ON	OFF	OFF
OFF	ON	ON
OFF	ON	OFF
OFF	OFF	ON
OFF	OFF	OFF

Which of the following statements must be true if an outcome is chosen at random?

A. The probability that all of the switches will be on is the same as the probability that all of the switches will be off.

B. The probability that one light switch is on is higher than the probability that two light switches are on.

C. The probability that exactly two switches have the same outcome is $\frac{1}{2}$.

D. The probability of having at least one light switch on is higher than the probability of having at least one light switch off.

2. The probability that Mirajanee gets a hit in softball is $\frac{3}{5}$. How many hits would you expect her to get in her next 60 at bats?

F. 50

G. 36

H. 30

I. 24

3. ✎ **GRIDDED RESPONSE** During the annual food drive, 60% of the students at Blackfoot Middle School donated at least one canned food item. If there are 660 students at Blackfoot Middle School, how many students did NOT participate in the food drive?

4. Of the 32 students surveyed in J.T.'s homeroom, 14 recycle at home. How many students would you expect to recycle at home if a total of 880 students were surveyed?

A. 495   **C.** 281

B. 385   **D.** 123

5. What is the value of x if $-5x - 4 = -34$?

F. -7 **H.** -6

G. 6   **I.** 7

6. A circle with a radius of 4 units has its center at $(1, -2)$ on a coordinate plane. If the circle is translated 5 units up and 4 units left, what will be the coordinates of the new center?

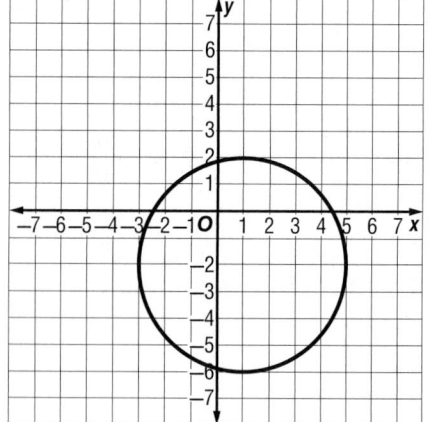

A. $(-5, 2)$ **C.** $(-3, 3)$

B. $(-4, 2)$ **D.** $(5, 3)$

7. A drawer contained two blue, three black, and four white socks. Michael removed one blue sock from the drawer and did NOT put the sock back in the drawer. He then randomly removed another sock from the drawer. What is the probability that the second sock Michael removed was blue?

    F. $\dfrac{1}{18}$ **H.** $\dfrac{1}{8}$

    G. $\dfrac{1}{9}$ **I.** $\dfrac{1}{4}$

8. **THINK SOLVE EXPLAIN** **SHORT RESPONSE** Funtime Party Rentals charges $75 to rent a party room, plus $12.50 per person for food.

    a. Write an equation in slope-intercept form to show the total cost y of renting a party room with x guests.

    b. Suppose Carly rented a party room and her total cost was $400. How many guests did she have at her party?

9. If three coins are tossed, what is the probability that they all show tails?

    A. 6.25%

    B. 12.5%

    C. 25%

    D. 50%

10. **GRIDDED RESPONSE** Dannie can make 3 bracelets in 55 minutes. At this rate, how many hours will it take her to make 18 bracelets?

11. Suppose you know the side lengths of each figure below. Which one would contain enough information to let you find the length of diagonal d?

    F. 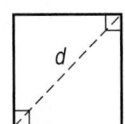   **H.**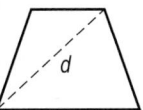

    G. 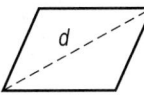   **I.**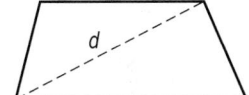

12. What is the solution of the inequality $3x + 10 \geq 5x - 16$?

    A. $x \leq -13$

    B. $x \geq -13$

    C. $x \leq 13$

    D. $x \geq 13$

13. **THINK SOLVE EXPLAIN** **EXTENDED RESPONSE** In a cookie jar, there are 5 chocolate chip, 5 peanut butter, and 5 oatmeal cookies. Two cookies are pulled out of the jar.

    a. What is the probability of choosing 2 chocolate chip cookies?

    b. From the cookies left, what is the probability of choosing an oatmeal cookie next?

NEED EXTRA HELP?													
If You Missed Question...	1	2	3	4	5	6	7	8	9	10	11	12	13
Go to Chapter-Lesson...	11-1A	11-2C	1-2C	11-3D	3-2B	8-3A	11-2A	6-2A	11-2A	1-2C	8-2B	4-2D	11-2A

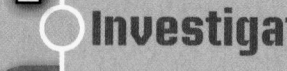
connectED.mcgraw-hill.com

Investigate

 Animations

 Vocabulary

 Multilingual eGlossary

Learn

 Personal Tutor

 Virtual Manipulatives

Graphing Calculator

 Audio

 Foldables

Practice

 Self-Check Practice

Worksheets

Assessment

Area and Volume

The ☆BIG Idea

How does knowing the areas of polygons help to find the surface area and volume of three-dimensional figures?

 FOLDABLES Study Organizer

Make this Foldable to help you organize your notes.

Review Vocabulary

perimeter   perímetro the distance around any closed geometric figure

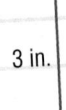

5 in.

3 in.

4 in.

$P = 3 + 4 + 5$ or 12 inches

Key Vocabulary

English	Español
circumference	circunferencia
pi	pi
total surface area	área de superficie total
volume	volumen

When Will I Use This?

Your Turn!

You will solve this problem in the chapter.

Are You Ready for the Chapter?

You have two options for checking prerequisite skills for this chapter.

Text Option Take the Quick Check below. Refer to the Quick Review for help.

QUICK Check

Find the area of each figure.

1.
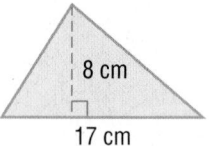
8 cm

17 cm

2.
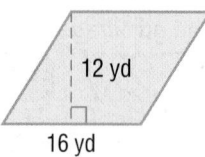
12 yd

16 yd

3.
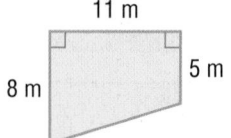
11 m

8 m

5 m

4.
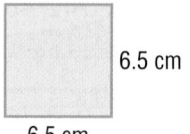
6.5 cm

6.5 cm

5.
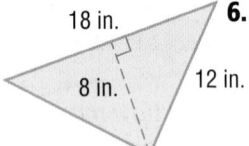
18 in.

8 in.

12 in.

6.
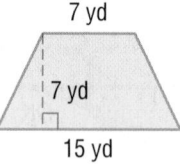
7 yd

7 yd

15 yd

7. GARDENS Marisol is making a rectangular flower bed along the side of her house. The flower bed will be 2.5 feet wide and cover an area of 55 square feet. How long is the flower bed?

Find the value of each expression. Use 3.14 for π. Round to the nearest tenth.

8. $\pi \cdot 15$

9. $2 \cdot \pi \cdot 3.2$

10. $\pi \cdot 7^2$

11. $\pi \cdot (19 \div 2)^2$

12. PIZZA The distance, in inches, around a circular pizza with diameter 14 inches is given by the expression $\pi \cdot 14$. Evaluate this expression. Round to the nearest tenth.

QUICK Review

EXAMPLE 1

Find the area of the triangle.

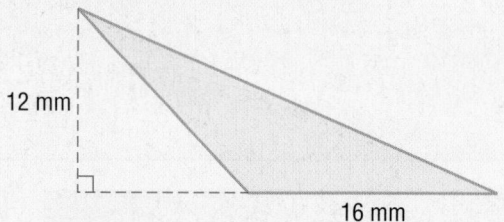
12 mm

16 mm

$A = \frac{1}{2}bh$ Formula for area of a triangle.

$A = \frac{1}{2} \cdot 16 \cdot 12$ Replace b with 16 and h with 12.

$A = 96$ Simplify.

The area is 96 square millimeters.

EXAMPLE 2

Evaluate $\pi \cdot 16^2$. Use 3.14 for π. Round to the nearest tenth.

$\pi \cdot 16^2 \approx 3.14 \cdot 256$ Evaluate 16^2.

        ≈ 803.8 Multiply.

Online Option Take the Online Readiness Quiz.

 **Perimeter and Area**

CRAFTS Kira is making a scrapbook page of her camp memories. She has 40 inches of material to use for the border around the outside. She wants the page to have the largest possible area with only whole number dimensions. Find the dimensions of the page.

ACTIVITY Use a Spreadsheet

If ℓ represents the length of the page, then $20 - \ell$ will represent the width of the page. These values are listed in column B. The areas are listed in column C.

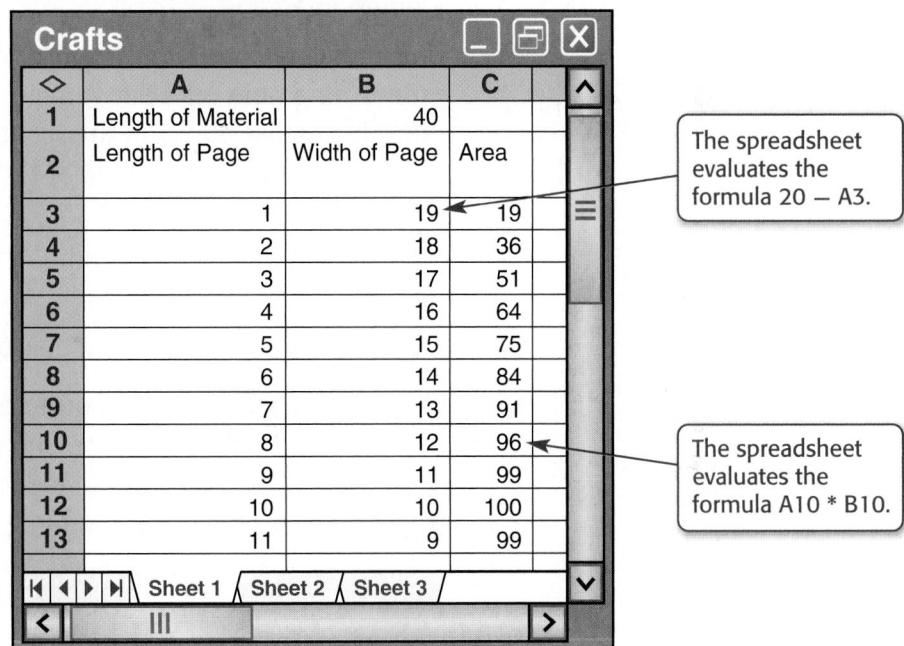

	A	B	C
1	Length of Material	40	
2	Length of Page	Width of Page	Area
3	1	19	19
4	2	18	36
5	3	17	51
6	4	16	64
7	5	15	75
8	6	14	84
9	7	13	91
10	8	12	96
11	9	11	99
12	10	10	100
13	11	9	99

The spreadsheet evaluates the formula 20 − A3.

The spreadsheet evaluates the formula A10 * B10.

The largest area is 100 square inches. This is when the length and width of the page are 10 inches.

Analyze the Results

1. What happens to the area of the page when the length is greater than 10 inches?

2. Suppose Kira has 50 inches of material. Which formula would need to change?

3. Use a spreadsheet to find the largest page area she could have with 30, 50, or 60 inches of material.

Main Idea

Find the circumference and area of circles.

Vocabulary

circle
center
radius
chord
diameter
circumference
pi
secant
tangent

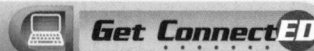

Circumference and Area of Circles

Explore

Step 1   Use a ruler to measure the distance across the center of a circular object like a CD. Record the length in a table.

Object	Distance Across (cm)	Distance Around (cm)

Step 2   Make a small mark at the edge of the circular object. Place a measuring tape on a flat surface. Place the mark you made on the circular object at the beginning of the measuring tape. Roll the object along the tape for one revolution, until you reach the mark again.

Step 3   Record the length in the table. This is the distance around the object.

Step 4   Repeat this activity with circular objects of various sizes.

1. Find the ratio of the distance around each object to the distance across each object. What do you notice?

2. MAKE A CONJECTURE Write a rule describing how you would find the distance around any circle if you know the distance across.

A **circle** is the set of all points in a plane that are the same distance from a given point in the plane.

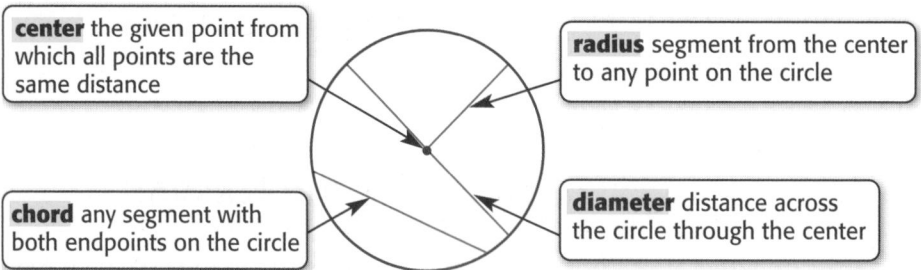

center the given point from which all points are the same distance

radius segment from the center to any point on the circle

chord any segment with both endpoints on the circle

diameter distance across the circle through the center

The distance around the circle is called the **circumference**. The ratio of the circumference of a circle to its diameter is always 3.1415926... .
It is represented by the Greek letter **π (pi)**. The numbers 3.14 and $\frac{22}{7}$ are often used as approximations for π. So, $\frac{C}{d} = \pi$. This can also be written as $C = \pi d$ or $C \approx 3.14d$.

Key Concept — Circumference of a Circle

Words The circumference C of a circle is equal to its diameter d times π, or 2 times its radius r times π.

Model

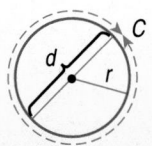

Symbols $C = \pi d$ or $C = 2\pi r$

EXAMPLES Find the Circumferences of Circles

Find the circumference of each circle. Round to the nearest tenth.

 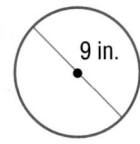

$C = \pi d$ Circumference of a circle

$C = \pi \cdot 9$ Replace d with 9.

$C = 9\pi$ This is the *exact* circumference.

Use a calculator to find 9π.   9 ☒ 2nd [π] ENTER   28.27433388

The circumference is about 28.3 inches.

Study Tip

Calculating with π When evaluating expressions involving π, use the π key on a calculator to obtain the most accurate result. However, using 3.14 for π will result in a close approximation.

 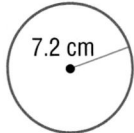

$C = 2\pi r$ Circumference of a circle

$C = 2 \cdot \pi \cdot 7.2$ Replace r with 7.2.

$C \approx 45.2$ Use a calculator.

The circumference is about 45.2 centimeters.

CHECK Your Progress

a. diameter: 6 ft   **b.** radius: 11.7 m   **c.** diameter: 7.5 in.

A circle can be decomposed into congruent wedge-like pieces that can be rearranged to form a figure that resembles a parallelogram.

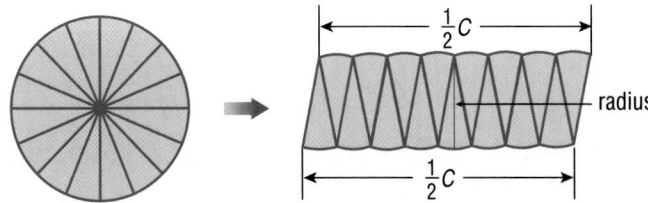

Since the circle has an area that is relatively close to the area of the parallelogram-shaped figure, you can use the formula for the area of a parallelogram to find the formula for the area of a circle.

$A = bh$ Area of a parallelogram

$A = \left(\dfrac{1}{2} \cdot C\right)r$ The base of the parallelogram is one half the circumference and the height is the radius.

$A = \left(\dfrac{1}{2} \cdot 2\pi r\right)r$ Replace C with $2\pi r$.

$A = \pi \cdot r \cdot r$ or πr^2 Simplify.

Key Concept   **Area of a Circle**

Words   The area A of a circle is equal to   **Model**
π times the square of the radius r.

Symbols $A = \pi r^2$

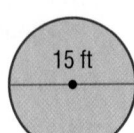 **EXAMPLE** **Find the Areas of Circles**

③ **Find the area of the circle. Round to the nearest tenth.**

15 ft

$A = \pi r^2$	Area of a circle
$A = \pi (7.5)^2$	Replace r with half of 15 or 7.5.
$A = \pi \cdot 56.25$	Evaluate 7.5^2. This is the *exact* area.
$A \approx 176.7$	Use a calculator.

The area is about 176.7 square feet.

 CHECK Your Progress

Find the area of each circle. Round to the nearest tenth.

d. The radius is 11 inches.   **e.** The diameter is 5 meters.

REAL-WORLD EXAMPLE

④ **STATE PARKS** Suppose you walk around the edge of the circular Point State Park fountain and estimate its circumference to be 470 feet. Based on your estimate, what is the approximate diameter of the fountain?

$C = \pi d$	Circumference of a circle
$470 = \pi d$	Replace C with 470.
$\dfrac{470}{\pi} = d$	Divide each side by π.
$149.6 \approx d$	Use a calculator.

The diameter of the fountain is about 150 feet.

 CHECK Your Progress

f. **HOME DECOR** A catalog states that a circular area rug covers 19.5 square feet. What is the approximate diameter of the rug?

Study Tip

Estimation To estimate the area of a circle, square the radius and then multiply by 3.

🌐 **Real-World Link** · · · ·
Point State Park in Pittsburgh, Pennsylvania, is located where the Allegheny and Monongahela rivers meet to form the Ohio River.

Examples 1 and 2 **Find the circumference of each circle. Round to the nearest tenth.**

1.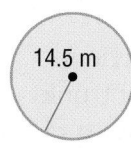
18 cm

2.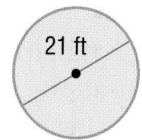
12 yd

3.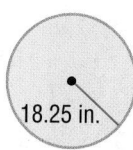
2.5 mi

Example 3 **Find the area of each circle. Round to the nearest tenth.**

4.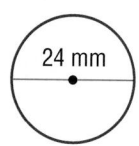
14.5 m

5.
21 ft

6.
18.25 in.

Example 4 **7. BRACELETS** When Cammie finished making a friendship bracelet, it was 7.9 inches long. What was the diameter of the bracelet?

Practice and Problem Solving

⬤ = **Step-by-Step Solutions** begin on page R1.
Extra Practice begins on page EP2.

Examples 1 and 2 **Find the circumference of each circle. Round to the nearest tenth.**

8.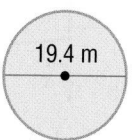
24 mm

9.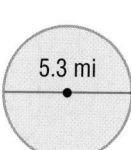
38 mi

10.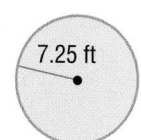
10 in.

11.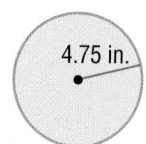
17 km

Example 3 **Find the area of each circle. Round to the nearest tenth.**

12.
19.4 m

13.
5.3 mi

14.
7.25 ft

15.
4.75 in.

Example 4 **16. PETS** Simone purchased a circular exercise pen with a radius of 2.5 feet to keep her new puppy safe. Find the area inside the pen.

17 MEASUREMENT A circular table top has a radius of $2\frac{1}{4}$ feet. A decorative trim is placed along the outside edge of the table. How long is the trim?

18. SAFETY A light in a parking lot illuminates a 175-square-meter circular area. What is the approximate radius of the area covered by the light?

19. BICYCLES The tires on Jarrod's mountain bike have a circumference of 80 inches. What is the approximate diameter of the tires?

Find the exact circumference and area of each circle.

20. The radius is 3.5 centimeters.

21. The diameter is 8.6 kilometers.

22. The diameter is 9 inches.

23. The radius is 0.6 mile.

24. SPORTS Three tennis balls are packaged one on top of the other in a can. Which measure is greater, the can's height or circumference? Explain.

25 BAKING Joaquin is making giant cookies for the school bake sale. They will be sold for $20 for one large cookie or $20 for three smaller cookies. Which offer is the better buy? Explain your reasoning.

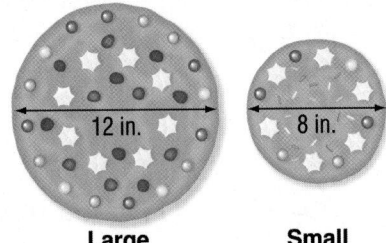

Large   Small

26. TREES During a construction project, barriers are placed around trees. For each inch of trunk diameter, the protection zone should have a radius of $1\frac{1}{2}$ feet. Find the approximate area of this zone for a tree with a trunk circumference of 63 inches.

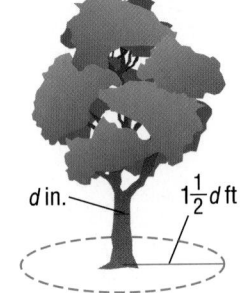

27. MATH IN THE MEDIA Find a photograph of a circle in a newspaper or magazine. Write a real-world problem in which you would determine the circumference of the circle.

d in. $1\frac{1}{2}d$ ft

H.O.T. Problems

28. FIND THE ERROR Carmen is finding the area, to the nearest tenth, of a circle with a diameter of 8 inches. Find her mistake and correct it.

$A = \pi r^2$
$A = \pi \cdot 8^2$
$A = \pi \cdot 64$
$A \approx 201.1 \text{ ft}^2$

29. NUMBER SENSE If the radius of a circle is halved, how will this affect its circumference and its area? What happens to the circumference and area if the radius is doubled or tripled? Explain your reasoning. (*Hint:* Find the circumference and area for each circle and organize the data in a table.)

CHALLENGE Find the area of each shaded region.

30.

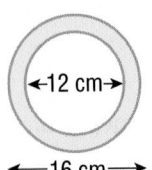

←12 cm→
←16 cm→

31.

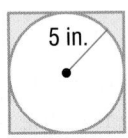

5 in.

32.

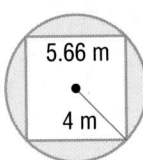

5.66 m
4 m

33. ✏️ **WRITE MATH** Explain how the circumference and area of a circle are related or different.

34. Mr. Rodgers is building a square patio with a circular rug in the middle.

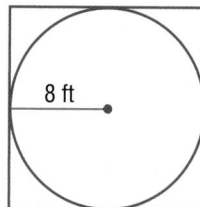

The radius of the rug is 8 feet. What is the perimeter of the patio?

A. 16π ft   C. 32 ft

B. 64 ft   D. 64π ft

35. Berto is painting a sign for a new coffee shop. On the sign, he drew a circle with a radius of 2 feet. He then drew another circle with a radius 1.5 times larger. How many times greater is the area of the second circle than the area of the first?

F. 1.5 times   H. 2.25 times

G. 2 times   I. 2.5 times

36. Using the two circles shown below, what is $\frac{\text{circumference of circle } x}{\text{circumference of circle } y}$?

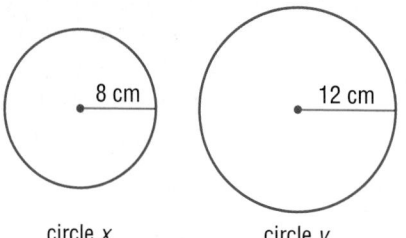

circle x   circle y

A. $\frac{3\pi}{4}$ C. $\frac{2}{3}$

B. $\frac{4\pi}{3}$ D. $\frac{4}{3}$

37. A circle that has a radius of 7 inches has an area of 49π square inches. If the radius is doubled, what is the area of the new circle?

F. 14π square inches

G. 98π square inches

H. 196π square inches

I. 392π square inches

More About Circles

The intersection of a line and a circle can be described in three different ways.

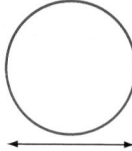

Zero points

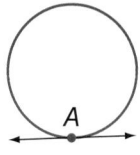

One point

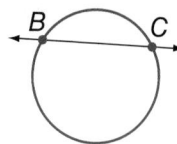

Two points

When a circle and a line intersect at exactly one point, the line is called a **tangent.** The line is *tangent* to the circle at the point of intersection. When a line intersects a circle at exactly two points, the line is called a **secant.**

Use the figure at the right to name each of the following.

38. a secant

39. a tangent

40. a chord

41. two different points of tangency

42. Draw a circle and a secant so that the diameter of the circle is part of the secant.

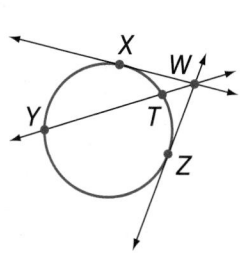

Arcs and Angles

Main Idea

Find measures of arcs and inscribed angles.

 Vocabulary

central angle
arc
minor arc
major arc
semicircle
inscribed angle

Get ConnectED

A **central angle** is an angle that intersects a circle in two points and has its vertex at the center of the circle. It separates the circle into two parts, each of which is an **arc**.

The measure of a central angle is equivalent to the measure of its corresponding arc. You can classify arcs by their measure.

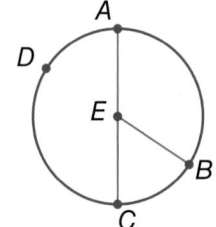

Arc	Measure	Example
minor arc	<180°	$\overset{\frown}{AB}$
major arc	>180°	$\overset{\frown}{ADB}$
semicircle	=180°	$\overset{\frown}{ABC}$

An **inscribed angle** is an angle that has its vertex on the circle, and its sides contain chords of the circle.

ACTIVITY   **Measure of Inscribed Angles**

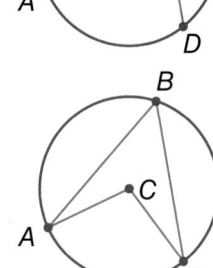

1 **STEP 1** Use a compass to draw circle C.

 STEP 2 Use a straightedge to draw chords BA and BD that do not go through the center of the circle.

 STEP 3 Use a straightedge to draw $\overline{AC}$ and $\overline{CD}$.

 STEP 4 Measure ∠ABD and ∠ACD.

Analyze the Results

Read Math

Arcs and Segments The symbol $\overset{\frown}{AB}$ is read *arc AB*. The symbol $\overline{AB}$ is read *segment AB*.

1. What seems to be the relationship between $m\angle ABD$ and $m\angle ACD$?

2. Repeat Steps 1–4 with several different inscribed angles.

3. **MAKE A PREDICTION** Circle A has a central angle that measures 60° and an inscribed angle that intercepts the same arc. Predict the measure of the inscribed angle.

ACTIVITY Angles Inscribed in a Semicircle

2 **STEP 1** Use a compass to draw a circle with center X and diameter $\overline{YZ}$.

STEP 2 Draw and label any point R on $\widehat{YZ}$. Use a straightedge to draw $\overline{RY}$ and $\overline{RZ}$.

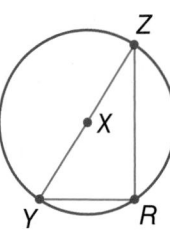

Analyze the Results

4. What shape is formed by $\overline{RY}$, $\overline{RZ}$, and $\overline{YZ}$?

5. Find $m\angle YRZ$. What kind of triangle is triangle YRZ?

6. Draw and label another point T on $\widehat{YZ}$. Draw $\overline{TY}$ and $\overline{TZ}$. Find $m\angle YTZ$.

7. **MAKE A CONJECTURE** What is true about inscribed angles that intercept a semicircle?

8. Find the measures of the missing angles and arcs in the figure at the right.

    a. $\widehat{DB}$ **b.** $\angle 1$ **c.** $\widehat{ECA}$ **d.** $\angle 2$

    e. $\angle ECB$ **f.** $\widehat{BA}$ **g.** $\widehat{DC}$ **h.** $\widehat{CB}$

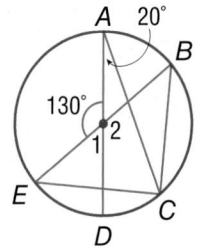

ACTIVITY Chords and Diameters

3 **STEP 1** Use a compass to draw a circle and label the center P. Draw a chord that is not a diameter. Label it $\overline{EF}$.

STEP 2 Construct a line segment through P that is perpendicular to $\overline{EF}$ with endpoints on the circle. Label this as diameter $\overline{GH}$.

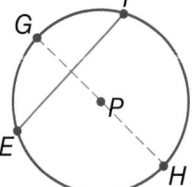

Analyze the Results

9. Compare the lengths of $\widehat{EG}$ and $\widehat{FG}$. Then compare the lengths of $\widehat{EH}$ and $\widehat{FH}$.

10. What is the relationship between diameter $\overline{GH}$ and chord $\overline{EF}$?

11. **MAKE A CONJECTURE** What is the relationship among a diameter, a chord, and its arc if the diameter is perpendicular to the chord?

Problem-Solving Investigation

Main Idea Solve problems by making a model.

P.S.I. TEAM +

Make a Model

ALICIA: I have a portrait that is 10 inches by 13 inches. I want to put it in a frame that is $2\frac{1}{4}$ inches wide on each side.

YOUR MISSION: How much wall space will I need to hang the portrait?

Understand	You know the size of the portrait and the size of the frame. You need to find the area of the framed portrait.
Plan	Make a model to find the area.
Solve	The portrait is 10 inches by 13 inches. The frame is $2\frac{1}{4}$ inches wide on each side. Find the dimensions of the framed portrait. length: $13 + 2\frac{1}{4} + 2\frac{1}{4}$ or $17\frac{1}{2}$ in. width: $10 + 2\frac{1}{4} + 2\frac{1}{4}$ or $14\frac{1}{2}$ in. To find the area, multiply length times width. $17\frac{1}{2} \times 14\frac{1}{2} = 253\frac{3}{4}$ The framed portrait will take up $253\frac{3}{4}$ square inches of wall space.
Check	Estimate. $17\frac{1}{2} \approx 15$ and $14\frac{1}{2} \approx 15$. $15 \times 15 = 225$. Since $225 \approx 253\frac{3}{4}$, the answer is reasonable. ✔

Analyze the Strategy

1. Describe a method you could use to find the area of just the frame border.

2. **WRITE MATH** Describe when you should use the *make a model* problem-solving strategy to solve a problem.

Mixed Problem Solving

 = **Step-by-Step Solutions** begin on page R1.
Extra Practice begins on page EP2.

- Make a model.
- Draw a diagram.
- Guess, check, and revise.
- Choose an operation.

Use the *make a model* strategy to solve
Exercises 3–5.

3. ART Jeffrey is making a model of his
backyard for an art class. His backyard
measures 72 feet by 96 feet. If he uses a
scale of 8 feet = $1\frac{1}{2}$ inches, what are the
dimensions of the backyard on the model?

4. PATTERNS Mrs. Padilla is making a quilt
using the following pattern. How many
squares would be in the 20th figure in this
pattern?

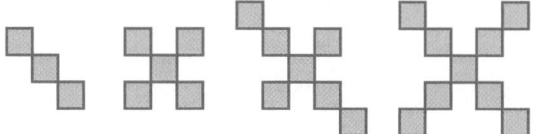

Figure 1   Figure 2   Figure 3   Figure 4

5. POPCORN A popcorn box is made from
rectangular sheets of cardboard measuring
$8\frac{1}{2}$ inches by 11 inches. To make the box,
each corner of the cardboard has a cut-out
square measuring $1\frac{1}{2}$ inches on each side.
Find the volume of the box of popcorn.
(*Hint:* The volume is found by multiplying
length, width, and height.)

Use any strategy to solve Exercises 6–11.

6. VOLLEYBALL A total of 8 players attended
volleyball practice on Monday. How many
different teams of 3 players can be made
from the total number of players that came
to practice?

7. TABLES Members of Student Council are
setting up tables end-to-end for an awards
banquet. How many square tables will they
need to put together for 32 people? Each
table will seat one person on each side.

8. MONEY Ken borrowed $250 from his
parents for a camping trip. He has already
repaid them $82. If he plans to pay them
$14 each week, how many weeks will it
take Ken to repay his parents?

9. POSTERS Jacinda has 3 posters that she
wants to hang on her bedroom wall. Each
poster measures 2 feet wide. She wants the
gaps between each poster and the ends of
the wall to be the same distance. If her wall
measures 18 feet, how wide should each
gap be?

10. TILE The diagram below shows the design
of a tile border around a rectangular
swimming pool that measures 7 feet by
4 feet. Each tile is a square measuring
1 foot on each side.

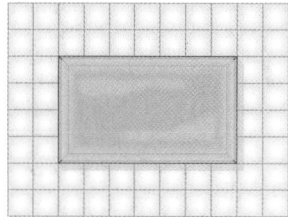

a. What is the area of the pool?

b. What is the area of both the pool and
the tiles?

c. Explain a method you could use to find
the area of just the tiles.

d. Using the model above, how many tiles
are needed if the pool is 18 feet long
and 12 feet wide?

e. How many tiles are needed for a pool
32 feet long and 20 feet wide?

11. LAUNDRY You need two clothespins to
hang one towel on a clothesline. One
clothespin can be used on a corner of one
towel and a corner of the towel next to it.
What is the least number of clothespins
you need to hang 8 towels?

Main Idea

Find the area of composite figures.

Vocabulary
composite figure

Get Connect**ED**

Area of Composite Figures

SPEEDWAY A diagram of the Indianapolis Motor Speedway is shown.

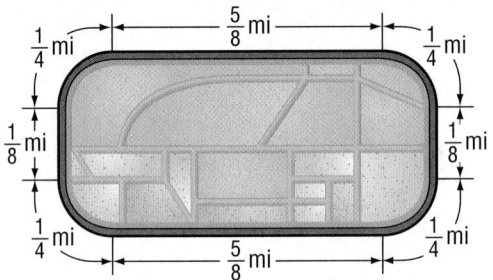

1. Identify some of the polygons that make up the infield of the speedway.

2. How can the polygons be used to find the total area of the infield?

A **composite figure** is made up of two or more shapes.

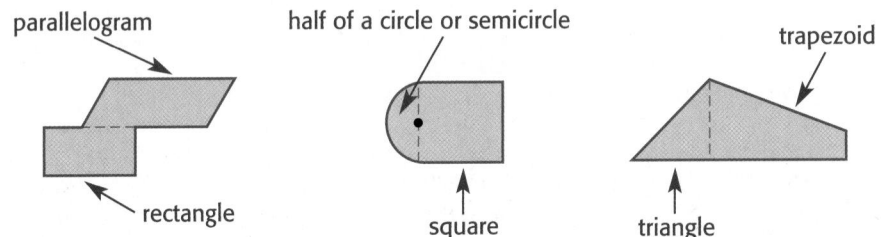

parallelogram

half of a circle or semicircle

trapezoid

rectangle

square

triangle

To find the area of a composite figure, decompose the figure into shapes with areas you know how to find. Then find the sum of these areas.

Key Concept	**Area Formulas**	
Shape	**Words**	**Formula**
Parallelogram	The area A of a parallelogram is the product of any base b and its height h.	$A = bh$
Triangle	The area A of a triangle is half the product of any base b and its height h.	$A = \frac{1}{2}bh$
Trapezoid	The area A of a trapezoid is half the product of the height h and the sum of the bases, b_1 and b_2.	$A = \frac{1}{2}h(b_1 + b_2)$
Circle	The area A of a circle is equal to π times the square of the radius r.	$A = \pi r^2$

EXAMPLE Find the Area of a Composite Figure

Study Tip

Semicircle Since a semicircle is half a circle, its area is $\frac{1}{2}\pi r^2$.

(1) **Find the area of the composite figure.**

The figure can be separated into a semicircle and a triangle.

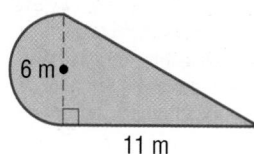

Area of semicircle

$A = \frac{1}{2}\pi r^2$

$A = \frac{1}{2} \cdot \pi \cdot 3^2$

$A \approx 14.1$

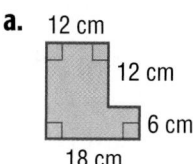

Area of triangle

$A = \frac{1}{2}bh$

$A = \frac{1}{2} \cdot 11 \cdot 6$

$A = 33$

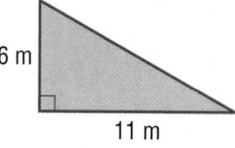

The area of the figure is about $14.1 + 33$ or 47.1 square meters.

CHECK Your Progress

Find the area of each figure. Round to the nearest tenth if necessary.

a. 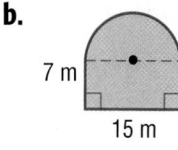 12 cm, 12 cm, 6 cm, 18 cm

b. 7 m, 15 m

c. 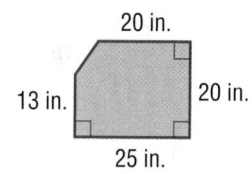 20 in., 13 in., 20 in., 25 in.

REAL-WORLD EXAMPLE

Real-World Link · · ·
There are 336 dimples on a regulation golf ball.

(2) **GOLF** The plan for one hole of a miniature golf course is shown. It is composed of a trapezoid and a parallelogram. How many square feet of turf will be needed for this plan?

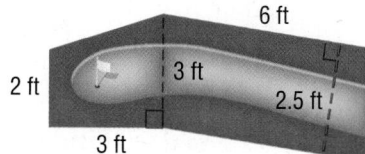
6 ft, 2 ft, 3 ft, 2.5 ft, 3 ft

Area of trapezoid

$A = \frac{1}{2}h(b_1 + b_2)$

$A = \frac{1}{2}(3)(2 + 3)$

$A = 7.5$

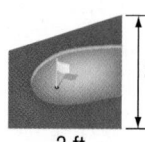
2 ft, 3 ft, 3 ft

Area of parallelogram

$A = bh$

$A = 6 \cdot 2.5$

$A = 15$

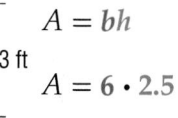
6 ft, 3 ft, 2.5 ft

So, $7.5 + 15$ or 22.5 square feet of turf will be needed.

CHECK Your Progress

d. SHEDS Pedro's father is building a shed. How many square feet of wood are needed to build the back of the shed shown at the right?

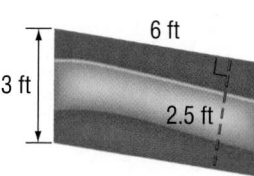

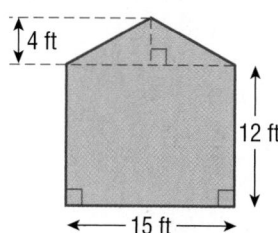
4 ft, 12 ft, 15 ft

 **EXAMPLE** Find the Area of a Shaded Region

Study Tip

Congruent Triangles
Congruent triangles have corresponding sides and angles that are congruent.

③ In the figure at the right, four congruent triangles are cut from a rectangle. Find the area of the shaded region. Round to the nearest tenth if necessary.

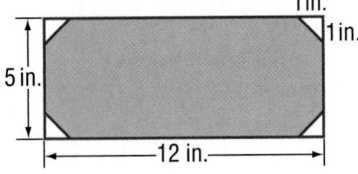

Find the area of the rectangle and subtract the area of the four triangles.

Area of rectangle	**Area of triangles**
$A = \ell w$	$A = 4 \cdot \left(\frac{1}{2}bh\right)$
$A = 12 \cdot 5$ $\ell = 12, w = 5$	$A = 4 \cdot \frac{1}{2} \cdot 1 \cdot 1$ $b = 1, h = 1$
$A = 60$ Simplify.	$A = 2$ Simplify.

The area of the shaded region is $60 - 2$ or 58 square inches.

CHECK Your Progress

e. Two rectangles are cut from a larger rectangle. Find the area of the shaded region. Round to the nearest tenth if necessary.

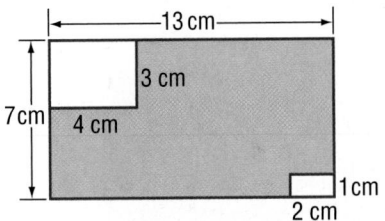

✓ CHECK Your Understanding

Example 1   Find the area of each figure. Round to the nearest tenth if necessary.

1.

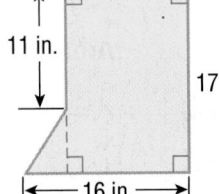

2.

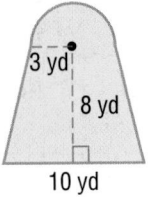

Examples 2 and 3   **③** **WINDOWS** The Lunas installed the window shown below. How many square feet is the window?

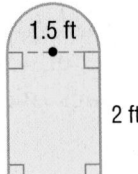

4. A triangle is cut from a rectangle. Find the area of the shaded region.

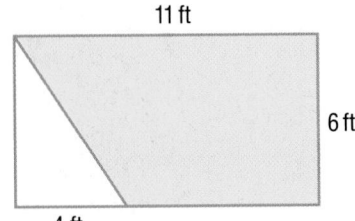

= **Step-by-Step Solutions** begin on page R1.
Extra Practice begins on page EP2.

Example 1 **Find the area of each figure. Round to the nearest tenth if necessary.**

5.
12 cm
4.5 cm
2 cm
5 cm

6.
6 yd   6 yd
16 yd   8 yd
24 yd

7.
15 cm
8 cm

8.
7 m
7 m

9.
6.4 ft   7 ft   3.6 ft
9 ft

10.
10 cm
6 cm   10 cm
20 cm

Example 2

11. CARPENTRY Daniel is constructing a deck like the one shown. What is the area of the deck?

5 ft
3.5 ft
12 ft

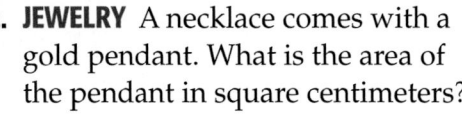

12. JEWELRY A necklace comes with a gold pendant. What is the area of the pendant in square centimeters?

1 cm
2 cm
3 cm
1 cm

Example 3 **Find the area of the shaded region. Round to the nearest tenth if necessary.**

13.
20 m
22 m
25 m
42 m

14.
10 yd
6 yd
9 yd
15 yd

15.
2 cm
2 cm
8 cm
16 cm

16.
5 ft
12 ft
25 ft

17 CARPETING Zoe's mom is carpeting her bedroom and needs to know the amount of floor space. How many square feet of carpeting are needed for the room? If she is also installing baseboards, how many feet of baseboards are needed?

10 ft
6 ft
8 ft
10 ft
12 ft
11 ft

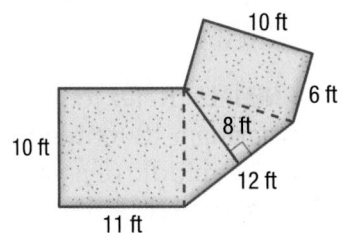

18. **CHALLENGE** The composite figure shown is made from a rectangle and a triangle. The area of the rectangle is 32 square feet. Find the area and perimeter of the entire figure.

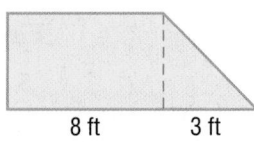

8 ft   3 ft

19. **REASONING** The side length of the square in the figure at the right is x units. Write expressions that represent the perimeter and area of the figure.

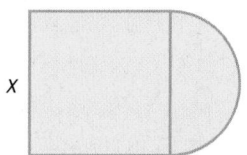

x

20. **CHALLENGE** In the diagram at the right, a 2-foot-wide flower border surrounds the pond. What is the area of the border?

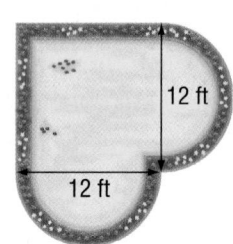

12 ft

12 ft

21. **WRITE MATH** Explain at least two different ways of finding the area of a hexagon. Include a drawing with your answer.

Test Practice

22. What is the total area of the figure shown?

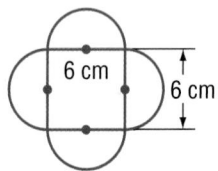

6 cm

6 cm

 A. 92.5 cm^2

 B. 64.3 cm^2

 C. 56.5 cm^2

 D. 36.0 cm^2

23. The Patels' backyard has a rectangular vegetable garden and a triangular pet exercise area.

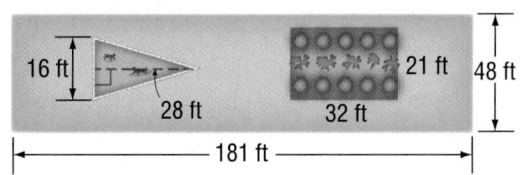

16 ft   28 ft   32 ft   21 ft   48 ft

181 ft

How many square feet of the backyard is NOT in one of these areas?

 F. 8,688 ft^2 H. 4,887 ft^2

 G. 7,792 ft^2 I. 896 ft^2

24. **MODELS** Suppose you had 100 cubes. Use the *make a model* strategy to determine the largest cube you could build with the cubes. (Lesson 1D)

25. **ANIMALS** Refer to the diagram at the right. A dog is wearing a 10-foot-long leash. If the leash is tied to a stake in the middle of a yard, how much space does the dog have to play? Round to the nearest tenth. (Lesson 1B)

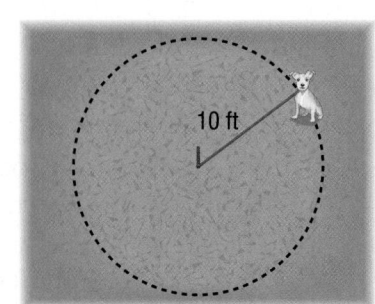

10 ft

Main Idea
Find the volumes of prisms and cylinders.

 Vocabulary
volume
composite solid

 Get ConnectED

 8.G.8

Volume of Prisms and Cylinders

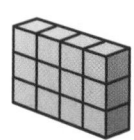

Explore The rectangular prism at the right has a volume of 12 cubic units. Model three other rectangular prisms with a volume of 12 cubic units. Then copy and complete the following table.

Prism	Length (units)	Width (units)	Height (units)	Area of Base (units²)
A	4	1	3	4
B				
C				
D				

1. Describe how the volume V of each prism is related to its length ℓ, width w, and height h.

2. Describe how the area of the base B and the height h of each prism is related to its volume V.

Volume is the measure of the space occupied by a solid. Volume is measured in cubic units.

Key Concept — Volume of a Prism

Words The volume V of a prism is the area of the base B times the height h.

Symbols $V = Bh$

Model

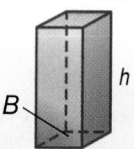

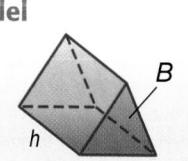

EXAMPLES Find the Volumes of Prisms

1 Find the volume of the rectangular prism.

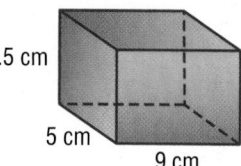

$V = Bh$ Volume of a prism

$V = (\ell \cdot w)h$ The base is a rectangle, so $B = \ell \cdot w$.

$V = (9 \cdot 5)6.5$ $\ell = 9, w = 5, h = 6.5$

$V = 292.5$ Simplify.

The volume is 292.5 cubic centimeters.

Study Tip

Common Error Remember that the bases of a triangular prism are triangles. In Example 2, these bases are not on the top and bottom of the figure, but on its sides.

② **Find the volume of the triangular prism.**

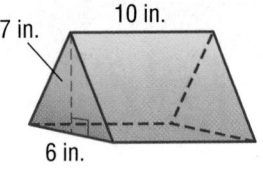

$V = Bh$ Volume of a prism

$V = \left(\frac{1}{2} \cdot 6 \cdot 7\right)h$ The base is a triangle, so $B = \frac{1}{2} \cdot 6 \cdot 7$.

$V = \left(\frac{1}{2} \cdot 6 \cdot 7\right)10$ The height of the prism is 10.

$V = 210$ Simplify.

The volume is 210 cubic inches.

✓ CHECK Your Progress

Find the volume of each prism.

a.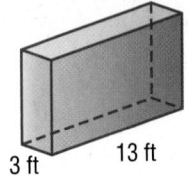

b.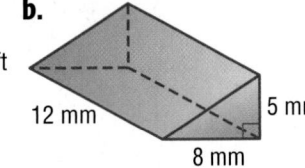

c.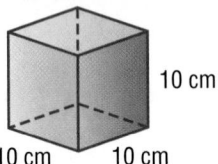

You can use the formula $V = Bh$ to find the volume of a cylinder.

Key Concept   **Volume of a Cylinder**

Words   The volume V of a cylinder with the area of the base B times the height h.

Model

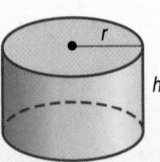

Symbols $V = Bh$

Study Tip

Cylinders The base of a cylinder is a circle. Use the formula $A = \pi r^2$ for B.

✎ EXAMPLE   **Find the Volume of a Cylinder**

③ **Find the volume of the cylinder. Round to the nearest tenth.**

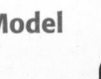

Since the diameter is 13 feet, the radius is 6.5 feet.

$V = \pi r^2 h$ Volume of a cylinder

$V = \pi \cdot 6.5^2 \cdot 20$ Replace r with 6.5 and h with 20.

$V \approx 2{,}654.6$ Simplify. Use a calculator.

The volume is about 2,654.6 cubic feet.

✓ CHECK Your Progress

Find the volume of each cylinder. Round to the nearest tenth.

d. radius, 2 in.; height, 7 in.   **e.** diameter, 18 cm; height, 5 cm

Recall that a composite figure is made up of two or more shapes. Objects that are made up of more than one type of *solid* are called **composite solids**. To find the volume of a composite solid, decompose the figure into solids whose volumes you know how to find.

Study Tip

Estimation You can check the reasonableness of your result in Example 4 by estimating the volume. The volume should be slightly less than 12 • 12 • 12 or 1,728 mm³.

🏃 ✏️ **REAL-WORLD EXAMPLE** **Find the Volume of a Composite Solid**

④ **CRAFTS** Tanya uses cube-shaped beads to make jewelry. Each bead has a circular hole through the middle. Find the volume of each bead.

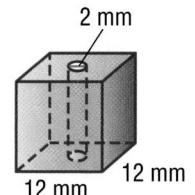

The bead is made of one rectangular prism and one cylinder. Find the volume of each solid. Then subtract to find the volume of the bead.

Rectangular Prism	**Cylinder**
$V = Bh$	$V = Bh$
$V = (12 \cdot 12)12$ or $1,728$	$V = (\pi \cdot 1^2)12$ or 37.7

The volume of the bead is $1,728 - 37.7$ or $1,690.3$ cubic millimeters.

✓ **CHECK Your Progress**

f. BIRDS The Ecology Club is building birdhouses, similar to the one shown at the right, to put in a nature preserve. Find the volume of the birdhouse.

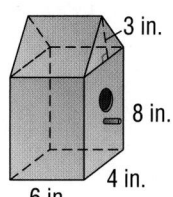

✓ **CHECK Your Understanding**

Examples 1 and 2 **Find the volume of each prism. Round to the nearest tenth if necessary.**

1.
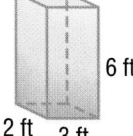

2.
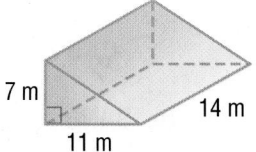

Example 3 **Find the volume of each cylinder. Round to the nearest tenth.**

3.
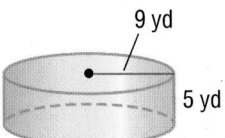

4.
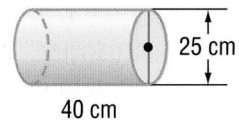

Example 4 ⑤ **TOYS** Lorenzo's younger sister Selma received the toy house shown as a gift. What is the volume of the toy house?

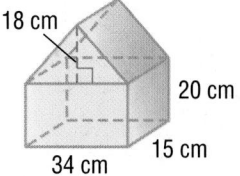

Practice and Problem Solving

= Step-by-Step Solutions begin on page R1.
Extra Practice begins on page EP2.

Examples 1–3 **Find the volume of each solid. Round to the nearest tenth if necessary.**

6.
4 in.
5 in.
$1\frac{1}{2}$ in.

7.
6 mm
6 mm
6 mm

8.
10 yd
7 yd
15 yd

9.
8 m
12 m
16 m

10.
7.4 cm
14 cm

11.
2.8 m
9 m

12. rectangular prism: length, 4 in.; width, 6 in.; height, 17 in.

13. triangular prism: base of triangle, 5 ft; altitude, 14 ft; height of prism, $8\frac{1}{2}$ ft

14. cylinder: radius, 25 m; height, 20 m

15. cylinder: diameter, 7.2 cm; height, 5.8 cm

Example 4 **16. MAILBOXES** The Klines' mailbox is shown below. Find the volume of the mailbox.

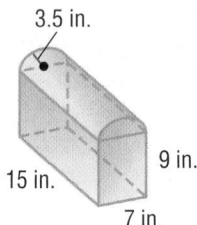

3.5 in.
15 in.
9 in.
7 in.

17. TOWELS An unused roll of paper towels has the dimensions shown. What is the volume of the unused roll?

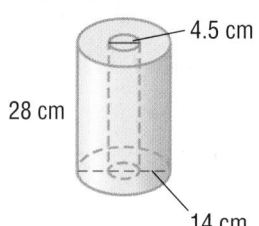

4.5 cm
28 cm
14 cm

18. Find the height of a rectangular prism with a length of 6.8 meters, a width of 1.5 meters, and a volume of 91.8 cubic meters.

19 Find the height of a cylinder with a radius of 4 inches and a volume of 301.6 cubic inches.

20. BAND The Band Boosters buy popcorn in large bags that have a volume of 2,500 cubic inches. They make individual boxes to sell that are 2 inches by 6 inches by 8 inches. If they sold 20 boxes, how much of the original bag is left?

21. BUSINESS The original package for the Crackle Cereal Company's cereal is shown. They want to design a new package with the same volume but in the shape of a cylinder. If the height remains the same as the original, what is the least value the company should use for the diameter of the new package? Explain your reasoning.

9 in.
3 in.
8 in.

22. POOLS A wading pool is to be 20 feet long, 11 feet wide, and 1.5 feet deep. The excavated dirt is to be hauled away by wheelbarrow. If the wheelbarrow holds 9 cubic feet of dirt, how many wheelbarrows of dirt must be hauled away from the site?

23 GEOMETRY Explain how you would find the volume of the hexagonal prism shown at the right. Then find its volume.

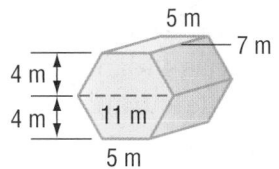

24. ⟳ **MULTIPLE REPRESENTATIONS** The formula for the volume of a cube is $V = s^3$.

  a. TABLES Copy and complete the table. It shows the volume V of a cube with side length s.

  b. GRAPHS Graph the ordered pairs (side length, volume) on a coordinate grid. Then connect the points.

Side Length (units)	Volume (units³)
1	1
2	▦
3	▦
4	▦
5	▦

  c. WORDS Describe the shape of the graph.

25. GRAPHIC NOVEL Refer to the graphic novel frame below for Exercises a–c.

  a. Find the volume of the bag and candle. Round to the nearest tenth.

  b. How much packing material is needed to fill the empty space in the bag after the candle is placed in the bag?

  c. There are 70 teachers in the school. If each package of packing material contains 575 cubic inches of material, how many packages do they need to buy to fill all of the gift bags?

H.O.T. Problems

26. CHALLENGE Does doubling the height of a cylinder have the same effect on the volume as doubling the radius? Explain.

27. 📝 **WRITE MATH** How are the formulas for the volume of prisms and cylinders similar?

28. ▦✎ **GRIDDED RESPONSE** A new television is packaged in a box that measures 2 feet on each edge. The box has 4 foam packaging cubes each measuring 0.5 foot on an edge.

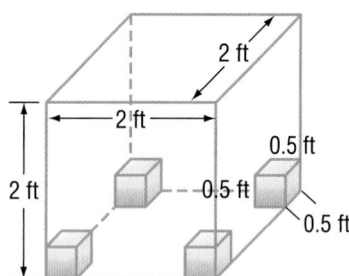

What is the volume, in cubic feet, of the empty space in the box when the 4 foam cubes are inside the box?

29. An ice cream company sells a container of ice cream that measures 5.5 inches in height and 4.2 inches in diameter.

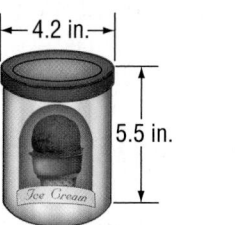
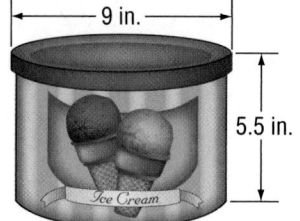

Another container is also 5.5 inches in height, but the diameter is 9 inches. About how many times more ice cream does the larger container hold?

A. 1.8 times   **C.** 4.6 times

B. 2.3 times   **D.** 5.2 times

More About Volume

The storage cubes shown can be arranged in several different ways. How does the volume change with each arrangement?

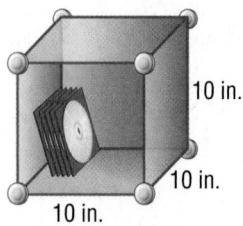

$V = 1,000 \text{ in}^3$

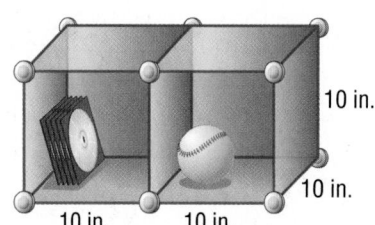

$V = 2,000 \text{ in}^3$

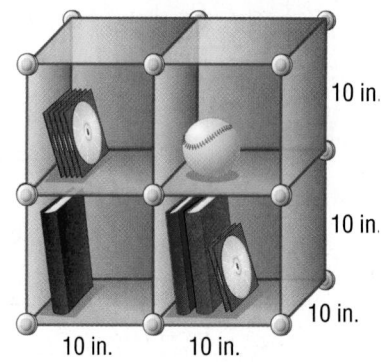

$V = 4,000 \text{ in}^3$

So, if you double the length of a cube, the volume doubles. If you double both the length and height of the cube, the volume is four times greater.

30. Refer to the rectangular prism at the right.

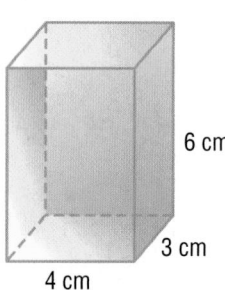

  a. Without calculating, describe how the volume would change if you double the width of the prism.

  b. Without calculating, describe how the volume would change if you double all three dimensions of the prism.

  c. *True or False?* Doubling the height will have a greater effect on the volume than doubling the width because the height is greater than the width. Explain your reasoning.

31. **REASONING** What effect does changing one measurement of a prism by a scale factor have on the volume of the prism?

Main Idea
Find the volumes of pyramids, cones, and spheres.

Vocabulary
sphere

*Get Connect**ED***

CCSS 8.G.8

Volume of Pyramids, Cones, and Spheres

Explore In this activity, you will investigate the relationship between the volumes of a pyramid and prism with the same base area and height.

Draw and cut out 5 squares.

2 in.

Tape together as shown.

Fold and tape to form a cube with an open top.

Draw and cut out 4 isosceles triangles.

1 in. $2\frac{1}{4}$ in.
2 in.

Tape together as shown.

Fold and tape to form an open square pyramid.

1. Compare the base areas and the heights of the two solids.

2. Fill the pyramid evenly with rice. Pour the rice into the cube. Repeat until the prism is filled. How many times did you fill the pyramid in order to fill the cube?

3. What fraction of the cube's volume does one pyramid fill?

The volume of a pyramid is one third the volume of a prism with the same base area and height.

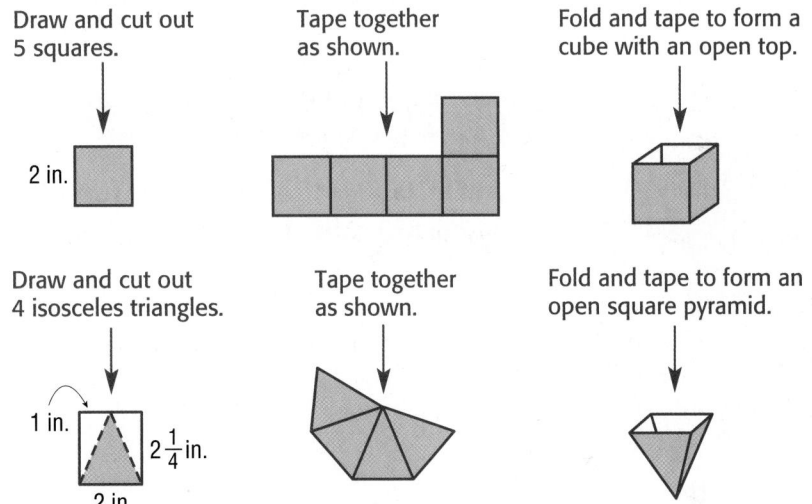

Key Concept Volume of a Pyramid

Words	The volume V of a pyramid is one-third the area of the base B times the height h.	Model
Symbols	$V = \frac{1}{3}Bh$	

The height of a pyramid or cone is the distance from the vertex, perpendicular to the base.

EXAMPLE Find the Volume of a Pyramid

① Find the volume of the pyramid. Round to the nearest tenth.

$V = \frac{1}{3}Bh$ Volume of a pyramid

$V = \frac{1}{3}\left(\frac{1}{2} \cdot 8.1 \cdot 6.4\right)11$ $B = \frac{1}{2} \cdot 8.1 \cdot 6.4, h = 11$

$V = 95.04$ Simplify.

The volume is about 95.0 cubic meters.

11 m, 6.4 m, 8.1 m

CHECK Your Progress

a. Find the volume of a pyramid that has a height of 5 yards and a square base with sides 2 yards long.

 REAL-WORLD EXAMPLE

② **ARCHITECTURE** Refer to the information at the left. If the volume is 296,875 cubic feet, find the height of the pyramid.

$V = \frac{1}{3}Bh$ Volume of a pyramid

$296,875 = \frac{1}{3} \cdot 12,544 \cdot h$ Replace V with 296,875 and B with 112 · 112 or 12,544.

$296,875 = \frac{12,544}{3}h$ Multiply.

$\frac{3}{12,544} \cdot 296,875 = \frac{3}{12,544} \cdot \frac{12,544}{3}h$ Multiply each side by $\frac{3}{12,544}$.

$71 \approx h$ Simplify.

The height of the pyramid is about 71 feet.

Real-World Link · · · ·
The glass pyramid which serves as an entrance to the Louvre museum in Paris has a square base with sides 112 feet long.

CHECK Your Progress

b. MODELS A pyramid-shaped model has a volume of 864 cubic inches. If its base has an area of 144 square inches, how high is the model?

The volumes of a cone and a cylinder are related in the same way as those of a pyramid and prism.

 Key Concept   Volume of a Cone

Words   The volume V of a cone with radius r is one third the area of the base B times the height h.

Model

Symbols $V = \frac{1}{3}Bh$ or $V = \frac{1}{3}\pi r^2 h$

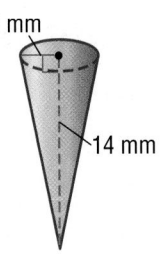

EXAMPLE   Find the Volume of a Cone

3 Find the volume of the cone.

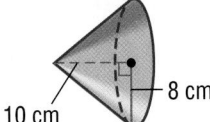

$V = \frac{1}{3}\pi r^2 h$ Volume of a cone

$V = \frac{1}{3} \cdot \pi \cdot 3^2 \cdot 14$ Replace r with 3 and h with 14.

$V \approx 131.9$ Simplify. Use a calculator.

The volume is about 131.9 cubic millimeters.

✓ CHECK Your Progress

Find the volume of each cone. Round to the nearest tenth.

c.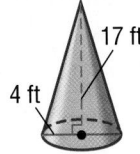

d.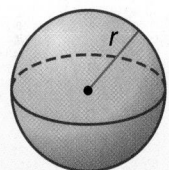

A **sphere** is the set of all points in space that are a given distance from a given point called the center. A sphere does not contain any bases. To find the volume of a sphere, use the following formula.

Key Concept   Volume of a Sphere

Words   The volume V of a sphere is four thirds the product of π and the cube of the radius r.

Model

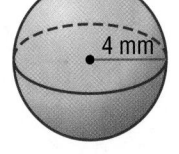

Symbols $V = \frac{4}{3}\pi r^3$

Study Tip

Exact and Approximate Whenever you round or use 3.14 for π, you are finding the approximate value. If an answer is left in the form $\frac{256}{3}\pi$, this is the exact value.

EXAMPLE   Find the Volume of a Sphere

4 Find the volume of the sphere.

$V = \frac{4}{3}\pi r^3$ Volume of a sphere

$V = \frac{4}{3} \cdot \pi \cdot 4^3$ Replace r with 4.

$V \approx 268.1$ Simplify. Use a calculator.

The volume of the sphere is about 268.1 cubic millimeters.

✓ CHECK Your Progress

Find the volume of each sphere. Round to the nearest tenth.

e. radius: 6 ft   **f.** diameter: 18 in.

Example 1 Find the volume of each pyramid. Round to the nearest tenth.

1. 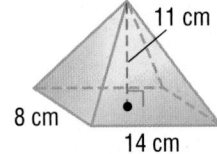 11 cm
8 cm
14 cm

2. 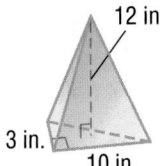 12 in.
3 in.
10 in.

3. 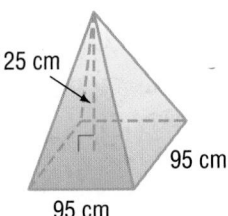 125 cm
95 cm
95 cm

Example 2 **4. ARCHAEOLOGY** El Castillo, the pyramid at Chichen Itza in Mexico, is 30 meters tall with a volume of about 30,580 cubic meters. What is the length of each side of the square base?

Example 3 Find the volume of each cone. Round to the nearest tenth.

5. 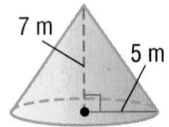 7 m
5 m

6. 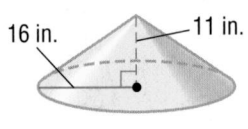 16 in. 11 in.

7. 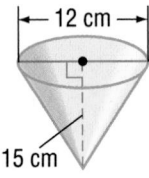 12 cm
15 cm

Example 4 Find the volume of each sphere. Round to the nearest tenth.

8. 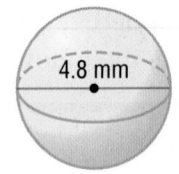 4.8 mm

9. 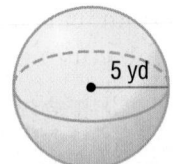 5 yd

10. 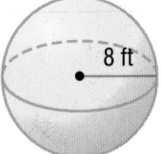 8 ft

Practice and Problem Solving

● = Step-by-Step Solutions begin on page R1.
Extra Practice begins on page EP2.

Example 1 Find the volume of each pyramid. Round to the nearest tenth.

11. 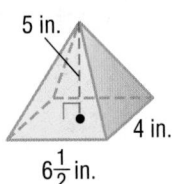 5 in.
4 in.
$6\frac{1}{2}$ in.

12. 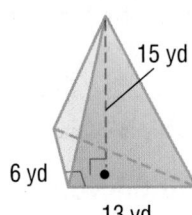 15 yd
6 yd
13 yd

13. 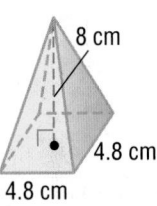 8 cm
4.8 cm
4.8 cm

14. triangular pyramid: triangle base, 10 cm; triangle height, 7 cm; pyramid height, 15 cm

15. square pyramid: base length, 22 ft; pyramid height, 17 ft

16. triangular pyramid: triangle base, 12 m; triangle height, 5 m; pyramid height, 22 m

Example 2 **17 SCIENCE** A model of a volcano constructed for a science project is cone-shaped with a diameter of 8 inches. If the volume of the model is about 201 cubic inches, how tall is the model?

18. ICE CREAM An ice cream cone can hold about 10 cubic inches of ice cream. If the cone has a height of 6 inches, what is the diameter of the cone?

Example 3 Find the volume of each cone. Round to the nearest tenth.

19.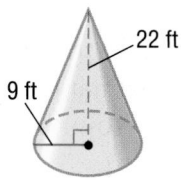
22 ft
9 ft

20.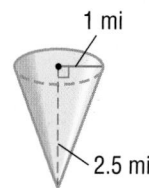
1 mi
2.5 mi

21.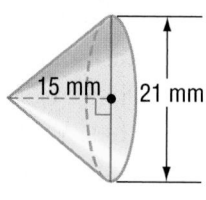
15 mm   21 mm

22. diameter, 12 m; height, 5 m

23. radius, 3.7 in.; height, 8.2 in.

Example 4 Find the volume of each sphere. Round to the nearest tenth.

24.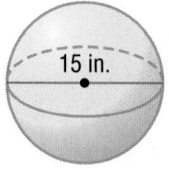
15 in.

25.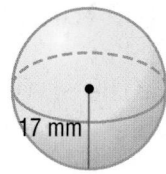
17 mm

26.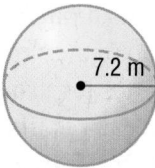
7.2 m

27. NATURE The eggs of a hummingbird are nearly spherical. If an egg is 1 centimeter in diameter, find its volume. Round to the nearest tenth.

28. SPORTS A soccer ball is a sphere with a radius of 11 centimeters. What is the volume of a soccer ball? Round to the nearest tenth.

Find the volume of each solid. Round to the nearest tenth if necessary.

29.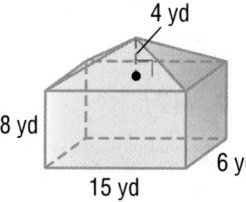
4 yd
8 yd
6 yd
15 yd

30.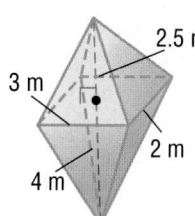
2.5 m
3 m
2 m
4 m

31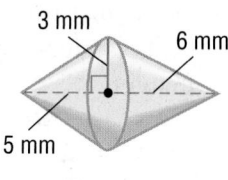
3 mm
6 mm
5 mm

H.O.T. Problems

32. CHALLENGE How could you change the height of a cone so that its volume would remain the same when its radius was tripled?

33. FIND THE ERROR Jacob is finding the volume of the square pyramid shown. Find his mistake and correct it.

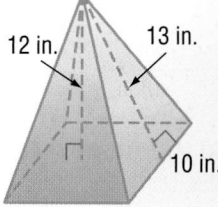
12 in.   13 in.
10 in.
10 in.

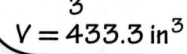
$V = \frac{1}{3}Bh$
$V = \frac{1}{3} \cdot 10 \cdot 10 \cdot 13$
$V = 433.3 \text{ in}^3$

34. ✍ **WRITE MATH** How are the formulas for the volume of cones and pyramids similar?

35. A rectangular pyramid has a base 18 inches by 30 inches and a height of 36 inches. Which is closest to the volume of the pyramid in cubic feet?

A. 2.5 ft³   **C.** 4 ft³

B. 3 ft³   **D.** 5.5 ft³

36. [THINK SOLVE EXPLAIN] **SHORT RESPONSE** The bag below is used as a party favor. The left part is the shape of a cylinder with a height of 3 inches and radius of 2 inches. The right part is in the shape of a cone with a height of 5 inches.

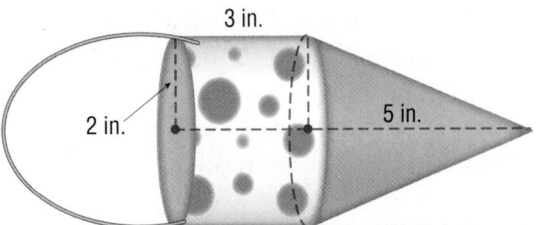

Determine the volume of the bag. Show all work necessary to justify your answer.

37. Which of the following is the **best** approximation for the volume of the cone below?

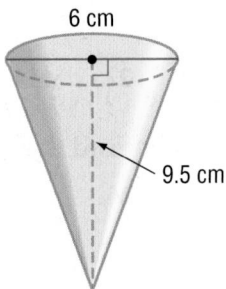

6 cm

9.5 cm

F. 1,074 cm³   **H.** 269 cm³

G. 358 cm³   **I.** 90 cm³

38. What is the **exact** volume of the sphere shown below?

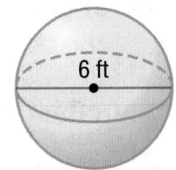

6 ft

A. 8π ft³   **C.** 36π ft³

B. 12π ft³   **D.** 288π ft³

Spiral Review

39. DISPENSER Find the volume of the soap dispenser at the right. (Lesson 2A)

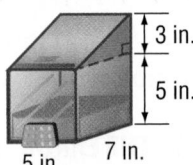

3 in.
5 in.
5 in.
7 in.

40. Find the area of the figure at the right. Write in simplest form. (Lesson 1E)

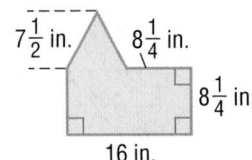

$7\frac{1}{2}$ in. $8\frac{1}{4}$ in.

$8\frac{1}{4}$ in.

16 in.

Find the circumference and area of each circle. Round to the nearest tenth.
(Lesson 1B)

41.

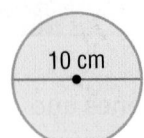

10 cm

42.

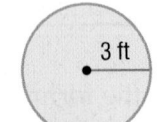

3 ft

43.

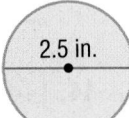

2.5 in.

Mid-Chapter Check

Find the circumference and area of each circle. Round to the nearest tenth. (Lesson 1B)

1.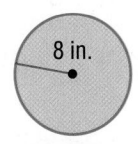
8 in.

2.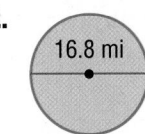
16.8 mi

3.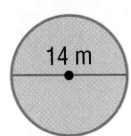
14 m

4.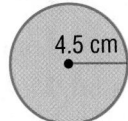
4.5 cm

5. MEASUREMENT A shot-putter must stay inside the circle shown when throwing the shot. (Lesson 1B)

7 ft

a. What is the area of the region in which the athlete is able to move?

b. What is the circumference of the circular region?

6. ART Leah is creating a model of her kitchen. The kitchen measures 18 feet by 12 feet. If she uses a scale of 2 feet = $1\frac{1}{2}$ inches, what are the dimensions of her kitchen on the model? Use the *make a model* strategy. (Lesson 1D)

7. GEOMETRY Two right triangles are side by side such that they form a larger isosceles triangle. The two right triangles are congruent, and each have angle measures of 90°, 45°, and 45°. What type of triangle will the new isosceles triangle be? Use the *make a model* strategy. (Lesson 1D)

8. MULTIPLE CHOICE What is the area of the figure below? Round to the nearest tenth. (Lesson 1E)

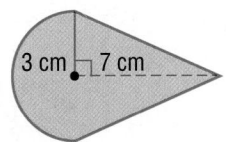
3 cm 7 cm

A. 30.4 cm^2
B. 35.1 cm^2
C. 39.8 cm^2
D. 49.3 cm^2

9. GAMES Find the volume of the puzzle cube. (Lesson 2A)

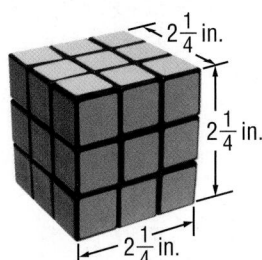
$2\frac{1}{4}$ in.
$2\frac{1}{4}$ in.
$2\frac{1}{4}$ in.

10. Find the width of a rectangular prism with a length of 7.6 meters, a height of 8 meters, and a volume of 88.4 cubic meters. Round to the nearest tenth. (Lesson 2A)

Find the volume of each solid. Round to the nearest tenth if necessary. (Lessons 2A and 2B)

11.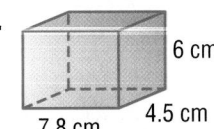
6 cm
7.8 cm   4.5 cm

12.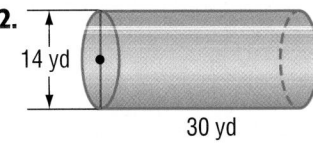
14 yd
30 yd

13.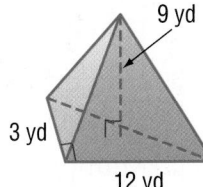
9 yd
3 yd
12 yd

14.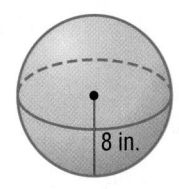
8 in.

15. MULTIPLE CHOICE Find the volume of a cone with a diameter of 6 centimeters and a height of 13 centimeters. Round to the nearest tenth. (Lesson 2B)

F. 122.5 cm^3
G. 367.6 cm^3
H. 452.4 cm^3
I. 1,470.3 cm^3

Surface Area of Cylinders

Main Idea

Find the surface area of cylinders using models and nets.

Vocabulary

net

Get Connect ED

Nets are two-dimensional patterns of three-dimensional figures. When you construct a net, you are decomposing the three-dimensional figure into separate shapes. You can use a net to find the area of each surface of a three-dimensional figure such as a cylinder.

ACTIVITY

STEP 1 Use an empty cylinder-shaped container that has a lid. Measure and record the height of the container.

STEP 2 Take off the lid of the container and make 2 cuts as shown. Next, cut off the sides of the lid. Finally, lay the lid, the curved side, and the bottom flat to form the net of the container.

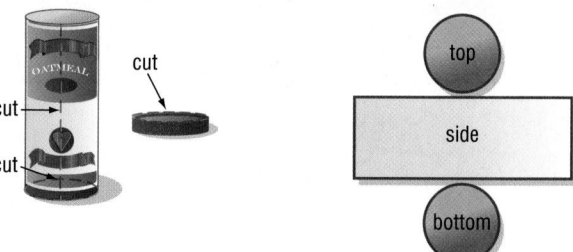

Analyze the Results

1. Classify the figures that make up the net of the container.

2. Find the area of each shape. Then find the sum of these areas.

3. Find the diameter of the top of the container and use it to find the circumference of that face.

4. Multiply the circumference by the height of the container. What does this product represent?

5. Add the product from Exercise 4 to the sum of the areas of the two circular bases.

6. Compare your answers from Exercises 2 and 5.

7. **MAKE A CONJECTURE** Write a method for finding the area of all of the surfaces of a cylinder given the measures of its height and the diameter of one of its bases.

Main Idea

Find the lateral and total surface area of prisms and cylinders.

Vocabulary

lateral face
lateral surface area
total surface area

Surface Area of Prisms and Cylinders

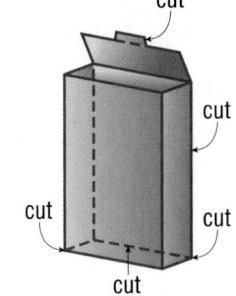

Explore

Step 1 Use an empty box with a tuck-in lid. Measure and record the height of the box and the perimeter of the top or bottom face.

Step 2 Label the top, bottom, front, back, and side faces using a marker.

Step 3 Open the lid and make 5 cuts as shown. Then open the box and lay it flat to form a net of the box. Measure and record the dimensions of each face.

1. Find the area of each face. Then find the sum of these areas.

2. Multiply the perimeter of a base by the height of the box. What does this product represent?

3. Add the product from Exercise 2 to the sum of the areas of the two bases.

4. Compare your answers from Exercises 1 and 3.

In the activity, you found the area of each surface, or face, of a box. A **lateral face** of a solid is any flat surface that is *not* a base. The **lateral surface area** of a solid is the sum of the areas of its lateral faces. The **total surface area** of a solid is the sum of the areas of all its surfaces.

Key Concept   Surface Area of a Prism

Lateral Area

Words   The lateral area *L.A.* of a prism is the perimeter *P* of the base times the height *h* of the prism.

Model

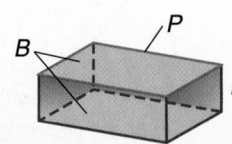

Symbols $L.A. = Ph$

Total Surface Area

Words   The total surface area *S.A.* of a prism is the lateral surface area *L.A.* plus the area of the two bases 2*B*.

Symbols $S.A. = L.A. + 2B$ or $S.A. = Ph + 2B$

---

1 Find the lateral and total surface areas of the rectangular prism.

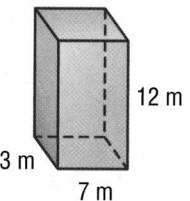

Begin by finding the perimeter and area of one base.

Perimeter of Base	Area of Base
$P = 2\ell + 2w$	$B = \ell w$
$P = 2(7) + 2(3)$ or 20	$B = 7(3)$ or 21

Use this information to find the lateral and total surface areas.

Lateral Surface Area	Total Surface Area
$L.A. = Ph$	$S.A. = L.A. + 2B$
$L.A. = 20(12)$ or 240	$S.A. = 240 + 2(21)$ or 282

The lateral surface area is 240 square meters and the total surface area of the prism is 282 square meters.

2 **WATER SKIING** The ramp for competitive water skiing is a wedge-shaped ramp that is covered in wax or fiberglass. Find the total surface area of the ramp.

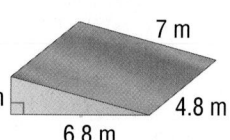

Estimate $S.A. = (2 + 7 + 7)5 + 7(2)$ or 94 m²

The bases of the prism are triangles with side lengths of 1.8 meters, 6.8 meters, and 7 meters. Find the perimeter and area of one base.

Perimeter of Base	Area of Base
$P = 1.8 + 6.8 + 7$	$B = \frac{1}{2}bh$
$P = 15.6$	$B = \frac{1}{2}(6.8)(1.8)$ or 6.12

Total Surface Area

$S.A. = Ph + 2B$ Total surface area of prism

$S.A. = 15.6(4.8) + 2(6.12)$ $P = 15.6, h = 4.8,$ and $B = 6.12.$

$S.A. = 87.12$ Simplify.

The surface area is 87.12 square meters.   Compare to the estimate.

✓ **CHECK Your Progress**

Find the lateral and total surface areas of each prism.

a.
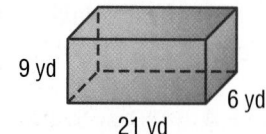

b.
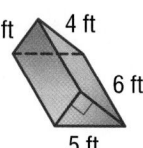

You can find the total surface area of a cylinder by finding the area of its two bases and adding the area of the curved surface. The lateral area of a cylinder is the area of the curved surface. If you unfold a cylinder, its net is two circles and a rectangle.

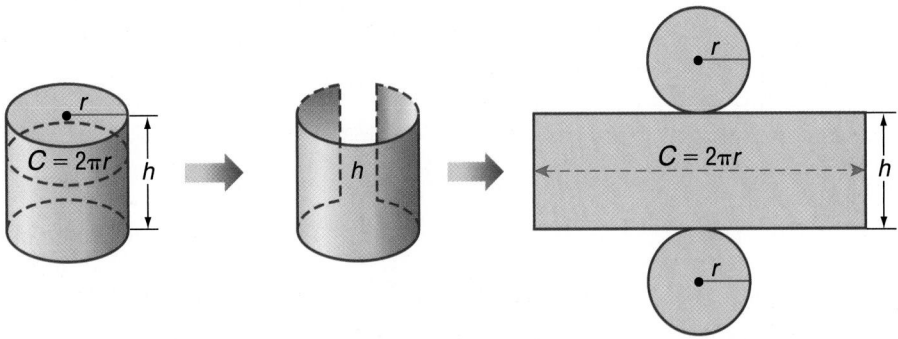

Model	Net	Area
2 circular bases	2 congruent circles with radius r	$2(\pi r^2)$ or $2\pi r^2$
1 curved surface	1 rectangle with width h and length $2\pi r$	$2\pi r \cdot h$ or $2\pi rh$

Key Concept Surface Area of a Cylinder

Lateral Area

area of a base $= \pi r^2$

Words   The lateral area L.A. of a cylinder with height h and radius r is the circumference of the base times the height.

Model

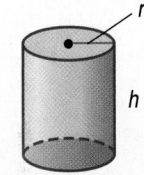

Symbols $L.A. = 2\pi rh$

Total Surface Area

Words   The surface area S.A. of a cylinder with height h and radius r is the lateral area plus the area of the two bases.

Symbols $S.A. = L.A. + 2\pi r^2$ or $S.A. = 2\pi rh + 2\pi r^2$

EXAMPLES Surface Area of a Cylinder

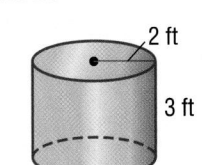

3 Find the lateral area and the total surface area of the cylinder. Round to the nearest tenth.

Lateral Surface Area	Total Surface Area
$L.A. = 2\pi rh$	$S.A. = L.A. + 2\pi r^2$
$L.A. = 2\pi(2)(3)$	$S.A. = 37.7 + 2\pi(2)^2$
$L.A. \approx 37.7$	$S.A. \approx 62.8$

The lateral area is about 37.7 square feet and the surface area of the cylinder is about 62.8 square feet.

4 **LABELS** Find the area of the label on the can of vegetables.

1.75 in.

5 in.

Since the label covers the lateral surface of the can, you only need to find the can's lateral surface area.

Estimate $L.A. = 2\pi rh$

            $L.A. \approx 2(3)(2)(5)$ $\pi \approx 3, r = 1.75 \approx 2, h = 5$

            $L.A. \approx 60 \text{ in}^2$

$L.A. = 2\pi rh$ Lateral surface area of cylinder

$L.A. = 2\pi(1.75)(5)$ $r = 1.75, h = 5$

$L.A. \approx 55.0$ Simplify.

The area of the label is about 55 square inches. Compare to the estimate.

CHECK Your Progress

Find the lateral and total surface areas of each cylinder. Round to the nearest tenth.

c.
5 mm
10 mm

d.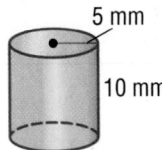
7 cm
14.8 cm

✓ CHECK Your Understanding

Examples 1 and 2   Find the lateral and total surface areas of each solid. Round to the nearest tenth if necessary.

1.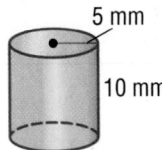
4 yd
5 yd
3 yd

2.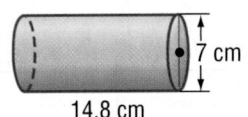
10 in.
6 in.
7 in.
8 in.

Example 3   **3.**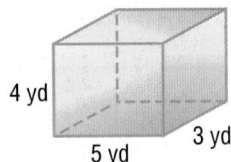
8 m
9.4 m

4.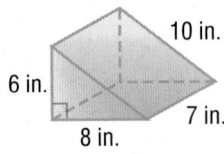
16 yd
25 yd

Example 4   **5** **CONTAINERS** Frozen orange juice often comes in cylindrical cardboard containers with metal lids. Find the area of the cardboard portion of the orange juice container shown.

2 in.
Orange Juice
6.5 in.

Practice and Problem Solving

● = **Step-by-Step Solutions** begin on page R1.
Extra Practice begins on page EP2.

Examples 1–3 **Find the lateral and total surface areas of each solid. Round to the nearest tenth if necessary.**

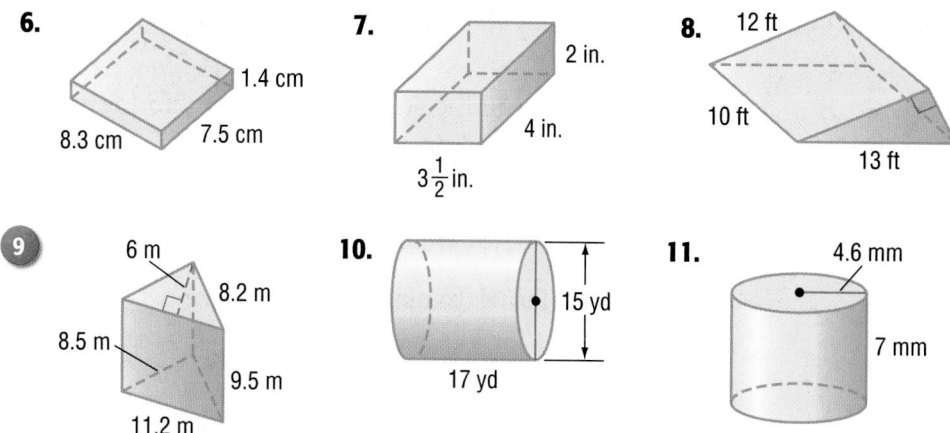

6. 1.4 cm   8.3 cm   7.5 cm

7. 2 in.   4 in. $3\frac{1}{2}$ in.

8. 12 ft   10 ft   5 ft   13 ft

9. 6 m   8.2 m   8.5 m   9.5 m   11.2 m

10. 15 yd   17 yd

11. 4.6 mm   7 mm

Example 4 **12. ART** Sabrina made a plant pot in ceramics class. A glaze will go on the outside and the bottom of the pot. How many square inches of surface will be glazed?

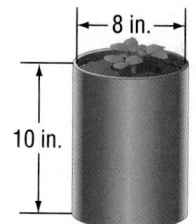

8 in.
10 in.

13 PACKAGING Two possible designs for a new cereal are shown. The volumes are approximately equal. Which design would use less material to produce? Explain.

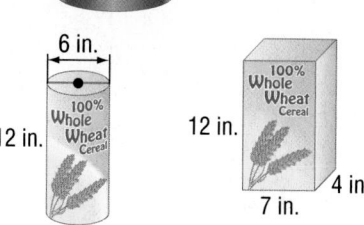

6 in.
12 in.   100% Whole Wheat Cereal
12 in.   100% Whole Wheat Cereal
7 in.   4 in.

14. GRAPHIC NOVEL Refer to the graphic novel frame below for Exercises a–b.

I think we should wrap these in gift paper before we put them in the bags.

You can review the dimensions of the candles in Lesson 2A.

a. What is the least amount of paper that will be needed to wrap one candle with no overlap?

b. How many square feet of wrapping paper will be needed to wrap all 70 candles?

H.O.T. Problems

15. **REASONING** Determine whether the following statement is *true* or *false*. If *false*, give a counterexample.

> *If two rectangular prisms have the same volume,*
> *then they also have the same surface area.*

16. **CHALLENGE** Will the surface area of a cylinder increase more if you double the height or double the radius? Explain your reasoning.

17. **NUMBER SENSE** If you triple the radius of a cylinder, explain how this affects the lateral area of the cylinder.

18. **WRITE MATH** Explain the difference between lateral area and surface area.

Test Practice

19. Oliver is painting the rectangular toy chest shown below.

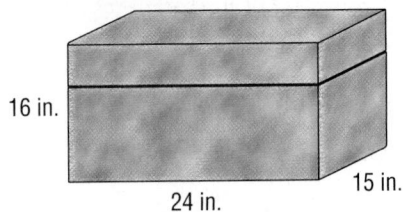

16 in.

24 in.   15 in.

If Oliver paints only the outside of the toy chest, what is the total surface area, in square inches, he will paint?

   A. 330 in^2 **C.** 1,968 in^2

   B. 399 in^2 **D.** 5,760 in^2

20. A roller like the one shown is used for painting.

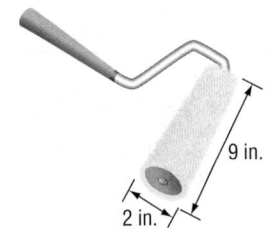

9 in.

2 in.

To the nearest tenth, how many square inches does a single rotation of the paint roller cover?

   F. 18.0 in^2 **H.** 56.5 in^2

   G. 28.3 in^2 **I.** 113.1 in^2

Spiral Review

21. Find the volume of a cone with a diameter of 22 inches and a height of 24 inches. Round to the nearest tenth. (Lesson 2B)

22. **HEALTH** The inside of a refrigerator in a medical laboratory measures 17 inches by 18 inches by 42 inches. You need at least 8 cubic feet to refrigerate some samples from the lab. Is the refrigerator large enough for the samples? Explain your reasoning. (Lesson 2A)

23. **DECORATING** A circular rug has an area of approximately 25 square feet. What is the diameter of the rug? Round to the nearest tenth.
(Lesson 1B)

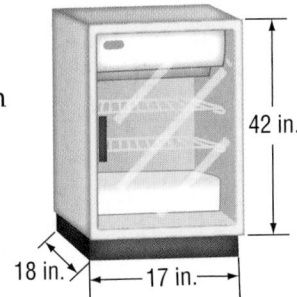

42 in.

18 in.   17 in.

Explore Nets of Cones

Main Idea

Make a net of a cone.

The lateral surface of a cone is part of a larger circle. So that the edges match, the circumference of the base is equal to *part* of the circumference of the larger circle.

ACTIVITY   Make a Net of a Cone

STEP 1 Use a compass to draw two circles slightly touching, one with a radius of 17 centimeters and one with a radius of 8 centimeters.

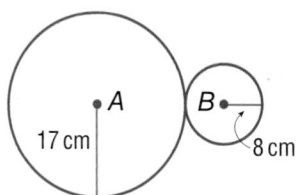

STEP 2 Think: What part of the circumference of *A* is equal to the circumference of *B*? Let *x* represent the part.

$x(34\pi) = 16\pi$ The circumference of *A* is 34π.
The circumference of *B* is 16π.

$\dfrac{x \cdot 34\pi}{34\pi} = \dfrac{16\pi}{34\pi}$ Divide each side by 34π.

$x \approx 0.47$ Simplify.

You need 0.47 of the circumference of *A*.

STEP 3 Find the size of the central angle to be cut from *A*.

$0.47 \cdot 360° \approx 170°$

Cut a central angle of 170° from circle *A* and make a cone.

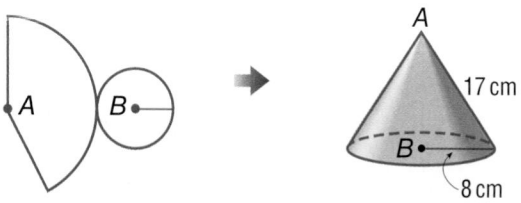

Analyze the Results

Find the central angle of each cone and then draw a net of the cone.

1.

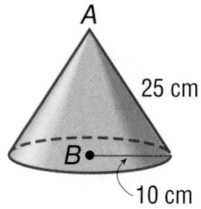

2.

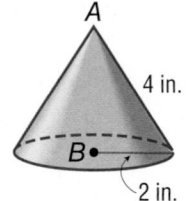

Main Idea

Find the lateral and total surface areas of pyramids and cones.

Vocabulary

regular pyramid
slant height

Get ConnectED

Surface Area of Pyramids and Cones

MUSEUMS The Rock and Roll Hall of Fame and Museum opened in Cleveland, Ohio, in 1995.

1. Not including the base, how many faces does this pyramid have? What shape are they?

2. How could you find the total area of the glass used for the building?

A **regular pyramid** is a pyramid with a base that is a regular polygon. The lateral faces of a regular pyramid are congruent isosceles triangles. At the top of the pyramid, these triangles meet at a common point called the vertex. The altitude or height of each lateral face is called the **slant height** of the pyramid.

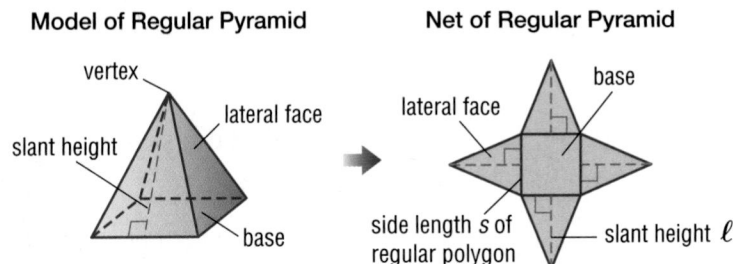

To find the lateral area *L.A.* of a regular pyramid, look at its net. The lateral area of a pyramid is the sum of the areas of its lateral faces, which are all triangles.

The net of a square pyramid is a square and four triangles as shown above.

$L.A. = 4\left(\frac{1}{2}s\ell\right)$ Area of the lateral faces

$L.A. = \frac{1}{2}(4s)\ell$ Commutative Property of Multiplication

$L.A. = \frac{1}{2}P\ell$ The perimeter of the base *P* is 4*s*.

The total surface area of a regular pyramid is the lateral surface area plus the area of the base.

Lateral Area

Words — The lateral surface area *L.A.* of a regular pyramid is half the perimeter *P* of the base times the slant height *ℓ*.

Model

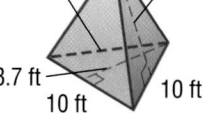

slant height *ℓ*

area of base *B*

perimeter of base *P*

Symbols — $L.A. = \frac{1}{2}P\ell$

Total Surface Area

Words — The total surface area *S.A.* of a regular pyramid is the lateral area *L.A.* plus the area of the base *B*.

Symbols — $S.A. = L.A. + B$ or $S.A. = \frac{1}{2}P\ell + B$

QUICK Review

Perimeter of a Square

The perimeter *P* of a square is four times the measure of any of its sides *s*.

$P = 4s$

 EXAMPLE — **Surface Areas of a Pyramid**

1 Find the lateral and total surface areas of the triangular pyramid.

10 ft   12 ft   8.7 ft   10 ft   10 ft

$L.A. = \frac{1}{2}P\ell$ $S.A. = L.A. + B$

$L.A. = \frac{1}{2} \cdot 30 \cdot 12$ $S.A. = 180 + 43.5$ $B = \frac{1}{2} \cdot 10 \cdot 8.7$

$L.A. = 180$ $S.A. = 223.5$

The lateral and total surface areas are 180 and 223.5 square feet.

 CHECK Your Progress

a. Find the lateral and total surface areas of a pyramid with a slant height of 18 meters and a square base with 11-meter sides.

Real-World Link . . . . .

The Pyramid of the Sun in Teotihuacán, Mexico, was built in the second century, A.D. It is about 71 meters tall, and its square base has side lengths of 223.5 meters.

REAL-WORLD EXAMPLE

2 **ARCHITECTURE** Use the information at the left to find the lateral surface area of the pyramid if it has a slant height of 132.5 meters.

$L.A. = \frac{1}{2}P\ell$ Lateral surface area of a pyramid

$L.A. = \frac{1}{2} \cdot 894 \cdot 132.5$ $P = 223.5(4)$ or 894 and $\ell = 132.5$

$L.A. = 59{,}227.5$ Simplify.

The lateral area of the pyramid is 59,227.5 square meters.

 CHECK Your Progress

b. **AWARDS** A music award is a square pyramid with a 6-inch-long base and a 13-inch slant height. Find the award's total surface area.

You can also find the surface area of a cone using a net. The surface area of a cone is the sum of its lateral area and the area of its base.

Model of Cone

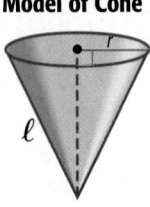

Net of Cone

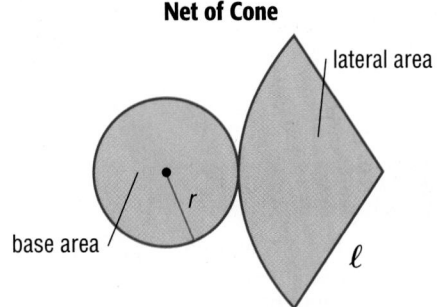

lateral area

base area

r

ℓ

Study Tip

Lateral Area of a Cone
The lateral area of a cone is one-half the circumference of the base times the slant height.
$L.A. = \frac{1}{2}(2\pi n)\ell$
$L.A. = \pi r\ell$

Key Concept   Surface Area of a Cone

Lateral Area

Words   The lateral area *L.A.* of a cone is π times the radius times the slant height ℓ.

Model

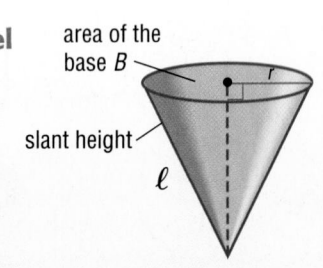

area of the base *B*

r

slant height

ℓ

Symbols $L.A. = \pi r\ell$

Total Surface Area

Words   The surface area *S.A.* of a cone with slant height ℓ and radius *r* is the lateral area plus the area of the base.

Symbols $S.A. = L.A. + \pi r^2$ or $S.A. = \pi r\ell + \pi r^2$

EXAMPLE   Surface Area of a Cone

3   Find the lateral and total surface areas of the cone. Round to the nearest tenth.

$L.A. = \pi r\ell$ Lateral area of a cone

$L.A. = \pi \cdot 5 \cdot 13$ Replace *r* with 5 and ℓ with 13.

$L.A. \approx 204.2$ Simplify.

13 mm

5 mm

Find the surface area.

$S.A. = L.A. + \pi r^2$ Surface area of a cone

$S.A. = 204.2 + \pi \cdot 5^2$ Replace *L.A.* with 204.2 and *r* with 5.

$S.A. \approx 282.7$ Simplify.

The lateral and total surface areas of the cone are about 204.2 and 282.7 square millimeters.

✓ CHECK Your Progress

c. Find the lateral and total surface areas of a cone with a radius of 9.5 inches and a slant height of 4 inches. Round to the nearest tenth.

Example 1 Find the lateral and total surface areas of each regular pyramid. Round to the nearest tenth if necessary.

1.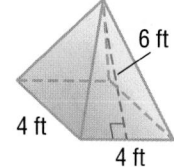

2.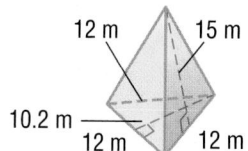

Example 2 **3. EVENTS** The Pyramid Arena in Memphis is a regular square pyramid. Each face of the arena has a base of 600 feet and a height of about 477 feet. Find the lateral surface area of the pyramid.

Example 3 Find the lateral and total surface areas of each cone. Round to the nearest tenth.

4.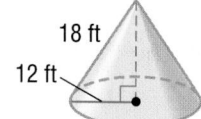

5.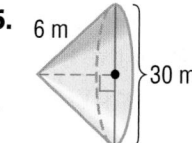

Practice and Problem Solving

● = Step-by-Step Solutions begin on page R1.
Extra Practice begins on page EP2.

Example 1 Find the lateral and total surface areas of each regular pyramid. Round to the nearest tenth if necessary.

6.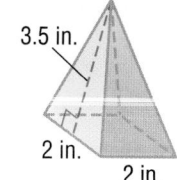

7.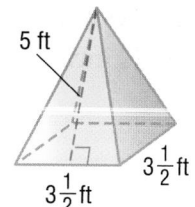

8.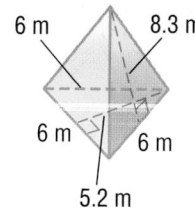

9.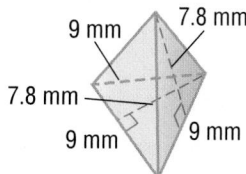

10.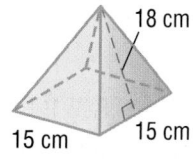

11.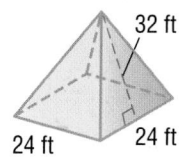

Example 2 **12. ARCHITECTURE** The Transamerica Pyramid in San Francisco is shaped like a square pyramid. It has a slant height of 856.1 feet and each side of its base is 145 feet long. Find the lateral area of the building.

13 ROOFS A pyramid-shaped roof has a slant height of 16 feet and its square base is 40 feet wide. How much roofing material is needed to cover the roof?

Example 3 **Find the lateral and total surface areas of each cone. Round to the nearest tenth.**

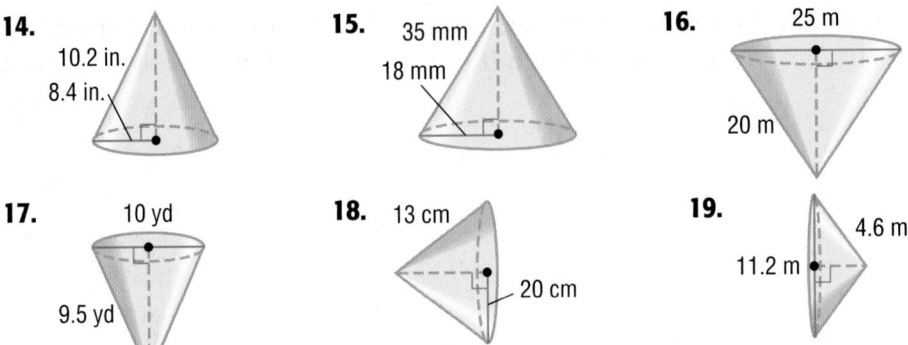

14. 10.2 in. 8.4 in.

15. 35 mm 18 mm

16. 25 m 20 m

17. 10 yd 9.5 yd

18. 13 cm 20 cm

19. 4.6 m 11.2 m

20. A square pyramid has a lateral area of 107.25 square centimeters and a slant height of 8.25 centimeters. Find the length of each side of its base.

21 **ARCHAEOLOGY** The Pyramid of Khafre in Egypt stands 471 feet tall. The sides of its square base are 705 feet in length. Find the lateral surface area of the Pyramid of Khafre. (*Hint*: Use the Pythagorean Theorem to find the pyramid's slant height ℓ.)

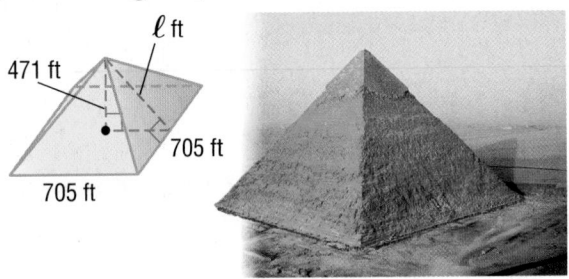

ℓ ft
471 ft
705 ft
705 ft

H.O.T. Problems

22. **CHALLENGE** Use the drawings of the figure shown. The total height of the figure is 20 inches.

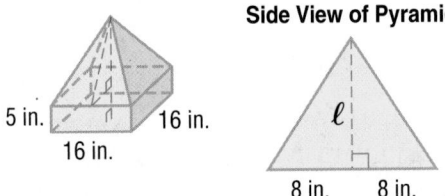

Side View of Pyramid

5 in. 16 in. 16 in.

ℓ

8 in. 8 in.

a. Find the height *h* of the pyramid.

b. Use the height of the pyramid to find the slant height, ℓ.

c. Which has a greater surface area, the prism or the pyramid? Explain your reasoning.

23. **OPEN ENDED** A pyramid has a base that is 3 inches square and a slant height of 4 inches. A rectangular prism has the same surface area. Give possible side lengths of the prism.

24. **WRITE MATH** How does the volume of a three-dimensional figure differ from its surface area?

25. Which is the **best** estimate for the surface area of the pyramid?

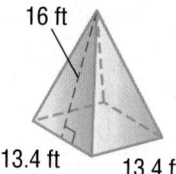

16 ft

13.4 ft   13.4 ft

A. 107 ft²

B. 180 ft²

C. 429 ft²

D. 608 ft²

26. **THINK SOLVE EXPLAIN** **SHORT RESPONSE** The lateral area of the cone is 351.9 square centimeters.

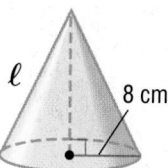

ℓ 8 cm

What is the slant height of the cone? Round to the nearest tenth.

27. The net of a paperweight is shown below. Which is **closest** to the lateral surface area of the paperweight?

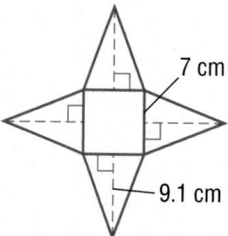

7 cm

9.1 cm

F. 32 cm²   **H.** 127 cm²

G. 49 cm²   **I.** 176 cm²

28. What is the total surface area of a cone with a diameter of 18 inches and a slant height of 12 inches? Round to the nearest tenth.

A. 1,696.5 in²

B. 678.6 in²

C. 593.8 in²

D. 339.3 in²

More About Surface Area

To find the surface area *S.A.* of a sphere, use the formula $S.A. = 4\pi r^2$, where r is the radius of the sphere.

EXAMPLE

Find the surface area of the sphere.

$S.A. = 4\pi r^2$ Surface area of a sphere

$S.A. = 4 \cdot \pi \cdot 9^2$ Replace r with 9.

$S.A. \approx 1,017.9$ Simplify. Use a calculator.

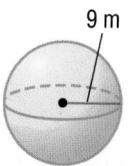

9 m

The surface area of the sphere is about 1,017.9 square meters.

Find the surface area of each sphere. Round to the nearest tenth.

29.

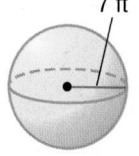

7 ft

30.

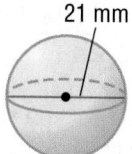

21 mm

31.

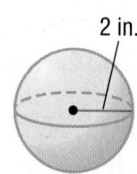

2 in.

Problem Solving in Architecture

OUT OF THIS WORLD ARCHITECTURE

Do you like building things? Are you an excellent problem solver? If so, you have what it takes to be a space architect. Space architects use principles from architecture, design, engineering, and science to create places for people to live and work in outer space. Their designs include transfer vehicles, lunar habitats, and Martian greenhouses. Because of the limitations, space architecture must be very efficient and functional. Every square inch of surface and every cubic inch of space must have a purpose.

21ˢᵗ Century Careers

Are you interested in a career as a space architect? Take some of the following courses in high school.

- Aerospace Technology
- Calculus
- Geometry
- Introductory Space Planning
- Intro to CAD

Get Connect**ED**

DESTINY

8.5 m

4.3 m

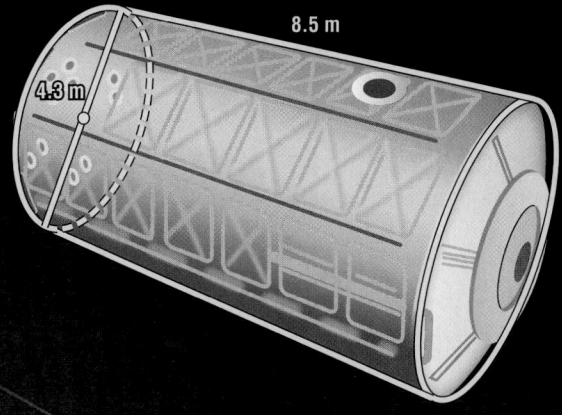

COLUMBUS

4.5 m

6.9 m

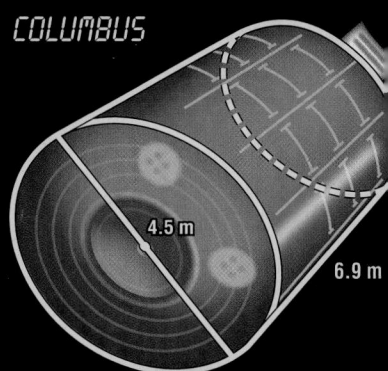

Real-World Math

Use the space laboratories to solve each problem. Round to the nearest tenth.

1. *Destiny* has one round window that is 20 inches in diameter. What is the circumference and area of the window?

2. What is the volume of *Destiny*?

3. The internal volume of *Columbus*, or the space where the astronauts live and work, is about 34.7 cubic meters less than the total volume. What is the internal volume of *Columbus*?

4. Find the surface area of *Destiny*.

5. Without calculating, predict whether *Destiny* or *Columbus* has a greater surface area. Then test your prediction by calculating the solution.

6. *Kibo* is a Japanese laboratory on the International Space Station. It is a cylinder 11.2 meters long with a radius of 2.2 meters. Compare its volume to the volumes of *Destiny* and *Columbus*.

FOLDABLES Study Organizer

Be sure the following Key Concepts are noted in your Foldable.

Key Concepts

Circles (Lesson 1)
- Circumference: $C = \pi d$ or $C = 2\pi r$
- Area: $A = \pi r^2$

Volume (Lesson 2)
- Prism: $V = Bh$
- Cylinder: $V = \pi r^2 h$
- Pyramid: $V = \frac{1}{3}Bh$
- Cone: $V = \frac{1}{3}\pi r^2 h$
- Sphere: $V = \frac{4}{3}\pi r^3$

Surface Area (Lesson 3)
- Prism
  Lateral Surface Area: $L.A. = Ph$
  Total Surface Area: $S.A. = L.A. + 2B$
- Cylinder
  Lateral Surface Area: $L.A. = 2\pi rh$
  Total Surface Area: $S.A. = L.A. + 2\pi r^2$
- Pyramid
  Lateral Surface Area: $L.A. = \frac{1}{2}P\ell$
  Total Surface Area: $S.A. = L.A. + B$
- Cone
  Lateral Surface Area: $L.A. = \pi r\ell$
  Total Surface Area: $S.A. = L.A. + \pi r^2$

Key Vocabulary

center	net
chord	pi
circle	radius
circumference	regular pyramid
composite figure	slant height
composite solid	sphere
diameter	surface area
lateral face	volume
lateral surface area	

Vocabulary Check

State whether each sentence is *true* or *false*. If *false*, replace the underlined word or number to make a true sentence.

1. A composite figure contains at least <u>three</u> or more shapes.

2. <u>Circumference</u> is the distance around a circle.

3. The measure of the space occupied by a solid is called the <u>total surface area</u>.

4. A <u>sphere</u> does not contain any bases.

5. The <u>lateral surface area</u> is the sum of the areas of all of a solid's faces.

6. A <u>lateral face</u> of a solid is any flat surface that is not a base.

7. The <u>radius</u> is the distance across a circle through its center.

8. A <u>circle</u> is the set of all points in a plane that are the same distance from a given point in the plane.

9. The lateral faces of a <u>net</u> are congruent isosceles triangles.

Multi-Part Lesson Review

Lesson 1   Circumference and Area

Circumference and Area of Circles (Lesson 1B)

Find the circumference and area of each circle. Round to the nearest tenth.

10. radius: 18 in.   **11.** diameter: 6 cm

12. LANDSCAPING Bill is planting a circular flower bed. What is the area of the flower bed if the diameter is 30 feet?

EXAMPLE 1   Find the circumference and area of the circle.

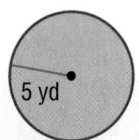

The radius is 5 yards.

$C = 2\pi r$ $A = \pi r^2$

$C = 2 \cdot \pi \cdot 5$ $A = \pi \cdot 5^2$

$C \approx 31.4$ yd $A \approx 78.5$ yd^2

PSI: Make a Model (Lesson 1D)

Solve the problem by using the *make a model* strategy.

13. MEASUREMENT Rosina has a postcard that measures 5 inches by 3 inches. She decides to frame it using a frame that is $1\frac{3}{4}$ inches wide. What is the perimeter of the framed postcard?

14. MAGAZINES A book store arranges magazines in the front window. In how many different ways can five magazines be arranged in a row?

EXAMPLE 2   Cans of oil are displayed in the shape of a pyramid. The top layer has 2 cans in it. One more can is added to each layer, and there are 4 layers in the pyramid. How many cans are there in the display?

Based on the model, there are 14 cans.

Area of Composite Figures (Lesson 1E)

Find the area of each figure. Round to the nearest tenth if necessary.

15.   **16.**

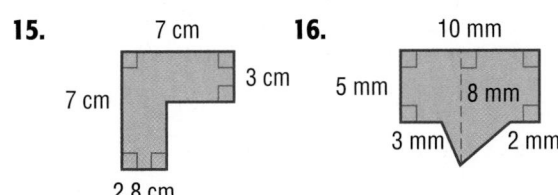

17. BASKETBALL Travis is going to paint part of a basketball court as shown. How many square feet will be painted?

EXAMPLE 3   Find the area of the figure. Round to the nearest tenth.

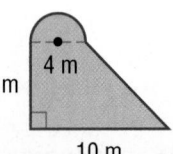

Area of semicircle   Area of trapezoid

$A = \frac{1}{2} \cdot \pi \cdot 2^2$ $A = \frac{1}{2}(6)(4 + 10)$

$A \approx 6.3$ $A = 42$

The area is about $6.3 + 42$ or 48.3 square meters.

Lesson 2 **Volume**

Volume of Prisms and Cylinders (Lesson 2A)

Find the volume of each solid. Round to the nearest tenth if necessary.

18.

7.2 mm
3 mm
4.3 mm

19.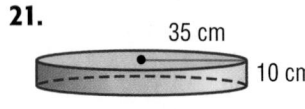
15 yd
8 yd
11 yd
17 yd

20.
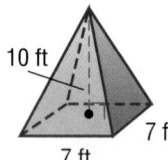
5 in.
8 in.

21.
35 cm
10 cm

22. MUSIC A drum is shown below. Find its volume to the nearest tenth.

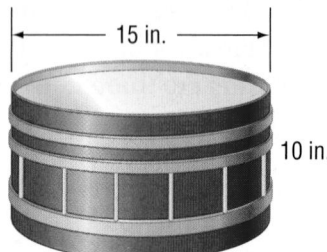
15 in.
10 in.

EXAMPLE 4 Find the volume of the solid.

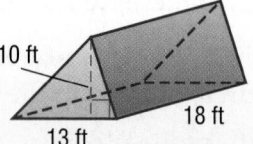
10 ft
13 ft
18 ft

The base of this prism is a triangle.

$V = Bh$

$V = \left(\frac{1}{2} \cdot 13 \cdot 10\right)18$

$V = 1{,}170 \text{ ft}^3$

EXAMPLE 5 Find the volume of the cylinder. Round to the nearest tenth.

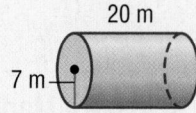
20 m
7 m

The radius is 7 meters. The height is 20 meters.

$V = \pi r^2 h$

$V = \pi \cdot 7^2 \cdot 20$

$V = \pi \cdot 980$

$V \approx 3{,}078.8 \text{ m}^3$

Volume of Pyramids, Cones, and Spheres (Lesson 2B)

Find the volume of each solid. Round to the nearest tenth if necessary.

23.
10 ft
7 ft
7 ft

24.
9 cm
5 cm
12 cm

25. cone: diameter, 9 yd; height, 21 yd

26. TOYS Find the volume of a spherical beach ball with a diameter of 15 inches. Round to the nearest tenth.

27. DECORATIONS A cone-shaped holiday decoration has a height of 12 inches. Find the volume of the decoration if it has a radius of 2.5 inches. Round to the nearest tenth.

EXAMPLE 6 Find the volume of the pyramid.

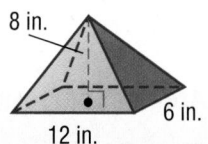
8 in.
6 in.
12 in.

The base B of the pyramid is a rectangle.

$V = \frac{1}{3}Bh$

$V = \frac{1}{3}(12 \cdot 6)8$

$V = 192 \text{ in}^3$

Surface Area of Prisms and Cylinders (Lesson 3B)

Find the lateral and total surface areas of each solid. Round to the nearest tenth if necessary.

28.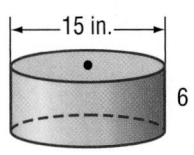
15 in.
6 in.

29.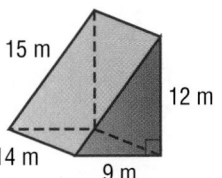
15 m
12 m
14 m
9 m

30.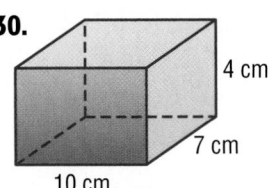
4 cm
7 cm
10 cm

31.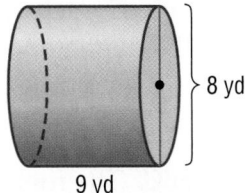
8 yd
9 yd

32. SPORTS The bicycle ramp shown needs to be painted. How many square feet will be painted, not including the bottom?

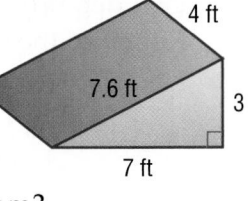
4 ft
7.6 ft
3 ft
7 ft

EXAMPLE 7 Find the total surface area of the prism.

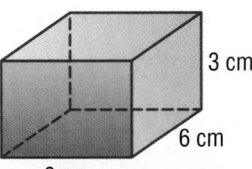
3 cm
6 cm
8 cm

$S.A. = Ph + 2B$
$S.A. = (28)(3) + 2(48)$
$S.A. = 180 \text{ cm}^2$

EXAMPLE 8 Find the total surface area of the cylinder.

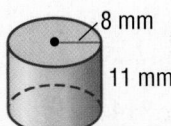
8 mm
11 mm

$S.A. = 2\pi r^2 + 2\pi rh$
$S.A. = 2\pi(8)^2 + 2\pi(8)(11)$
$S.A. \approx 955.0 \text{ mm}^2$

Surface Area of Pyramids and Cones (Lesson 3D)

Find the lateral and total surface areas of each solid. Round to the nearest tenth if necessary.

33.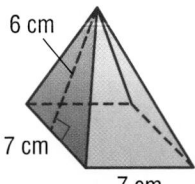
6 cm
7 cm
7 cm

34.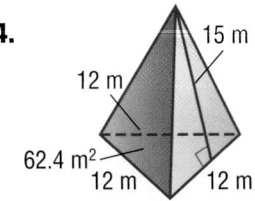
15 m
12 m
62.4 m²
12 m
12 m

35.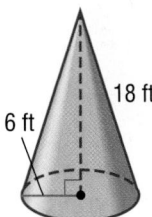
18 ft
6 ft

36.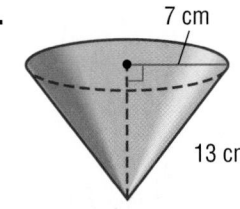
7 cm
13 cm

37. PYRAMIDS A square pyramid has a slant height of 92.5 meters and each side of its base is 183.5 meters long. What is the lateral surface area of the pyramid?

EXAMPLE 9 Find the total surface area of the square pyramid.

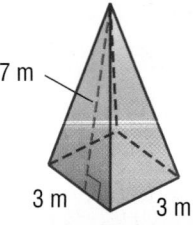
7 m
3 m
3 m

$S.A. = \frac{1}{2}P\ell + B$
$S.A. = \frac{1}{2}(12)(7) + 9$
$S.A. = 51 \text{ m}^2$

EXAMPLE 10 Find the total surface area of the cone.

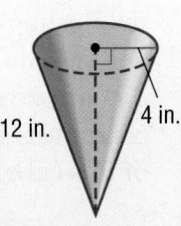
12 in.
4 in.

$S.A. = \pi r\ell + \pi r^2$
$S.A. = \pi(4)(12) + \pi(4)^2$
$S.A. \approx 201.1 \text{ in}^2$

Find the circumference and area of each figure. Round to the nearest tenth if necessary.

1.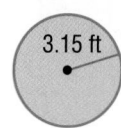
3.15 ft

2.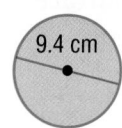
9.4 cm

3. MULTIPLE CHOICE A jogger ran around a circular track two times. If the track has a radius of 25 yards, about how far did the jogger run?

A. 314 yd

B. 157 yd

C. 78.5 yd

D. 50 yd

Find the area of each figure. Round to the nearest tenth if necessary.

4.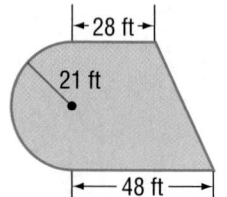
28 ft
21 ft
48 ft

5.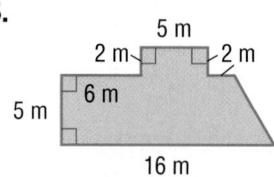
5 m
2 m
2 m
6 m
5 m
16 m

6. INVITATIONS Abigail made 3-inch-by-5-inch invitations for her birthday party. She added a $1\frac{1}{2}$-inch-wide red border on all sides. What is the area of the border? Use the *make a model* strategy.

7. CAKE DECORATION Mrs. Lee designed the flashlight birthday cake shown below. One container of frosting covers 250 square inches of cake. How many containers will she need to frost the top of this cake? Explain.

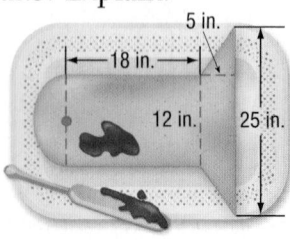
5 in.
18 in.
12 in.
25 in.

Find the volume of each solid. Round to the nearest tenth.

8.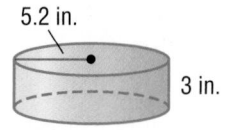
5.2 in.
3 in.

9.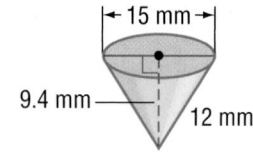
15 mm
9.4 mm
12 mm

10. CIRCUS A clown is juggling four balls. If the diameter of each ball is 2 inches, what is the volume of one ball? Round to the nearest tenth.

11. FUEL The fuel tank is made up of a cylinder. What is the volume of the tank? Round to the nearest tenth.

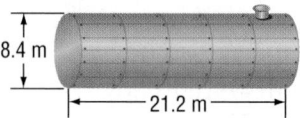
8.4 m
21.2 m

Find the lateral and total surface areas of each solid. Round to the nearest tenth if necessary.

12.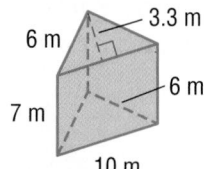
6 m
3.3 m
7 m
6 m
10 m

13.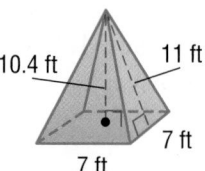
10.4 ft
11 ft
7 ft
7 ft

14. **THINK SOLVE EXPLAIN** **EXTENDED RESPONSE** Refer to the figure below.

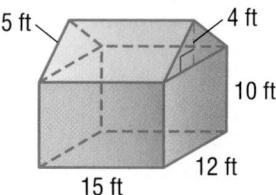
5 ft
4 ft
10 ft
15 ft
12 ft

Part A Describe a method you could use to find the volume of the figure.

Part B Use the method from **Part A** to find the volume.

Part C What is the total surface area of the figure?

Preparing for Standardized Tests

 **Gridded Response: Rounding**

Sometimes it is necessary to round the answer to a gridded-response question. In a situation like this, there may be more than one correct response.

TEST EXAMPLE

Two identical right square pyramids are placed together at their bases to create a party decoration. The base length of each pyramid is 8 centimeters and the total height is 28 centimeters. What is the volume, in cubic centimeters, of the decoration?

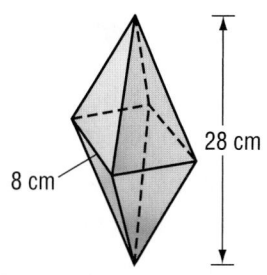

The height of each pyramid is 28 ÷ 2 or 14 centimeters. Find the volume of one pyramid and double it.

$V = \frac{1}{3}\ell wh$

$V = \frac{1}{3}(8)(8)(14)$

$V \approx 298.6666$

The volume is about 298.7 · 2 or 597.4 cubic centimeters. Other correct responses include 597 or 597.0 cubic centimeters, as shown at the right.

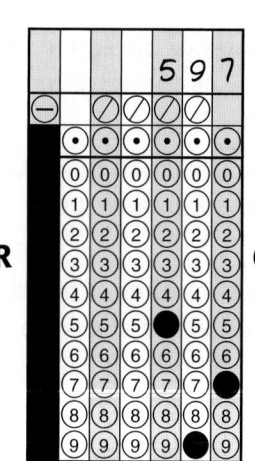 **OR**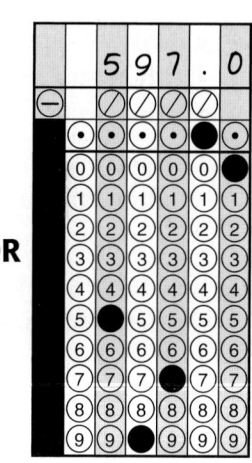

Work on It

Refer to the decoration shown above. The slant height of each pyramid is about 14.6 centimeters. What is the area of the outside surface, in square centimeters, of the decoration? (*Hint:* The bases of the pyramids are not included in this area.) Fill in your answer on an answer grid.

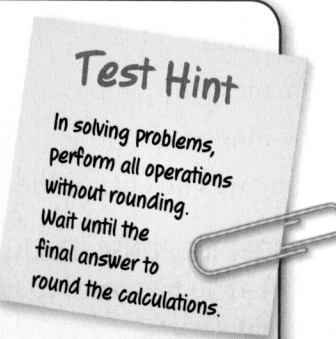

Test Hint

In solving problems, perform all operations without rounding. Wait until the final answer to round the calculations.

Read each question. Then fill in the correct answer on the answer document provided by your teacher or on a sheet of paper.

1. The figure shows a circle inside a square.

Which procedure should be used to find the area of the shaded region?

A. Find the area of the square and then subtract the area of the circle.

B. Find the area of the circle and then subtract the area of the square.

C. Find the perimeter of the square and then subtract the circumference of the circle.

D. Find the circumference of the circle and then subtract the perimeter of the square.

2. The table shows the number of hours students have volunteered at a community center over several months. If the students volunteer 290 hours during the month of September, which measure of data will change the most?

Student Volunteer Hours						
Month	Jan	Feb	Mar	Apr	May	Jun
Hours	145	150	125	165	160	155

F. the mean

G. the median

H. the mode

I. They will all change the same amount.

3. ![THINK SOLVE EXPLAIN icon] **SHORT RESPONSE** The hypotenuse of a right triangle measures 15 feet. If one of the legs measures 9 feet, what is the length of the other leg?

4. ![gridded response icon] **GRIDDED RESPONSE** A rectangular prism has a length of 7.5 inches, a width of 1.4 inches, and a volume of 86.4 cubic inches. What is the height, in inches, of the rectangular prism? Round to the nearest tenth.

5. Which of the following conclusions about the number of rebounds per game and the height of a player is **best** supported by the scatter plot below?

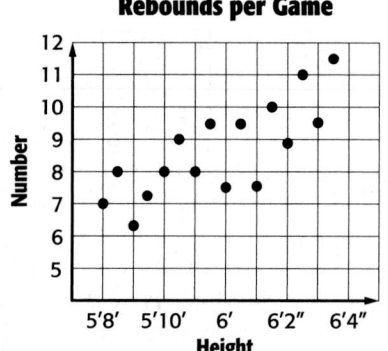

A. The number of rebounds increases as the player's height decreases.

B. The number of rebounds is unchanged as the player's height increases.

C. The number of rebounds increases as the player's height increases.

D. There is no relationship between the number of rebounds and the player's height.

6. A car tire travels about 100 inches in 1 full rotation. What is the radius of the tire, to the nearest inch?

F. 32 inches

G. 28 inches

H. 24 inches

I. 16 inches

7. Which point on the number line below is closest to $\sqrt{50}$?

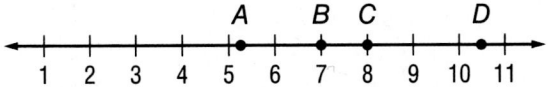

A. point A C. point C

B. point B D. point D

8. ✏ **GRIDDED RESPONSE** In the figure below, every angle is a right angle. What is the area of the figure in square units?

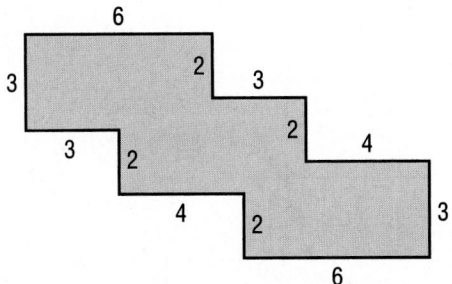

9. Mr. Brauen's farm has a square cornfield. Which of the following is a possible area for the cornfield if the sides are measured in whole numbers?

F. 164,000 ft^2 H. 170,586 ft^2

G. 170,150 ft^2 I. 174,724 ft^2

10. Allison, Carl, and Theo drove from Austin, Texas, to Los Angeles, California, a distance of 1,224 miles. Allison drove $\frac{1}{3}$ of the total distance, Carl drove 40%, and Theo drove the remainder. How many miles were driven by the person who drove the greatest distance?

A. 326.4 mi   C. 489.6 mi

B. 408 mi   D. 897.6 mi

11. The back-to-back stem-and-leaf plot shows the amount of protein in certain foods.

Amount of Protein (g)

Dairy Products	Stem	Legumes, Nuts, Seeds
9 8 8 7 7 6 2 2	0	5 6 9
0	1	4 5 8
6	2	
	3	9

$6|2 = 26$ grams $3|9 = 39$ grams

Which of the following is a true statement?

F. The median amount of protein in dairy products is 9 grams.

G. The average amount of protein in legumes, nuts, and seeds is more than the average amount in dairy products.

H. The difference between the greatest and least amount of protein in dairy products is 28 grams.

I. The greatest amount of protein in legumes, nuts, and seeds is 93 grams.

12. 📋 **EXTENDED RESPONSE** A movie theater has two different containers for popcorn.

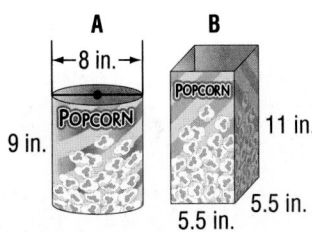

Part A Which container will hold more popcorn? Justify your selection.

Part B Which container requires less packaging to construct? Explain your reasoning.

NEED EXTRA HELP?												
If You Missed Question...	1	2	3	4	5	6	7	8	9	10	11	12
Go to Chapter-Lesson...	12-1E	10-1C	8-2B	12-2A	10-3C	12-1B	2-3C	12-1E	2-3A	1-2B	10-1A	12-3B

Problem-Solving Projects

Design That Bridge

Web Design 101

Basketball All-Star

When will I ever use this?

Have you ever said that? Did you wonder when you would use the math you are learning?

The Problem-Solving Projects apply the math you have learned so far in school. You'll see math in everyday events. Try them!

Green Thumb

Music to My Ears

Web Design 101

Have you ever wondered what it would be like to create a Web page? You have been hired to create a Web page about another country. As a Web designer, you'll research the country to find information about its population, geography, government, and economy. Your Web page will include tables and graphs that can be used to predict future data. You are going to need your algebra skills to design a successful Web page, so grab your mouse and get ready to go live!

What You'll Do

Select and research information about a country. Design a Web page that contains the information you found about the country. Then share your Web page with the class.

Materials:

- Internet

Procedure

1. Choose any country and research the history, government, language, and economy. Record this information with detail.

2. **Get ConnectED** Locate the longitude and latitude of your chosen country. This should be cross-checked on more than one Web site. You can go to connectED.mcgraw-hill.com/ to use the given map as a second source.

3. Research the population of your country over the past 20 years. Create a graph that will predict the population in 20 years.

4. Design your Web page about your country. Include tables, graphs, and photos to help make it visually appealing. Be careful not to overload your page. Share your Web page with the class.

Making the Connection

Use the information collected about your chosen country as needed to help in these investigations.

Language Arts

Write a travel brochure describing the country you are researching. Show your brochure to a person who has never been to the country. Include illustrations and graphs as well as written text.

Art

Create a design for the background of your Web page. Use colors and symbols that are meaningful to the country you are researching.

Physical Education

Research a sport that is popular in the country you chose. Write a few paragraphs describing the sport, including the participants, rules, equipment, and current champions.

Congratulations!

Nice work as a Web designer! We hope you enjoyed getting a taste of an ever-growing career field and learning the mathematical talent it requires.

Design That Bridge

Are you ready to become a civil engineer? You've been one of three civil engineers selected to propose a drawing of a new bridge to be built in your city. Along the way, you'll determine the size, dimensions, material, and layout of your bridge. Your journey will begin soon, so pack your geometry tool kit.

What You'll Do

Create a detailed drawing of a bridge. Use information about similar triangles, parallel lines cut by a transversal, and the Pythagorean Theorem to label all lengths and angles of your bridge. Then present your drawing to the class as if you were presenting it to a committee in your city.

Materials:

• Internet

Procedure

1. Research information about the construction of bridges, including information about the different types. Then write 1 or 2 paragraphs

describing the type of bridge you will design. Also include information about the purpose your bridge will serve in the city.

2. Research information about the different types of material used for bridges. Write 1-2 paragraphs about the material you would like to use for your bridge and why you chose that material.

3. Create a detailed drawing of your bridge, including the front, side, and top views of the bridge. You can use a computer program or make your drawing by hand. Use your geometry skills to label all dimensions and angle measures of your bridge.

4. Create a visual presentation that you would use to help your city choose your bridge design. Include your detailed drawing as well as information about the type of bridge and the materials that would be used.

▶ **Technology Tips**

• Use **geometry software** to help measure angles in your bridge design.

• Use **presentation software** to share your bridge design.

Making the Connection ························

Use the information collected about bridges as needed to help in these investigations.

Language Arts

Select a famous bridge in your state. Research information about that bridge and write an expository paper about it.

Social Studies

Select one bridge style and research its history. How were mathematics involved in the first bridges of this type? How have the bridges improved over time?

Art

Select a famous bridge and draw it. The drawing doesn't necessarily have to be a scale drawing; it can be an artistic interpretation.

Congratulations!

Congratulations on a job well done! Perhaps someday you will become a civil engineer and will use geometry in designing bridges, buildings, and other structures. We hope you enjoyed your task and were able to put your geometry skills to good use.

PROJECT 3 Basketball All-Star

How much do you know about basketball? The Women's National Basketball Association (WNBA) has asked you to analyze several seasons of their basketball data. You'll be required to organize the data in a variety of different representations and analyze the data using the measures of central tendency and measures of variation. Tip-off will soon begin. Let's see if you can make a slam dunk!

What You'll Do

A sports statistician organizes and analyzes data. You will research and find data for given seasons of the WNBA. Then you will analyze the data and create different graphs of it. Finally, you will create your own team of six WNBA players.

Materials:

- Internet

Procedure

1. **Get ConnectED** Go to <u>connectED.mcgraw-hill.com/</u> to get the recording sheet to compute the average points per game for 12 different WNBA teams. Use the Internet to find this information, and then

complete the recording sheet. Find the measures of central tendency and measures of variation for this data set. Then create a box-and-whisker plot of the data.

2. **Get Connect ED** Go to connectED.mcgraw-hill.com/
to get the recording sheet to compute the average points per game for one particular team over the past 10 seasons. Use the Internet to find this information, and then complete the recording sheet. Make a scatter plot of the data. Use the plot to make a prediction of the average number of points the team will score in the next season.

3. Create a fantasy team of six players by researching the statistics of different WNBA players. Once you have chosen your team, create two or three different data displays showing such information as the average number of points per game, average number of rebounds per game, or the average number of assists per game for the players of your team.

4. Make a poster that displays your fantasy team. Include your data displays as well as facts and information about each of your six players.

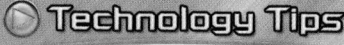

Technology Tips

- Use a **spreadsheet** to create different data displays for your project.

- Use **publishing software** to help in the design of your poster board.

Projects

Making the Connection

Use the information collected about the WNBA as needed to help in these investigations.

Physical Education

Select one of your school's sports and research information about how to play or participate in that sport. Describe the sport, including offense, defense, rules, and ways to win.

Language Arts

Attend a sporting event for your school and assume the role of a journalist. Afterwards, write an article for the sports section of a local newspaper that will highlight the event.

Congratulations!

Great work! Your statistics tool kit likely came in handy with your task. Did you find out some interesting things about the world of basketball? We hope you enjoyed making up your own dream team and saw the large role mathematics plays in sports.

Green Thumb

Do you have a green thumb for gardening? Do you or your family have a garden? If so, what grows in it? You have been chosen to create a new garden. You'll research soil, fruits, vegetables, and flowers for your garden. You will also need to consider the size, weather, and climate. Grab your gardening and math tools as you get ready to break ground.

What You'll Do

Design a garden of your own. Research and include information about the type and amount of soil and seeds you will plant. Create a blueprint of your garden that has dimensions labeled and gives its area and volume.

Materials:

- Internet

Procedure

1. Research the fruits, vegetables, and/or flowers you would like to grow in your garden. Make a table that lists each item, the type of soil needed, planting width, and the ideal temperature for the plant to grow. List the temperature in both degrees Fahrenheit and Celsius.

2. Based on the plants you want to grow and how many of each plant, determine the size of your garden. Once you determine the size, calculate how much soil you need. Give the amount in cubic feet and cubic meters.

3. Create a detailed blueprint of your garden. Include dimensions in feet and meters, and carefully draw and label the location of each type of plant. You need to include a legend that describes each type of plant.

4. Create a visual presentation that includes the following:
   • detailed blueprint of your garden
   • table with specifics about plants in your garden
   • information about the soil in your garden
   • area and volume of your garden

▶ Technology Tips

• Use **presentation software** to share information about your plants with the class.

• Use a **computer projector** to share your garden blueprint with the class.

Making the Connection

Use the information collected about gardening as needed to help in these investigations.

Language Arts

Suppose your garden produces wonderful fruits and vegetables. You decide to sell some of the produce. Create a flyer that advertises what produce you have available, as well as prices. You may even include a recipe that can be used with the produce.

Health

Research the nutritional value of 10 different fruits or 10 different vegetables. Make a table that shows your findings.

Science

Continue your study of the Earth's soil. Write 1-2 paragraphs that describe the qualities of a type of soil that allow for plants to grow well.

Congratulations!

Great job as a gardener! We hope you enjoyed the opportunity to study and learn about gardening. Hopefully someday you can test your gardening skills and math skills and plant a garden of your own.

PROJECT 5 — Music to My Ears

Grab some sheet music. You're about to be a composer! In this project, you'll learn about the connection between math and music. Along the way, you'll research Pythagoras' findings about music and learn how to make harmony. You'll need your problem-solving skills and an ear for music. This is one project that's sure to top the charts!

What You'll Do

Research Pythagoras' findings about music, notes and frequency, and harmony. At the end, you will gather all of the knowledge you have gained about notes and frequency to write your own music. Then write a report explaining how your music is harmonious.

Materials:

- Internet

Procedure

1. Use the Internet to research how Pythagoras' findings affected music. Include information about frequencies, tones, and harmony. Remember to think of the math aspect of the music while you are researching.

2. [**Get ConnectED**] Create your own piece of music based on your knowledge of notes and frequency. It should be around 1-2 minutes long. Be creative when writing your own music, but keep in mind what you learned. You can go to connectED.mcgraw-hill.com/ to obtain a blank piece of sheet music to record your music.

3. Prepare and give a presentation that includes your piece of music and a one-page report of how it is harmonious.

Making the Connection

Use the information collected about Pythagoras and music as needed to help in these investigations.

Science

Research the human ear and how it interprets music. Why are you able to listen to music? What causes your emotions to change while listening? Do different types of music cause different reactions?

Health

Many studies have been done that show a positive connection between music and good health. Research the Internet to find information about one such study. Create a presentation to share with the class to inform them about the study you researched.

Language Arts

Think about different events you attend in which you listen to music. Then research a place or event that centers on music. Write an essay that describes a certain event and discuss the type of entertainment offered and what music is played. Include an explanation of how the music may affect those at the event.

Congratulations!

We hope you enjoyed your journey through music. Now that you are finally back in the moment, what were your impressions of Pythagoras? He was quite a remarkable man! And to think that he and other ancient Greeks discovered so much of the mathematics we know today without the aid of a computer!

Student Handbook

How to Use the Student Handbook

The Student Handbook is the additional skill and reference material found at the end of books. The Student Handbook can help answer these questions.

What if I need more practice?

You, or your teacher, may decide that working through some additional problems would be helpful. The **Extra Practice** section provides these problems for each lesson so you have ample opportunity to practice new skills.

What if I forget a vocabulary word?

The **English/Spanish Glossary** provides a list of new vocabulary words used throughout the textbook. It provides a definition in English and Spanish.

What if I need to check a homework answer?

The answers to the odd-numbered problems are included in **Selected Answers and Solutions**. Check your answers to make sure you understand how to solve all of the assigned problems. Fully worked out solutions to selected problems are also included in this section.

What if I need to find something quickly?

The **Index** alphabetically lists the subjects covered throughout the entire textbook and the pages on which each subject can be found.

What if I forget a formula?

Inside the back cover of your math book is a **Quick Reference** that lists formulas that are used in the book.

Additional Lessons

Extend Scientific Notation Using Technology

Main Idea

Interpret scientific notation when using technology.

 8.EE.3, 8.EE.4

SOLAR SYSTEM The table below shows the mass of some planets in our solar system.

Planet	Mass (kg)
Earth	5,973,700,000,000,000,000,000,000
Mars	641,850,000,000,000,000,000,000
Saturn	568,510,000,000,000,000,000,000

What is the mass of Earth written in scientific notation?

In this activity, you will explore how scientific notation is displayed on a graphing calculator.

ACTIVITY

 What do you need to find? the mass of Earth in scientific notation

 STEP 1 Press CLEAR to clear the home screen.

 **STEP 2** Enter the value for Earth's mass. Press ENTER.

Analyze the Results

1. What are the similarities and differences between 5,973,700,000,000,000,000,000,000 written in scientific notation and the calculator notation shown on your screen?

2. Repeat Steps 1 and 2 for the mass of Mars. What is the mass of Mars in calculator notation?

3. Based on your answer for Exercise 2, what is the mass of Mars in scientific notation?

4. What does the E symbol represent on the calculator screen? What does the value after the E symbol represent?

5. **MAKE A CONJECTURE** Without entering the value in your calculator, predict how the mass of Saturn will be displayed on the calculator screen.

6. The following expressions are written in calculator notation. Write each expression in scientific notation and in standard form.

   a. 3.1E7   **b.** 6.39E10   **c.** 1.7E−11

ACTIVITY

2 **MEASUREMENT** A human blood cell is about 0.000001 meter in diameter. The Moon is about 3,476,000 meters in diameter. How many times greater is the diameter of the Moon than the diameter of a blood cell?

STEP 1 Press CLEAR to clear the home screen.

STEP 2 Enter the values into the calculator.

```
3476000/.000001
           3.476E12
```

KEYSTROKES: 3476000 ÷ 0.000001 ENTER

Analyze the Results

7. Write 3.476E12 in scientific notation. Interpret the meaning of the answer displayed on the calculator screen. Write this value in standard form.

Practice and Apply

8. Refer to the information at the beginning of the lesson. When Saturn's mass is divided by Mars' mass, suppose the calculator displays **8.857365428E2**. What does this value represent?

9. A *micrometer* is 0.000001 meter. Use your calculator to determine how many micrometers are in each of the following. Write your answer in both calculator and scientific notation.

    a. 5,000 meters   **b.** 4.08E14 meters   **c.** 2.9E⁻10 meters

Put your calculator in *scientific* mode by pressing MODE ▶ ENTER. Then press CLEAR to return to the home screen.

10. **MEASUREMENT** The approximate areas of several states are given in the table.

    a. Enter the area of Alaska on your calculator. Press ENTER. What is displayed on the screen? What does this value represent?

    b. Using your calculator, find the area of the remaining states in scientific notation.

    c. About how many times greater is the area of Alaska than the area of New Jersey?

State	Area (mi²)
Alaska	656,000
Texas	269,000
California	164,000
Michigan	97,000
Pennsylvania	46,000
New Jersey	9,000

Lesson 2

Solve Multi-Step Equations

Main Idea

Use Properties of Equality to solve multi-step equations.

New Vocabulary

null set

identity

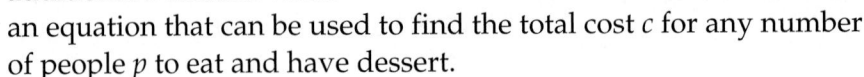

FOOD An all-you-can-eat buffet costs $15 per person.

1. Write an equation that can be used to find the total cost c for any number of people p.

2. In order to have dessert, each person must pay an additional d dollars. Write an equation that can be used to find the total cost c for any number of people p to eat and have dessert.

3. Suppose the total cost for 5 people to eat and have dessert is $90. Write an equation to show the total cost of the buffet if all 5 people order dessert.

CCSS 8.EE.7, 8.EE.7a, 8.EE.7b

To find the cost of the dessert in the above example, you can solve the equation $5(15 + d) = 90$. First, you can use the Distributive Property to remove the grouping symbols. Then solve the equation using Properties of Equality.

EXAMPLE Solve Multi-Step Equations

① Solve $5(15 + d) = 90$.

$5(15 + d) =$	90	Write the equation.
$75 + 5d =$	90	Distributive Property
$-75 = -75$		Subtraction Property of Equality
$5d = 15$		Simplify.
$\dfrac{5d}{5} = \dfrac{15}{5}$		Division Property of Equality
$d = 3$		Simplify.

 CHECK Your Progress

Solve each equation. Check your solution.

a. $-3(9 + x) = 33$ **b.** $5(a - 7) = 24$

c. $2(g + 8) = 4(g - 3)$ **d.** $-6(n + 9) = 4(5n - 7)$

2 **MONEY** At the fair, Hunter bought 3 snacks and 10 ride tickets. Each ride ticket costs $1.50 less than a snack. If he spent a total of $24.00, what was the cost of each snack?

Use a bar diagram.

$24
snack	snack	snack
$--s--$	$--s--$	$--s--$

ticket	ticket	ticket	ticket	ticket	ticket	ticket	ticket	ticket	ticket
$s-1.5$	$s-1.5$	$s-1.5$	$s-1.5$	$s-1.5$	$s-1.5$	$s-1.5$	$s-1.5$	$s-1.5$	$s-1.5$

Write an equation to represent the bar model.

$24 = 3s + 10(s - 1.5)$ Write the equation.

$24 = 3s + 10s - 15$ Distributive Property

$24 = 13s - 15$ Simplify.

$\underline{+15 = \qquad +15}$ Addition Property of Equality

$39 = 13s$ Simplify.

$\dfrac{39}{13} = \dfrac{13s}{13}$ Division Property of Equality

$3 = s$ Simplify.

The cost of each snack was $3.

Check If a snack costs $3, then a ride ticket costs $3 − $1.50 or $1.50. Hunter bought three snacks for $9 and 10 ride tickets for $15. Since $15 + $9 = $24, the answer is correct. ✓

 CHECK Your Progress

e. PETS Deandra's dog weighs fifteen pounds more than Ruby's dog. Jennifer's dog weighs twice the amount of Deandra's dog. If the dogs weigh 91 pounds altogether, how many pounds does Deandra's dog weigh?

Some equations have no solution. When this occurs, the solution is the **null or empty set** and is shown by the symbol ∅ or { }. Other equations may have every number as their solution. An equation that is true for every value of the variable is called an **identity**.

No solution	**All numbers**
$3x + 4 = \quad 3x$	$4x + 2 = 4x + 2$
$\underline{-3x \qquad = -3x}$	$\underline{\quad -2 = \qquad -2}$
$4 = 0$	$4x \qquad = 4x$
	$x = x$
Since $4 \neq 0$, there is no solution.	Since $x = x$, the solution is all numbers.

Solve each equation.

3 $6(x - 3) + 10 = 2(3x - 4)$

$6(x - 3) + 10 = 2(3x - 4)$	Write the equation.
$6x - 18 + 10 = 6x - 8$	Distributive Property
$6x - 8 = 6x - 8$	Simplify.
$\underline{+ 8 = \quad + 8}$	Addition Property of Equality
$6x = 6x$	Simplify.
$\dfrac{6x}{6} = \dfrac{6x}{6}$	Division Property of Equality
$x = x$	Simplify.

The statement $x = x$ is *always* true. The equation is an identity and the solution set is all numbers.

Check	$6(x - 3) + 10 = 2(3x - 4)$	Write the original equation.
	$6(5 - 3) + 10 \stackrel{?}{=} 2[3(5) - 4]$	Substitute any value for x.
	$6(2) + 10 \stackrel{?}{=} 2(15 - 4)$	Simplify.
	$22 = 22 ✓$	

4 $8(4 - 2x) = 4(3 - 5x) + 4x$

$8(4 - 2x) = 4(3 - 5x) + 4x$	Write the equation.
$32 - 16x = 12 - 20x + 4x$	Distributive Property
$32 - 16x = 12 - 16x$	Simplify.
$\underline{+ 16x = \quad + 16x}$	Addition Property of Equality
$32 = 12$	Simplify.

The statement $32 = 12$ is *never* true. The equation has no solution and the solution set is Ø.

Check	$8(4 - 2x) = 4(3 - 5x) + 4x$	Write the equation.
	$8[4 - 2(2)] \stackrel{?}{=} 4[3 - 5(2)] + 4(2)$	Substitute any value for x.
	$8(0) \stackrel{?}{=} 4(-7) + 8$	Simplify.
	$0 \neq -20 ✓$	Since 0 ≠ −20, the equation has no solution.

CHECK Your Progress

Solve each equation. Check your solution.

f. $3(6 - 4x) = -2(6x - 9)$ **g.** $4(5 + 2x) - 5 = 3(3x + 5) - x$

h. $6(4x - 8) + 3 = 15 + 8(3x - 2)$ **i.** $2(3x + 5) = 5(2x - 4) - 4x$

✓ **CHECK Your Understanding**

Examples 1–4 **Solve each equation. Check your solution.**

1. $5(a - 4) = 30$

2. $-8(w - 6) = 32$

3. $4(t - 9) = 6(t + 7)$

4. $5(2d + 8) = 7(2d + 8)$

5. $9(g - 10) - 4g = 8g + 27$

6. $6(r - 4) = 2(r - 8) + 3r$

7. $12(x + 3) = 4(2x + 9) + 4x$

8. $8z - 22 = 3(3z + 11) - z$

Example 2 **9. CHARITY** Mr. Richards's class is holding a canned food drive for charity. Juliet collected 10 more cans than Rosana. Santiago collected twice as many cans as Juliet. If they collected 130 cans altogether, how many cans did Juliet collect?

Practice and Problem Solving

Examples 1–4 **Solve each equation. Check your solution.**

10. $9(j - 4) = 81$

11. $-12(k + 4) = 60$

12. $-5(3m + 6) = -3(4m - 2)$

13. $8(3a + 6) = 9(2a - 4)$

14. $\frac{1}{2}r + 2\left(\frac{3}{4}r - 1\right) = \frac{1}{4}r + 6$

15. $\frac{1}{3}h - 4\left(\frac{2}{3}h - 3\right) = \frac{2}{3}h - 6$

16. $8(4q - 5) - 7q = 5(5q - 8)$

17. $8(t + 2) - 3(t - 4) = 6(t - 7) + 8$

18. $-7(k + 9) = 9(k - 5) - 14k$

19. $-10y + 18 = -3(5y - 7) + 5y$

20. $10p - 2(3p - 6) = 4(3p - 6) - 8p$ **21.** $8(c - 9) = 6(2c - 12) - 4c$

Example 2 **22. PARTIES** The school has budgeted $2,000 for an end-of-year party at the local park. The cost to rent the park shelter is $150. How much can the student council spend per student on food if each of the 225 students receives a $3.50 gift?

23. SCHOOL The table shows the number of students in each homeroom.

a. Write an equation to find the number of students in Mr. Boggs's homeroom if the total number of students is 90.

b. Solve the equation from part **a** to find the number of students in Mr. Boggs's homeroom.

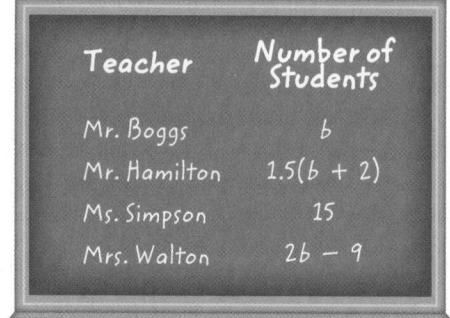

Teacher	Number of Students
Mr. Boggs	b
Mr. Hamilton	$1.5(b + 2)$
Ms. Simpson	15
Mrs. Walton	$2b - 9$

H.O.T. Problems

24. **REASONING** Does a multi-step equation *always, sometimes,* or *never* have a solution? Explain your reasoning.

25. **CHALLENGE** The perimeter of a rectangle is $8(2x + 1)$ inches. If the length of the sides of the rectangle are $3x + 4$ inches and $4x + 3$ inches, what is the length of each side of the rectangle?

26. **WRITE MATH** Write about a real-world situation that could be represented by the equation $5(x + 2) = 5x$.

Test Practice

27. The Yeoman family spent a total of $26.75 on lunch. They bought 5 drinks and 3 sandwiches. Each drink costs $2.50 less than a sandwich. Which of the following equations could be used to find the cost of each sandwich?

   A. $\$26.75 = 5(\$2.50) + 3s$

   B. $\$26.75 = 3(\$2.50) + 5s$

   C. $\$26.75 = 5s + 3(s + 2.50)$

   D. $\$26.75 = 3s + 5(s - \$2.50)$

28. What value of x makes the perimeters of the figures below equal?

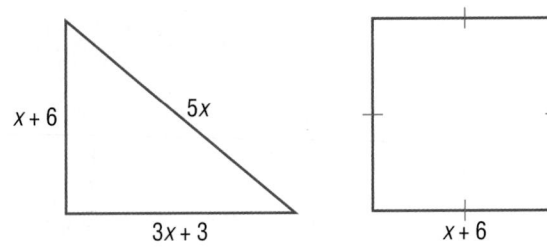

   F. 2

   G. 3

   H. 4

   I. 5

Lesson 3

Solve Equations with Rational Coefficients

Main Idea

Solve equations with rational coefficients.

New Vocabulary

multiplicative inverse

Get Connect**ED**

CCSS 8.EE.7, 8.EE.7b

SOCIAL NETWORKS The bar diagram shows that 18 is the number of students in Naveed's science class that belong to a social network.

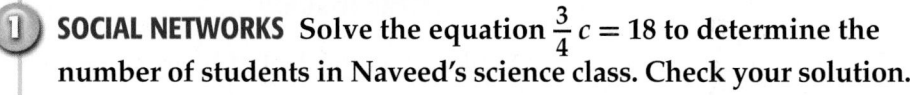

1. What does each section in the diagram represent?
2. What part of Naveed's class belongs to a social network?
3. Based on the diagram, write an equation that can be used to find the number of students in Naveed's science class.

In the equation $\frac{3}{4}c = 18$, the coefficient of c is a rational number. To solve an equation when the coefficient is a fraction, multiply each side by the reciprocal of the fraction.

REAL-WORLD EXAMPLE

1 **SOCIAL NETWORKS** Solve the equation $\frac{3}{4}c = 18$ to determine the number of students in Naveed's science class. Check your solution.

$$\frac{3}{4}c = 18 \qquad \text{Write the equation.}$$

$$\left(\frac{4}{3}\right) \cdot \frac{3}{4}c = \left(\frac{4}{3}\right) \cdot 18 \qquad \text{Multiply each side by the reciprocal of } \frac{3}{4}, \frac{4}{3}.$$

$$\frac{\cancel{4}}{\cancel{3}} \cdot \frac{\cancel{3}}{\cancel{4}}c = \frac{4}{\cancel{3}} \cdot \frac{\cancel{18}^{6}}{1} \qquad \text{Write 18 as } \frac{18}{1}. \text{ Divide by common factors.}$$

$$c = 24 \qquad \text{Simplify.}$$

There are 24 students in Naveed's science class.

Check $\qquad \frac{3}{4}c = 18 \qquad$ Write the original equation.

$$\frac{3}{4}(24) \stackrel{?}{=} 18 \qquad \text{Replace } c \text{ with 24.}$$

$$\frac{3}{\cancel{4}} \left(\frac{\cancel{24}^{6}}{1}\right) \stackrel{?}{=} 18 \qquad \text{Write 24 as } \frac{24}{1}. \text{ Divide by common factors.}$$

$$18 = 18 \checkmark \qquad \text{This sentence is true.}$$

CHECK Your Progress

Solve each equation. Check your solution.

a. $\frac{1}{5}x = 12$ **b.** $-\frac{2}{9}d = 4$ **c.** $15 = \frac{5}{3}n$

In Example 1, you multiplied each side of the equation by $\frac{4}{3}$ because $\frac{3}{4} \cdot \frac{4}{3}$ is 1. Two numbers with a product of 1, such as $\frac{3}{4}$ and $\frac{4}{3}$, are called reciprocals or **multiplicative inverses**.

Key Concept **Inverse Property of Multiplication**

Words	The product of a number and its multiplicative inverse is 1.
Numbers	$\frac{7}{8} \times \frac{8}{7} = 1$ $-\frac{3}{2} \times -\frac{2}{3} = 1$
Symbols	$\frac{a}{b} \cdot \frac{b}{a} = 1$, where a and $b \neq 0$

REAL-WORLD EXAMPLE

2 **CARPENTRY** Mallory has $16\frac{1}{2}$ feet of boards. She is making shelves from the boards that are $1\frac{1}{2}$ feet long. Define a variable. Then write and solve an equation to find the number of shelves Mallory can make.

Mallory has $16\frac{1}{2}$ feet of boards. Each shelf is $1\frac{1}{2}$ feet long.

Let s represent the number of shelves. Write and solve an equation.

$$1\frac{1}{2}s = 16\frac{1}{2}$$ Write the equation.

$$\frac{3}{2}s = \frac{33}{2}$$ Rename $1\frac{1}{2}$ as $\frac{3}{2}$ and $16\frac{1}{2}$ as $\frac{33}{2}$.

$$\left(\frac{2}{3}\right) \cdot \frac{3}{2}s = \left(\frac{2}{3}\right) \cdot \frac{33}{2}$$ Multiply each side by the multiplicative inverse of $\frac{3}{2}$, $\frac{2}{3}$.

$$\frac{\overset{1}{\cancel{2}}}{\underset{1}{\cancel{3}}} \cdot \frac{\overset{1}{\cancel{3}}}{\underset{1}{\cancel{2}}}s = \frac{\overset{1}{\cancel{2}}}{\underset{1}{\cancel{3}}} \cdot \frac{\overset{11}{\cancel{33}}}{\underset{1}{\cancel{2}}}$$ Divide by common factors.

$$s = 11$$ Simplify.

Mallory can make 11 shelves.

CHECK Your Progress

d. **JEWELRY** Juan is cutting pieces of wire from a 10-yard roll to make necklaces for a craft fair. Each piece of wire is $2\frac{1}{2}$ feet long. Define a variable. Then write and solve an equation to determine the number of necklaces Juan can make from the 10-yard roll of wire. (*Hint:* 3 feet = 1 yard)

Real-World Link
Beads date back as far as 40,000 years. Earlier civilizations would trade beads for items of greater value. Today, teens use beads to make one-of-a-kind jewelry pieces.

Sometimes the rational coefficient is a decimal. In this case, divide each side of the equation by the coefficient.

EXAMPLE **Decimal Coefficients**

3 Solve $3.15 = 0.45n$. Check your solution.

$$3.15 = 0.45n \qquad \text{Write the equation.}$$

$$\frac{3.15}{0.45} = \frac{0.45n}{0.45} \qquad \text{Division Property of Equality}$$

$$7 = n \qquad \text{Simplify.}$$

Check $3.15 = 0.45n \qquad$ Write the original equation.

$\, 3.15 = 0.45(7) \qquad$ Replace n with 7.

$\, 3.15 = 3.15\ \checkmark \qquad$ The sentence is true.

QUICK Review

Division

$$0.45\overline{)3.15}$$
$$\underline{-3.15}$$
$$0$$

with result 7

✓ **CHECK Your Progress**

Solve each equation. Check your solution.

e. $4.9 = 0.7t$ **f.** $-1.4m = 2.1$ **g.** $-5.6k = -12.88$

REAL-WORLD EXAMPLE

4 **SPORTS** Latoya's softball team won 75%, or 18, of its games. Define a variable. Then write and solve an equation to determine the number of games the team played.

Latoya's softball team won 18 games, which was 75% of the games played. Let n represent the number of games played. Write and solve an equation.

$$0.75n = 18 \qquad \text{Write the equation. Write 75\% as 0.75.}$$

$$\frac{0.75n}{0.75} = \frac{18}{0.75} \qquad \text{Division Property of Equality}$$

$$n = 24 \qquad \text{Simplify.}$$

Latoya's softball team played 24 games.

QUICK Review

To change a percent to a decimal, move the decimal point two places to the left. Add zeros, if necessary. For example, 3% = 0.03 and 75% = 0.75.

✓ **CHECK Your Progress**

For each situation, define a variable. Then write and solve an equation.

h. SCHOOL Mrs. Henderson has decorated 30%, or 15, of the bulletin boards in the school hallways. How many bulletin boards have been decorated?

i. AREA The land area of Calfornia is about 1% of the land area of the United States. If the land area of California is about 36,000 square miles, what is the approximate land area of the United States?

✓ CHECK Your Understanding

Examples 1 and 3 **Solve each equation. Check your solution.**

1. $\frac{1}{7}n = 20$ **2.** $60 = \frac{3}{4}p$ **3.** $-\frac{27}{25}x = -\frac{9}{5}$

4. $2.5m = 17.5$ **5.** $-1.08 = -0.9y$ **6.** $-2.7t = 810$

Examples 2 and 4 **Define a variable. Then write and solve an equation for each situation.**

7. SAVINGS Demitrius deposited 60% of his paycheck into his savings account. What was the amount of his paycheck?

Savings Deposit Slip	
Demetrius Matthews	
Name	
Amount Deposited	$41.67

8. BOOKS Paula has read $\frac{7}{10}$ of the total pages in a book she is reading for English class. If Paula has read 84 pages, how many pages are in the book?

Practice and Problem Solving

Examples 1 and 3 **Solve each equation. Check your solution.**

9. $6 = \frac{1}{12}v$ **10.** $-\frac{2}{3}w = 60$ **11.** $-\frac{7}{8}k = -21$

12. $\frac{1}{10}s = -13$ **13.** $\frac{20}{9} = -\frac{4}{5}m$ **14.** $-\frac{14}{5} = -\frac{2}{7}n$

15. $9.6 = 1.2b$ **16.** $0.75a = -9$ **17.** $-413.4 = -15.9n$

18. $0.6w = 0.48$ **19.** $-226.8 = 21.6y$ **20.** $-30 = 1.25c$

Examples 2 and 4 **Define a variable. Then write and solve an equation for each situation.**

21. FOOD One third of the bagels in a bakery are sesame bagels. If there are 72 sesame bagels, how many bagels are there?

22. RUNNING To train for a marathon, Uyen ran a total of 71 miles in one month. This distance is $2\frac{1}{2}$ times the distance that she ran in the first week. How many miles did Uyen run in the first week?

23. SCHOOL José correctly answered 80% of the questions on a language arts quiz. If he answered 16 questions correctly, how many questions were on the language arts quiz?

24. TRAVEL The Parker family drove a total of 180 miles on their road trip. This distance is 1.5 times the distance they drove on the first day. How many miles did the Parker family drive on the first day?

25. SURVEY A principal proposed a change in the school dress code. She surveyed students about the proposed change. Yes votes came from 90% of sixth graders, 80% of seventh graders, and 50% of eighth graders. About what percent of the total students voted in favor of the change?

Class	Number of Yes Votes
Grade 6	198
Grade 7	204
Grade 8	91

26. OPEN ENDED Write a real-world problem that can be represented by the equation $\frac{3}{4}c = 20$.

CHALLENGE Determine whether each statement is *true* or *false*. Explain your reasoning.

27. The product of a fraction and its multiplicative inverse is 1.

28. To change a percent to a decimal, move the decimal place two places to the right.

29. To solve an equation with a coefficient that is a fraction, divide each side of the equation by the reciprocal of the fraction.

30. REASONING Complete the statement: If $10 = \frac{1}{5}x$, then $x + 3 = $ ■. Explain your reasoning.

31. WRITE MATH Explain how a multiplicative inverse can be used to solve an equation which has a rational coefficient that is a fraction.

Test Practice

32. THINK SOLVE EXPLAIN **SHORT RESPONSE** What is the reciprocal of $-\frac{4}{3}$?

33. An airplane travels 100 miles in 0.4 hour. Which speed represents the rate of the airplane?

   A. 50 miles per hour

   B. 100 miles per hour

   C. 250 miles per hour

   D. 500 miles per hour

34. A store is having a sale on notebook computers.

Which equation can be used to find the regular price x of a notebook computer that is on sale for $799?

   F. $799x = 0.5$

   G. $0.5x = 799$

   H. $\frac{1}{799}x = 0.5$

   I. $\frac{1}{0.5}x = 799$

Extend Investigating Linear Equations

Main Idea

Solve real-world mathematical problems using two linear equations in two variables.

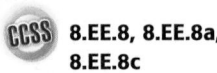

CCSS 8.EE.8, 8.EE.8a, 8.EE.8c

MAPS A map uses a coordinate grid to show the locations of cities and towns. The map locations for four towns are shown in the table. Suppose Brent travels from Town A to Town B and Maria travels from Town C to Town D. Make a graph to determine if Brent and Maria's routes pass through a common location.

Town	Location
A	(0, 6)
B	(5, 1)
C	(0, 4)
D	(4, 8)

ACTIVITY

STEP 1 Copy the coordinate grid shown. Plot the points of each town.

STEP 2 Draw line segments to represent Brent's route and Maria's route.

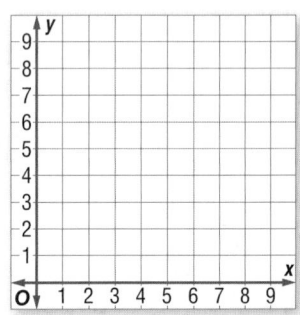

Analyze the Results

1. Find the slope of the lines that represent Brent's route and Maria's route.

2. How can you determine if the lines representing each route intersect using the slope?

3. What do you know about the slopes of parallel lines?

4. Write an equation for the lines that represent Brent's route and Maria's route.

5. Solve the system of equations algebraically.

6. **WRITE MATH** Explain how the solution to Exercise 5 determines if the lines intersect.

Practice and Apply

Write an equation for the line that passes through each pair of points. Then solve the system of equations to find the point of intersection of the lines.

7. (0, −1) and (4, 3); (2, 1) and (0, 3)   8. (6, 7) and (0, 9); (0, 2) and (3, 8)

9. **WRITE MATH** Describe a situation similar to the one above where the routes do not intersect.

Lesson 5

Compare Properties of Functions

Main Idea

Compare properties of two functions represented in different ways.

CCSS 8.F.2, 8.EE.5

TEXTING Carlos and Stephanie have different monthly texting plans. Carlos' plan can be represented by the function $c = 9.99$ where c represents the cost in dollars. Stephanie's plan is shown in the table.

Number of Texts	Cost ($)
1	0.20
2	0.40
3	0.60
4	0.80

1. Describe the rate of change for each function.

2. Who pays more for 50 text messages in one month? Explain.

Functions can be represented by a table, graph, equation, or words. You can compare two functions represented in different forms.

REAL-WORLD EXAMPLE   Compare Two Functions

1 **ANIMALS** A zebra's main predator is a lion. Lions can run at a speed of 53 feet per second over short distances. The graph at the right shows the speed of a zebra. Compare their speeds.

Compare the rates of change.

A lion can travel at a rate of 53 feet per second.

To find the rate of change for a zebra, choose two points on the line and find the rate of change between them.

$$\frac{\text{Change in distance}}{\text{Change in time}} = \frac{(118 - 59)}{(2 - 1)} \text{ or } \frac{59}{1}$$

A zebra can travel at a rate of 59 feet per second. Since $59 > 53$, the speed of a zebra is greater than the speed of a lion.

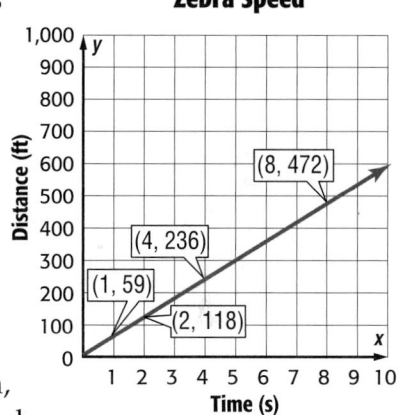

Zebra Speed

CHECK Your Progress

a. **CARS** A certain car has a gas mileage of 22 miles per gallon. The gas mileage of a certain sport utility vehicle is represented by the function shown. Compare their gas mileage.

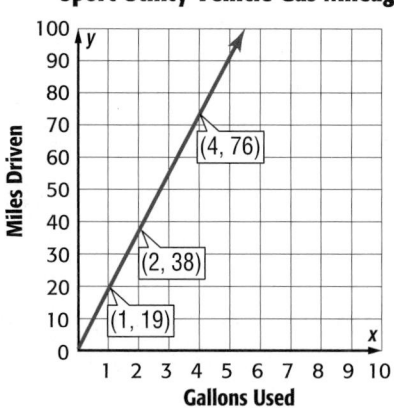

Sport Utility Vehicle Gas Mileage

REAL-WORLD EXAMPLE

2 **TRAINS** The function $m = 140h$, where m is the miles traveled in h hours, represents the speed of the first Japanese high speed train. The speed of a high speed train operating today in China is shown in the table. Assume the relationship between the two quantities is linear.

Train Rate in China	
Hours	Miles
1	217
2	434
3	651

a. Compare the functions' *y*-intercepts and rates of change.

Compare the *y*-intercepts.

At 0 hours, no distance has been covered. So, the *y*-intercepts are the same, 0.

Compare the rates of change.

The speed of the Japanese train is 140 miles per hour.

Use the table to find the speed of the Chinese train.

Train Rate in China	
Hours	Miles
1	217
2	434
3	651

+1 between hours; +217 between miles

The speed of the Chinese train is $\dfrac{217\text{ miles}}{1\text{ hour}}$, or 217 miles per hour.

Since $217 > 140$, the function representing the Chinese high speed train has a greater rate of change than the function representing the Japanese high speed train.

b. If you ride each train for 5 hours, how far will you travel on each?

Find the distance on the Japanese train.

$m = 140h$ Write the function.

$m = 140(5)$ Replace *h* with 5.

$m = 700$ Simplify.

You will travel 700 miles in 5 hours on the Japanese train.

Find the distance on the Chinese train by extending the table.

Train Rate in China	
Hours	Miles
1	217
2	434
3	651
4	868
5	1,085

+1 between hours; +217 between miles

You will travel 1,085 miles in 5 hours on the Chinese train.

 **CHECK Your Progress**

MOVIES The number of new movies a store receives can be represented by the equation $m = 7w + 2$, where m represents the number of movies and w represents the number of weeks. The number of games the same store receives is shown in the table.

Week	Number of New Games
1	3
2	6
3	9

b. Compare the functions' y-intercepts and rates of change.

c. How many new movies and games will the store have in Week 6?

REAL-WORLD EXAMPLE

3 **SAVINGS** Jesse and Juan each open savings accounts. The amounts in each account each week are shown below. Who will save more in 8 weeks? Justify your response.

Jesse's Savings	
Week	Amount Saved ($)
1	5
2	10
3	15
4	20
5	25

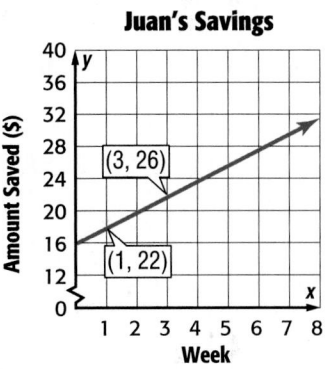

Juan's Savings

Jesse's savings can be represented by the function $s = 5w$, where s represents the amount saved and w represents the week. After Week 8, Jesse will have 5(8) or $40.

Juan's savings can be represented by the function $s = 2w + 20$. After Week 8, Juan will have 2(8) + 20 or $36. So, Jesse will save more after 8 weeks than Juan.

 CHECK Your Progress

d. MOVING The cost to rent a truck from two different companies is shown. Which company should you use if you plan to rent the truck for 20 miles? Justify your response.

Ron's Rentals	
Miles	Cost ($)
10	25
20	50
30	75

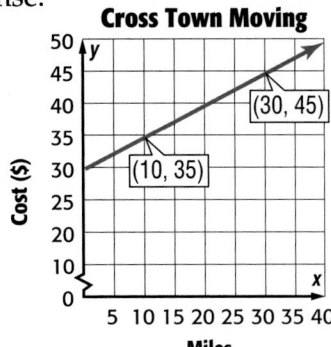

Cross Town Moving

CHECK Your Understanding

Example 1

1. ZOOS A tiger in captivity is fed 13.5 pounds of food a day. The graph shows the pounds of food an elephant in captivity eats per day. Compare the functions by comparing their rates of change.

Elephants

(5, 625)
(4, 500)
(3, 375)
(2, 250)
(1, 125)

Time (days)

Example 2

2. JEWELRY One person's profit at a craft fair is represented by the function $p = 5b - 15$, where p is the profit and b is the number of bracelets sold. A second person's profit is shown in the table.

a. Compare the functions by comparing their y-intercepts and rates of change.

b. How much will the first person make if he sells 30 bracelets?

Bracelets Sold	Profit ($)
1	5
2	10
3	15
4	20

Example 3

3. RAFTING The cost to rent a raft from two different companies is shown. Which company should you use if you plan have the raft for 9 hours? Justify your response.

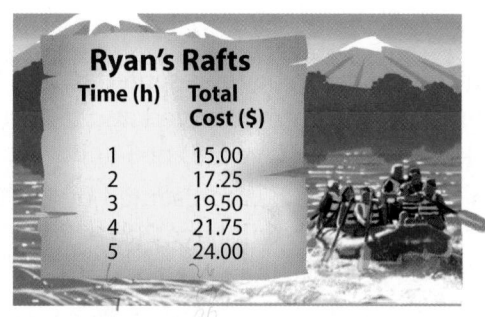

Ryan's Rafts

Time (h)	Total Cost ($)
1	15.00
2	17.25
3	19.50
4	21.75
5	24.00

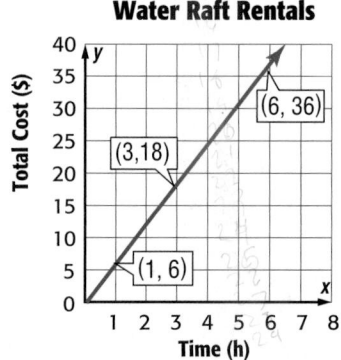

Water Raft Rentals

Total Cost ($)

(6, 36)
(3,18)
(1, 6)

Time (h)

Practice and Problem Solving

Example 1

4. ROAD TRIP For the first leg of the Ramirez family's trip, their speed averages 68 miles per hour. The second leg is shown in the graph. Compare the functions for each part of their trip by comparing the speeds.

Second Leg

Miles Driven

(6, 330)
(1, 55)

Time (h)

5. FABRIC A fabric store sells cotton for $7.00 a yard. The price of special occasion fabric is shown in the graph. Compare the functions' rates of change.

Special Occasion Fabric

(graph showing Cost ($) vs Yards with points (1, 9) and (2, 18))

Example 2 **6. BOOKS** The late fees for a school library are represented by the function $c = 0.25d$, where c is the total cost and d is the number of days a book is late. The fees charged by a city library are shown in the table.

Days Late	1	2	3
Cost ($)	1.25	1.50	1.75

a. Compare the functions' y-intercepts and rates of change.

b. Shamar checks out one book at each library and returns both books 3 days late. What is the total Shamar will have to pay?

7. PATIO The Shaw family is building a patio. One person can lay pavers at a rate of 4.5 per hour. The equation $p = 11h$ represents the number of pavers p that two people can lay for the number of hours h.

a. Compare the functions' y-intercepts and rates of change.

b. How many more pavers can 2 people lay in 3 hours than one person?

Example 3 **8. HOBBIES** Matt and Seth purchase baseball cards each week. The amount of cards they each have in their collection is shown in the graph and table. Who will have more cards in Week 20? Justify your response.

Seth's Collection	
Week	Number of Cards
1	4
2	8
3	12

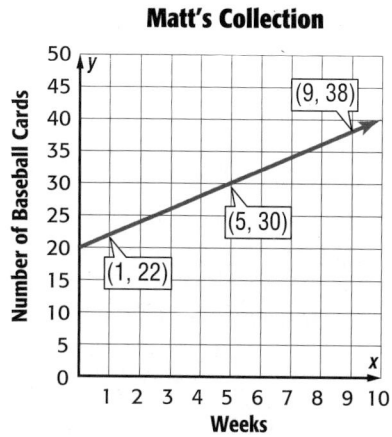

Matt's Collection

(graph showing Number of Baseball Cards vs Weeks with points (1, 22), (5, 30), (9, 38))

9. SPORTS Canada Olympic Park features sports training and entertainment facilities. The Monster zip line produces average speeds of 120 kilometers per hour. A smaller line produces speeds represented by the function $d = 50h$ where d is the distance in kilometers after h hours. How much farther could you travel on the Monster zip line in 0.25 hours? Justify your answer.

10. **MEASUREMENT** Refer to the conversions in the tables below.

Cups	Ounces
1	8
2	16
3	24
4	32

Pints	Ounces
1	16
2	32
3	48
4	64

Quarts	Ounces
1	32
2	64
3	96
4	128

   a. Write an equation for each table.

   b. If you graph the points, the graph for which equation would have the steepest slope? Justify your response.

   c. Which function has the least rate of change? Explain.

11. **RACE** Raj starts running 20 minutes before Jacinda. If Raj runs at a rate of 4.5 miles per hour and Jacinda's progress is represented by a graph that goes through the points (1, 10), (2, 20), and (3, 30), how long will Jacinda need to run to catch up with Raj?

H.O.T. Problems

12. **OPEN ENDED** Write a real-world problem where you would want to compare rates of change for two different functions.

13. **CHALLENGE** Explain why the graph of the function $y = 3x + 40$ will never intersect the graph of the function $y = 3x + 35$.

14. **WRITE MATH** Explain an advantage and disadvantage to representing a function as an equation instead of a graph.

Test Practice

15. A museum charges $12.50 per adult ticket. The price of a student ticket is represented in the table.

Student Ticket Price			
Tickets	1	2	3
Price ($)	8.50	17	25.5

Which statement is not true?

  A. The adult ticket price has a greater rate of change.

  B. Both functions have the same y-intercept.

  C. The student ticket price has a greater rate of change.

  D. Both functions show a direct variation.

16. The exchange rate to convert U.S. dollars to British pounds is represented by the function $p = 1.54d$ where p is the amount in pounds and d is the number of dollars. One U.S. dollar can also be exchange for 0.77 European euro. If you exchange $250 for pounds and $250 for euro, which of the following is true?

  F. You will receive 385 pounds and 192.5 euro.

  G. You will recieve about 162 pounds and about 325 euro.

  H. You will receive 250 dollars.

  I. You will receive the same amount of pounds and euros.

Construct Functions

PARTIES Grace is planning to have her birthday party at a skating rink. The rink charges a party fee plus an additional charge for each guest.

1. Write a function to represent this situation.

2. Use the function to find the amount the skating rink charges for the party fee.

Number of Guests, x	Total Cost ($), y
1	53
2	56
3	59
4	62
5	65
6	68

In the example above, the party fee is the initial cost. The *initial value of a function* is the corresponding y-value when x equals 0. You can find the initial value of a function from tables, graphs, and words.

REAL-WORLD EXAMPLE **Analyze Tables**

1 **MONEY** The table shows how much money Ava has saved. Determine the initial amount of money Ava had in her savings account. Assume the relationship between the two quantities is linear.

Number of Months, x	Money Saved ($), y
3	110
4	130
5	150
6	170

Choose two points from the table and find the rate of change.
We chose (3, 110) and (5, 150).

$$\frac{\text{change in money saved}}{\text{number of months}} = \frac{\$150 - \$110}{(5 - 3) \text{ months}}$$

$$= \frac{\$40}{2 \text{ months}} \text{ or } \frac{\$20}{1 \text{ month}}$$

Ava saves $20 each month. Use this information to determine the initial amount of money Ava had in her savings account. The initial amount of money corresponds to an x-value of 0. Complete the table to find the initial value.

	−1	−1	−1	−1	−1	−1	
Number of Months, x	0	1	2	3	4	5	6
Money Saved ($), y	50	70	90	110	130	150	170
		−20	−20	−20	−20	−20	−20

So, Ava initially had $50 in her savings account.

Real-World Link . . . . .

Approximately 45% of teens use a cell phone and 33% use text messaging.

✓ **CHECK** Your Progress

. . . . . **a. TEXT MESSAGING** The table shows the monthly cost of sending text messages. Determine the initial cost of the phone plan. Assume the relationship between the two quantities is linear.

Number of Messages, x	Cost ($), y
5	10.50
6	10.60
7	10.70

REAL-WORLD EXAMPLE **Analyze Graphs**

② **REWARDS** A new shoe store is offering free points to customers who sign up for their rewards card. Then for each pair of shoes a customer buys they earn an additional number of points. The graph shows the total number of points earned. Determine the number of points initially earned.

Membership Points

Choose two points from the graph and find the rate of change. We chose (2, 60) and (4, 90).

$$\frac{\text{change in points}}{\text{change in pairs}} = \frac{(90 - 60) \text{ points}}{(4 - 2) \text{ pairs}}$$

$$= \frac{30 \text{ points}}{2 \text{ pairs}} \text{ or } \frac{15 \text{ points}}{1 \text{ pair}}$$

For each pair purchased, a customer earns 15 points.

Use the rate of change and the graph to find the y-values when $x = 1$ and 0.

When $x = 1$, $y = 45$. When $x = 0$, $y = 30$.

Membership Points

So, the initial amount of points earned is 30.

✓ **CHECK** Your Progress

b. MUSIC Music Inc. charges a yearly subscription fee plus a monthly fee. The total cost for different numbers of months, including the yearly fee, is shown in the graph. Determine the amount Music Inc. charges for the yearly subscription fee.

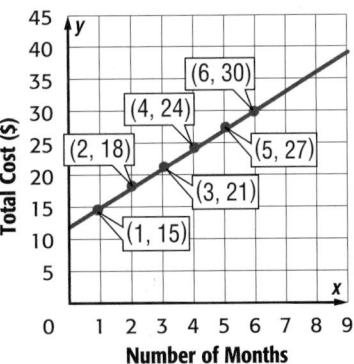

Music Inc. Charges

(6, 30)
(4, 24)
(2, 18)
(5, 27)
(3, 21)
(1, 15)

3 **PHOTOS** Joan has some photos in her photo album. Each week she plans to add an additional amount of photos. The total number of photos in her album for Weeks 2, 4, 8, and 16 respectively are 48, 72, 120, and 216. What was the initial number of photos Joan had in her album? Assume the relationship is linear.

Write the week number and number of photos as ordered pairs, (week number, number of photos): (2, 48), (4, 72), (8, 120), and (16, 216).

Choose two points and find the rate of change of the function. We chose (2, 48) and (4, 72).

$$\frac{\text{change in number of photos}}{\text{change in weeks}} = \frac{(72 - 48) \text{ photos}}{(4 - 2) \text{ weeks}}$$

$$= \frac{24 \text{ photos}}{2 \text{ weeks}} \text{ or } \frac{12 \text{ photos}}{1 \text{ week}}$$

Use the rate of change to find the corresponding y-values when x equals 1 and 0. When $x = 1$, $y = 48 - 12$ or 36. When $x = 0$, $y = 36 - 12$ or 24. So, the initial number of photos is 24.

 CHECK Your Progress

c. **ZOOS** The zoo charges a rental fee plus an hourly usage fee for an audio tour guide. The total costs of 2, 3, 4, and 5 hours respectively are $7, $9, $11, and $13. What is the rental fee? Assume the relationship is linear.

CHECK Your Understanding

Example 1
1. MEMBERSHIP A science center charges an initial membership fee and then the total cost of the membership depends on the number of people included. Use the table that shows the cost of memberships to determine the initial fee. Assume the relationship between the two quantities is linear.

Number of People, x	2	3	4	5
Cost ($), y	65	80	95	110

Examples 2 and 3
2. GAMES As part of a grand opening, an arcade gave out free tokens to the first 100 customers. The graph shows the number of tokens customers received for each dollar spent at the Play More Arcade. Find the initial number of free tokens.

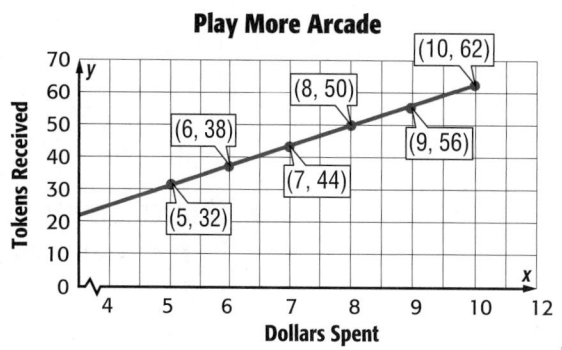

Example 1

3. **CUPCAKES** Melissa is frosting cupcakes for a birthday party. Use the table that shows the number of cupcakes remaining to determine the initial number of cupcakes she had to frost. Assume the relationship between the two quantities is linear.

Time (min), x	5	10	15	20
Remaining Cupcakes, y	28	24	20	16

4. **DVDS** Jonas has a certain number of DVDs in his collection and he decides that each month he will add DVDs to his collection. Use the table to determine the initial number of DVDs he had. Assume the relationship between the two quantities is linear.

Month, x	3	6	9	12
Number of DVDs, y	18	27	36	45

Example 2

5. **READING** A teacher has already read an initial number of pages of a book to a class. The graph shows the number of pages read by the teacher over the next several days. Determine the number of pages initially read by the teacher.

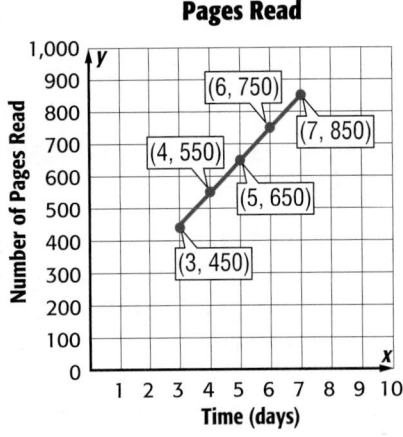

6. **SNOWBOARDING** A ski instructor charges an initial fee and then an hourly rate for private ski lessons. The graph shows the total cost of snowboarding lessons for certain numbers of hours. Determine the initial fee.

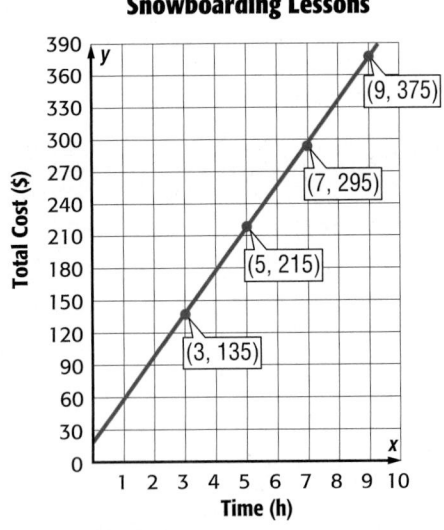

Example 3

7. **WATER PARKS** A water park charges a rental fee plus an hourly fee for inflatable rafts. The total costs of 3, 4, 5, and 6 hours respectively are $8.50, $9, $9.50, and $10. What is the cost of the rental fee? Assume the relationship is linear.

8. **CANNED GOODS** A class has collected a certain number of canned goods. Each day the class plans to bring in the same number of canned goods. The total number of canned goods for Days 5 and 10 respectively are 165 and 205. How many canned goods did the class initially collect?

9. **MULTIPLE REPRESENTATIONS** The Coughlin family is driving from Boston to Chicago. The total distance of the trip is 986 miles and each hour they will drive 65 miles.

   a. **ALGEBRA** Write an equation to represent the number of remaining miles y after driving any number of hours x.

   b. **GRAPHS** Graph the equation from part **a** on a coordinate plane.

   c. **NUMBERS** What is the rate of change and y-intercept of the line?

   d. **WORDS** Explain why the line *slopes down* by 65 for each hour.

   e. **WORDS** Why does the line cross the y-axis at 986?

H.O.T. Problems

10. **OPEN ENDED** Write and solve a real-world problem in which you need to find the initial value of a function. Then explain how you solved your problem.

11. **WRITE MATH** Describe two different methods for finding the initial value of a function. Which method do you prefer to use? Explain your reasoning.

Test Practice

12. The graph shows the number of gallons of water in a bathtub after filling it for a certain number of minutes.

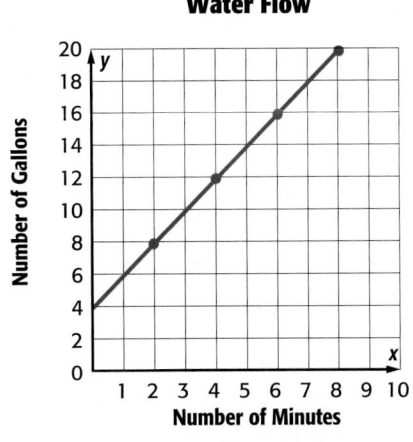

Water Flow

How many gallons of water did the bathtub originally contain?

  A. 2 gallons

  B. 4 gallons

  C. 6 gallons

  D. 8 gallons

13. **EXTENDED RESPONSE** Coach Keller is having the last names of his players printed on their shirts. The lettering company charges a flat fee and a charge per letter. The table shows the total costs.

Number of Letters, x	15	20	25	30
Cost ($), y	7	8.25	9.5	10.75

  PART A What is the rate of change? Explain its meaning.

  PART B What is the initial fee? Explain how you solved.

Lesson 7

Main Idea

Sketch and describe qualitative graphs.

New Vocabulary

qualitative graph

 8.F.5

Qualitative Graphs

Time (s)	Percent Downloaded
0	0
2	15
4	30
6	30
8	64
10	64
12	82
14	100

DOWNLOADS Emily is downloading photos from her digital camera to her computer. The table shows the percent of photos downloaded for several seconds.

1. During which period(s) of time did the percent downloaded not change?

2. During which period of time did the percent downloaded change the most?

The graph below displays the relationship from the table.

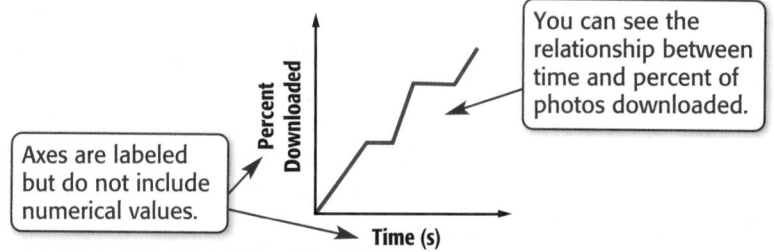

Axes are labeled but do not include numerical values.

You can see the relationship between time and percent of photos downloaded.

The graph above is a qualitative graph. **Qualitative graphs** are graphs used to represent situations that may not have numerical values or graphs in which numerical values are not included. Qualitative graphs represent the essential elements of a situation in a graphical form.

REAL-WORLD EXAMPLE **Analyze Qualitative Graphs**

1. **BATHTUBS** The graph at the right displays the water level in a bathtub. Describe the change in the water level over time.

At time zero, the water level in the bathtub is zero. The water level in the bathtub increases at a constant rate. Then the water is turned off and the water level does not change. Finally, the drain plug is pulled and the water level decreases at a constant rate until the water level is zero.

CHECK Your Progress

a. **WEATHER** The graph at the right displays the temperature throughout the day. Describe the change in the temperature over time.

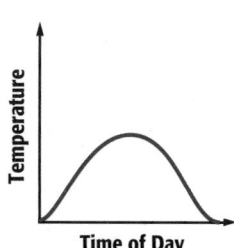

REAL-WORLD EXAMPLE **Sketch Qualitative Graphs**

 SPORTS A tennis ball is dropped onto the floor. On each successive bounce, it rebounds to a height less than its previous bounce height until it comes to rest on the floor. Sketch a qualitative graph to represent the situation.

Step 1 Draw the axes. Label the vertical axis "Distance from Floor." Label the horizontal axis "Time."

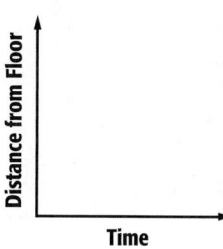

Step 2 Sketch the shape of the graph. The distance from the floor starts out at a high value. The ball falls to the floor, bounces, and rebounds to a height less than its drop height. This pattern is repeated several times until the ball comes to rest on the floor.

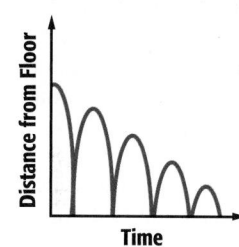

Real-World Link · · · ·
The bounce height of a standard tennis ball is between 53 and 58 inches when dropped onto concrete from a height of 100 inches.

✓ **CHECK Your Progress**

b. DRIVING A car is traveling at a constant speed. The car slows down steadily to come to rest at a stop light. Sketch a qualitative graph to represent the situation.

CHECK Your Understanding

Example 1

1. FLYING The graph below displays the height of an airplane. Describe the change in the airplane's height over time.

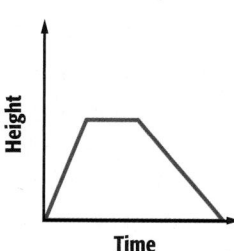

Time

2. MONEY The graph below displays the amount of money in Jared's account. Describe the change in the amount over time.

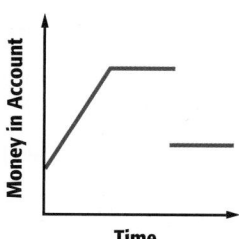

Time

Example 2

3. BICYCLES Tamar rides her bicycle at a steady rate. She coasts downhill which increases her speed at increasing rates. Sketch a qualitative graph to represent the situation.

4. ACTION FIGURES Jamaal purchased the same number of action figures daily for one week. Over the next week, he sold most of them on the Internet. Sketch a qualitative graph to represent the situation.

Real-World Link . . . .

Hot chocolate and other hot liquids do not cool down at a constant rate of change. The temperature drops at a greater rate in the first few minutes and then gradually reaches room temperature.

Example 1

5. WALKING The graph below displays the distance from Luis' home as he walks his dog in his neighborhood. Describe the change in the distance from his home over time.

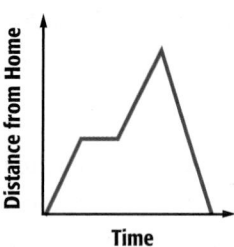

6. BUSES The graph below displays the speed of a city bus as it stops frequently to pick up passengers. Describe the change in the speed over time.

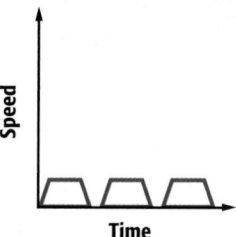

7. TEMPERATURE The graph below displays the temperature of a cup of hot chocolate. Describe the change in the temperature over time.

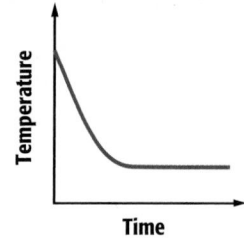

8. BILLS The graph below displays the amount of Mrs. Fraser's electric bill throughout the year. Describe the change in the amount of the bill over time.

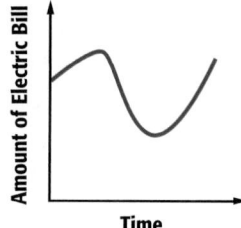

Example 2

9. HAIRCUTS At the beginning of the year, Kristin had medium length hair. She grew her hair out for most of the year. At the end of the year, she got a short haircut. Sketch a qualitative graph to represent the situation.

10. RUNNING Tobias started his morning run slowly. He sped up to his maximum running speed, stopped to drink some water, then increased his speed to his maximum running speed again. Sketch a qualitative graph to represent the situation.

11. WEATHER The outside temperature rises throughout the day at varied rates, then drops at night. Sketch a qualitative graph to represent the situation.

12. LIONS A lion cub is resting in the grass. He sees another lion cub nearby and races after it, picking up speed as it runs. Sketch a qualitative graph to represent the situation.

HIKING For Exercises 13–15, use the graph at the right which displays the rate at which Hector hiked along a path.

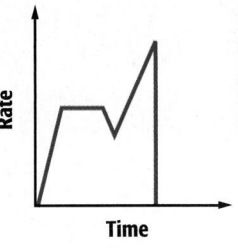

13. What situation could the horizontal line segment represent?

14. What situation could the vertical line segment represent?

15. Did Hector's rate increase or decrease during the first portion of his hike? Explain your reasoning.

H.O.T. Problems

CHALLENGE For Exercises 16 and 17, use the graph at the right which displays the speed of a car as time increases.

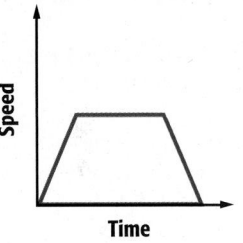

16. Draw a qualitative graph that represents the distance the car travels as time increases.

17. Describe how the distance changes as time passes.

18. **REASONING** A tree grows steadily. When it reaches a specific height, it stops growing. Which graph displays this relationship? Explain.

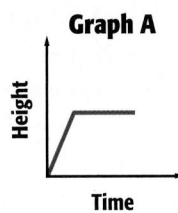
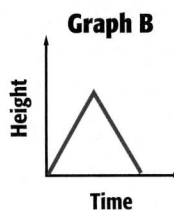
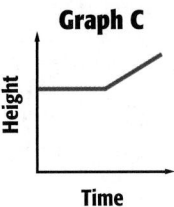

19. **WRITE MATH** Explain some advantages of displaying the relationship between two quantities using a qualitative graph.

Test Practice

20. The graph displays the number of pies sold by a bakery.

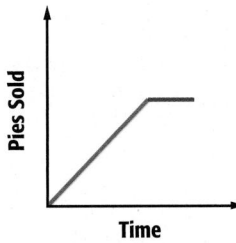

What does the red line segment represent?

A. The number of pies sold increases at a constant rate.

B. The number of pies sold decreases at a constant rate.

C. The number of pies sold decreases, but not at a constant rate.

D. The number of pies sold increases, but not at a constant rate.

Congruence and Transformations

Explore You can compare figures to determine if they are the same size and shape.

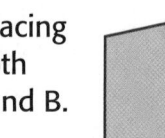

STEP 1 Copy the figure shown on tracing paper two times. Cut out both figures. Label the figures A and B.

STEP 2 Place Figure B on top of Figure A. Are the figures the same?

STEP 3 Slide Figure B up and over on your desk. Can you move Figure A on top of Figure B so all sides and angles match?

STEP 4 Flip Figure B over. Can you move Figure A on top of Figure B so all sides and angles match?

In the activity above, you matched Figure A to Figure B by a translation and a reflection. Two figures are congruent if the second can be obtained from the first by a series of rotations, reflections, and/or translations.

EXAMPLE Identify Congruency

1. **Determine if the two figures are congruent by using transformations.**

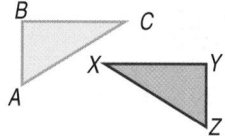

Step 1 Reflect $\triangle ABC$ over a vertical line. Label the vertices of the image A', B', and C'.

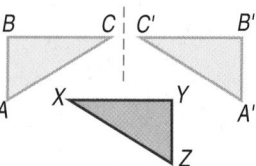

Step 2 Translate $\triangle A'B'C'$ until all sides and angles match $\triangle XYZ$.

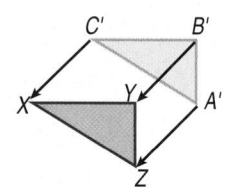

So, the two triangles are congruent because a reflection followed by a translation will map $\triangle ABC$ onto $\triangle XYZ$.

CHECK Your Progress

a. Determine if the two figures are congruent by using transformations. Explain your reasoning.

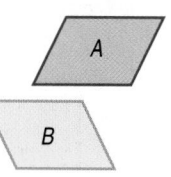

Read Math

Transformations
Translations, reflections and rotations are called *isometries*. In an isometry, the distance between two points in an image is the same as the distance in the preimage.

iso / metry
↓ ↓
same distance

EXAMPLE

2 Determine if the two figures are congruent by using transformations.

Reflect the red figure over a vertical line.

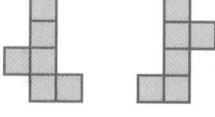

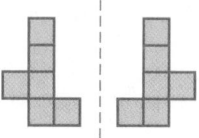

Even if the reflected figure is translated up and over, it will not match the green figure exactly. The two figures are not congruent.

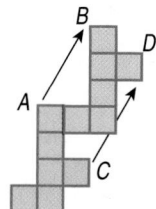

 CHECK Your Progress

b. Determine if the two figures are congruent by using transformations. Explain your reasoning.

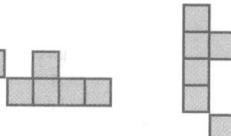

If you have two congruent figures, you can determine the transformation, or series of transformations, that maps one figure onto the other by analyzing the orientation or relative position of the figures.

Translation	Reflection	Rotation
• length is the same • orientation is the same	• length is the same • orientation is reversed	• length is the same • orientation is changed
Notice the segments are facing the same way.	Notice the segments are facing the opposite way.	Notice the segments are facing a different way.

REAL-WORLD EXAMPLE

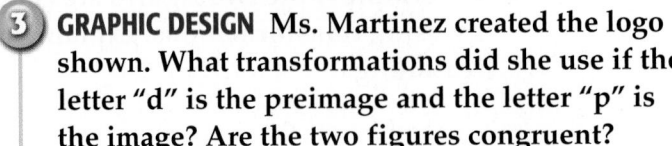

3 **GRAPHIC DESIGN** Ms. Martinez created the logo shown. What transformations did she use if the letter "d" is the preimage and the letter "p" is the image? Are the two figures congruent?

Step 1 Start with the preimage. Determine which transformation will change the orientation of the letter.

Step 2 Rotations or reflections change orientation. Rotate the letter "d" 180° about point A.

Step 3 Translate the new image up.

So, Ms. Martinez used a rotation and translation to create the logo. The letters are congruent because images produced by a rotation and translation have the same shape and size.

Check Trace the letter "d" with tracing paper. Rotate the letter 180° around Point A. Slide it up to line up with the letter "p." The letters are the same shape and size. They are congruent ✓

✅ CHECK Your Progress

c. GRAPHIC DESIGN What transformations could be used if the letter "W" is the preimage and the letter "M" is the image in the logo shown?

CHECK Your Understanding

Examples 1 and 2 Determine if the two figures are congruent by using transformations. Explain your reasoning.

1.

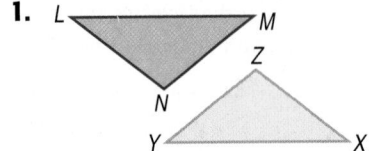

2.

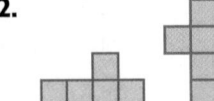

Example 3 **3. LOGOS** The Boyd Box Company uses the logo shown. What transformations could be used if the top, red trapezoid is the preimage and the bottom, blue trapezoid is the image?

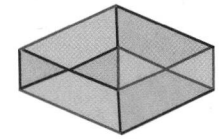

Examples 1 and 2   **Determine if the two figures are congruent by using transformations. Explain your reasoning.**

4.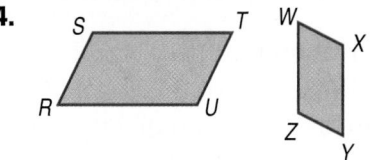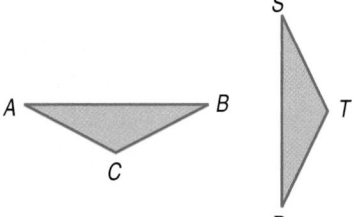

5.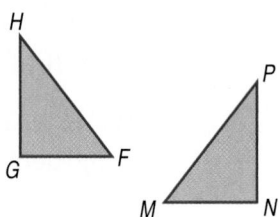

6.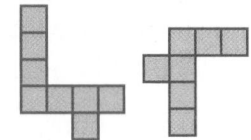

7.

Example 3   **8. STATIONERY** Nilda purchased some custom printed stationery with her initials. What transformations could be used if the letter "Z" is the preimage and the letter "N" is the image in the design shown?

9. ART Simon is illustrating a graphic novel for a friend. He is using the two thought bubbles shown. What transformations did he use if Figure A is the preimage and Figure B is the image?

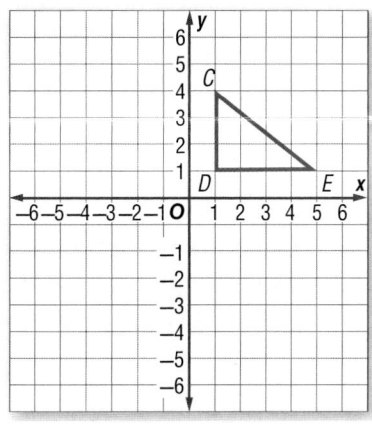

10. One way to identify congruent triangles is to prove their matching sides have the same measure. Triangle CDE has vertices at (1, 4), (1, 1), and (5, 1).

  a. Find the lengths of the sides of $\triangle CDE$.

  b. Reflect $\triangle CDE$ over the y-axis, then translate it 2 units left. Label the vertices of the image $C'D'E'$.

  c. Find the lengths of the sides of $\triangle C'D'E'$.

  d. Are the two triangles congruent? Justify your response.

Find the lengths of the sides of the preimage with the given vertices and the image after the transformations are performed. Then determine if the two figures are congruent.

11. preimage: (0, 1), (4, 0), (4, 1)
    transformations: translate 3 units up then reflect over the y-axis

12. preimage: (0, 0), (4, 0), (0, 4)
    transformations: reflect over the x-axis then dilate by a scale factor of 2

13. **OPEN ENDED** Create a design using a series of transformations that produce congruent figures. Exchange designs with a classmate and determine what transformations were used to create their design.

14. **CHALLENGE** Angle ABC has points $A(-3, 4)$, $B(-2, 1)$ and $C(2, 2)$. Find the coordinates of the image of the angle after a 90° clockwise rotation about the origin, a translation of 2 units up, and a reflection over the y-axis.

15. **CHALLENGE** Line segment XY has endpoints at $X(3, 1)$ and $Y(-2, 0)$. Its image after a series of transformations has endpoints at $X'(0, 1)$ and $Y'(5, 0)$. Find the series of transformations that maps $\overline{XY}$ onto $\overline{X'Y'}$. Then find the exact length of both segments.

16. **WRITE MATH** Explain why rotations, reflections, and translations create congruent images.

Test Practice

17.  **SHORT RESPONSE** Gregory is creating a mosaic for art class. He started by using triangular tiles as shown.

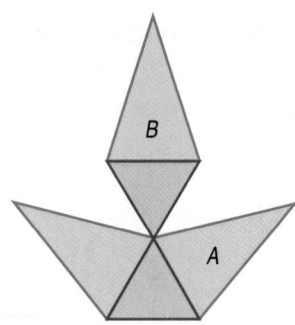

What are possible transformations he used if Figure A is the preimage and Figure B is the image?

18. Triangle MNO is congruent to triangle RST.

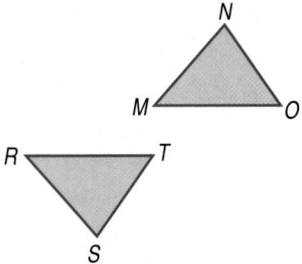

Which series of transformations maps $\triangle MNO$ onto $\triangle RST$?

A. 90° clockwise rotation about M then reflection

B. translation then dilation

C. 90° clockwise rotation about M then translation

D. reflection then translation

Lesson 9

Similarity and Transformations

 In the figure, $\triangle MNP$ is a dilation of $\triangle ABC$.

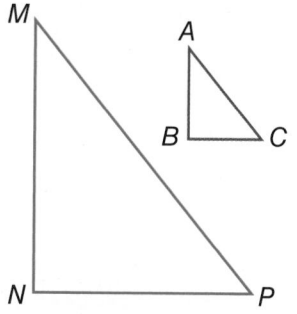

STEP 1 Using a centimeter ruler, measure the side lengths of each triangle. Copy and complete the table.

Figure	Side Length		
$\triangle ABC$	AB	BC	CA
$\triangle MNP$	MN	NP	PM

STEP 2 Copy $\triangle MNP$ on tracing paper. Lay the copy of the triangle so that $\angle M$ is on top of $\angle A$.

STEP 3 Repeat Step 2 for the other two angles. What do you notice about the sides of the triangles? the angles?

Recall that a dilation changes the size of a figure by a scale factor. Since the size of a figure is changed, the image and the preimage are not congruent. Two figures are **similar** if the second can be obtained from the first by a sequence of transformations and dilations.

EXAMPLES Identify Similarity

1. **Determine if the two triangles are similar by using transformations.**

   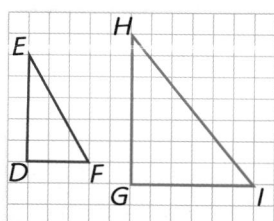

   Since the orientation of the figures is the same, one of the transformations is a translation.

   STEP 1 Translate $\triangle DEF$ down 1 unit and 5 units to the right so D maps onto G.

   STEP 2 Write ratios comparing the lengths of each side.

   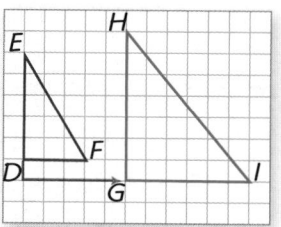

   $$\frac{HG}{ED} = \frac{8}{4} \text{ or } \frac{2}{1}; \frac{GI}{DF} = \frac{6}{3} \text{ or } \frac{2}{1}; \frac{IH}{FE} = \frac{10}{5} \text{ or } \frac{2}{1}$$

   Since the ratios are equal, $\triangle HGI$ is the dilated image of $\triangle EDF$. So, the two triangles are similar because a translation and a dilation maps $\triangle EDF$ onto $\triangle HGI$.

Determine if the two rectangles are similar by using transformations.

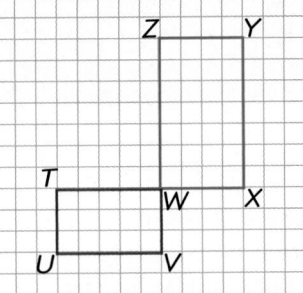

The orientation of the figures is different, so one of the transformations is a rotation.

STEP 1 Rotate rectangle *VWTU* 90° clockwise about W so that it is oriented the same way as rectangle *WXYZ*.

STEP 2 Write ratios comparing the lengths of each side.

$$\frac{WT}{XY} = \frac{5}{7}; \frac{TU}{YZ} = \frac{3}{4};$$

$$\frac{UV}{ZW} = \frac{5}{7}; \frac{VW}{WX} = \frac{3}{4}$$

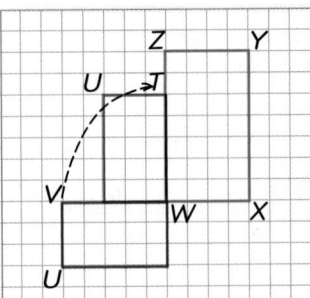

The ratios are not equal. So, the two rectangles are not similar since a dilation did not occur.

 CHECK Your Progress

Determine if the two figures are similar by using transformations. Explain your reasoning.

a.

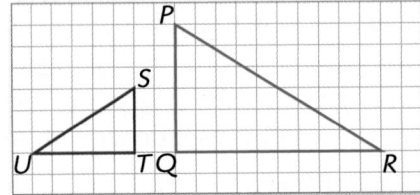

b.

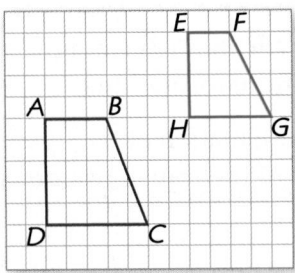

Similar figures have the same shape, but may have different sizes. The sizes of the two figures are related to the scale factor of the dilation.

If the scale factor of the dilation is ...	then the dilated figure is ...
between 0 and 1	smaller than the original
equal to 1	the same size as the original
greater than 1	larger than the original

REAL-WORLD EXAMPLE

PHOTOGRAPY Ken enlarges the photo shown by a scale factor of 2 for his webpage. He enlarges the webpage photo by a scale factor of 1.5 to print. What are the dimensions of the print? Are the enlarged photos similar to the original?

3 in.

2 in.

Multiply each dimension of the original photo by 2 to find the dimensions of the webpage photo.

2 in. × 2 = 4 in.   3 in. × 2 = 6 in.

So, the webpage photo will be 4 inches by 6 inches. Multiply the dimensions of that photo by 1.5 to find the dimensions of the print.

4 in. × 1.5 = 6 in.   6 in. × 1.5 = 9 in.

The printed photo will be 6 inches by 9 inches. All three photos are similar since each enlargement was the result of a dilation.

CHECK Your Progress

c. ART An art show offers different sized prints of the same painting. The original print measures 24 centimeters by 30 centimeters. A printer enlarges the original by a scale factor of 1.5, and then enlarges the second image by a scale factor of 3. What are the dimensions of the largest print? Are both of the enlarged prints similar to the original?

CHECK Your Understanding

Examples 1 and 2   Determine if the two figures are similar by using transformations. Explain your reasoning.

1.
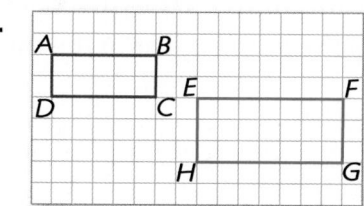

2.
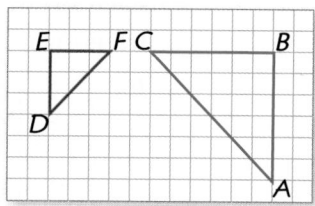

Example 3   **3. DESIGN** A T-shirt iron-on measures 2 inches by 1 inch. It is enlarged by a scale factor of 3 for the back of the shirt. The second iron-on is enlarged by a scale factor of 2 for the front of the shirt. What are the dimensions of the largest iron-on? Are both of the enlarged iron-ons similar to the original?

Practice and Problem Solving

Examples 1 and 2 Determine if the two figures are similar by using transformations. Explain your reasoning.

4.

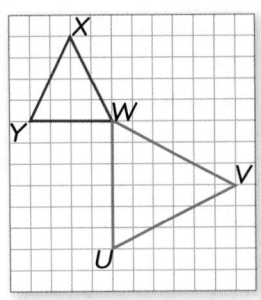

5.

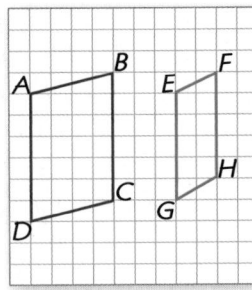

6.

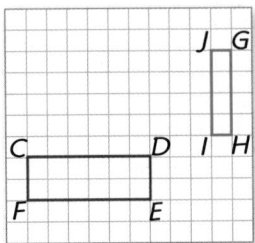

7.

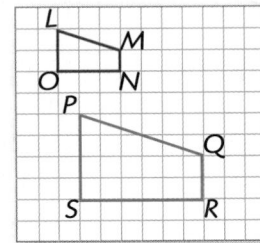

Example 3 **8. SCRAPBOOKING** Felisa is creating a scrapbook of her family. A photo of her grandmother measures 3 inches by 5 inches. She enlarges it by a scale factor of 1.5 to place in the scrapbook. Then she enlarges the second photo by a scale factor of 1.5 to place on the cover of the scrapbook. What are the dimensions of the photo for the cover of the scrapbook? Are all of the photos similar?

9. BLANKETS Shannon is making three different sizes of blankets from the same material. The first measures 2.5 feet by 2 feet. She wants to enlarge it by a scale factor of 2 to make the second blanket. Then she will enlarge the second one by a scale factor of 1.5 to make the third blanket. What are the dimensions of the third blanket? Are all of the blankets similar?

Each preimage and image are similar. Describe a sequence of transformations that maps the preimage onto the image.

10.

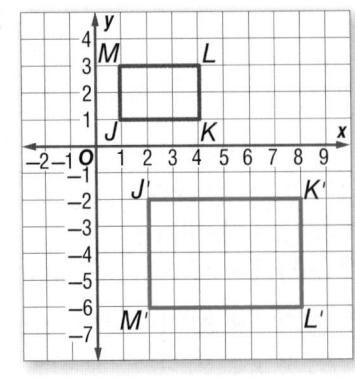

11.

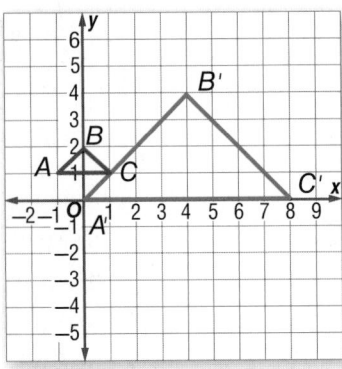

12. In the figure shown, △A′B′C′ is the image of △ABC after a dilation followed by a translation.

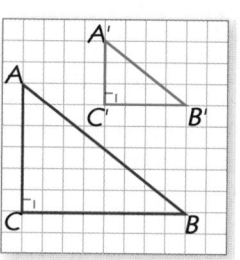

   a. Find the lengths of the sides of △ABC.

   b. Find the lengths of the sides of △A′B′C′.

   c. Write and simplify the following ratios: $\frac{AC}{A'C'}$, $\frac{CB}{C'B'}$, and $\frac{AB}{A'B'}$.

   d. What do you notice about the ratios?

13. In the figure shown, trapezoid RSTU has vertices R(1, 3), S(4, 3), T(3, 1), and U(2, 1).

   a. Draw RSTU on a coordinate grid. Then draw the image of the trapezoid after a translation of 2 units down followed by a dilation with a scale factor of 2. Label the vertices ABCD.

   b. On a different coordinate grid, draw the image of RSTU after a dilation with a scale factor of 2, followed by a translation of 2 units down. Label the vertices EFGH.

   c. Which figures are similar? Which figures are congruent?

   d. Are ABCD and EFGH in the same location? If they are not, what transformation would map ABCD onto EFGH?

H.O.T. Problems

14. **OPEN ENDED** Using at least one dilation, describe a series of transformations where the image is congruent to the preimage.

15. **CHALLENGE** The image of △DEF after two transformations has vertices at D′(3, 3), E′(6, 3) and F′(3, −6). If the two triangles are similar, determine what two transformations map △DEF onto △D′E′F′.

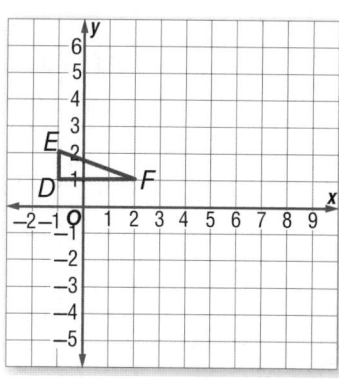

16. **REASONING** *True* or *false*. If a dilation is in a composition of transformations, the order in which you perform the composition does not matter. Explain your reasoning.

17. **WRITE MATH** Explain the difference between using transformations to create similar and congruent figures.

18. Triangle *DEF* is the image of △*ABC* after a sequence of transformations.

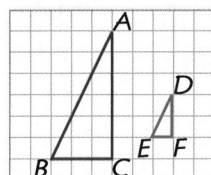

What is the scale factor of the dilation in the sequence?

A. 3

B. $\frac{1}{3}$

C. $-\frac{1}{3}$

D. -3

19. Which transformation produces similar figures that are enlargements or reductions?

F. translation

G. rotation

H. reflection

I. dilation

20. Trapezoid *ABCD* is similar to trapezoid *PQRS*.

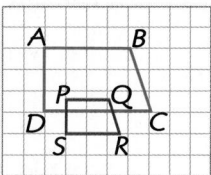

Which series of transformations maps point *C* onto point *R?*

A. rotation then a dilation

B. reflection then a dilation

C. translation then a dilation

D. two dilations

21. ✎ **GRIDDED RESPONSE** Figure B is produced after Figure A is reflected over the *y*-axis and then dilated by a scale factor of $\frac{3}{4}$. What is the ratio comparing the lengths of the sides of Figure A to Figure B?

Lesson 10

Main Idea

Prove the Pythagorean Theorem and its converse.

 Get Connect**ED**

 8.G.6

Extend Proofs About the Pythagorean Theorem

You have used the Pythagorean Theorem and its converse working with specific side measures of triangles. In this activity, you will construct a *geometric proof* of the side relationships of any right triangle. A geometric proof uses definitions or properties to show a general statement is true. This is called *deductive reasoning*.

 ACTIVITY Prove the Pythagorean Theorem

① **STEP 1** Draw and cut out 8 copies of a right triangle. Label each pair of legs *a* and *b*, and each hypotenuse *c*.

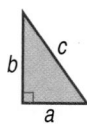

STEP 2 Arrange four of the triangles in a square as shown on a piece of paper. Trace the figure formed by the hypotenuses.

> The length of each side of the large square is $a + b$, so the area of the large square is $(a + b)^2$.

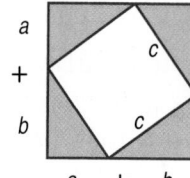

STEP 3 Arrange the remaining triangles as shown. Draw the two figures shown by the dashed lines.

> The length of each side of the large square is $a + b$, so the area of the large square is also $(a + b)^2$.

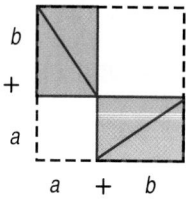

STEP 4 Since the area of each of the two composite figures you created is $(a + b)^2$, the areas are equal.

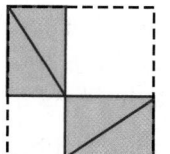 =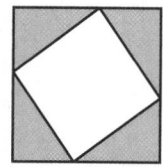

STEP 5 Remove the triangles from each side.

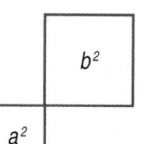

Analyze the Results

1. In Step 2, is the figure formed by the hypotenuses of the triangles a square? Explain.

2. Write an expression for the area of the figure from Exercise 1.

3. In Step 3, are the two figures represented by dashed lines squares? Explain.

4. Find the areas of the two figures from Exercise 3.

5. What property would justify removing four triangles from each side of the diagram in Step 5?

6. Write an algebraic equation that represents the relationship between the figures shown in Step 5.

7. Describe the relationship between the three sides of a right triangle measuring a units, b units, and c units.

The converse of the Pythagorean Theorem states if a triangle has side lengths a, b, and c units such that $a^2 + b^2 = c^2$, then the triangle is a right triangle. In this activity, you will prove the converse of the Pythagorean Theorem.

You'll start with a triangle with sides a and b units long, and then draw a *right* triangle with legs a and b units long. If you can show that the triangles are the same, the first triangle is also a right triangle.

ACTIVITY **Prove the Converse of the Pythagorean Theorem**

2 **GIVEN:** $\triangle ABC$ such that $a^2 + b^2 = c^2$.

PROVE: $\triangle ABC$ is a right triangle.

Complete the table.

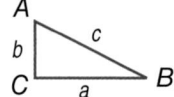

Statement	Reason
Draw a right triangle *DEF* so that side *DE* is a units long, and side *DF* is b units long. Label side *FE* as d.	
Write an equation that describes the relationship between the side lengths of $\triangle DEF$. State the theorem that allows you to make that statement.	

You will finish this proof in Exercise 8.

Analyze the Results

8. Fill in the missing justifications to complete the proof of the converse of the Pythagorean Theorem.

Statement	Reason
You know from the given triangle, that a2 + $b^2 = c^2$.	Given
If $a^2 + b^2 = c^2$ and $a^2 + b^2 = d^2$, what allows you to state that $d^2 = c^2$?	
If $d^2 = c^2$, what allows you to state that $d = c$?	
If $d = c$, what allows you to say that they are the same length?	
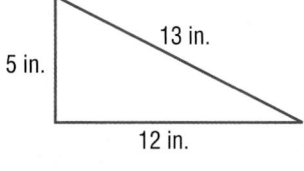 In the two triangles, $AC = FD$, $CB = DE$, and $AB = FE$. So, $\triangle ABC$ and $\triangle FED$ are the same shape and size.	If three sides of a triangle are the same length as the corresponding sides of another triangle, the triangles are the same shape and size.
$\angle C$ is the same as $\angle D$	Corresponding parts of triangles with the same shape and size are the same.
$\angle C$ is a right angle.	
$\triangle ABC$ is a right triangle.	

So, if a triangle has side lengths a, b, and c units such that $a^2 + b^2 = c^2$, then the triangle is a right triangle.

Determine whether the following figures are right triangles. Justify your answer.

9.
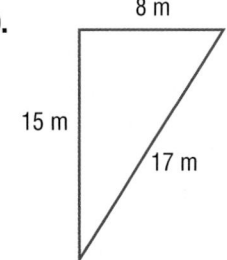
5 in.
13 in.
12 in.

10.
8 m
15 m
17 m

11.
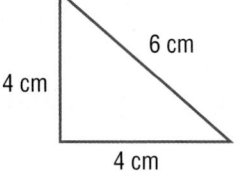
4 cm
6 cm
4 cm

12.
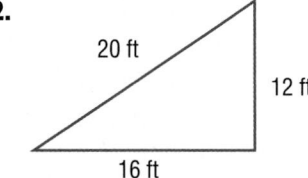
20 ft
12 ft
16 ft

Lesson 11

Main Idea

Use technology to describe nonlinear patterns in scatter plots.

New Vocabulary

nonlinear association

CCSS 8.SP.1

Extend Linear and Nonlinear Association

You can use a graphing calculator to display a scatter plot and find a line of best fit for data that show a *linear association*, or linear relationship.

ACTIVITY

1 **LEISURE** The table shows the weekly number of hours spent watching television and the weekly number of hours spent exercising. Construct a scatter plot of the data. Then find and graph a line of best fit.

Weekly Television (h)	17	20	11	10	15	38	5	25
Weekly Exercise (h)	5	4.5	7.5	8	6.5	1	7.5	3

Weekly Television (h)	25	32	5	17	40	28	20	30
Weekly Exercise (h)	2.5	3.5	6	7	0.5	5	4	1.5

STEP 1 Clear the existing data by pressing [STAT] [ENTER] [▲] [CLEAR] [ENTER]. Then enter the data. Input the number of weekly hours spent watching television in L$_1$ and press [ENTER]. Then enter the weekly hours spent exercising in L$_2$.

STEP 2 Turn on the statistical plot by pressing [2nd] [STAT PLOT] [ENTER] [ENTER]. Select the scatter plot and confirm L$_1$ as the Xlist, L$_2$ as the Ylist, and the square as the mark.

STEP 3 Graph the data by pressing [ZOOM] 9. Use the Trace feature and the left and right arrow keys to move from one point to another.

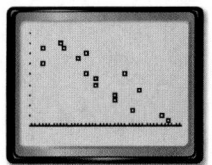

STEP 4 Access the CALC menu by pressing [STAT] [▶]. Select 4 to find a line of best fit in the form $y = ax + b$. Press [2nd] [L$_1$] [,] [2nd] [L$_2$] [ENTER] to find a line of best fit for the data in lists L$_1$ and L$_2$.

STEP 5 Graph the line of best fit in Y₁ by pressing $\boxed{Y=}$ and then $\boxed{VARS}$ 5 to access the Statistics... menu. Use the $\boxed{\blacktriangleright}$ and $\boxed{ENTER}$ keys to select EQ and then press 1 to select RegEQ, the line of best fit equation. Finally, press $\boxed{GRAPH}$.

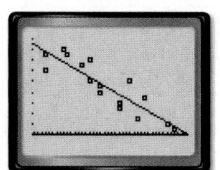

Analyze the Results

1. **MAKE A PREDICTION** Use the Trace feature to predict the average number of hours of exercise someone who watches 35 hours of television would get.

2. **COLLECT THE DATA** Collect a set of data that can be represented in a scatter plot. Use a graphing calculator to determine whether the data have a *positive, negative,* or *no* relationship. Then use the calculator to find a line of best fit and to make a prediction.

There are times when a scatter plot will show an association that is not linear. A **nonlinear association** is one where the pattern does not follow a linear trend.

ACTIVITY

2 The table shows the side lengths and the corresponding areas for various squares. Construct a scatter plot of the data to determine what kind of relationship, if any, exists between the side length of a square and its area.

Side Length (cm)	Area (cm²)
0.5	0.25
1	1
1.5	2.25
2	4
2.5	6.25
3	9
3.5	12.25

STEP 1 Clear the equation from Y₁ by pressing $\boxed{Y=}$ $\boxed{CLEAR}$. Clear data from L₂ in the same manner. Clear the existing data by pressing $\boxed{STAT}$ $\boxed{ENTER}$ $\boxed{\blacktriangle}$ $\boxed{CLEAR}$ $\boxed{ENTER}$.

STEP 2 Next, enter the data. Input the side lengths in L₁ and press $\boxed{ENTER}$. Then enter the areas in L₂.

STEP 3 Turn on the statistical plot by pressing $\boxed{2nd}$ [STAT PLOT] $\boxed{ENTER}$ $\boxed{ENTER}$. Select the scatter plot and confirm L₁ as the Xlist, L₂ as the Ylist, and the square as the mark.

STEP 4 Graph the data by pressing $\boxed{ZOOM}$ 9. Use the Trace feature and the left and right arrow keys to move from one point to another.

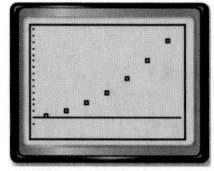

Analyze the Results

3. Does the scatter plot show a *linear association* or a *nonlinear association*? Explain.

4. Find the rate of change between each pair of points. What do you notice about the rates of change as x increases?

5. Write an equation to represent the data.

6. How are the rates of change from this scatter plot different from the rates of change from a scatter plot showing a linear association?

7. In the same viewing window, plot the ordered pairs (side length, perimeter). How is this scatter plot different from the nonlinear scatter plot?

ACTIVITY

3 **ENERGY** Camryn lives in Missouri. She kept track of how much her energy bill was every month for one year. She displayed it in the table shown at the right. Construct and describe a scatter plot of the data.

Month	Bill ($)
January	146
February	138
March	116
April	84
May	72
June	73
July	94
August	114
September	92
October	91
November	126
December	139

STEP 1 Clear the existing data by pressing [STAT] [ENTER] [▲] [CLEAR] [ENTER]. Clear data from L_2 in the same manner.

STEP 2 Next, enter the data. Input the month numbers in L_1 and press [ENTER]. Then enter the amounts of the electric bill in L_2.

STEP 3 Turn on the statistical plot by pressing [2nd] [STAT PLOT] [ENTER] [ENTER]. Select the scatter plot and confirm L_1 as the Xlist, L_2 as the Ylist, and the square as the mark.

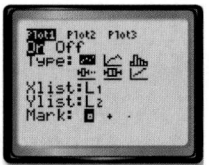

STEP 4 Graph the data by pressing [ZOOM] 9. Use the Trace feature and the left and right arrow keys to move from one point to another.

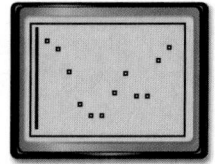

Analyze the Results

8. How is the scatter plot for Activity 3 different from the scatter plot for Activity 2?

9. For Activity 3, what does a negative rate of change mean in the problem's context?

10. **WRITE MATH** For Activity 3, describe how the rates of change are affected as the x-values increase.

Two-Way Tables

Main Idea

Construct and interpret two-way tables.

New Vocabulary
two-way table
relative frequency

CCSS 8.SP.4

SCHOOL The data from a survey of 50 students is shown in the Venn diagram. The students were asked whether or not they were taking a foreign language and whether or not they played a sport.

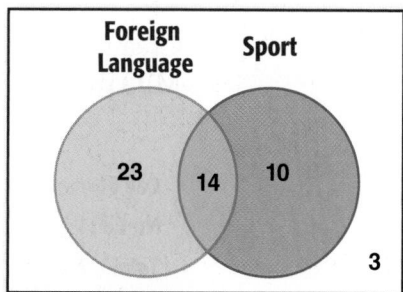

1. How many students are taking a foreign language?
2. How many students play a sport?
3. How many students do both?
4. How many students do not play a sport and do not take a foreign language?
5. How many students play a sport but do not take a foreign language?

A two-way table is similar to a Venn diagram. A **two-way table** shows data that pertain to two different categories. The data from one sample group is shown as it relates to two different categories.

The same information from the Venn diagram above is shown below as a two-way table, where one category is represented by rows and the other category is represented by columns.

	Play a Sport	Do Not Play a Sport	Total
Take a Foreign Language	14	23	14 + 23 or 37
Do Not Take a Foreign Language	10	3	10 + 3 or 13
Total	14 + 10 or 24	23 + 3 or 26	50

The totals shown are for the corresponding row or column with a grand total of 50 students in the data set.

In this lesson, you will learn how to construct and analyze two-way tables from words and diagrams.

EXAMPLE Construct a Two-Way Table

1. **TECHNOLOGY** Felipe surveyed students at his school. He found that 78 students own a cell phone and 57 of those students own an MP3 player. There are 13 students that do not own a cell phone, but own an MP3 player. Nine students do not own either device. Construct a two-way table summarizing the data.

Step 1 Create a table using the two categories: cell phones and MP3 players.

	MP3 Player	No MP3 Player	Total
Cell Phone			
No Cell Phone			
Total			

Step 2 Use the values given to fill in the table.

	MP3 Player	No MP3 Player	Total
Cell Phone	57		78
No Cell Phone	13	9	
Total			

Step 3 Use reasoning to complete the table. Remember, the totals are for each row and column. The column labeled "Total" should have the same sum as the row labeled "Total."

	MP3 Player	No MP3 Player	Total
Cell Phone	57	21	78
No Cell Phone	13	9	22
Total	70	30	100

✔ CHECK Your Progress

a. **SUMMER CAMP** There are 150 children at summer camp and 71 signed up for swimming. There were a total of 62 children that signed up for canoeing and 28 of them also signed up for swimming. Construct a two-way table summarizing the data.

Real-World Link · · · ·
There are about 12,000 summer camps across the United States.

A two-way table can also show relative frequencies. **Relative frequency** is the ratio of the value of a subtotal to the value of the total. In Example 1, the relative frequency of students who own a cell phone who also own an MP3 player is $\frac{57}{78}$ or about 0.73.

$\frac{57}{78}$ ← number of students who own a cell phone and an MP3 player
← total number of students

A two-way table can show relative frequencies for rows or for columns, rather than the actual values.

EXAMPLES **Relative Frequencies in a Two-Way Table**

SCHOOL Using the two-way table from the beginning of the lesson, find the relative frequencies by row and then by column.

2 What is the relative frequency of students that take a foreign language and play a sport to all students taking a foreign language?

To find the relative frequencies by row, write the ratios of each value to the total in that row. Round to the nearest hundredth if necessary.

Frequency by Row	Play a Sport	Do Not Play a Sport	Total
Take a Foreign Language	$\frac{14}{37} \approx 0.38$	$\frac{23}{37} \approx 0.62$	1.00
Do Not Take a Foreign Language	$\frac{10}{13} \approx 0.77$	$\frac{3}{13} \approx 0.23$	1.00
Total	$\frac{24}{50} = 0.48$	$\frac{26}{50} = 0.52$	1.00

> **QUICK Review**
>
> When the numerator and denominator of a fraction are equal, the decimal equivalent is 1.00.

The relative frequency is $\frac{14}{37}$ or about 0.38.

3 What is the relative frequency of students that neither play a sport nor take a foreign language to all students that do not play a sport?

When creating a two-way table with relative frequencies by column, use the total of the columns when writing the ratios.

Frequency by Column	Play a Sport	Do Not Play a Sport	Total
Take a Foreign Language	$\frac{14}{24} \approx 0.58$	$\frac{23}{26} \approx 0.88$	$\frac{37}{50} = 0.74$
Do Not Take a Foreign Language	$\frac{10}{24} \approx 0.42$	$\frac{3}{26} \approx 0.12$	$\frac{13}{50} = 0.26$
Total	1.00	1.00	1.00

The relative frequency is $\frac{3}{26}$ or about 0.12.

 CHECK Your Progress

b. TRAVEL A class was surveyed about whether they have been to Canada or Mexico. Find the relative frequencies by row and then by column for the two-way table shown. Round to the nearest hundredth if necessary. What is the relative frequency of a student who has been to both Canada and Mexico to all students that have been to Mexico?

	Have Been to Canada	Have Not Been to Canada	Total
Have Been to Mexico	6	3	9
Have Not Been to Mexico	5	11	16
Total	11	14	25

CHECK Your Understanding

Example 1   Use the information to construct a two-way table.

1. **SURVEY** Eloise surveyed the students in her cafeteria and found that 38 males agree with the new cafeteria rules while 70 do not. There were 92 females surveyed and 41 of them agree with the new cafeteria rules.

Examples 2 and 3   2. **NEWS** The two-way table shows how some students get their news.

	TV	Internet
7th grade	13	49
8th grade	20	68

   a. How many students were surveyed?

   b. What is the relative frequency of students that responded TV to the total number of students surveyed? Round to the nearest hundredth if necessary.

   c. Do a higher percent of 7th graders or 8th graders get their news from the Internet? Justify your response.

Practice and Problem Solving

Use the information to construct a two-way table.

Example 1   3. **FOOD** There were 100 customers in a restaurant that were asked whether they liked chicken or beef and whether they liked rice or pasta. Out of 30 customers that liked rice, 20 liked chicken. There were 60 customers that liked chicken.

4. **MOVIES** As each person entered the theater, Aaron counted how many of the 105 people had popcorn and how many had a drink. He found that out of 84 people that had popcorn, only 10 did not have a drink. Six people walked in without popcorn or a drink.

Examples 2 and 3   5. **ALLOWANCE** The two-way table shows the number of students that do or do not do chores at home and whether they receive an allowance or not.

	Allowance	No Allowance
Do Chores	13	3
Do Not Do Chores	5	4

   a. How many total students do chores?

   b. What is the relative frequency of students that do chores and get an allowance to the number of students that do chores? Round to the nearest hundredth if necessary.

   c. What is the relative frequency of students that do not do chores nor get an allowance to the total number of students? Round to the nearest hundredth if necessary.

6. SCHOOL The two-way table shows the number of Sasha's soccer teammates that are in her Math class and English class.

	Math Class	Not in Math Class
English Class	4	2
Not in English Class	1	3

a. How many teammates does Sasha have?

b. What is the relative frequency of teammates that are in both of Sasha's classes to all of her teammates?

c. Of the teammates in her math class, which percentage is higher: the percentage of teammates that are in her English class or the percentage of teammates that are not in her English class?

7. MESSAGING The results of a survey show the number of 7th graders and the number of 8th graders that message on a daily basis. Find the relative frequencies by row and then by column. Round to the nearest hundredth if necessary.

	Text Message	Instant Message
7th graders	46	38
8th graders	59	41

Real-World Link · · · ·
A total of 63.4 million volunteers contributed 8.1 billion hours of service in 2009.

8. VOLUNTEERING The two-way table shows the places that males and females volunteered in the past month. Do a higher percentage of males or females volunteer at the animal shelter? Justify your response.

	Males	Females
Animal Shelter	26	21
Hospital	13	17
Library	9	14

9. EXERCISE The Venn diagram shows the number of students that exercise in different ways. Construct a two-way table that displays the data. What is the relative frequency of the number of students that jog and do aerobics to the total number of students? Round to the nearest hundredth if necessary.

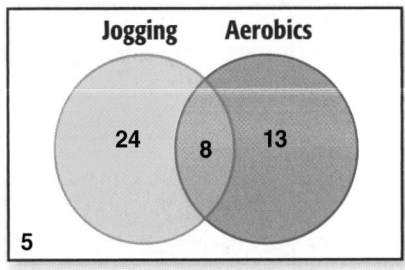

10. LUNCH Cali surveyed the students in the cafeteria about the number of times they bring their lunch to school per month. The table shows her findings. Construct a two-way table that shows the relative frequencies by columns. What is the relative frequency of the number of girls that bring their lunch to school less than 6 times a month to the total number of students surveyed? Round to the nearest hundredth if necessary.

Number of Times per Month	Males	Females
0–5	35	25
6–10	23	16
11–15	22	13
16–20	18	8

H.O.T. Problems

11. **OPEN ENDED** Survey your classmates to find out what kinds of after school jobs they prefer. Make a two-way table that displays your results.

12. **CHALLENGE** The two-way table below shows the number of students with each hair color and eye color.

		Hair Color				
		Black	Brown	Red	Blond	Total
Eye Color	Brown	7	12	3	1	23
	Blue	2	8	2	9	21
	Hazel	2	5	1	1	9
	Green	1	3	1	2	7
	Total	12	28	7	13	60

Which is greater: the percentage of the brown-haired students with blue eyes or the percentage of the red-haired students with brown eyes?

13. **WRITE MATH** Refer to Example 2. Explain how to find the relative frequency of students that do not take a foreign language but play a sport to the students that do not take a foreign language.

Test Practice

14. The two-way table below shows the number of hours students studied and whether they studied independently or with a study group.

	Studied Less Than 2 Hours	Studied More Than 2 Hours
Studied Independently	12	4
Studied with a Study Group	8	11

What is the relative frequency of students that studied independently for more than 2 hours to the total number of students that studied independently?

A. 0.4

B. 0.33

C. 0.25

D. 0.11

15. **SHORT RESPONSE** The Pep Club was asked to vote for which dinner they would like for their banquet. Construct a two-way table for the information shown in the Venn diagram below.

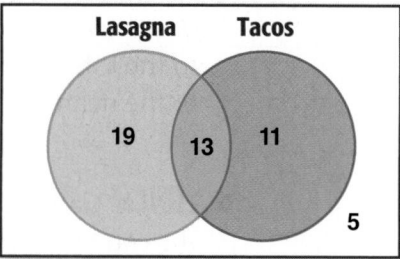

Extra Practice

Multi-Part Lesson 1-1: Rational Numbers

PART A   PAGES 28–32

Write each fraction or mixed number as a decimal.

1. $\frac{2}{5}$

2. $2\frac{3}{11}$

3. $-\frac{3}{4}$

4. $\frac{5}{7}$

5. $\frac{3}{4}$

6. $-\frac{2}{3}$

7. $\frac{7}{11}$

8. $\frac{1}{2}$

9. $\frac{5}{6}$

10. $1\frac{3}{5}$

11. $-2\frac{1}{4}$

12. $\frac{8}{9}$

Write each decimal as a fraction or mixed number in simplest form.

13. 0.5

14. $0.\overline{8}$

15. 0.32

16. -0.75

17. $2.\overline{2}$

18. $0.\overline{38}$

19. -0.486

20. 20.08

21. -9.36

22. $10.1\overline{8}$

23. 1.24

24. $-5.\overline{7}$

PART B   PAGES 33–38

Add or subtract. Write in simplest form.

1. $\frac{17}{21} + \left(-\frac{13}{21}\right)$

2. $\frac{5}{11} + \frac{6}{11}$

3. $-\frac{8}{13} + \left(-\frac{11}{13}\right)$

4. $-\frac{7}{12} + \frac{5}{12}$

5. $\frac{13}{28} - \frac{9}{28}$

6. $-1\frac{2}{9} - \frac{7}{9}$

7. $\frac{15}{16} + \frac{13}{16}$

8. $2\frac{1}{3} - \frac{2}{3}$

9. $-\frac{29}{9} - \left(-\frac{26}{9}\right)$

10. $2\frac{3}{5} + 7\frac{3}{5}$

11. $\frac{5}{18} - \frac{13}{18}$

12. $-2\frac{2}{7} + \left(-1\frac{6}{7}\right)$

13. $\frac{7}{12} + \frac{7}{24}$

14. $-\frac{3}{4} + \frac{7}{8}$

15. $\frac{2}{5} + \left(-\frac{2}{7}\right)$

16. $-\frac{3}{5} - \left(-\frac{5}{6}\right)$

17. $\frac{5}{24} - \frac{3}{8}$

18. $-\frac{7}{12} - \frac{3}{4}$

19. $-\frac{3}{8} + \left(-\frac{4}{5}\right)$

20. $\frac{2}{15} + \left(-\frac{3}{10}\right)$

21. $-\frac{2}{9} - \left(-\frac{2}{3}\right)$

22. $-\frac{7}{15} - \frac{5}{12}$

23. $\frac{3}{8} + \frac{7}{12}$

24. $-2\frac{1}{4} + \left(-1\frac{1}{3}\right)$

25. $3\frac{2}{5} - 3\frac{1}{4}$

26. $\frac{3}{4} + \left(-\frac{4}{15}\right)$

27. $-1\frac{2}{3} + 4\frac{3}{4}$

28. $-\frac{1}{8} - 2\frac{1}{2}$

29. $3\frac{2}{5} - 1\frac{1}{3}$

30. $5\frac{1}{3} + \left(-8\frac{3}{7}\right)$

31. $\frac{3}{5} - \frac{2}{3}$

32. $1\frac{1}{3} - 2\frac{5}{6}$

PART C   PAGES 39–43

Multiply. Write in simplest form.

1. $\frac{2}{11} \cdot \frac{3}{4}$

2. $4\left(-\frac{7}{8}\right)$

3. $-\frac{4}{7} \cdot \frac{3}{5}$

4. $\frac{6}{7}\left(-\frac{7}{12}\right)$

5. $\frac{7}{8} \cdot \frac{1}{3}$

6. $\frac{3}{4} \cdot \frac{4}{5}$

7. $-1\frac{1}{2} \cdot \frac{2}{3}$

8. $\frac{5}{6} \cdot \frac{6}{7}$

9. $8\left(-2\frac{1}{4}\right)$

10. $-3\frac{3}{4} \cdot \frac{8}{9}$

11. $\frac{10}{21}\left(-\frac{7}{8}\right)$

12. $-1\frac{4}{5}\left(-\frac{5}{6}\right)$

13. $5\frac{1}{4} \cdot 6\frac{2}{3}$

14. $-8\frac{3}{4} \cdot 4\frac{2}{5}$

15. $6 \cdot 8\frac{2}{3}$

16. $\left(\frac{3}{5}\right)\left(\frac{3}{5}\right)$

17. $-4\frac{1}{5}\left(-3\frac{1}{3}\right)$

18. $-8\left(\frac{3}{4}\right)$

19. $3\frac{2}{3}\left(-3\frac{1}{2}\right)$

20. $\left(-\frac{2}{5}\right)\left(-\frac{2}{5}\right)$

21. $4\frac{1}{2}\left(-1\frac{1}{3}\right)$

22. $-5\left(-3\frac{1}{5}\right)$

23. $4\frac{1}{3} \cdot 1\frac{1}{2}$

24. $-5\left(3\frac{1}{3}\right)$

Multi-Part Lesson 1-1 (continued)

PART D

PAGES 44–48

Write the multiplicative inverse of each number.

1. 3   **2.** -5 **3.** $\frac{2}{3}$ **4.** $2\frac{1}{8}$

5. $\frac{1}{15}$ **6.** -8 **7.** $1\frac{1}{3}$ **8.** $-\frac{4}{5}$

Divide. Write in simplest form.

9. $\frac{2}{3} \div \frac{3}{4}$ **10.** $-\frac{4}{9} \div \frac{5}{6}$ **11.** $\frac{1}{3} \div 4$ **12.** $5\frac{1}{4} \div \left(-2\frac{1}{2}\right)$

13. $-6 \div \left(-\frac{4}{7}\right)$ **14.** $-6\frac{3}{8} \div \frac{1}{4}$ **15.** $\frac{6}{7} \div \frac{3}{5}$ **16.** $3\frac{1}{3} \div (-4)$

17. BUILDING Mr. Thompson and his two children built a tree house in their backyard. It took them 15 days to complete the project. How long would it take Mr. Franklin and 4 children to build a similar tree house?

Multi-Part Lesson 1-2: Percents

PART A

PAGES 50–51

Use the *look for a pattern* strategy to solve Exercises 1 and 2.

1. NUMBERS Find the next two integers in the pattern 48, 36, 25, 15, 6, ▪, ▪.

2. MONEY A car rental company charges a flat rate of $24.95 and $0.12 per mile. If the total cost of renting a car was $60.95, how many miles were driven?

Number of Miles	Charges	Cost ($)
0	24.95 + 0(0.12)	24.95
50	24.95 + 50(0.12)	30.95
100	24.95 + 100(0.12)	36.95
150	24.95 + 150(0.12)	42.95

PART B

PAGES 52–57

Write each percent as a decimal.

1. 2%   **2.** 25%   **3.** 29%   **4.** 6.2%

5. 16.8%   **6.** 14%   **7.** 23.7%   **8.** 42%

Write each decimal as a percent.

9. 6.21   **10.** 0.08   **11.** 0.036   **12.** 2.34

13. 0.4   **14.** 0.75   **15.** 0.125   **16.** 0.01

Write each fraction as a percent.

17. $\frac{2}{5}$ **18.** $\frac{1}{3}$ **19.** $\frac{2}{25}$ **20.** $\frac{9}{75}$

21. $\frac{7}{3}$ **22.** $\frac{14}{25}$ **23.** $\frac{11}{40}$ **24.** $\frac{9}{20}$

Multi-Part Lesson 1-2 (continued)
PART C
PAGES 58–63

Solve each problem using a percent proportion or equation.

1. 39 is 5% of what number? **2.** What is 19% of 200?

3. 6 is what percent of 30? **4.** 24 is what percent of 72?

5. 9 is $33\frac{1}{3}$% of what number? **6.** Find 55% of 134.

7. 8 is what percent of 32? **8.** What is 35% of 215?

9. 62 is 50% of what number? **10.** 93 is what percent of 186?

11. 90 is 36% of what number? **12.** 15 is 60% of what number?

13. What is 15% of 60? **14.** 15 is 20% of what number?

15. 66 is 75% of what number? **16.** 31 is what percent of 155?

17. 22 is 25% of what number? **18.** What is 65% of 150?

19. 6 is 75% of what number? **20.** 27 is what percent of 100?

Multi-Part Lesson 1-3: Apply Percents
PART A
PAGES 64–68

Find the sale price or total cost of each item to the nearest cent.

1. piano: $4,220, 35% off **2.** scissors: $14, 10% off

3. book: $29, 40% off **4.** sweater: $38, 25% off

5. jeans: $45, 6% tax **6.** skateboard: $105, 7.5% tax

7. gloves: $49.95, 5.25% tax **8.** coat: $145, 6.25% tax

Find the selling price for each item given the cost to the store and the markup.

9. golf clubs: $250, 30% markup **10.** compact disc: $17, 15% markup

11. shoes: $57, 45% markup **12.** necklace: $85, 35% markup

PARTS B C
PAGES 69–73

Find the simple interest to the nearest cent.

1. $500 at 7% for 2 years **2.** $2,500 at 6.5% for 36 months

3. $8,000 at 6% for 1 year **4.** $1,890 at 9% for 42 months

5. $760 at 4.5% for $2\frac{1}{2}$ years **6.** $12,340 at 5% for 6 months

Find the total amount in each account to the nearest cent, if the interest is compounded annually.

7. $300 at 10% for 3 years **8.** $3,200 at 8% for 2 years

9. $20,000 at 14% for 2 years **10.** $4,000 at 12.5% for 4 years

11. $450 at 11% for 2 years **12.** $17,000 at 15% for 3 years

Multi-Part Lesson 1-3 (continued)
PART D

PAGES 74–77

Find each percent of change. Round to the nearest tenth if necessary. State whether the percent of change is an *increase* or a *decrease*.

1. original: 450 centimeters
   new: 675 centimeters

2. original: 77 million
   new: 200.2 million

3. original: 500 albums
   new: 100 albums

4. original: 350 yards
   new: 420 yards

5. original: 3.25 meters
   new: 2.95 meters

6. original: $65
   new: $75

7. original: 180 dishes
   new: 160 dishes

8. original: 450 pieces
   new: 445.5 pieces

9. original: 700 grams
   new: 910 grams

10. original: 55 women
    new: 11 women

Multi-Part Lesson 2-1: Laws of Exponents
PART A

PAGES 91–96

Write each expression using exponents.

1. $4 \cdot 4 \cdot 4 \cdot 4$

2. $\frac{3}{4} \cdot \frac{3}{4}$

3. $7 \cdot 7 \cdot 7 \cdot 7 \cdot 7 \cdot 7$

4. $4 \cdot 4 \cdot 4 \cdot 4 \cdot 4 \cdot 5 \cdot 5 \cdot 5 \cdot 5 \cdot 5 \cdot 5 \cdot 5 \cdot 5$

5. $3 \cdot 2 \cdot \frac{5}{6} \cdot \frac{5}{6} \cdot \frac{5}{6} \cdot 2 \cdot 2 \cdot 2 \cdot 3 \cdot \frac{5}{6}$

6. $b \cdot b \cdot b \cdot b \cdot c \cdot c \cdot c \cdot c \cdot c \cdot c$

7. $3 \cdot 2 \cdot 5 \cdot 5 \cdot 5 \cdot 2 \cdot 2 \cdot 2 \cdot 3 \cdot 5$

Evaluate each expression.

8. 4^3

9. 6^2

10. $\left(\frac{2}{5}\right)^3$

11. $5^2 \cdot 6^2$

12. $3 \cdot 2^4$

13. $10^4 \cdot 3^2$

14. $5^3 \cdot 1^9$

15. $2^2 \cdot 2^4$

16. $2 \cdot 3^2 \cdot 4^2$

17. 7^3

18. $\left(\frac{1}{2}\right)^3 \cdot 4^5$

19. $3^5 \cdot 4^2$

20. $7^2 \cdot 3^4$

21. $\left(\frac{1}{3}\right)^3$

22. $(-2)^4$

23. $(-5)^3$

PART B

PAGES 97–101

Simplify. Express using exponents.

1. $2^3 \cdot 2^4$

2. $5^6 \cdot 5$

3. $t^4 \cdot t^2$

4. $y^5 \cdot y^3$

5. $(-3x^3)(-2x^2)$

6. $b^{12} \cdot b$

7. $3^5 \cdot 3^8$

8. $(-2y^3)(5y^7)$

9. $(6a^5)(-3a^6)$

10. $(-x)(-6x^3)$

11. $(3x^2)(2x^5)$

12. $(-6y^2)(-2y^5)$

13. $\frac{x^{11}}{x^2}$

14. $\frac{a^6}{a^3}$

15. $\frac{7^9}{7^6}$

16. $\frac{2^5}{2^2}$

17. $\frac{16x^3}{4x^2}$

18. $\frac{25y^5}{5y^2}$

19. $\frac{-48y^3}{-8y}$

20. $\frac{12y^5}{3y^2}$

21. $\frac{39x^7y^5}{3x^3y}$

Multi-Part Lesson 2-1 (continued)

PART C

PAGES 102–105

Simplify.

1. $(2^3)^2$

2. $(4^3)^3$

3. $(6^2)^4$

4. $(a^4)^3$

5. $(m^7)^8$

6. $(k^5)^7$

7. $[(3^2)^2]^3$

8. $[(4^2)^2]^2$

9. $[(2^3)^2]^3$

10. $(6z^4)^5$

11. $(8c^8)^3$

12. $(-3a^5b^{12})^5$

PART D

PAGES 106–107

Use the *act it out* strategy to solve Exercises 1 and 2.

1. **BRIDGES** One third of a bridge support is underground, another one sixth of it is covered by water, and 325 feet are out of the water. What is the total height of the bridge support?

2. **GAMES** Frederico is playing a game with his little sister that requires him to arrange cards face down into an array of columns and rows. When he puts 4 cards in each row, he has 3 left over. When he puts 5 cards in each row, he has 1 left over. Give two possible numbers of cards Frederico might have.

Multi-Part Lesson 2-2: Scientific Notation

PART A

PAGES 108–112

Write each expression using a positive exponent.

1. 5^{-3}

2. 6^{-10}

3. $(-2)^{-5}$

4. $(-3)^{-2}$

5. m^{-6}

6. g^{-2}

7. n^{-9}

8. r^{-8}

9. h^{-7}

Write each fraction as an expression using a negative exponent.

10. $\dfrac{1}{4^5}$

11. $\dfrac{1}{3^4}$

12. $\dfrac{1}{(-3)^3}$

13. $\dfrac{1}{(-6)^5}$

14. $\dfrac{1}{64}$

15. $\dfrac{1}{49}$

16. $\dfrac{1}{243}$

17. $\dfrac{1}{625}$

18. $\dfrac{1}{216}$

Simplify. Express using positive exponents.

19. $3^{-2} \cdot 3^7$

20. $5^{-3} \cdot 5^{-4}$

21. $x^{-5} \cdot x^{-3}$

22. $a^4 \cdot a^{-7}$

23. $a^{-2}b^3 \cdot a^{-5}b$

24. $x^3y^{-2} \cdot x^{-5}y^3$

25. $\dfrac{7^{-2}}{7^{-6}}$

26. $\dfrac{x^{-4}}{x^5}$

27. $\dfrac{24a^3}{-6a^2}$

28. $\dfrac{18y^{-4}}{3y^{-10}}$

29. $\dfrac{4^2x^5}{4^5x^{-2}}$

30. $\dfrac{6^{-2}a^4}{6^{-3}a^{-2}}$

Multi-Part Lesson 2-2 (continued)

PART B

Write each number in standard form.

1. 4.5×10^3 **2.** 2×10^4 **3.** 1.725896×10^6

4. 9.61×10^2 **5.** 1×10^7 **6.** 8.256×10^8

7. 5.26×10^4 **8.** 3.25×10^2 **9.** 6.79×10^5

10. 3.1×10^{-4} **11.** 2.51×10^{-2} **12.** 6×10^{-1}

Write each number in scientific notation.

13. 720   **14.** 7,560   **15.** 892

16. 1,400   **17.** 91,256   **18.** 51,000

19. 0.012   **20.** 0.0002   **21.** 0.054

22. 0.231   **23.** 0.0000056   **24.** 0.000123

PART C

Evaluate each expression. Express the result in scientific notation.

1. $(3.2 \times 10^4)(1.4 \times 10^2)$ **2.** $(6.1 \times 10^5)(8.2 \times 10^4)$

3. $(5.2 \times 10^{-3})(7.4 \times 10^5)$ **4.** $(4.8 \times 10^6)(3.9 \times 10^{-8})$

5. $(9.3 \times 10^{-5})(2.7 \times 10^{-2})$ **6.** $(4.3 \times 10^{-2})(5.6 \times 10^{-4})$

7. $\dfrac{4.55 \times 10^7}{1.3 \times 10^4}$ **8.** $\dfrac{8.84 \times 10^{-5}}{3.4 \times 10^{-2}}$ **9.** $\dfrac{8.05 \times 10^4}{2.3 \times 10^{-2}}$

10. $\dfrac{7.56 \times 10^6}{4.2 \times 10^7}$ **11.** $\dfrac{2.016 \times 10^7}{8.4 \times 10^3}$ **12.** $\dfrac{1.175 \times 10^{-3}}{1.25 \times 10^{-6}}$

13. $(2.7 \times 10^3) + (3.4 \times 10^2)$ **14.** $(7.2 \times 10^6) + (1.25 \times 10^5)$

15. $(8.4 \times 10^5) - (7.9 \times 10^3)$ **16.** $(9.2 \times 10^3) - (9.6 \times 10^2)$

17. $(6.5 \times 10^{12}) + (3.1 \times 10^{11})$ **18.** $(2.6 \times 10^9) - (7.4 \times 10^7)$

Multi-Part Lesson 2-3: Square Roots and Cube Roots

PART A

Find each square root.

1. $\sqrt{9}$ **2.** $\sqrt{81}$ **3.** $-\sqrt{625}$

4. $\sqrt{36}$ **5.** $-\sqrt{169}$ **6.** $\sqrt{144}$

7. $\sqrt{961}$ **8.** $\sqrt{324}$ **9.** $-\sqrt{225}$

10. $-\sqrt{4}$ **11.** $\sqrt{529}$ **12.** $-\sqrt{484}$

13. $\sqrt{0.04}$ **14.** $\sqrt{2.25}$ **15.** $\sqrt{0.01}$

16. $-\sqrt{0.09}$ **17.** $\sqrt{0.49}$ **18.** $\sqrt{1.69}$

19. $-\sqrt{\dfrac{4}{9}}$ **20.** $-\sqrt{\dfrac{81}{64}}$ **21.** $\sqrt{\dfrac{25}{81}}$

Multi-Part Lesson 2-3 (continued)

Estimate to the nearest whole number.

1. $\sqrt{229}$ 2. $\sqrt{63}$ 3. $\sqrt{290}$ 4. $\sqrt{27}$

5. $\sqrt{333}$ 6. $\sqrt{23}$ 7. $\sqrt{96}$ 8. $\sqrt{200}$

9. $\sqrt{117}$ 10. $\sqrt{47}$ 11. $\sqrt{1.30}$ 12. $\sqrt{8.4}$

13. $\sqrt{18.35}$ 14. $\sqrt{25.70}$ 15. $\sqrt{14.1}$ 16. $\sqrt{15.3}$

Name all sets of numbers to which each real number belongs.

1. 6.5 2. $\sqrt{25}$ 3. $\sqrt{3}$

4. -7.2 5. $-0.\overline{61}$ 6. $\frac{1}{2}$

7. $\frac{16}{4}$ 8. -102.1 9. $\sqrt{29}$

Order each set of numbers from least to greatest.

10. $\sqrt{12}, 3, 3\frac{1}{2}, 3.\overline{5}$

11. $\sqrt{2}, 140\%, 1.45, 1.\overline{4}$

12. $-\sqrt{10}, -\sqrt{11}, -3.5, -3\frac{1}{3}$

Replace each ● with <, >, or = to make a true statement.

13. $\sqrt{7}$ ● 2.8 14. $2\frac{1}{3}$ ● $2.\overline{3}$

15. $\sqrt{121}$ ● 11 16. 5.6 ● $\sqrt{30}$

17. 9.45 ● $9.\overline{4}$ 18. $\sqrt{5}$ ● 2.23

19. $\sqrt{6.25}$ ● $2\frac{1}{2}$ 20. $5\frac{1}{3}$ ● $\sqrt{30}$

21. $4\frac{2}{3}$ ● $\sqrt{22}$ 22. $\sqrt{8}$ ● 2.9

Multi-Part Lesson 3-1: One-Step Equations

Use the *work backward* strategy to solve Exercises 1–3.

1. **SCHEDULE** The closing day activities at camp must be over by 2:45 P.M. Trent needs $1\frac{1}{2}$ hours to hold the field competitions, 45 minutes for the awards ceremony, and an hour and 15 minutes for the cookout. Then everyone will need an hour to pack and check out. What time will Trent need to start the camp activities?

2. **BOWLING** Alexia's bowling scores are 166, 176, 172, 171, and 159. What is the minimum score she can bowl in her next game to maintain an average of at least 170?

3. **STOCK** A share of stock increased in value by 25%. Then it decreased in value by $4 and then it doubled. If the stock is now worth $32, how much was the stock worth originally?

Multi-Part Lesson 3-1 (continued)

PART **B**

Define a variable. Then write an equation to model each situation.

1. When the marbles were divided among the 3 players, each player received 48 marbles.

2. The low temperature of −2°F was 20 degrees less than the high temperature.

3. A team of 84 football players separated into equal-size groups results in 12 players per group.

Define a variable. Then write an equation that could be used to solve each problem.

4. **TICKETS** The total cost of concert tickets is equally divided among 4 friends. If the cost of 1 ticket is $56, what was the total cost of the 4 concert tickets?

5. **READING** Sonja has read 163 pages in her book. If the book has a total of 395 pages, how many more pages does Sonja have to read?

6. **SCORES** Dale's score of −12 is one fifth of Tabitha's score. What is Tabitha's score?

7. **BASEBALL CARDS** Abe's baseball card collection has 76 more cards than Kerry's baseball card collection. If Kerry has 349 baseball cards, how many baseball cards does Abe have?

PART **C**

Solve each equation. Check your solution.

1. $g - 3 = 10$

2. $b + 7 = 12$

3. $a + 3 = 15$

4. $r - 3 = 4$

5. $t + 3 = 21$

6. $s + 10 = 23$

7. $9 + n = 13$

8. $13 + v = 31$

9. $-4 + b = 12$

10. $z - 10 = -8$

11. $-7 = x + 12$

12. $a + 6 = -9$

PART **D**

Solve each equation. Check your solution.

1. $4x = 36$

2. $39 = 3y$

3. $4z = 16$

4. $\frac{t}{5} = 6$

5. $100 = 20b$

6. $8 = \frac{w}{8}$

7. $10a = 40$

8. $\frac{s}{9} = 8$

9. $420 = 5s$

10. $8k = 72$

11. $2m = 18$

12. $\frac{m}{8} = 5$

13. $\frac{r}{7} = -8$

14. $\frac{w}{7} = 8$

15. $18q = 36$

16. $9w = 54$

17. $4 = p \div 4$

18. $14 = 2p$

19. $12 = 3t$

20. $\frac{m}{4} = 12$

21. $6h = 12$

22. $-2a = -8$

23. $0 = 6r$

24. $\frac{y}{12} = -6$

25. $3m = -15$

26. $\frac{c}{-4} = 10$

27. $-6f = -36$

Multi-Part Lesson 3-2: Two-Step Equations

PARTS A B

PAGES 171–176

Solve each equation. Check your solution.

1. $2x + 4 = 14$
2. $5p - 10 = 0$
3. $5 + 6a = 41$
4. $\frac{x}{3} - 7 = 2$
5. $18 = 6q - 24$
6. $18 = 4m - 6$
7. $3r - 3 = 9$
8. $2x + 3 = 5$
9. $0 = 4x - 28$
10. $3x - 1 = 5$
11. $3z + 5 = 14$
12. $3x - 15 = 12$
13. $9a - 8 = 73$
14. $2x - 3 = 7$
15. $3t + 6 = 9$
16. $2y + 10 = 22$
17. $15 = 2y - 5$
18. $3c - 4 = 2$
19. $6 + 2p = 16$
20. $8 = 2 + 3x$
21. $4b + 24 = 24$
22. $5x - 6 = 19$
23. $-2x - 6 = 14$
24. $3x - 9 = -18$
25. $-a + 1 = 15$
26. $2x + 6 = -10$
27. $3a + 2 = 11$
28. $2a - 4 = -10$
29. $3 = 2a + 1$
30. $8y - 1 = 15$

PART C

PAGES 177–181

Translate each sentence into an equation. Then find each number.

1. Seven more than three times a number is 16.
2. Seven more than the quotient of a number and -2 is 6.
3. Six more than twice a number is 20.
4. Two less than five times a number is equal to 8.
5. Twice a number plus 5 is -3.
6. The product of a number and 3 plus 1 is 19.
7. The product of a number and 4 plus 2 is 14.
8. Eight less than the quotient of a number and 3 is 5.
9. The difference of twice a number and 3 is 11.
10. The sum of 3 times a number and 7 is 25.

Multi-Part Lesson 3-3: One-Step Inequalities

PART A

PAGES 183–186

Write an inequality for each sentence.

1. A number is less than 10.
2. A number is greater than or equal to -7.
3. A number is less than -2.
4. A number is more than 5.
5. A number is less than or equal to 11.
6. A number is no more than 8.

Graph each inequality on a number line.

7. $x > 5$
8. $y > 0$
9. $z < -2$
10. $a \geq 6$
11. $b \leq 2$
12. $x \geq 1$
13. $a \leq 3$
14. $b \geq 1$
15. $x < -2$
16. $n \geq -3$
17. $t > -1$
18. $y \leq -5$

Multi-Part Lesson 3-3 (continued)

PART B

Solve each inequality. Graph the solution set on a number line.

1. $y + 3 > 7$ **2.** $c - 9 < 5$ **3.** $x + 4 \geq 9$

4. $y - 3 < 15$ **5.** $t - 13 \geq 5$ **6.** $x + 3 < 10$

7. $y - 6 \geq 2$ **8.** $x - 3 \geq -6$ **9.** $a + 3 \leq 5$

10. $c - 2 \leq 11$ **11.** $a + 15 \geq 6$ **12.** $y + 3 \geq 18$

13. $y + 16 \geq -22$ **14.** $x - 3 \geq 17$ **15.** $y - 6 > -17$

16. $y - 11 < 7$ **17.** $a + 5 \geq 21$ **18.** $c + 3 > -16$

19. $x - 12 \geq 12$ **20.** $x + 5 \geq 5$ **21.** $y - 6 > 31$

22. $a - 6 > 17$ **23.** $y + 7 > 3$ **24.** $a + 13 \geq -16$

25. $y - 6 > 5$ **26.** $y + 6 < -5$ **27.** $x - 17 \geq 34$

28. $y + 1 \leq 16$ **29.** $a - 14 \geq 16$ **30.** $x + 14 \leq 20$

PART C

Solve each inequality. Graph the solution set on a number line.

1. $5p \geq 25$ **2.** $4x < 12$ **3.** $15 \leq 3m$

4. $\frac{d}{3} > 15$ **5.** $8 < \frac{r}{7}$ **6.** $9g < 27$

7. $4p \geq 24$ **8.** $5p > 25$ **9.** $-4 > \frac{-k}{3}$

10. $\frac{-z}{5} > 2$ **11.** $-3x \leq 9$ **12.** $-5x > -35$

13. $\frac{a}{-6} < 1$ **14.** $\frac{x}{-5} \leq -2$ **15.** $-2x < 16$

Multi-Part Lesson 3-4: Two-Step Inequalities

PART A

Solve each inequality. Graph the solution set on a number line.

1. $2x - 3 > 11$ **2.** $6x + 5 \leq 23$ **3.** $12 \leq 3x - 6$

4. $-3 < 4x + 1$ **5.** $-8x + 4 \leq -12$ **6.** $-5x - 6 > 19$

7. $\frac{x}{4} + 2 > -3$ **8.** $-18 < -3x + 6$ **9.** $4 \leq 6 + \frac{x}{3}$

10. $-7 + 2x \leq 5$ **11.** $-4x - 5 > 7$ **12.** $-10.4 > 0.8 + 1.6x$

13. $2.1x - 3 > 5.4$ **14.** $\frac{x}{-5} + 2 < 4$ **15.** $80 \leq -15x + 5$

16. $3 \geq 5 + \frac{x}{-3}$ **17.** $-0.6x + 1.3 < 2.5$ **18.** $-3.7 \geq -1.2x + 1.1$

Multi-Part Lesson 3-4 (continued)

PART B

PAGES 202–206

Write a compound inequality to represent each situation.

1. **TICKETS** Tickets for a concert cost at least $31 and no more than $58.

2. **AGE** A person applying to be a Federal Law Enforcement Officer must be at least 21 years of age but no older than 37 at the time of his or her appointment.

Graph the solution set of each inequality.

3. $m > 3$ and $m < 9$

4. $r \geq -6$ or $r \leq -10$

5. $a < -2$ or $a > 3$

6. $k > -3$ and $k < 0$

7. $d < -6$ or $d \geq -2$

8. $n > 2$ and $n \leq 7$

9. $x < -4$ and $x \geq -10$

10. $y \geq 3$ or $y \leq -1$

11. $p > 6$ or $p \leq 4$

Write a compound inequality to represent each situation. Then graph the solution set on a number line.

12. Negative 4 is less than a number, which is less than or equal to 2.

13. A number is greater than or equal to 5 but less than 9.

Multi-Part Lesson 4-1: Properties of Mathematics

PART A

PAGES 220–224

Name the property shown by each statement.

1. $1 \cdot 4 = 4$

2. $6 + (b + 2) = (6 + b) + 2$

3. $9(6n) = (9 \cdot 6)n$

4. $8t \cdot 0 = 0 \cdot 8t$

5. $0(13n) = 0$

6. $7 + t = t + 7$

Simplify each expression.

7. $(12 + x) + 9$

8. $31 + (15 + c)$

9. $d + (8 + 19)$

10. $2 \cdot (6 \cdot m)$

11. $(5 \cdot p) \cdot 3$

12. $9(4f)$

PART B

PAGES 225–230

Use the Distributive Property to evaluate each expression.

1. $2(4 + 5)$

2. $4(5 + 3)$

3. $3(7 - 6)$

4. $(2 + 5)9$

5. $(10 - 4)3$

6. $-6(1 + 3)$

Use the Distributive Property to rewrite each expression.

7. $3(m + 4)$

8. $(y + 7)5$

9. $-6(x + 3)$

10. $(p - 4)5$

11. $-3(s - 9)$

12. $5(x + y)$

13. $b(c + 3d)$

14. $(a - b)(-5)$

15. $-6(v - 3w)$

16. $5(x + 12)$

17. $(m - 6)(4)$

18. $-2(a - b)$

19. $(8 - m)(-3)$

20. $8(p - 3q)$

21. $(2x + 3y)(4)$

22. $2(x + 3)$

23. $3(a + 7)$

24. $3(g - 6)$

25. $-2(a + 3)$

26. $-1(x - 6)$

27. $4(a - 5)$

Multi-Part Lesson 4-1 (continued)
PART C

PAGES 231–235

Identify the terms, like terms, coefficients, and constants in each expression.

1. $8b + 7b - 4 - 6b$ **2.** $9 + 8z - 3 + 5z$ **3.** $11q - 5 + 2q - 7$

4. $a + 1 + 2a + 8a$ **5.** $1 - 2c - 3c + 100$ **6.** $14j - 6 + 8j - 5$

Write each expression in simplest form.

7. $3x + 2x$ **8.** $6x - 3x$ **9.** $2a - 5a$

10. $5x - 6x$ **11.** $8a - 3a$ **12.** $a - 4a$

13. $3a + 2a - 6$ **14.** $6x + 2x - 3$ **15.** $5a - 3 + 2a$

16. $3x + 7 - 5x$ **17.** $x - 3 + 5x$ **18.** $6x - 3x - 2$

19. $a - 2a + 5$ **20.** $6x - 2 + 7x$ **21.** $5a - 7a + 2$

22. $4a + 2 - 7a - 5$ **23.** $3a - 2 + 5a - 7$ **24.** $5x - 3x + 2 - 5$

PART D

PAGES 236–237

Use the *solve a simpler problem* strategy to solve Exercises 1–3.

1. **MEASUREMENT** Describe a method that could be used to determine the thickness of one sheet of paper in a textbook.

2. **HEALTH** A human heart beats an average of 72 times in one minute. Estimate the number of times a human heart beats in one year.

3. **GIFT WRAPPING** During the holidays, Marcos and Renee earn extra money by wrapping gifts at a department store. Marcos wraps 8 packages an hour and Renee wraps 10 packages an hour. Working together, about how long will it take them to wrap 40 packages?

Multi-Part Lesson 4-2: Multi-Step Equations and Inequalities
PARTS A B

PAGES 239–245

Solve each equation. Check your solution.

1. $6x + 10 = 1x$ **2.** $2a - 5 = -3a$ **3.** $7a - 5 = 2a$

4. $4a + 7 = 10 + a$ **5.** $8x + 3 = 2x$ **6.** $6x - 3 = -18 + x$

7. $3a - 1 = 2a$ **8.** $8a - 2 = 12 + a$ **9.** $3x + 6 = x$

10. $2x + 7 = 11 - 2x$ **11.** $8x + 10 = 3x$ **12.** $7a + 4 = 3a$

13. $7x + 8 = 11x$ **14.** $21x + 11 = 10x$ **15.** $5x + 5 = 14 + 2x$

16. $7b - 4 = 2b + 16$ **17.** $2y - 3 = 5 - 2y$ **18.** $3m = 2m + 7$

19. $9t + 1 = 4t - 9$ **20.** $-2a + 3 = a - 12$ **21.** $3x = 9x - 12$

22. $2c + 3 = 3c - 4$ **23.** $s - 3 = 5 - s$ **24.** $3w - 5 = 5w - 7$

25. $4x - 7 = 11 + x$ **26.** $5x + 2 = 10 + x$ **27.** $3x + 2 = 2x + 5$

Multi-Part Lesson 4-2 (continued)

PART C PAGES 246–249

Solve each equation. Check your solution.

1. $6(m - 2) = 12$

2. $4(x - 3) = 4$

3. $5(2d + 4) = 35$

4. $w + 6 = 2(w - 6)$

5. $3(b + 1) = 4b - 1$

6. $7w - 6 = 3(w + 6)$

7. $4(k - 6) = 6(k + 2)$

8. $3(x - 0.8) = 4x + 4$

9. $\frac{5}{9}(g + 18) = \frac{1}{6}g + 3$

10. $4(c + 12) = 2c + 18$

11. $7(d - 2) = 5(d + 2)$

12. $5p - 17 = 2(2p - 7)$

13. $4(3z - 2) = 9z - 7$

14. $7s + 2 = 4(s + 1)$

15. $6(k + 1) = 2k + 7$

16. $6(n - 1) + 2(n + 1)$
    $= 10 + 7n$

17. $\frac{1}{4}(x + 8) = 8$

18. $\frac{2}{3}(3q + 6) = 8$

PART D PAGES 250–253

Solve each inequality. Graph the solution set on a number line.

1. $3(y - 2) > 2(y - 1)$

2. $4(1 - a) \leq 28$

3. $3(2 - x) + 2 < 2(2 - x)$

4. $3(z + 3) < 2z + 5$

5. $5(p - 1) \geq p - 6$

6. $\frac{1}{3}(3r + 6) > 2$

7. $\frac{7}{8}(-16t + 24) < 7$

8. $3b - 1 > 2(2 - b)$

9. $2(3c - 5) - 4(c + 1)$
    $\leq c - 12$

10. $2(m + 1) < 10$

11. $3(k + 2) \geq 24$

12. $-2 > 2(5 - x)$

13. $2(-3a + 1) \geq 14$

14. $3(y + 3) < 0$

15. $2(\frac{d}{4} + 3) \geq -22$

16. $3(\frac{x}{3} - 5) < 18$

17. $-5g + 6 < 2(1.5g + 13)$

18. $3(n + 1) \geq 5(n + 5)$

Multi-Part Lesson 5-1: Expressions

PART A PAGES 268–269

Use the *make a table* strategy to solve Exercises 1 and 2.

1. RENTALS Mr. Nelson wants to rent a house boat. The prices to rent the boat from two different companies are shown. For how many days must he rent the boat for the cost from each place to be the same?

Company	Rental Fee ($)	Cost per Day ($)
Boat Magic	150	75
Seaside Rentals	75	100

2. RIDES There are 75 people standing in line for a ride at an amusement park. Every four minutes, 15 people get on the ride, but 20 more people get in line. After one hour, how many people are standing in line?

Multi-Part Lesson 5-1 (continued)

PART B

PAGES 270–274

Translate each phrase into an algebraic expression.

1. twenty-four football cards divided equally among c collectors
2. eleven points less than twice the amount scored in the last game
3. seven times as many DVDs
4. six more laps than Dora swam on Monday

Evaluate each expression if $a = 3$, $b = -2$, $c = 5$, and $d = -4$.

5. $3a + b$
6. $4c - 2d$
7. $ac + bd$
8. $7c + 2a - b$
9. $ad - bc$
10. $abc - cd$
11. $\dfrac{a + 2c}{3d}$
12. $\dfrac{a + d}{b}$
13. $\dfrac{ac}{b + 10}$

PART C

PAGES 275–280

Graph each ordered pair on a coordinate plane.

1. $A\left(4\frac{1}{2}, 2\frac{2}{3}\right)$
2. $B(-2.75, 3.5)$
3. $C(-1.5, 0.25)$
4. $D\left(3\frac{1}{4}, \frac{3}{4}\right)$
5. $E(2.4, -1.75)$
6. $F\left(-2\frac{1}{3}, -3\frac{2}{5}\right)$

Express each relation as a table and a graph. Then state the domain and range.

7. $\{(3, 2), (-2, 4), (4, -4), (4, 0), (-1, -3)\}$
8. $\{(-1, -5), (2, -3), (3, -2), (5, 1), (-4, 2)\}$
9. $\{(0, 3), \left(1\frac{1}{2}, -2\right), \left(-3\frac{1}{2}, 2\frac{1}{2}\right), (-4, -3), (-2, 0)\}$
10. $\{(1.5, -2), (-2, 3.5), (-2.5, -2.5), (0, 4.5), (3.25, 1.5)\}$

Multi-Part Lesson 5-2: Translate Among Words, Tables, Graphs, and Equations

PARTS A B

PAGES 281–286

Write an expression that can be used to find the nth term of each sequence. Then use the expression to find the next three terms.

1.

Term Number (n)	1	2	3	4
Term	-3	-6	-9	-12

2.

Term Number (n)	1	2	3	4
Term	2	3	4	5

3.

Term Number (n)	1	2	3	4
Term	5	7	9	11

4.

Term Number (n)	1	2	3	4
Term	-1	3	7	11

5. $\dfrac{1}{3}, \dfrac{2}{3}, 1, 1\frac{1}{3}, \dots$
6. $4, 3, 2, 1, \dots$
7. $9, 11, 13, 15, \dots$
8. $2, 8, 14, 20, \dots$
9. $-5, -8, -11, -14, \dots$
10. $\dfrac{1}{8}, \dfrac{1}{4}, \dfrac{3}{8}, \dfrac{1}{2}, \dots$

Multi-Part Lesson 5-2 (continued)

PART C

PAGES 287–291

Write an algebraic expression to represent the data in each graph.

1.

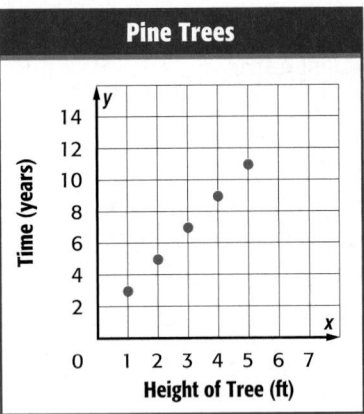

2.

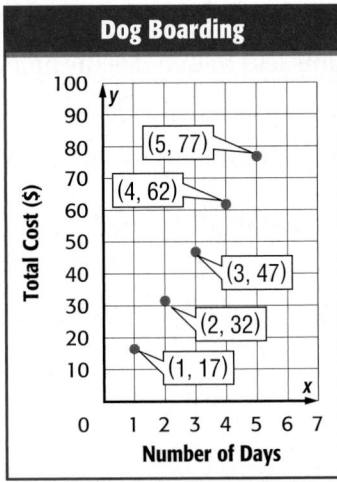

3.

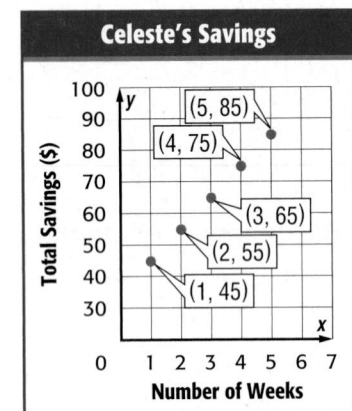

4.

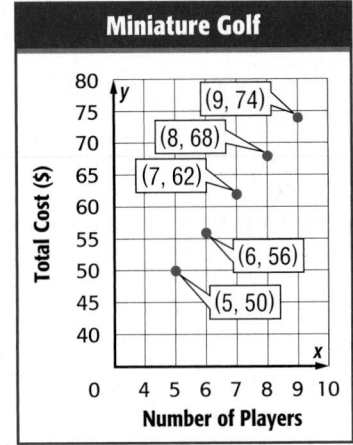

PART D

PAGES 292–297

1. RECYCLING The zoo is holding a recycling day. It deducts $0.10 from the price of admission for each recyclable item donated at the zoo. Admission is $15. The table shows the reduced price for admission if 10, 11, 12, or 13 items are donated.

Number of Items Donated	Price ($)
10	14.00
11	13.90
12	13.80
13	13.70
d	?

a. Write an equation to determine the price of admission p if any number of items are donated d.

b. Use the equation to find the cost of admission if 25 items are donated.

2. RENTALS Jen is renting a mountain bike. The bike rental shop charges a $55 deposit for the bike and helmet plus an additional $15 per day.

a. Write an equation to find the cost c of renting a mountain bike for any number of days d.

b. Make a table to find the cost for 4, 5, 6, and 7 days. Then graph the ordered pairs.

Multi-Part Lesson 5-3: Relations and Functions

PARTS Ⓐ Ⓑ

PAGES 299–305

Find each function value.

1. $f\left(\frac{1}{2}\right)$ if $f(x) = 2x - 6$

2. $f(-4)$ if $f(x) = -\frac{1}{2}x + 4$

3. $f(1)$ if $f(x) = -5x + 1$

4. $f(6)$ if $f(x) = \frac{2}{3}x - 5$

5. $f(0)$ if $f(x) = 1.6x + 4$

6. $f(2)$ if $f(x) = 2x - 8$

Copy and complete each function table. Then state the domain and range of the function.

7. $f(x) = -4x$

x	−4x	f(x)
−2		
−1		
0		
1		

8. $f(x) = x + 6$

x	x + 6	f(x)
−6		
−4		
−2		
0		

9. $f(x) = 3x + 2$

x	3x + 2	f(x)
−3		
−2		
−1		
0		

PART Ⓒ

PAGES 306–311

Graph each function.

1. $y = 6x + 2$

2. $y = -2x + 3$

3. $y = -5x$

4. $y = 10x - 2$

5. $y = -2.5x - 1.5$

6. $y = 7x + 3$

7. $y = \frac{x}{4} - 8$

8. $y = 3x + 1$

9. $y = 25 - 2x$

10. $y = \frac{x}{6}$

11. $y = -2x + 11$

12. $y = 7x - 3$

13. SALES The school spirit club is selling T-shirts for $10.50 each.

   a. Write a function to represent this situation.

   b. Make a function table to find the total cost of 1, 2, 3, 4, or 5 T-shirts.

   c. Graph the function. Is the function continuous or discrete? Explain.

14. DISTANCE The table shows the distance a car travels at a constant rate in a certain amount of time.

   a. Write a function to represent this situation.

   b. Make a function table to find the distance that can be traveled in 5, 5.5, 6, or 6.5 hours.

   c. Is this function continuous or discrete? Explain.

   d. At this speed, how far can the car travel in 12.5 hours?

Time (h)	Distance (mi)
1	65
1.5	97.5
2	130
2.5	162.5
3	195

Multi-Part Lesson 5-4: Nonlinear Functions

PART A

PAGES 314–319

Determine whether each table represents a *linear* or *nonlinear* function. Explain.

1.

x	−1	0	1	2
y	2	0	2	8

2.

x	−1	0	1	2
y	−1	0	1	8

3.

x	−1	0	1	2
y	−3	0	3	6

4.

x	−3	−2	0	1
y	−5	−3	−2	−1

5.

x	2	5	8	11
y	−4	−2	0	2

6.

x	−6	−2	2	6
y	0	−1	−2	−3

PARTS B C

PAGES 320–325

Graph each function.

1. $y = x^2 - 1$

2. $y = 1.5x^2 + 3$

3. $y = 2x^2 - 2$

4. $y = 2x^2$

5. $y = x^2 + 3$

6. $y = -3x^2 + 4$

7. $y = -x^2 + 7$

8. $y = 3x^2$

9. $y = -3x^2$

10. $y = -x^2$

11. $y = \frac{1}{2}x^2 + 1$

12. $y = 5x^2 - 4$

13. $y = -x^2 + 3$

14. $y = 2.5x^2$

15. $y = -2x^2$

16. $y = 8x^2 + 3$

17. $y = -2x^2 + 5$

18. $y = -4x^2 + 4$

19. $y = 4x^2 + 3$

20. $y = -4x^2 + 1$

21. $y = 2x^2 + 1$

Multi-Part Lesson 6-1: Slope

PART A

PAGES 337–342

Determine whether the relationship between the two quantities described in each table is linear. If so, find the constant rate of change. If not, explain your reasoning.

1.

Calories Burned				
Time (min)	1	2	3	4
Calories	4.3	8.6	12.9	16.3

2.

Punch Recipe				
Soda (c)	2	4	6	8
Juice (c)	$1\frac{1}{4}$	$2\frac{1}{2}$	$3\frac{3}{4}$	5

Find the constant rate of change for each graph and interpret its meaning.

3.

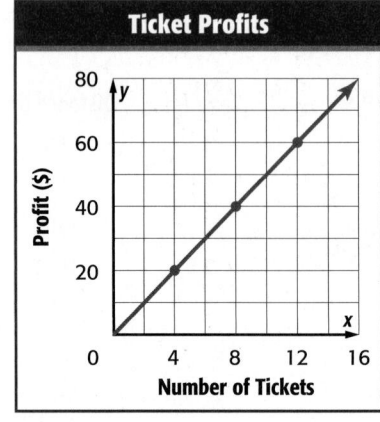

4.

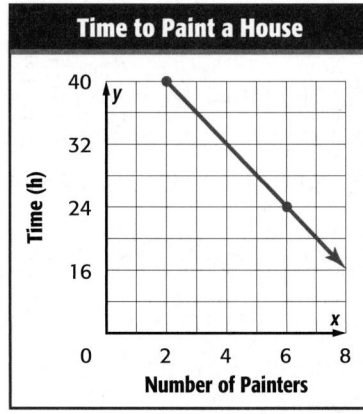

Multi-Part Lesson 6-1 (continued)
PARTS B C

PAGES 343–349

Find the slope of the line that passes through each pair of points.

1. $A(2, 3)$, $B(1, 5)$

2. $C(-6, 1)$, $D(2, 1)$

3. $E(3, 0)$, $F(5, 0)$

4. $G(-1, -3)$, $H(-2, -5)$

5. $I(6, 7)$, $J(11, 1)$

6. $K(5, 3)$, $L(5, -2)$

7. $M(10, 2)$, $N(-3, 5)$

8. $O(6, 2)$, $P(1, 7)$

9. $Q(5, 8)$, $R(-3, -2)$

10. $S(-1, 7)$, $T(3, 8)$

11. $U(4, -1)$, $V(-5, -2)$

12. $W(3, -2)$, $X(7, -1)$

13. $Y(0, 5)$, $Z(2, 1)$

14. $A(6, 5)$, $B(-3, -5)$

15. $C(2, 1)$, $D(7, -1)$

PARTS D E

PAGES 350–356

1. TRAVEL Refer to the graph.

   a. The number of miles traveled varies directly with the number of hours traveled. What is the rate of speed in miles per hour?

   b. Going at the rate shown, what distance would one travel in 39 hours?

2. GAS MILEAGE Mira's car can travel about 100 miles on 3 gallons of gas. Assuming that the distance traveled remains constant to the amount of gas used, how many gallons of gas would be needed to travel 650 miles?

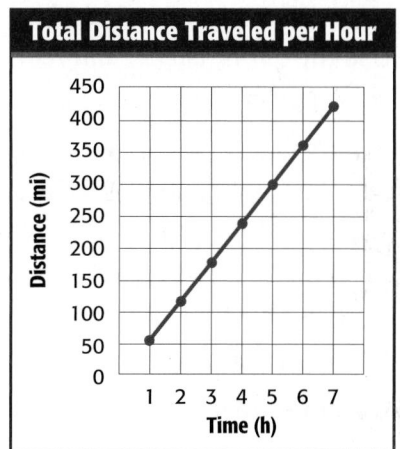

Total Distance Traveled per Hour

3. MONEY Determine whether the linear function shown is a direct variation. If so, state the constant of variation.

Savings, x	$2,154	$3,231	$4,308	$5,385
Years, y	2	3	4	5

Multi-Part Lesson 6-2: Intercepts
PART A

PAGES 357–362

State the slope and y-intercept for the graph of each equation.

1. $y = 3x - 5$

2. $y = 2x - 6$

3. $y = -6x + \frac{1}{2}$

4. $y = -7x + \frac{5}{2}$

5. $y = \frac{1}{2}x + 7$

6. $y = \frac{3}{4}x + 8$

7. $y = -\frac{2}{3}x - \frac{1}{3}$

8. $y = -\frac{1}{8}x - \frac{3}{8}$

9. $y = \frac{2}{3}x + 5$

10. $y = -\frac{2}{7}x - 1$

11. $3x + y = 6$

12. $y - 4x = 7$

Graph each equation using the slope and the y-intercept.

13. $y = -2x + 5$

14. $y = -3x + 1$

15. $y = -x + 1$

16. $y = -x + 3$

17. $y = x - 3$

18. $y = x - 5$

19. $y = 3x - 6$

20. $y = \frac{5}{2}x - 1$

21. $y = \frac{1}{2}x + 3$

22. $y = -2x - 2$

23. $y - 4x = -1$

24. $2x + y = 3$

Multi-Part Lesson 6-2 (continued)

PARTS B C

PAGES 363–368

State the *x*- and *y*-intercepts of each function. Then graph the function.

1. $3x + y = 9$

2. $-2x + y = 6$

3. $3x - 2y = 12$

4. $2x - y = 6$

5. $2x - 4y = 8$

6. $-5x + y = -10$

7. $6x + 2y = 12$

8. $5x - 2y = 15$

9. $4.5x + 2y = 9$

10. $\dfrac{x}{-2} + 3y = 12$

11. $-4x - 3y = 24$

12. $8x + 5y = 20$

Multi-Part Lesson 6-3: Systems of Equations

PART A

PAGES 370–371

Use the *guess, check, and revise* strategy to solve Exercises 1–4.

1. NUMBER THEORY A number squared is 529. Find the number.

2. VOLUME The volume of a rectangular prism is 216 cubic centimeters. The length of the prism is two thirds the height of the prism. If the width is 9 centimeters, find the length and height. (*Hint*: $V = \ell wh$)

3. MONEY Edwin has $2.35 in quarters, dimes, nickels, and pennies. He has 31 coins. The number of pennies is twice as great as the number of quarters and $\dfrac{5}{6}$ as great as the number of nickels. He has fewer dimes than any other coin. How many of each coin does he have?

4. LEGS On a farm, there are 100 total legs among the people and animals. Each cow and horse has 4 legs, and each person has 2 legs. There are 4 more than ten times as many cows and horses than people. How many cows and horses and how many people are there?

PARTS B C

PAGES 372–377

Write a system of equations that represents each situation. Use a table or graph to solve. Interpret the solution.

1. BAND The school band has a total of 125 students. There are 25 more girls than boys. How many boys and girls are in the band?

2. PIZZA Pearson Middle School ordered a total of 75 cheese and pepperoni pizzas. The total cost of the pizzas was $445. Each cheese pizza cost $5, and each pepperoni pizza cost $7. How many of each type did they order?

Solve each system of equations by graphing.

3. $y = x - 1$
$y = -x + 11$

4. $y = -x$
$y = 2x$

5. $y = -x + 3$
$y = x + 3$

6. $y = x - 3$
$y = 2x$

7. $y = -x + 6$
$y = x + 2$

8. $y = -x + 2$
$y = x - 4$

9. $y = -3x + 6$
$y = x - 2$

10. $y = 3x - 4$
$y = -3x - 4$

11. $y = 2x + 1$
$y = 3x$

12. $y = -x + 4$
$y = x - 10$

13. $y = -x + 6$
$y = 2x$

14. $y = x - 4$
$y = -2x + 5$

Multi-Part Lesson 6-3 (continued)
PART D
PAGES 378–381

Solve each system of equations by substitution.

1. $y = x - 2$
$y = 4$

2. $y = x + 8$
$y = -2$

3. $y = x - 5$
$y = 6$

4. $y = x + 6$
$y = -4$

5. $y = x - 9$
$y = 3$

6. $y = x + 4$
$y = -x$

7. $y = x - 5$
$y = 2x$

8. $y = x + 12$
$y = \frac{1}{2}x$

9. $y = x - 8$
$y = 3x$

Multi-Part Lesson 7-1: Angle Measure
PARTS A B
PAGES 395–400

Name each angle in four ways. Then classify each angle as *acute, right, obtuse,* or *straight.*

1.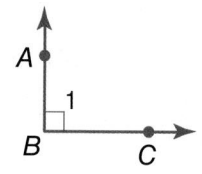

2.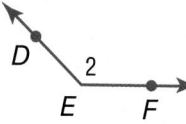

3.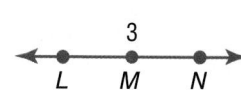

4.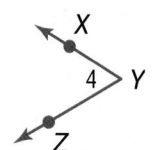

5. Identify a pair of vertical angles in the diagram at the right.

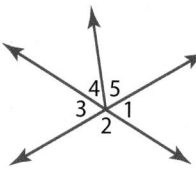

PART C
PAGES 401–405

Identify each pair of angles as *complementary, supplementary,* or *neither.*

1.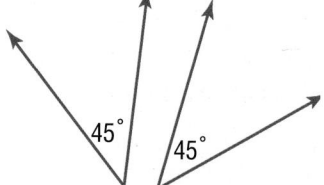

2.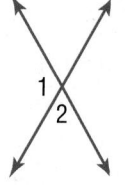

3.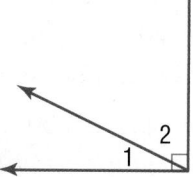

ALGEBRA Find the value of x in each figure.

4.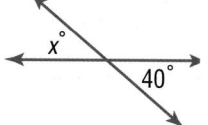

5.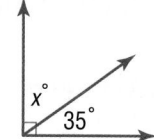

6.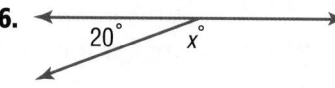

Multi-Part Lesson 7-1 (continued)

PART D

PAGES 406–407

Use *logical reasoning* to solve Exercises 1–3.

1. **GEOMETRY** Can a polygon containing two right angles be a triangle? Explain your reasoning. Can it be a quadrilateral? Explain.

2. **HEIGHT** Kristina is $\frac{2}{3}$ the height of Jodi, who is $\frac{3}{4}$ as tall as Destini. If Destini is 6 feet tall, how tall are the others?

3. **SPELLING** The top four finishers in the spelling bee were Kina, Niko, Gia, and Martez. Niko and the first place winner studied with Kina for the spelling bee. Gia is not the first place winner. Who is the first place winner?

Multi-Part Lesson 7-2: Lines

PARTS A B

PAGES 408–414

Classify each pair of angles as *alternate interior, alternate exterior,* or *corresponding.*

1. $\angle 3$ and $\angle 6$

2. $\angle 7$ and $\angle 3$

3. $\angle 5$ and $\angle 4$

4. $\angle 8$ and $\angle 1$

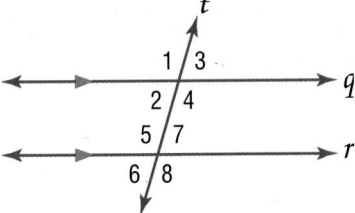

Refer to the figure at the right. Line a is parallel to line b and $m\angle 2$ is 145°. Find each given angle measure. Justify your answer.

5. $m\angle 9$

6. $m\angle 7$

7. $m\angle 3$

8. $m\angle 4$

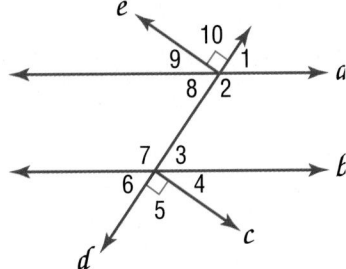

Multi-Part Lesson 7-3: Angle Relationships in Polygons

PARTS A B

PAGES 416–423

Find the value of x in each triangle with the given angle measures.

1. 80°, 40°, $x°$

2. 30°, 80°, $x°$

3. 10°, 46°, x

4. 25°, 27°, $x°$

5. 60°, 25°, $x°$

6. 50°, 50°, $x°$

Classify each triangle with the given angles and side measures.

7. angles: 40°, 50°, 90°
   sides: 21 cm, 25 cm, 32.7 cm

8. angles: 20°, 20°, 140°
   sides: 26.5 in., 26.5 in., 50 in.

9. angles: 30°, 70°, 80°
   sides: 8.5 mm, 16 mm, 16.8 mm

10. angles: 35°, 55°, 90°
    sides: 7 m, 10 m, 12.2 m

11. angles: 30°, 40°, 110°
    sides: 116 cm, 130 cm, 200 cm

12. angles: 60°, 60°, 60°
    sides: 8 ft, 8 ft, 8 ft

13. The measures of the angles of $\triangle PQR$ are in the ratio 2:5:5. What are the measures of the angles?

14. The measures of the angles of $\triangle KLM$ are in the ratio 2:3:4. What are the measures of the angles?

Multi-Part Lesson 7-3 (continued)
PARTS C D

PAGES 424–430

Classify each quadrilateral with the name that best describes it.

1.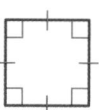

2.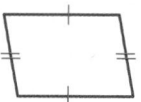

3.

ALGEBRA Find the missing angle measure in each quadrilateral.

4.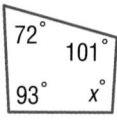

5.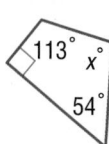

6.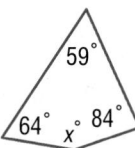

7. **ALGEBRA** Find $m\angle A$ in quadrilateral $ABCD$ if $m\angle B = 75°$, $m\angle C = 125°$, and $m\angle D = 35°$.

8. **ALGEBRA** What is $m\angle E$ in quadrilateral $DEFG$ if $m\angle D = 53°$, $m\angle F = 136°$, and $m\angle G = 84°$?

PARTS E F

PAGES 431–437

Determine whether each figure is a polygon. If it is, classify the polygon. If it is not a polygon, explain why.

1.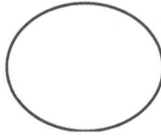

2.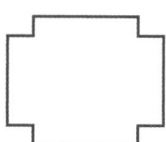

3.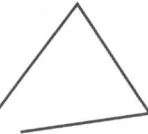

Find the sum of the interior angle measures of each polygon.

4. dodecagon (12-gon) 5. 17-gon 6. 21-gon

Find the measure of one interior angle in each regular polygon. Round to the nearest tenth if necessary.

7. 18-gon 8. 22-gon 9. octagon

Multi-Part Lesson 7-4: Three-Dimensional Figures
PARTS A B C

PAGES 440–448

Identify each figure. Then name the bases, faces, edges, and vertices.

1.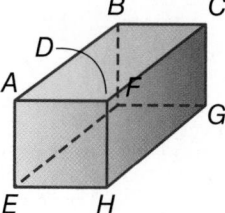

2.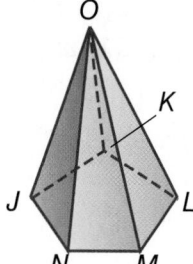

3.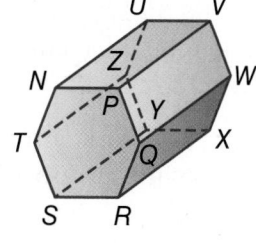

Multi-Part Lesson 8-1: Similar Triangles

PART A

PAGES 462–463

Use the *draw a diagram* strategy to solve Exercises 1–3.

1. **PICTURE FRAMES** Mr. Francisco has 4 picture frames that he wants to hang on the wall. In how many different ways can he hang the picture frames on the wall?

2. **PONDS** Carter is filling the pond in his backyard. After 2 minutes and 20 seconds, the pond is only $\frac{1}{7}$ full. If the pond can hold 280 gallons, how much longer will it take to fill the pond?

3. **MARCHING BAND** The marching band is in formation on the field. In the first row, there are 10 band members. Each additional row has 6 more members in it. If there are a total of 6 rows, how many band members are there?

PARTS B C

PAGES 464–470

Determine whether each pair of polygons is similar. Explain.

1.

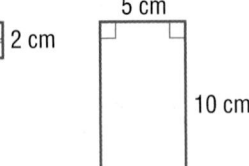

2.

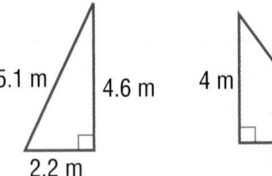

Each pair of polygons is similar. Find each missing side measure.

3.

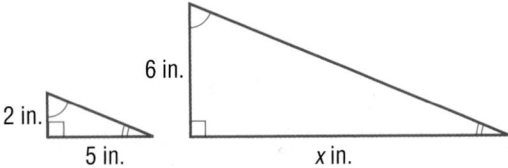

4.

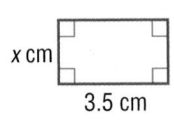

PART D

PAGES 471–474

Write a proportion and solve the problem.

1. A road sign casts a shadow 14 meters long, while a tree nearby casts a shadow 27.8 meters long. If the road sign is 3.5 meters high, how tall is the tree?

2. Use the diagram to find the distance across Catfish Lake. Assume the triangles are similar.

3. A 7-foot tall golf flag casts a shadow 21 feet long. A golfer standing nearby casts a shadow 16.5 feet long. How tall is the golfer?

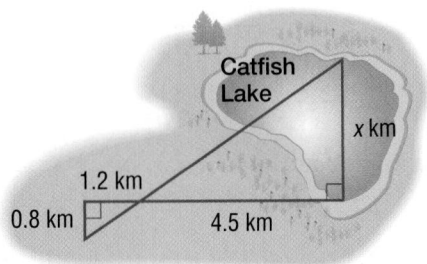

Multi-Part Lesson 8-1 (continued)

PART E

PAGES 475–479

Find the tangent of each acute angle. Explain its meaning.

1.
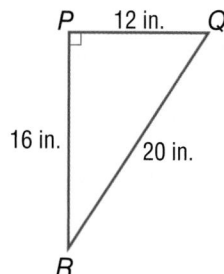

2.
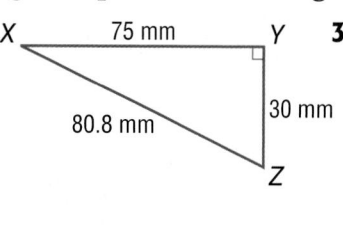

3.
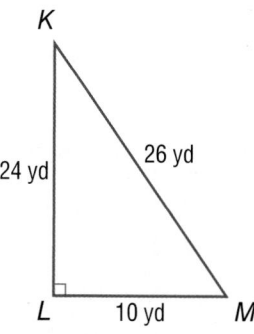

4.
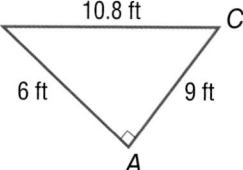

5.
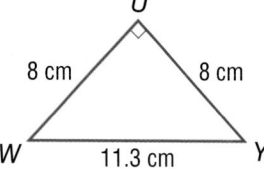

6.
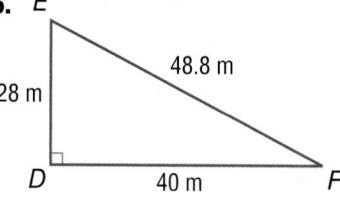

7. HOUSES Mason is looking at the top of his house at a 30° angle. He is standing 40 feet from the house. How tall is the house? Round to the nearest tenth.

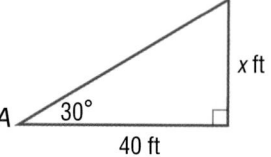

Multi-Part Lesson 8-2: The Pythagorean Theorem

PARTS A B

PAGES 481–486

Write an equation you could use to find the length of the missing side of each right triangle. Then find the missing length. Round to the nearest tenth if necessary.

1.
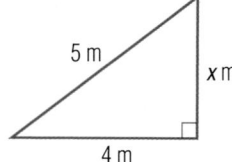

2.
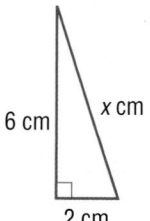

3.
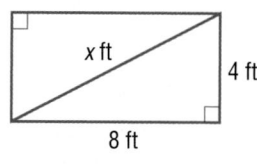

4. a, 6 cm; b, 5 cm

5. a, 12 ft; b, 12 ft

6. a, 8 in.; b, 6 in.

7. a, 20 m; c, 25 m

8. a, 9 mm; c, 14 mm

9. b, 15 m; c, 20 m

Determine whether each triangle with sides of given lengths is a right triangle.

10. 15 m, 8 m, 17 m

11. 7 yd, 5 yd, 9 yd

12. 5 in., 12 in., 13 in.

13. 9 in., 12 in., 16 in.

14. 10 ft, 24 ft, 26 ft

15. 2 ft, 2 ft, 3 ft

Multi-Part Lesson 8-2 (continued)

PART C

PAGES 487–492

**Write an equation that can be used to answer the question. Then solve.
Round to the nearest tenth if necessary.**

1. How far apart are the boats?

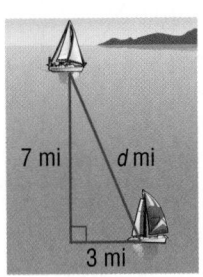

7 mi d mi 3 mi

2. How high does the ladder reach?

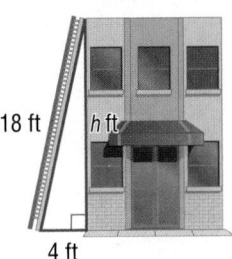

18 ft h ft 4 ft

3. How long is each rafter?

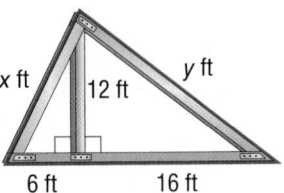

x ft 12 ft y ft 6 ft 16 ft

PARTS D E

PAGES 493–499

**Graph each pair of ordered pairs. Then find the distance between the points.
Round to the nearest tenth if necessary.**

1. $(-4, 2), (4, 17)$

2. $(5, -1), (11, 7)$

3. $(-3, 5), (2, 7)$

4. $(7, -9), (4, 3)$

5. $(5, 4), (-3, 8)$

6. $(-8, -4), (-3, 8)$

7. $(2, 7), (10, -4)$

8. $(9, -2), (3, 6)$

9. $(2, 3), (-1, 6)$

10. $(-5, 1), (2, -3)$

11. $(0, 1), (5, 2)$

12. $(-1, 2), (-2, 3)$

PART F

PAGES 500–505

Find each missing measure.

1.
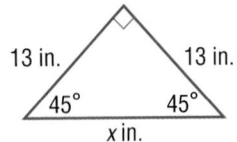

13 in. 13 in. 45° 45° x in.

2.
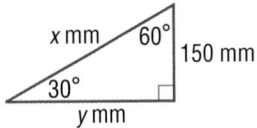

x mm 60° 150 mm 30° y mm

3.
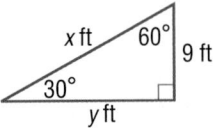

x ft 60° 9 ft 30° y ft

4.
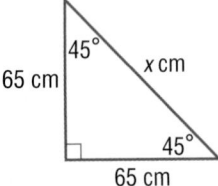

45° x cm 65 cm 45° 65 cm

5.
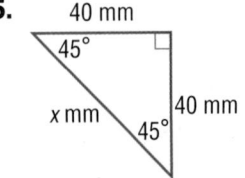

40 mm 45° x mm 40 mm 45°

6.
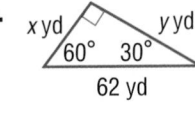

x yd y yd 60° 30° 62 yd

7. In a 45°-45°-90° triangle, a leg is 11.3 feet long. Find the exact length of the hypotenuse.

8. In a 30°-60°-90° triangle, the shorter leg is 12 centimeters long. Find the exact lengths of the hypotenuse and the longer leg.

9. In a 30°-60°-90° triangle, the hypotenuse is 36 millimeters long. Find the exact lengths of the shorter and longer legs.

Multi-Part Lesson 8-3: Transformations

PART A

PAGES 508–511

Graph each figure with the given vertices. Then graph the image of the figure after the indicated translation, and write the coordinates of its vertices.

1. rectangle $PQRS$ with vertices $P(-7, 6)$, $Q(-5, 6)$, $R(-5, 2)$, and $S(-7, 2)$ translated 9 units right and 1 unit down

2. pentagon $DGLMR$ with vertices $D(1, 3)$, $G(2, 4)$, $L(4, 4)$, $M(5, 3)$ and $R(3, 1)$ translated 5 units left and 7 units down

3. triangle TRI with vertices $T(2, 1)$, $R(0, 3)$, and $I(-1, 1)$ translated 2 units left and 3 units down

4. quadrilateral $QUAD$ with vertices $Q(3, 2)$, $U(3, 0)$, $A(6, 0)$ and $D(6, 2)$, translated 3 units left and 1 unit down

PART B

PAGES 512–516

Graph each figure and its reflection over the given axis.

1. y-axis

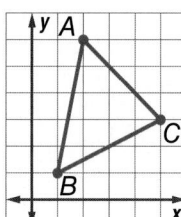

2. x-axis

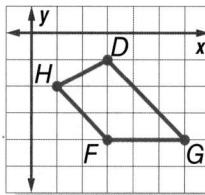

Graph each figure and its reflection over the given axis. Then find the coordinates of the reflected image.

3. triangle CAT with vertices $C(2, 3)$, $A(8, 2)$, and $T(4, -3)$; x-axis

4. trapezoid $TRAP$ with vertices $T(-2, 5)$, $R(1, 5)$, $A(4, 2)$, and $P(-5, 2)$; y-axis

PARTS C D

PAGES 517–522

Graph $\triangle XYZ$ and its image after each rotation. Then give the coordinates of the vertices for $\triangle X'Y'Z'$.

1. 180° counterclockwise about vertex X

2. 90° clockwise about vertex Z

3. 180° clockwise about the origin

4. 270° counterclockwise about the origin

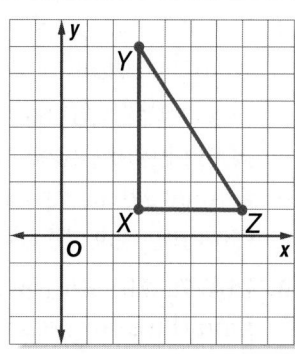

Multi-Part Lesson 8-3 (continued)

PARTS E F

PAGES 523–528

Find the coordinates of the vertices of each figure after a dilation with the given scale factor k. Then graph the original image and the dilation.

1. $F(-4, -6)$, $G(-4, 4)$, $H(4, -6)$; $k = \frac{1}{2}$

2. $R(-3, -1)$, $S(0, 4)$, $T(4, -1)$; $k = 3$

3. $A(-3, 2)$, $B(2, 2)$, $C(-3, -3)$, $D(2, -3)$; $k = 2$

4. $W(4, 8)$, $X(8, 8)$, $Y(8, 4)$, $Z(4, 4)$; $k = \frac{1}{4}$

Multi-Part Lesson 9-1: Literal Equations

PART A

PAGES 541–544

Solve each equation for the indicated variable.

1. $A = bh$, for h
2. $k = mp + 4$, for p
3. $R = 3k - q$, for k
4. $\pi = \frac{C}{2r}$, for C
5. $a^2 + b^2 = c^2$, for b
6. $a = \frac{x}{2} + y$, for x
7. $2a + 3b = 4c$, for b
8. $V = \pi r^2 h$, for h
9. $a = b + \frac{c}{2}$, for b
10. $E = mc^2$, for c
11. $F = \frac{9}{5}C + 32$, for C
12. $y = \frac{d}{m} + 15$, for m

PART B

PAGES 545–549

Complete each conversion. Round to the nearest hundredth if necessary.

1. $22°C = \blacksquare °F$
2. $100°F = \blacksquare °C$
3. $-21°F = \blacksquare °C$
4. $-13.4°C = \blacksquare °F$
5. $0.5°C = \blacksquare °F$
6. $32°F = \blacksquare °C$
7. $-16°C = \blacksquare °F$
8. $112°C = \blacksquare °F$
9. $98.6°F = \blacksquare °C$
10. $-0.8°F = \blacksquare °C$
11. $-115°C = \blacksquare °F$
12. $-17°F = \blacksquare °C$
13. $100°C = \blacksquare °F$
14. $-6.4°C = \blacksquare °F$
15. $10°F = \blacksquare °C$
16. $112°F = \blacksquare °C$
17. $0°C = \blacksquare °F$
18. $143°F = \blacksquare °C$

PART C

PAGES 550–551

Determine reasonable answers to solve Exercises 1–4.

1. **TIME** Sondra estimates that she spends 30% of her work day answering E-mails. If she worked 8.7 hours in one day, did she answer E-mails for 2, 3, or 4 hours?

2. **SEWING** Norma is making scarves for the craft show. Each scarf needs 48 inches of material. If she has 676 inches of material, about how many scarves can she make: 14, 16, or 18?

3. **POPCORN** The cost of renting a popcorn machine is $134.99. If the student council sells bags of popcorn for $0.85, should they sell 125, 140, or 150 bags of popcorn to pay for the popcorn machine?

4. **TRAVEL** The Wright family has traveled 310 miles. If this is 77% of the trip, would the number of miles left to travel be about 70, 90, or 240?

Multi-Part Lesson 9-2: Convert Units of Measure

PARTS (A) (B)

PAGES 553–560

Complete.

1. 4,000 lb = ■ T
2. 5 T = ■ lb
3. 5 lb = ■ oz
4. 12,000 lb = ■ T
5. $\frac{1}{4}$ lb = ■ oz
6. 12 pt = ■ c
7. 3 gal = ■ pt
8. 24 fl oz = ■ c
9. 8 pt = ■ c
10. 10 pt = ■ qt
11. $2\frac{1}{4}$ c = ■ fl oz
12. 6 lb = ■ oz
13. 10 gal = ■ qt
14. 4 qt = ■ fl oz
15. 4 pt = ■ c

Complete each conversion. Round to the nearest hundredth if necessary.

16. 662 m = ■ km
17. 5,283 mL = ■ L
18. 0.24 cm = ■ mm
19. 380 kL = ■ L
20. 10.8 g = ■ mg
21. 83,000 mL = ■ L
22. 56 in. ≈ ■ cm
23. 32.8 ft ≈ ■ m
24. 609 yd ≈ ■ m
25. 21.78 mi ≈ ■ km
26. 48 lb ≈ ■ g
27. 2.3 T ≈ ■ kg
28. 8.5 c ≈ ■ mL
29. 33 gal ≈ ■ L
30. 1.8 qt ≈ ■ mL

PART (C)

PAGES 561–564

Complete each conversion. Round to the nearest hundredth if necessary.

1. 62 mi/h = ■ ft/s
2. 500 gal/h ≈ ■ L/min
3. 100 mL/min = ■ L/h
4. 100 m/min ≈ ■ ft/s
5. 2 lb/min ≈ ■ kg/h
6. 55 mi/h = ■ ft/min
7. 18 km/L ≈ ■ mi/gal
8. 200 gal/yr ≈ ■ L/mo
9. 5 fl oz/s = ■ qt/min
10. 180 m/min ≈ ■ ft/s
11. 3 qt/min = ■ gal/h
12. 220 kg/day ≈ ■ oz/min

PART (D)

PAGES 565–569

Complete each conversion.

1. 420 mm^2 = ■ cm^2
2. 2.5 yd^2 = ■ ft^2
3. 3.25 ft^2 = ■ in^2
4. 0.5 m^2 = ■ cm^2
5. 2 cm^3 = ■ mm^3
6. 459 ft^3 = ■ yd^3
7. 8,640 in^3 = ■ ft^3
8. 0.3 m^3 = ■ cm^3

Complete each conversion. Round to the nearest hundredth if necessary.

9. 2 m^2 ≈ ■ ft^2
10. 13 yd^2 ≈ ■ m^2
11. 12 in^2 ≈ ■ cm^2
12. 5.5 km^2 ≈ ■ mi^2
13. 4 m^3 ≈ ■ yd^3
14. 25 cm^3 ≈ ■ in^3
15. 17 m^3 ≈ ■ ft^3
16. 115 ft^3 ≈ ■ m^3

Multi-Part Lesson 10-1: Analyze Data

PARTS (A) (B)

PAGES 583–588

Find the mean, median, and mode of each data set. Round to the nearest tenth if necessary.

1. 2, 7, 9, 12, 5, 14, 4, 8, 3, 10

2. 58, 52, 49, 60, 61, 56, 50, 61

3. 122, 134, 129, 140, 125, 134, 137

4. 36, 41, 43, 45, 48, 52, 54, 56, 56, 57, 60, 64, 65

5. 11, 15, 21, 11, 6, 10, 11

6. 21, 20, 19, 20, 18, 21, 23, 25

7. 1, 3, 2, 1, 1, 2, 2, 2, 3

8. 23, 35, 42, 26, 27, 29, 31, 29, 27

PART (C)

PAGES 589–592

Describe how the mean, median, and mode are affected if the indicated value is removed from the data set.

1. minutes Olympics were watched: 75, 92, 80, 160, 125, 80

2. prices of a CD at 6 stores: 18, 19, 18, 20, 17, 21

3. scores on math quizzes: 10, 10, 6, 8, 7, 10, 9, 8

4. hours spent exercising per week: 5, 5, 8, 9, 4, 3, 7

5. cost for a gallon of gasoline: 2.85, 2.93, 3.41, 3.76, 4.01, 3.76

6. number of ambulance runs per month: 26, 53, 44, 31, 62, 44

7. graduates in the senior class: 412, 292, 226, 378, 402, 402

8. miles driven per week: 220, 365, 220, 180, 512, 260

9. MONEY The hourly wages of employees at a restaurant are: $6, $5, $5.75, $6.50, $20, $5.75, and $7. Which measure of central tendency will change the most if the highest amount is removed?

Multi-Part Lesson 10-2: Box-and-Whisker Plots

PART (A)

PAGES 593–598

Find the measures of variation and any outliers for each data set.

1. 15, 12, 21, 18, 25, 11, 17, 19, 20

2. 2, 24, 6, 13, 8, 6, 11, 4

3. 189, 149, 155, 290, 141, 152

4. 451, 501, 388, 428, 510, 480, 390

5. 22, 18, 9, 26, 14, 15, 6, 19, 28

6. 245, 218, 251, 255, 248, 241, 250

7. 46, 45, 50, 40, 49, 42, 64

8. 128, 148, 130, 142, 164, 120, 152, 202

Multi-Part Lesson 10-2 (continued)

PART B
PAGES 599–603

Construct a box-and-whisker plot for each data set.

1. 2, 3, 5, 4, 3, 3, 2, 5, 6

2. 6, 7, 9, 10, 11, 11, 13, 14, 12, 11, 12

3. 15, 12, 21, 18, 25, 11, 17, 19, 20

4. 2, 24, 6, 13, 8, 6, 11, 4

5. ZOOS Use the following box-and-whisker plot.

Area (acres) of Major Zoos in the United States

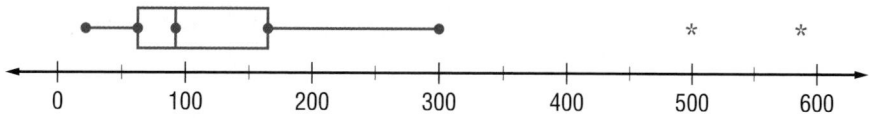

a. How many outliers are in the data?

b. Describe the distribution of the data. What can you say about the areas of the major zoos in the United States?

PARTS C D
PAGES 604–608

1. LUNCH The table shows the results of a school survey about types of lunch purchased.

Types of Lunch Purchased							
Full Lunch				**À la Carte**			
152	202	107	97	91	103	281	257
89	110	101	269	175	223	135	82

   a. Construct a double box-and-whisker plot for the data.

   b. Compare the data for each line.

2. CARS Refer to the double box-and-whisker plot below that shows the number of car shows attended by two car clubs each year.

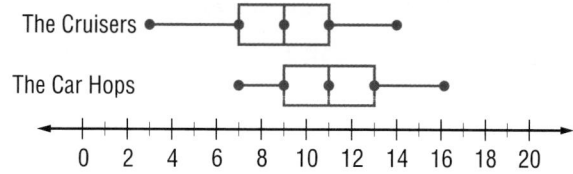

   a. Compare the number of shows for The Cruisers to The Car Hops.

   b. About what percent of The Cruisers' attendance and what percent of The Car Hops' attendance are 11 or more?

Multi-Part Lesson 10-3: Scatter Plots

PART A
PAGES 610–611

1. FOOD DRIVES The table shows the number of canned goods collected in Mrs. Michel's class.

   a. Make a graph of the data.

   b. Describe how the number of canned goods changed from 2008 to 2011.

Year	Number of Canned Goods
2008	30
2009	32
2010	41
2011	49

Multi-Part Lesson 10-3 (continued)

PARTS (B) (C)

PAGES 612–618

Would a scatter plot of the data for each of the following show a *positive*, *negative*, or *no* relationship?

1. height and hair color
2. hours spent studying and test scores
3. income and month of birth
4. child's age and height

Determine whether each graph shows a *positive*, *negative*, or *no* relationship.

5.
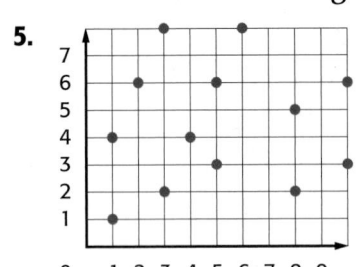

6.
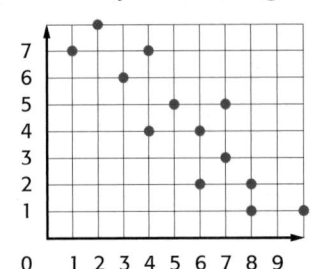

7.
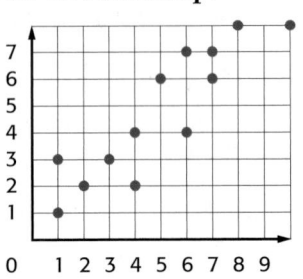

PARTS (D) (E) (F)

PAGES 619–626

1. **SILVER** The table shows the average price of an ounce of silver to the nearest ten cents for certain years.

Years Since 1975	0	5	10	15	20	25	30
Cost per Ounce	4.4	21.0	6.1	4.8	5.2	5.0	7.2

  a. Construct a scatter plot of the data. Then draw a line that best represents the data.

  b. Use the line of best fit to make a conjecture about the cost of an ounce of silver in the year 2015.

2. **GOLD** The table shows the average price of an ounce of gold to the nearest ten dollars for certain years.

Years Since 1975	0	5	10	15	20
Cost per Ounce	350	420	390	270	520

  a. Construct a scatter plot of the data. Then draw a line that best represents the data.

  b. Write an equation in slope-intercept form for the line of best fit.

  c. Use the equation to predict the cost of an ounce of gold in the year 2015.

PART (G)

PAGES 627–631

Select an appropriate display for each situation. Justify your reasoning.

1. the amount of each flavor of ice cream sold relative to the total sales

2. the number of people attending a fair for specific intervals of ages

3. **RADIO** Construct an appropriate display of the data below.

Adult Audience of Oldies Radio				
Age	18 to 24	25 to 34	35 to 44	45 to 54	55 or older
Percent of Audience	10%	14%	29%	33%	14%

Multi-Part Lesson 11-1: Outcomes

PART A

PAGES 645–649

Draw a tree diagram to determine the number of possible outcomes.

1. A car comes in white, black, or red with standard or automatic transmission and with a 4-cylinder or 6-cylinder engine.

2. A customer can buy roses or carnations in red, yellow, pink, or white.

3. A pizza can be ordered with a regular or deep dish crust and with a choice of one topping, two toppings, or three toppings.

Use the Fundamental Counting Principle to find the number of possible outcomes.

4. A woman's shoe comes in red, white, blue, or black with a choice of high, medium, or low heels.

5. Sugar cookies, chocolate chip, or oatmeal raisin cookies can be ordered either with or without icing.

PARTS B C

PAGES 650–654

1. **CIRCUS** In how many different ways can a family of four be seated on a bleacher at the circus?

2. **LETTERS** In how many different ways can you arrange the letters in the word *orange* if you take the letters five at a time?

3. **MUSIC** In how many different ways can Kevin listen to each of his ten CDs once?

Find each value.

4. $P(7, 4)$ 5. $P(4, 3)$ 6. $P(5, 5)$ 7. $P(3, 1)$

8. $P(9, 4)$ 9. $P(6, 2)$ 10. $P(10, 3)$ 11. $P(12, 4)$

12. $P(1, 1)$ 13. $P(12, 5)$ 14. $P(10, 2)$ 15. $P(6, 4)$

PART D

PAGES 655–658

1. How many combinations of four songs can you pick from a CD with 8 songs on it?

2. How many combinations of eight colors can you pick from a box that has 30 different-colored crayons?

3. How many combinations of nine friends can you pick from a group of 10 people?

4. How many combinations of three instruments can you pick from a band that has 7?

Determine whether each situation is a *permutation* or a *combination*. Then find the number of possible outcomes.

5. choosing a committee of 3 from a class of 25

6. placing 6 different math books in a line

7. choosing 4 flavors of ice cream from 20 different flavors of ice cream

Multi-Part Lesson 11-2: Probability

PART A PAGES 659–664

1. **COINS** Two evenly balanced nickels are tossed. Find the probability that one head and one tail result.

2. **MONEY** A wallet contains four $5 bills, two $10 bills, and eight $1 bills. A bill is randomly selected. Find the probability of selecting a $5 bill or a $1 bill.

3. Two chips are selected from a box containing 6 blue chips, 4 red chips, and 3 green chips. The first chip selected is replaced before the second is drawn. Find P(red and green).

4. A bag contains 7 blue, 4 orange, 8 red, and 5 purple marbles. Suppose one marble is chosen and not replaced. A second marble is then chosen. Find P(purple and red).

PARTS B C D PAGES 665–672

1. **FOOD** Use the survey results at the right.

   a. What is the probability that a person's favorite pizza topping is pepperoni?

   b. Out of 280 people, how many would you expect to have pepperoni as their favorite pizza topping?

   c. What is the probability that a person's favorite pizza topping is pepperoni or sausage?

Favorite Pizza Topping	
Topping	Number
pepperoni	45
sausage	25
green pepper	15
mushrooms	5
other	10

2. **SURVEYS** In a survey, 53 out of 110 people said their favorite season was summer. Thirty-two people said spring, 17 people chose autumn, and 8 people said winter was their favorite season. What is the experimental probability that a person chose spring as their favorite season?

Multi-Part Lesson 11-3: Data Collection

PART A PAGES 674–675

Use the *act it out* strategy to solve Exercises 1–3.

1. **PICTURES** Gabriella, Taylor, and Kelsi are standing next to each other for a yearbook picture. In how many different ways can they stand next to each other for the picture?

2. **BASKETBALL** There are six boys on the basketball team. How many combinations of a captain and co-captain are possible?

3. **CLOTHES** Ryan is going to a dance. He can choose a blue, black, or white sports coat. He can also choose from a green, black, blue, or red tie. How many coat and tie combinations can Ryan wear?

Multi-Part Lesson 11-3 (continued)

PARTS B C

PAGES 676–681

1. **PRIZES** A restaurant offers six kids-meal prizes. The prizes are placed in the meals at random. Describe a model that could be used to simulate selecting one of the prizes.

2. **FOOD** A pizza parlor offers three different types of crust. Each crust type is equally likely to be ordered. Describe a model that could be used to simulate this situation. Based on your simulation, how many customers must order a pizza in order to sell all possible combinations?

3. **FORECAST** A weather forecaster has predicted a 25% chance of precipitation for the next 4 days. Describe a model that could be used to find the experimental probability of rain all 4 days.

4. **CHARITY** Fifty percent of the clothes a local charity receives are coats. Describe a model that could be used to find the experimental probability of the charity receiving coats during the next 10 donations.

PARTS D E

PAGES 682–689

Determine whether each conclusion is valid. Justify your answer.

1. To determine whether most students participate in after school activities, the principal of Humberson Middle School randomly surveyed 75 students from each grade level. Of these, 34% said they participate in after school activities. The principal concluded that about a third of the students at Humberson Middle School participate in after school activities.

2. To evaluate their product, the manager of an assembly line inspected the first 100 watches produced on Monday. Of these, 2 were defective. The manager concluded that about 2% of all watches produced are defective.

3. A television program asked its viewers to dial one of two phone numbers indicating their preference for one of two brands of shampoo. Of those that responded, 76% said they prefer Brand A. The program concluded that Brand A was the most popular brand of shampoo.

Multi-Part Lesson 12-1: Circumference and Area

PARTS A B C

PAGES 703–711

Find the circumference and area of each circle. Round to the nearest tenth.

1.
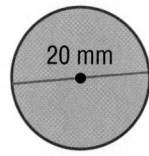
20 mm

2.
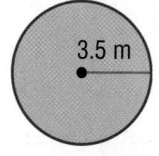
3.5 m

3.
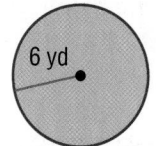
6 yd

4.
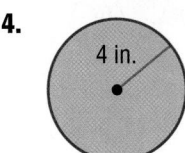
4 in.

5.
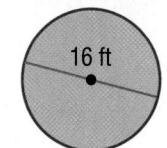
16 ft

6.
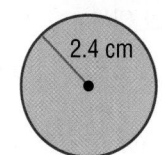
2.4 cm

Multi-Part Lesson 12-1 (continued)

PART D

PAGES 712–713

Use the *make a model* strategy to solve Exercises 1 and 2.

1. **GEARS** The set of gears shown has diameters of 10 inches, 12 inches, 12 inches, and 20 inches. After how many complete revolutions of the smallest gear will the largest gear make one complete revolution?

2. **PACKAGING** Cecil needs packing tape to ship a number of identically shaped packages. The packages are cubes. He needs to tape all the way around the box, in both directions, and have 2 inches of overlap in each direction. Write an expression that can be used to find the amount of tape needed to wrap *p* such packages.

PART E

PAGES 714–718

Find the area of the shaded region. Round to the nearest tenth if necessary.

1.
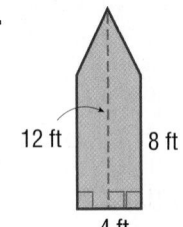
12 ft   8 ft
4 ft

2.
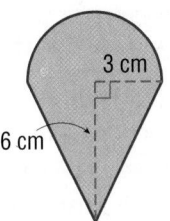
3 cm
6 cm

3.
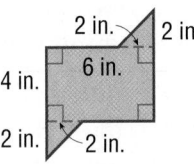
2 in.   2 in.
4 in.   6 in.
2 in.   2 in.

4.
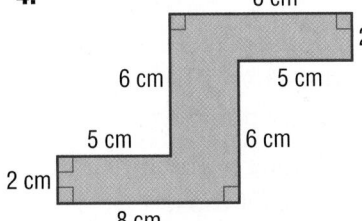
8 cm
2 cm
6 cm   5 cm
5 cm   6 cm
2 cm
8 cm

5.
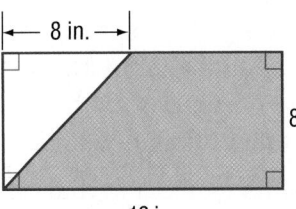
8 in.
8 in.
18 in.

6.
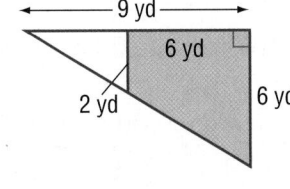
9 yd
6 yd
2 yd   6 yd

Multi-Part Lesson 12-2: Volume

PART A

PAGES 719–724

Find the volume of each solid. Round to the nearest tenth if necessary.

1.
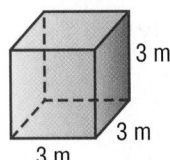
3 m
3 m
3 m

2.
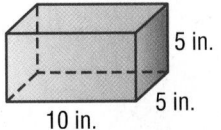
5 in.
5 in.
10 in.

3.
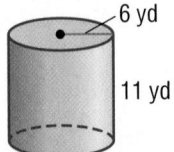
6 yd
11 yd

4.
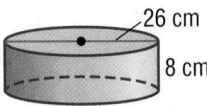
26 cm
8 cm

5.
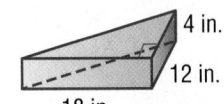
4 in.
12 in.
18 in.

6.
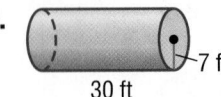
7 ft
30 ft

7. triangular prism: base of triangle, 7 yd; altitude, 18 yd; height of prism, $5\frac{1}{3}$ yd

Multi-Part Lesson 12-2 (continued)
PART B

PAGES 725–730

Find the volume of each solid. Round to the nearest tenth if necessary.

1.
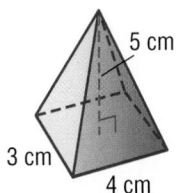
5 cm
3 cm
4 cm

2.
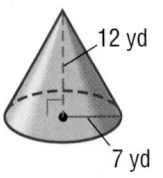
12 yd
7 yd

3.
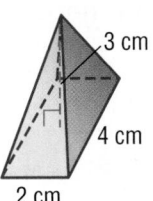
3 cm
4 cm
2 cm

4.
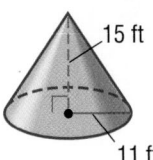
15 ft
11 ft

5.
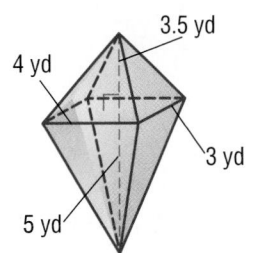
3.5 yd
4 yd
3 yd
5 yd

6.
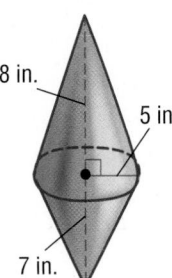
8 in.
5 in.
7 in.

Find the volume of each sphere. Round to the nearest tenth.

7.

3.6 mm

8.
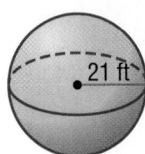
21 ft

9.
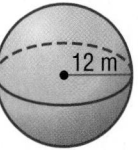
12 m

Multi-Part Lesson 12-3: Surface Area
PARTS A B

PAGES 732–738

Find the lateral and total surface areas of each solid. Round to the nearest tenth if necessary.

1.

2 ft
2 ft
2 ft

2.
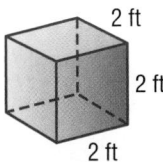
3 ft
4 ft
6 ft

3.
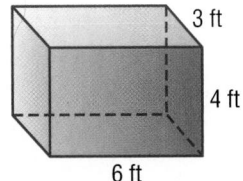
4 cm
8 cm
5 cm
3 cm

4.
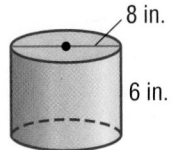
8 in.
6 in.

5.
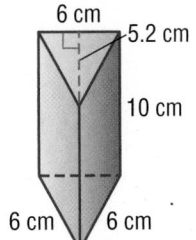
6 cm
5.2 cm
10 cm
6 cm   6 cm

6.
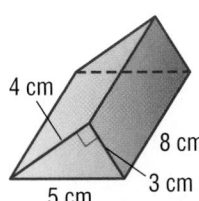
14 cm
3 cm

7. PLUMBING A hollow piece of a cylindrical pipe is shown. Find the total surface area of the pipe, including the interior. Round to the nearest tenth.

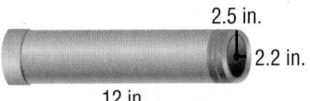
2.5 in.
2.2 in.
12 in.

Multi-Part Lesson 12-3 (continued)

PARTS C D

PAGES 739–745

Find the lateral and total surface areas of each regular pyramid. Round to the nearest tenth if necessary.

1.
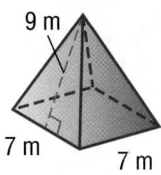
9 m
7 m   7 m

2.
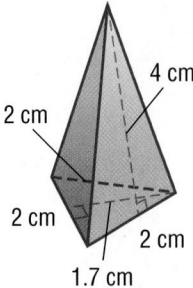
4 cm
2 cm
2 cm
2 cm
1.7 cm

3.
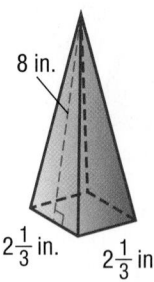
8 in.
$2\frac{1}{3}$ in. $2\frac{1}{3}$ in.

Find the lateral and total surface areas of each cone. Round to the nearest tenth.

4.
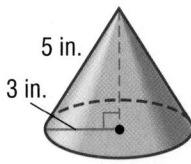
5 in.
3 in.

5.
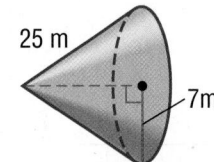
25 m
7m

6.
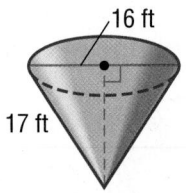
16 ft
17 ft

7. HISTORY The great pyramid of Khufu in Egypt was originally 481 feet high with a square base measuring 756 feet on a side and slant height of about 611.8 feet. What was its lateral surface area? Round to the nearest tenth.

Go to Hotmath.com for step-by-step solutions of most odd-numbered exercises free of charge.

Chapter 1 Rational Numbers and Percent

p. 26   Are You Ready?
1. -9 **3.** -14 **5.** -84 **7.** -9 **9.** $17°F$ **11.** 6
13. 14.8 **15.** 2 **17.** 6 eggs

p. 30–32   Lesson 1-1A
1. 0.8 **3.** -1.725 **5.** $4.8\overline{3}$ **7.** 0.438

⑨ $0.32 = \dfrac{32}{100}$ 0.32 is 32 hundredths.
$= \dfrac{8}{25}$ Simplify.

11. $-\dfrac{5}{9}$ **13.** $2\dfrac{5}{33}$ **15.** 0.4 **17.** 0.825
19. -0.15625 **21.** 5.3125

㉓ $-\dfrac{6}{11}$ means $-6 \div 11$.

$$11\overline{)6.0000}$$
$$\begin{array}{r} 0.5454... \\ \underline{-55} \\ 50 \\ \underline{-44} \\ 60 \\ \underline{-55} \\ 50 \\ \underline{-44} \\ 6 \end{array}$$

Divide 6 by 11 and add a negative sign.

The fraction $-\dfrac{6}{11}$ can be written as $-0.\overline{54}$.

25. $-7.1\overline{7}$ **27.** $-\dfrac{2}{5}$ **29.** $5\dfrac{11}{20}$ **31.** $\dfrac{2}{9}$ **33.** $-3\dfrac{1}{11}$

㉟ Rainfall for Friday is 0.08 inch.

$0.08 = \dfrac{8}{100}$ 0.08 is 8 hundredths.
$= \dfrac{2}{25}$ Simplify.

37. $\dfrac{7}{200}$ **39.** $1\dfrac{1}{16}$ in.; 1.0625 in.   **41.** Sample answer: $0.\overline{12}$; Since $0.\overline{12} = \dfrac{4}{33}$, it is a rational number.
43. Sample answer: When dividing, there are two possibilities for the remainder. If the remainder is 0, the decimal terminates. If the remainder is not 0, then at the point where the remainder repeats or equals the original dividend the decimal begins to repeat.   **45.** D   **47.** I

pp. 36–38   Lesson 1-1B
1. $-\dfrac{2}{5}$ **3.** $-1\dfrac{3}{5}$ **5.** $\dfrac{7}{12}$ **7.** $\dfrac{1}{8}$ **9.** $3\dfrac{2}{9}$

⑪ $-3\dfrac{2}{5} + 1\dfrac{5}{6}$
$= \dfrac{-17}{5} + \dfrac{11}{6}$ Write as improper fractions.
$= \dfrac{-102}{30} + \dfrac{55}{30}$ $\dfrac{-17}{5} \cdot \dfrac{6}{6} = \dfrac{-102}{30}$ and $\dfrac{11}{6} \cdot \dfrac{5}{5} = \dfrac{55}{30}$
$= \dfrac{-102 + 55}{30}$ Add the numerators.
$= \dfrac{-47}{30}$ Simplify.
$= -1\dfrac{17}{30}$

13. $2\dfrac{1}{2}$ h   **15.** $-\dfrac{5}{7}$ **17.** $\dfrac{1}{3}$ **19.** $\dfrac{3}{8}$ **21.** $-\dfrac{2}{3}$ **23.** $\dfrac{11}{24}$
25. $\dfrac{13}{72}$ **27.** $\dfrac{14}{15}$ **29.** $\dfrac{1}{75}$ **31.** $2\dfrac{1}{6}$ c   **33.** $5\dfrac{1}{5}$
35. $-3\dfrac{23}{24}$ **37.** $1\dfrac{3}{5}$ **39.** $-13\dfrac{1}{6}$

㊶ $42\dfrac{6}{10} - 14\dfrac{3}{10}$
$= \dfrac{426}{10} - \dfrac{143}{10}$ Write as improper fractions.
$= \dfrac{426 - 143}{10}$ Subtract the numerators.
$= \dfrac{283}{10}$ Simplify.
$= 28\dfrac{3}{10}$ in.

43. $2\dfrac{11}{24}$ **45.** $-\dfrac{1}{2}$ **47.** $-3\dfrac{11}{24}$ **49.** 10 h 5 min
51. $2\dfrac{1}{2}$ **55.** Greater than; since both fractions are greater than $\dfrac{1}{2}$, the sum will be greater than $\dfrac{1}{2} + \dfrac{1}{2}$ or 1.   **57.** Sample answer: You are 2 miles away from your destination. You travel $1\dfrac{3}{10}$ miles. How far are you from your destination? $\dfrac{7}{10}$ mi
59. I   **61.** 0.7 **63.** $0.\overline{81}$ **65.** 2.25 **67.** $4.58\overline{3}$
69. 0.875 in.   **71.** $1\dfrac{3}{5}$ **73.** $\dfrac{47}{50}$ **75.** $-2\dfrac{2}{9}$ **77.** $4\dfrac{65}{99}$

pp. 41–43   Lesson 1-1C
1. $\dfrac{3}{7}$ **3.** $-\dfrac{1}{18}$ **5.** $3\dfrac{1}{2}$ **7.** $\dfrac{1}{8}$

⑨ $P(\text{greater than } 2) = \dfrac{2}{3}$ and $P(\text{yellow}) = \dfrac{3}{8}$
$P(\text{greater than 2 and yellow})$
$= \dfrac{2}{3} \cdot \dfrac{3}{8}$ Multiply.
$= \dfrac{6}{24}$ or $\dfrac{1}{4}$ Simplify.

11. $\dfrac{2 \text{ dollars}}{1 \text{ pound}} \cdot 2\dfrac{5}{8} \text{ pounds} = \$5.25$ **13.** $\dfrac{1}{48}$ **15.** $\dfrac{3}{5}$
17. $-\dfrac{9}{40}$ **19.** $\dfrac{1}{35}$

21 $4\frac{1}{4} \cdot 3\frac{1}{3} = \frac{17}{4} \cdot \frac{10}{3}$ Rename $4\frac{1}{4}$ as $\frac{17}{4}$ and $3\frac{1}{3}$ as $\frac{10}{3}$.

$= \frac{17}{\cancel{4}_2} \cdot \frac{\cancel{10}^5}{3}$ Divide out common factors.

$= \frac{17 \cdot 5}{2 \cdot 3}$ Multiply.

$= \frac{85}{6} = 14\frac{1}{6}$ Simplify.

23. $1\frac{1}{2}$ **25.** $\frac{7}{20}$ **27.** $\frac{1}{4}$

29. $\frac{150,000 \text{ people}}{1 \text{ square mile}} \cdot 2.25 \text{ square miles} =$ 337,500 people **31.** $\frac{2}{9}$

33 rtv

$= \frac{1}{4} \cdot \frac{8}{9} \cdot \left(\frac{-2}{3}\right)$ $r = \frac{1}{4}, t = \frac{8}{9}, v = -\frac{2}{3}$

$= \frac{1 \cdot 8 \cdot (-2)}{4 \cdot 9 \cdot 3}$ Multiply.

$= \frac{(-16)}{108}$ Simplify.

$= -\frac{4}{27}$

35. $-\frac{1}{10}$ **37.** 25 **39.** $-\frac{2}{27}$ **41.** In order to multiply mixed numbers, you must first rename them as improper fractions.

$2\frac{1}{2} \cdot 3\frac{1}{4} = \frac{5}{2} \cdot \frac{13}{4}$

$= \frac{65}{8}$

$= 8\frac{1}{8}$

43. $\frac{6}{7}$ **45.** B **47.** $\frac{13}{42}$ **49.** $-12\frac{3}{10}$ **51.** $\frac{11}{100}$ oz

---

pp. 46–48 Lesson 1-1D

1. $\frac{7}{5}$ **3.** $-\frac{4}{11}$ **5.** $1\frac{1}{4}$ **7.** $\frac{1}{2}$ **9.** $\frac{3}{10}$

11 $-3\frac{7}{12} \div 6\frac{5}{6}$

$= -\frac{43}{12} \div \frac{41}{6}$ Write as improper fractions.

$= -\frac{43}{12} \cdot \frac{6}{41}$ The multiplicative inverse of $\frac{41}{6}$ is $\frac{6}{41}$.

$= -\frac{43}{\cancel{12}_2} \cdot \frac{\cancel{6}^1}{41}$ Divide 12 and 6 by their GCF, 6.

$= -\frac{43}{82}$ Multiply.

13. $-\frac{9}{7}$

15 Since $15\left(\frac{1}{15}\right) = 1$, the multiplicative inverse of 15 is $\frac{1}{15}$. **17.** $\frac{5}{17}$ **19.** $\frac{8}{15}$ **21.** $\frac{4}{5}$

23. $-1\frac{1}{15}$ **25.** $\frac{5}{6}$ **27.** $\frac{1}{10}$ **29.** $\frac{2}{15}$ **31.** $1\frac{1}{2}$

33. $-2\frac{5}{8}$ **35a.** $3\frac{2}{3}$ **35b.** $1\frac{5}{6}$

---

37 Divide 22 centimeters, the size of the large hummingbird, by $5\frac{1}{2}$ centimeters, the size of the small hummingbird.

$22 \div 5\frac{1}{2}$

$= \frac{22}{1} \div \frac{11}{2}$ Write as improper fractions.

$= \frac{22}{1} \cdot \frac{2}{11}$ The multiplicative inverse of $\frac{11}{2}$ is $\frac{2}{11}$.

$= \frac{22}{1} \cdot \frac{2}{\cancel{11}_1}$ Divide 22 and 11 by their GCF, 11.

$= \frac{4}{1}$ or 4 Multiply. Then simplify.

4 hummingbirds

39. 4 batches, $\frac{7}{12}$ cup left over **41.** $30 \div \frac{3}{4}$; 30 times a number less than 1 will be less than 30. However, 30 divided by a number less than 1 will be greater than 30. **43.** $\frac{53}{72}$ **45.** B **47.** 15 **49.** $\frac{2}{3}$ **51.** $2\frac{1}{6}$ **53.** 6.75 lb

---

pp. 50–51 Lesson 1-2A PSI

1. Sample answer: Each day he volunteers for $2\frac{2}{3}$ hours. After 10 days, he will have $26\frac{2}{3}$ hours. Multiply $2\frac{2}{3}$ by 5 to find the additional hours. Then add that to $26\frac{2}{3}$ to find the number of hours Drew could have after 15 days. **3.** 6th bounce **5.** 8 lawns **7.** 5:40 P.M.

9 Convert each shutter speed value to the same format and compare them.

$\frac{1}{125} = 0.008, 0.0\overline{6}, \frac{1}{60} = 0.01\overline{6}, 0.125, 0.004, \frac{1}{4} = 0.25$

Reorder the decimals from smallest to largest. The fastest shutter speed would be the smallest decimal quantity.

$0.004, 0.008, 0.01\overline{6}, 0.0\overline{6}, 0.125, 0.25$

The fastest shutter speed is 0.004 second.

11. 1,500 times longer

---

pp. 54–57 Lesson 1-2B

1. 0.4 **3.** 0.003 **5.** 123%

7 $\frac{11}{25} = 0.44,$

$= 44\%$

9. $83.\overline{3}\%$ **11.** $0.062, 60\%, \frac{13}{20}, \frac{17}{25}$ **13.** 0.9 **15.** 1.72

17. 0.004 **19.** 0.07 **21.** 0.11 **23.** 62% **25.** 47.5%

27. 0.7% **29.** 275% **31.** 21% **33.** 85% **35.** 160%

37. 2.5% **39.** $44.\overline{4}\%$

41 Write 6 in 30 people as a fraction.

$\frac{6}{30} = 0.20,$

$= 20\%$

43a. $33\frac{1}{3}\%$ **43b.** 17% **45.** $8\%, \frac{7}{10}, \frac{3}{4}, 0.8$

47. $\frac{1}{20}, 7\%, \frac{2}{25}, 0.09$ **49.** = **51.** >

53 To find the percent of trips that involve air travel, write a fraction.

$\frac{14,000,000}{50,000,000} = 0.28$,

$= 28\%$

55. 160%;

$1\frac{3}{5} = 1 + \frac{3}{5}$ Definition of mixed number

$= 1 + 0.6$ $3 \div 5 = 0.6$

$= 1.6$ Add.

$= 160\%$ Multiply by 100 and add % symbol.

57. Less than; by multiplying by 100 and adding the percent symbol, 0.04 = 4%. Since 4% < 40%, 0.04 < 40%. **59.** H **61.** 12 mi **63.** $-4\frac{11}{16}$

65. $-2\frac{1}{4}$ **67.** $4\frac{5}{6}$ **69.** $1.06 = 1\frac{3}{50}$; $0.24 = \frac{6}{25}$;

$-2.72 = -2\frac{18}{25}$; $-3.40 = -3\frac{2}{5}$

pp. 61–63 Lesson 1-2C

1 $\frac{70}{280} = \frac{n}{100}$ Write the percent proportion.

$70 \cdot 100 = 280 \cdot n$ Find cross products.

$7,000 = 280n$ Multiply.

$\frac{7,000}{280} = \frac{280n}{280}$ Divide each side by 280.

$25 = n$ Simplify.

70 is 25% of 280.

3. $\frac{n}{19} = \frac{118}{100}$; 22.42 **5.** $\frac{151.5}{n} = \frac{75}{100}$; 202 **7.** 4%

9. 782 **11.** 2,000 **13.** $7,420 **15.** $\frac{120}{360} = \frac{n}{100}$; 33.3%

17. $\frac{n}{350} = \frac{17}{100}$; 59.5 **19.** $\frac{95}{n} = \frac{95}{100}$; 100 **21.** $\frac{n}{57} = \frac{250}{100}$;

142.5 **23.** 36 **25.** 30% **27.** 200 **29.** 20.16

31 Each week Michaela will increase her study time by 40%. Make a table to determine in what week she will begin studying for at least 40 minutes per day.

Week	Study Time
Week 1	15 minutes
Week 2	15 + (15 × 0.4) = 15 + 6 or 21 minutes
Week 3	21 + (21 × 0.4) = 21 + 8.4 or 29.4 minutes
Week 4	29.4 + (29.4 × 0.4) = 29.4 + 11.76 or 41.16 minutes

41.16 > 40. So, in week 4, Michaela will begin studying for at least 40 minutes per day.

33. $\frac{4}{550} = \frac{n}{100}$; 0.7% **35.** $\frac{n}{42} = \frac{5.8}{100}$; 2.4

37. $\frac{57}{n} = \frac{13.5}{100}$; 422.2

39 $\frac{p}{40} = \frac{87.5}{100}$ Write the percent proportion.

$p \cdot 100 = 40 \cdot 87.5$ Find cross products.

$100p = 3,500$ Multiply.

$\frac{100p}{100} = \frac{3,500}{100}$ Divide each side by 100.

$p = 35$ Simplify.

The team won 35 games.

41. Sample answer: Let x equal 2 and y equal 5. 2% of 5 is 0.1 and 5% of 2 is 0.1. The result will always be the same for any two numbers x and y. x% of $y = x(0.01) \cdot y$ and y% of $x = y(0.01) \cdot x$. By the Commutative Property of Multiplication, $x(0.01) \cdot y = y(0.01) \cdot x$. **43.** D **45.** 90.93 **47.** 0.875 **49.** $0.\overline{2}$ **51.** University of Virginia

pp. 66–68 Lesson 1-3A

1. $84 **3.** $115.71 **5.** $10.57

7 Let d represent the total discount.

$d = 0.10 \cdot 19$ Write the percent equation.

$= 1.9$ Multiply.

Subtract the discount from the original price to find the sale price.

$19 − $1.90 = $17.10.

Let t represent the sales tax.

$t = 0.075 \cdot 17.10$ Write the percent equation.

≈ 1.28 Multiply.

Add the sales tax to the sale price to find the total price.

$17.10 + $1.28 = $18.38.

9. $117.96 **11.** $339.15

13 Find the price of the earrings after the discount. If the amount of the discount is 35%, the percent paid for the earrings is 100% − 35% or 65%. Find 65% of $19.50. Let s represent the sale price.

$s = 0.65 \cdot 19.50$ Write the percent equation.

$s \approx 12.68$ Multiply.

The sale price of the earrings is $12.68.

15. $118.64 **17.** $235.73 **19.** $910 **21.** $36.25

23 $\frac{p}{p + 6.80} = \frac{20}{100}$ Write the percent equation.

$p \cdot 100 = 20(p + 6.80)$ Find cross products.

$100p = 20p + 136$ Multiply and distribute.

$\underline{− 20p = −20p}$ Subtract 20p from each side.

$80p = 136$ Subtract.

$\frac{80p}{80} = \frac{136}{80}$ Divide each side by 80.

$p = 1.70$

Adult tickets cost $1.70 more than the student tickets. The price of the adult ticket is $8.50.

25. 25% **27.** 0.7242x **29.** Sample answer: Discount is when a percentage is taken off an item, resulting in a lower price. Markup is when a percentage is applied to an item, resulting in a higher price. Markup is usually used by businesses to make a profit. **31.** G **33.** 18 **35.** 150 **37a.** 0.35; $\frac{7}{20}$

37b. more

pp. 71–72 Lesson 1-3B

1. $112.50 **3.** 5.25% **5.** $618.88 **7.** $45

9. $10.08 **11.** about 39.77%

13 Determine the simple interest for each year separately.

$I_{(year\ 1)} = prt$ Write the simple interest formula.

$= 2{,}250 \cdot 0.05 \cdot 1$ Replace p with 2,250, r with 0.05, and t with 1.

$= 112.5$ Multiply.

At the end of the first year, there is $2,250 + $112.50 or $2,362.50.

$I_{(year\ 2)} = 2{,}362.50 \cdot 0.05 \cdot 1$ Substitution

$= 118.13$ Multiply.

At the end of the second year, there is $2,362.50 + $118.13 or $2,480.63.

$I_{(year\ 3)} = 2{,}480.63 \cdot 0.05 \cdot 1$ Substitution

$= 124.03$ Multiply.

So, at the end of the third year, there is $2,480.63 + $124.03 or $2,604.66.

15. $612.18

17 Determine the simple interest for each year separately.

$I_{(year\ 1)} = prt$ Write the simple interest formula.

$= 15{,}000 \cdot 0.11 \cdot 1$ Substitution

$= 1{,}650$ Multiply.

At the end of the first year, there is $15,000 + $1,650 or $16,650.

$I_{(year\ 2)} = 16{,}650 \cdot 0.11 \cdot 1$ Substitution

$= 1{,}831.50$ Multiply.

At the end of the second year, there is $16,650 + $1,831.50 or $18,481.50.

$I_{(year\ 3)} = 18{,}481.50 \cdot 0.11 \cdot 1$ Substitution

$= 2{,}032.97$ Multiply.

So, at the end of the third year, there is $18,481.50 + $2,032.97 or $20,514.47.

$I_{(year\ 4)} = 20{,}514.47 \cdot 0.11 \cdot 1$ Substitution

$= 2{,}256.59$ Multiply.

So, at the end of the fourth year, there is $20,514.47 + $2,256.59 or $22,771.06.

$I_{(year\ 5)} = 22{,}771.06 \cdot 0.11 \cdot 1$ Substitution

$= 2{,}504.82$ Multiply.

So, at the end of the fifth year, Felicia will pay $22,771.06 + $2,504.82 or $25,275.88.

19. $1,029.60   **21.** 25-year mortgage loan **23.** Sample answer: $1,000 principal with a 22% simple interest rate   **25.** B   **27.** 23%

29. $0.016, 16\%, \frac{1}{6}$ **31.** $-5\frac{6}{25}$ **33.** $-12\frac{25}{33}$

pp. 76–77   *Lesson 1-3D*

1. −20%; decrease

3 The amount of change is $400 - 325 = 75$.

percent of change $= \frac{75}{325}$ Percent of change equation

≈ 0.231 Divide.

$= 23.1\%$ Express the ratio as a percent.

There was a 23.1% increase in the number of miles.

---

5. 50%; increase   **7.** −20%; decrease
9. −25%; decrease   **11.** −6.7%

13 Percent of change

$= \dfrac{\text{amount of change}}{\text{original amount}}$ Definition of percent of change

$0.35 = \dfrac{x}{8}$ Substitution.

$2.8 = x$ Multiply.

The decrease in time is 2.8 seconds. Faster speed is $8 - 2.8$ or 5.2 seconds.

15. Sample answer: Increase in attendance from one year to the next. **17.** Sample answer: Both are percents of change however a percent of increase shows a growing change in quantity while a percent of decrease shows a lessening change in quantity. **19.** H   **21.** −15%   **23.** $180   **25.** $7\frac{1}{6}$ in.

pp. 80–83   *Chapter Study Guide and Review*

1. true   **3.** false; rational numbers   **5.** true

7. false; discount   **9.** $\frac{3}{10}$ **11.** $4\frac{1}{3}$ **13.** $\frac{2}{5}$

15. $-\frac{3}{4}$ **17.** $2\frac{1}{2}$ h   **19.** $\frac{4}{9}$ **21.** $5\frac{1}{4}$ c   **23.** $-\frac{7}{8}$

25. −4   **27.** 15.2 min or 15 min 12 s   **29.** 0.043

31. 0.007   **33.** 1.5%   **35.** Andrea: $\frac{7}{8} = 87.5\%$, $87.5\% < 88\%$ **37.** 8.1   **39.** 16 attempts   **41.** $5.28

43. $10.40   **45.** $405.17   **47.** $1,962.50

49. −20%; decrease

Chapter 2 Real Numbers and Monomials

p. 90   *Are You Ready?*

1. 1,296   **3.** 256   **5.** $2,048   **7.** $2 \times 2 \times 3 \times 3$
9. $2 \times 3 \times 3$ **11.** 11×11 **13.** $-1 \times 2 \times 2 \times 2 \times 2 \times 2 \times 2$ **15.** cheetah: 1×71; wildebeest: $2 \times 5 \times 5$; brown hare: $2 \times 2 \times 2 \times 2 \times 3$; horse: $3 \times 3 \times 5$

pp. 94–96   *Lesson 2-1A*

1. 4^4 **3.** a^4 **5.** $r^5 \cdot s^3$

7 $2 \times 2 \times 2 \times 2 \times 2 \times 2$ or 64   **9.** 256   **11.** $\frac{1}{343}$

13. $\frac{81}{256}$ **15.** black bear: 350 lb; key deer: 75 lb; panther: 120 lb   **17.** 2,400   **19.** 2,744   **21.** $\left(\frac{5}{6}\right)^3$

23. $3^2 \cdot 5 \cdot q^3$ **25.** $4^2 \cdot b^4$ **27.** $2^2 \cdot d^3 \cdot k^2$ **29.** 8

31. $\frac{1}{81}$ **33.** 24   **35.** 432   **37.** 8,000,000,000 or 8 billion

39 $g^5 - h^3 = (2)^5 - (7)^3$

$= 2 \times 2 \times 2 \times 2 \times 2 - 7 \times 7 \times 7$

$= 32 - 343$

$= -311$

41. 16   **43.** 10

45 a. $9.3 \times 10^7 = 9.3 \times 10{,}000{,}000$ or 93,000,000 mi

b. $8.87 \times 10^8 = 8.87 \times 100{,}000{,}000$ or 887,000,000 mi

c. $2.8 \times 10^9 = 2.8 \times 1{,}000{,}000{,}000$ or 2,800,000,000 mi

d. $2{,}800{,}000{,}000 - 887{,}000{,}000 = 1{,}913{,}000{,}000$ mi

47. =   **49.** <   **51.** =   **53a.** 10^6 **53b.** 10^9 **53c.** 10^{15}

55. $1; \frac{1}{3}, \frac{1}{9}, \frac{1}{27}$ **57.** B   **59.** 1331
61. -121 **63.** 529

pp. 100–101   Lesson 2-1B

1. 4^8 **3.** r^{10} **5.** $-6a^5$ **7.** 7^5

(9) $\frac{y^8}{y^5} = y^{8-5}$
$\quad\quad = y^3$

11. $3c^5$ **13.** 2^4 or 16 times   **15.** 2,352   **17.** $(-6)^{13}$
19. $15x^9$ **21.** $-8w^{11}$ **23.** $40y^9$ **25.** 8^{11} **27.** h
29. $6d^5$ **31.** x^2y^5
(33) $10^{11} \cdot 10^3 = 10^{11+3}$ or 10^{14} instructions
35. 120   **37.** 540   **39.** Sample answer: $5^{10} \cdot 5^3$
41. 2^{31} **43.** Sample answer: The Quotient of Powers Rule can only be used when the bases are the same. In the expression $\frac{x^2}{y^2}$ the bases x and y are different.   **45.** G   **47.** C

pp. 104–105   Lesson 2-1C

1. 3^{10} or 59,049   **3.** 2^{18} or 262,144   **5.** $625g^{32}k^{48}$

(7) Volume = side³ or side · side · side
$\quad\quad = 3c^3d^2 \cdot 3c^3d^2 \cdot 3c^3d^2$
$\quad\quad = (3c^3d^2)^3$
$\quad\quad = 3^3(c^3)^3(d^2)^3$
$\quad\quad = 3^3c^{3 \cdot 3}d^{2 \cdot 3}$
$\quad\quad = 27c^9d^6$ cubic units

9. 2^{14} or 16,384   **11.** 3^8 or 6,561   **13.** m^{40} **15.** z^{55}
17. 4^{12} or 16,777,216   **19.** 2^{18} or 262,144
21. $32,768v^{45}$ **23.** $38,416y^4$ **25.** $64m^{30}n^{66}$
27. $625r^{16}s^{48}$ **29.** $144d^{12}e^{14}$ units²   **31.** $343m^{18}n^{27}$ units³   **33.** $0.027p^{21}$ **35.** $\frac{9}{25}a^{12}b^{18}$

(37) $(-2v^7)^3(-4v^2)^4 = (-2)^3(v^7)^3(-4)^4(v^2)^4$
$\quad\quad = -8 \cdot v^{(7 \cdot 3)} \cdot 256 \cdot v^{2 \cdot 4}$
$\quad\quad = -2,048v^{(7 \cdot 3) + (2 \cdot 4)}$
$\quad\quad = -2,048v^{21+8}$
$\quad\quad = -2,048v^{29}$

39. $25(2^{2x}); 125(2^{3x})$ **41.** 5   **43.** Sample answer: To simplify $(2a^3)(4a^6)$, multiply 2 by 4. Then add the exponents 3 and 6 and write this sum as the final exponent on a. To simplify $(2a^3)^6$, evaluate 2^6. Then multiply the exponents 3 and 6 and write this product as the final exponent on a.   **45.** G
47. 6^{11} **49.** $18x^{14}$ **51.** Bridaveil: 620 ft; Fall Creek: 256 ft; Shoshone: 212 ft

pp. 106–107   Lesson 2-1D   PSI

1. Sample answer: You were able to count just the number of counters in the 4th row.   **3.** Sample answer: In how many different ways can four people be seated in a car if there are 2 front seats and 2 back seats and only 3 of the four people are able to drive? There are 18 possible ways.
5. 6 times   **7.** $64s^5t$ ft²

(9) Use the work backward strategy.
Caroline ended up with $11. Before that, she spent $9 at the movies. To undo spending $9, add $9; $11 + $9 = $20.
Before the movie, she received $10. To undo receiving $10, subtract $10; $20 − $10 = $10.
Before she received the $10, she spent half of what she had. To undo spending half, multiply by 2; $10 · 2 = $20.
Before she spent half of the money, she lent $5 to her sister. To undo giving her sister $5, add $5; $20 + $5 = $25.
So, Caroline received $25 for the birthday gift.
11. $256^2 = 65,536$ **13.** 81 friends

pp. 110–112   Lesson 2-2A

1. $\frac{1}{2^4}$ **3.** $\frac{1}{a^4}$ **5.** $\frac{1}{729}$ **7.** $\frac{1}{64}$ **9.** 3^{-4} **11.** 4^{-2} or 2^{-4}
13. 10^{-5}

(15) $3^{-3} \cdot 3^{-2} = 3^{-3+(-2)}$
$\quad\quad = 3^{-5}$
$\quad\quad = \frac{1}{3^5}$

17. $\frac{1}{m^7}$ **19.** 12^2 **21.** p^5 **23.** $\frac{1}{7^{10}}$ **25.** $\frac{1}{(-5)^4}$
27. $\frac{1}{g^7}$ **29.** $\frac{1}{w^{13}}$ **31.** $\frac{1}{343}$ **33.** $\frac{1}{1,728}$ **35.** $-\frac{1}{729}$
37. $-\frac{1}{32,768}$ **39.** 12^{-4} **41.** $(-5)^{-7}$ **43.** 5^{-3}
45. 2^{-10} or 4^{-5} **47.** 10^{-10} **49.** $\frac{1}{2^7}$

(51) $y^{-1} \cdot y^4 = y^{(-1+4)}$
$\quad\quad = y^3$

53. $\frac{1}{s^7}$ **55.** $-\frac{12b^4}{a^2}$ **57.** 3^4 **59.** a^2 **61.** y^4 **63.** z^4

(65) $\frac{10^{-18}}{10^{-23}} = 10^{-18-(-23)}$
$\quad\quad = 10^5$ or 100,000 times

67. 12   **69.** -6 **71.** $11^{-3}, 11^0, 11^2$; Sample answer: The exponents in order from least to greatest are $-3, 0, 2$.   **73.** Sample answer: $\left(\frac{1}{2}\right)^{-1} = 2$, $\left(\frac{34}{43}\right)^{-1} = \left(\frac{43}{34}\right)$, $\left(\frac{56}{65}\right)^{-1} = \left(\frac{65}{56}\right)$; When you raise a fraction to the -1 power, it is the same as finding the reciprocal of the fraction.   **75.** A   **77.** I   **79.** 6^{15} **81.** $16a^{12}b^8$
83. 2^3 or 8 times

pp. 115–117   Lesson 2-2B

1. 73,200   **3.** 0.455   **5.** 2.77×10^5 **7.** 4.955×10^{-5}
9. year 4, year 3, year 2, year 1   **11.** 3,160
13. 4,265,000   **15.** 0.00011   **17.** 0.0000252
19. 4.3×10^4 **21.** 1.47×10^8 **23.** 7.2×10^{-3}
25. 9.01×10^{-5}

(27) Rewrite each of the areas to the same power of 10. Then compare the numbers.

Ocean	Area (mi²)
Atlantic	29.6×10^6
Arctic	5.43×10^6
Indian	26.5×10^6
Pacific	60.0×10^6
Southern	7.85×10^6

Since $5.43 < 7.85 < 26.5 < 29.6 < 60.0$, the oceans in order from least to greatest area are Arctic, Southern, Indian, Atlantic, and Pacific.
29. $-4.56 \times 10^2, -4.56 \times 10^{-3}, 4.56 \times 10^{-2}, 4.56 \times 10^2$

33 Rewrite each of the numbers in standard notation to compare them. $6,250 \circ 6,300 \rightarrow 6,250 < 6,300$; So $6.25 \times 10^3 < 6.3 \times 10^3$.
35. $<$ **37a.** Jacob: 1,700,000,000 nanometers, Sarah: 1,520,000,000 nanometers **37b.** Jacob: 1.7×10^9 nanometers, Sarah: 1.52×10^9 nanometers

39a. $\dfrac{(1.3 \times 10^5)(5.7 \times 10^{-3})}{4 \times 10^{-4}} = 1.8525 \times 10^6$

39b. $\dfrac{(9 \times 10^4)(1.6 \times 10^{-3})}{(2 \times 10^5)(3 \times 10^4)(1.2 \times 10^{-4})} = 2 \times 10^{-4}$

41. D **43.** F **45.** $\dfrac{1}{5^4}$ **47.** $\dfrac{1}{7^2}$ **49.** 66 ice cream cones
51. $9g^{10}h^2$ **53.** $50x^7$ **55.** 6^3 **57.** $5^4 \cdot t^3$

---

pp. 120–122   Lesson 2-2C

1. 4.94×10^7 **3.** 4.44×10^1 **5.** 3.1×10^5
7. 5.4×10^{-14} **9.** 4.5 times **11.** 8.9042×10^9
13. 9.5871×10^8 **15.** 8.97×10^8 **17.** 3.762×10^{-9}
19. 8.19×10^{-2} **21.** 6.3×10^4 **23.** 3.75×10^{-19}

25 $\dfrac{1.14 \times 10^6}{4.8 \times 10^{-3}} = \left(\dfrac{1.14}{4.8}\right)\left(\dfrac{10^6}{10^{-3}}\right)$
$= (0.2375)(10^9)$
$= 2.375 \times 10^8$

27. about 1.39×10^{14} tons **29.** 9.563×10^{11}
31. 1.334864×10^{10} **33.** 9.83×10^8
35. 8.70366×10^4

37 $\dfrac{(1.37 \times 10^4)(2.64 \times 10^2)}{4.356 \times 10^4} = \dfrac{3.6168 \times 10^6}{4.356 \times 10^4}$
$= \left(\dfrac{3.6168}{4.356}\right)\left(\dfrac{10^6}{10^4}\right)$
$\approx 0.8303 \times 10^2$ or
8.303×10^1
about 83.03 acres

39. 1.25×10^4 in. **41.** 1.4×10^9; Sample answer: All the other expressions are equivalent. **43.** D
45. B **47.** $\dfrac{1}{5^4}$ **49.** $\dfrac{1}{3^5}$ **51.** $64r^9s^3$ **53.** 3^2 or 9 times

---

pp. 126–128   Lesson 2-3A

1. 5 **3.** -1.3 **5.** ± 10 **7.** no real solution
9. 6 or -6 **11.** 2.5 or -2.5 **13.** 6 **15.** -2 **17.** 4
19. -22 **21.** $\dfrac{11}{18}$ **23.** $\pm\dfrac{3}{7}$ **25.** -1.6 **27.** no real solution **29.** ± 9 **31.** ± 12 **33.** $\pm\dfrac{3}{5}$ **35.** ± 0.13

37 $\sqrt{169} = \pm 13$; The answer has to be $+13$ because there cannot be a negative number of students.

39. 11 **41.** 13 **43.** 12 **45.** 15 **47.** 25
49 $z = (10.5)^2 = 110.25$
51. 20 ft **53.** 7 in. **55.** 16 ft **57a.** 36 **57b.** $\dfrac{25}{81}$
57c. 199 **57d.** x **59.** $\sqrt{1,296} = 36$ **61.** I
63. 1.428×10^7 **65.** 696,000,000 meters **67.** 280
69. $-15,552$

---

pp. 132–134   Lesson 2-3C

1. 5 **3.** 12 **5.** 6 **7.** about 62.5 swings **9.** 2 **11.** 5
13. 7 **15.** 5 **17.** 14 **19.** 5 **21.** 6

23 $t = \dfrac{\sqrt{h}}{4}$
$= \dfrac{\sqrt{125}}{4}$
$\approx \dfrac{11}{4}$ or 2.75
about 2.75 seconds

25. 3 **27.** 3 **29.** 10 **31.** 6 **33.** 5 **35.** 7, $\sqrt{50}$, 9, $\sqrt{85}$ **37.** $\sqrt{34}$, 6, $\sqrt{62}$, 8

39 95 is rounded to 100 as the nearest integer. So, $\sqrt{95}$ is close to $\sqrt{100} = 10$ or -10.
41. 6 in. **43.** No; 12 inches = 1 foot and the cube root of $4 > 1$. **45.** 10; Since 94 is less than 100, $\sqrt{94}$ is less than 10. **47.** She incorrectly estimated. She found half of 200, not the square root.

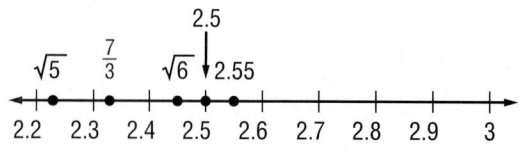

49. Since $\sqrt{64} < \sqrt{78} < \sqrt{81}$, the square root of 78 is between 8 and 9. Since 78 is closer to 81 than 64, graph $\sqrt{78}$ closer to 9 than 8. **51.** B **53.** 7.29×10^4
55. 5^{-3} **57.** 6^{-4}

---

pp. 138–139   Lesson 2-3D

1. rational **3.** irrational **5.** $>$
7 Write $\sqrt{5.2}$ as a decimal and compare.
$\sqrt{5.2} \approx 2.28$; since $2.\overline{21} < 2.28$, $2.\overline{21} < \sqrt{5.2}$.
9. about 30.6 cm^2 **11.** rational **13.** irrational
15. rational **17.** whole, integer, rational **19.** $<$
21. $=$ **23.** $>$

25 Write each number as a decimal and compare them.

$\sqrt{5}$	$\sqrt{6}$	2.5	2.55	$\dfrac{7}{3}$
2.24	2.45	2.5	2.55	$2.\overline{3}$

In order from least to greatest:
$\sqrt{5}, \dfrac{7}{3}, \sqrt{6}, 2.5, 2.55$

27. about 1.9 m^2 **29.** Sample answer: $\sqrt{4}$; $\sqrt{4} = 2$ and 2 is a rational number. **31.** $<$ **33.** always
35. Sometimes; sample answer: The product of the rational number 0 and any irrational number is the rational number 0. **37.** D **39.** F **41.** 5 or -5
43. 0.8 or -0.8 **45.** 5.736×10^5

---

pp. 142–145   Chapter Study Guide and Review

1. false; scientific notation **3.** true
5. false; Sample answer: 9 **7.** true **9.** $x^4 \cdot y$

11. 216 calls **13.** $36y^{11}$ **15.** 3^1 or 3 times
17. 5^4y^{20} **19.** $125x^6t^{12}$ **21.** No; sample answer: She needs $14\frac{3}{5}$ feet for 5 shelves and she only has $14\frac{1}{4}$ feet. **23.** $\frac{81}{2^7}$ **25.** 6^3 **27.** 9^5 **29.** 0.0032
31. 6.4×10^{-5} **33.** 0.004375 lb **35.** 7.425×10^{-8}
37. 1.11905×10^3 **39.** 9 **41.** -8 **43.** 16 squares
45. 6 **47.** 15 **49.** 5 **51.** 4 **53.** ± 8 **55.** rational
57. integer, rational **59.** whole, integer, rational
61. $2\frac{1}{5}, 2.\overline{2}, \sqrt{5}, 2.25$

Chapter 3 Equations and Inequalities

p. 152 Are You Ready?

1. 51 **3.** 152 **5.** -16 **7.** -8 **9.** 260 mi **11.** true
13. true **15.** true **17.** Des Moines; $-5 > -7$

pp. 154–155 Lesson 3-1A PSI

1. Sample answer: The students have an ending number and the operations that lead to that number. They need to work their way back to the beginning number. **3.** Sample answer: Jacob spent a third of his money as a deposit on a campsite. Then he bought sports equipment that cost $21. Finally, he spent $16 at the grocery store. How much money did Jacob have initially if he now has $41 left?
Start with 41 and add 16. $\rightarrow 41 + 16 = 57$
Add 21. $\rightarrow 57 + 21 = 78$
Divide by $\frac{2}{3}$. $\rightarrow 78 \div \frac{2}{3} = 117$; So, Jacob initially had $117. **5.** 6:15 A.M. **7.** Week 12
9 The total cost of the sofa is $150 plus the monthly charge of $37.50 for 12 months. The expression to represent this is $150 + 37.50(12)$ or $600. **11.** $1,238.50

pp. 158–160 Lesson 3-1B

1. $s =$ Corey's score; $20 = 4s$ **3.** $d =$ original depth; $d - 75 = -600$ **5.** $r = m + 4$ **7.** $a =$ class average; $a - 5 = 82$

9 Let d equal number of days. Write an expression for the depth over several days. That expression would be -75 divided by number of days. Set this expression equal to the average dirt removal, -15.
$\frac{-75}{d} = -15$
11. $m =$ amount of money; $\frac{m}{4} = 235$
13. $h =$ height; $15 = \frac{h}{4}$ **15.** $s =$ leader's score; $s + 5 = -3$ **17.** $d = 24g$ **19.** $t = m + \frac{1}{2}$
21 Determine the relationship between the values in the table. Each of the f-values equals 3 times the y-value. So, $f = 3y$.

23.

Map Distance, m (inches)	Actual Distance, a (miles)
1	20
2	40
3	60
4	80
m	$20m$

$a = 20m$

25. Sample answer: A number in the sequence is 2 times its position number. **27.** Sample answer: A number would now be 2 less than 2 times its position number; $n = 2p - 2$. **29.** C **31.** Sample answer: The guests at a party were divided into 7 equal-size groups, resulting in 4 guests per table. How many guests were at the party? **33.** Sample answer: A swimming competition had a total of 60 swimmers. There were 4 teams, each with an equal number of swimmers. How many swimmers were on each team?

pp. 164–165 Lesson 3-1C

1. 6 **3.** -12 **5.** 7 **7.** 3
9 $n + 3 = 20$ Write the equation.
$n + 3 - 3 = 20 - 3$ Subtract 3 from both sides.
     $n = 17$
To check your solution, substitute 17 into the original equation.
 $n + 3 = 20$ Write the equation.
$17 + 3 \overset{?}{=} 20$ Substitute.
  $20 = 20$ ✓
11. -6 **13.** -14 **15.** 1 **17.** -10 **19.** 7
21. $b - 50 = 124$; $174

23 The hole is 18 inches deep. This is represented by -18. The planted tree is 54 inches tall. Write an equation to find the total height. $(-18) + h = 54$. Then solve.
     $(-18) + h = 54$ Write the equation.
$(-18) + h + 18 = 54 + 18$ Add 18 to both sides.
         $h = 72$
The height of the tree is 72 inches or 6 feet.
25. Sample answer: $n + 5 = 2$; $n - 6 = -9$
27. ± 2; Subtract 5 from each side to get $|x| = 2$. If $x = -2$, $|-2| = 2$ and if $x = 2$, $|2| = 2$. **29.** C
31. G **33.** $h =$ height of Lindsay's sister; $h - 5 = 59$
35. 20

pp. 168–170 Lesson 3-1D

1. 8 **3.** 6 **5.** -36 **7.** $\frac{3}{4}x = 1,350$; $1,800
9 $9b = 72$ Write the equation.
 $\frac{9b}{9} = \frac{72}{9}$ Divide each side by 9.
  $b = 8$
Now check your solution.
 $9b = 72$ Write the equation.
$9 \cdot 8 \overset{?}{=} 72$ Substitute.
  $72 = 72$ ✓
11. -2 **13.** -7 **15.** 54 **17.** -100 **19.** -72

21. $1{,}200t = 6{,}000$; \$5   **23.** $16c = 64$; 4 c   **25.** $8g = 120$; 15 gal   **27.** -8 **29.** 6

31 Use the distance formula, $d = rt$.

$d = rt$	Distance formula
$245 = 70 \cdot t$	Replace d with 245 and r with 70.
$\dfrac{245}{70} = \dfrac{70t}{70}$	Divide each side by 70.
$3.5 = t$	$245 \div 70 = 3.5$

So, the car traveled 3.5 hours.

33 The number of times more visitors t can be found by writing an equation. Compare the values for the Golden Gate National Recreation Area, 13.9, and the Great Smoky Mountains National Park, 9.4. So, $9.4t = 13.9$. Solve this equation by dividing both sides by 9.4 to get $t = 1.5$. There are 1.5 times more visitors to the Golden Gate National Recreation Area than the Great Smoky Mountains National Park.

35. 31.823 m^2 **37.** $d = \dfrac{W}{F}$ **39.** B   **41.** D

43. $x + 4 = 58$; 54 in.   **45.** $t =$ temperature at 6 A.M.; $t - 28 = 17$ **47.** $a =$ Malik's age; $a + 13 = 26$

49. $p =$ total slices of pizza; $\dfrac{p}{5} = 3$

pp. 174–176   Lesson 3-2B

1. 4   **3.** 28   **5.** 8   **7.** -1 **9.** $2\dfrac{5}{8}$

11 Solve the equation for p.

$63 = 6.50p + 17.50$	Write the equation.
$-17.50 = -17.50$	Subtract 17.50 from both sides.
$45.50 = 6.50p$	Simplify.
$\dfrac{45.40}{6.50} = \dfrac{6.50p}{6.50}$	Divide each side by 6.50.
$7 = p$	Simplify.

7 people went to the movies.

13. -3 **15.** -2 **17.** 5   **19.** -64 **21.** 7

23

$15 - \dfrac{w}{4} = 28$	Write the equation.
$-15 = -15$	Subtract 15 from both sides.
$-\dfrac{w}{4} = 13$	Simplify.
$-\dfrac{w}{4}(-4) = 13(-4)$	Multiply each side by -4.
$w = -52$	Simplify.

Check the solution by replacing w with -52.

$15 - \dfrac{w}{4} = 28$	Write the equation.
$15 - \dfrac{-52}{4} \stackrel{?}{=} 28$	Replace w with -52.
$15 + 13 \stackrel{?}{=} 28$	Divide.
$28 = 28$ ✓	The statement is true.

25. 65   **27.** -10 **29.** -12 **31.** 5 bracelets   **33.** 15 rounds   **35.** -35

37 Use the information in the figure to find the area of the floor. One side is labeled as 14 feet and the opposite side is $(5 + 3c)$ feet. Set these expressions equal to each other and solve for c. $14 = 5 + 3c$, so $c = 3$. Substitute the value of c into each expression to find the lengths of the two sides. Then multiply to find the area.

Side 1	Side 2
$6c - 8$	$5 + 3c$
$6(3) - 8$	$5 + 3(3)$
$18 - 8$ or 10	$5 + 9$ or 14

The area of the room is $10 \cdot 14$ or 140 square feet.

39. $13 + 3x = 25$; 4   **41.** -12 and 2   **43.** B   **45.** 40   **47.** -22 **49.** -4 **51.** $m = 0.62k$

pp. 179–181   Lesson 3-2C

1. $3n + 1 = 7$ **3.** $\dfrac{n}{5} - 10 = 3$ **5.** $121 = s + (s + 45)$; \$38   **7.** $2n + 15 = 9$

9 Translate the sentence into an equation.

Six less than seven times a number is equal to -20. This means 6 subtracted from $7 \cdot n$ is 20. $7n - 6 = -20$.

11. $3x + 1.99 = 55.99$; \$18   **13.** $13 + 1.50r = 35.50$; 15 rides   **15a.** 65 mph   **15b.** 34 mph   **15c.** 23 mph

17 $550 =$ lessons + junior season pass

A junior season pass is \$315. Therefore, Elsie has \$235 that she can spend on lessons. For that amount of money, Elsie can take $\dfrac{235}{45}$ or 5 semi-private lessons or $\dfrac{235}{60}$ or 3 private lessons. So, she can take two more semi-private lessons than private lessons.

19a. Sample answer: Hunter; although he has money saved, he makes considerably less per hour than Amado. So, he will have to work longer.   **19b.** $7.50h + 150 = 600$; 60 h; $12h = 600$; 50 h

21. Sample answer: You and your friend spent \$25 at the mall. You spent \$6 less than your friend. How much did your friend spend?   **23.** $n + 2n + (n + 3) = 27$; 6, 9, 12   **25.** D   **27.** D   **29.** 3   **31.** -56 **33.** 154   **35.** 300   **37.** -648 **39.** $p - 14 = 17$; 31 points   **41.** $n =$ number of buses; $\dfrac{200}{n} = 50$

pp. 185–186   Lesson 3-3A

1. $s \le 55$ **3.** false   **5.** true

7 The $\le$ tells you to place a closed circle at 2. Since p is less than or equal to 2, draw a line and an arrow to the left.

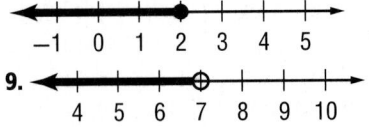

9.

11. $t \ge 10$ **13.** $c \le 25$ **15.** $w > 200$

17

$15 - k > 6$	Write the inequality.
$15 - 8 \stackrel{?}{>} 6$	Substitute.
$7 > 6$ ✓	Simplify.

The inequality is true for the given value of k.

19. true   **21.** false

23.

25.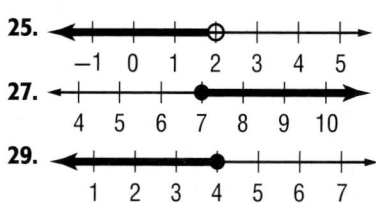

27.

29.

31. Roberto used the "less than or equal to" symbol when he should have used the "greater than or equal to" symbol. $h \geq 2$ **33.** Sample answer: $a < c$; $a = 2, b = 4, c = 6$: $2 < 4, 4 < 6$, and $2 < 6$; $a = -10$, $b = -5, c = -1$: $-10 < -5, -5 < -1$, and $-10 < -1$
35. G **37.** $4b + \$2.35 = \$34.15$; $7.95 per book
39. -5 **41.** 4

pp. 189–190   Lesson 3-3B

1. $b > 4$

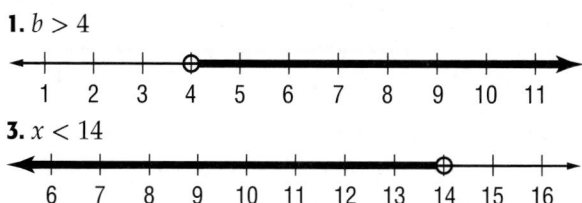

3. $x < 14$

5 $5 + x \leq 18$ Write the inequality.
$\underline{-5 \qquad -5}$ Subtraction Property of Inequality
$\qquad x \leq 13$

7. $k > -10$

9. $c < -1$

11. $m \geq 1.5$

13. $v > 8.7$

15. $d \leq -1\frac{1}{6}$

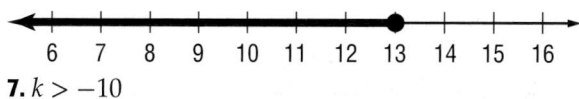

17 Amos is 15 and has to be at least 18 in order to join the City Basketball League. Write an inequality using y to represent the number of years until he is able to join the league. $15 + y \geq 18$. Solve for y. $y \geq 3$. Amos needs to wait for at least three years before he can join the league.

19b. $4.5 < 8$ **19c.** $4.5 + 4.5 < 8 + 8$ or $9 < 16$
19d. $4.5 + 4.5x < 8 + 8x$ **21.** more than one
23. more than one **25.** Sample answer: $x + 4 < 13$, $x - 6 < 3$ **27.** $m \leq 55$ **29.** I **31.** false
33. $7t + 6 = 27$; 3 touchdowns **35.** -12
37. $a + 244.72 = 874$; $629.28

pp. 194–195   Lesson 3-3C

1. $x > 4$

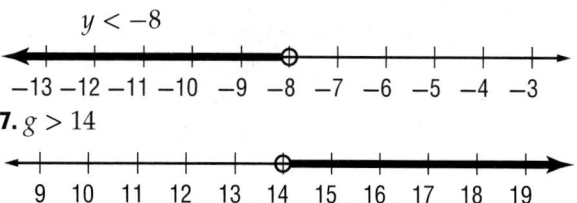

3. $x \leq -9$

5 $-4y > 32$ Write the inequality.
$\dfrac{-4y}{-4} < \dfrac{32}{-4}$ Division Property of Inequality
Remember to reverse the inequality symbol when dividing by a negative number.

$y < -8$

7. $g > 14$

9. $1.5d \geq 6$; $d \geq 4$; She will have to practice the piano for at least 4 days.

11. $n \leq 5$

13. $g < -4$

15. $y < -11$

17. $n < -98$

19. $t \geq 10$

21. $k < 20$

23 Max makes $6 per hour or $6h$. He wants to make at least $89. So, $6h$ is greater than or equal to 89 can be written as the inequality $6h \geq 89$. Solve for h. $h \geq 14.8\overline{3}$
Max must work at least $14.8\overline{3}$ hours. So, Max must work at least 15 hours to buy the shoes.

25. $\dfrac{x}{-6} + 5 \leq 9$; $x \geq -24$; Sample answers: $-12, 0$, and 12, because they are all greater than -24.
27. $4a < -16$; Sample answer: All of the others involve dividing by a negative integer and then reversing the inequality symbol. **29.** A **31.** B

33. $y < 2$

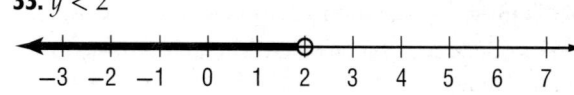

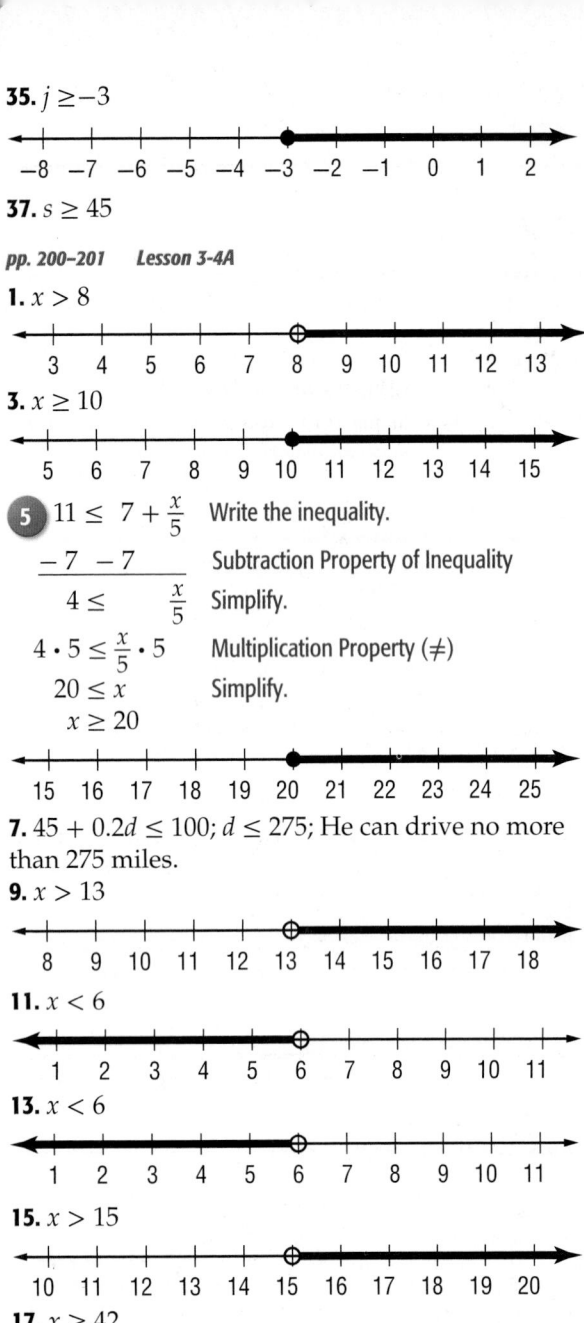

35. $j \geq -3$

(number line from −8 to 2, closed dot at −3, shaded right)

37. $s \geq 45$

pp. 200–201   Lesson 3-4A

1. $x > 8$

(number line from 3 to 13, open dot at 8, shaded right)

3. $x \geq 10$

(number line from 5 to 15, closed dot at 10, shaded right)

5 $11 \leq 7 + \dfrac{x}{5}$ Write the inequality.

$\dfrac{-7 \quad -7}{}$ Subtraction Property of Inequality

$4 \leq \qquad \dfrac{x}{5}$ Simplify.

$4 \cdot 5 \leq \dfrac{x}{5} \cdot 5$ Multiplication Property ($\neq$)

$20 \leq x$ Simplify.

$x \geq 20$

(number line from 15 to 25, closed dot at 20, shaded right)

7. $45 + 0.2d \leq 100;\ d \leq 275;$ He can drive no more than 275 miles.

9. $x > 13$

(number line from 8 to 18, open dot at 13, shaded right)

11. $x < 6$

(number line from 1 to 11, open dot at 6, shaded left)

13. $x < 6$

(number line from 1 to 11, open dot at 6, shaded left)

15. $x > 15$

(number line from 10 to 20, open dot at 15, shaded right)

17. $x \geq 42$

(number line from 37 to 47, closed dot at 42, shaded right)

19. $x < 12$

(number line from 7 to 17, open dot at 12, shaded left)

21. $x \leq -8$

(number line from −13 to −3, closed dot at −8, shaded left)

23. $x \leq -15$

(number line from −20 to −10, closed dot at −15, shaded left)

25. $x \leq -12$

(number line from −17 to −7, closed dot at −12, shaded left)

27 Catie spent $26 to advertise. Let h be the number of hours she works. Write an expression for her earnings with an initial fee of $5 and $3 per

hour, $5 + 3h$. Write an inequality to show when she will make a profit.

$5 + 3h > 26$ Write an inequality.

$\quad 3h > 21$ Subtraction Property of Inequality

$\quad\ \ h > 7$ Division Property of Inequality

Catie will have to babysit more than 7 hours to make a profit.

29. $g \leq 4$

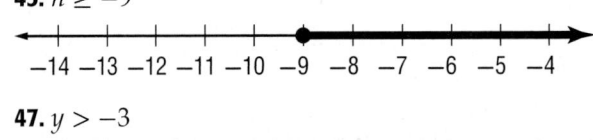

(number line from −1 to 9, closed dot at 4, shaded left)

31. $3x + 4 < -62;\ x < -22$

(number line from −29 to −19, open dot at −22, shaded left)

33. $\dfrac{x}{3} - 2 \geq -12;\ x \geq -30$

(number line from −33 to −23, closed dot at −30, shaded right)

35. Sample answer: Felicia plans on spending at least $32 on a new purse. She has already saved $8. If she earns $4 an hour tutoring, how many hours will she need to tutor to save at least $32?; $x \geq 6$

37. $x > 5;\ x \leq 5;$ Sample answer: Both inequalities contain the same terms. However, one is greater than and the other is less than or equal to. The solution sets of the inequalities differ as well. Five is not included in the first solution set, but is included in the second.   **39.** A   **41.** D   **43.** $26x > 2{,}600{,}000;\ x > 100{,}000;$ There were more than 100,000 spectators per mile.

45. $n \geq -9$

(number line from −14 to −4, closed dot at −9, shaded right)

47. $y > -3$

(number line from −8 to 2, open dot at −3, shaded right)

pp. 204–206   Lesson 3-4B

1 The inequality needs to show that the pressure is between 28 and 35. So, show p as the pressure between 28 and 35. $28 \leq p \leq 35$

3. (number line from 5 to 15, open dot at 8, closed dot at 13, shaded between)

5. (number line from −3 to 7, closed dot at −2, open dot at 6, shaded between)

7.

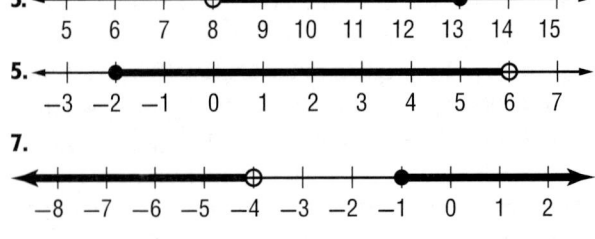

(number line from −8 to 2, open dot at −4, closed dot at −1, shaded between)

9. $150 < j < 275$

11. $w \geq 131$

13 Since the two inequalities are joined by the word "and", you want to find the intersection of each solution set. Graph both simple inequalities to find the intersection.

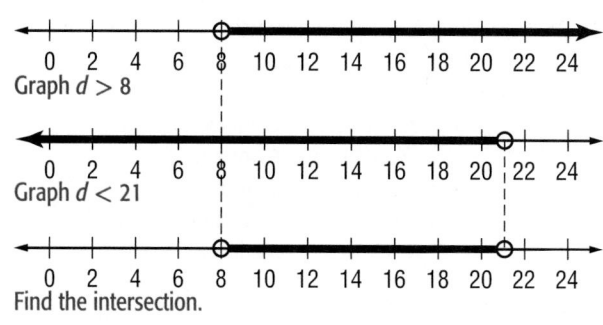

Graph $d > 8$

Graph $d < 21$

Find the intersection.

The intersection is all real numbers greater than 8 and less than 21.

15.

17.

19.

21.

23.

25. $-6 \leq x < 2$ **27.** $x \leq -12$ or $x \geq -8$

29. $19 \leq n < 25$

31 **a.** Let w represent weight. If a loggerhead turtle weighs 200 to 350 pounds, this can be represented by $200 \leq w \leq 350$. If a green turtle weighs 300 to 350 pounds, this can be represented by $300 \leq w \leq 350$.

loggerhead turtle: $200 \leq w \leq 350$

green turtle: $300 \leq w \leq 350$

b. The union of the two graphs includes all of the possible range of weights, $200 \leq w \leq 350$. And the intersection of the two graphs includes the weights that occur in both species, $300 \leq w \leq 350$.

33. $-4 < p \leq 7$

35. $-6 \leq s \leq 1$

37. Sample answer: An intersection includes only the values that are solutions of both inequalities. A union includes all values that are solutions of either

inequality. **39.** H

41. $m < -4$

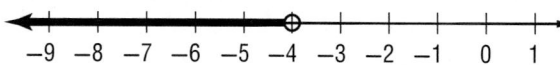

43. $5.25h \geq 42; h \geq 8$

pp. 207–211 *Chapter Study Guide and Review*

1. false; inverse operations **3.** true **5.** true

7. true **9.** 25 tickets **11.** $n = \frac{1}{4}\ell$ **13.** 13 **15.** -22

17. -5 **19.** 294 **21.** $28x = 168$; 6 months **23.** 6

25. 200 **27.** $9.25 **29.** $\frac{n}{8} - 2 = 5$; 56 **31.** $p \geq 12$

33.

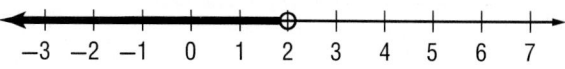

35. $t \leq 3$

37. $n < 12$

39. $k > -9$

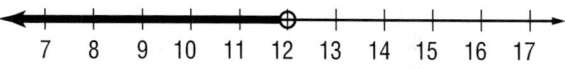

41. $n < 24$

43. $x > 3$

45. 3 boxes

47. $b < -5$

49. $y \geq 20$

51. $s < 1$

53.

55.

57.

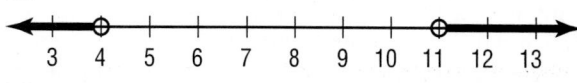

59.

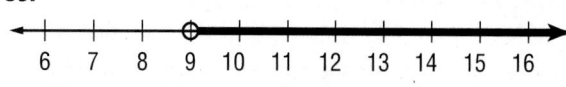

61. $14 \leq w \leq 16$

Chapter 4 Multi-Step Equations and Inequalities

p. 218 **Are You Ready?**

1. -17 **3.** 19 **5.** -6 **7.** -32 **9.** 520 **11.** 55
13. $18 + h = 92$; 74 marbles
15. $b < -7$

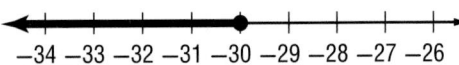

17. $d > -23$

19. $f > -6$

21. $n > -5$

23. $h \leq -30$

25. $s \leq 40$

pp. 223–224 **Lesson 4-1A**

1 The Multiplicative Property of Zero states that when any number is multiplied by 0, the product is 0. $3m$ and $5m$ are being multiplied by 0 in this problem.
3. false; Sample answer: $(8 - 5) - 3 \neq 8 - (5 - 3)$
5.

$$
\begin{aligned}
9c + (8 + 3c) &= 9c + (3c + 8) & &\text{Commutative } (+)\\
&= (9c + 3c) + 8 & &\text{Associative } (+)\\
&= 12c + 8 & &\text{Simplify.}
\end{aligned}
$$

7.

$$
\begin{aligned}
5 \cdot (7h \cdot 4) &\\
&= 5 \cdot (4 \cdot 7h) & &\text{Commutative } (\times)\\
&= (5 \cdot 4) \cdot 7h & &\text{Associative } (\times)\\
&= 20 \cdot 7h & &\text{Simplify.}\\
&= (20 \cdot 7) \cdot h & &\text{Associative } (\times)\\
&= 140h & &\text{Simplify.}
\end{aligned}
$$

9. Identity $(+)$ **11.** Multiplicative (0)
13. Commutative $(\times)$ **15.** false; Sample answer: $10 - 4 \neq 4 - 10$

17 Sample answer: When you use mental math, you want to group numbers so that they add together easily. Look at the cents part of each number. The 0.75 and 0.50 add easily to 1.25. The 0.85 and 0.15 add easily to 1.00. Then add the 2 and 8 to get 10. You are left with 3, 1, and the 2.25 from adding the cents part, which leaves you with $16.25.
19.

$$
\begin{aligned}
(22 + 19b) + 7 &= (19b + 22) + 7 & &\text{Commutative } (+)\\
&= 19b + (22 + 7) & &\text{Associative } (+)\\
&= 19b + 29 & &\text{Simplify.}
\end{aligned}
$$

21.

$$
\begin{aligned}
11s(4) &= 11 \cdot 4 \cdot s & &\text{Commutative } (\times)\\
&= (11 \cdot 4) \cdot s & &\text{Associative } (\times)\\
&= 44s & &\text{Simplify.}
\end{aligned}
$$

23.

$$
\begin{aligned}
3x \cdot (7 \cdot x) &= 3x \cdot (x \cdot 7) & &\text{Commutative } (\times)\\
&= (3x \cdot x) \cdot 7 & &\text{Associative } (\times)\\
&= 3x^2 \cdot 7 & &\text{Simplify.}\\
&= 3 \cdot 7 \cdot x^2 & &\text{Commutative } (\times)\\
&= (3 \cdot 7) \cdot x^2 & &\text{Associative } (\times)\\
&= 21x^2 & &\text{Simplify.}
\end{aligned}
$$

27. Brian incorrectly multiplied both the 5 and m by 4. He should have used the Associative Property to group the 5 and 4 together, simplify and then multiply by m.

$$
\begin{aligned}
4 \cdot (5 \cdot m) &\\
&= (4 \cdot 5) \cdot m & &\text{Associative } (\times)\\
&= 20 \cdot m & &\text{Simplify.}
\end{aligned}
$$

29. C **31a.** $2.29 + 2.50 + 2.21$
31b.

$$
\begin{aligned}
2.29 + 2.50 + 2.21 &\\
&= 2.29 + 2.21 + 2.50 & &\text{Commutative } (+)\\
&= (2.29 + 2.21) + 2.50 & &\text{Associative } (+)\\
&= 4.50 + 2.50 & &\text{Simplify.}\\
&= 7.00 & &\text{Simplify.}
\end{aligned}
$$

The total cost for all three items is $7.00.

pp. 227–230 **Lesson 4-1B**

1. $7 \cdot 5 + 7 \cdot 4$; 63 **3.** $9 \cdot 10 - 9 \cdot 6$; 36

5 Multiplying $6.85 times 4 is difficult. So, rewrite $6.85 as ($7.00 − $0.15). Then multiply by 4.

$$
\begin{aligned}
4(7.00 - 0.15) & &\text{Set up an expression to make the}\\
& &\text{multiplication simpler.}\\
= 4 \cdot 7 - 4 \cdot 0.15 & &\text{Distributive Property}\\
= 28 - 0.60 & &\text{Multiply.}\\
= 27.40 & &\text{Subtract.}
\end{aligned}
$$

So, $4 \cdot \$6.85 = \$27.40$.
7. $5x + 20$ **9.** $3y + 18$ **11.** $24 - 6k$
13. $3 \cdot 5 + 3 \cdot 6$; 33 **15.** $3 \cdot (-8) + 6 \cdot (-8)$; -72
17. $4 \cdot 8 - 4 \cdot 7$; 4 **19.** $(-6) \cdot 9 - (-6) \cdot 4$; -30
21. $12 \cdot (-5) - 4 \cdot (-5)$; -40 **23.** $31.96;
$4(\$8.00 - \$0.01) = 4 \cdot 8 - 4 \cdot 0.01$ **25.** $-8a - 8$
27. $-2p - 14$ **29.** $30 - 6q$ **31.** $-15 + 3b$
33a. $3(7 + 11 + 8)$ and $3(7) + 3(11) + 3(8)$
33b. $78 **35.** $6y + 3$ **37.** $-72 + 48n$
39. $-6a + 4b$ **41.** $5xy - 5xz$

43 Use the Distributive Property.

$$
\begin{aligned}
-4m(3n - 6p) &\\
&= -4m(3n) - (-4m)(6p) & &\text{Distributive Property}\\
&= -12mn - (-24mp) & &\text{Multiply.}\\
&= -12mn + 24mp & &\text{Simplify.}
\end{aligned}
$$

45. $10(x + 5)$; $10x + 50$

47 The area of a triangle is $\frac{1}{2}bh$.

$$
\begin{aligned}
A &= \frac{1}{2}bh & &\text{Area of a triangle}\\
&= \frac{1}{2}(16)(x + 4) & &\text{Replace } b \text{ with 16 and } h \text{ with } x + 4.\\
&= 8(x + 4) & &\text{Multiply.}\\
&= 8(x) + 8(4) & &\text{Distributive Property.}\\
&= 8x + 32 & &\text{Simplify.}
\end{aligned}
$$

49. 315; $9(30 + 5) = 9(30) + 9(5) = 270 + 45$
51. 672; $(100 + 12)6 = 100(6) + 12(6) = 600 + 72$
53. 488; $4(120 + 2) = 4(120) + 4(2) = 480 + 8$
55. 756; $(100 + 8)7 = 100(7) + 8(7) = 700 + 56$
57a. Jacob: $50 = 20 + 0.15(54 + m)$; Roberto: $50 = 30 + 0.1(65 + m)$ **57b.** Jacob: 146 messages; Roberto: 135 messages   **59.** $7b(x + y)$ **61.** Sample answer: The perimeter of a rectangle can be found in two ways. You can find the sum of the base and height, and then multiply by two, $2(b + h)$. You can also find the sum of twice the base and twice the height, $(2b + 2h)$. These two expressions are equivalent and an example of the Distributive Property.   **63.** H
65. False; Sample answer: $5 - 7 = -2$ and -2 is not a whole number.   **67.** False; Sample answer: $1 \div 5 = 0.2$ and 0.2 is not a whole number.   **69.** Yes; the sum of two even numbers is always even.

pp. 233–235   Lesson 4-1C

1. terms: $5n$, $-2n$, -3, n; like terms: $5n$, $-2n$, and n; coefficients: 5, -2, 1; constant: -3 **3.** terms: 7, $-3d$, -8, d; like terms: $-3d$ and d, 7 and -8; coefficients: -3, 1; constants: 7, -8 **5.** 5   **7.** $3x + 4.50$
9. terms: 7, $-5x$, 1; like terms: 7, 1; coefficient: -5; constants: 7, 1 **11.** terms: n, $4n$, $-7n$, -1; like terms: n, $4n$, $-7n$; coefficients: 1, 4, -7; constant: -1
13. terms: 9, $-z$, 3, $-2z$; like terms: 9 and 3, $-z$ and $-2z$; coefficients: -1, -2; constants: 9, 3 **15.** $11c$
17. $2 + 4d$ **19.** $-8j + 5$

21 Set up a table showing the number of minutes watched each day.

Day	Minutes
Monday	x
Wednesday	x
Friday	30

Add the totals for each day. $(x) + (x) + 30 = 2x + 30$

23. $2y - 5$ **27.** Sample answer: You are 14 years younger than 6 times your brother's age, a.

29 a. Write the information as it is given.
$7 to get in, 5 hot dogs for x dollars each ($5x$), 4 popcorns for y dollars each ($4y$), 2 pretzels for z dollars each ($2z$); $7 + 5x + 4y + 2z$
b. $7 + 5x + 4y + 2z$ Write the expression.
$= 7 + 5(4) + 4(3) + 2(2)$ Replace x with 4, y with 3, z with 2.
$= \$43$ Simplify.
31. $16a + 8b + 4$ **33.** $20x + 9$ **35.** $38g + 36h - 38$
37. $14m + 20n + 12$ **39.** $7m - 20$ **41.** $4(x - 2)$; Sample answer: $4(x - 2)$ is equivalent to $4x - 8$, while the other three expressions are equivalent to $4x - 2$.   **43.** yes; $2(x - 1) + 3(x - 1) = 2x - 2 + 3x - 3$ or $5x - 5$ which is equivalent to $5(x - 1)$.
45. G   **47.** D   **49.** $2p^2 + 5$ **51.** $8x^2 + 12x$
53. $11m^2 + 6m$

pp. 236–237   Lesson 4-1D   PSI

1. There would be too many squares to count in the 5×5 square.   **3.** 210 chairs   **5.** 3 packages of 30 and 2 packages of 80   **7.** 165 hours   **9.** 22%
11. 16 pieces

13
| $25 + 0.03x =$ | 35.38 | Write an equation to represent the situation. |

$\dfrac{-25 \qquad = -25}{\dfrac{0.03x}{0.03} = \dfrac{10.38}{0.03}}$ Subtract 25 from each side.

Divide each side by 0.03.

$x = 346$ Simplify.
Cora used the phone for 346 minutes.

pp. 243–245   Lesson 4-2B

1. -3 **3.** -4 **5.** 5   **7.** 75 mi   **9.** 8   **11.** -9

13 Use Properties of Equality to solve the equation.

$8y - 3 = 6y + 17$ Write the equation.
$-6y \qquad = -6y$ Subtraction Property of Equality
$2y - 3 = 17$ Simplify.
$+3 = +3$ Addition Property of Equality
$2y = 20$ Simplify.
$\dfrac{2y}{2} = \dfrac{20}{2}$ Division Property of Equality
$y = 10$ Simplify.
15. 1   **17.** 5   **19.** 3.6   **21.** Let $n =$ the number; $4n + 11 = n - 7$; -6 **23.** Let $n =$ the number; $0.5n - 9 = 4n + 5$; -4

25
Name	Points per Game	Number of Games	Score	Ongoing Total
Will	18	p	483	$483 + 18p$
Tom	21	p	462	$462 + 21p$

To determine when the two players will have scored the same number of points, set the two expressions equal and solve for p.
$483 + 18p = 462 + 21p$ Write the equation.
$-462 \qquad = -462$ Subtraction Property of Equality
$21 + 18p = 21p$ Simplify.
$-18p = -18p$ Subtraction Property of Equality
$21 = 3p$ Simplify.
$\dfrac{21}{3} = \dfrac{3p}{3}$ Division Property of Equality
$7 = p$ Simplify.
Will and Tom will have scored the same number of points after 7 games.
27. $60x = 8x + 26$; 0.5   **29.** -4 **31.** -2.75

33 Solve for x.

$-19.7x - 12.4 = -8.5x + 15.6$ Write the equation.
$+12.4 = +12.4$ Addition ($=$)
$-19.7x = -8.5x + 28$ Simplify.
$+8.5x = +8.5x$ Addition ($=$)
$-11.2x = 28$ Simplify.
$\dfrac{-11.2x}{-11.2} = \dfrac{28}{-11.2}$ Division ($=$)
$x = -2.5$ Simplify.

For Homework Help, go to Hotmath.com

Selected Answers and Solutions   **R13**

Selected Answers and Solutions

35. $10 + 0.07(15x) + 9x = 15x$; 3 sweatshirts
37. Sample answer: You have 20 crafts made and continue to make crafts at the rate of 3 per hour. How many hours will it take you and your friend to make the same amount of crafts, if she makes crafts at a rate of 5 per hour? **39.** Sample answer: $2 - 4x + 4x = 6x - 8 + 4x$ [Addition (=)]; $2 = 10x - 8$ (Simplify); $2 + 8 = 10x - 8 + 8$ [Addition (=)]; $10 = 10x$ (Simplify); $\frac{10}{10} = \frac{10x}{10}$ [Division (=)]; $1 = x$ (Simplify.) **41.** F **43.** G **45.** $4x + 6$ **47.** -9
49. $14a + 10$ **51.** $3h + 4j + 5$ **53.** $7 \cdot 9 - 7 \cdot 4$; 35
55. $12 \cdot (-7) - 8 \cdot (-7)$; -28 **57.** $-8 \cdot 10 - (-8) \cdot 5$; -40

pp. 247–249   Lesson 4-2C

1. 10 **3.** -39 **5.** -39 **7.** 6

9 Let c represent the number of cans Rosana collected, $c + 10$ represent the cans Juliet collected, and $2(c + 10)$ represent the cans Santiago collected. Since they collected 130 cans altogether, $c + (c + 10) + 2(c + 10) = 130$.

$c + (c + 10) + 2(c + 10) = 130$	Write the equation.
$c + (c + 10) + 2c + 20 = 130$	Distributive Property
$4c + 30 = 130$	Simplify.
$\underline{-30 = -30}$	Subtraction (=)
$4c\ \ = 100$	Simplify.
$\frac{4c}{4} = \frac{100}{4}$	Division (=)
$c = 25$	Simplify.

Since Rosana collected 25 cans of food, Juliet collected $25 + 10$ or 35 cans.
11. -9 **13.** -14

15

$\frac{1}{3}h - 4\left(\frac{2}{3}h - 3\right) = \frac{2}{3}h - 6$	Write the equation.
$\frac{1}{3}h - 4 \cdot \frac{2}{3}h - 4(-3) = \frac{2}{3}h - 6$	Distributive Property
$-\frac{7}{3}h + 12 = \frac{2}{3}h - 6$	Simplify.
$-\frac{2}{3}h - \frac{7}{3}h + 12 = \frac{2}{3}h - 6 - \frac{2}{3}h$	Subtraction Property of Equality
$-3h + 12 = -6$	Simplify.
$-3h + 12 - 12 = -6 - 12$	Subtraction Property of Equality
$-3h = -18$	Simplify.
$h = 6$	Division Property of Equality

17. 62 **19.** -1 **21.** $\frac{3}{5}$

23 **a.** Add the number of students in each class and set it equal to 90.

$90 = b + 1.5(b + 2) + 15 + (2b - 9)$	Write the equation.
$= 4.5b + 9$	Combine like terms.

The equation that represents the number of students in Mr. Bogg's homeroom is $90 = 4.5b + 9$.

b. Solve for b in the equation from part **a.**

$90 = 4.5b + 9$	Write the equation.
$\underline{-9 = \qquad -9}$	Subtraction Property of Equality
$81 = 4.5b$	Simplify.
$\frac{81}{4.5} = \frac{4.5b}{4.5}$	Division Property of Equality
$18 = b$	Simplify.

So, there are 18 students in Mr. Boggs' homeroom.
25. Sometimes; sample answer: When you solve the equation $3(2x - 4) = 4(x + 5) + 2x$, all of the x's are eliminated and you are left with $12 = 20$. Since this is not true, there is no solution to this equation.
27. Sample answer: Mrs. Fraser spent the same amount on snacks as drinks at the movie theater for her son and his four friends. If drinks cost $2 more than snacks, how much does each item cost?
29. G **31.** Sample answer: about 25% of 200 or $50
33. Additive Identity Property **35.** Commutative Property (+)

pp. 251–253   Lesson 4-2D

1. $v \geq 3$

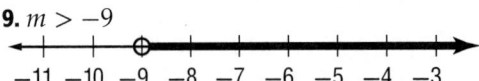

3. $w \leq 9$

5. $e > 8$

7 "No more than" means that Elise must spend $10 or less. Each picture costs $(c + 0.03)$. There are 40 pictures. Also, there is a shipping fee of $3.50.

$40(c + 0.03) + 3.5 \leq 10$	Write the inequality.
$40c + 1.2 + 3.5 \leq 10$	Distributive Property
$40c + 4.7 \leq 10$	Simplify.
$\underline{\quad -4.7 \quad -4.7}$	Subtraction Prop. ($\neq$)
$\frac{40c}{40} \leq \frac{5.3}{40}$	Division Prop. ($\neq$)
$c \leq 0.1325$	Simplify.

9. $m > -9$

11

$-12(g + 8) > 24$	Write the inequality.
$-12g - 96 > 24$	Distributive Property
$-12g - 96 + 96 > 24 + 96$	Addition Property of Inequality
$-12g > 120$	Simplify.
$\frac{-12g}{-12} < \frac{120}{-12}$	Division Property of Inequality
$g < -10$	

13. $b < 9$

15. $g \leq -3$

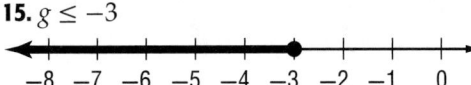

17. $x < 8$

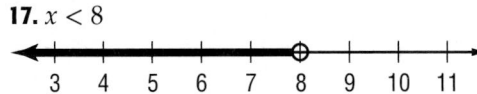

19. $a > -4$

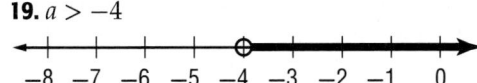

21. no more than 2 hours and 48 minutes

23 The perimeter is the sum of the three sides of the triangle.

$$(x + 4) + (4x - 8) + (2x + 8) \geq 88 \qquad \text{Write the inequality.}$$
$$x + 4x + 2x + 4 - 8 + 8 \geq 88 \qquad \text{Commutative (+)}$$
$$7x + 4 \geq 88 \qquad \text{Combine like terms.}$$
$$\underline{-4 \quad -4} \qquad \text{Subtraction } (\neq)$$
$$\frac{7x}{7} \geq \frac{84}{7} \qquad \text{Division Prop. } (\neq)$$
$$x \geq 12 \qquad \text{Simplify.}$$

Since $x \geq 12$, we can determine the minimum length of each side by using $x = 12$.

Expression	$x = 12$	Side Length
$x + 4$	$12 + 4$	16
$4x - 8$	$4(12) - 8$	40
$2x + 8$	$2(12) + 8$	32

The sides of the triangle are 16, 40, and 32 units.
25. $2(2x + 9) \geq 6$; Sample answer: The solution set of all the other inequalities is $x \geq 3$. The solution set of $2(2x + 9) \geq 6$ is $x \geq -3$.
27. Yes; sample answer:
$$3(x - 2) \leq 4(x - 4) - x$$
$$3x - 6 \leq 4x - 16 - x$$
$$3x - 6 + 6 \leq 3x - 16 + 6$$
$$3x \leq 3x - 10$$
$$3x - 3x \leq 3x - 10 - 3x$$
$$0 \leq -10$$
There is no solution.
29. H **31.** 4 **33.** -4 **35.** terms: $6n, -3n, -4, n$; like terms: $6n, -3n, n$; coefficients: $6, -3, 1$; constant: -4 **37.** terms: $4, 8k, 9k$; like terms: $8k, 9k$; coefficients: $8, 9$; constant: 4 **39.** terms: $j, 15j, -8j, -3$; like terms: $j, 15j, -8j$; coefficients: $1, 15, -8$; constant: -3
41a. $2(2.75 + 1 + 3.5)$ and $2(2.75) + 2(1) + 2(3.5)$
41b. $14.50 **43.** Commutative (+)
45. Commutative ($\times$) **47.** Associative ($\times$)

pp. 256–259 *Chapter Study Guide and Review*

1. false; like **3.** false; property **5.** true **7.** false; counterexample **9.** Associative Property (+)
11. Identity ($\times$) **13.** $3 \cdot 8 + 3 \cdot 7$; 45
15. $9 \cdot 3 + 9 \cdot 9$; 108 **17.** $4a + 52$ **19.** $8r - 72$
21. $13; $0.20(60 + 5) = 0.2 \cdot 60 + 0.2 \cdot 5$

23. $13b + 2$ **25.** s **27.** $8{,}123.5$ mi^2 **29.** -2
31. CA: 840 mi; LA: 397 mi **33.** -12 **35.** $9.17
37. $p < 2$

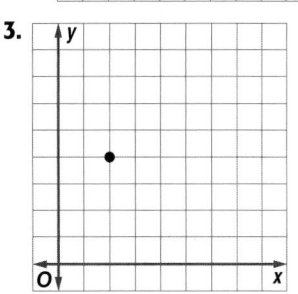

39. $a < -13$

Chapter 5 Expressions and Functions

p. 266 *Are You Ready?*

1. $(3, 5)$ **3.** $(0, 3)$ **5.** $(2, 2)$
7–12.

13.

15. 15 **17.** 11 **19.** 2

p. 268–269 *Lesson 5-1A PSI*

1. Sample answer: You can use guess, check, and revise to find the distance by picking various times and finding how far each has traveled. **5.** 1 h 20 min
7 Set up a table to determine which person is from which state.

	Ohio	Idaho	Colorado	Arizona
William	no		no	no
Scott	no	no	no	yes
Sophia	yes	no	no	no
Christina	no	no	yes	no

Fill in the boxes based on the information given. William lives in Idaho. **9.** 240 episodes **11.** about 14.6 trillion

1. $12p$

③ $3\left(a + \dfrac{c}{a}\right) - b$

$= 3\left(2 + \dfrac{4}{2}\right) - 7$ Replace a with 2, b with 7, and c with 4.

$= 3(2 + 2) - 7$ Perform operations in the parentheses first.

$= 3 \cdot 4 - 7$ Multiply first. Then subtract.

$= 5$

5. 14   **7a.** $4.25 + 0.15p$ **7b.** \$5.23

⑨ Translate *the number of pieces of candy divided equally among 6 friends* into an algebraic expression.

Words	the number of pieces of candy divided equally among 6 friends
Variable	Let c represent the number of pieces of candy.
Expression	$\dfrac{c}{6}$

11. $2c - 35$ **13.** -19 **15.** 23   **17.** -7 **19.** -1

㉑ a. For Ryan's Rafts, the cost to rent for one hour is \$15. Write the expression for h hours. The cost increases \$2.25 for each additional hour. The expression is $15 + 2.25(h - 1)$. For Water Raft Rentals, the cost to rent for one hour is \$6. Write the expression for h hours. The cost increases \$3 for each additional hour. The expression is $6 + 3(h - 1)$.
b. The cost to rent from Ryan's Rafts for 8 hours is $15 + 2.25(8 - 1)$. Simplify using the order of operations to get \$30.75. The cost to rent from Water Raft Rentals for 8 hours is $6 + 3(8 - 1)$. Simplify using the order of operations to get \$27.
23. Sample answer: The expression simplifies to $-19 + n$. If n is a negative integer or a positive integer less than 19, the value will be negative. If n is 19, the value will be 0. If n is greater than 19, the value will be positive.   **25.** $3n + 4; 34$ **27.** D
29. B   **31a.** $20 + 5h$

31b.

Hours	Total Cost ($)
2	30
3	35
4	40
5	45

1. $\left(-1\dfrac{1}{3}, 1\dfrac{1}{3}\right)$ **3.** $\left(1, -\dfrac{2}{3}\right)$

5–8.

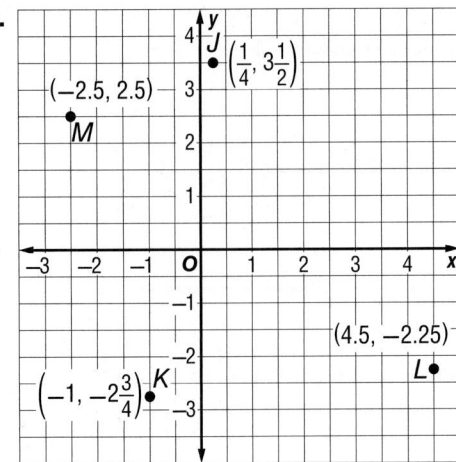

⑨ $\{(-4, 3), (2, 1), (0, 3), (-3, -2)\}$
List x-values and y-values in a table with the x-values in the left column and the y-values in the right column. Then graph the ordered pairs.

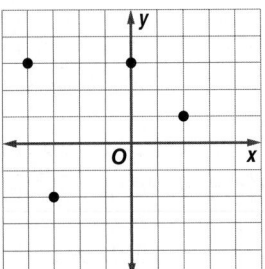

x	y
−4	3
2	1
0	3
−3	−2

The domain is the set of x-values in order from least to greatest. $\{-4, -3, 0, 2\}$; The range is the set of y-values in order from least to greatest. $\{-2, 1, 3\}$

11a.

x	y
1	20
2	40
3	60
4	80

11b.

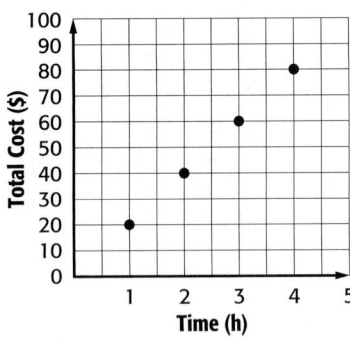

Watercraft Rentals

13. $\left(\dfrac{3}{4}, \dfrac{1}{2}\right)$ **15.** $\left(1, -\dfrac{3}{4}\right)$ **17.** $\left(-\dfrac{1}{2}, -\dfrac{1}{2}\right)$

⑲ Name the ordered pair for point W. Start at the origin. Move left to find the x-coordinate, which is -1. Move up to find the y-coordinate, which is $\dfrac{1}{4}$. So, the ordered pair for point W is $\left(-1, \dfrac{1}{4}\right)$.

20–25.

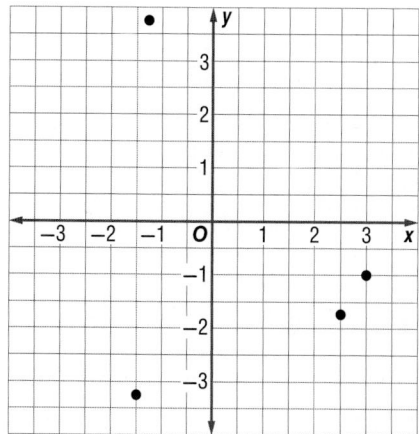

27.

x	y
9	4
5	−7
−3	−4
−8	7

D: {−8, −3, 5, 9}
R: {−7, −4, 4, 7}

29.

x	y
−1.25	3.75
2.5	−1.75
3	−1
−1.5	−3.25

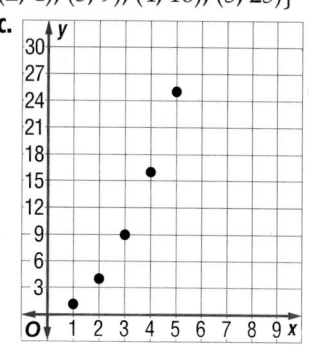

D: {−1.5, −1.25, 2.5, 3}
R: {−3.25, −1.75, −1, 3.75}

31a.

x	y
30	1.25
60	2.5
90	3.75
120	5

31b.

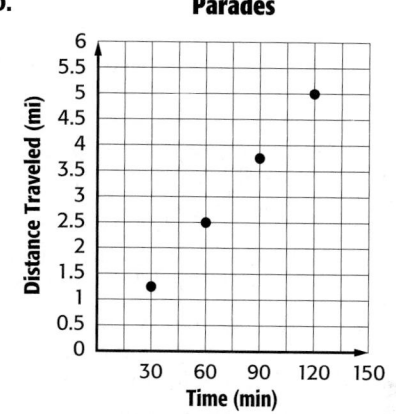

Parades

33 a. Look at the table to compare the *x*-values with the *y*-values. To get the *y*-value, the *x*-value was multiplied by itself. $1 \cdot 1 = 1$. $2 \cdot 2 = 4$. $3 \cdot 3 = 9$. $4 \cdot 4 = 16$. $5 \cdot 5 = 25$.
b. List the ordered pairs in the form (x, y). {(1, 1), (2, 4), (3, 9), (4, 16), (5, 25)}

c.

d. Sample answer: This graph curves upward. The points in all of the other graphs in the lesson lie in a straight line.
37. H   **39.** 14   **41.** $2b − 3$

pp. 284–286   Lesson 5-2B

1. $3n$; 15, 18, 21   **3.** $\frac{1}{6}n$; $\frac{5}{6}$, 1, $1\frac{1}{6}$ **5a.** $8w$ **5b.** 80 min

7 Look at the numbers in the table.
The sequence is 12, 24, 36, 48. The difference between each term is 12. Term 1 is $12 \cdot 1$. Term 2 is $12 \cdot 2$. Term 3 is $12 \cdot 3$. Term 4 is $12 \cdot 4$. So, the expression is $12n$.

Term Number (*n*)	5	6	7
Term	60	72	84

9. $3n − 2$; 13, 16, 19   **11.** $−2n + 27$; 17, 15, 13
13. $\frac{1}{10}n$; $\frac{1}{2}$, $\frac{3}{5}$, $\frac{7}{10}$ **15.** $−\frac{3}{4}n + 2\frac{3}{4}$; −1, $−1\frac{3}{4}$, $−2\frac{1}{2}$
17a. $1.5m − 25$ **17b.** $170

19 a. Set up the table as shown:

Figure	Number of Tiles
1	3
2	5
3	7

b. The difference between each term is 2. Start with $2f$.

$2(1) + \blacksquare = 3$
$2(2) + \blacksquare = 5$
$2(3) + \blacksquare = 7$

Since you need to add 1 to get the number of tiles, the expression is $2f + 1$.
c. Substitute 18 for f in the expression, $2(18) + 1 = 37$ tiles   **d.** No; sample answer: If the number of tiles in each figure is proportional to the number of the figure, each ratio of f to $2f + 1$ would be equal. They are not. So, the number of tiles in each figure is not proportional to the number of the figure.
21. (2, 5); Sample answer: Each ordered pair follows the pattern 3 times the term number plus one. (2, 5) does not fit the pattern.   **25.** 12   **27.** Arithmetic; the terms have a common difference of 3; 16, 19, 22.

pp. 289–291   Lesson 5-2C

1. $3p + 12$ **3a.** $6d + 2$ **3b.** 182 tokens

5 Make a table of the data in the graph. Compare the x- and y-values to determine the expression.

Number of Lawns	5	6	7	8	9
Money Earned ($)	100	120	140	160	180

The expression is $20m$.
7. $100d + 150$

9 a. Make a table of the data in the graph. Compare the x- and y-values to determine the expression.

Time (min)	1	2	3	4	5	6
Rotations	110	220	330	440	550	660

The expression is $110m$.
b. Convert 1.5 hours into 90 minutes. Substitute 90 for m. A tire will make 110 · (90) or 9,900 rotations in 1.5 hours.
11. Sample answer: Mandar's expression $2n + 2$ would be correct if the first ordered pair was (1, 3). Since the first ordered pair is (2, 3), the expression is $2n - 1$.

13. Sample answer: It is easier to make a table because you can more easily see a pattern in the values.   **15.** 342

17–20.

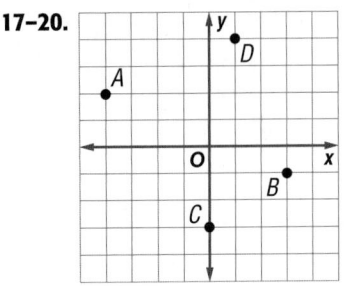

pp. 295–297   Lesson 5-2D

1a. $m = 50d$; Brad sends an average of 50 messages every day.   **1b.** 1,500 messages   **3a.** $m = 7w$

3b.

w	7w	m
4	7(4)	28
5	7(5)	35
6	7(6)	42
7	7(7)	49

New Movies

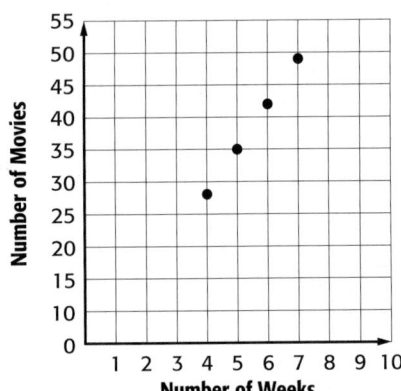

5 a. Compare the values in the table. There are 45 baskets made each day. The equation is $b = 45d$.
b. The number of days in a non-leap year is 365. Therefore, there are $45(365) = 16,425$ baskets made in one non-leap year.

7a. $f = 3.5 + 0.15d$

7b.

d	3.5 + 0.15d	f
10	3.5 + 0.15(10)	5.00
15	3.5 + 0.15(15)	5.75
20	3.5 + 0.15(20)	6.50
25	3.5 + 0.15(25)	7.25

Library Fees

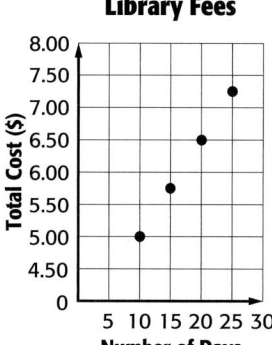

9 **a.**

Week Number	Kara's Savings ($)	Mandy's Savings ($)
1	20	50
2	35	60
3	50	70
4	65	80
5	80	90
6	95	100
7	110	110
8	125	120
9	140	130
10	155	140

b. **Money Saved**

c. Sample answer: Each graph is linear. Kara's savings increase at a quicker rate than Mandy's savings. The overlap point is the week that the girls have saved the same amount of money.   **d.** Kara started with $20 and saved $15 each week after that. Mandy started with $50 and saved $10 each week after that.   **e.** Kara: $S = 5 + 15w$; Mandy: $S = 40 + 10w$ **f.** It is the point when they have both saved the same amount after the same amount of time.
11. $A = \pi \cdot r \cdot r$ or $A = \pi r^2$ **13.** B   **15a.** $28.4z$
15b. 4,260 g   **17.** $2n - 8$; 4, 6, 8

pp. 302–305   Lesson 5-3B
1. -2
3 Set up a table and choose four whole number values to substitute for x. Sample answer:

x	$8 - x$	$f(x)$
-3	$8 - (-3)$	11
-1	$8 - (-1)$	9
2	$8 - 2$	6
4	$8 - 4$	4

D: $\{-3, -1, 2, 4\}$
R: $\{11, 9, 6, 4\}$

5. Sample answer:

x	$3x - 2$	$f(x)$
-5	$3(-5) - 2$	-17
-2	$3(-2) - 2$	-8
2	$3(2) - 2$	4
5	$3(5) - 2$	13

D: $\{-5, -2, 2, 5\}$
R: $\{-17, -8, 4, 13\}$

7. $d(t) = 55t$; 275 miles   **9.** 22   **11.** 15   **13.** -9
15. Sample answer:

x	$5 - 2x$	$f(x)$
-2	$5 - 2(-2)$	9
0	$5 - 2(0)$	5
3	$5 - 2(3)$	-1
5	$5 - 2(5)$	-5

D: $\{-2, 0, 3, 5\}$
R: $\{9, 5, -1, -5\}$

17. Sample answer:

x	$x - 9$	$f(x)$
-2	$-2 - 9$	-11
-1	$-1 - 9$	-10
7	$7 - 9$	-2
12	$12 - 9$	3

D: $\{-2, -1, 7, 12\}$
R: $\{-11, -10, -2, 3\}$

19. Sample answer:

x	$4x + 3$	$f(x)$
-4	$4(-4) + 3$	-13
-2	$4(-2) + 3$	-5
3	$4(3) + 3$	15
5	$4(5) + 3$	23

D: $\{-4, -2, 3, 5\}$
R: $\{-13, -5, 15, 23\}$

21 **a.** Since the total pictures depends on the number of sessions, the number of pictures p is the dependent variable and the number of sessions s is the independent variable.   **b.** Only whole numbers make sense for the domain because you cannot have a fraction of a session. The range values depend on the domain values, so the range will be multiples of 15.
23. $m(s) = 5 + 0.50s$; $20   **25.** $2\frac{1}{4}$
27. $p(d) = \frac{49}{110}d + 14.7$; 92.7 lb/in^2

For Homework Help, go to Hotmath.com

29. A function requires that there is only one y-value for each x-value. Put the ordered pairs into a table and see if there are any duplicate x-values.

x	−12	−8	8	12
y	−7	5	5	7

There are no duplicate x-values, so this relation is a function.

31a. $c(p) = 0.15p + 2.99$

31b.

p	c(p) = 0.15p + 2.99	c
25	0.15(25) + 2.99	6.74
50	0.15(50) + 2.99	10.49
75	0.15(75) + 2.99	14.24
100	0.15(100) + 2.99	17.99

31c.

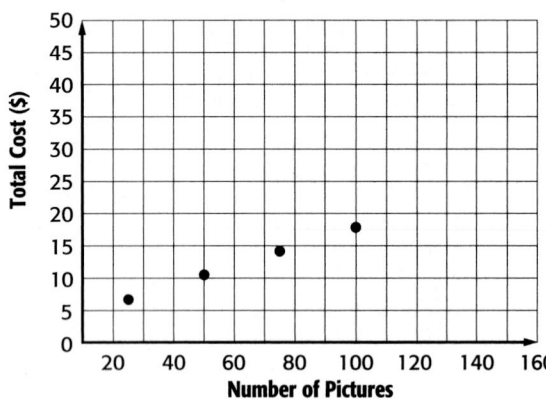

Brian's Pictures

It appears that he will be able to print about 145 pictures for $25.   **33a.** $f(x) = 10x$ **33b.** $f(x) = x - 4$
33c. $y = 2x + 1$ **33d.** $y = 2x - 1$ **35.** B   **37a.** $m = 85w$

37b.

m	85w	w
3	85(3)	255
4	85(4)	340
5	85(5)	425
6	85(6)	510

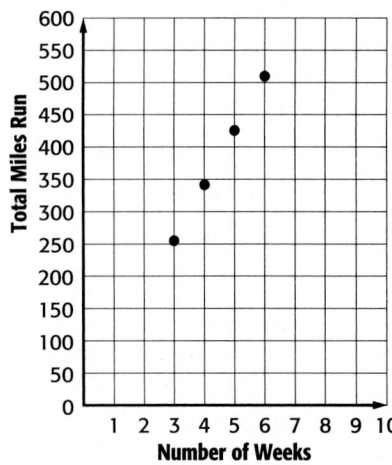

Marathon Training

39. 3   **41.** 8   **43.** 1

pp. 309–311   Lesson 5-3C

1.

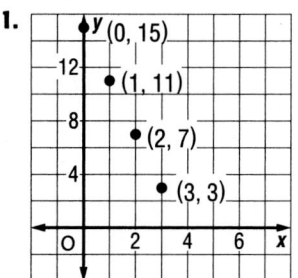

3.

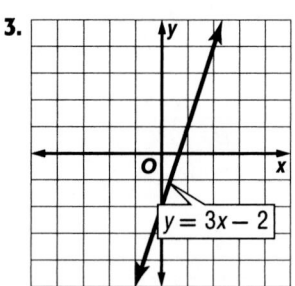

$y = 3x - 2$

5. a. Writing the function puts the numbers into an algebraic equation. The company charges $50 upfront and $39.95 every month. So, let c be the total cost after m months. $c = 50 + 35.95m$.

b.

m	50 + 35.95m	c
1	50 + 35.95(1)	85.95
2	50 + 35.95(2)	121.90
3	50 + 35.95(3)	157.85
4	50 + 35.95(4)	193.80
5	50 + 35.95(5)	229.75

c.

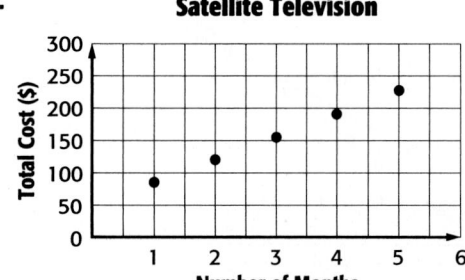

Satellite Television

This situation is discrete because you cannot pay for a partial month of service.

7.

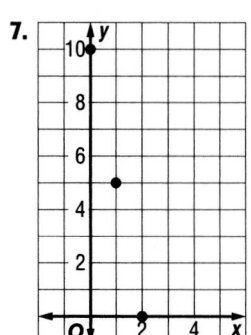

Selected Answers and Solutions

9.

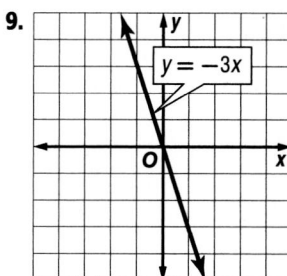

11.

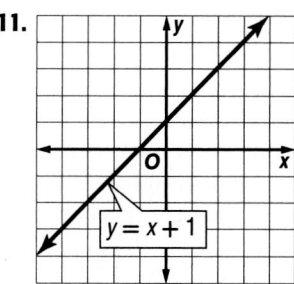

13.

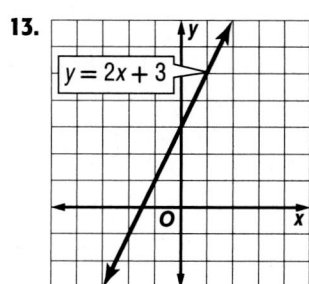

15.

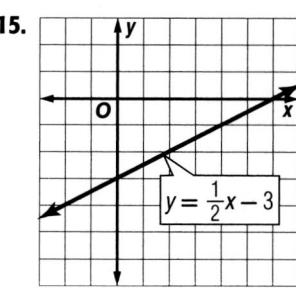

17a. bike: $c = 15 + 4.25h$; scooter: $c = 25 + 2.5h$

17b.

Mountain Bike Rental		
h	$15 + 4.25h$	c
2	$15 + 4.25(2)$	23.50
3	$15 + 4.25(3)$	27.75
4	$15 + 4.25(4)$	32.00
5	$15 + 4.25(5)$	36.25

Scooter Rental		
h	$25 + 2.5h$	c
2	$25 + 2.5(2)$	30.00
3	$25 + 2.5(3)$	32.50
4	$25 + 2.5(4)$	35.00
5	$25 + 2.5(5)$	37.50

17c.

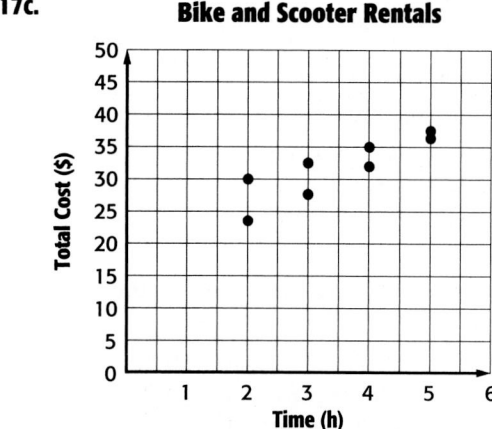

Both situations are discrete because you cannot rent either piece of equipment for a partial hour.

17d. mountain bike

19 a. You may want to draw a picture of what is being asked. The relationship compares yards and meters. Neither of these values can be negative. You could not have a negative distance.

b.

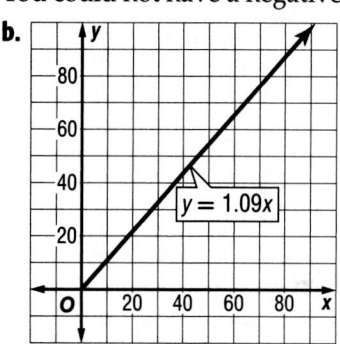

c. Substitute 40 for y in the equation $y = 1.09x$; about 36.7 meters.

21.

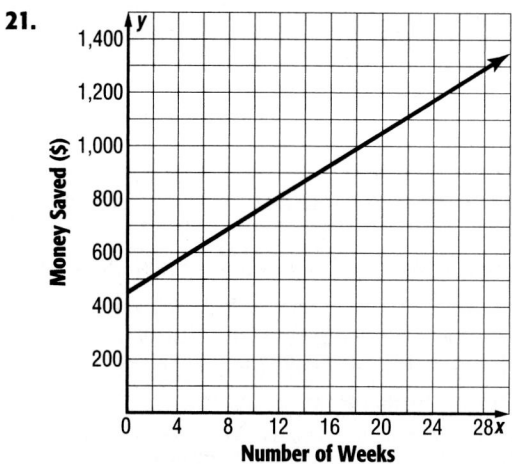

25 weeks

23. $(2, 5)$; $5 \neq -4(2) + 3$ or $5 \neq -5$.
25. Make ordered pairs using the x-value and its corresponding y-value. Then graph the ordered pairs on a coordinate plane. Draw a line that the points suggest.   **27.** no; $(0, -4)$ **29.** 39   **31.** -1
33. $15n$; 75, 90, 105

1. Nonlinear; as *x* increases by 1, *y* increases by a greater amount each time.

3 A function is linear if the rate of change between any two points is constant.

Change in *x*	2	2	2
Change in *y*	1	1	1

Since the rates of change are constant, the function is linear.

5. No; the rate of change is not constant.   **7.** Linear; rate of change is constant; as *x* increases by 3, *y* decreases by 2.   **9.** Linear; rate of change is constant; as *x* increases by 5, *y* increases by 15.
11. Nonlinear; rate of change is not constant.

13 Yes; the rate of change is constant; as the time increases by 1 hour, the distance increases by 65 miles.

15. Linear; sample answer: If you graph the function, the ordered pairs (hours, seconds) lie on a straight line.

17.

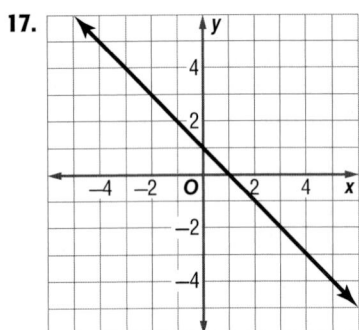

Linear; sample answer: The points lie on a straight line.

19.

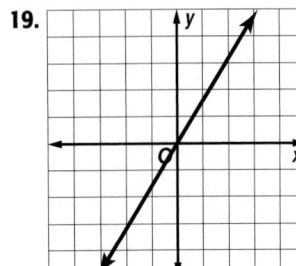

Linear; sample answer: The points lie on a straight line.

21 The formula $P = 4s$ relates the perimeter of a square to the side of a square. The formula $A = s^2$ relates the area of a square to the side of a square. Set up a table showing the side length, the perimeter and the area of a square.

Side length (*s*)	Perimeter (4*s*)	Area (*s*²)
1	4	1
2	8	4
3	12	9
4	16	16

Graph the point (perimeter, area) for each side length.

Perimeter and Area of Square

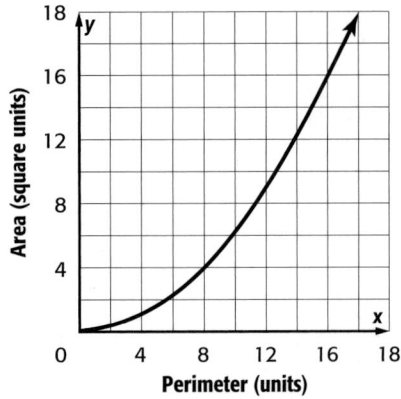

Sample answer: The function is nonlinear because the graph of the function is not a straight line.
25. No; sample answer: the graphs of vertical lines are not functions because there is more than one value of *y* that corresponds to *x* = 2.   **27.** Sample answer: The table of values indicates a proportional rate of change between the *x*- and *y*-values; the graph of the function is a straight line that is not vertical.

29a.

Month	Jung	Miguel
0	200	200
1	210	220
2	220	242
3	230	266.20
4	240	292.82
5	250	322.10
6	260	354.31
7	270	389.74

29b.   **Savings**

Miguel's Savings

Jung's Savings

Total Savings ($)

Number of Months

29c. Jung's savings represent a linear function because the rate of change is constant. As the months increase by 1, the savings increase by 10. Miguel's savings represent a nonlinear function because as the months increase by 1, the total savings increases at a different rate.

31.

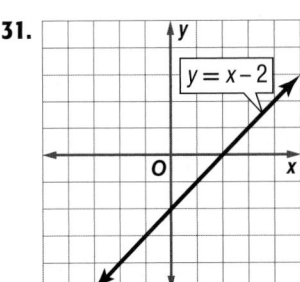

33.

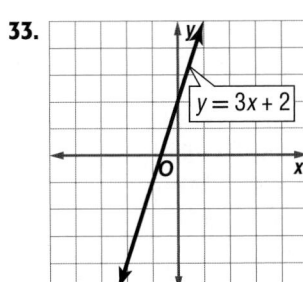

35. −14   **37a.** $c = 5d$; Riley makes an average of 5 phone calls per day.   **37b.** 35 phone calls
39. (3, −4)   **41.** (3, 0)   **43.** (4, 4)

pp. 321–323   Lesson 5-4B

1.

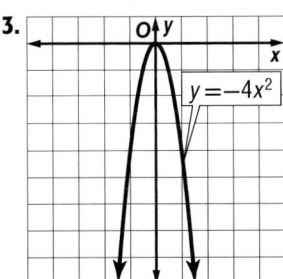

3.

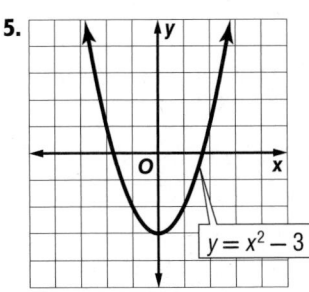

5.

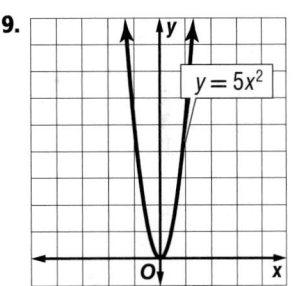

s	$d = 0.006s^2$	(s, d)
0	$0.006(0)^2 = 0$	(0, 0)
4	$0.006(4)^2 = 0.096$	(4, 0.096)
8	$0.006(8)^2 = 0.384$	(8, 0.384)
12	$0.006(12)^2 = 0.864$	(12, 0.864)
16	$0.006(16)^2 = 1.536$	(16, 1.536)
20	$0.006(20)^2 = 2.4$	(20, 2.4)

Plot these points on a grid and connect the points with a curve. Compare your graph to the answer. Extend the line to where d is approximately 12; s would be approximately 45 kilometers per second.

Braking Distance

9.

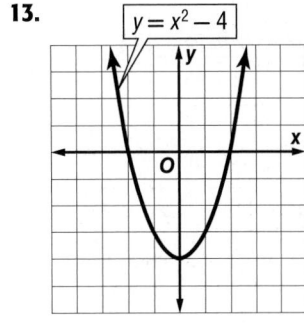

11.

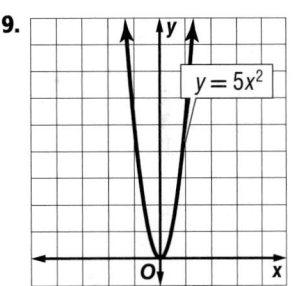

13.

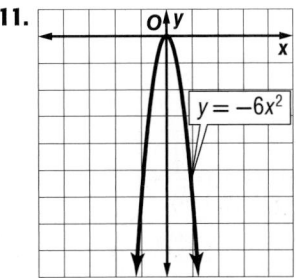

7. Set up a table to determine the values for your graph. Distance and speed can only be positive, so include only positive values of s.

For Homework Help, go to Hotmath.com

15.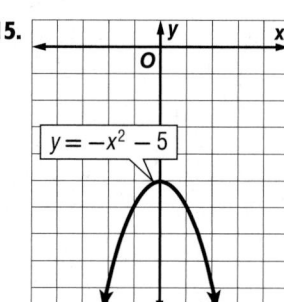

$y = -x^2 - 5$

17.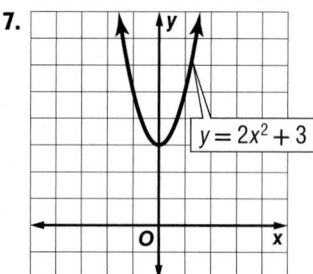

$y = 2x^2 + 3$

19.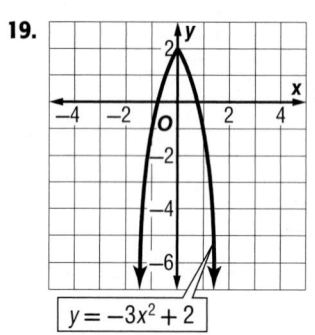

$y = -3x^2 + 2$

21 Set up a table to determine ordered pairs to graph the function. Since the distance is from the surface of the water, the graph will curve downward to the x-axis. The penny will reach the water when $d = 0$.

t	$d = -16t^2 + 196$	(t, d)
0.0	$-16(0.0)^2 + 196 = 196$	(0, 196)
0.5	$-16(0.5)^2 + 196 = 192$	(0.5, 192)
1.0	$-16(1.0)^2 + 196 = 180$	(1.0, 180)
1.5	$-16(1.5)^2 + 196 = 160$	(1.5, 160)
2.0	$-16(2.0)^2 + 196 = 132$	(2.0, 132)
2.5	$-16(2.5)^2 + 196 = 96$	(2.5, 96)
3.0	$-16(3.0)^2 + 196 = 52$	(3.0, 52)
3.5	$-16(3.5)^2 + 196 = 0$	(3.5, 0)

Use these points to draw the graph. Draw a curve to connect the points. The curve will cross the x-axis at 3.5. The penny will reach the water after about 3.5 seconds.

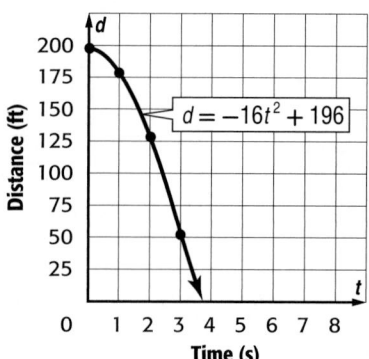

23. nonlinear; The function is quadratic.   **25.** linear; When graphed, the function is a straight line.
27. nonlinear; The function is quadratic.
29. nonlinear; The function is quadratic.

31a. $A = x^2 + 4x$

31b.

Fabric Memo Board

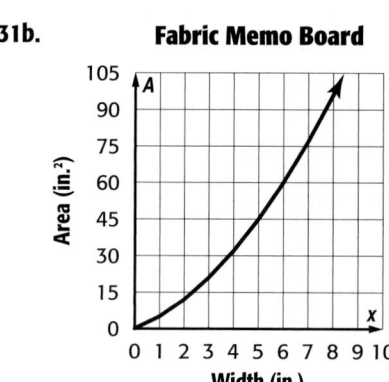

31c. 96 in^2

33.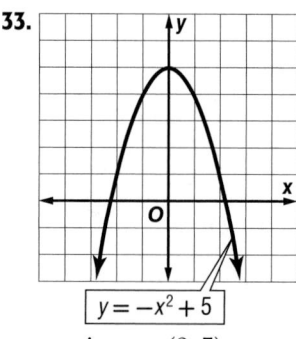

$y = -x^2 + 5$

maximum; (0, 5)

35. Sample answer: $y = x^2 - 3.5$

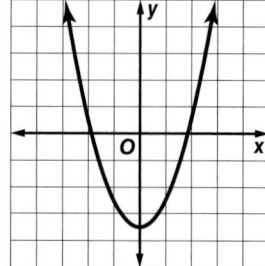

37. A

39.

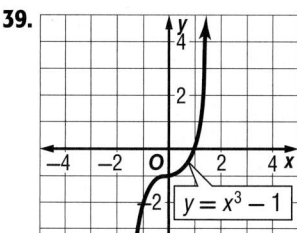

41.

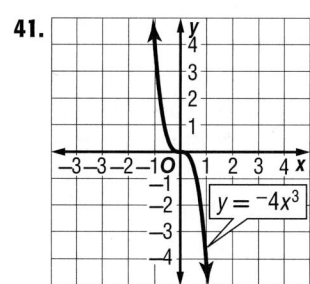

pp. 326–329 *Chapter Study Guide and Review*

1. domain   **3.** dependent   **5.** function
7. relation   **9.** 350 people   **11.** 3 mi

12–13.

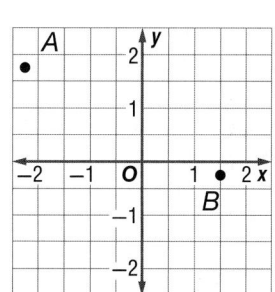

15.

x	y
7	2
−3	9
0	4
−3	2
7	8

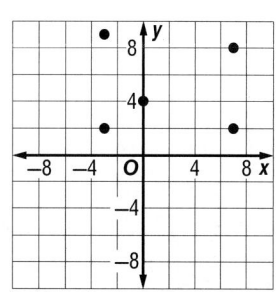

D: $\{-3, 0, 7\}$; R: $\{2, 4, 8, 9\}$

17a. $2h + 3$ **17b.** $19   **19.** 10   **21.** −3

23.

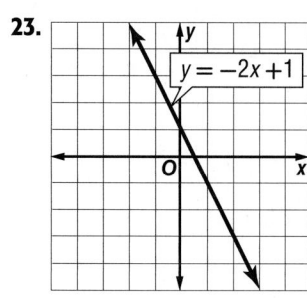

25.

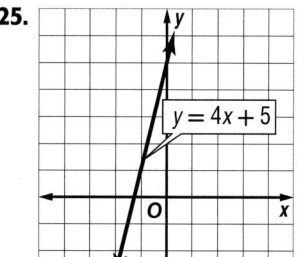

27. Nonlinear; sample answer: As x increases by 1, y increases by a different amount each time.

29.

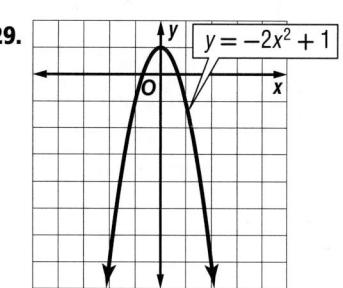

31.

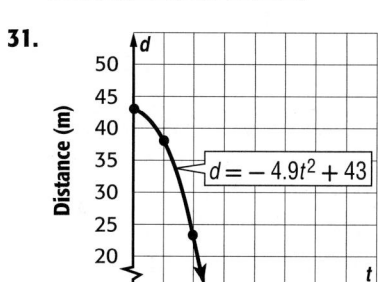

Chapter 6 Linear Functions and Systems of Equations

p. 336   *Are You Ready?*

1. 9   **3.** −7   **5.** 6   **7.** −15   **9.** −6   **11.** −13°F
13. $\frac{2}{5}$ **15.** $\frac{1}{5}$ **17.** $-\frac{2}{5}$ **19.** 2   **21.** $-\frac{5}{6}$

pp. 340–342   *Lesson 6-1A*

1. No; the rate of change from 2 to 3 cm, $\frac{27 - 8}{3 - 2}$ or 19 cm³ per cm, is not the same as the rate of change from 3 to 4 cm, $\frac{64 - 27}{4 - 3}$ or 37 cm³ per cm, so the rate of change is not constant.

3 The constant rate of change is the change in actual distance divided by the change in map distance.
$\frac{(45 - 15)}{(6 - 2)} = \frac{30}{4}$ or 7.5 miles for each inch on the map.

5. Yes; the graph is a line, so the relationship is linear. The ratio of actual distance to map distance is a constant 7.5 mi/in., so the relationship is proportional.   **7.** Yes; the rate of change between cost and time for each hour is a constant 3¢ per hour.   **9.** No; the rate of change from 1 to 2 meters, $\frac{19.6 - 4.9}{2 - 1}$ or 14.7 m/s, is not the same as the rate of change from 2 to 3 meters, $\frac{44.1 - 19.6}{3 - 2}$ or 24.5 m/s,

so the rate of change is not constant.   **11.** 2 in./min; the level of the aquarium went up 2 inches every minute.   **13.** −250 ft/min; a decrease of 250 feet each minute   **15.** 0.5; $\frac{1}{2}$ of retail price   **17.** Yes; the graph is a line, so the relationship is linear. The ratio of water level to time is a constant 2 inches per minute, so the relationship is proportional.
19. No; the graph is a line, so the relationship is linear. However, the ratios of altitude to time for 2 and 6 minutes are $\frac{3,000}{2}$ or 1,500 and $\frac{2,000}{6}$ or 333.$\overline{3}$, respectively. Since these ratios are not the same, the relationship is not proportional.
21. Yes; the graph is a line, so the relationship is linear. The ratio of sale price to retail price is a constant 0.5, so, the relationship is proportional.
23 **a.** Tyrell; he is spending $0.50 per minute, while Miriam is only paying about $0.17 per minute.
b. Tyrell; the ratio of cost to time for Tyrell's plan is a constant $0.50 per minute, while this ratio for Miriam's plan is not constant, with a ratio of $\frac{4.50}{3}$ or 1.5 for 3 minutes and a ratio of $\frac{5.00}{6}$ or 0.$\overline{83}$ for 6 minutes.
25. the origin, (0, 0)   **27.** D   **29.** 15; Will saves $15 each week

pp. 347–349   Lesson 6-1C

1. $\frac{1}{5}$ or $-\frac{1}{5}$ **3.** $-\frac{1}{3}$ **5.** $\frac{3}{4}$ **7.** $-\frac{8}{9}$

9 slope = $\frac{\text{rise}}{\text{run}}$ Definition of slope

$\quad = \frac{-15}{24}$ rise = −15 ft, run = 24 ft

$\quad = \frac{-5}{8}$ Simplify.

The slope of the ski run is $-\frac{5}{8}$.

11. $\frac{1}{2}$ **13.** −3

15. $-\frac{5}{2}$

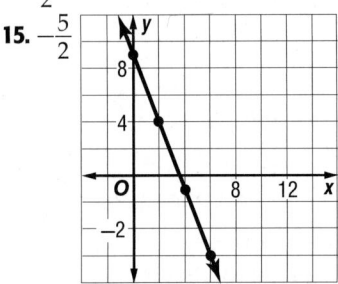

17. 3   **19.** $\frac{5}{3}$ **21.** $-\frac{2}{11}$

23 **a.**   **Work**

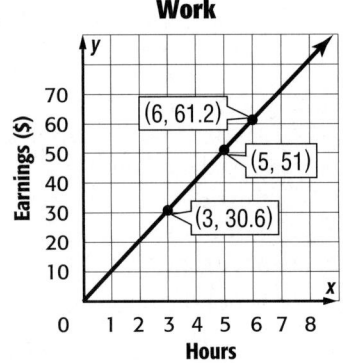

b. The slope of the line is the rise divided by the run. Slope = $\frac{(51 - 30.6)}{5 - 3} = \frac{20.4}{2}$ or 10.2   **c.** The slope of the line represents how much money Sofia makes each hour of work or $10.20 per hour.
25. yes; $\frac{1}{15} < \frac{1}{12}$ **27.** 2π **29.** 2   **31.** I

pp. 354–356   Lesson 6-1E

1. 25 computers per hour   **3.** yes; 58

5 The slope of the line relating earnings to newspapers delivered is the constant rate of change. It is also the amount Dusty earns for each newspaper delivery. Choose two points on the line and find the slope.
$\frac{(6 - 2)}{(12 - 4)} = \frac{4}{8} = \frac{1}{2} = \$0.50$ per newspaper
7. $3.49/DVD   **9.** $53\frac{1}{3}$ lb   **11.** $6\frac{7}{8}$ c   **13.** no
15. yes; 0.07   **17.** $y = \frac{2}{5}x; 4$ **19.** $y = \frac{7}{8}x; 28\frac{4}{7}$

21 If the number of centimeters varies directly with the number of inches, write a proportion to find the number of centimeters in 50 inches. Use one pair of numbers that form a ratio.

$\frac{6}{15.24} = \frac{50}{x}$

$\quad 6x = 15.24 \cdot 50$

$\quad 6x = 762$

$\quad \frac{6x}{6} = \frac{762}{6}$

$\quad\quad x = 127$

The object is 127 centimeters long.
23. Sample answer: $x = 3, y = 1\frac{11}{16}$ **25.** Sample answer: $y = 6x$; If you multiply x by 3, then $y = 6 \cdot 3x$ or $18x$.   **27.** 36 pages   **29.** $\frac{2}{1}$ or 2   **31.** $-\frac{7}{3}$

pp. 359–362   Lesson 6-2A

1. −2; 3   **3.** $y = 4x + 2$

5 The equation is written in slope-intercept form, so the slope of the line is $\frac{1}{3}$ and the y-intercept is −2. Place the point (0, −2) on a coordinate plane. Then move up 1 unit and right 3 units to graph the next point. Draw a line to connect the points.

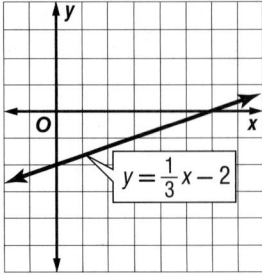

7. 3; 4   **9.** $\frac{1}{2}$; −6   **11.** 2; 8   **13.** $y = -\frac{3}{4}x - 2$
15. $y = -\frac{2}{3}x - 10$ **17.** $y = -\frac{1}{3}x + 5$ **19.** $y = \frac{5}{4}x - 12$

21.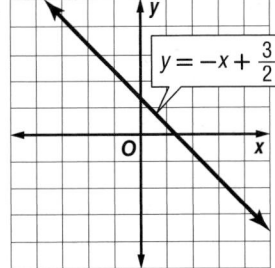

$$y = -x + \frac{3}{2}$$

23.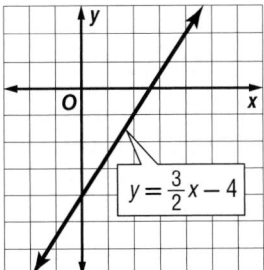

$$y = \frac{3}{2}x - 4$$

25.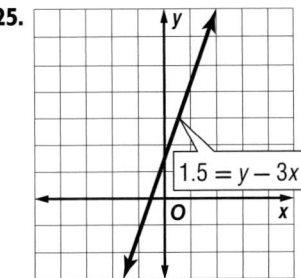

$$1.5 = y - 3x$$

27a.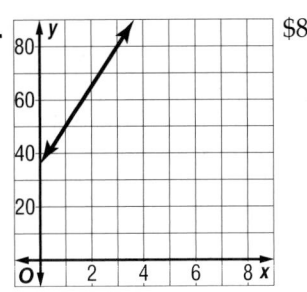

$80

27b. the hourly rental charge, $15, and the base rental fee, $35

29 Make a table based on the equation.

x (chirps)	15x + 37	y (temperature)
0.5	15(0.5) + 37	44.5
1	15(1) + 37	52
1.5	15(1.5) + 37	59.5
2	15(2) + 37	67
2.5	15(2.5) + 37	74.5
3	15(3) + 37	82
3.5	15(3.5) + 37	89.5

Use the ordered pairs from the table to graph the equation.

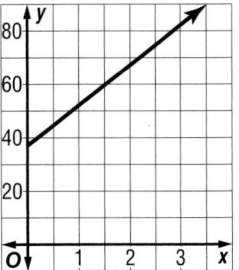

From the graph and the data table, we find that a cricket chirps about 3 times in 15 seconds at 80 degrees Fahrenheit.

31 **a.** Determine the slope between any two sets of points.
$$\frac{y_2 - y_1}{x_2 - x_1} = \frac{(9 - 5)}{(2 - 1)} = \frac{4}{1} = 4$$
Since one of the points of the line is (0, 1), the line crosses the y-axis at 1. So, the y-intercept is 1.

b. The slope of the graph rises to the right. The line intercepts the y-axis at +1.

c. Substitute the slope and y-intercept into the slope-intercept equation.
$$y = mx + b \quad \text{Slope-intercept form}$$
$$y = 4x + 1 \quad \text{Replace } m \text{ with 4 and } b \text{ with 1.}$$

33. Sample answer: 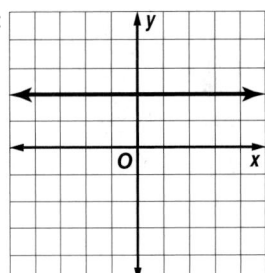 0

35. Sample answer: Cell phone charge of $10 plus $0.05 per minute. In a table, the y-intercept would be the amount charged for 0 minutes, $10, and the slope would be the amount added for each minute, $0.05. In an equation, the y-intercept is the constant value, 10, and the slope is the coefficient of x, 0.05. In a graph, the y-intercept is the y-coordinate where the line crosses the y-axis, (0, 10), and the slope reflects the direction and steepness of the line.

37.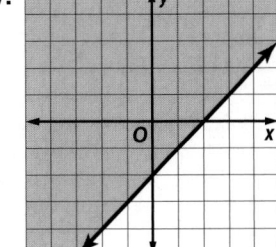

39.

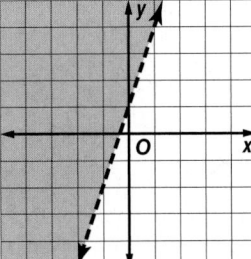

41.

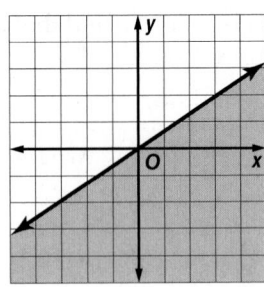

pp. 365–366   Lesson 6-2B

1. x-intercept: 4; y-intercept: 5

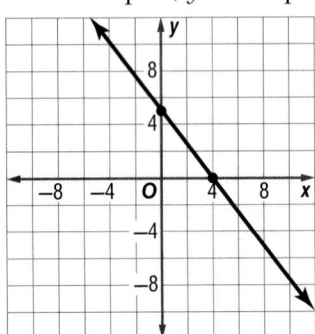

3. x-intercept: -4; y-intercept: 2

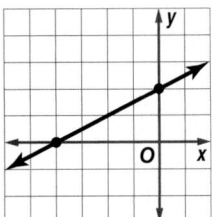

5. x-intercept: 10; y-intercept: 1.5

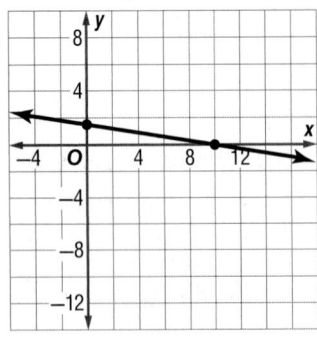

7.   **Juice Boxes**

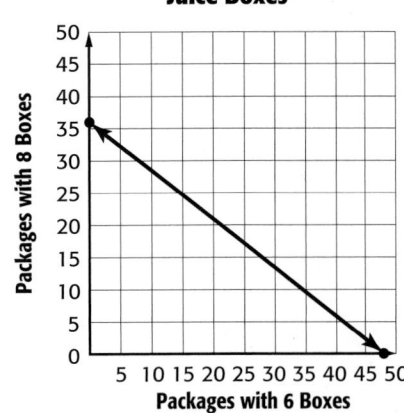

The x-intercept is at the point (48, 0). This means that the store has 48 packages of 6 boxes and 0 packages of 8 boxes to have a total of 288 boxes. The y-intercept is at the point (0, 36). This means that the store has 36 packages of 8 boxes and 0 packages of 6 boxes to have a total of 288 boxes.

9 To determine the x-intercept, substitute 0 for y and solve for x. To determine the y-intercept, substitute 0 for x and solve for y.

x-intercept	y-intercept
$12x + 9(0) = 15$	$12(0) + 9y = 15$
$12x = 15$	$9y = 15$
$x = 1\frac{1}{4}$	$y = 1\frac{2}{3}$

x-intercept: $1\frac{1}{4}$; y-intercept: $1\frac{2}{3}$

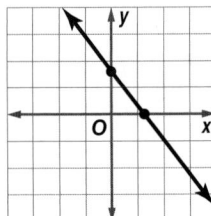

11. x-intercept: $-\frac{1}{2}$; y-intercept: $\frac{2}{3}$

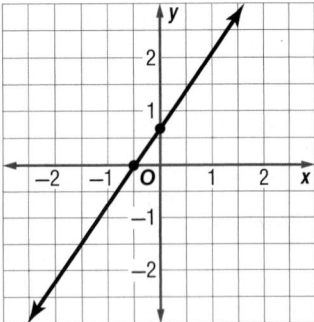

13. x-intercept: 5; y-intercept: -3

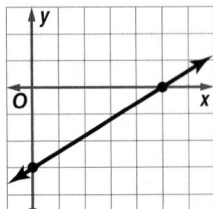

15. x-intercept: $\frac{1}{5}$; y-intercept: $-\frac{2}{3}$

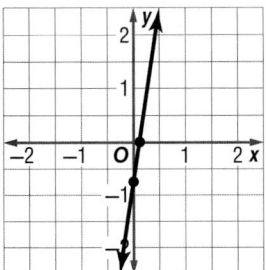

17. x-intercept: -18; y-intercept: -16

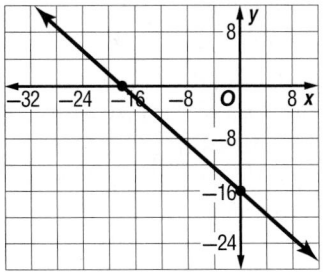

19. x-intercept: 12; y-intercept: 7.5

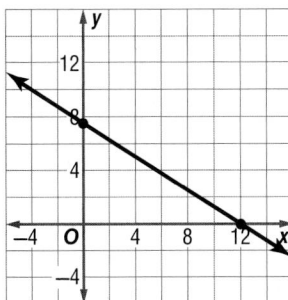

21.

Animals at the Zoo

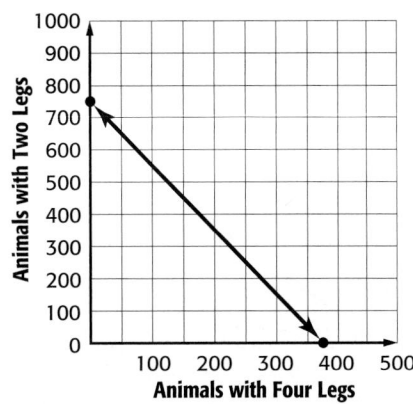

The x-intercept is at the point (375, 0). This means that if the zoo had only four-legged animals, there would be 375 of them for a total of 1,500 legs. The y-intercept is at the point (0, 750). This means that if the zoo had only two-legged animals, there would be 750 of them for a total of 1,500 legs.

23 a. Rewrite the perimeter equation specifically for this exercise. $2x + 2y = 24$
b. To find the x-intercept, let $y = 0$. To find the y-intercept, let $x = 0$.

x-intercept	y-intercept
$2x + 2y = 24$	$2x + 2y = 24$
$2x + 2(0) = 24$	$2(0) + 2y = 24$
$2x = 24$	$2y = 24$
$x = 12$	$y = 12$

The x-intercept is 12.   The y-intercept is 12.
Sample answer: The x-intercept is at the point (12, 0) and the y-intercept is at the point (0, 12). These points are not solutions in this situation because the length or width of the rectangle cannot be 0.
25. Sample answer: You can graph a function by making a function table and plotting ordered pairs. You can also find the x- and y-intercepts of the function and graph those points.   **27.** I   **29.** 3; 4
31. $\frac{2}{3}$; -1 **33a.** 6.5   **33b.** the amount she made per hour, $6.50

pp. 370–371   Lesson 6-3A PSI

1. There are 30 people going to the museum, not $23 + 5$ or 28 people.   **3.** 24 or -24
5 Use the *guess, check,* and *revise* strategy. You know that Shyla bought a total of 8 gifts for a total of $53. The rings r cost $6 each and the toys t cost $7 each. Make a table to organize the information.

Number of Rings, r	Number of Toys, t	$6r + 7t$	
4	4	$6(4) + 7(4) = 52$	too low
2	6	$6(2) + 7(6) = 54$	too high
3	5	$6(3) + 7(5) = 53$	correct

So, Shyla bought 3 rings and 5 toys.
7. Sample answer: 3, 8, 12   **9.** Sample answer: 8 in. by 5 in. by 10 in.; 10 in. by 10 in. by 4 in.
11. Sample answer: 32, 36, 40   **13.** $\frac{1}{100}$

pp. 375–377   Lesson 6-3C

1. $(-2, 1)$

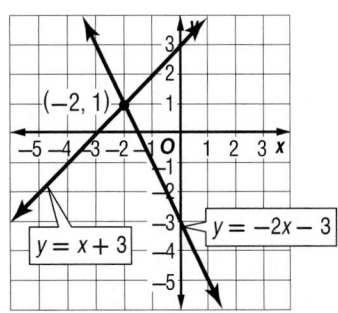

3 Graph each equation. The solution is where the two graphs intersect. Because both equations can be rewritten as $y = 2x + 6$, we know that the two equations are the same. Therefore, there are infinitely many solutions because the two graphs intersect everywhere.

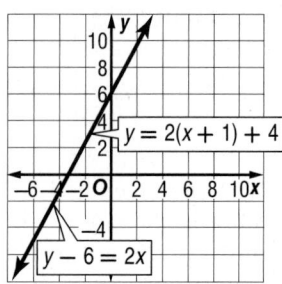

5. (4, 4)

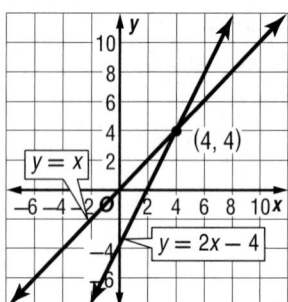

7. no solution

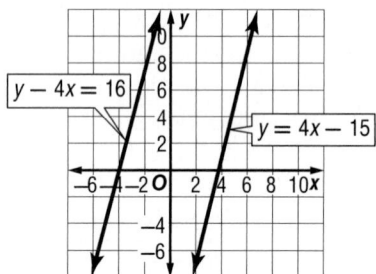

9. (3, 4)

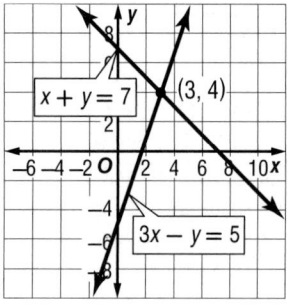

11. infinitely many solutions

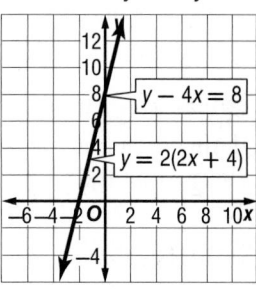

13. (4, 2)

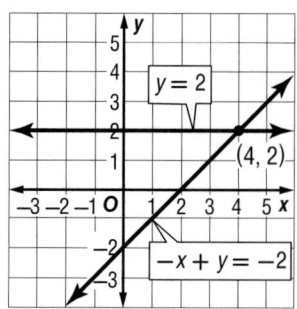

15 Sample answer: Let x equal the number of beads purchased and y equal the total cost.

$y = 0.12x + 3.25$ Babs' Beads charges \$0.12 per bead with a shipping cost of \$3.25.

$y = 0.25x$ Jewels by Jo charges \$0.25 per bead with no shipping costs.

Graph each equation on the same coordinate plane.

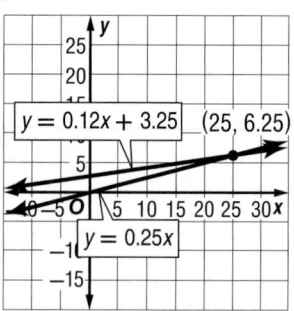

The graphs appear to intersect at (25, 6.25). So, at 25 beads, the costs would be the same.

17 a. Graph $y = 0.71x$ and $y = 25$ on the same coordinate plane.

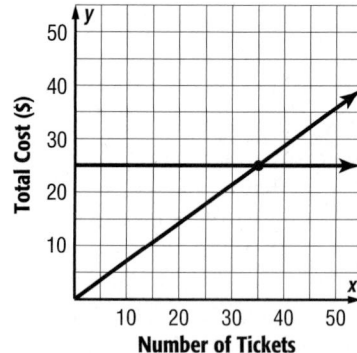

b. Find the intersection point of the two lines to find the place where the cost is the same for individual tickets and for the wristband. The two lines intersect at 35 tickets. If each ride uses 2 tickets, then $35 \div 2 = 17.5$ rides. So, the wristband is a better deal at 18 rides.

21. G   **23.** x-intercept: 9; y-intercept: 6

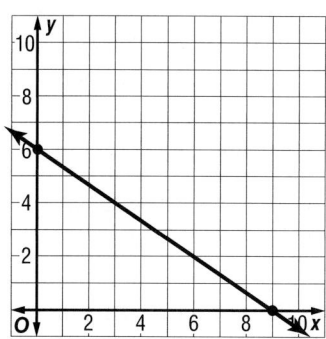

25. x-intercept: -6; y-intercept: -4

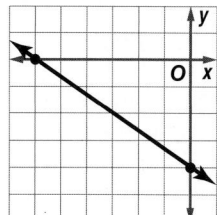

27. -1

pp. 380–381   Lesson 6-3D

1. $(-3, 4)$ **3.** $(2.5, 7.5)$ **5.** Sample answer:
$y = x + 1$; $y + x = 7$; $(3, 4)$; Four adults and three
children went to the movies.   **7.** $(-30, -18)$

9 Set the two equations equal and solve for x.
Then substitute the value of x into the equation to
find y.

$-x - 14 = 15$	Set the two equations equal.
$x = -29$	Solve for x.
$y = -(-29) - 14$	Substitute.
$= 15$	Solve.

The solution of the system of equations is $(-29, 15)$.

11. $\left(3\frac{2}{3}, 18\frac{1}{3}\right)$ **13.** $(6, -18)$ **15.** Sample answer:
$p + h = 49$; $h = p + 11$; $(19, 30)$; Preston has
19 games and Horatio has 30 games.

17 Sample answer: If Brad has 3 more dimes than
nickels, write an expression for the amount of
dimes Brad has. Let n be the number of nickels.
Three more dimes, d, would be $n + 3$. Now write
an equation for the number of nickels and the
number of dimes, $d = n + 3$. If the coins total
90 cents, write an equation using the value of a
dime and the value of a nickel, $90 = 10d + 5n$.
Solve the system of equations.
$$d = n + 3$$
$$10d + 5n = 90$$
Substitute for d in the second equation.

$10(n + 3) + 5n = 90$	Write the equation.
$10n + 30 + 5n = 90$	Simplify.
$15n + 30 = 90$	Simplify.
$\underline{-30 = -30}$	Subtract 30 from each side.
$15n = 60$	Simplify.

$\dfrac{15n}{15} = \dfrac{60}{15}$	Divide each side by 15.
$n = 4$	Simplify.

Now substitute $n = 4$ into the first equation to find
the number of dimes.
$$d = n + 3$$
$$d = 4 + 3 \text{ or } 7 \text{ dimes}$$
Brad has 4 nickels and 7 dimes.

19. $(7, -1)$ **21.** C   **23.** $(-3, -8)$ **25.** $(-4, -5)$

pp. 384–387   *Chapter Study Guide and Review*

1. y-intercept   **3.** rise   **5.** standard   **7.** constant
9. Yes; sample answer: The rate of change between
cost and time for each minute is a constant $7 per
minute.   **11.** $-\dfrac{5}{6}$ **13.** $5.20   **15.** $2; 5$ **17.** $4; 7$
19. $y = -3x - 2$
21. x-intercept: 8; y-intercept: 4

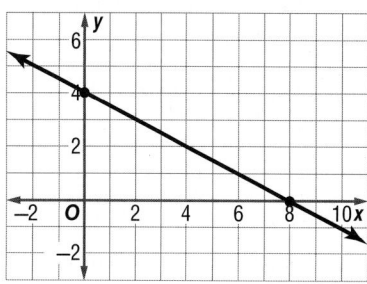

23a.

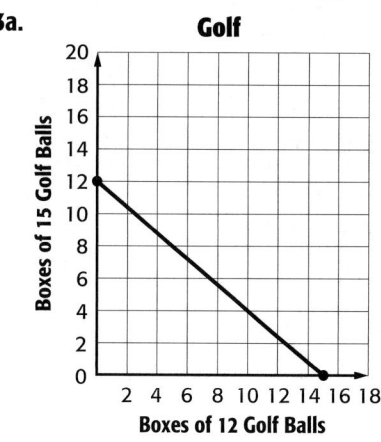

Golf

23b. The x-intercept is at the point $(15, 0)$. This
means that Coach Taylor bought 15 boxes of
12 golf balls and 0 boxes of 15 golf balls for a total
of 180 golf balls. The y-intercept is at the point
$(0, 12)$. This means that Coach Taylor bought
12 boxes of 15 golf balls and 0 boxes of 12 golf balls
for a total of 180 golf balls.
25. 3 bags of apples and 1 bag of oranges
27. $(-2, -2)$

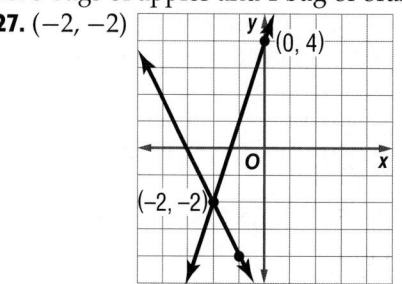

29. Sample answer: Let $x =$ the number that

preferred steak and y = the number that preferred pizza; $x + y = 25$, $y = x + 5$; (10, 15); 10 students preferred steak and 15 students preferred pizza.

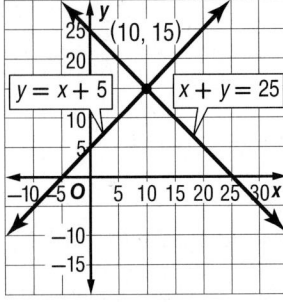

31. (2, −2)   **33.** (−2, −8)

Chapter 7 Two- and Three-Dimensional Geometry

p. 394   Are You Ready?

1. 86   **3.** 98   **5.** 120   **7.** 120 tickets   **9.** 900
11. 1,620   **13.** 1,800   **15.** 4,140

pp. 398–400   Lesson 7-1B

1 An angle can be named by the rays that form it ∠MNP or ∠PNM, by the number in it, ∠1, or the letter at the vertex ∠N. Since the angle is between 90° and 180°, it is obtuse.
3. ∠ABC, ∠CBA, ∠B, ∠1; obtuse   **5.** Sample answer: ∠1 and ∠3 are vertical angles because they are opposite angles formed by the intersection of two lines. ∠2 and ∠3 are adjacent angles because they share a common vertex and a common side, and they do not overlap.
7. 52   **9.** 131   **11.** ∠DEF, ∠FED, ∠E, ∠5; right
13. ∠MNP, ∠PNM, ∠N, ∠7; straight
15. ∠RTS, ∠STR, ∠T, ∠9; acute

17 The angles are not across from each other nor do they share a ray. Therefore, they are neither vertical nor adjacent.
19. adjacent   **21.** vertical   **23.** 75°   **25.** 55°
27. 125°
29 **a.** Since ∠1 and ∠2 share a common vertex and a common side and do not overlap, they are adjacent angles.
b. Since ∠2 and ∠4 are opposite angles formed by the intersection of two lines, they are vertical angles.
c. Since ∠3 and ∠4 share a common vertex and a common side and do not overlap, they are adjacent angles.
d. Since ∠1 and ∠3 are opposite angles formed by the intersection of two lines, they are vertical angles.
e. ∠2 and ∠4 are vertical angles because they are opposite angles formed by the intersection of two lines. Since vertical angles are congruent, the measure of ∠4 is 66°. ∠1 and ∠2 form a straight

angle, so the sum of their measures is 180°. So, the measure of ∠1 is 180° − 66° or 114°. Since ∠1 and ∠3 are vertical angles, and vertical angles are congruent, then ∠3 also measures 114°.
33. true; Sample answer:

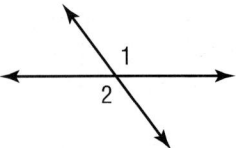

35. Sample answer: Vertical angles are opposite angles formed by the intersection of two lines, share a common vertex, and do not share a common side. Adjacent angles can be formed by the intersection of two lines but must share a common side and a common vertex.   **37.** F
39a. $x + 102 = 180$; 78°   **39b.** acute; The measure of ∠OLN is 102° since it is a vertical angle with ∠PLR. The measure of ∠OLM plus the measure of ∠MLN is 102°. The measure of ∠OLM plus 60° is 102°, so the measure of ∠OLM is 42°. Since the measure of ∠OLM is 42°, it is acute.   **39c.** No; the vertex of ∠OLP is at point L with sides $\overrightarrow{LO}$ and $\overrightarrow{LP}$. The vertex of ∠OPL is at point P with sides $\overrightarrow{PL}$ and $\overrightarrow{PO}$.

pp. 403–405   Lesson 7-1C

1. supplementary   **3.** 135   **5.** supplementary
7. supplementary   **9.** neither

11 We are given the fact that the angles are supplementary and we are given the measure of one of the angles. We can use that information to find the missing angle value.
$m∠J + m∠K = 180°$ Definition of supplementary angles
$m∠J + 115° = 180°$ Substitution
$m∠J = 65°$ Subtract 115 from both sides.

13 The given angle and the missing angle form supplementary angles. Use this information to find the measure of the missing angle.
$x + 43° = 180°$ Definition of supplementary angles
$x = 137°$ Subtract 43 from each side.
15. Sample answer: ∠CGK, ∠KGJ   **17a.** adjacent; adjacent; vertical   **17b.** $m∠1 + m∠2 = 180°$; $m∠2 + m∠3 = 180°$ **17c.** $m∠1 = 180° − m∠2$; $m∠3 = 180° − m∠2$; Sample answer: $m∠1$ and $m∠3$ are equal.
17d. Sample answer: Vertical angles are congruent.
19. never; Sample answer: One right angle measures 90°.

21 This statement would sometimes be true.
The sum of complementary angles is 90°. Vertical angles are congruent. Therefore, $2x = 90$. If the vertical angles measured 45°, then they would be complementary.
23. 55°   **25.** Sample answer: Adjacent angles are two angles that share a common side and vertex. Vertical angles will never be adjacent. Complementary and supplementary angles may or may not be adjacent.   **27.** 45

29. Given: $\angle 1 \cong \angle 2$, $\angle 1$ and $\angle 2$ are supplementary
Prove: $\angle 1$ and $\angle 2$ are right angles
Proof: $m\angle 1 + m\angle 2 = 180°$ since they are supplementary angles. Since $\angle 1 \cong \angle 2$ then $m\angle 1 + m\angle 2 = 180°$ by substitution. Solving the equation gives $m\angle 1 = 90°$. Since $\angle 1 \cong \angle 2$, the $m\angle 2$ is also 90°. Therefore, $\angle 1$ and $\angle 2$ are right angles.

pp. 406–407   Lesson 7-1D   PSI

1. Yes, Chris used inductive reasoning. Chris observed that the acute angles of several different examples of right triangles were complementary to decide that the acute angles of all right triangles are complementary.   **3.** The diagonals of a rectangle are congruent.   **5.** $0.\overline{09}, 0.\overline{36}, 0.\overline{72}; 0.\overline{27}, 0.\overline{54}, 0.\overline{81}$
7. Sample answer: Paulo: 10 h; Clarissa: 20 h
9. One 1-lb package and three $2\frac{1}{2}$-lb packages

11 If the arctic tern flies 21,750 miles per year for 20 years, it has flown $21{,}750 \cdot 20$ by the end of its lifespan. $21{,}750 \text{ miles/year} \cdot 20 \text{ years} = 435{,}000$ miles

pp. 411–414   Lesson 7-2B

1. alternate exterior   **3.** alternate interior

5 145°; Sample answer: $\angle 7$ and $\angle 5$ are supplementary. So, $m\angle 5 = 180 - 35$ or 145°. $\angle 5$ and $\angle 1$ are corresponding angles. Since corresponding angles are congruent, $m\angle 1 = 145°$.
7. 35°; Sample answer: $\angle 7$ and $\angle 3$ are corresponding angles. So, $m\angle 3 = 35°$.   **9.** 45°; Sample answer: $\angle 2$ and angles 9 and 10 are vertical angles. So, $m\angle 9 + m\angle 10 = 135°$. So, $m\angle 9 = 135 - 90$ or 45°.   **11.** 45°; Sample answer: $\angle 1$ and $\angle 2$ are supplementary. So, $m\angle 1 = 45°$. $\angle 1$ and $\angle 3$ are corresponding angles, so, $m\angle 3 = 45°$.
13. corresponding   **15.** corresponding
17. alternate exterior

19 $m\angle 2 = 120°$, $m\angle 3 = 60°$; Sample answer: $\angle 1$ and $\angle 2$ are alternate exterior angles, so they are congruent. $\angle 2$ and $\angle 3$ are supplementary. So, $m\angle 3 = 180° - 120°$ or 60°.
21. 70°; Sample answer: $\angle 2$ and $\angle 6$ are congruent corresponding angles. $\angle 6$ and $\angle 7$ are supplementary. So, $m\angle 7 = 180 - 110$ or 70°.
23. 110°; Sample answer: $\angle 6$ and $\angle 8$ are congruent vertical angles.   **25.** 137°; Sample answer: $\angle 11$ and $\angle 3$ are congruent corresponding angles.   **27.** $\angle 1$ and $\angle 2$ are alternate interior angles. $m\angle 2 = 22°$ because $\angle 1 \cong \angle 2$.   **29a.** 20   **29b.** 40

33 **a.** The top and bottom of the ramp are parallel. The slanted part of the ramp can be considered a transversal. You can use angle relationships of parallel lines to find the measure of the missing angle.   **b.** Since the 152° angle and the exterior angle of the missing angle are alternate interior angles, the measure of the missing angle is $180° - 152°$ or 28°.

35. Sample answer: $\angle DAB$ and $\angle ADC$ are supplementary. Extend the sides as shown. Since the lines are parallel, $\angle DAB \cong \angle ADE$ (alternate interior angles are $\cong$). Since $\angle ADE$ and ADC lie on the same line, they are supplementary, and $m\angle ADE + m\angle ADC = 180°$. Substitute $\angle DAB$ for $\angle ADE$. Therefore, $m\angle ADC + m\angle DAB = 180°$.   **37.** D
39. A   **41.** 116   **43.** adjacent   **45.** neither

pp. 420–423   Lesson 7-3B

1 We know that the sum of the interior angles of a triangle is 180 degrees. Use that to find the missing measure.
$x + 75 + 60 = 180$ Sum of the angles of a triangle
$\qquad\quad x = 45$ Subtract 75 and 60 from each side.
3. 68   **5.** 22.5°, 45°, 112.5°   **7.** obtuse scalene   **9.** 30
11. 65   **13.** 95   **15.** 65   **17.** 24°, 48°, 108°   **19.** 45°, 45°, 90°   **21.** obtuse isosceles   **23.** obtuse scalene
25. right isosceles

27 We know that the sum of the interior angles of a triangle is 180 degrees. Use that to find the missing measure.
$x + 25 + 50 = 180$ Sum of the angles of a triangle
$\qquad\quad x = 105$ Subtract 25 and 50 from each side.
29. $m\angle A = 47°$, $m\angle B = 90°$, $m\angle C = 43°$
31. Not possible; sample answer: An equilateral triangle has three congruent sides and three congruent angles that each measure 60°. Therefore, it cannot be obtuse.   **33.** isosceles or equilateral
35. scalene

37 We know that the sum of the interior angles of a triangle is 180 degrees. Since each of the given angles is a multiple of x, we can add them to find the value of x.
$2x + 3x + x = 180$ Sum of the angles of a triangle
$\qquad\quad 6x = 180$ Combine like terms.
$\qquad\quad\; x = 30$ Divide both sides by 6.
Since $x = 30$, $2(30) = 60$, and $3(30) = 90$. So, the angles measure 30°, 60°, and 90°.
39. 53°, 55°, 72°   **41.** $x = 62$; $y = 40$
43. She incorrectly simplified the equation. $x + 3x + 5x$ simplifies to $9x$, not $8x$.
$x + 3x + 5x = 180$
$\qquad\quad 9x = 180$
$\qquad\quad\; x = 20$
The angles measure 20°, 3(20°) or 60°, and 5(20°) or 100°.   **45.** D   **47.** 66°   **49.** Yes, the two corners at the intersection have measures of 108° and 72°. Therefore, it is within the safety limit.   **51.** 51°
53. $\angle 2$; $\angle S$; $\angle RST$; $\angle TSR$; right

pp. 427–430   Lesson 7-3D

1. rectangle   **3.** parallelogram

5 We know that the sum of the interior angles of a quadrilateral is 360 degrees. Therefore, we can add the angles and solve for the missing angle measure.

$m\angle D + m\angle E + m\angle F + m\angle G = 360°$ Sum of the angles of a quadrilateral

$57° + 78° + m\angle F + 105° = 360°$ Substitution

$240° + m\angle F = 360°$ Simplify.

$m\angle F = 120°$ Subtract 240 from each side.

7. 68   **9.** square   **11.** quadrilateral   **13.** trapezoid
15. 62   **17.** 67   **19.** 90   **21.** 116°   **23.** triangles, quadrilaterals, trapezoid, and rectangle

25 We know that the sum of the angles in a quadrilateral is 360°. Write and solve an equation to find x.

$2x + 2x + 2x + 2x = 360$

$8x = 360$

$x = 45$

27. 80

29 No; a quadrilateral with three right angles will have both pairs of opposite sides parallel. So, it cannot be a trapezoid.

31. Yes; a square is a rhombus and a rectangle.

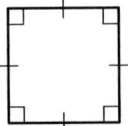

33. Sometimes; sample answer: A quadrilateral can also be a rectangle.   **35.** Always; sample answer: A square has all the properties of a rectangle.
37. Since a square has all the properties of a rectangle and a rhombus, the diagonals of a square must be congruent and perpendicular. Nothing can be concluded about the diagonals of a parallelogram unless more information is provided. If a quadrilateral is a parallelogram, it is not necessarily a rectangle or a rhombus. So, it would not necessarily have the properties of a rectangle or a rhombus.
39. I   **41.** H   **43.** right scalene   **45.** 115°; Sample answer: $\angle 3$ and $\angle 6$ are alternate interior angles.
47. 115°; Sample answer: $\angle 1$ and $\angle 6$ are vertical angles.   **49.** Nina

pp. 433–435   *Lesson 7-3E*

1. The figure is not a polygon because it is an open figure.   **3.** The figure has ten sides that intersect only at their endpoints. It is a decagon.

5 Use the equation for the Interior Angle Sum of a Polygon to find the interior angle measure for the nonagon.

$S = (n - 2)180$ Interior Angle Sum of a Polygon

$S = (9 - 2)180$ Substitution

$S = 1,260$ Simplify.

7. 60°   **9.** The figure is not a polygon because it is an open figure.   **11.** The figure is not a polygon because all the segments do not intersect at their endpoints.
13. The figure has four sides that intersect only at their endpoints. It is a quadrilateral.   **15.** 900°
17. 2,160°   **19.** 3,960°   **21.** 140°

23 To find the measure of one interior angle of a 13-gon, we have to find the interior angle sum and then divide by the number of angles.

$S = (n - 2)180$ Interior Angle Sum of a Polygon

$S = (13 - 2)180$ Substitution

$S = 1,980$ Simplify.

$\dfrac{1,980}{13} \approx 152.3$ Divide the sum by the number of angles to determine the measure of the individual angle.

25. 90°, 120°, 150°, 360°

27 The sum of the measures of the interior angles of a polygon is $(n - 2)180$, where n represents the number of sides. An equilateral triangle has 3 sides. Replace n with 3 and simplify.

$S = (n - 2)180$

$S = (3 - 2)180$

$S = (1)180$ or 180

The sum of the measures of the interior angles of an equilateral triangle is 180°. Divide 180 by 3, the number of interior angles, to determine the measure of one interior angle of the equilateral triangle. So, the measure of one interior angle of an equilateral triangle is 180° ÷ 3 or 60°.
The measure of each angle in each outlined triangle is 60°. If a triangle is equilateral, the measure of each angle will be 60° regardless of the size of the triangle.
29a. Sample answer: The sum of the interior angles will still be 720° because even though the figures are not regular, they are still hexagons.   **29b.** 720°
31. D   **33.** H   **35.** isosceles

pp. 445–447   *Lesson 7-4B*

1. rectangular prism; bases: *ABCD* and *EFGH*, *ABFE* and *DCGH*, *ADHE* and *BCGF*; faces: *ABCD*, *EFGH*, *ABFE*, *DCGH*, *ADHE*, *BCGF*; edges: $\overline{AB}$, $\overline{BC}$, $\overline{CD}$, $\overline{AD}$, $\overline{EF}$, $\overline{FG}$, $\overline{GH}$, $\overline{EH}$, $\overline{AE}$, $\overline{BF}$, $\overline{CG}$, $\overline{DH}$; vertices: *A, B, C, D, E, F, G, H*
3. cylinder; bases: *A* and *B*; faces: none; edges: none; vertices: none   **5.** circle   **7.** parallelogram

9 The figure has one base that is a hexagon, so it is a hexagonal pyramid.

base: *MNOPQR*

faces: *LMN, LNO, LOP, LPQ, LQR, LRM*

edges: $\overline{LM}$, $\overline{LN}$, $\overline{LO}$, $\overline{LP}$, $\overline{LQ}$, $\overline{LR}$, $\overline{MN}$, $\overline{NO}$, $\overline{OP}$, $\overline{PQ}$, $\overline{RM}$

vertices: *L, M, N, O, P, Q, R*

11.

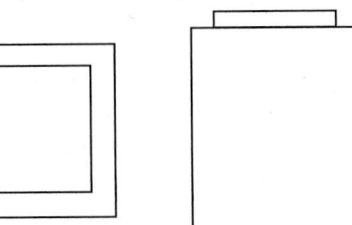

top view   front and side views

13. rectangle **15.** triangle

17 The shape of the cross-section of the cone would be a curve. Think of a snow cone cup and how a piece of paper would have to be shaped to rest against the edge of the cup.

19. False; two planes intersect at a line, which is an infinite number of points. **23.** Sometimes; a rectangular prism has 2 bases and 4 faces, but a triangular prism has 2 bases and 3 faces.

25. Sample answer: a top-front-side view diagram is not always sufficient to draw a figure. It is possible to draw several figures for the same top-front-side view diagram.

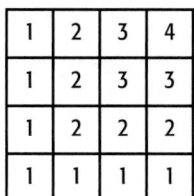

top   front   side

The following top-count views can all represent the figure.

1	2	3	4
1	1	1	3
1	1	1	2
1	1	1	1

1	2	3	4
1	2	2	3
1	2	2	2
1	1	1	1

1	2	3	4
1	2	3	3
1	2	2	2
1	1	1	1

27. F **29.** F **31.** 108° **33.** 140° **35.** trapezoid

---

pp. 449–453 Chapter Study Guide and Review

1. true **3.** false; supplementary **5.** false; acute angle **7.** false; parallelogram **9.** ∠1 and ∠4 **11.** 79 **13.** supplementary **15.** 60 **17.** Riley is first, Hazen is second, and Alyca is third. **19.** alternate exterior **21.** alternate interior **23.** 49 **25.** right isosceles **27.** rhombus **29.** trapezoid **31.** 64 **33.** 1,440° **35.** 128.6° **37.** 135° **39.** hexagonal pyramid; base: $MNOPQR$; faces: LMN, LNO, LOP, LPQ, LQR, LMR; edges: $\overline{LM}$, $\overline{LN}$, $\overline{LO}$, $\overline{LP}$, $\overline{LQ}$, $\overline{LR}$, $\overline{MN}$, $\overline{NO}$, $\overline{OP}$, $\overline{PQ}$, $\overline{QR}$, $\overline{RM}$; vertices: L, M, N, O, P, Q, R **41.** triangle

Chapter 8 Triangles and Transformations

p. 460 Are You Ready?

1. 20 **3.** 89 **5.** 95 **7.** 64 **9.** 394

10–17.

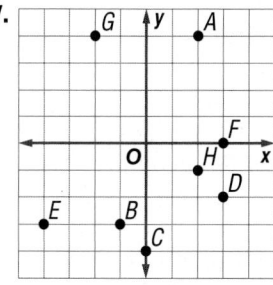

pp. 462–463 Lesson 8-1A PSI

1. Sample answer: You could act it out by placing counters around a triangle.

3 Mrs. Rogers has driven 45 miles of her trip. Draw a diagram to represent the total distance of her trip.

Total distance of the trip

← 45 miles →	← remaining distance →	
← $\frac{5}{6}$ of the trip →	← $\frac{1}{6}$ of the trip →	

$45 \div \frac{5}{6} = 54$ miles for the total trip.

$54 - 45$ is 9 miles remaining.

5. 55 baseballs **7.** $95,310 **9.** 7.5 gal **11.** 2 quarters, 3 dimes, and 4 nickels **13.** 18 ft, 27 ft, 36 ft

pp. 467–469 Lesson 8-1B

1. No; $\frac{5}{3} \neq \frac{13}{5}$. **3.** $GH = 12$; $KL = 4.5$ **5.** No; The corresponding angles are congruent, but $\frac{3}{7} \neq \frac{4}{8}$.

7. Yes; the corresponding angles are congruent and $\frac{20}{15} = \frac{16}{12} = \frac{24}{18}$.

9 Find the scale factor from one parallelogram to the other by setting up the scale factor.

scale factor: $\frac{3}{12}$

$x = \frac{3}{12}(8)$ Write the equation.

$x = 2$ Multiply.

11. 20

13 The actual length is proportional to the length on the drawing with a ratio of $\frac{20\text{ ft}}{4\text{ in.}}$. To find the scale factor, convert feet to inches and divide out units, $\frac{20\text{ ft}}{4\text{ in.}} = \frac{240\text{ in.}}{4\text{ in.}}$ or $\frac{60}{1}$. To find the perimeter of the actual garden, multiply the perimeter of the garden on the drawing by the scale factor. $14 \cdot 60 = 840$ in. or 70 ft

15a. Figure 1: 96 cm^2; Figure 2: 294 cm^2

15b. Sample answer: The scale factor of the side lengths is $\frac{14}{8}$ or $\frac{7}{4}$. The ratio of the areas is $\frac{49}{16}$. The ratio of the areas is the scale factor of the side lengths squared. **17.** Roberto incorrectly found the scale factor. The scale factor should be greater than 1.

scale factor $= \dfrac{15}{12}$ or 1.25

$x = 15(1.25)$ or 18.75

19. Sometimes; counterexample: All corresponding angles are congruent among the three rectangles.

Rectangle A is similar to Rectangle C, since $\dfrac{6}{9} = \dfrac{4}{6}$,

but Rectangle A is not similar to Rectangle B, since $\dfrac{4}{4} \neq \dfrac{6}{12}$.

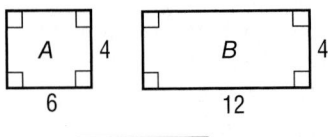

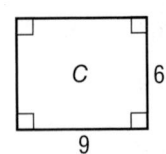

21. A **23.** D **25.** 24 in^2

---

pp. 472–474 Lesson 8-1D

1 Set up a proportion comparing the base and height of each triangle.

$\dfrac{0.45}{0.3} = \dfrac{h}{2.2}$ Write the proportion.

$0.45 \cdot 2.2 = 0.3h$ Find the cross products.

$0.99 = 0.3h$ Multiply.

$\dfrac{0.99}{0.3} = \dfrac{0.3h}{0.3}$ Divide each side by 0.3.

$3.3 = h$

So, the height of the tree is 3.3 meters.

3. 200 ft

5 The distance from the Ferris wheel to the pirate ship corresponds to the distance from the tent to the pirate ship. The distance from the rollercoaster to the pirate ship corresponds to the distance from the log ride to the pirate ship. Set up a proportion to find the missing length.

$$\dfrac{\text{distance from Ferris wheel to pirate ship}}{\text{distance from tent to pirate ship}} =$$
$$\dfrac{\text{distance from roller coaster to pirate ship}}{\text{distance from the log ride to pirate ship}}$$

Let x represent the distance from the log ride to the pirate ship.

$\dfrac{8}{25} = \dfrac{12}{x}$ Write the proportion.

$8x = 25 \cdot 12$ Find the cross products.

$8x = 300$ Multiply.

$\dfrac{8x}{8} = \dfrac{300}{8}$ Divide each side by 8.

$x = 37.5$

The log ride is 37.5 meters from the pirate ship.

7. 4.2 ft

---

9 Draw a picture to help you set up the proportion. Simplify the picture into two similar triangles.

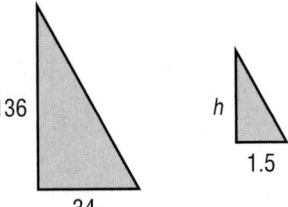

$\dfrac{136}{34} = \dfrac{h}{1.5}$ Write the proportion.

$34h = 204$ Multiply.

$h = 6$

So, the man is 6 feet tall.

11. Sample answer: At the same time a baby giraffe casts a 3.2-foot shadow, a 15-foot adult giraffe casts an 8-foot shadow. How tall is the baby giraffe? Solving the proportion $\dfrac{x}{3.2} = \dfrac{15}{8}$ gives a height of 6 ft for the baby giraffe. **13.** Sample answer: The length of the tall object's shadow, the length of the shadow of a nearby object with a height that is directly measurable, and the height of the nearby object. **15.** H **17.** No; the corresponding angles are not congruent and $\dfrac{3}{5} \neq \dfrac{5}{9}$.

---

pp. 477–479 Lesson 8-1E

1. $\tan B = 2.33$; $\tan C = 0.43$; The side opposite $\angle B$ is 2.33 times as long as the side adjacent to $\angle B$. The side opposite $\angle C$ is 0.43 times as long as the side adjacent to $\angle C$. **3.** $\tan S = 0.8$; $\tan T = 1.25$; The side opposite $\angle S$ is 0.8 times as long as the side adjacent to $\angle S$. The side opposite $\angle T$ is 1.25 times as long as the side adjacent to $\angle T$. **5.** $\tan X = 1.67$; $\tan Z = 0.6$; The side opposite $\angle X$ is 1.67 times as long as the side adjacent to $\angle X$. The side opposite $\angle Z$ is 0.6 times as long as the side adjacent to $\angle Z$.

7 The tangent of an angle is the ratio of the measure of the leg opposite the angle and the measure of the leg adjacent to the angle.

$\tan X = \dfrac{WY}{XY}$ Write the ratio.

$\tan X = \dfrac{7}{5} = 1.4$ Substitution

The side opposite $\angle X$ is 1.4 times as long as the side adjacent to $\angle X$.

$\tan W = \dfrac{XY}{WY}$ Write the ratio.

$\tan W = \dfrac{5}{7} \approx 0.71$ Substitution

The side opposite $\angle W$ is 0.71 times as long as the side adjacent to $\angle W$.

9. $\tan J = 1.44$; $\tan L = 0.69$; The side opposite $\angle J$ is 1.44 times as long as the side adjacent to $\angle J$. The side opposite $\angle L$ is 0.69 times as long as the side

---

adjacent to $\angle L$.   **11.** 13,826.17 ft   **13.** 2.01 ft

15 Since you are being asked to find the inverse tangent of the ratio, set up each ratio and then find the inverse tangent using your calculator.

$\tan P = \frac{8}{6}$ $\tan R = \frac{6}{8}$

$\angle P = \tan^{-1}\left(\frac{8}{6}\right) = 53$ $\angle R = \tan^{-1}\left(\frac{6}{8}\right) = 37$

Therefore, $\angle P = 53°$ and $\angle R = 37°$.

17a.

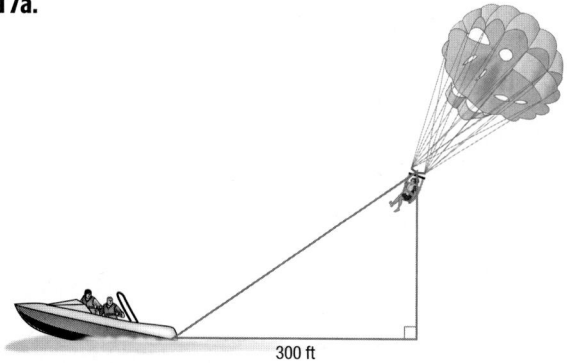

300 ft

17b. 210.06 ft   **19.** 15.2   **21.** B   **23.** 62.5 ft
25. 30 ft up the wall

pp. 484–486   Lesson 8-2B

1 The Pythagorean Theorem tells us that the sum of the square of the lengths of the two sides of a right triangle gives us the square of the length of the hypotenuse. So, we need to find the sum of the squares of the two given sides ($12^2 + 16^2$) and set it equal to the square of the missing side (c^2). Write an equation and solve for c.

$12^2 + 16^2 = c^2$ Write the equation.
$144 + 256 = c^2$ $12^2 = 144; 16^2 = 256$
$400 = c^2$ Add.
$\pm 20 = c$ Take the square root of each side.

Since length cannot be negative, take the positive square root. The length of the hypotenuse is 20 m.
3. $7^2 + b^2 = 12^2$; about 9.7 in.   **5.** yes; $9^2 + 40^2 = 41^2$
7. $5^2 + 12^2 = c^2$; 13 in.   **9.** $a^2 + 51^2 = 60^2$; 31.6 yd
11. $8^2 + b^2 = 18^2$; 16.1 m   **13.** no; $30^2 + 122^2 \neq 125^2$
15. no; $135^2 + 140^2 \neq 175^2$ **17.** no; $44^2 + 55^2 \neq 70^2$

19 Use the Pythagorean Theorem.

$a^2 + b^2 = c^2$ Pythagorean Theorem
$365^2 + 275^2 = c^2$ Replace a with 365 and
b with 275.
$133,225 + 75,625 = c^2$ Evaluate 365^2 and 275^2.
$208,850 = c^2$ Add 133,225 and 75,625.
$\pm\sqrt{208,850} = c$ Definition of square root
$c \approx 457.0$ or -457.0 Simplify. Use a calculator.

The equation has two solutions, approximately 457.0 and -457.0. However, the length of a side must be positive. So, the diagonal of the state of Wyoming is about 457.0 miles.

21 The Pythagorean Theorem tells us that the sum of the square of the lengths of the two sides of a right triangle gives us the square of the length of the hypotenuse. So, we need to find the sum of the squares of the two given sides ($48^2 + 55^2$) and set it equal to the square of the missing side (c^2). Write out this equation and solve for c.

$48^2 + 55^2 = c^2$ Write the equation.
$2,304 + 3,025 = c^2$ $48^2 = 2,304; 55^2 = 3,025$
$5,329 = c^2$ Add.
$\pm 73 \approx c$ Take the square root of each side.

Since length cannot be negative, take the positive square root. The length of the hypotenuse is 73 yd.
23. $23^2 + 18^2 = c^2$; 29.2 in.   **25.** $a^2 + 5.1^2 = 12.3^2$; 11.2 m   **27.** Sample answer: 3, 4, 5; $3^2 + 4^2 = 5^2$; $9 + 16 = 25$; $25 = 25$ **29.** Sample answer: 6, 8, 10; 5, 12, 13; 10, 24, 26   **31.** C   **33.** $\tan A = 2.25$; $\tan C = 0.44$; The side opposite $\angle A$ is 2.25 times as long as the side adjacent to $\angle A$. The side opposite $\angle C$ is 0.44 times as long as the side adjacent to $\angle C$.
35. 18 ft

pp. 489–492   Lesson 8-2C

1. $3^2 + h^2 = 5^2$; 4 ft   **3.** about 5.7 in.

5 We are given one side and the hypotenuse. Substitute the values into the Pythagorean Theorem and solve for h, the missing side.

$5^2 + h^2 = 12^2$
$25 + h^2 = 144$ $5^2 = 25; 12^2 = 144$
$\underline{-25 \qquad = -25}$ Subtraction Property of Equality
$h^2 = 119$ Take the square root of each side.
$h \approx 10.9$

The cat is 10.9 feet up the tree.
7. $70^2 + 20^2 = x^2$; 72.8 feet   **9.** 2 blocks

11 We assume that the box is a rectangle with right angle corners. Then we need to determine the diagonal length of the box. Using the Pythagorean Theorem, we can find that length.

$4^2 + 4^2 = c^2$ Write the equation.
$16 + 16 = c^2$ $4^2 = 16$
$32 = c^2$ Add.
$5.66 \approx c$ Take the square root of each side.

The diagonal of the box is approximately 5.66 feet long, which is 5 feet 8 inches. The fishing pole is 5.5 feet long. There will actually be about 2 inches extra on the diagonal.
13a. They form a right triangle.   **13b.** 316.2 ft

15 Within the three-dimensional figure, there is a triangle with the measure of two legs given and the hypotenuse marked with an x. Use the Pythagorean Theorem to determine the value of x.
$x^2 = 11^2 + 4^2$ Write the equation.
$x^2 = 121 + 16$ or 137   Perform the operations.
$x \approx 11.7$ Take the square root of each side.
The height of the pyramid is about 11.7 cm.
17. 13.9 mm   **21.** 3–5–7; $3^2 + 5^2 \neq 7^2$
23. Use the Pythagorean Theorem to set up the equation $a^2 + b^2 = (\sqrt{288})^2$. Since the legs in an isosceles triangle are equal, you can substitute a for b and solve for a.

$$a^2 + a^2 = (\sqrt{288})^2$$
$$2a^2 = 288$$
$$a^2 = 144$$
$$a = 12$$

The length of the legs are 12 units.

25. H   **27a.** $5^2 + 12^2 = x^2$ **27b.** 13 in.

27c. Sample answer: 10 in., 24 in., and 26 in.

29. 8.4 in.

pp. 496–498   Lesson 8-2D

1. 4.5 units

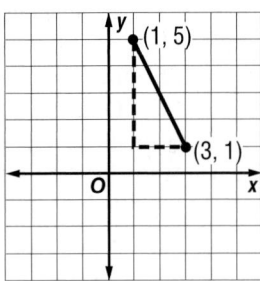

3. 9.4 units

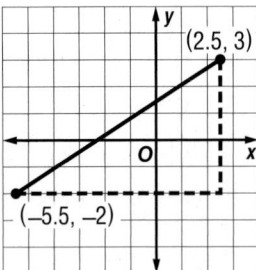

5. about 1.0 mi

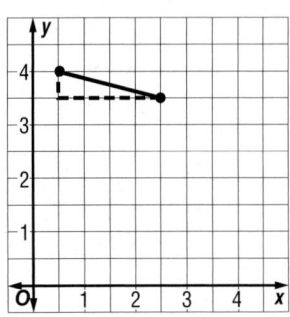

7. 7.6 units

9. 3.6 units

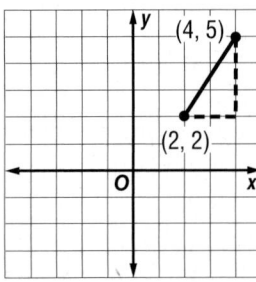

11. 4.1 units

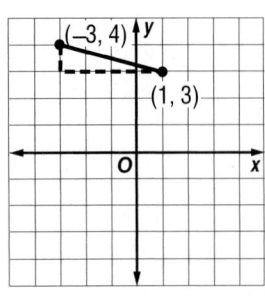

13. 7.2 units

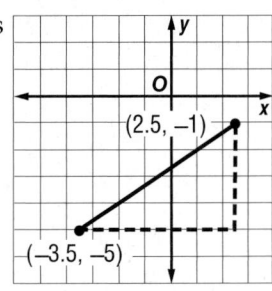

15 After reading through the exercise, we determine that we need to find the distance between the two given points, (4, 12) and (6, 2).

$d = \sqrt{(x_2 - x_1)^2 + (y_2 - y_1)^2}$ Distance Formula

$d = \sqrt{(6 - 4)^2 + (2 - 12)^2}$ Substitution. $(x_1, y_1) =$ (4, 12), $(x_2, y_2) = $ (6, 2)

$d = \sqrt{(2)^2 + (-10)^2}$ Simplify.

$d = \sqrt{4 + 100}$ Evaluate 2^2 and $(-10)^2$.

$d = \sqrt{104}$ Add 4 and 100.

$d \approx 10.2$ Use a calculator.

The distance between the island and Ferry Landing *B* is about 10.2 units. The grid is 0.5 mile per unit. Multiply the number of units by the miles per unit to get the miles. $10.2 \cdot 0.5 = 5.1$; The distance between the island and Ferry Landing *B* is about 5.1 miles.

17. 1.4 units   **19.** 7.6 units   **21.** 15.9 units

23 To find the perimeter and the area, we need to find the length of each side. Since *ABCD* is a rectangle, we only need to find the length of *AB* and *BC*. Then we can determine the area as $AB \cdot BC$ and the perimeter as $2AB + 2BC$. First determine the locations of *A, B, C*: $A(3, 8), B(5, 4), C(-4, -1)$.

$AB = \sqrt{(5 - 3)^2 + (4 - 8)^2}$ Set up the equation.

$AB = \sqrt{20}$ Simplify by performing operations.

$AB \approx 4.5$ Find the square root.

The length of side *AB* is 4.5 units.

$BC = \sqrt{(-4 - 5)^2 + (-1 - 4)^2}$ Set up the equation.

$BC = \sqrt{106}$ Simplify by performing operations.

$BC \approx 10.3$ Find the square root.

The length of side *BC* is 10.3 units.

Area of $ABCD = AB \cdot BC = 4.5 \cdot 10.3 = 46.4$ units^2

Perimeter of $ABCD = 2AB + 2BC = 2(4.5) + 2(10.3) = 29.6$ units

25. about 150 mi   **27a.** The sum of the areas of the two smaller squares equals the area of the larger square.   **27b.** about 4.5 units   **29.** Sample answer: (1, 2) and (4, 6)   **31.** D   **33.** 4.5 spaces   **35.** 9.6 in.

37. 37.5 ft

pp. 503–505   Lesson 8-2F

1. $40\sqrt{2}$ in.   **3.** $x = 4.5$ m, $y = 4.5\sqrt{3}$ m

5. $30\sqrt{2}$ cm   **7.** $x = 8$ ft, $y = 6.9$ ft

9 In a 45°-45°-90° triangle, the length of the hypotenuse is $\sqrt{2}$ times the length of a leg. The triangle given is 8 feet on each side. Therefore, the length of the hypotenuse is $8\sqrt{2}$ feet.
11. $x = 14$ cm, $y = 7\sqrt{3}$ cm **13.** $220\sqrt{2}$ cm
15. shorter leg: 34 cm, longer leg: $34\sqrt{3}$ cm **17.** 14 ft

19 We are given a square with an area of 1,600 square feet. This means that the square is 40 feet on each side. The diagonal of the floor mat would be the hypotenuse of a triangle formed when the diagonal is drawn. This triangle would be a 45°-45°-90° triangle. We know that the hypotenuse is $\sqrt{2}$ times the length of a leg. Therefore, the measure of the hypotenuse is $40\sqrt{2}$ feet.
21. 18 cm **23.** 6 m **25.** 52.4 cm^2 **27.** sometimes; Sample answer: In order for a 30°-60°-90° triangle to be formed, the diagonal of the rectangle must be twice its width. **29.** B **31.** G
33. 5 units

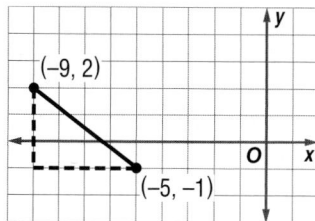

35. 8 ft **37.** tan $X = 3.43$; tan $Z = 0.29$; The side opposite $\angle X$ is 3.43 times as long as the side adjacent to $\angle X$. The side opposite $\angle Z$ is 0.75 times as long as the side adjacent to $\angle Z$.

pp. 510–511 Lesson 8-3A

1 Graph points $X(-4, -4)$, $Y(-3, -1)$ and $Z(2, -2)$. The translation is 3 units to the right and 4 units up. Start with X. Count 3 units to the right then 4 units up.

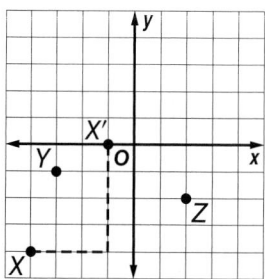

Repeat with points Y and Z. The image has coordinates $X'(-1, 0)$, $Y'(0, 3)$ and $Z'(5, 2)$.

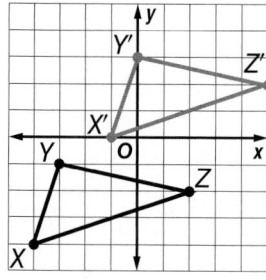

3. $(x + 2, y + 5)$
5. $J'(-2, -2)$, $K'(4, 1)$, $L'(5, -1)$, $M'(-1, -4)$

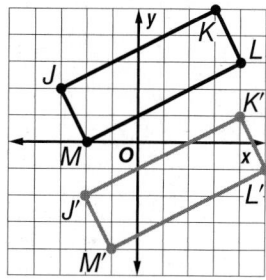

7. $(x - 2, y - 4)$

9 First translate $KLMN$ by the first translation $(x + 2, y - 1)$. Add 2 to each x-coordinate and -1 to each y-coordinate.

$K(-2, -2) \rightarrow (-2 + 2, -2 + (-1))$ or $(0, -3)$
$L(1, 1) \rightarrow (1 + 2, 1 + (-1))$ or $(3, 0)$
$M(0, 4) \rightarrow (0 + 2, 4 + (-1))$ or $(2, 3)$
$N(-3, 5) \rightarrow (-3 + 2, 5 + (-1))$ or $(-1, 4)$

Next translate $K'L'M'N'$ by the next translation $(x - 3, y + 4)$. Add -3 to each x-coordinate and 4 to each y-coordinate.

$K'(0, -3) \rightarrow (0 +(-3), -3 + 4)$ or $(-3, 1)$
$L'(3, 0) \rightarrow (3 + (-3), 0 + 4)$ or $(0, 4)$
$M'(2, 3) \rightarrow (2 + (-3), 3 + 4)$ or $(-1, 7)$
$N'(-1, 4) \rightarrow (-1 + (-3), 4 + 4)$ or $(-4, 8)$

The coordinates after both translations are $K''(-3, 1)$, $L''(0, 4)$, $M''(-1, 7)$, $N''(-4, 8)$.
11. The final position of the figure is the same as the original position of the figure; Sample answer: translating a figure by $(x - 5, y + 7)$ means that the figure is translated 5 units left and 7 units up. Translating this image by $(x + 5, y - 7)$ means that the figure is translated 5 units right and 7 units down, reversing the effects of the first translation. The final image is in the same position as the original figure. **15.** H **17.** $x = 5\sqrt{3}$ cm; $y = 10$ cm

19. 8.6 units

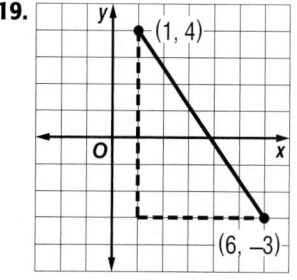

21. 4.5 units

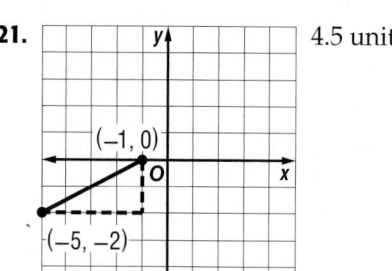

pp. 514–516   **Lesson 8-3B**

1.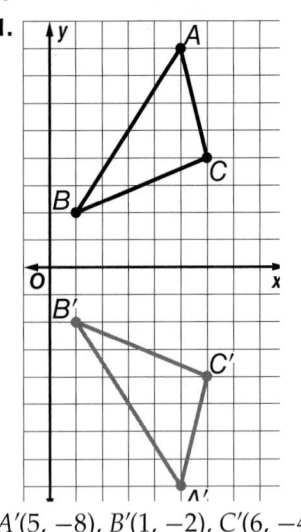

$A'(5, -8)$, $B'(1, -2)$, $C'(6, -4)$

3.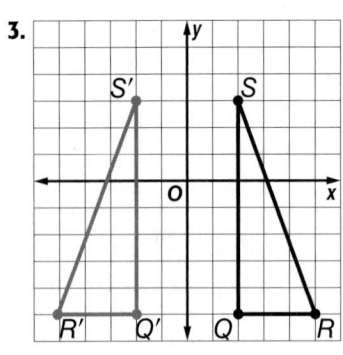

$Q'(-2, -5)$, $R'(-5, -5)$, $S'(-2, 3)$

5.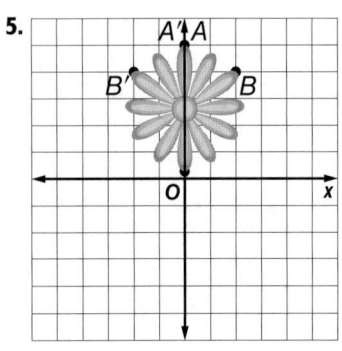

$A'(0, 5)$, $B'(-2, 4)$

7.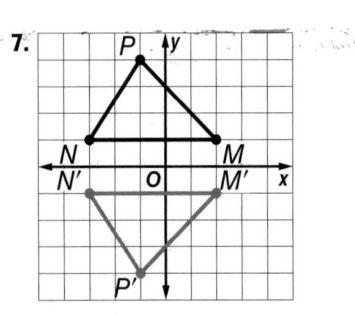

$M'(2, -1)$, $N'(-3, -1)$, $P'(-1, -4)$

9 First graph $WXYZ$.

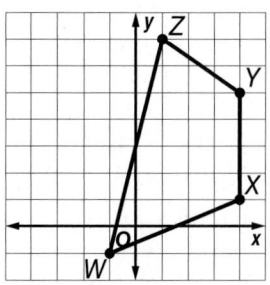

Then plot each vertex of $W'X'Y'Z'$ the same distance from the x-axis as its corresponding vertex on $WXYZ$.

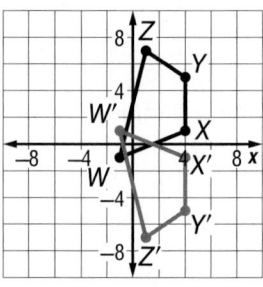

$W'(-1, 1)$, $X'(4, -1)$, $Y'(4, -5)$, $Z'(1, -7)$

11.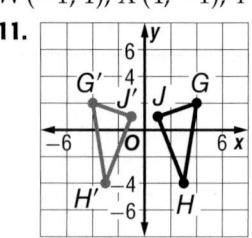

$G'(-4, 2)$, $H'(-3, -4)$, $J'(-1, 1)$

13.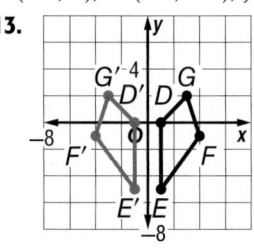

$D'(-1, 0)$, $E'(-1, -5)$, $F'(-4, -1)$, $G'(-3, 2)$

15. $A'(-3, -3)$, $B'(3, -3)$

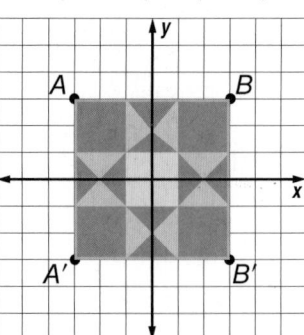

17 In the reflection $M(3, 3) \rightarrow M'(3, -3)$, the x-coordinate stays the same and the y-coordinate changes from 3 to -3. This indicates a reflection over the x-axis.

19. *y*-axis   **21.** *J*′(7, −4), *K*′(−7, −1), *L*′(−2, 2)   **23.** C
25. yes   **27.** no symmetry

pp. 520–522   *Lesson 8-3D*

1.

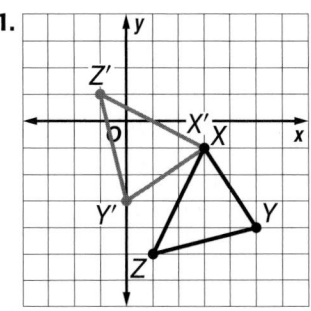

X′(3, −1), *Y*′(0, −3), *Z*′(−1, 1)

3.

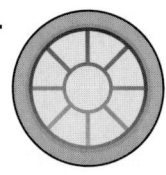

5.

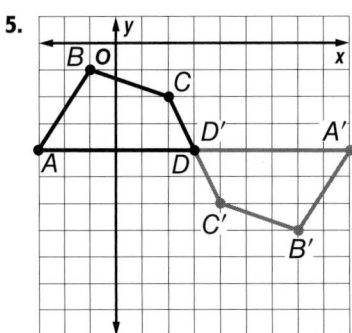

A′(9, −4), *B*′(7, −7), *C*′(4, −6), *D*′(3, −4)

7. Graph △*RST* on a coordinate plane. Sketch segment $\overline{TO}$ connecting point *T* to the origin. Sketch another segment, $\overline{T'O}$ so that the angle between point *T*, *O*, and *T*′ measures 180°, or is a straight line. The segment must be congruent to $\overline{TO}$.

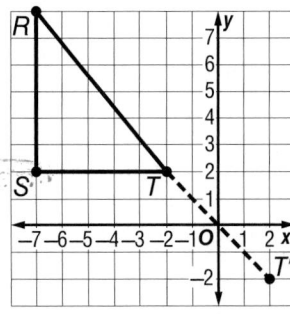

Repeat Step 2 for points *R* and *S*. Then connect the vertices to form △*R*′*S*′*T*′. So, the points of the image after the rotation are *R*′(7, −8), *S*′(7, −2), and *T*′(2, −2).

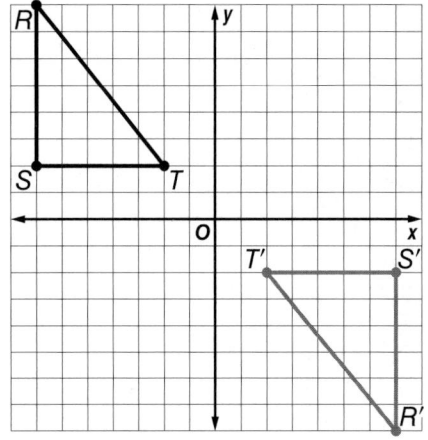

9.

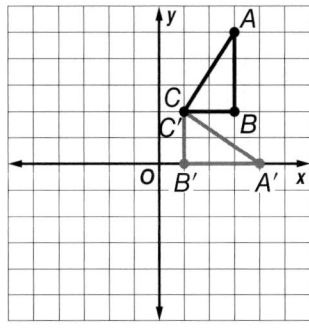

6 slices

11. Rotate the word VIRGINIA 180° or $\frac{1}{2}$ of a full rotation: VINIϽЯIΛ. The letters I and N appear the same as the original.

13. Sample answer:

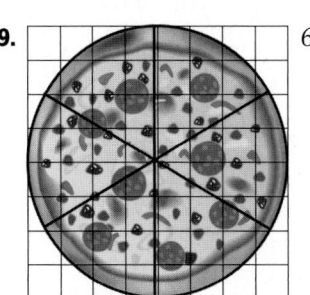

15. *Q*(−6, 6), *R*(−6, 0), *S*(0, 0)   **17.** B   **19a.** *A*′(−2, 2), *B*′(−1, −2), *C*′(1, 0)

19b.

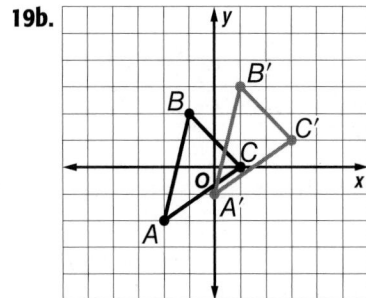

pp. 525–526   Lesson 8-3E

1. $A'(6, 10)$, $B'(0, 8)$, $C'(-4, -4)$

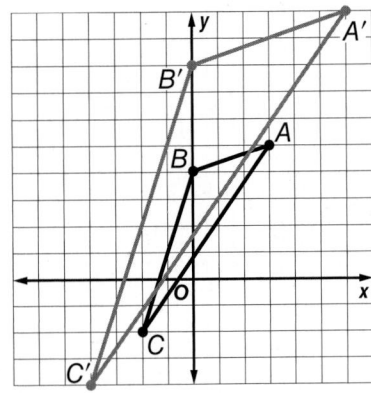

3 Write a ratio comparing the sizes of the photograph.

$$\frac{\text{width in reduction}}{\text{width in original}} = \frac{320}{480} \text{ or } \frac{2}{3}$$

$$\frac{\text{length in reduction}}{\text{length in original}} = \frac{720}{1,080} \text{ or } \frac{2}{3}$$

So, the scale factor of the reduction is $\frac{2}{3}$.

5. $V'(-9, 12)$, $X'(-6, 0)$, $W'(3, 6)$

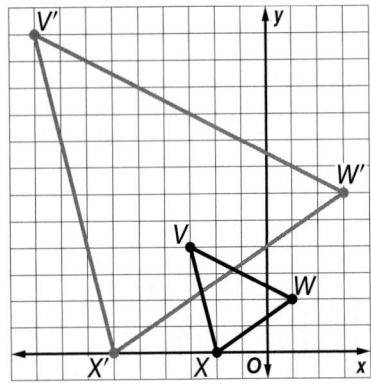

7. $R'(2, 2)$, $S'(2, 4)$, $T'(4, 4)$, $U'(4, 2)$

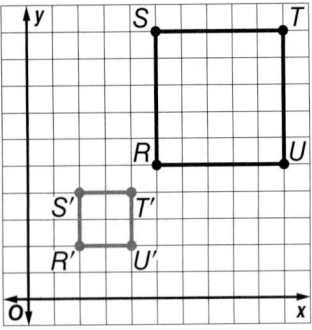

9. 3

11 a. The triangle is reflected over the x-axis so the x-coordinate stays the same and the y-coordinate changes to its opposite.
$A(-2, 3) \rightarrow A'(-2, -3)$
$B(0, 0) \rightarrow B'(0, 0)$
$C(1, 1) \rightarrow C'(1, -1)$
When you dilate that image by a scale factor of 3, multiply the x- and y-coordinates by 3.

$A'(-2, -3) \rightarrow A''(-6, -9)$
$B'(0, 0) \rightarrow B''(0, 0)$
$C'(1, -1) \rightarrow C''(3, -3)$
b. Dilate ABC by a scale factor of 3. Multiply each x- and y-coordinate by 3.
$A(-2, 3) \rightarrow A'(-6, 9)$
$B(0, 0) \rightarrow B'(0, 0)$
$C(1, 1) \rightarrow C'(3, 3)$
When the triangle is reflected over the x-axis, the x-coordinate stays the same and the y-coordinate changes to its opposite.
$A'(-6, 9) \rightarrow A''(-6, -9)$
$B'(0, 0) \rightarrow B''(0, 0)$
$C'(3, 3) \rightarrow C''(3, -3)$
c. The two transformations are commutative since the coordinates of the answers to Exercises a and b are the same. The order in which you perform them does not matter.

13. No; Sample answer: both coordinates of all the points must be multiplied by the same scale factor. The x-coordinates are multiplied by 4, but the y-coordinates are only multiplied by 2.   **15.** C

17. H: 180°; I: 180°; N: 180°; O: 180°; S: 180°; X: 180°; Z: 180°   **19.** 4.3 m

pp. 529–533   *Chapter Study Guide and Review*

1. is   **3.** longest   **5.** Indirect measurement
7. $\frac{3}{4}$ ft   **9.** $\frac{49}{64}$ **11.** about 26 ft   **13.** $\tan T = 1.79$; $\tan U = 0.56$ **15.** $5^2 + b^2 = 6^2$; 3.3 in.   **17.** 15 in.
19. $25^2 + x^2 = 47^2$; 79.6 cm
21. 7.8 units

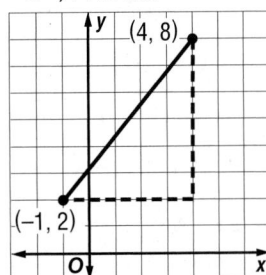

23. 3.6 units

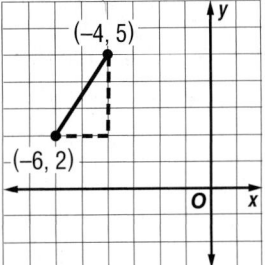

25. 3.2 units

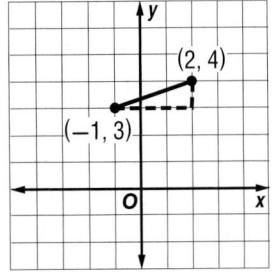

27. $12\sqrt{2}$ ft   **29.** longer leg: $12\sqrt{3}$ in., hypotenuse: 24 in.

31.

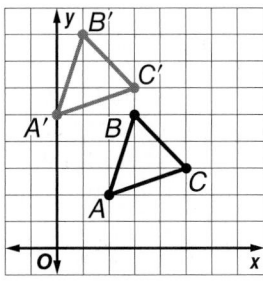

$A'(0, 5)$, $B'(1, 8)$, $C'(3, 6)$

33.

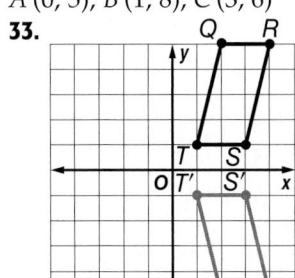

$Q'(2, -5)$, $R'(4, -5)$, $S'(3, -1)$, $T'(1, -1)$

35.

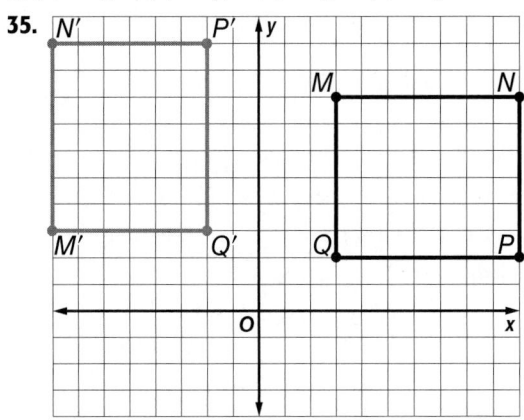

$M'(-8, 3)$, $N'(-8, 10)$, $P'(-2, 10)$, $Q'(-2, 3)$
37. $A'(-6, -9)$, $B'(-6, 9)$, $C'(9, -9)$

Chapter 9 Units of Measure

p. 540   Are You Ready?

1. 13   **3.** -1 **5.** -3 **7.** 14   **9.** 3 touchdowns
11. 2.745   **13.** 15.12   **15.** $23.\overline{3}$ **17.** 57.6

pp. 542–544   Lesson 9-1A

1. $w = \dfrac{V}{\ell h}$

3 Use properties of equality solve for t.

$I = prt$	Write the equation.
$\dfrac{I}{pr} = \dfrac{prt}{pr}$	Division Property of Equality
$\dfrac{I}{pr} = t$	Simplify.

So, $t = \dfrac{I}{pr}$.

5. $s = \sqrt{A}$ **7.** $v = \dfrac{m}{D}$ **9.** $s = \sqrt{\dfrac{S}{6}}$

11 a. | $I = prt$ | Write the equation. |
$I = pt \cdot r$	Commutative Property
$\dfrac{I}{pt} = \dfrac{pt \cdot r}{pt}$	Division Property of Equality
$\dfrac{I}{pt} = r$	Simplify.

So, $r = \dfrac{I}{pt}$.

b. Use the equation from **part a**. Substitute 2,500 for p, 2 for t, and 362.50 for I.

$$r = \frac{I}{pt}$$
$$= \frac{362.50}{2,500 \cdot 2}$$
$$= \frac{362.50}{5,000}$$
$$= 0.0725$$

So, $r = 0.0725$. To turn 0.0725 into a percent, multiply by 100 or move the decimal point two places to the right.
The interest rate is 7.25%.

13 a. Use properties of equality to solve the equation for r.

$V = \dfrac{1}{3}\pi r^2 h$	Write the equation.
$3 \cdot V = \dfrac{1}{3}\pi r^2 h \cdot 3$	Multiplication Property of Equality
$3V = \pi r^2 h$	Simplify.
$\dfrac{3V}{\pi h} = r^2$	Division Property of Equality
$\sqrt{\dfrac{3V}{\pi h}} = \sqrt{r^2}$	Take the square root of each side.
$\sqrt{\dfrac{3V}{\pi h}} = r$	

b.

Bag	Height (in.)	Volume (in³)	Radius (in.)
A	6	6.3	$\sqrt{\dfrac{3 \cdot 6.3}{6\pi}} \approx 1.00$
B	8	33.5	$\sqrt{\dfrac{3 \cdot 33.5}{8\pi}} \approx 1.99$
C	12	113.0	$\sqrt{\dfrac{3 \cdot 113.0}{12\pi}} \approx 2.99$

Bag C; Sample answer: Bag A has a radius of 1 inch, Bag B has a radius of 2 inches, and Bag C has a radius of 3 inches. Since Bag C has the largest radius, it is the widest bag.

15. Sample answer: $\ell = \dfrac{\dfrac{S.A.}{2} - wh}{h + w}$ **17.** C   **19.** A
21. $y = \dfrac{3}{4}x + 4$ **23.** $y = 5x - 4$

pp. 547–549   Lesson 9-1B

1. 50   **3.** 21.2   **5.** -1.11

7 The two temperatures shown are 79°F and 68°F. Use the equation to convert them to degrees Celsius.

$C = \frac{5}{9}(F - 32)$ Write the equation.

$C = \frac{5}{9}(79 - 32)$ Replace F with 79.

$C \approx 26.11$ Simplify.

$C = \frac{5}{9}(68 - 32)$ Replace F with 68.

$C = 20$ Simplify.

So, 79°F = 26.11°C and 68°F = 20°C.

9. 113   **11.** 45

13 Use the formula $F = \frac{9}{5}C + 32$ to convert from degrees Celsius to degrees Fahrenheit.

$F = \frac{9}{5}C + 32$

$F = \frac{9}{5} \cdot 0 + 32$

$F = 0 + 32$ or 32

So, 0°C is the same as 32°F.

15. −26.11   **17.** 61.7

19 Convert 40°C to degrees Fahrenheit.

$F = \frac{9}{5}C + 32$ Write the equation.

$F = \frac{9}{5}(40) + 32$ Replace C with 40.

$F = 72 + 32$ Multiply.

$F = 104$ Add.

A computer's fan turns on when the hard drive reaches a temperature of 104°F.

21. <

23a.

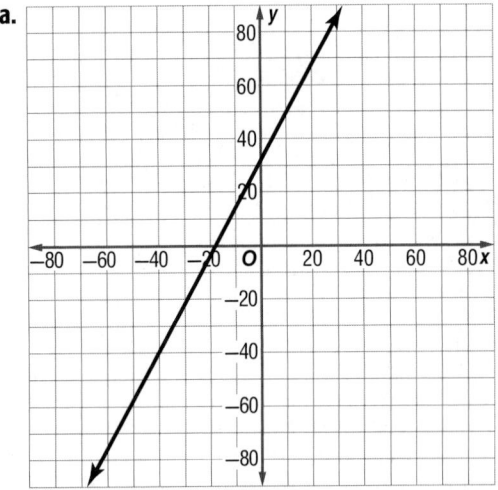

23b. Sample answer: The y-intercept is (0, 32); this means that when it is 0°C, it is 32°F. The x-intercept is about (−18, 0); this means that when it is −18°C, it is 0°F. The slope is $\frac{9}{5}$. This means that for every 9° change on the Fahrenheit scale, the temperature changes 5° on the Celsius scale.   **23c.** about −5°
25. degrees Fahrenheit; Sample answer: 135°C is about 275°F which is well over boiling point. So, 136°F is the reasonable temperature.   **27.** 15°F; Sample answer: Since 15°C is about 59°F, 15°F is the

more reasonable temperature.   **29.** H   **31.** 373.15 K
33. 268.15 K   **35.** 274.26 K   **37.** 253.71 K

pp. 550–551   Lesson 9-1C PSI

1. An exact temperature was not needed.   **3.** $90; $30 is less than 50% of $129, and $60 is about 50% of $129. The sale price must be $90 because it is greater than 50%.   **5.** $12; Sample answer: $0.454 \approx 0.5$, $5 \div 0.5 = 10$. Since the package is about 10 lb, it will cost about 10 × $1.20 or $12 to ship.
7. 1,234,567,654,321

9 Let the money Emmett had at the beginning of the shopping trip be x.

$\frac{3}{4}x$	Emmett bought invitations which cost $\frac{1}{4}$ of the money he had. This leaves $\frac{3}{4}$ of the money.
$\frac{1}{2}\left(\frac{3}{4}x\right)$	Then he bought decorations, which cost $\frac{1}{2}$ of what remained.
$\frac{1}{2}\left(\frac{3}{4}x\right) = 15$	He had $15 left for a cake. This gives us an equation. Now solve for x.
$\frac{3}{8}x = 15$	Simplify.
$x = \$40$	Multiply both sides by $\frac{8}{3}$.

11. about 36,000,000   **13.** 1,875 boxes

pp. 556–558   Lesson 9-2A

1. 80   **3.** 10.5   **5.** 6.86   **7.** 2.5 lb   **9.** 0.63   **11.** 1.69

13 To convert 19 kilograms to pounds, use the conversion factor 1 kg ≈ 2.203 lb.

$19 \text{ kg} \approx 19 \text{ kg} \cdot \dfrac{2.203 \text{ lb}}{1 \text{ kg}}$ Since 1 kg ≈ 2.203 lb, multiply by $\frac{2.203 \text{ lb}}{1 \text{ kg}}$.

$\approx 19 \text{ kg} \cdot \dfrac{2.203 \text{ lb}}{1 \text{ kg}}$ Divide out common units, leaving the desired unit, pounds.

$\approx 19 \text{ kg} \cdot 2.203$ or 41.86 lb   Multiply.

So, 19 kilograms is about 41.86 pounds.
15. 10,560   **17.** 8,250   **19.** 140   **21.** 6   **23.** 4.57
25. 2   **27.** 4.75 lb   **29.** 7.62   **31.** 6.69   **33.** 4.23
35. 1,816   **37.** 18.93   **39.** 3,216.4   **41.** 1,362 kg

43 **a.** Convert 37.18 kilometers to yards.

$\dfrac{37.18 \text{ km}}{1} \cdot \dfrac{1000 \text{ m}}{1 \text{ km}} \cdot \dfrac{1 \text{ yd}}{0.914 \text{ m}} = 40{,}678.34 \text{ yd}$

b. Convert 34.9 cm to inches.

$\dfrac{34.9 \text{ cm}}{1} \cdot \dfrac{0.394 \text{ in.}}{1 \text{ cm}} = 13.75 \text{ in.}$

c. Convert 23,487.5 kilometers to meters.

$\dfrac{23{,}487.5 \text{ km}}{1} \cdot \dfrac{1000 \text{ m}}{1 \text{ km}} = 23{,}487{,}500 \text{ m}$

45. <   **47.** >   **49.** =

51 To be able to order the measurements, they need to have the same units. We will convert all the units to feet.

10 ft = 10 ft

$$\frac{3.1 \text{ m}}{1} \cdot \frac{3.279 \text{ ft}}{1 \text{ m}} = 10.16 \text{ ft}$$

$$\frac{3 \text{ yd}}{1} \cdot \frac{3 \text{ ft}}{1 \text{ yd}} = 9 \text{ ft}$$

$$\frac{300 \text{ cm}}{1} \cdot \frac{0.394 \text{ in.}}{1 \text{ cm}} \cdot \frac{1 \text{ ft}}{12 \text{ in.}} = 9.85 \text{ ft}$$

The order is 9, 9.85, 10, and 10.16. Then list the numbers in their original units: 3 yd, 300 cm, 10 ft, 3.1 m

53. 2,350 oz, 67,000 g, 67.9 kg, 150 lb **55a.** yes; Sample answer: Her luggage weighs 40 pounds, which is about 18.16 kilograms. Since 18.16 < 30, her luggage is under the weight limit. **55b.** no; Sample answer: Before the trip, her luggage weighed 40 pounds or about 18.16 kilograms. 18.16 kg + 13 kg = 31.16 kg. Since 31.16 > 30, her luggage is no longer under the weight limit.
57. Sample answer: To compare the distances, convert one measurement to the other unit of measure. Jesse Owens' rate in the 100-yard dash was faster at about 31.9 ft/s and his rate in the 100-meter dash was about 31.8 ft/s. **59a.** 6,670 km
59b. 255,200 yd **61.** I **63.** 68 **65.** 23 **67.** −22.22
69a. $h = \dfrac{3V}{s^2}$ **69b.** $h = \dfrac{V}{\pi r^2}$ **69c.** cylinder; The height of the pyramid is 7.5 inches, which is too tall. The height of the cylinder is about 5 inches.

pp. 563–564 Lesson 9-2C

1. 73.33 **3.** 559.35

5 Convert 7.5 pounds in 12 minutes to ounces per second.

$$\frac{7.5 \text{ lb}}{12 \text{ min}} \cdot \frac{1 \text{ min}}{60 \text{ s}} \cdot \frac{16 \text{ oz}}{1 \text{ lb}} = 0.17 \text{ oz/s}$$

7. 102.67 **9.** 62.5 **11.** 1.39 **13.** 4.29
15. 2,304.54 mi/h

17 Convert 3 pounds per year to milligrams per day.

$$\frac{3 \text{ lb}}{1 \text{ year}} \cdot \frac{1 \text{ year}}{365 \text{ days}} \cdot \frac{0.454 \text{ kg}}{1 \text{ lb}} \cdot \frac{1000 \text{ g}}{1 \text{ kg}} \cdot \frac{1000 \text{ mg}}{1 \text{ g}} =$$

3,731.51 mg/day

19. 40 kg/h, 25 oz/min, 95 lb/h **23.** No; Sample answer: The speed is only 70.71 miles per hour.
25. B **27a.** Car B: 20.01 mi/gal; Car C: 33.90 mi/gal
27b. Car B, Car A, Car C **29.** 21-oz box; the 21-oz box costs about 27.1¢/oz and the 17-oz box costs about 28.8¢/oz

pp. 567–569 Lesson 9-2D

1. 432 **3.** 116.1

5 Convert 270 square feet to square yards.

$$270 \times \text{ft} \times \text{ft} \times \frac{1 \text{ yd}}{3 \text{ ft}} \times \frac{1 \text{ yd}}{3 \text{ ft}} = 30 \text{ yd}^2$$

7. 929.03 **9.** 7.93 **11.** 14.4 **13.** 28,000 **15.** 3,096 in²

17 Convert 5 cubic yards to cubic feet to determine how much stone is needed.

$$5 \times \text{yd} \times \text{yd} \times \text{yd} \times \frac{3 \text{ ft}}{1 \text{ yd}} \times \frac{3 \text{ ft}}{1 \text{ yd}} \times \frac{3 \text{ ft}}{1 \text{ yd}} = 135 \text{ ft}^3$$

No, 100 cubic feet would not be enough stone. 135 ft³ is needed.
19. 1,548.38 **21.** 0.28 **23.** 0.5 cm³

25 Use dimensional analysis and the conversion factor 0.305 m = 1 ft.

$$400 \text{ ft}^2 \approx 400 \times \text{ft} \times \text{ft} \times \frac{0.305 \text{ m}}{1 \text{ ft}} \times \frac{0.305 \text{ m}}{1 \text{ ft}}$$

$$\approx 37.21 \text{ m}^2$$

One gallon of paint can cover 400 ft² of wall. This is equivalent to approximately 37.21 m².
27. 11.5 in. **29.** 1,667.75 **31.** 2,669.85 **33.** about 13 acres **35.** C **37.** 2884.1 **39.** $7\frac{1}{3}$ **41.** 600,000

pp. 572–575 Chapter Study Guide and Review

1. derived unit **3.** literal equation **5.** literal equation **7.** $m = Dv$ **9.** $b_1 = \dfrac{2A}{h} - b_2$

11a. $r = \sqrt{\dfrac{V}{\pi h}}$ **11b.** 1.0 ft **13.** 5 **15.** −6.67

17. 38.78 **19.** 84 consumers; Sample answer: 10% of 1,413 is about 140, and 6% is a little more than half of 140 or 70. Since 84 is slightly greater than 70, it is a reasonable answer. **21.** 330°F; Sample answer: $\frac{9}{5} \approx 2$, 160°C ≈ 150°C, 32 ≈ 30. 2 × 150 + 30 = 300 + 30 or 330 **23.** 18,480 **25.** 26,000
27. 15.54 **29.** 850.5 **31.** 9.46 **33.** 4-mile race
35. 56.97 **37.** 1,743.08 **39.** 23.07 mi/h **41.** 3,240
43. 172.0 **45.** 7.32 **47.** 34,884.38 m²

Chapter 10 Data Analysis and Statistics

p. 582 Are You Ready?

1.

Life Spans of Certain Mammals (Years)

Stem	Leaf
0	9
1	0 1 2
2	0 4 5
3	5 1|2 = 12 years

3. 16 **5.** 16.5

pp. 585–587 Lesson 10-1A

1. 9; 9; no mode **3.** 12; 9; 8

5 Find the mean, median, and mode of the data. Mean is the sum of the data divided by the number of data items; $\dfrac{1,222}{29} = 42.1$. Median means the middle number when the data are ordered from least to greatest. Since there are 29 data points, the median is the 15th number, which is 40. The mode is the most common number in the list, which is 50.

7. The mean, 103.1, median, 100, and mode, 100, equally represent the data as most of the data is near 100.

9 Gregory needs an 85% average to go on the field trip. His current scores are given. If x is the test score he has yet to take, set up the equation that represents the situation and solve for x.

$$\frac{94 + 82 + 78 + 80 + x}{5} = 85 \quad \text{Average equation}$$

$$\frac{334 + x}{5} = 85 \quad \text{Simplify.}$$

$$334 + x = 425 \quad \text{Multiply each side by 5.}$$

$$x = 91 \quad \text{Subtract 334 from both sides.}$$

Gregory must receive at least a 91% on his 5th test to be able to attend the class trip.
11. Sample answer: 4, 6, 7, 10, 10 **13.** Never; the mode must always be a member of the data set, but the mean and median may or may not be a member of the data set. **17.** G **19.** Misleading; the mean is 76%, and the median is 79%. The mode is 85% which is the 2nd highest score.

pp. 590–592 **Lesson 10-1C**

1 To determine how the measures would change, find each of the measures for each situation.
First situation: 1, 3, 2, 2 over four days.

mean: $\dfrac{1 + 3 + 2 + 2}{4} = 2$

median: 1, 2, 2, 3
$\uparrow$
2

mode: 2

Second situation: 2, 3, 2, 2 over four days.

mean: $\dfrac{2 + 3 + 2 + 2}{4} = 2.25$

median: 2, 2, 2, 3
$\uparrow$
2

mode: 2

The mean increases from 2 to 2.25 hours per day. The median and mode are not affected.
3. If the data for Yankee Stadium are not included, the mean decreases from about 43,505 to 41,741. The median decreases from 42,099 to 40,793. There still is no mode.

5 To see which measure will change the most, we have to determine the mean, median, and mode before the greatest and least price are excluded and after.

mean: $\dfrac{4{,}120}{18} \approx 228.89$

median: 90, 100, 100, 130, 130, 150, 180, 180, 180, 200, 200, 250, 250, 280, 300, 300, 350, 750
Since there are an even number of data, find the mean of the middle two numbers.
$\dfrac{180 + 200}{2} = 190$

mode: 180

Exclude the greatest and least priced cameras.

mean: $\dfrac{3{,}280}{16} = 205$

median: 100, 100, 130, 130, 150, 180, 180, 180, 200, 200, 250, 250, 280, 300, 300, 350
Since there are an even number of data, find the mean of the middle two numbers.
$\dfrac{180 + 200}{2} = 190$

mode: 180

The median and mode are not affected because the highest and lowest prices were removed, so the middle number did not change and neither of the numbers removed were any of the modes. The mean was reduced some. The high price was much higher than most of the prices, which would have skewed the initial mean higher than it should have been to represent the data.
7. The mean increases from 95 to 106. **9.** The mean decreases from 28 to 22.67.

11 Calculate each measure with and without the two lowest essay scores.

With the lowest scores:
Mean: $\dfrac{8 + 6 + 5 + 8 + 9 + 10 + 5 + 7}{8} = \dfrac{58}{8}$ or 7.25 points
Median: 5, 5, 6, 7, 8, 8, 9, 10
There is no middle number. The median is the mean of the two numbers in the middle, 7 and 8.
$\dfrac{7 + 8}{2} = \dfrac{15}{2}$ or 7.5 points
Modes: 5 and 8

Without the lowest scores:
Mean: $\dfrac{8 + 6 + 8 + 9 + 10 + 7}{6} = \dfrac{48}{6}$ or 8 points
Median: 6, 7, 8, 8, 9, 10
The median is 8.
Mode: 8
The mean increases from 7.25 to 8. The difference is 0.75. The median increases from 7.5 to 8. The difference is 0.5. The mode changes from being 8 and 5 to just 8.
13. mean; Sample answer: In general, the mean will be affected the most because it is found using all of the values in the data set. The median and mode might change a little, but the mean will typically change the most. **15.** 90 **17.** A **19.** A

pp. 596–598 **Lesson 10-2A**

1. range: 21; median: 82; upper quartile: 90.5; lower quartile: 78; interquartile range: 12.5; outliers: none **3.** range: 371,000; median: 210,000; upper quartile: 347,500; lower quartile: 67,500; interquartile range: 280,000; no outliers; Sample answer: The spread of the data is 371,000 gallons. The middle number is 210,000 gallons. About one fourth of the states had a maple syrup production at or above 347,500 gallons, and about one fourth of the states had a maple syrup production at or below 67,500 gallons. The number of gallons of maple syrup produced by half of the states was in the interval 67,500–347,500.

5 Finding the measures of variation means finding the range, median, upper quartile, lower quartile, and interquartile range. You also need to find any outliers. The data are listed in order from greatest to least.

The range is the difference between the greatest and least data points.

$1{,}100{,}000 - 5{,}000 = 1{,}095{,}000$.

Median is the middle number or the average of the two middle numbers if there are an even number of data.

$\dfrac{9{,}000 + 9{,}000}{2} = 9{,}000$

Upper quartile is the median of the upper half of the data, 24,500.

Lower quartile is the median of the lower half of the data, 8,000.

The interquartile range is the difference between the upper and lower quartiles.

$24{,}500 - 8{,}000 = 16{,}500$.

An outlier of the data is one that is more than 1.5 times the value of the interquartile range beyond either quartile.

The interquartile range is 16,500.

$16{,}500 \times 1.5 = 24{,}750$

So, any values that are greater than $24{,}500 + 24{,}750$ or 49,250 and less than $8{,}000 - 24{,}750$ or $-16{,}750$ are outliers. Since $1{,}100{,}000 > 49{,}250$, it is an outlier.

Sample answer: The spread of the data is 1,095,000 species. The middle number is 9,000 species. About one fourth of the animal groups had 24,500 or more species and about one fourth of the animal groups had 8,000 or fewer species. The number of species for half of the animal groups was in the interval 8,000–24,500.

7. Brandon: range: 10; median: −1; lower quartile: −3; upper quartile: 3; interquartile range: 6. Rashan: range: 10; median: −1; lower quartile: −4; upper quartile: 4; interquartile range: 8. Sample answer: The spread of both data sets is 10 and the middle score of both data sets is −1.

9 **a.** The range of the data is $3.6 - 1.5 = 2.1$.

The mean of the data is

$$\dfrac{3.6 + 2.9 + 2.7 + 2.5 + 2.3 + 2 \cdot 2.1 + 2 \cdot 1.9 + 1.7 + 7 \cdot 1.5}{17} = 2.01$$

The median of the data is 1.9.

The mode is 1.5.

The upper quartile is $\dfrac{2.3 + 2.5}{2}$ or 2.4.

The lower quartile is $\dfrac{1.5 + 1.5}{2}$ or 1.5.

The interquartile range is $2.4 - 1.5$ or 0.9.

b. There are no outliers because no values are more than 1.5 times the value of the interquartile range beyond either quartile. **c.** Sample answer: The spread of the data is 2.1. The middle of the data is

1.9. About one fourth of the data lies at or below 1.5 and about one fourth of the data lies at or above 2.4. Half of the data lies between 1.5 and 2.4.

13. Same range but different interquartile ranges: Sample answer: {1, 1, 2, 2, 2, 5, 9, 9, 9, 10, 10} and {1, 4, 4, 4, 4, 5, 5, 5, 9, 10, 10}; Same medians and same quartiles but different ranges: Sample answer: {1, 2, 5, 7, 9, 10, 12, 14, 15, 17, 22} and {0, 2, 5, 7, 9, 10, 12, 14, 15, 17, 27} **15.** True; Sample answer: The mean involves the sum of all the values. Since an outlier is an extreme value, the sum of the data values is affected, so the mean is the most affected measure of central tendency.

17. C **19.** 36.4; 33; 28, 33

pp. 601–603   Lesson 10-2B

1.

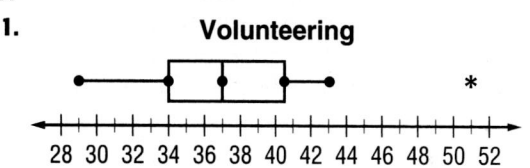

Volunteering

3a. 75   **3b.** 125 fish

5.

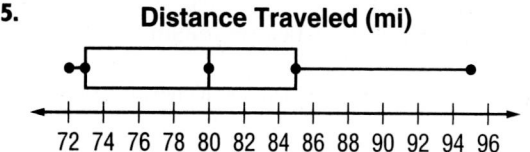

Distance Traveled (mi)

7.

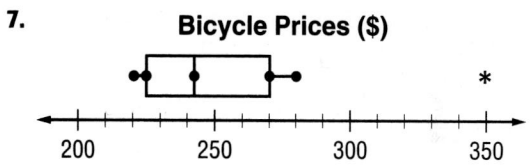

Bicycle Prices ($)

9 **a.** In box-and-whisker plots, outliers are the points not included in the plot and represented by an asterisk (*). There are two outliers.   **b.** Sample answer: The top half of the data is much more spread out than the bottom half of the data. Most major zoos are considerably smaller in area than the few zoos that have very large areas.

11a.

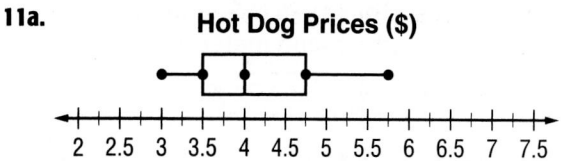

Hot Dog Prices ($)

11b. the prices above the median

13 **a.** Draw a number line that includes the least and greatest numbers in the data. Check for outliers. The outliers for this set of data are 2,261 and 9,406. Mark the following values above the number line: minimum non-outlier: 4,097; lower quartile: 4,362; median: 4,659; upper quartile: 5,538; maximum non-outlier: 5,685. Place an asterisk at 2,261 and 9,406 to indicate they are outliers. Draw the box and the whiskers.

Acreage of Lakes

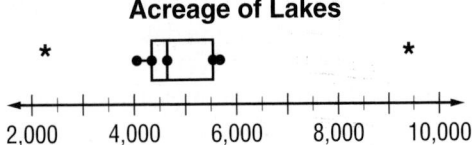

b. Sample answer: The data between the median and the upper quartile are more spread out than the data between the median and the lower quartile. The whisker at the left is longer than the whisker at the right, so the data below the lower quartile are more spread out than the data above the upper quartile.

15a. Sample answer: $x = 60$, $y = 65$ **15b.** Sample answer: $x = 64$, $y = 66$ **17.** A   **19.** range: 45; lower quartile: 38.5; median: 49; upper quartile: 56; interquartile range: 17.5; no outliers

---

pp. 605–607   Lesson 10-2C

1 **a.** West: minimum: 0.5, lower quartile: 1.2, median: 2.6, upper quartile: 5.6, maximum: 6.4, outlier: 36.5
Southeast: minimum: 1.8, lower quartile: 3.55, median: 4.45, upper quartile: 8.25, maximum: 9.4, outlier: 18.1

Regional Populations (millions)

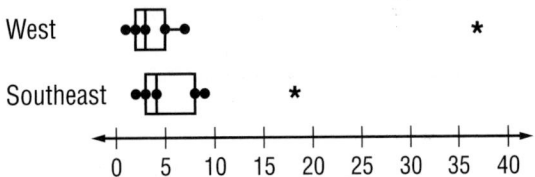

b. Sample answer: In general, the southeastern states have a greater population than the western states. The median population for the southeastern states is nearly double the median population of the western states.

3a.   **Longest Bridges in U.S. and Europe (thousand meters)**

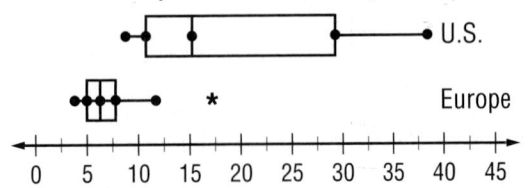

3b. Sample answer: All of the longest bridges in the U.S. are longer than 75% of the longest bridges in Europe.

5 **a.** 67 miles per hour is the upper quartile of the box-and-whisker plot for the wood roller coasters. By definition, 75% of wood roller coasters have slower speeds and 25% have faster speeds.

b. For the steel roller coasters, 82 miles per hour is the lower quartile. By definition, 75% of steel roller coasters have speeds faster than 82 miles per hour. **c.** Faster; 100% of the steel roller coasters travel faster than all of the wood roller coasters. The maximum speed of a wood roller coaster is slower than the minimum speed of a steel roller coaster.

7. Sample answer: Set A: 20, 21, 24, 25, 27, 28, 29, 29, 30, 30, 31, 32, 32, 33, 34, 34, 34, 37, 38; Set B: 16, 18, 18, 19, 20, 20, 21, 22, 24, 25, 27, 28, 30, 30, 30, 30, 31, 34, 35   **9.** C   **11.** Store A   **13a.** range: 1,367 mi; median: 1,454.5 mi; UQ: 1,900 mi; LQ: 1,243 mi; interquartile range: 657 mi   **13b.** no outliers

13c. Sample answer: The range of the data is 1,367 miles. The median is 1,454.5 miles. One fourth of the rivers are 1,243 or fewer miles and one fourth of the rivers are 1,900 or more miles.

---

pp. 610–611   Lesson 10-3A PSI

1. The most popular Web site has an average download time of 1.4 seconds, which is not the fastest download time. This ordered pair is an exception to the general statement that the most popular Web sites are faster than the less popular Web sites.   **3.** Company A

5 Find the next two numbers in the sequence 4, 0, −4, −8, …. First, determine the common difference between each term in the sequence. The common difference is −4. Apply that to continue the sequence and find the next two numbers. −8 + (−4) = −12 and −12 + (−4) = −16

7. 38%   **9.** Yes

---

pp. 616–618   Lesson 10-3C

1 **a.** Let the x-axis represent the time in hours. Let the y-axis represent the number of units produced. Then graph the ordered pairs (time, units produced).

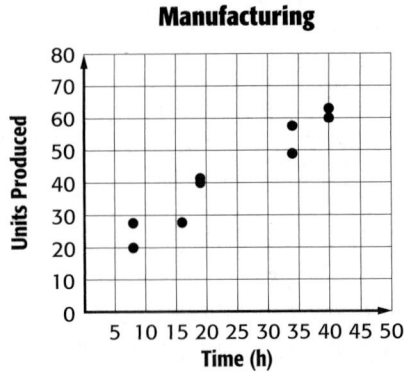

Manufacturing

b. As the time increases, the units produced increases. A positive relationship exists between the time and the units produced.

c. Look at the graph and estimate a continuation of the trend of the data. At 50 hours, there will be approximately 70 units produced.

3.

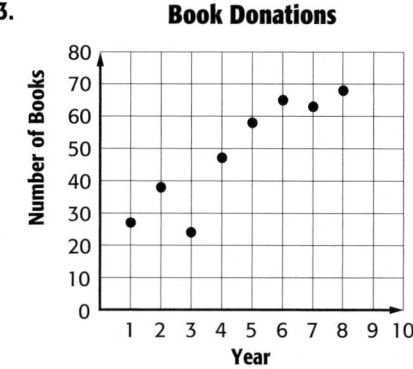

Book Donations

5 Look at the graph to see if there is a trend in the data. The data is all over the place. There is no noticeable trend so the scatter plot shows no relationship.

7 a. Let the *x*-axis represent the day. Let the *y*-axis represent the number of E-mails. Graph the ordered pairs (day, number of E-mails).

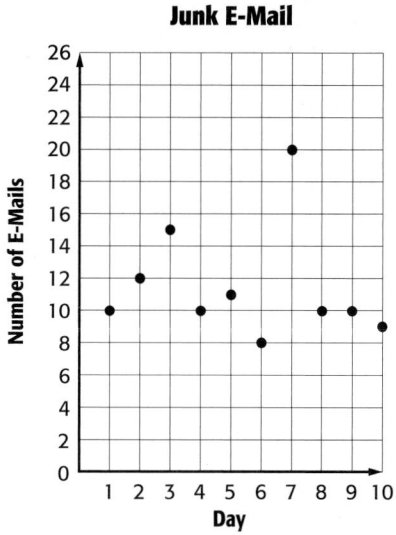

Junk E-Mail

b. The day does not affect the number of E-mails received. So, the scatter plot shows no relationship.

c. Since the scatter plot shows no relationship between the data, it is not possible to predict how many E-mails will be received on Day 15.

11. Sometimes; sample answer: The price per gallon of gasoline would increase proportionally as the number of gallons bought increases. But, as the level of education increases, salary may or may not increase proportionally.

13. A   **15.** 0.75; strong positive relationship

pp. 622–624   Lesson 10-3E

1a.

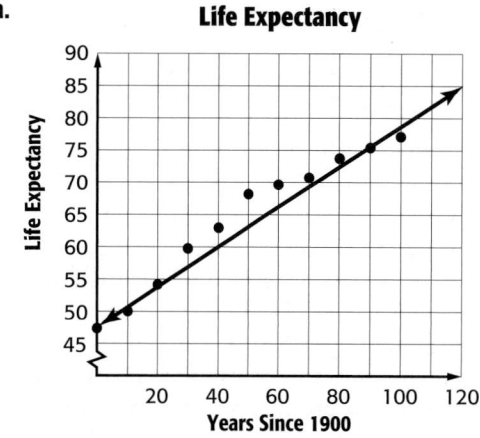

Life Expectancy

1b. Sample answer: 85 years

3 a. Let the *x*-axis represent the shoe size. Let the *y*-axis represent the person's height in inches. Then graph the ordered pairs (shoe size, height) and draw a line that fits the data.

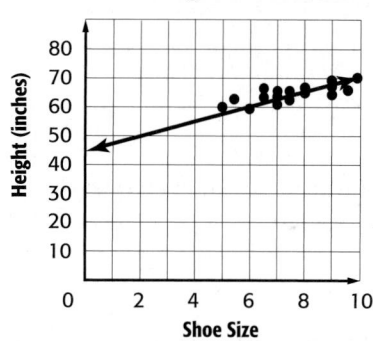

Female Height and Shoe Size

b. Sample answer: Look at the line of best fit to see where it intersects with a shoe size of 5. From the line, a reasonable estimate for the height of a woman with a shoe size of 5 is 57.5 inches.

5 a. Sample answer: Determine the slope of the line by choosing two points that the line passes through.

$$m = \frac{6{,}000 - 4{,}000}{7.5 - 3.5} = 500$$

Then use the slope and a point on the line to find the *y*-intercept.

$y = mx + b$	Slope-intercept equation
$6{,}000 = 500(7.5) + b$	Substitute in a point and the slope.
$2{,}250 = b$	Solve for *b*.

Therefore, a sample equation for the line of best fit is $y = 500x + 2{,}250$.

b. In 2020, *x* will equal 24. Evaluate the equation for $x = 24$.

$y = 500(24) + 2{,}250$	Substitute.
$y = 14{,}250$	Simplify.

There will be approximately 14,250 girls participating in ice hockey in 2020.

7a.

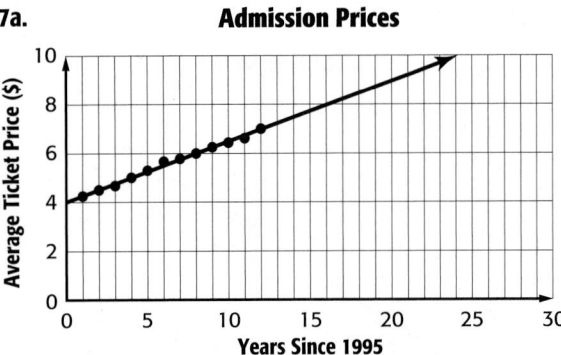

Admission Prices

7b. Sample answer: $y = 0.2x + 4$; about $9

9. Sample answer: A line of best fit helps in making interpretations and predictions about the situation modeled in the data set. **11.** C

13a.

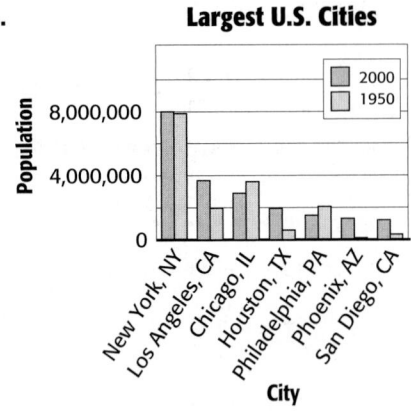

Largest U.S. Cities

13b. Sample answer: It increased by about 300%.
13c. Phoenix

pp. 629–631 Lesson 10-3G

1 Sample answer: A bar graph would be the most appropriate display because it shows the number of items in specific categories.

3. Sample answer: Box-and-whisker plot; it shows how the data are separated into 4 equal sets.

Test Scores Period 4

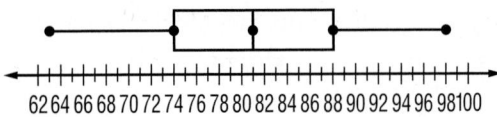

5. Sample answer: Bar graph; it shows the number of items in specific categories.   **7.** Sample answer: Box-and-whisker plot; it shows the measures of variation for a set of data.

9 Sample answer: The type of display that would best display the number of Americans who speak Spanish, French, and/or German would have to allow for an overlap of data. Some people may speak two or even three of the languages. Therefore, a Venn diagram would be the best display.

11. Sample answer: Line graph; it shows change over a period of time.

Average Height of Females

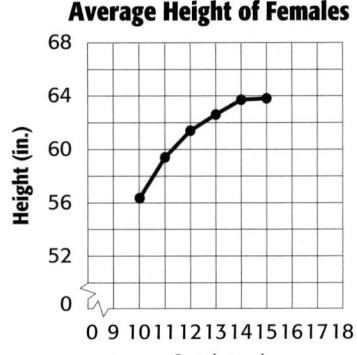

13. Sample answer:

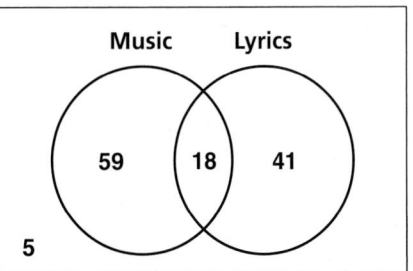

19. Never; sample answer: A line graph shows change over a period of time while a Venn diagram shows how elements of a set are related. These two types of displays show completely different data.   **21.** Sample answer: Both bar graphs and histograms use bars to show how many things are in each category. A histogram shows the frequency of data that has been organized into equal intervals. There is no space between the bars in a histogram. It would be appropriate to use a histogram instead of a bar graph when the data can be organized into equal intervals.

pp. 634–637 Chapter Study Guide and Review

1. false; scatter plot   **3.** false; measures of variation   **5.** true   **7.** true   **9.** 15; 15; 15   **11.** If the height is 24 inches instead of 42 inches, both the median and the mode stay the same. The mean decreases from 11.84 to 11.12 inches.   **13.** range: 11; median: 3; upper quartile: 5; lower quartile: 2; interquartile range: 3; outlier: 12

15.   **Time Working (h)**

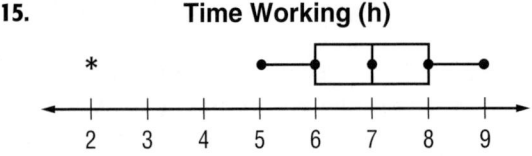

17a.   **400-Meter-Dash Times (s)**

Team A

Team B

60   64   68   72   76   80   84

17b. Sample answer: All of the times for Team A are less than 50% of the times for Team B.

19.

Shopping

21a.

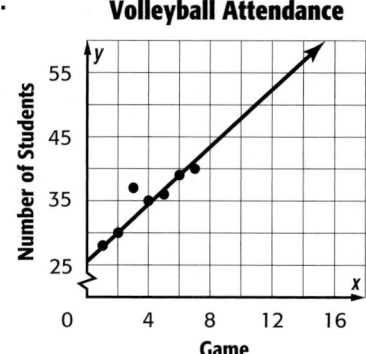

Volleyball Attendance

21b. Sample answer: $y = \frac{9}{4}x + 27$ **21c.** 54
students   **23.** Yes; a circle graph compares parts of
the data to the whole.

Chapter 11   Probability and Combinations

p. 644   Are You Ready?

1. $\frac{2}{3}$ **3.** $\frac{7}{33}$ **5.** $\frac{5}{8}$ **7.** $\frac{2}{9}$ **9.** 142.8   **11.** 3.6   **13.** 45%
15. 60%

pp. 647–649   Lesson 11-1A

1.

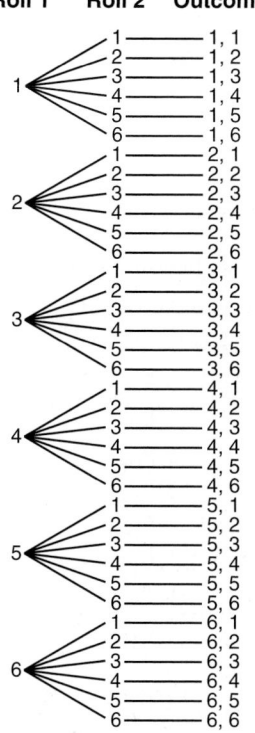

| Roll 1 | Roll 2 | Outcome |

36 outcomes

3. $\frac{1}{10,000}$

5. Number Cube

	Penny	Outcome
1	Heads	1, Heads
	Tails	1, Tails
2	Heads	2, Heads
	Tails	2, Tails
3	Heads	3, Heads
	Tails	3, Tails
4	Heads	4, Heads
	Tails	4, Tails
5	Heads	5, Heads
	Tails	5, Tails
6	Heads	6, Heads
	Tails	6, Tails

12 outcomes

7.

Flavor	Cone	Outcome
Chocolate	Regular	Chocolate, Regular
	Sugar	Chocolate, Sugar
Vanilla	Regular	Vanilla, Regular
	Sugar	Vanilla, Sugar
Strawberry	Regular	Strawberry, Regular
	Sugar	Strawberry, Sugar

6 outcomes   **9.** 216 outcomes

11 There are 4 choices for each of the 5 questions
on the test. The Fundamental Counting Principle
tells us that we can multiply to find out the total
number of outcomes. $4 \cdot 4 \cdot 4 \cdot 4 \cdot 4 = 1{,}024$
outcomes

13. 1,757,600   **15.** $\frac{1}{10,000}$

17 a. Draw a tree diagram. Circle the branches
where at least one of the marbles is blue.

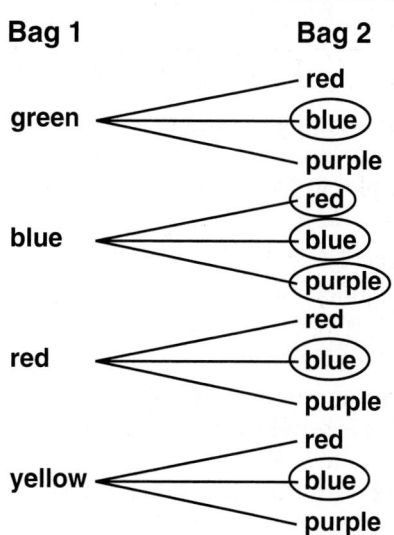

Bag 1 / Bag 2

There are 12 possible outcomes and 6 of them have at least one blue marble. P(at least one blue marble) $= \frac{6}{12}$ or $\frac{1}{2}$.

b. Using the same tree diagram, circle all the branches where at least one marble is yellow.

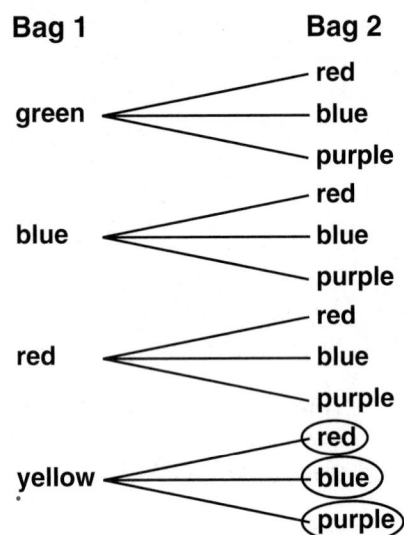

Bag 1   Bag 2

green — red / blue / purple
blue — red / blue / purple
red — red / blue / purple
yellow — (red) / (blue) / (purple)

There are 12 possible outcomes and 3 of them have at least one yellow marble. P(at least one yellow marble) $= \frac{3}{12}$ or $\frac{1}{4}$.

19. Sample answer: Maria can choose from 5 flavors of ice cream and 3 different toppings. How many desserts can Maria make with one flavor of ice cream and one topping? **21.** 6^x **23.** C **25.** B **27.** 1:1 **29.** 3:1

pp. 653–654   **Lesson 11-1C**

1. 504 ways

3 $P(7, 4)$ means a permutation of 7 things taken 4 at a time. Start with 7 and use 4 factors.
$7 \cdot 6 \cdot 5 \cdot 4 = 840$

5. 40,320 **7.** 5,040 codes **9.** 1,320 ways

11. 120 **13.** 120 **15.** 240,240 **17.** 303,600 **19.** $\frac{1}{5}$

21 There are 9 squares on a tic-tac-toe board. Since the order of the marks matters, this is a permutation of 9 things taken 3 at a time.
$P(9, 3) = 9 \cdot 8 \cdot 7$ or 504 ways.

23. $P(7, 3)$ means to start with 7 and use 3 factors. $P(7, 3) = 7 \cdot 6 \cdot 5$ or 210. **25.** They are the same; $P(n, n) = n \cdot (n - 1) \cdot (n - 2) \cdot \ldots 2 \cdot 1$ and $P(n, n - 1) = n \cdot (n - 1) \cdot (n - 2) \cdot \ldots 2$, which are equal. **27.** B **29.** C

pp. 657–658   **Lesson 11-1D**

1 Since the order of the endpoints is not important, this is a combination. $C(8, 2) = \frac{P(8, 2)}{2 \cdot 1} = \frac{8 \cdot 7}{2 \cdot 1}$ or 28. So, there are 28 segments that can be drawn.

3. combination; 35 **5.** 210 squads **7.** combination; 91

9 Since the 4 photographs will be placed at specific locations, order matters so this is a permutation. $P(12, 4) = 12 \cdot 11 \cdot 10 \cdot 9 = 11,880$. So, there are 11,880 different ways to display 4 photographs out of 12 in specific locations.

11. Sample answer: the number of five-person committees that could be formed from a group of 15 people **13.** Sometimes; they are equal if $y = 1$.

15. B **17.** 42 **19.** 1,860,480 **21.** 12 lunches

pp. 662–664   **Lesson 11-2A**

1. $\frac{1}{30}$ **3.** $\frac{1}{10}$

5 Find the probability that she will choose a pair of black pants and a white shirt.
P(black pants and white shirt)
$= P$(black pants) $\cdot P$(white shirt)
$= \frac{2}{6} \cdot \frac{4}{6}$ 2 out of 6 pants are black. 4 out of 6 shirts are white.
$= \frac{8}{36}$ Multiply.
$= \frac{2}{9}$ Simplify.

So, the probability she will choose a pair of black pants and a white shirt is $\frac{2}{9}$.

7. $\frac{1}{90}$ **9.** $\frac{2}{3}$ **11.** $\frac{1}{2}$ **13.** $\frac{1}{6}$

15 List the sample space for choosing one card from each set of cards. In this example, R_1 and R_2 will represent the two red cards in the second deck.

1, R_1	2, R_1	3, R_1	4, R_1	5, R_1
1, R_2	2, R_2	3, R_2	4, R_2	5, R_2
1, P_1	2, P_1	3, P_1	4, P_1	5, P_1
1, P_2	2, P_2	3, P_2	4, P_2	5, P_2
1, P_3	2, P_3	3, P_3	4, P_3	5, P_3
1, G_1	2, G_1	3, G_1	4, G_1	5, G_1
1, G_2	2, G_2	3, G_2	4, G_2	5, G_2
1, G_3	2, G_3	3, G_3	4, G_3	5, G_3
1, G_4	2, G_4	3, G_4	4, G_4	5, G_4
1, G_5	2, G_5	3, G_5	4, G_5	5, G_5

There are 50 outcomes when choosing the two cards. Six of those are successes.

$1, R_1$ $1, R_2$
$3, R_1$ $3, R_2$
$5, R_1$ $5, R_2$

P(odd number and a red card) $=$
$\dfrac{\text{number of times an odd card and a red card occurs}}{\text{number of possible outcomes}}$

So, the probability of choosing an odd card and a red card is $\dfrac{6}{50}$ or $\dfrac{3}{25}$.

17. 0   **19.** $\dfrac{3}{16}$ **21.** $\dfrac{15}{92}$ **23.** $\dfrac{3}{7}$ **25.** $\dfrac{4}{7}$

27 The coin is tossed twice and a letter is chosen from the word *event*. Since one event does not influence the other events, the events are independent. Multiply the probabilities of each event.

P(two tails and a vowel) $= P$(tail) $\cdot P$(tail) $\cdot P$(vowel)
$= \dfrac{1}{2} \cdot \dfrac{1}{2} \cdot \dfrac{2}{5}$ or $\dfrac{1}{10}$

29. He found the probability of two disjoint events. He should have found the probability of two independent events. P(two heads) $= \dfrac{1}{2} \cdot \dfrac{1}{2}$ or $\dfrac{1}{4}$

31. $\dfrac{1}{6}$; the probability decreases to $\dfrac{1}{9}$ **33.** A

35. 6 burritos

pp. 668–670   Lesson 11-2C

1a. $\dfrac{3}{8}$ **1b.** $\dfrac{1}{2}$ **1c.** equally likely; half the tosses have exactly two heads.   **3a.** The experimental probability $\dfrac{1}{5}$ is greater than the theoretical probability $\dfrac{1}{8}$.

3b. 36   **3c.** The prediction is not reasonable; sample answer: the experimental probability of landing on a 4 or 8 is $\dfrac{7}{50}$. It is much more likely to land on one of the other numbers.

5 a. Based on the fact that Jeanette won 24 out of 30 matches, the probability she will win her next match is $\dfrac{24}{30}$ or $\dfrac{4}{5}$.
b. Write a proportion to find how many of the next 50 matches she can expect to win.

She wins 4 out of 5. → $\dfrac{4}{5} = \dfrac{x}{50}$ ← She can expect to win x out of 50 matches.

Solve the proportion.
$$\dfrac{4}{5} = \dfrac{x}{50}$$
$$200 = 5x$$
$$40 = x$$

So, Jeanette should expect to win 40 out of the next 50 matches.

7a. $\dfrac{10}{79}$ **7b.** about 5

9 a. The experimental probability of getting out is $\dfrac{120}{200}$ or $\dfrac{3}{5}$. Since $\dfrac{3}{5}$ or 60% is greater than 50%, it is likely that he will be out.

b. The experimental probability of the player hitting a single or a double is $\dfrac{32}{200} + \dfrac{18}{200}$ or $\dfrac{1}{4}$. This is less than 50% so it is somewhat likely he will hit a single or a double.

13. $\dfrac{21}{80}$ **15.** C   **17.** The theoretical probability of Player 1 winning is $\dfrac{2}{5}$. The theoretical probability of Player 2 winning is $\dfrac{3}{5}$.

pp. 674–675   Lesson 11-3A PSI

1. Sample answer: You can make a prediction about what will actually happen in the problem.

3. 24 ways   **5.** math, science, reading, art; math, science, art, reading; math, reading, art, science; math, reading, science, art; math, art, science, reading; math, art, reading, science   **7.** about 233 students   **9.** 1,140 combinations

11 Make a table showing ratios of $\dfrac{\text{green jelly beans}}{\text{red jelly beans}}$ equivalent to $\dfrac{3}{4}$.

$\dfrac{3}{4}$	$\dfrac{3-4}{4} = -\dfrac{1}{4}$	No
$\dfrac{6}{8}$	$\dfrac{6-4}{8} = \dfrac{1}{4}$	No
$\dfrac{9}{12}$	$\dfrac{9-4}{12} = \dfrac{5}{12}$	No
$\dfrac{12}{16}$	$\dfrac{12-4}{16} = \dfrac{1}{2}$	Yes

So, Danielle started with 12 green jelly beans.

13. 4 blue marbles, 3 red marbles, 2 green marbles, and 1 yellow marble

pp. 678–679   Lesson 11-3B

1 Sample answer: Toss a coin. Repeat the simulation until all possible cones are obtained.

3. Sample answer: Spin a spinner with 4 equal-size sections 50 times.   **5.** Sample answer: Spin a spinner divided into 3 equal sections and roll a number cube. Repeat the simulation until all types of cookies are obtained.   **7.** Sample answer: Draw 1 marble from 10 marbles, 3 red to represent winning and 7 blue to represent losing, 4 times, replacing the marble each time.

9 Sample answer: Use a coin to decide which way the mouse will go at each intersection. Record whether the mouse goes out the Out opening, goes out the In opening, or comes to a dead end. Repeat the experiment numerous times.

11. 5 times; Sample answer: A number should begin with 5 about half of the time. Half of 10 is 5.

13. D   **15.** The experimental probability is $\dfrac{1}{5}$ and the theoretical probability is $\dfrac{1}{4}$. Currently, the experimental probability is less than the theoretical probability. **17a.** $\frac{1}{18}$ **b.** $\frac{2}{3}$

pp. 684–687 Lesson 11-3D

1 The researcher divided the company into floors, and then randomly chose 10 people from each floor. This is an unbiased stratified random sample so the conclusion is valid.
3. This is an unbiased simple random sample, so the sample is valid; about 102 students.
5 The restaurant asked its customers to call a number to complete the survey. This is a voluntary response sample and it is biased. The conclusion is not valid.
7. The conclusion is valid. This is an unbiased stratified random sample. **9.** Yes; this is an unbiased simple random sample; about 4,600 people. **11.** The sample is a convenience sample. Therefore, no conclusion can be made.
13 Sample answer: Pedro could use a random sample, asking every 10th sixth grader who enters the school. This would be a systematic random sample and it is unbiased.
17. Yes; Sample answer: Every 10th person at a basketball game is asked whether they prefer basketball or baseball. This survey is systematic because every 10th person is surveyed. It is also a convenience sample because people attending a basketball game probably prefer basketball.
19. C **21.** Sample answer: On a number cube, designate 2 numbers to represent winning a prize, and 4 numbers to represent not winning a prize. Then roll the number cube 50 times and record the results in a table. **23.** about 292 students

pp. 692–695 Chapter Study Guide and Review

1. sample space **3.** compound event
5. Theoretical
7.

18 outcomes
9. $\frac{1}{6}$ **11.** 6 **13.** 60 **15.** 720 **17.** 120 numbers
19. 4 **21.** 126 **23.** 21 **25.** $\frac{1}{15}$ **27.** $\frac{1}{15}$ **29a.** $\frac{1}{5}$
29b. 5 words **31.** 5 **33.** Sample answer: Use a

spinner divided into 4 equal sections where each section represents a different biscuit. Repeat the simulation 10 times.

Chapter 12 Area and Volume

p. 702 Are You Ready?

1. 68 cm^2 **3.** 71.5 m^2 **5.** 72 in^2 **7.** 22 ft
9. 20.1 **11.** 283.4

pp. 707–709 Lesson 12-1B

1. 56.5 cm **3.** 7.9 mi **5.** 346.4 ft^2 **7.** about 2.5 inches **9.** 119.4 mi **11.** 106.8 km
13. 22.1 mi^2 **15.** 70.9 in^2

17 The decorative trim is around the edge of the table so we need to find the circumference of the table. We are given the radius, so we use $C = 2\pi r$, where r is $2\frac{1}{4}$ feet.

$C = 2\pi r$	Circumference of a circle
$C = 2 \cdot \pi \cdot 2\frac{1}{4}$	Replace r with $2\frac{1}{4}$.
$C \approx 14.137$	Use a calculator.

So, the trim is about 14.1 feet long.
19. 25.5 in. **21.** 8.6π km; 18.49π km^2
23. 1.2π mi; 0.36π mi^2

25 To determine the better offer, find the cost per square inch for each.
First Offer (one large cookie for \$20)

$A = \pi r^2$	Area of a circle
$A = \pi(6)^2$	Replace r with 6.
$A = 36\pi$	Evaluate 6^2.
$A \approx 113.1$	Use a calculator.

The area of one large cookie is about 113 square inches. Find the unit rate to find the cost per square inch.
$\frac{\$20}{113 \text{ in}^2} \approx \$0.18$ per square inch
Second Offer (three small cookies for \$20)

$A = \pi r^2$	Area of a circle
$A = \pi(4)^2$	Replace r with 4.
$A = 16\pi$	Evaluate 4^2.
$A \approx 50.3$	Use a calculator.

Since the area of one small cookie is about 50 square inches, three small cookies have an area of 150 square inches. Find the unit rate to find the cost per square inch.
$\frac{\$20}{150 \text{ in}^2} \approx \$0.13$ per square inch
Since \$0.13 < \$0.18, 3 small cookies for \$20 is the better offer.
29. If the radius is halved, the circumference will be halved since $2\pi\left(\frac{1}{2}r\right) = \pi r$ and the area will be one fourth of the original area since $\pi\left(\frac{1}{2}r\right)^2 = \frac{1}{4}(\pi r^2)$. If the radius is doubled, the circumference will double since $2\pi(2r) = 2(2\pi r)$, and the area will quadruple since $\pi(2r)^2 = 4(\pi r^2)$. If the radius is

tripled, the circumference will triple since $2\pi(3r) = 3(2\pi r)$, and the area will be nine times the original since $\pi(3r)^2 = 9(\pi r^2)$. **31.** 21.5 in^2 **33.** Sample answer: They are related in that they are both measurements of a circle. They are different in that circumference is measured in linear units while area is measured in square units. **35.** H **37.** H **39.** Sample answer: $\overleftrightarrow{WZ}$ **41.** X and Z

pp. 712–713   Lesson 12-1D PSI

1. Sample answer: Subtract the area of just the portrait from the area of the framed portrait. **3.** 13.5 in. by 18 in. **5.** 66 in^3 **7.** 15 tables

9 You may want to sketch the situation to visualize the distances. To determine the amount of space left over after the posters are hung, subtract the width of all the posters from the width of the wall. $18 - 3(2) = 12$ feet. There are four gaps that need to be of equal width. $\frac{12}{4} = 3$; Each gap should be 3 feet wide. **11.** 9 clothespins

pp. 716–718   Lesson 12-1E

1. 216 in^2

3 To find the area of the window, we need to find the sum of the areas of the rectangle and semicircle.

$A = \frac{1}{2}\pi r^2$ Area of a semicircle

$A = \frac{1}{2}\pi\left(\frac{1.5}{2}\right)^2$ Substitute $\frac{1.5}{2}$ for r.

$A \approx 0.88$ Use a calculator.
$A = \ell w$ Area of a rectangle
$A = 1.5 \cdot 2$ Substitute.
$A = 3.0$ Multiply.

Add the two areas to find total area. $0.88 + 3.0 = 3.9$ square feet.
5. 64 cm^2 **7.** 220.5 cm^2

9 To find the area of the composite figure, find the sum of the areas of the triangle and the trapezoid.

Area of the triangle
The height of the triangle is the overall height of the figure minus the height of the trapezoid. So, the height of the triangle is $6.4 - 3.6$ or 2.8 feet.

$A = \frac{1}{2}bh$ Area of a triangle
$A = \frac{1}{2} \cdot 7 \cdot 2.8$ Substitute 7 for b and 2.8 for h.
$A = 9.8$ Simplify.

Area of the trapezoid
$A = \frac{1}{2}h(b_1 + b_2)$ Area of a trapezoid

$A = \frac{1}{2} \cdot 3.6(7 + 9)$ Substitute 7 for b_1, 9 for b_2, and 3.6 for h.

$A = 1.8(16)$ Simplify.
$A = 28.8$ Multiply.

So, the area of the composite figure is $9.8 + 28.8$ or 38.6 square feet.
11. 119.5 ft^2 **13.** 610 m^2 **15.** 120 cm^2

17 To find the area of the bedroom, we need to find the sum of the areas of the two rectangles and the triangle.

$A = \ell w$ Area of a rectangle
$A = (10)(11)$ Substitute for rectangle 1.
$A = 110$ Multiply.
$A = (10)(6)$ Substitute for rectangle 2.
$A = 60$ Multiply.
$A = \frac{1}{2}bh$ Area of a triangle

$A = \frac{1}{2}(12)(8)$ Substitute.

$A = 48$ Multiply.
Total area is $110 + 60 + 48$ or 218 square feet. Since the baseboard goes around the bedroom, find the perimeter of the room. Add the lengths of all the sides of the room.
$P = 10$ ft $+ 11$ ft $+ 12$ ft $+ 6$ ft $+ 10$ ft $+ 6$ ft $+ 11$ ft
$P = 66$ ft
So, Zoe's mom will need 218 square feet of carpet and 66 feet of baseboards.

19. perimeter: $3x + \frac{1}{2}\pi x$; area: $x^2 + \frac{1}{2}\pi\left(\frac{x}{2}\right)^2$

21. Divide the composite figure horizontally into two trapezoids, find the area of each, and then find the sum of their areas. Divide the composite figure up vertically into two triangles and a rectangle, find the area of each figure, then find the sum of their areas. **23.** G **25.** 314.2 ft^2

pp. 721–724   Lesson 12-2A

1. 36 ft^3 **3.** 1,272.3 yd^3

5 To find the volume of the toy house, find the sum of the volumes of the rectangular prism and the triangular prism.

$V = Bh$ Volume of a prism
$V = (\ell \cdot w)h$ The base is a rectangle, so $B = \ell w$.
$V = (34 \cdot 15)20$ $\ell = 34, w = 15, h = 20$
$V = 10,200$ Simplify.
$V = Bh$ Volume of triangular prism
$V = \left(\frac{1}{2} \cdot 34 \cdot 18\right)h$ The base is a triangle, so $B = \frac{1}{2} \cdot 34 \cdot 18$.

$V = \left(\frac{1}{2} \cdot 34 \cdot 18\right)15$ The height of the prism is 15.

$V = 4,590$ Simplify.
The total volume of the toy house is 10,200 cubic centimeters plus 4,590 cubic centimeters or 14,790 cm^3.

7. 216 mm^3 **9.** 768 m^3 **11.** 55.4 m^3
13. 297.5 ft^3 **15.** 236.1 cm^3 **17.** 3,864.9 cm^3
19 We are given the radius and the volume, so, we need to solve for the height.

$V = \pi r^2 h$ Volume of a cylinder

$301.6 = \pi(4)^2 h$ Replace V with 301.6 and r with 4.

$301.6 = 16\pi h$ $4^2 = 16$

$\dfrac{301.6}{16\pi} = \dfrac{16\pi h}{16\pi}$ Divide each side by 16π.

$6.0 \approx h$ Use a calculator.

The height of the cylinder is 6 inches.

21. Sample answer: 5.5 inches; the height remains the same, so πr^2 must equal 24. Solve the equation for r, then double it to get the diameter.

23 To find the volume of the hexagonal prism, split the prism in half so that there are two trapezoidal prisms. The volume of a prism is the area of the base times the height. The area of a trapezoid is $\frac{1}{2}h(b_1 + b_2)$, where b_1 and b_2 are the lengths of the two bases and h is the height of the trapezoid.

$V = Bh$ Volume of a prism

$V = \left(\frac{1}{2}h(b_1 + b_2)\right)h$ The base is a trapezoid, so $B = \frac{1}{2}h(b_1 + b_2)$.

$V = \left(\frac{1}{2}(4)(5 + 11)\right)7$ Replace h with 4, b_1 with 5, b_2 with 11, and h with 7.

$V = (32)7$ Simplify.

$V = 224$ Multiply.

The volume of the trapezoidal prism is 224 cubic meters. So, the volume of the hexagonal prism is 224(2) or 448 cubic meters.

25a. bag: 132 in^3; candle: 29.5 in^3 **25b.** 102.5 in^3

25c. 13 packages   **27.** Sample answer: For both formulas, you are finding the area of the base and multiplying that by the height.   **29.** C   **31.** Sample answer: When you change only one measurement by a scale factor, the volume is changed by the same scale factor.

pp. 728–730   **Lesson 12-2B**

1. 410.7 cm^3 **3.** 376,041.7 cm^3 **5.** 183.3 m^3
7. 565.5 cm^3 **9.** 523.6 yd^3 **11.** 43.3 in^3
13. 61.4 cm^3 **15.** 2,742.7 ft^3

17 The model is shaped like a cone with a diameter of 8 in. and volume of 201 in^3. Since d is 8, r is 4. Solve the volume equation for height.

$V = \frac{1}{3}\pi r^2 h$ Volume of a cone

$201 = \frac{1}{3}\pi(4)^2 h$ Replace V with 201 and r with 4.

$201 = \frac{16}{3}\pi h$ $4^2 = 16$

$603 = 16\pi h$ Multiply each side by 3.

$\dfrac{603}{16\pi} = \dfrac{16\pi h}{16\pi}$ Divide each side by 16π.

$12.0 \approx h$ Use a calculator.

The height of the volcano is approximately 12 in.

19. 1,866.1 ft^3 **21.** 1,731.8 mm^3 **23.** 117.6 in^3
25. 20,579.5 mm^3 **27.** 0.5 cm^3 **29.** 840 yd^3

31 This solid is made up of two cones put together. So, we need to find the sum of the volumes of the cones.

$V_1 = \frac{1}{3}\pi r_1^2 h_1$ Volume of a cone

$V_1 = \frac{1}{3}\pi(3)^2(5)$ Replace r_1 with 3 and h_1 with 5.

$V_1 \approx 47.1$ Simplify. Use a calculator.

$V_2 = \frac{1}{3}\pi r_2^2 h_2$ Volume of a cone

$V_2 = \frac{1}{3}\pi(3)^2(6)$ Replace r_2 with 3 and h_2 with 6.

$V_2 \approx 56.5$ Simplify. Use a calculator.

$V_1 + V_2 \approx 103.7$ Sum of the volumes.

The volume of this solid is approximately 103.7 mm^3.

33. Jacob used the height of the triangle, not the height of the pyramid.

$V = \frac{1}{3}Bh$

$V = \frac{1}{3} \cdot 10 \cdot 10 \cdot 12$

$V = 400$ in^3

35. C   **37.** I   **39.** 227.5 in^3 **41.** 31.4 cm; 78.5 cm^2
43. 7.9 in.; 4.9 in^2

pp. 736–738   **Lesson 12-3B**

1. 64 yd^2; 94 yd^2 **3.** 236.2 m^2; 336.7 m^2

5 The cardboard portion of the orange juice container is a rectangle. The height of the cylinder is the height of the rectangle and the circumference of the top is the base of the rectangle.

$C = 2\pi r$ Circumference of a circle

$C = 2\pi(2)$ Replace r with 2.

$C \approx 12.6$ Simplify. Use a calculator.

$A = \ell w$ Area of a rectangle.

$A = (12.6)(6.5)$ Replace ℓ wiith 12.6 and w with 6.5.

$A \approx 81.7$ Simplify.

The area of the cardboard portion of the orange juice container is approximately 81.7 in^2.

7. 30 in^2; 58 in^2

9 The solid is a triangular prism. Find the perimeter and area of one base.

Perimeter

$P = 8.2 + 8.5 + 11.2$

$P = 27.9$

Area of the Base

$B = \frac{1}{2}bh$ Area of a triangle

$B = \frac{1}{2} \cdot 11.2 \cdot 6$ Substitute 11.2 for b and 6 for h.

$B = 33.6$ Simplify.

Use this information to find the lateral and total surface areas.

Lateral Area

$L.A. = Ph$

$L.A. = 27.9 \cdot 9.5$ or 265.1

Total Surface Area

$S.A. = L.A. + 2B$

$S.A. = 265.1 + (2 \cdot 33.6)$

$S.A. = 265.1 + 67.2$ or 332.3

The lateral area is 265.1 square meters and the total surface area is 332.3 square meters.
11. 202.3 mm^2; 335.3 mm^2

13 To compare the material used for the two designs, we need to compare the total surface area. The first container is a cylinder. The second container is a prism.

$$S.A. = 2\pi rh + 2\pi r^2 \qquad \text{Surface area of a cylinder}$$
$$= 2\pi(3)(12) + 2\pi(3)^2 \qquad \text{Replace } r \text{ with 3 and } h \text{ with 12.}$$
$$\approx 282.7 \qquad \text{Simplify. Use a calculator.}$$
$$S.A. = Ph + 2B \qquad \text{Surface area of a prism}$$
$$= [2(4) + 2(7)]12 + 2(7 \cdot 4) \qquad P = 2(4) + 2(7),$$
$$h = 12, B = (7 \cdot 4)$$
$$= 22(12) + 2(28) \qquad \text{Simplify.}$$
$$= 264 + 56 \text{ or } 320 \qquad \text{Add.}$$

Since 282.7 < 320, the cylinder uses less material.
15. False; a rectangular prism 2 ft long, 4 ft wide, and 6 feet high has the same volume as a prism 2 ft long, 2 ft wide, and 12 ft high, 48 ft^3. The surface area of the first prism is 88 ft^2, but the surface area of the second prism is 104 ft^2. **17.** The lateral area triples. The original lateral area is $2\pi rh$; tripling the radius gives a lateral area of $2\pi(3r)h$ or $6\pi rh$, which is three times the original. **19.** C **21.** 3,041.1 in^3
23. 5.6 ft

pp. 743–745   Lesson 12-3D

1. 48 ft^2; 64 ft^2 **3.** 572,400 ft^2
5. 282.7 m^2; 989.6 m^2 **7.** 35 ft^2; 47.3 ft^2
9. 105.3 mm^2; 140.4 mm^2 **11.** 1,536 ft^2; 2,112 ft^2

13 We only need the lateral area of the roof because that is the only part that will have roofing materials on it.

$$L.A. = \frac{1}{2}P\ell \qquad \text{Lateral area of a regular pyramid}$$
$$= \frac{1}{2}(4 \cdot 40)(16) \qquad \text{Replace } P \text{ with the perimeter of the square base, which is } 4 \cdot 40, \text{ and replace } \ell \text{ with 16, the slant height.}$$
$$= 1,280 \qquad \text{Simplify.}$$

So, the roof needs 1,280 ft^2 of roofing material.

15. 1,979.2 mm^2; 2,997.1 mm^2
17. 149.2 yd^2; 227.7 yd^2 **19.** 80.9 m^2; 179.4 m^2

21 The information that is missing is the slant height, ℓ. Use the Pythagorean Theorem to find the slant height.

$$a^2 + b^2 = c^2 \qquad \text{Pythagorean Theorem}$$
$$(471)^2 + (352.5)^2 = \ell^2 \qquad \text{Replace } c \text{ with } \ell, a \text{ with 471, and } b \text{ with } \frac{705}{2}.$$
$$346,097.3 = \ell^2 \qquad \text{Simplify.}$$
$$588.3 = \ell \qquad \text{Take the square root of each side.}$$

Now use the slant height to find the lateral surface area of the Pyramid.

$$L.A. = \frac{1}{2}P\ell \qquad \text{Lateral area of a regular pyramid}$$
$$= \frac{1}{2}(4 \cdot 705)(588.3) \qquad \text{Replace } P \text{ with } (705 \cdot 4) \text{ which is the perimeter of the base. Replace } \ell \text{ with 588.3 for the slant height.}$$
$$= 829,503 \qquad \text{Simplify.}$$

So, the lateral surface area of the Pyramid of Khafre is 829,503 ft^2.
23. Sample answer: 2 in. by 3 in. by 2.1 in. **25.** D
27. H **29.** 615.8 ft^2 **31.** 50.3 in^2

pp. 748–751   Chapter Study Guide and Review

1. false; two **3.** false; volume **5.** false; total surface area **7.** false; diameter **9.** false; regular pyramid **11.** 18.8 cm; 28.3 cm^2 **13.** 30 in.
15. 32.2 cm^2 **17.** 200.5 ft^2 **19.** 660 yd^3
21. 38,484.5 cm^3 **23.** 163.3 ft^3 **25.** 445.3 yd^3
27. 78.5 in^3 **29.** 504 m^2; 612 m^2
31. 226.2 yd^2; 326.7 yd^2 **33.** 84 cm^2; 133 cm^2
35. 339.3 ft^2; 452.4 ft^2 **37.** 33,947.5 m^2

Photo Credits

Glossary/Glosario

 Click on the eGlossary link to find out more about these words in the following 13 languages.

Arabic	Cantonese	Hmong	Spanish	Urdu
Bengali	English	Korean	Tagalog	Vietnamese
Brazilian Portuguese	Haitian Creole	Russian		

English

Spanish
(Español)

Aa

accuracy   The degree of closeness of a measurement to the true value.

acute angle   An angle whose measure is less than 90°.

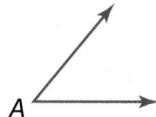

acute triangle   A triangle with all acute angles.

Addition Property of Equality   If you add the same number to each side of an equation, the two sides remain equal.

adjacent angles   Angles that share a common vertex, a common side, and do not overlap. In the figure, the adjacent angles are ∠5 and ∠6.

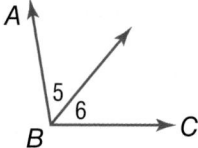

algebra   A branch of mathematics that involves expressions with variables.

algebraic expression   A combination of variables, numbers, and at least one operation.

---

exactitud   Cercanía de una medida a su valor verdadero.

ángulo agudo   Ángulo que mide menos de 90°.

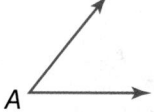

triángulo acutángulo   Triángulo con todos los ángulos agudos.

propiedad de adición de la igualdad   Si sumas el mismo número a ambos lados de una ecuación, los dos lados permanecen iguales.

ángulos adyacentes   Ángulos que comparten un vértice, un lado común y no se traslapan. En la figura, los ángulos adyacentes son ∠5 y ∠6.

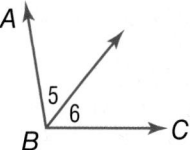

álgebra   Rama de las matemáticas que trabaja con expresiones con variables.

expresión algebraica   Una combinación de variables, números y por lo menos una operación.

alternate exterior angles   Exterior angles that lie on opposite sides of the transversal. In the figure, transversal *t* intersects lines ℓ and *m*. $\angle 1$ and $\angle 7$, and $\angle 2$ and $\angle 8$ are alternate exterior angles. If line ℓ and *m* are parallel, then these pairs of angles are congruent.

ángulos alternos externos   Ángulos externos que se encuentran en lados opuestos de la transversal. En la figura, la transversal *t* interseca las rectas ℓ y *m*. $\angle 1$ y $\angle 7$, y $\angle 2$ y $\angle 8$ son ángulos alternos externos. Si las rectas ℓ y *m* son paralelas, entonces estos ángulos son pares de ángulos congruentes.

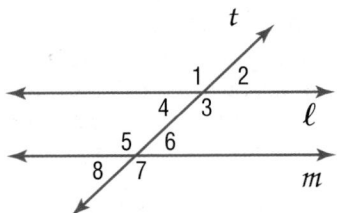

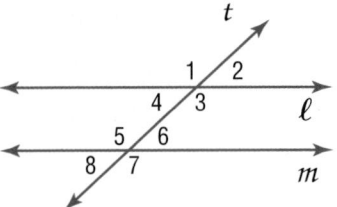

alternate interior angles   Interior angles that lie on opposite sides of the transversal. In the figure below, transversal *t* intersects lines ℓ and *m*. $\angle 3$ and $\angle 5$, and $\angle 4$ and $\angle 6$ are alternate interior angles. If lines ℓ and *m* are parallel, then these pairs of angles are congruent.

ángulos alternos internos   Ángulos internos que se encuentran en lados opuestos de la transversal. En la figura, la transversal *t* interseca las rectas ℓ y *m*. $\angle 3$ y $\angle 5$, y $\angle 4$ y $\angle 6$ son ángulos alternos internos. Si las rectas ℓ y *m* son paralelas, entonces estos ángulos son pares de ángulos congruentes.

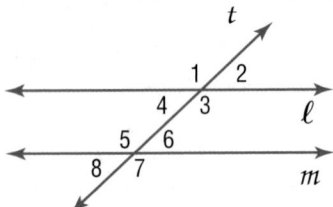

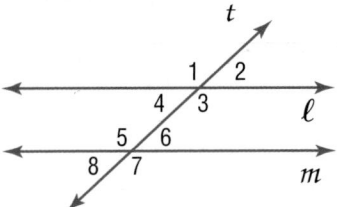

arc   One of two parts of a circle separated by a central angle.

arco   Una de dos partes de un círculo separadas por un ángulo central.

arithmetic sequence   A sequence in which the difference between any two consecutive terms is the same.

sucesión aritmética   Sucesión en la cual la diferencia entre dos términos consecutivos es constante.

Associative Property   The way in which three numbers are grouped when they are added or multiplied does not change their sum or product.

propiedad asociativa   La forma en que se agrupan tres números al sumarlos o multiplicarlos no altera su suma o producto.

Bb

base   In a power, the number that is the common factor. In 10^3, the base is 10. That is, $10^3 = 10 \times 10 \times 10$.

base   En una potencia, número que es el factor común. En 10^3, la base es 10. Es decir, $10^3 = 10 \times 10 \times 10$.

base   The bases of a prism are the two parallel congruent faces.

base   Las bases de un prisma son las dos caras congruentes paralelas.

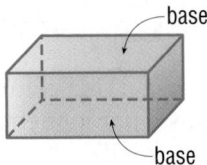

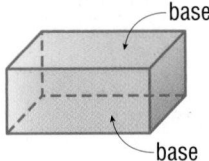

biased sample   A sample drawn in such a way that one or more parts of the population are favored over others.

muestra sesgada   Muestra en que se favorece una o más partes de una población.

boundary   A line that defines the edge of a graph of a linear inequality.

frontera   Recta que define el límite de una gráfica de una desigualdad lineal.

box-and-whisker plot   A diagram that summarizes data using five values: the median, the upper and lower quartiles, and the extreme values. A box is drawn around the quartile values and whiskers extend from each quartile to the extreme data points.

diagrama de caja y patillas   Diagrama que resume información usando cinco valores: la mediana, los cuartiles superior e inferior y los valores extremos. Se dibuja una caja alrededor de los cuartiles y se trazan patillas que los unan a los valores extremos respectivos.

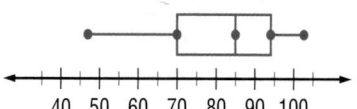

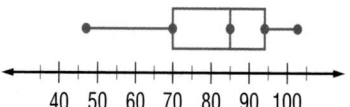

Celsius (°C)   A unit used to measure temperature in the Celsius temperature scale.

Celsius (°C)   Unidad que se usa para medir la temperatura en la escala Celsius.

center   The given point from which all points on a circle are the same distance.

centro   Un punto dado del cual equidistan todos los puntos de un círculo.

center of rotation   A fixed point around which shapes move in a circular motion to a new position.

centro de rotación   Punto fijo alrededor del cual se giran las figuras en movimiento circular alrededor de un punto fijo.

central angle   An angle that intersects a circle in two points and has its vertex at the center of the circle.

ángulo central   Ángulo que interseca un círculo en dos puntos y cuyo vértice es el centro del círculo.

circle   The set of all points in a plane that are the same distance from a given point called the center.

círculo   Conjunto de todos los puntos en un plano que equidistan de un punto dado llamado centro.

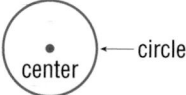

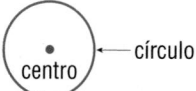

circumference   The distance around a circle.

circunferencia   La distancia alrededor de un círculo.

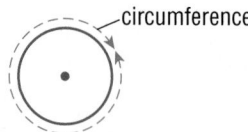

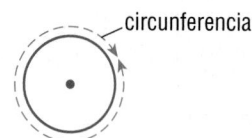

chord   A segment with endpoints that are on a circle.

cuerda   Segmento cuyos extremos están sobre un círculo.

coefficient   The numerical factor of a term that contains a variable.

coeficiente   Factor numérico de un término que contiene una variable.

combination An arrangement or listing in which order is not important.

combinación Arreglo o lista en que el orden no es importante.

common difference The difference between any two consecutive terms in an arithmetic sequence.

diferencia común La diferencia entre cualquier par de términos consecutivos en una sucesión aritmética.

Commutative Property The order in which two numbers are added or multiplied does not change their sum or product.

propiedad conmutativa La forma en que se suman o multiplican dos números no altera su suma o producto.

complementary angles Two angles are complementary if the sum of their measures is 90°.

ángulos complementarios Dos ángulos son complementarios si la suma de sus medidas es 90°.

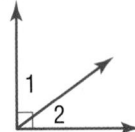

∠1 and ∠2 are complementary angles.

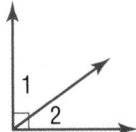

∠1 y ∠2 son complementarios.

composite figure A figure that is made up of two or more shapes.

figura compleja Figura compuesta de dos o más formas.

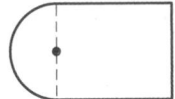

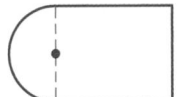

composite solid An object made up of more than one type of solid.

sólido complejo Cuerpo compuesto de más de un tipo de sólido.

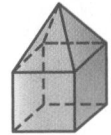

composition of transformations The resulting transformation when a transformation is applied to a figure and then another transformation is applied to its image.

composición de transformaciones Transformación que resulta cuando se aplica una transformación a una figura y luego se le aplica otra transformación a su imagen.

compound event An event that consists of two or more simple events.

evento compuesto Evento que consta de dos o más eventos simples.

compound inequality Two inequalities connected by the words *and* or *or.*

desigualdad compuesta Dos desigualdades conectadas por las palabras *y* u *o.*

compound interest Interest paid on the initial principal and on interest earned in the past.

interés compuesto Interés que se paga por el capital inicial y sobre el interés ganado en el pasado.

cone   A three-dimensional figure with one circular base and a vertex connected by a curved side.

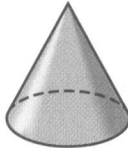

cono   Figura tridimensional con una base circular y un vértice conectado por un lado curvo.

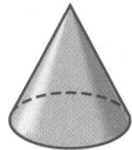

congruent   Having the same measure.

congruente   Que tienen la misma medida.

constant   A term without a variable.

constante   Término sin variables.

constant rate of change   The rate of change between any two points in a linear relationship is the same or *constant*.

tasa constante de cambio   La tasa de cambio entre dos puntos cualesquiera en una relación lineal permanece igual o *constante*.

constant of variation   A constant ratio in a direct variation.

constante de variación   Razón constante en una relación de variación directa.

continuous data   Data that can take on any value. There is no space between data values for a given domain. Graphs are represented by solid lines.

datos continuos   Datos que pueden tomar cualquier valor. No hay espacio entre los valores de los datos para un dominio dado. Las gráficas se representan con rectas sólidas.

convenience sample   A sample which includes members of the population that are easily accessed.

muestra de conveniencia   Muestra que incluye miembros de una población fácilmente accesibles.

converse   The converse of a theorem is formed when the parts of the theorem are reversed. The converse of the Pythagorean Theorem can be used to test whether a triangle is a right triangle. If the sides of the triangle have lengths a, b, and c, such that $c^2 = a^2 + b^2$, then the triangle is a right triangle.

recíproco   El recíproco de un teorema se forma cuando se invierten las partes del teorema. El recíproco del teorema de Pitágoras puede usarse para averiguar si un triángulo es un triángulo rectángulo. Si las longitudes de los lados de un triángulo son a, b y c, tales que $c^2 = a^2 + b^2$, entonces el triángulo es un triángulo rectángulo.

coordinate plane   A coordinate system in which a horizontal number line and a vertical number line intersect at their zero points.

plano de coordenadas   Sistema de coordenadas en que una recta numérica horizontal y una recta numérica vertical se intersecan en sus puntos cero.

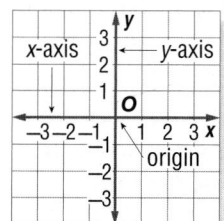

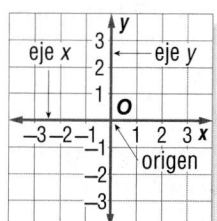

coplanar   Lines that lie in the same plane.

coplanario   Rectas que yacen en el mismo plano.

corresponding angles   Angles that are in the same position on two parallel lines in relation to a transversal.

ángulos correspondientes   Ángulos que están en la misma posición sobre dos rectas paralelas en relación con la transversal.

corresponding parts   Parts of congruent or similar figures that match.

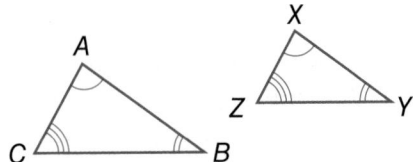

counterexample   A statement or example that shows a conjecture is false.

cross section   The intersection of a solid and a plane.

cube root   One of three equal factors of a number. If $a^3 = b$, then a is the cube root of b. The cube root of 64 is 4 since $4^3 = 64$.

cubic function   A nonlinear function in which the greatest power of the variable is 3.

cylinder   A three-dimensional figure with congruent, parallel bases that are circles connected with a curved side.

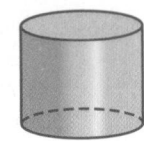

partes correspondientes   Partes de figuras congruentes o semejantes que coinciden.

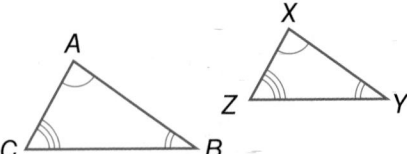

contraejemplo   Ejemplo o enunciado que demuestra que una conjetura es falsa.

sección transversal   Intersección de un sólido y un plano.

raíz cúbica   Uno de tres factores iguales de un número. Si $a^3 = b$, entonces a es la raíz cúbica de b. La raíz cúbica de 64 es 4, dado que $4^3 = 64$.

función cúbica   Función no lineal en la cual la mayor potencia de la variable es 3.

cilindro   Figura tridimensional con bases circulares congruentes y paralelas unidas por un lado curvo.

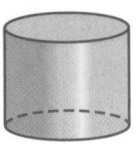

Dd

defining a variable   Choosing a variable and a quantity for the variable to represent in an expression or equation.

degree   A unit used to measure angles.

degree   A unit used to measure temperature.

dependent events   Two or more events in which the outcome of one event does affect the outcome of the other event or events.

dependent variable   The variable in a relation with a value that depends on the value of the independent variable.

derived unit   A unit that is derived from a measurement system base unit, such as length, mass, or time.

definir una variable   El elegir una variable y una cantidad que esté representada por la variable en una expresión o en una ecuación.

grado   Unidad que se usa para medir ángulos.

grado   Unidad que se usa para medir la temperatura.

eventos dependientes   Dos o más eventos en que el resultado de uno de ellos afecta el resultado de los otros eventos.

variable dependiente   La variable en una relación cuyo valor depende del valor de la variable independiente.

unidad derivada   Unidad derivada de una unidad básica de un sistema de medidas como por ejemplo, la longitud, la masa o el tiempo.

diagonal A line segment whose endpoints are vertices that are neither adjacent nor on the same face.

diameter The distance across a circle through its center.

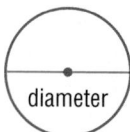

dilation A transformation that enlarges or reduces a figure by a scale factor.

dimensional analysis The process of including units of measurement when you compute.

direct variation A relationship between two variable quantities with a constant ratio.

discount The amount by which a regular price is reduced.

discrete data Data with space between possible data values. Graphs are represented by dots.

disjoint events Events that cannot happen at the same time.

Distance Formula The distance d between two points with coordinates (x_1, y_1) and (x_2, y_2) is given by the formula
$$d = \sqrt{(x_1 - x_2)^2 + (y_1 - y_2)^2}.$$

Distributive Property To multiply a sum by a number, multiply each addend by the number outside the parentheses.
$$5(x + 3) = 5x + 15$$

Division Property of Equality If you divide each side of an equation by the same nonzero number, the two sides remain equal.

domain The set of x-coordinates in a relation.

double box-and-whisker plot Two box-and-whisker plots graphed on the same number line.

diagonal Segmento de recta cuyos extremos son vértices que no son ni adyacentes ni yacen en la misma cara.

diámetro La distancia a través de un círculo pasando por el centro.

homotecia Transformación que produce la ampliación o reducción de una imagen por un factor de escala.

análisis dimensional Proceso que incorpora las unidades de medida al hacer cálculos.

variación directa Relación entre dos cantidades variables con una razón constante.

descuento La cantidad de reducción del precio normal.

datos discretos Datos con espacios entre posibles valores de datos. Las gráficas están representadas por puntos.

eventos disjuntos Eventos que no pueden ocurrir al mismo tiempo.

fórmula de la distancia La distancia d entre dos puntos con coordenadas (x_1, y_1) y (x_2, y_2) viene dada por la fórmula
$$d = \sqrt{(x_1 - x_2)^2 + (y_1 - y_2)^2}.$$

propiedad distributiva Para multiplicar una suma por un número, multiplica cada sumando por el número fuera de los paréntesis.
$$5(x + 3) = 5x + 15$$

propiedad de división de la igualdad Si cada lado de una ecuación se divide entre el mismo número no nulo, los dos lados permanecen iguales.

dominio Conjunto de coordenadas x en una relación.

diagrama de caja y patilla doble Dos diagramas de caja y patilla graficados sobre la misma recta numérica.

edge   The intersection of two faces of a three-dimensional figure.

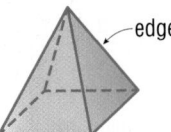

arista   La intersección de dos caras de una figura tridimensional.

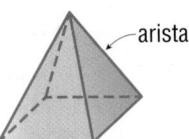

equation   A mathematical sentence stating that two quantities are equal.

ecuación   Enunciado matemático que establece que dos cantidades son iguales.

equiangular   A polygon in which all angles are congruent.

equiangular   Polígono en el cual todos los ángulos son congruentes.

equilateral triangle   A triangle with three congruent sides.

triángulo equilátero   Triángulo con tres lados congruentes.

equivalent expressions   Expressions that have the same value regardless of the value(s) of the variable(s).

expresiones equivalentes   Expresiones que poseen el mismo valor, sin importar los valores de la(s) variable(s).

event   An outcome is a possible result.

evento   Un resultado posible.

experimental probability   An estimated probability based on the relative frequency of positive outcomes occurring during an experiment.

probabilidad experimental   Probabilidad estimada que se basa en la frecuencia relativa de los resultados positivos que ocurren durante un experimento.

exponent   In a power, the number of times the base is used as a factor. In 10^3, the exponent is 3.

exponente   En una potencia, el número de veces que la base se usa como factor. En 10^3, el exponente es 3.

exponential function   A nonlinear function in which the base is a constant and the exponent is an independent variable.

función exponencial   Función no lineal en la cual la base es una constante y el exponente es una variable independiente.

exterior angles   The four outer angles formed by two lines cut by a transversal.

ángulo externo   Los cuatro ángulos exteriores que se forman cuando una transversal corta dos rectas.

Ff

face   Any surface that forms a side or a base of a prism.

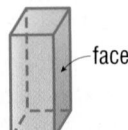

cara   Cualquier superficie que forma un lado o una base de un prisma.

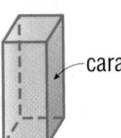

Fahrenheit (°F)   A unit used to measure temperature in the Fahrenheit temperature scale.

fair game   A game where each player has an equally likely chance of winning.

function   A relation in which each member of the domain (input value) is paired with exactly one member of the range (output value).

function table   A table organizing the domain, rule, and range of a function.

Fundamental Counting Principle   Uses multiplication of the number of ways each event in an experiment can occur to find the number of possible outcomes in a sample space.

Fahrenheit (°F)   Unidad que se usa para medir la temperatura en la escala Fahrenheit.

juego justo   Juego donde cada jugador tiene igual posibilidad de ganar.

función   Relación en la cual a cada elemento del dominio (valor de entrada) le corresponde exactamente un único elemento del rango (valor de salida).

tabla de funciones   Tabla que organiza la regla de entrada y de salida de una función.

principio fundamental de contar   Método que usa la multiplicación del número de maneras en que cada evento puede ocurrir en un experimento, para calcular el número de resultados posibles en un espacio muestral.

Gg

geometric sequence   A sequence in which each term after the first is found by multiplying the previous term by a constant.

sucesión geométrica   Sucesión en la cual cada término después del primero se determina multiplicando el término anterior por una constante.

Hh

half-plane   The part of the coordinate plane on one side of the boundary.

hypotenuse   The side opposite the right angle in a right triangle.

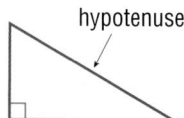

semiplano   Parte del plano de coordenadas en un lado de la frontera.

hipotenusa   El lado opuesto al ángulo recto de un triángulo rectángulo.

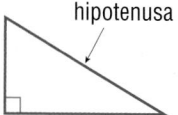

Ii

image   The resulting figure after a transformation.

independent events   Two or more events in which the outcome of one event does not affect the outcome of the other event(s).

independent variable   The variable in a function with a value that is subject to choice.

imagen   Figura que resulta después de una transformación.

eventos independientes   Dos o más eventos en los cuales el resultado de un evento no afecta el resultado de los otros eventos.

variable independiente   Variable en una función cuyo valor está sujeto a elección.

indirect measurement   A technique using properties of similar polygons to find distances or lengths that are difficult to measure directly.

medición indirecta   Técnica que usa las propiedades de polígonos semejantes para calcular distancias o longitudes difíciles de medir directamente.

inequality   A mathematical sentence that contains $<$, $>$, $\neq$, $\leq$, or $\geq$.

desigualdad   Enunciado matemático que contiene $<$, $>$, $\neq$, $\leq$, o $\geq$.

inscribed angle   An angle that has its vertex on the circle. Its sides contain chords of the circle.

ángulo inscrito   Ángulo cuyo vértice está en el círculo y cuyos lados contienen cuerdas del círculo.

interest   The amount of money paid or earned for the use of money.

interés   Cantidad que se cobra o se paga por el uso del dinero.

interior angle   An angle inside a polygon.

ángulo interno   Ángulo dentro de un polígono.

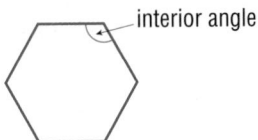

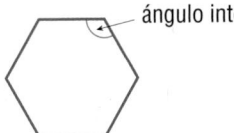

interior angles   The four inside angles formed by two lines cut by a transversal.

ángulo interno   Los cuatro ángulos internos formados por dos rectas intersecadas por una transversal.

interquartile range   The range of the middle half of a set of data. It is the difference between the upper quartile and the lower quartile.

rango intercuartílico   El rango de la mitad central de un conjunto de datos. Es la diferencia entre el cuartil superior y el cuartil inferior.

intersection   The overlapping of two graphs of a compound inequality.

intersección   Dos gráficas de una desigualdad compuesta que se traslapan.

inverse operations   Pairs of operations that undo each other. Addition and subtraction are inverse operations. Multiplication and division are inverse operations.

peraciones inversas   Pares de operaciones que se anulan mutuamente. La adición y la sustracción son operaciones inversas. La multiplicación y la división son operaciones inversas.

irrational number   A number that cannot be expressed as the quotient $\frac{a}{b}$, where a and b are integers and $b \neq 0$.

números irracionales   Número que no se puede expresar como el cociente $\frac{a}{b}$, donde a y b son enteros y $b \neq 0$

isosceles triangle   A triangle with at least two congruent sides.

triángulo isóceles   Triángulo con por lo menos dos lados congruentes.

Kk

kelvin (K)   A unit used to measure temperature in the Kelvin temperature scale.

kelvin (K)   Unidad que se usa para medir la temperatura en la escala Kelvin.

lateral face   Any flat surface that is not a base.

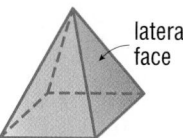

lateral
face

cara lateral   Cualquier superficie plana que no es la base.

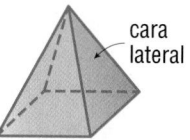

cara
lateral

lateral surface area   The sum of the areas of the lateral faces of a solid.

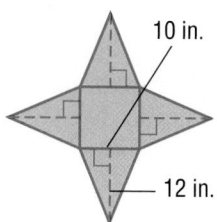

10 in.

12 in.

lateral area = $4\left(\frac{1}{2} \times 10 \times 12\right)$ = 240 square inches

área de superficie lateral   La suma de las áreas de las caras laterales de un sólido.

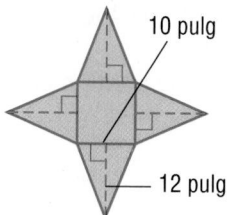

10 pulg

12 pulg

área lateral = $4\left(\frac{1}{2} \times 10 \times 12\right)$ = 240 pulgadas cuadradas

legs   The two sides of a right triangle that form the right angle.

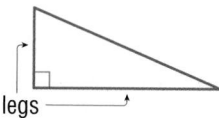

legs

catetos   Los dos lados de un triángulo rectángulo que forman el ángulo recto.

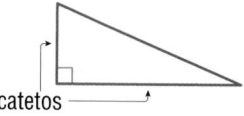

catetos

like fractions   Fractions that have the same denominators.

fracciones semejantes   Fracciones que tienen el mismo denominador.

like terms   Terms that contain the same variable(s) to the same powers.

términos semejantes   Términos que contienen la misma variable o variables elevadas a la misma potencia.

linear   To fall in a straight line.

lineal   Que cae en una línea recta.

linear equation   An equation with a graph that is a straight line.

ecuación lineal   Ecuación cuya gráfica es una recta.

linear function   A function in which the graph of the solutions forms a line.

función lineal   Función en la cual la gráfica de las soluciones forma un recta.

linear relationship   A relationship that has a straight-line graph.

relación lineal   Relación cuya gráfica es una recta.

line of best fit   A line that is very close to most of the data points in a scatter plot.

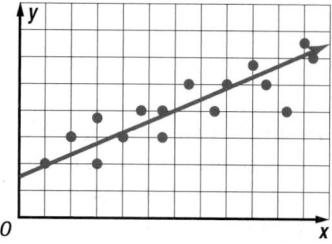

recta de mejor ajuste   Recta que más se acerca a la mayoría de puntos de los datos en un diagrama de dispersión.

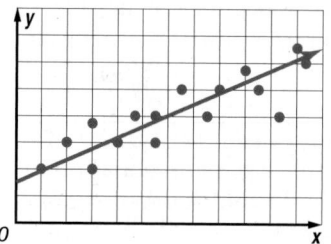

line of reflection   The line over which a figure is reflected.

line of symmetry   Each half of a figure is a mirror image of the other half when a line of symmetry is drawn.

line symmetry   A figure has line symmetry if a line can be drawn so that one half of the figure is a mirror image of the other half.

literal equation   An equation or formula that has more than one variable.

lower quartile   The median of the lower half of a set of data, represented by LQ.

línea de reflexión   Línea a través de la cual se refleja una figura.

eje de simetría   Recta que divide una figura en dos mitades especulares.

simetría lineal   Una figura tiene simetría lineal si se puede trazar una recta de manera que una mitad de la figura sea una imagen especular de la otra mitad.

ecuación literal   Ecuación o fórmula con más de una variable.

cuartil inferior   La mediana de la mitad inferior de un conjunto de datos, la cual se denota por CI.

Mm

major arc   An arc measuring more than 180°.

markup   The amount the price of an item is increased above the price the store paid for the item.

mean   The sum of the data divided by the number of items in the set.

measures of central tendency   Numbers that describe the center of a set of data.

measures of variation   Numbers used to describe the distribution or spread of a set of data.

median   The middle number of a data set ordered from least to greatest, or the mean of the middle two numbers.

minor arc   An arc measuring less than 180°.

mode   The number(s) or item(s) that appear most often in a set of data.

monomial   A number, a variable, or a product of a number and one or more variables.

Multiplication Property of Equality   If you multiply each side of an equation by the same number, the two sides remain equal.

arco mayor   Arco que mide más de 180°.

margen de utilidad   Cantidad de aumento en el precio de un artículo por encima del precio que paga la tienda por dicho artículo.

media   La suma de datos dividida entre el número total de artículos.

medidas de tendencia central   Números que describen el centro de un conjunto de datos.

medidas de variación   Números que se usan para describir la distribución o separación de un conjunto de datos.

mediana   El número central de los datos ordenados de menor a mayor o la media de los dos números centrales.

arco menor   Arco que mide menos de 180°.

moda   El número(s) o artículo(s) que aparece con más frecuencia en un conjunto de datos.

monomio   Un número, una variable o el producto de un número por una o más variables.

propiedad de multiplicación de la igualdad   Si cada lado de una ecuación se multiplica por el mismo número, los lados permanecen iguales.

multiplicative inverses Two numbers with a product of 1. The multiplicative inverse of $\frac{2}{3}$ is $\frac{3}{2}$.

inversos multiplicativo Dos números cuyo producto es 1. El inverso multiplicativo de $\frac{2}{3}$ es $\frac{3}{2}$.

Nn

net A two-dimensional pattern of a three-dimensional figure.

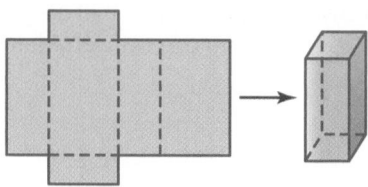

red Patrón bidimensional de una figura tridimensional.

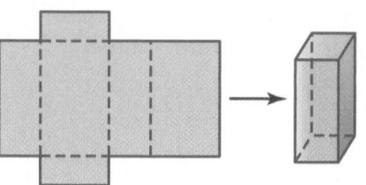

nonlinear function A function whose rate of change is not constant. The graph of a nonlinear function is not a straight line.

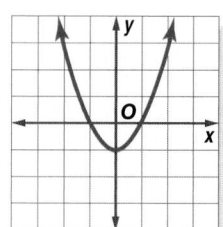

función no lineal Función cuya tasa de cambio no es constante. La gráfica de una función no lineal no es una recta.

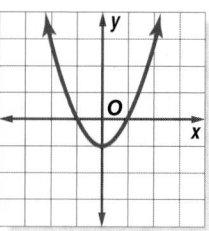

Oo

obtuse angle An angle whose measure is between 90° and 180°.

ángulo obtuso Ángulo cuya medida está entre 90° y 180°.

obtuse triangle A triangle with one obtuse angle.

triángulo obtusángulo Triángulo con un ángulo obtuso.

odds against The ratio that compares the number of ways the event cannot occur to the number of ways the event can occur.

posibilidades en contra Razón que compara el número de maneras en que puede ocurrir un evento no al número de maneras en que puede ocurrir el evento.

odds in favor The ratio that compares the number of ways the event can occur to the number of ways the event cannot occur.

posibilidades a favor Razón que compara el número de maneras en que puede ocurrir un evento al número de maneras en que no puede ocurrir el evento.

ordered pair A pair of numbers used to locate a point in the coordinate plane. The ordered pair is written in this form: (*x*-coordinate, *y*-coordinate).

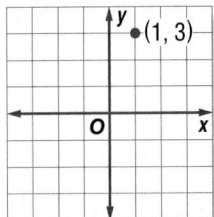

par ordenado Par de números que se utiliza para ubicar un punto en un plano de coordenadas. Se escribe de la siguiente forma: (coordenada *x*, coordenada *y*).

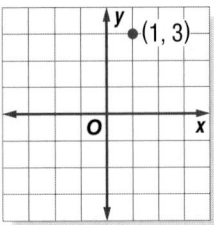

origin The point of intersection of the *x*-axis and *y*-axis in a coordinate plane.

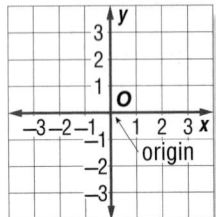

origen Punto en que el eje *x* y el eje *y* se intersecan en un plano de coordenadas.

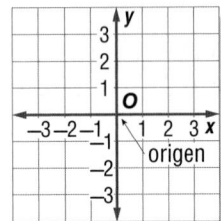

outcome One possible result of a probability event. For example, 4 is an outcome when a number cube is rolled.

resultado Una consecuencia posible de un evento de probabilidad. Por ejemplo, 4 es un resultado posible al lanzar un cubo numérico.

outlier Data that are more than 1.5 times the interquartile range from the upper or lower quartiles.

valor atípico Datos que distan de los cuartiles respectivos más de 1.5 veces la amplitud intercuartílica.

Pp

paragraph proof A paragraph that explains why a statement or conjecture is true.

prueba por párrafo Párrafo que explica por qué es verdadero un enunciado o una conjetura.

parallel Lines that never intersect no matter how far they extend.

paralelo Rectas que nunca se intersecan sea cual sea su extensión.

parallel lines Lines in the same plane that never intersect or cross. The symbol ∥ means parallel.

rectas paralelas Rectas que yacen en un mismo plano y que no se intersecan. El símbolo ∥ significa paralela a.

parallelogram A quadrilateral with both pairs of opposite sides parallel and congruent.

paralelogramo Cuadrilátero con ambos pares de lados opuestos, paralelos y congruentes.

percent equation An equivalent form of a percent proportion in which the percent is written as a decimal.

$$\text{part} = \text{percent} \cdot \text{whole}$$

ecuación porcentual Forma equivalente de proporción porcentual en la cual el por ciento se escribe como un decimal.

$$\text{parte} = \text{por ciento} \cdot \text{entero}$$

percent of change   A ratio that compares the change in quantity to the original amount.

$$\text{percent of change} = \frac{\text{amount of change}}{\text{original amount}}$$

percent of decrease   When the percent of change is negative.

percent of increase   When the percent of change is positive.

percent proportion   Compares part of a quantity to the whole quantity using a percent.

$$\frac{\text{part}}{\text{whole}} = \frac{\text{percent}}{100}$$

perfect cube   A rational number whose cube root is a whole number. 27 is a perfect cube because its cube root is 3.

perfect square   A rational number whose square root is a whole number. 25 is a perfect square because its square root is 5.

permutation   An arrangement or listing in which order is important.

perpendicular lines   Two lines that intersect to form right angles.

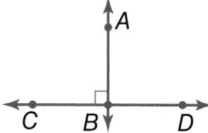

pi   The ratio of the circumference of a circle to its diameter. The Greek letter π represents this number. The value of pi is always 3.1415926… .

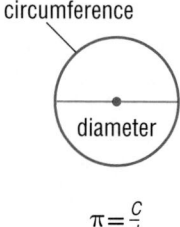

$$\pi = \frac{C}{d}$$

polygon   A simple, closed figure formed by three or more line segments.

porcentaje de cambio   cantidad de cambio/cantidad original

$$\text{procentaje de cambio} = \frac{\text{cantidad de cambio}}{\text{cantidad original}}$$

porcentaje de disminución   Cuando el porcentaje de cambio es negativo.

porcentaje de aumento   Cuando el porcentaje de cambio es positivo.

proporción porcentual   Compara parte de una cantidad con la cantidad total mediante un por ciento.

$$\frac{\text{parte}}{\text{entero}} = \frac{\text{por ciento}}{100}$$

cubo perfecto   Número racional cuya raíz cúbica es un número entero. 27 es un cubo perfecto porque su raíz cúbica es 3.

cuadrados perfectos   Número racional cuya raíz cuadrada es un número entero. 25 es un cuadrado perfecto porque su raíz cuadrada es 5.

permutación   Arreglo o lista donde el orden es importante.

rectas perpendiculares   Dos rectas que se intersecan formando ángulos rectos.

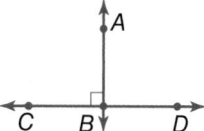

pi   Razón de la circunferencia de un círculo al diámetro del mismo. La letra griega π representa este número. El valor de pi es siempre 3.1415926… .

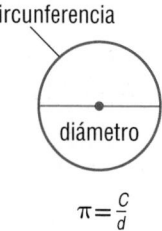

$$\pi = \frac{C}{d}$$

polígono   Figura simple y cerrada formada por tres o más segmentos de recta.

polyhedron   A solid with flat surfaces that are polygons.

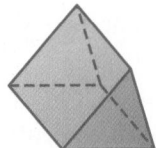

poliedro   Sólido cuyas superficies planas son polígonos.

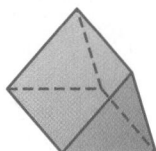

power   A product of repeated factors using an exponent and a base. The power 7^3 is read *seven to the third power,* or *seven cubed.*

potencia   Producto de factores repetidos con un exponente y una base. La potencia 7^3 se lee *siete a la tercera potencia* o *siete al cubo.*

precision   The ability of a measurement to be consistently reproduced.

precisión   Capacidad de una medida a ser reproducida consistentemente.

preimage   The original figure before a transformation.

preimagen   Figura original antes de una transformación.

principal   The amount of money invested or borrowed.

capital   Cantidad de dinero que se invierte o que se toma prestada.

prism   A polyhedron with two parallel, congruent faces called bases.

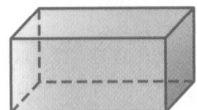

prisma   Poliedro con dos caras congruentes y paralelas llamadas bases.

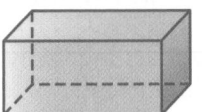

probability   The chance that some event will happen. It is the ratio of the number of ways a certain event can occur to the number of possible outcomes.

probabilidad   La posibilidad de que suceda un evento. Es la razón del número de maneras en que puede ocurrir un evento al número total de resultados posibles.

proof   A logical argument in which each statement that is made is supported by a statement that is accepted as true.

prueba   Argumento lógico en el cual cada enunciado hecho se respalda con un enunciado que se acepta como verdadero.

property   A statement that is true for any numbers.

propiedad   Enunciado que se cumple para cualquier número.

pyramid   A polyhedron with one base that is a polygon and faces that are triangles.

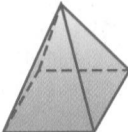

pirámide   Poliedro cuya base tiene forma de polígono y caras en forma de triángulos.

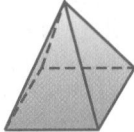

Pythagorean Theorem   In a right triangle, the square of the length of the hypotenuse c is equal to the sum of the squares of the lengths of the legs a and b. $a^2 + b^2 = c^2$

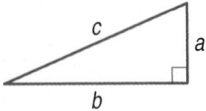

Teorema de Pitágoras   En un triángulo rectángulo, el cuadrado de la longitud de la hipotenusa es igual a la suma de los cuadrados de las longitudes de los catetos. $a^2 + b^2 = c^2$

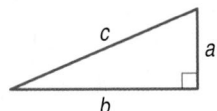

Qq

quadrants   The four sections of the coordinate plane.

```
              y-axis ▲
         ┌──────┬──────┐
         │Quadrant II│Quadrant I│
    ◄────┼──────O──────┼────► x-axis
         │Quadrant III│Quadrant IV│
         └──────┴──────┘
              ▼
```

quadratic function   A function in which the greatest power of the variable is 2.

quadrilateral   A closed figure with four sides and four angles.

qualitative graph   A graph used to represent situations that do not necessarily have numerical values.

quartiles   Values that divide a set of data into four equal parts.

cuadrantes   Las cuatro secciones del plano de coordenadas.

```
              eje y ▲
         ┌──────┬──────┐
         │Cuadrante II│Cuadrante I│
    ◄────┼──────O──────┼────► eje x
         │Cuadrante III│Cuadrante IV│
         └──────┴──────┘
              ▼
```

función cuadrática   Función en la cual la potencia mayor de la variable es 2.

cuadrilátero   Figura cerrada con cuatro lados y cuatro ángulos.

gráfica cualitativa   Gráfica que se usa para representar situaciones que no tienen valores numéricos necesariamente.

cuartiles   Valores que dividen un conjunto de datos en cuatro partes iguales.

Rr

radical sign   The symbol used to indicate a positive square root, $\sqrt{}$.

radius   The distance from the center of a circle to any point on the circle.

random   Outcomes occur at random if each outcome is equally likely to occur.

range   The set of y-coordinates in a relation.

range   The difference between the greatest number (maximum) and the least number (minimum) in a set of data.

rational number   Numbers that can be written as the ratio of two integers in which the denominator is not zero. All integers, fractions, mixed numbers, and percents are rational numbers.

signo radical   Símbolo que se usa para indicar una raíz cuadrada no positiva, $\sqrt{}$.

radio   Distancia desde el centro de un círculo hasta cualquier punto del mismo.

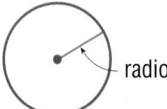

azar   Los resultados ocurren al azar si todos los resultados son equiprobables.

rango   Conjunto de coordenadas y en una relación.

rango   La diferencia entre el número mayor (máximo) y el número menor (mínimo) en un conjunto de datos.

número racional   Números que pueden escribirse como la razón de dos enteros en los que el denominador no es cero. Todos los enteros, fracciones, números mixtos y porcentajes son números racionales.

real numbers   The set of rational numbers together with the set of irrational numbers.

número real   El conjunto de números racionales junto con el conjunto de números irracionales.

reciprocals   The multiplicative inverse of a number. The product of reciprocals is 1.

recíproco   El inverso multiplicativo de un número. El producto de recíprocos es 1.

reflection   A transformation where a figure is flipped over a line. Also called a flip.

reflexión   Transformación en la cual una figura se voltea sobre una recta. También se conoce como simetría de espejo.

regular polygon   A polygon that is equilateral and equiangular.

polígono regular   Polígono equilátero y equiangular.

regular pyramid   A pyramid whose base is a regular polygon.

pirámide regular   Pirámide cuya base es un polígono regular.

relation   Any set of ordered pairs.

relación   Cualquier conjunto de pares ordenados.

relative frequency   The ratio of the number of experimental successes to the total number of experimental attempts.

frecuencia relativa   Razón del número de éxitos experimentales al número total de intentos experimentales.

repeating decimal   A decimal whose digits repeat in groups of one or more. Examples are 0.181818… and 0.8333… .

decimal periódico   Decimal cuyos dígitos se repiten en grupos de uno o más. Por ejemplo: 0.181818… y 0.8333… .

rhombus   A parallelogram with four congruent sides.

rombo   Paralelogramo con cuatro lados congruentes.

right angle   An angle whose measure is exactly 90°.

ángulo recto   Ángulo que mide exactamente 90°.

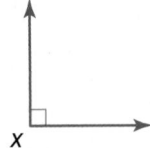

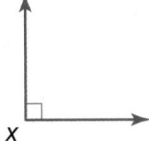

right triangle   A triangle with one right angle.

triángulo rectángulo   Triángulo con un ángulo recto.

rise   The vertical change between any two points on a line.

elevación   El cambio vertical entre cualquier par de puntos en una recta.

rotation   A transformation in which a figure is turned about a fixed point.

rotación   Transformación en la cual una figura se gira alrededor de un punto fijo.

run   The horizontal change between any two points on a line.

carrera   El cambio horizontal entre cualquier par de puntos en una recta.

sales tax   An additional amount of money charged on certain goods and services.

impuesto sobre las ventas   Cantidad de dinero adicional que se cobra por ciertos artículos y servicios.

sample   A randomly-selected group chosen for the purpose of collecting data.

muestra   Subconjunto de una población que se usa con el propósito de recoger datos.

sample space   The set of all possible outcomes of a probability experiment.

espacio muestral   Conjunto de todos los resultados posibles de un experimento de probabilidad.

scale factor   The ratio of the lengths of two corresponding sides of two similar polygons.

factor de escala   La razón de las longitudes de dos lados correspondientes de dos polígonos semejantes.

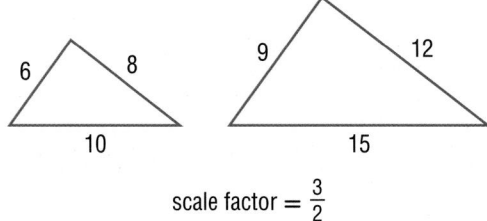

scale factor = $\frac{3}{2}$

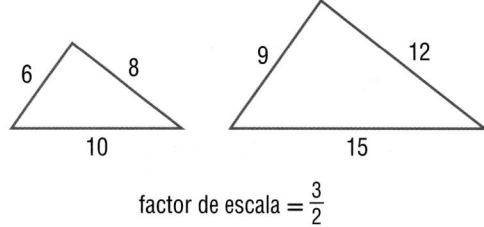

factor de escala = $\frac{3}{2}$

scalene triangle   A triangle with no congruent sides.

triángulo escaleno   Triángulo sin lados congruentes.

scatter plot   A graph that shows the relationship between a data set with two variables graphed as ordered pairs on a coordinate plane.

diagrama de dispersión   Gráfica que muestra la relación entre un conjunto de datos con dos variables graficadas como pares ordenados en un plano de coordenadas.

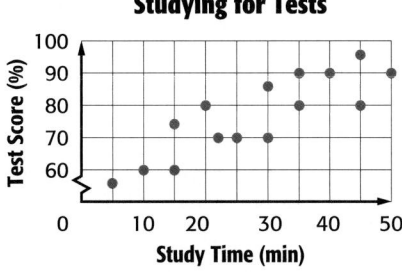

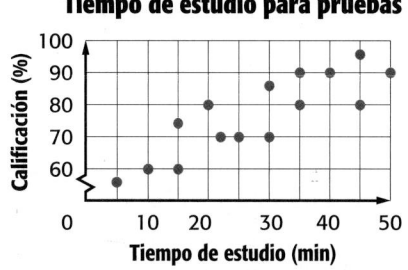

scientific notation   A compact way of writing numbers with absolute values that are very large or very small. In scientific notation, 5,500 is 5.5×10^3.

notación científica   Manera abreviada de escribir números con valores absolutos que son muy grandes o muy pequeños. En notación científica, 5,500 es 5.5×10^3.

secant   Any line that intersects a circle in exactly two points.

secante   Cualquier recta que interseca un círculo exactamente en dos puntos.

selling price   The amount the customer pays for an item.

precio de venta   Cantidad de dinero que paga un consumidor por un artículo.

semicircle   An arc measuring 180°.

semicírculo   Arco que mide 180°.

sequence An ordered list of numbers, such as 0, 1, 2, 3 or 2, 4, 6, 8.

similar polygons Polygons that have the same shape.

simple interest Interest paid only on the initial principal of a savings account or loan.

simple random sample A sample where each item or person in the population is as likely to be chosen as any other.

simplest form An algebraic expression that has no like terms and no parentheses.

simplify To perform all possible operations in an expression.

simulation An experiment that is designed to act out a given situation.

slant height The altitude or height of each lateral face of a pyramid.

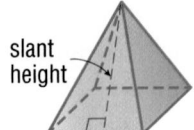

slope The rate of change between any two points on a line. The ratio of the rise, or vertical change, to the run, or horizontal change.

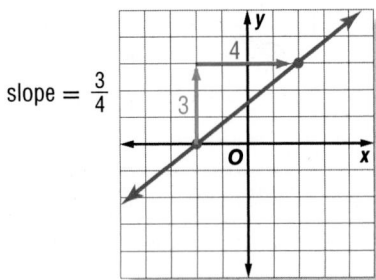

slope = $\frac{3}{4}$

slope-intercept form An equation written in the form $y = mx + b$, where m is the slope and b is the y-intercept.

solid A three-dimensional figure formed by intersecting planes.

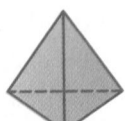

sucesión Lista ordenada de números, tales como 0, 1, 2, 3 o 2, 4, 6, 8.

polígonos semejantes Polígonos con la misma forma.

interés simple Interés que se paga sólo sobre el capital inicial de una cuenta de ahorros o préstamo.

muestra aleatoria simple Muestra de una población que tiene la misma probabilidad de escogerse que cualquier otra.

forma reducida Expresión algebraica que carece de términos semejantes y de paréntesis.

simplificar Realizar todas las operaciones posibles en una expresión.

simulacro Experimento diseñado para representar una situación dada.

altura oblicua La longitud de la altura de cada cara lateral de una pirámide.

pendiente Razón de cambio entre cualquier par de puntos en una recta. La razón de la altura, o cambio vertical, a la carrera, o cambio horizontal.

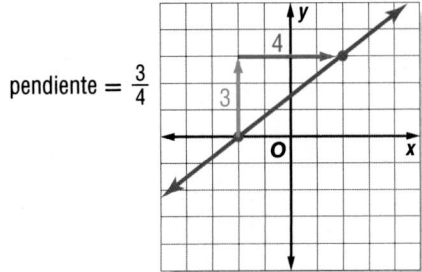

pendiente = $\frac{3}{4}$

forma pendiente intersección Ecuación de la forma $y = mx + b$, donde m es la pendiente y b es la intersección y.

sólido Figura tridimensional formada por planos que se intersecan.

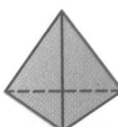

sphere   The set of all points in space that are a given distance from a given point called the center.

esfera   Conjunto de todos los puntos en el espacio que están a una distancia dada de un punto dado llamado centro.

square root   One of the two equal factors of a number. If $a^2 = b$, then a is the square root of b. A square root of 144 is 12 since $12^2 = 144$.

raíz cuadrada   Uno de dos factores iguales de un número. Si $a^2 = b$, la a es la raíz cuadrada de b. Una raíz cuadrada de 144 es 12 porque $12^2 = 144$.

standard form   An equation written in the form $Ax + By = C$.

forma estándar   Ecuación escrita en la forma $Ax + By = C$.

straight angle   An angle whose measure is exactly 180°.

ángulo llano   Ángulo que mide exactamente 180°.

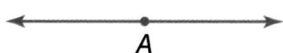

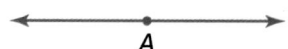

stratified random sample   A sample where the population is divided into similar non-overlapping groups.

muestra aleatoria estratificada   Muestra donde la polación se divide en grupos semejantes que no se sobreponen.

substitution   An algebraic model that can be used to find the exact solution of a system of equations.

sustitución   Modelo algebraico que se puede usar para calcular la solución exacta de un sistema de ecuaciones.

Subtraction Property of Equality   If you subtract the same number from each side of an equation, the two sides remain equal.

propiedad de sustracción de la igualdad   Si sustraes el mismo número de ambos lados de una ecuación, los dos lados permanecen iguales.

supplementary angles   Two angles are supplementary if the sum of their measures is 180°.

ángulos suplementarios   Dos ángulos son suplementarios si la suma de sus medidas es 180°.

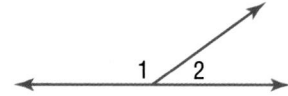

∠1 and ∠2 are supplementary angles.

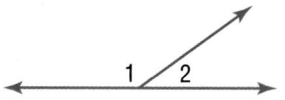

∠1 y ∠2 son ángulos suplementarios.

systemic random sample   A sample where the items or people are selected according to a specific time or item interval.

muestra aleatoria sistemática   Muestra en que los elementos o personas se eligen según un intervalo de tiempo o elemento específico.

system of equations   A set of two or more equations with the same variables.

sistema de ecuaciones   Sistema de ecuaciones con las mismas variables.

Tt

tangent   Any line that intersects a circle in exactly one point.

tangente   Cualquier recta que interseca un círculo exactamente en un sólo punto.

tangent ratio   A ratio that compares the measure of the leg opposite an angle with the measure of the leg adjacent to that angle.

razón tangente   Razón que compara la medida del cateto opuesto a un ángulo con la medida del cateto adyacente a ese ángulo.

term A number, a variable, or a product of numbers and variables.

term Each part of an algebraic expression separated by an addition or subtraction sign.

terminating decimal A decimal whose digits end. Every terminating decimal can be written as a fraction with a denominator of 10, 100, 1,000, and so on.

theoretical probability Probability based on known characteristics or facts.

total surface area The sum of the areas of the surfaces of a solid.

transformation An operation that maps a geometric figure, preimage, onto a new figure, image.

translation A transformation that slides a figure from one position to another without turning.

transversal A line that intersects two or more other lines.

trapezoid A quadrilateral with exactly one pair of parallel sides.

tree diagram A diagram used to show the total number of possible outcomes in a probability experiment.

triangle A figure formed by three line segments that intersect only at their endpoints.

trigonometric ratio A ratio of the lengths of two sides of a right triangle.

trigonometry The study of the properties of triangles.

two-step equation An equation that contains two operations.

two-step inequality An inequality that contains two operations.

término Un número, una variable o un producto de números y variables.

término Cada parte de un expresión algebraica separada por un signo adición o un signo sustracción.

decimal terminal Decimal cuyos dígitos terminan. Todo decimal terminal puede escribirse como una fracción con un denominador 10, 100, 1,000, etc.

probabilidad teórica Probabilidad que se basa en características o hechos conocidos.

área de superficie total La suma del área de las superficies de un sólido.

transformación Operación que convierte una figura geométrica, la pre-imagen, en una figura nueva, la imagen.

traslación Transformación en la cual una figura se desliza de una posición a otra sin hacerla girar.

transversal Recta que interseca dos o más rectas.

trapecio Cuadrilátero con exactamente un par de lados paralelos.

diagrama de árbol Diagrama que se usa para mostrar el número total de resultados posibles en un experimento de probabilidad.

triángulo Figura formada por tres segmentos de recta que se intersecan sólo en sus extremos.

razón trigonométrica Razón de las longitudes de dos lados de un triángulo rectángulo.

trigonometría Estudio de las propiedades de los triángulos.

ecuación de dos pasos Ecuación que contiene dos operaciones.

desigualdad de dos pasos Desigualdad que contiene dos operaciones.

Uu

unbiased sample A sample that is selected so that it is representative of the entire population.

union Everything shown in both graphs of a compound inequality.

muestra no sesgada Muestra que se selecciona de modo que sea representativa de la población entera.

unión Todo lo que muestran ambas gráficas de una desigualdad compuesta.

unit rate/ratio   A rate or ratio with a denominator of 1.

unlike fractions   Fractions whose denominators are different.

upper quartile   The median of the upper half of a set of data, represented by UQ.

tasa/razón unitaria   Una tasa o razón con un denominador de 1.

fracciones con distinto denominador   Fracciones cuyos denominadores son diferentes.

cuartil superior   La mediana de la mitad superior de un conjunto de números, denotada por CS.

variable   A symbol, usually a letter, used to represent a number in mathematical expressions or sentences.

vertex   The point where the sides of an angle meet.

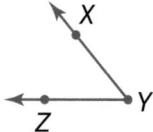

vertex   The vertex of a prism is the point where three or more planes intersect.

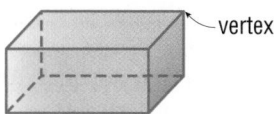

vertical angles   Opposite angles formed by the intersection of two lines. Vertical angles are congruent. In the figure, the vertical angles are $\angle 1$ and $\angle 3$, and $\angle 2$ and $\angle 4$.

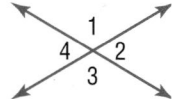

volume   The measure of the space occupied by a solid. Standard measures are cubic units such as in^3 or ft^3.

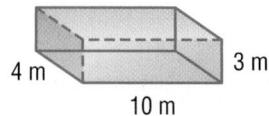

$V = 10 \times 4 \times 3 = 120$ cubic meters

voluntary response sample   A sample which involves only those who want to participate in the sampling.

variable   Un símbolo, por lo general, una letra, que se usa para representar números en expresiones o enunciados matemáticos.

vértice   Punto donde se encuentran los lados.

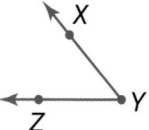

vértice   El vértice de un prisma es el punto en que se intersecan dos o más planos del prisma.

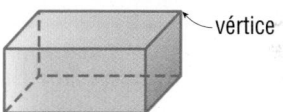

ángulos opuestos por el vértice   Ángulos congruentes que se forman de la intersección de dos rectas. En la figura, los ángulos opuestos por el vértice son $\angle 1$ y $\angle 3$, y $\angle 2$ y $\angle 4$.

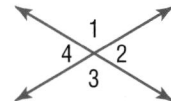

volumen   Medida del espacio que ocupa un sólido. Las medidas estándares son las unidades cúbicas, como pulg^3 o pies^3.

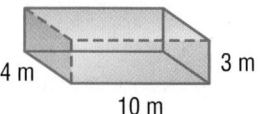

$V = 10 \times 4 \times 3 = 120$ metros cúbicos

muestra de respuesta voluntaria   Muestra que involucra sólo aquellos que quieren participar en el muestreo.

x-axis   The horizontal number line that helps to form the coordinate plane.

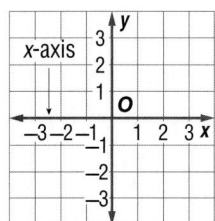

eje *x*   La recta numérica horizontal que ayuda a formar el plano de coordenadas.

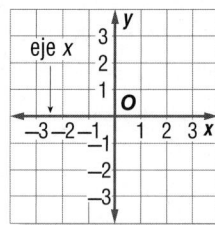

x-coordinate   The first number of an ordered pair.

x-intercept   The *x*-coordinate of the point where the line crosses the *x*-axis.

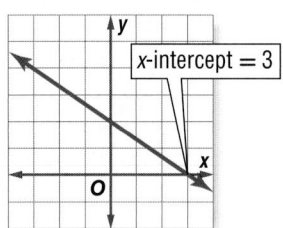

coordenada *x*   El primer número de un par ordenado.

intersección *x*   La coordenada *x* del punto donde cruza la gráfica el eje *x*.

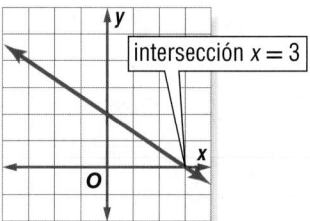

y-axis   The vertical number line that helps to form the coordinate plane.

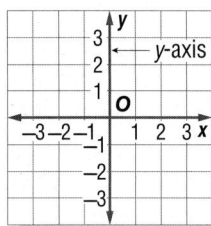

eje *y*   La recta numérica vertical que ayuda a formar el plano de coordenadas.

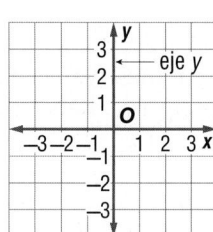

y-coordinate   The second number of an ordered pair.

y-intercept   The *y*-coordinate of the point where the line crosses the *y*-axis.

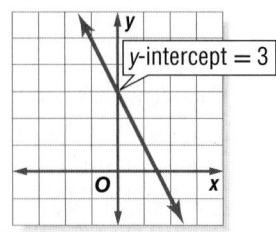

coordenada *y*   El segundo número de un par ordenado.

intersección *y*   La coordenada *y* del punto donde cruza la gráfica el eje *y*.

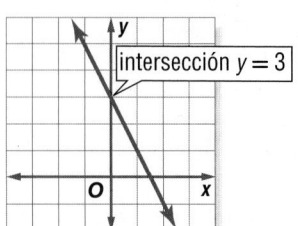

Index

Dd

Ee

Ss

Quick Reference

Pythagorean Theorem	Simple Interest Formula
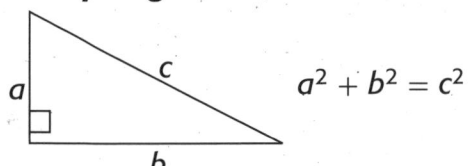 $a^2 + b^2 = c^2$	$I = prt$ where p = principal, r = rate, t = time
Slope-Intercept Form of an Equation of a Line $y = mx + b$ where m = slope and b = y-intercept	**Distance, Rate, Time Formula** $d = rt$ where d = distance, r = rate, t = time

Conversions within a System of Measure

1 yard = 3 feet
1 mile = 1760 yards = 5280 feet
1 acre = 43,560 square feet
1 minute = 60 seconds
1 hour = 60 minutes
1 year = 52 weeks = 365 days

1 cup = 8 fluid ounces
1 pint = 2 cups
1 quart = 2 pints
1 gallon = 4 quarts

1 liter = 1000 milliliters = 1000 cubic centimeters
1 meter = 100 centimeters = 1000 millimeters
1 kilometer = 1000 meters
1 gram = 1000 milligrams
1 kilogram = 1000 grams

1 pound = 16 ounces
1 ton = 2000 pounds

Conversions Between Systems of Measure

When converting from Customary to Metric, use these approximations.

1 inch ≈ 2.54 centimeters
1 foot ≈ 0.305 meter
1 mile ≈ 1.61 kilometers

1 cup ≈ 0.24 liter
1 gallon ≈ 3.785 liters
1 ounce ≈ 28.35 grams
1 pound ≈ 0.454 kilogram

When converting from Metric to Customary, use these approximations.

1 centimeter ≈ 0.39 inch
1 meter ≈ 3.28 feet
1 kilometer ≈ 0.62 mile

1 liter ≈ 4.23 cups
1 liter ≈ 0.26 gallon
1 gram ≈ 0.035 ounce
1 kilogram ≈ 2.21 pounds

Temperature Conversions between Celsius and Fahrenheit

$$C = (F - 32) \div 1.8$$
$$F = (C \times 1.8) + 32$$

Quick Reference

Area

	Rectangle	$A = \ell w$

	Parallelogram	$A = bh$

	Triangle	$A = \frac{1}{2}bh$

	Trapezoid	$A = \frac{1}{2}h(b_1 + b_2)$

	Circle	$A = \pi r^2$

Circumference

$C = \pi d$ or $C = 2\pi r$

Volume

 Rectangular Prism $V = \ell wh$
or $V = Bh$

   Surface Area

   S.A. = L.A. + 2B or $2\ell h + 2hw + 2\ell w$

 Right Circular Cylinder $V = \pi r^2 h$
or $V = Bh$

   S.A. = L.A. + 2B or $2\pi rh + 2\pi r^2$

 Right Square Pyramid $V = \frac{1}{3}Bh$

  Right Circular Cone $V = \frac{1}{3}Bh$

or $V = \frac{1}{3}\pi r^2 h$